Early Hominin Landscapes in Northern Pakistan

Investigations in the Pabbi Hills

R. W. Dennell

With contributions by: M. Anwar, M. Beech, R. Coard, L.M. Hurcombe, H.M. Rendell and A. Turner

BAR International Series 1265
2004

Published in 2019 by
BAR Publishing, Oxford

BAR International Series 1265

Early Hominin Landscapes in Northern Pakistan

ISBN 9781841713717 paperback
ISBN 9781407326832 e-book

DOI https://doi.org/10.30861/9781841713717

A catalogue record for this book is available from the British Library

This book is available at www.barpublishing.com

BAR Publishing is the trading name of British Archaeological Reports (Oxford) Ltd. British Archaeological Reports was first incorporated in 1974 to publish the BAR Series, International and British. In 1992 Hadrian Books Ltd became part of the BAR group. This volume was originally published by John and Erica Hedges in conjunction with British Archaeological Reports (Oxford) Ltd / Hadrian Books Ltd, the Series principal publisher, in 2004. This present volume is published by BAR Publishing, 2019.

BAR
PUBLISHING

BAR titles are available from:

BAR Publishing
122 Banbury Rd, Oxford, OX2 7BP, UK
EMAIL info@barpublishing.com
PHONE +44 (0)1865 310431
FAX +44 (0)1865 316916
www.barpublishing.com

Early Hominin Landscapes in Northern Pakistan

Table of Contents

List of Chapters	i - iii
List of Tables	iv - vii
List of Figures	viii - x
Acknowledgements	xi -xii
Introduction	xiii - xv

List of Chapters:

Part 1: Background to the Fieldwork in the Pabbi Hills

Chapter 1: The Palaeoanthropological Background to the Investigations in the Pabbi Hills.	1-12
1: An East African genesis?	2
1.1: How well documented is the East African record?	2
1.2: Where were those grasslands in which hominids evolved?	3
1.2.1. Hominids, grasslands and East Africa	3
1.2.2: Hominids and grasslands before 2.5 Mya ago	3
1.3: When did African hominids first live beyond the East African Rift Valley?	4
1.4: What does the distribution of chimpanzees and gorillas tell us?	5
2: Out of Africa?	5
2.1: *Homo erectus* and *H. ergaster*: a taxonomic digression	5
2.2 *Homo ergaster (*or *H. erectus)*: the first hominid outside Africa?	5
2.3: Asia, the unknown continent	6
2.4: The earliest Eurasian evidence	8
3: Discussion	12
Chapter 2: The Palaeontological Background to the Fieldwork in the Pabbi Hills.	13-23
2.1 Siwalik Biostratigraphy	13
2.1.1. History of research	14
2.1.2. Siwalik Zonation	17
2.1.3. The Upper Siwaliks: Tatrot, Pinjor, and Boulder Conglomerate "Faunal Stages"	18
2.1.4. Recent research into the Pinjor	18
2.1.5. The Boulder Conglomerate Faunal Stage	20
2.1.6. Summary	20
2. 2. Fossil hominid remains and fluvial landscapes	20
2. 3. Fossils, rivers and landscapes	22

Part 2: Results of the Surveys in the Pabbi Hills

Chapter 3: The Pabbi Hills: Introduction and Overview.	24-31
3. 1 Fieldwork problems	25
3. 2. Field procedures and survey methods	26
3.2.1. Collecting strategy	26
3.2.2. Time-banding and stratigraphic zonation	26
3. 2. Field procedures and survey methods	26
3.3. Recording methods	29
3.4. Relationship between fossil occurrences and palaeolandscapes	31
Chapter 4: Magnetic Polarity Stratigraphy of Upper Siwalik Sediments in the Pabbi Hills (Helen Rendell)	32-36
Chapter 5: The 1986 survey (fossil occurrences 14 - 156) of Sandstone 12 and adjacent units.	37-88
Set 1: The North-East Side of Kanarawala Kas: 14-25, 42-50 and 109.	39
Set 2: The West side of Kanarawala Kas (26 - 41).	40
Set 3: Fossil occurrences 51-63.	42
Set 4: Fossil occurrences 64 – 75.	42
Set 5: Fossil occurrences 74 - 85 (Site 73 to the G.T road), Sandstone 12.	43
Set 6: The badlands in front of Sandstone 12: (nos. 110 - 150, even numbers only).	45

Set 7: Nos. 111 - 137 (odd numbers only), Sandstone 12, south of the G.T. road.	46
Set 8: Nos. 139-153 (odd numbers only), 154, 156; older than Sandstone 12, south of the G.T.Road.	46
Set 9: The survey of Sandstone 14 (fossil occurrences 90 - 108).	47
Tables accompanying Chapter 5	48-88
Chapter 6: The 1987 survey of the Pabbi Hills	89-187
Set 1: Choawala Kas headwaters (1.7 - 1.9 Mya-old): 200-218 (even nos. only), 365, 367, 388.	89
Set 2: Choawala West (1.4 - 1.7 Mya-old): 222 – 232 (even numbers); also 332, 334, 336.	94
Set 3: Choawala West, 362 area (1.7 - 1.9 Mya-old): 338 - 366.	95
Set 4: Choawala East (1.7 - 1.9 Mya-old). (Even numbers 236-260 for Happy Valley, and 262, 264, 416, 418, 434 and 518, Death Valley).	96
Set 5: Bangial Kas and Choawala Kas (mostly 1.4 - 1. 7 Mya-old): 225-255, 270, 272.	97
Set 6: Sohawa Kas: the 203 horizon (1.9 - 2.2 Mya-old).	99
Set 7: Sohawa Kas headwaters (205, 207, 209) and Sohawa Kas West (217-223 (odd nos. only), 288, 316, and 318): 1.8 - 2.2 Mya-old.	100
Set 8: Sohawa Kas West (1.7 - 1.9 Mya-old): 276-286 (even nos. only).	101
Set 9: Sohawa East, 291 area (1.9 - 2.2 Mya-old): 237, 257, 263, 273-295 (odd. nos.), 296-308 (even nos.), 320-330 (even nos.), 315-319 (odd nos.), 376; 305, 309, 311,319.	102
Set 10: the Flag Hill area (1.7 - 1.9 Mya-old): 327-335 (odd nos.), 368-372 (even nos. only).	104
Set 11: West Branch of Kotha Kas (various numbers from 337 to 535).	106
Set 12: West Branch of Kotha Kas (397, 399, 477, 501- 511, odd. nos. only).	107
Set 13: West Branch of Kotha Kas (the 481 and 489 areas).	107
Set 14: West Branch of Kotha Kas (343-357, 383-395).	109
Set 15: The Viaduct area (1.9 - 2.2 Mya-old): 359-363, 369-381 (odd nos. only).	110
Set 16: East Branch of Kotha Kas (1.9-2.2 Mya-old): 407-425 (odd nos. only) and 499.	110
Set 17: Kotha Kas (1.9-2.2 Mya-old): 427-46 (odd nos. only).	111
Set 18: Head of Kotha Kas, between Happy and Death Valleys (?1.7 - 1.9 Mya-old): 398 – 414 (even nos. only).	112
Set 19: East Branch of Kotha Kas (453-475, odd. nos. only).	113
Set 20: Sandstones 14 - 16, Choawala Kas (0.9 - 1.2 Mya-old): 500-510, 516, 520-528 (even nos. only).	113
Set 21: Ratial and Dhorian Kas (382-386, 420-432).	114
Set 22: The badlands north of the G.T. Road (436-528 even nos. only).	114
Tables accompanying Chapter 6	117-187
Chapter 7: The 1989 and 1990 surveys of the Pabbi Hills.	188-221
Area 1: The Bulani area (Bandgwara Kas: 600 – 607)	188
Area 2: Besa Kas (608-615); Pir Jaffa (616-625)	188
Area 3. The Baroth area (622-640, odd and even numbers)	189
Area 4: Locality 642 to Ban Barilla (637-648)	192
Area 5: Ban Barilla - Doga (649-674)	192
Area 6: Hairpin bend (657-685)	192
Tables accompanying Chapter 7	193-221
Chapter 8 The stone artefacts from the Pabbi Hills (Linda Hurcombe)	222-292
Section A: The context of the survey and the lithic recording system	222
Stone collecting and recording systems	223-228
Section B: Results, analysis and discussion	229-238
Discussion: the overall nature and affinities of the Pabbi Hills lithic material	235-238
Section C: Taphonomic investigations of the distribution, movement and flaking of stone	238-249
The occurrence of natural stone in the Pabbi Hills	238-239
Pebble beds in modern rivers	240
Survey data (occurrences 346, 218, 387, 722, 625	240-243
Investigations at 269 and surrounding areas	243-244
Places with pottery	244

Experiments monitoring falling fractures, erosion, and tool movement 244-249
Falling fractures 244
Rates of erosion and stone artefact movement 245
Gully and slope experiments 246-249
Section D: Interpreting patterns within the assemblage: discussion and conclusions 249-252
Conclusions 251-252
Tables accompanying Chapter 8 253-268
Drawings of stone tools accompanying Chapter 8 269-281
Illustrations accompanying Chapter 8 282-292

Chapter 9: The excavations of fossil localities 73, 362 and 642. 293-371
Locality 73 (R.Dennell) 294-313
Locality 362 (M. Beech and M. Anwar) 314-327
Locality 642 (M. Beech and M. Anwar) 328-352
Taphonomic observations of localities 73, 362 and 642 (R. Coard) 353-371
Observations at a distance: an interpretation of locality 73 353-355
A taphonomic analysis of Locality 362 355-361
A taphonomic analysis of Locality 642 362-371

Part 3:Analaysis

Chapter 10: The fossil vertebrate record of the Pabbi Hills: palaeontology and biostratigraphy. 372-411
Part 1: Fossil remains from the Pabbi Hills: Artiodactyla 373-382
i) Bovid teeth and horns 373-377
ii) Cervid teeth and antler 377-378
Artiodactyla (Bovid/Cervid) post-cranial remains 379
iii) Giraffidae 380
iv) Suidae 381
v) Perissodactyla: Equidae 382
Hipparion or *Equus?* 383
Equid post-cranial specimens 384
vi) Rhinoceritidae and vii) Proboscidea 385
viii) other taxa: Anthracoceridae, Hippopotamidae, Chelonia, Crocodile, Mollusca, Primates, Coprolites 386-388
Part 2: an overview of the faunal remains from the Pabbi Hills 388
1) Continuity and extinction in the Pabbi Hills sequence 388
2) The *Elephas hysudricus* faunal zone 389
3) Opdyke et al's (1979) Upper Siwalik biostratigraphy reassessed 390
4) Faunal turnover in South Asia compared with that in East Africa 390
Tables accompanying Chapter 10 394-403
Appendix: Vertebrate Fossils: Carnivora (Alan Turner) 404-411

Chapter 11: The taphonomy of the fossil record of the Pabbi Hills. 412-442
1) What was found where: the distribution of fossils across the landscape 412
2) The sedimentary context of fossils 413
3) The main types of fossil occurrences 414-418
4) Type of animals 418
5) Information loss: weathering, fragmentation and disarticulation 422-428
6) Types of skeletal elements 428
7) Carcass dispersal 430
8) The sampling of rare taxa: needles and haystacks 430
9) The absence of hominins: real or a sampling problem? 433
Overview 435
Tables accompanying chapter 11 437-442

Summary and Post-script 443

Bibliography 444-454

List of Tables:

1	Table 1.1: The size (in square miles) of S.W. Asia, East Africa, the European Union (EU), and Europe west of the former Soviet Union	7
2	Table 2.1: Hominid skeletal remains from fluvial deposits	21
3	Table 3.1: Numbers of fossil specimens collected by field season and time-band in the Pabbi Hills	27
4	Table 3.2: The Chinji classification scheme for recording fossil provenance	30
5	Table 4.1: Magnetic polarity timescale for the last 3.22 million years	32
6	Table 4. 2: Details of palaeomagnetic sampling in the Pabbi Hills	33
7	Table 4. 3. Mean blocking temperature (Tb) components for sampled sections.	34
8	Table 5.1: Fossil occurrences of the 1986 survey grouped according to age, sandstone (SST) unit, and location north or south of the Grand Trunk (G.T.) Road	38
9	Table 5.2: Fossil occurrences from the 1986 field survey of the Pabbi Hills	48
10	Table 5.3. Contents of fossil locality 14, Sandstone 12	50
11	Table 5. 4: Contents of fossil occurrences 15-25 and 109, Sandstone 12, Kanarawala Kas North East, 1986	53
12	Table 5.5: Contents of fossil occurrences 42 – 50, Sandstone 12, Kanarawala Kas North East, north of site 14	55
13	Table 5.6: Contents of fossil occurrences 26 - 41, Sandstone 12, Kanarawala Kas West	57
14	Table 5.7: Contents of fossil occurrences 51-63 (Sandstone 12, south of occurrence 30, 1986)	60
15	Table 5. 8: Contents of fossil occurrences 64 – 75, Sandstone 12, excluding localities 68 and site 73	62
16	Table 5. 9: Contents of fossil locality 68, Sandstone 12	63
17	Table 5. 10: Contents of locality 73, zone A, Sandstone 12	66
18	Table 5. 11: Contents of fossil locality 73, zone B, Sandstone 12	67
19	Table 5. 12: Contents of fossil locality 73, zone C, Sandstone 12	69
20	Table 5. 13: Contents of fossil locality 73, zone X, Sandstone 12	70
21	Table 5. 14: Contents of fossil locality 73, zone Y, Sandstone 12	70
22	Table 5. 15: Contents of fossil locality 73, zone Z, Sandstone 12	71
23	Table 5. 16: Contents of fossil occurrences 74 – 85, Sandstone 12, between locality 73 and the GT road	72
24	Table 5. 17: Contents of locality 83, Sandstone 12	73
25	Table 5. 18: Contents of fossil occurrences in the "badlands" in front of Sandstone 12 north of the G.T. road	74
26	Table 5. 19: Contents of fossil locality 124, badlands in front of Sandstone 12	78
27	Table 5. 20. Contents of fossil locality 128, badlands in front of Sandstone 12	80
28	Table 5. 21: Contents of fossil occurrences along Sandstone 12 on the south side of the G.T. road, (excluding locality 133)	81
29	Table 5. 22: Fossil material from locality 133, Sandstone 12	83
30	Table 5. 23: Contents of fossil occurrences south of the G.T. Road and in front of Sandstone 12	84
31	Table 5. 24: Fossil material from locality 156	85
32	Table 5. 25: Contents of fossil occurrences from Sandstone 14, 1986	85
33	Table 6.1: Fossil sites found in 1987, listed in numerical order	117
34	Table 6.2: Total numbers of specimens per set and the average number per fossil occurrence from the 1987 survey	123
35	Table 6.3: Fossil material from Set 1, 1987, headwaters of Choawala Kas (200-220, even nos. only; 365, 367, 368)	123
36	Table 6. 4: Fossil material from Set 2, Choawala West (222-236, even nos. only, 332,334 and 336)	126
37	Table 6. 5: Fossil material from Set 3, Choawala West (338-366)	127
38	Table 6. 6: Fossil material from locality 362, Set 3, Choawala West	128
39	Table 6. 7: Fossil material from Set 4, Choawala East (Happy Valley 236-260 even nos. only; Death Valley 262, 264, 416, 418, 434 and 518).	130
40	Table 6. 8: Fossil material from Set 5, Bangial Kas and Choawala Kas (225-255 odd nos.), 270-272 (even nos.), 532 .	133
41	Table 6. 9: Fossil material from Set 6, the 203 horizon, Sohawa Kas (1.9 - 2.2 Mya-old	135
42	Table 6. 10: Fossil material from Set 7, Sohawa Headwaters (205, 207, 209) and Sohawa Kas West (217-223 (odd nos. only), 288, 316, and 318)	140
43	Table 6. 11: Fossil material from Set 8, Sohawa West (276-290)	142
44	Table 6. 12: Fossil material from Set 9, Sohawa Kas East (principally 273-295 odd. nos, 296-308 even nos.)	143
45	Table 6. 13: Fossil material from Set 10, Flag Hill area (327-335 (odd nos.), 368-372 even nos.)	147
46	Table 6. 14: Fossil material from Set 11, West Branch of Kotha Kas (337-341, 507-535, odd nos.)	148

47	Table 6. 15: Fossil material from Set 12, West Branch of Kotha Kas (397-399, 477, 501-511, odd nos.)	151
48	Table 6. 16: Fossil material from Set 13, West Branch of Kotha Kas (401-405, 479-497 odd nos.).	153
49	Table 6. 17: Fossil material from Locality 481, Set 13, West Branch of Kotha Kas	156
50	Table 6: 18: Fossil material from Locality 489, Set 13, West Branch of Kotha Kas	158
51	Table 6. 19: Fossil material from Locality 491, Set 13, West Branch of Kotha Kas	161
52	Table 6. 20: Fossil material from Set 14, West Branch of Kotha Kas (343-357, 383-395, odd nos.).	163
53	Table 6. 21: Fossil material from Set 15, the Viaduct area (359-381, odd. nos. only).	166
54	Table 6. 22: Fossil material from Set 16, the East Branch of Kotha Kas (407-425 odd nos. only) and 499.	168
55	Table 6. 23: Fossil material from Set 17, East Branch of Kotha Kas (427-441, odd nos. only).	170
56	Table 6. 24: Fossil material from Set 17, Kotha Kas (445-461, odd. nos. only).	175
57	Table 6. 25: Fossil material from Set 18, the area between Happy and Death Valleys (398-414 even nos.).	177
58	Table 6. 26: Fossil specimens from Set 19, Kotha Kas South East (497-535)	178
59	Table 6. 27: Fossil material from Set 20, Sandstone 14, Choawala Kas (500-510, 516, 520-528).	180
60	Table 6. 28: Fossil material from Set 21, Dhorian Kas and Ratial Kas.	182
61	Table 6. 29: Fossil material from Set 22, the badlands north of the G.T. road, 1987 (436-528 even nos.).	183
62	Table 7.1: Fossil occurrences found in 1989	193
63	Table 7:2. The Bulani area - Bandgwara Kas (600-607), Besa Kas (608-615) and Pir Jaffa (616-621)	194
64	Table 7.3. Fossil material from locality 610	197
65	Table 7: 4. Fossil specimens from Baroth area (622-640).	201
66	Table 7.5: Fossil specimens from locality 642, zone 1, 1989.	204
67	Table 7: 6 Fossil specimens from locality 642, zone 2, 1989.	206
68	Table 7: 7. Fossil specimens from locality 642, zone 3, 1989.	208
69	Table 7. 8: Fossil specimens from locality 642, zone X, 1989	211
70	Table 7: 9. Fossil material from locality 642, zone 1, clusters 1 and 2, 1989.	212
71	Table 7.10: Fossil specimens from find spots on locality 642, 1989	213
72	Table 7: 11. Fossil specimens from between locality 642 (Baroth area) and Ban Barilla.	215
73	Table 7: 12. Fossil specimens from between Ban Barilla and Doga (650-674).	216
74	Table 7: 13. Fossil specimens near hairpin bend, watershed track, Pabbi Hills north (657-685).	219
75	Table 8.1: Types of flakes, cores and polished/ground stonework	253
76	Table 8. 2: The number of stone artefacts found grouped by age of underlying deposit showing the low density overall.	263
77	Table 8.3: Summary of the lithic assemblage from the Pabbi Hills by type, retouch, and lithology	263
78	Table 8.4: Technological types for all quartzite flake artefacts	264
79	Table 8.5: The technotypes for all cores and core tools	264
80	Table 8.6: The types of artefacts grouped by the age of the geological strata on which they were found.	265
81	Table 8.7: The technotypes of flakes and cores grouped by the age of the geological strata on which they were found.	266
82	Table 8.8: The percentage of cortex remaining for flakes grouped by the age of the geological strata on which they were found.	267
83	Table 8.9: The percentage of cortex remaining for cores grouped by the age of the geological strata on which they were found.	267
84	Table 8. 10: The occurrence of stone in the 269 area as indicated on Figure 8.24.	268
85	Table 8. 11: Summary of the pieces created when 18 nodules fractured, showing the numbers that were potentially confusable with intentional flaking.	268
86	Table 8. 12: Details of erosion experiments of stone along gully bases.	268
87	Table 9. 1: Principal taxa represented in the excavated assemblage from locality 73.	295
88	Table 9. 2: Bone surface weathering at locality 73	297
89	Table 9. 3: Measurements of limb-bones from locality 73.	298
90	Table 9. 4: Fossil specimens from the excavation of locality 73, 1986.	303
91	Table 9. 5: Fossil specimens from the excavation of locality 73, 1987.	305
92	Table 9. 6: Fossil material collected in 1989 from the gully down slope of locality 362	322
93	Table 9. 7: Catalogue of site 362 fossil specimens, excavated 1989	324
94	Table 9. 8: Comparison of the surface and excavated fossil assemblages from site 362.	315
95	Table 9. 9: Fossil specimens from the surface of locality 642, zone 1, 1990	330
96	Table 9. 10: Fossil specimens from the surface of locality 642, zone 2, 1990	331

97	Table 9. 11: Fossil specimens from the surface of locality 642, zone 3, 1990	332
98	Table 9. 12: Fossil specimens from find spots on the surface of locality 642, 1990	333
99	Table 9. 13: Measurements (in mm.) of specimens from the surface collection of locality 642, 1990.	335
100	Table 9. 14: Catalogue of locality 642 fossil specimens, excavated in 1990.	336
101	Table 9. 15: Measurements (in mm.) of specimens from the excavation of locality 642, 1990.	346
102	Table 9. 16: Number and % of individual specimens present from the excavation and surface collections of locality 362	360
103	Table 9. 17: Number of individual specimens present (NISP)and minimum number of individuals (MNI) at locality 362	360
104	Table 9. 18: NISP for all mammal taxa at locality 362 (N=352) (Table 1, 362) – to come	361
105	Table 9. 19: Number of individual specimens present from the 1989 and 1990 surface collections and 1990 excavation at locality 642	368
106	Table 9. 20: Skeletal part analysis for all taxa at Locality 642	369
107	Table 9. 21: Skeletal part analysis for size classes and indeterminate material at Locality 642.	370
108	Table 9. 22: Number of individual specimens present and minimum number of individuals at locality 642	371
109	Table 10.1: The Pabbi Hills fossil material in terms of provenance, dating and grading	372
110	Table 10. 2: Size of Type 1 bovid teeth (*Damalops palaeindicus*)	373
111	Table 10. 3: Size of Type 1 third molars (*Damalops palaeindicus*)	373
112	Table 10. 4: Third molar measurements of Type 1B alcelaphine-type teeth	374
113	Table 10. 5: Size of Type 2 bovid (*Hemibos triqueticornis*) third molars	375
114	Table 10. 6: Size of Type 3 (Gazelle-like) teeth.	376
115	Table 10. 7: Size of Type A horns (*Damalops palaeindicus*)	376
116	Table 10. 8: Type B horn-cores (*Hemibos triqueticornis*)	377
117	Table 10. 9: Suggested linkage of bovid teeth and horn-cores (end of chapter)	390
118	Table 10. 10: Size of cervid teeth.	377
119	Table 10. 11: Types and provenance of cervid antler (end of chapter)	394
120	Table 10.12: Distal measurements for bovid and cervid post-cranial bones (end of chapter)	394
121	Table 10. 13: Bovid and cervid proximal post-cranial measurements (end of chapter)	396
122	Table 10. 14: Astraguli measurements from the Pabbi Hills (end of chapter)	398
123	Table 10. 15: Specimens from locality 642 attributed to *Gazella* (end of chapter)	399
124	Table 10. 16: *Sivatherium giganteum* tooth and post-cranial remains	380
125	Table 10: 17. Isolated suid tooth measurements	382
126	Table 10. 18: Details of equid mandibles	382
127	Table 10. 19: Equid lower molars/premolars from the Pabbi Hills (end of chapter)	400
128	Table 10. 20: Equid upper molars/premolars from the Pabbi Hills (end of chapter)	400
129	Table 10. 21: Comparative specimens of equid teeth.	383
130	Table 10.22: Equid complete post-cranial specimens (end of chapter)	401
131	Table 10. 23: Additional equid post-cranial measurements (end of chapter)	401
132	Table 10. 24: *Rhinoceros sivalensis* specimens from the Pabbi Hills (end of chapter)	402
133	Table 10. 25: Principal *Elephas hysudricus* specimens.	385
134	Table 10. 26. Hippopotamid specimens from the Pabbi Hills	387
	Table10. 27: Siwalik Hyaenidae: lower dentitions.	410
137	Table 10. 28: Siwalik Hyaenidae: upper dentitions.	410
138	Table 10. 29: Siwalik and Modern Felidae: lower dentitions.	411
139	Table 10. 30: Siwalik Felidae: metapodia.	411
140	Table 10. 31: Siwalik Canidae: lower dentitions.	411
141	Table 11. 1: The 20 most productive fossil occurrences in the Pabbi Hills	412
141	Table 11. 2: The number and % of fossils from the most productive localities in the Pabbi Hills	413
143	Table 11. 3: Size of fossil occurrences from each field season	413
144	Table 11.4: Matrix on fossil specimens by type of fossil occurrence and numbers of specimens (NISP) (in parentheses).	413
145	Table 11. 5: Combined totals of matrix adhering to fossils by numbers of occurrences and specimens.	414
146	Table 11. 6: The main types of fossil occurrences in the Pabbi Hills.	415
147	Table 11. 7: Taxa represented at large, concentrated (Type 1a) localities.	416

148	Table 11. 8: Taxa represented at land-based fossil occurrences >1.6 Mya old (end of chapter)	436
149	Table 11. 9: Taxa represented at land-based fossil occurrences from Sandstone 12 (1.2 - 1.4 Mya-old). (end of chapter)	437
150	Table 11. 10: Taxa represented at large, dispersed (Type 1b) localities.	416
151	Table 11. 11: Taxa represented at fluvial, Type 1b fossil occurrences >1.6 Mya old (end of chapter)	438
152	Table 11. 12: Taxa represented at fluvial, Type 1b fossil occurrences from Sandstone 12 (1.2 - 1.4 Mya-old) (end of chapter)	439
153	Table 11: 13: Overall composition of scatters	417
154	Table 11. 14: Overall composition of general collections .	418
155	Table 11. 15: Total number of specimens of each taxa represented in localities, scatters, and general collections in the Pabbi Hills. (end of chapter)	439
156	Table 11: 16: Probable body size of taxa represented in the Pabbi Hills	417
157	Table 11. 17: Immature specimens in the Pabbi Hills (end of chapter)	440
158	Table 11. 18: Taphonomic processes and bone fragmentation.	422
159	Table: 11. 19: Summary of surface weathering of specimens from the Pabbi Hills (end of chapter)	441
160	Table 11. 20: Proportions of diagnostic and indeterminate specimens from non-fluvial fossil occurrences	423
161	Table 11. 21 Proportions of diagnostic and indeterminate specimens from fluvial fossil occurrences.	424
162	Table 11. 22 Fragmentation at non-fluvial occurrences >1.6 Mya-old.	424
163	Table 11. 23: Fragmentation at non-fluvial fossil occurrences from Sandstone 12 (1.2 - 1.4 Mya-old)	424
164	Table 11. 24: Fragmentation at fluvial fossil occurrences from Sandstone 12 (1.2 - 1.4 Mya-old)	424
165	Table 11. 25: Fragmentation at fluvial fossil occurrences >1.6 Mya old	425
166	Table 11. 26: Limb-bone fragmentation in non-fluvial fossil occurrences (all surface data, 1986, 1987 and 1989).	425
167	Table 11. 27: Limb-bone fragmentation in fluvial fossil occurrences (all surface data, 1986, 1987 and 1989).	426
168	Table 11. 28: Jaw and tooth preservation in non-fluvial fossil occurrences (all surface data, 1986, 1987 and 1989).	426
169	Table 11. 29: Jaw and tooth preservation in fluvial fossil occurrences (all surface data, 1986, 1987 and 1989).	426
170	Table 11. 30: Tooth and vertebrae preservation in different types of fossil occurrences.	427
171	Table 11. 31: Carcass part representation in land-based fossil occurrences (all surface data from 1986, 1987 and 1989).	429
172	Table 11. 32: Carcass part representation in fluvial fossil occurrences (all surface data from 1986, 1987 and 1989).	429
173	Table 11. 33: Head and trunk parts in land-based occurrences.	429
174	Table 11. 34: Head and trunk parts in fluvial occurrences.	430
175	Table 11. 35: The relative abundance of various fossil taxa in the Pabbi Hills.	431

List of Figures:

1	Figure 1.1: The dispersal of *Homo erectus* out of Africa and into Asia and Europe, as commonly portrayed in the 1980's and early 1990's	1
2	Figure 1.2: Grasslands in the northern hemisphere now and in the Late Pliocene	4
3	Figure 1.3: The size of areas that need to be sampled for early hominid sites in Europe, Africa and Asia respectively	7
4	Figure 1.4: The earliest hominid and archaeological localities in southern Eurasia.	8
5	Figure 2.1: Map of Siwalik deposits across northern Pakistan and India.	13
6	Figure 2.2: The lithological and palaeontological zonation of the Siwaliks. (Source: Barry 1982).	14
7	Figure 2.3: Principal fossil localities in the Potwar Plateau and adjacent areas	15
8	Figure 2.4: Stratigraphic sections, age ranges, and occurrences of fossils in Siwalik sediments	16
9	Figure 2.5: The faunal zonation of the Siwalik deposits on the Potwar Plateau	18
10	Figure 2. 6: Correlations of the Faunal Interval Zones of Barry et al. (1982) with those proposed by Hussain et al. (1992) for the Mangla-Samwal anticline	19
11	Figure 3.1: The areas surveyed in the Pabbi Hills, 1986-1989	24
12	Figure 3.2: Cross-section and palaeomagnetic dating of the Pabbi Hills anticline	27
13	Figure 3.3: Correlation of palaeomagnetic sections in the Pabbi Hills	28
14	Figure 3.4: Keller et al.'s (1977) palaeomagnetic zonation of the Pabbi Hills.	28
15	Figure 3.5: The main type of fossil occurrences in the Pabbi Hills.	29
16	Figure 4.1: Combined palaeomagnetic stratigraphy of the Pabbi Hills	35
17	Figure 5.1: The main areas surveyed along Sandstone 12 in 1986	37
18	Figure 5.2: Sedimentary section along Sandstone 12 between Kanarawala Kas and the G.T. road.	39
19	Figure 5.3: Schematic view towards fossil occurrences 14-22, scarp face of Sandstone 12, north bank of Kanarawala Kas	40
20	Figure 5.4: Fossil occurrences 33-39, Sandstone 12	41
21	Figure 5.5: Schematic section at fossil occurrence 30	41
22	Figure 5. 6: Schematic view of locality 68	42
23	Figure 5. 7: Schematic map of locality 73	43
24	Figure 5. 8: Schematic section at locality 83	44
25	Figure 5. 9: Schematic map of localities 83, 124 and 128	45
26	Figure 6. 1: Areas surveyed in 1987 south of the G.T. road	90
27	Figure 6. 2: Principal collecting areas in Choawala Kas	91
28	Figure 6. 3: The stratigraphic position and age of the principal fossil occurrences found in 1987	92
29	Figure 6.4: Location of fossil sites in Choawala Kas, Set 1 (200-220, even nos. only; 365, 367, 368).	93
30	Figure 6. 5. Location of fossil sites in Choawala Kas West, Set 2 (222-236, even nos. only, 332,334 and 336).	94
31	Figure 6.6: Stratigraphic context of fossil occurrences in Sets 2 and 3, Choawala Kas West.	94
32	Figure 6. 7: Location of fossil sites in Choawala West (362 series), Set 3 (338-366).	95
33	Figure 6.8: Location of fossil sites in Choawala East, Set 4 (Happy Valley, Death Valley).	96
34	Figure 6.9: Principal collecting areas in Sohawa Kas.	98
35	Figure 6.10: Location of fossil sites in Sohawa Kas (in and below the 203 horizon).	99
36	Figure 6.11. Stratigraphic provenance of fossils from the 203 horizon, Sohawa Kas.	100
37	Figure 6. 12: Location of fossil sites in Sohawa Kas West.	101
38	Figure 6. 13: Location of fossil sites in Sohawa East, 291 area.	102
39	Figure 6. 14: Schematic section of fossil occurrences 293 and 295.	103
40	Figure 6. 15: Location of sites in the Flag Hill area	104
41	Figure 6. 16: Principal collecting areas in Kotha Kas.	105
42	Figure 6. 17: Schematic cross-section down the West branch of Kotha Kas.	106
43	Figure 6. 18: Plan of locality 481.	107
44	Figure 6. 19: Schematic section at locality 481	108
45	Figure 6. 20: Schematic section at localities 489 and 491	108
46	Figure 6. 21: Plan of the area around localities 489 and 491.	108

47	Figure 6. 22: Schematic section of fossil occurrence 349.	109
48	Figure 6. 23: Location of fossil sites in the Viaduct area (359-379, odd. nos.), and section at 361.	110
49	Figure 6. 24: Schematic section of localities 427, 435 and 437.	111
50	Figure 6. 25: Principal occurrences in the area of localities 427 and 437.	111
51	Figure 6. 26: Field plan of the 437 area.	112
52	Figure 6. 27: Schematic section of locality 445.	112
53	Figure 6.28: Schematic section of occurrences 400-412.	112
54	Figure 6. 29: A schematic view of the fault between localities 427 and 435, and the watershed track.	113
55	Figure 6. 30: Schematic section of stratigraphic relationships of fossil material at occurrences 427 and 465, East Branch of Kotha Kas.	113
56	Figure 6. 31. Schematic sections of occurrences 504 and 516.	114
57	Figure 6. 32: Principal collecting areas in the badlands near the brick weir north of the G.T. road.	115
58	Figure 6. 33: Principal collecting areas in the badlands near Bawli Kas, north of the G.T. road.	115
59	Figure 6. 34: Schematic section of occurrences 116, 120, 444, 452 and 456.	116
60	Figure 6. 35: Schematic section of occurrence 492.	116
61	Figure 6. 36: Schematic section of occurrence 488.	116
62	Figure 7.1: Plan of locality 610	189
63	Figure 7.2: Map of the area surveyed in 1989 around Baroth, northern part of the Pabbi Hills	190
64	Figure 7.3: Map of the area of locality 642, near Baroth, northern part of the Pabbi Hills	190
65	Figure 7.4: Plan of locality 642	191
66	Figure 7.5: Schematic section of profile near locality 642	191
67	Figure 7. 6: Plan of Proboscidean remains at occurrence 664	192
68	Figure 8.1: Flake techno-types defined by amounts of cortex as a percentage of total surface area and the presence or absence of cortex on the striking platform	269
69	Figure 8.2: Core technotypes defined by flaking strategy and sequence	270
70	Figure 8.3: Flint artefacts, including micro-cores, flakes and scraper	271
71	Figure 8.4: Ground and polished stone objects including axes, hammerstone and ringstones	271
72	Figure 8.5: Proto-biface	272
73	Figure 8.6: Straight-edged tools (SECT's), mostly made on cores	272
74	Figure 8.7: High-action core tools (HACT's) with use wear, and associated flakes	273
75	Figure 8.8: Flake-blades cores	274
76	Figure 8.9: Squared cores	274
77	Figure 8.10: Split cobbles, irregular and bi-polar cores as evidence of core reduction problems	275
78	Figure 8.11: Pebble cores showing uni- and bi-directional flaking, some forming chopper-like artefacts	276
79	Figure 8.12: Simple and complex disc cores, and polyhedral cores	277
80	Figure 8.13: Flake blades and straight-edged flakes	278
81	Figure 8.14: Simple and complex round flakes, and salami slices	279
82	Figure 8.15: Simple and double-bulb flakes as evidence of knapping problems, and angular flakes	280
83	Figure 8.16: Retouched flakes, including knife and scrapers	281
84	Figure 8.17: The conglomerate capping Sandstone 14 here exposed as a consolidated slab and showing the size of the pebbles	282
85	Figure 8.18: The sands with pebble stringers running along the top of the Sandstone 14 scarp slope	282
86	Figure 8.19: Laterally restricted conglomerate layer at the base of the sands in Sandstone 14 showing the 50 cm section cut back into the hillside (scale = 20cm)	283
87	Figure 8.20: The small spur of locality 218 being searched for stones, and the arrow marking where an artefact was recovered in 1987.	283
88	Figure 8.21: The conglomerate and pebble exposures in the Baroth area mapped against artefacts from the collecting zones.	284
89	Figure 8.22: The 269 stone and fossil bone in close proximity as found in 1987. The fossil refitted with fragments further upslope.	285
90	Figure 8.23: The general area of 269 looking North in 1989 showing the main scarp in the background and the 'horseshoe' containing the stone artefact, with person standing at th level of the path in the area of excavations.	285
91	Figure 8.24: Map of the 269 area, showing also collecting areas, and natural and flaked stone.	286
92	Figure 8.25: The 269 horseshoe area showing the level of the path with soft sands above, and steep slope below.	287
93	Figure 8.26: Contour map of the 269 horseshoe area and excavation trenches.	287
94	Figure 8.27: The North section of Trench A, site 269.	288

95	Figure 8.28: The East section of Trench A, site 269.	288
96	Figure 8.29: Detail (D), refits ®, and sequence (S) drawings of fractures resulting from vertical falling experiments of rock on rock.	289
97	Figure 8.30: Hillside on the west bank of the west fork of Kotha Kas (1999) indicating the layout of the erosion experiments. Flag Hill can be seen in the background.	290
98	Figure 8.31: The movement of stones from 1989 to 1990 and 1992 in two of the erosion experiment stations.	291
99	Figure 8.32: Erosion and movement of stones at experiment station 5 in 1999 after 10 years, showing the spread of material away from the original cluster.	292
100	Figure 8.33: Erosion and movement of stones at experiment station 4 in 1999 after 10 years, showing the thorn bush roots resisting erosion to form a cone above the present land surface.	292
101	Figure 9.1: Location of localities 73, 362 and 642.	294
102	Figure 9. 2: Schematic section of the local profile at locality 73.	295
103	Figure 9. 3: Plan of excavation of upper part of locality 73.	300
104	Figure 9. 4: Plan of excavation of middle part of locality 73.	301
105	Figure 9. 5: Plan of excavation of bottom part of locality 73.	302
106	Figure 9. 6: Section of locality 362.	316
107	Figure 9. 7: Plan of level 1, locality 362.	318
108	Figure 9. 8: Plan of level 2, locality 362.	319
109	Figure 9. 9: Plan of level 3, locality 362.	320
110	Figure 9. 10: Plan of level 4, locality 362.	321
111	Figure 9. 11: Collecting zones, datum point, and find-spots at locality 642, 1990.	328
112	Figure 9. 12: Excavation grid and areas excavated at locality 642.	329
113	Figure 9. 13: Plan of excavation of locality 642, area A, level 1.	347
114	Figure 9. 14: Plan of excavation of locality 642, area A, level 2.	348
115	Figure 9. 15: Plan of excavation of locality 642, area A, level 3.	349
116	Figure 9. 16: Plan of excavation of locality 642, area B, level 1.	350
117	Figure 9. 17: Plan of excavation of locality 642, area B, level 2.	350
118	Figure 9. 18: Plan of excavation of locality 642, area B, level 3.	351
119	Figure 9. 19: Plan of excavation of locality 642, area B, level 4.	351
120	Figure 9. 20: East section of locality 642, Area B	352
121	Figure 9. 21: Comparison of *Damalops paleindicus* (Falconer) GB642 EX1717 and *Proamphibos lachrymans* (reproduced from Pilgrim 1939).	366
122	Figure 10. 1: Cervid antler.	378
123	Figure 10. 2: Scattergram of bovid distal metacarpals	379
124	Figure 10. 3: Scattergram of bovid and cervid astraguli	380
125	Figure 10. 4: Hipparionine and equine teeth.	383
126	Figure 10. 5: Equine molar occlusal morphology from the Pabbi Hills (end of chapter).	390
127	Figure 10. 6: Scattergram of equid distal metapodia from the Pabbi Hills and Mirpur.	384
128	Figure 10. 7: *Elephas* molar plate details of 535GB1.	385
129	Figure 10.8: Detail of *Stegodon* tooth X507	386
130	Figure 10. 9: Overview of the main biostratigraphic features of the Pabbi Hills.	392
131	Figure 10. 10: Overview of the main biostratigraphic features of the Matuyama Chron (2.5 – 0.7 Mya) as suggested by Opdyke et al. (1979, Figure 21).	393
132	Figure 10. 11: Mandible of *Canis cautleyi* from locality 362.	409
133	Figure 11. 1: Fossil collecting along steep slopes of Sandstone 12	419
134	Figure 11. 2: Survey in an area of moderate slopes, high erosion rates and little ground cover	419
135	Figure 11.3: Sandstone 12, near locality 83	420
136	Figure 11.4: A general view from Kanarawala kas down Sandstone 12	420
137	Figure 11. 5: Sohawa kas headwaters (near the 203 horizon)	421
138	Figure 11. 6: Kotha kas, showing slope angles and modern erosion	421
139	Figure 11.7: The relative frequencies of skeletal parts in different modern situations in the Amboseli Park, East Africa. (After Behrensmeyer 1975: Figure 5).	428
140	Figure 11. 8: The effect of river-levels upon the availability of stone.	435

Acknowledgements

Many people in Britain and Pakistan need to be thanked for their contribution to the work described in this volume. The first are those who enabled it to happen in the first place. The most important in Britain were Dr. Bridget and Professor Raymond Allchin, the indefatigable Directors of the Ancient India and Iran Trust, Cambridge, and the Directors of the British Archaeological Mission to Pakistan, in whose name the permits for this fieldwork were granted by the Pakistani authorities. More than any others in the U.K., they have encouraged interest in the rich archaeological heritage of the Indian sub-continent through their own publications, and by organising conferences, academic exchanges, and fieldwork programmes such as the one in the Pabbi Hills. Throughout this work, they were unfailingly supportive, generous, and a ready source of advice and constructive criticism. The Leverhulme Trust is especially thanked for funding a three-year Senior Research Fellowship that allowed me to act as full-time Field Director of the British Archaeological Mission to Pakistan, and thus spend large parts of 1989-92 in Pakistan. The British Academy and South Asian Society are thanked as the main sources of funding. Other funding bodies that must be thanked for smaller grants to team members are the Ancient India and Iran Trust, the Boise Fund, the Antiquaries Society, the Charles Wallis (Pakistan) Trust, the Prehistoric Society, and the Harold Hyam Wingate Foundation for funding Dr. Coard's research.

Several members of the British Council and British High Commission should be thanked. The British Council representatives in Islamabad were unfailingly supportive. Roger Budd is thanked for arranging a Commonwealth Scholarship for the late Asif Jah (see below) of the Geological Survey of Pakistan. Les Phillips, his successor and a former graduate in archaeology, was exceptionally helpful, and a generous and convivial host. He is also thanked for enabling Mr. Saleem ul-Haq of the Dept. of Archaeology, Govt. of Pakistan, to study for an M.Sc. in Archaeology at Sheffield. Geoffrey Kaye, British Council Deputy, and his wife Beryl deserve a very special set of thanks. They were always an oasis of calm, willing to help, and generous with their time. Beryl especially is thanked for her nursing skills, thankfully not often required, but indispensable when they were. Two members of the British High Commission are also singled out for thanks. The first is Sir Nicholas Barrington, the High Commissioner during our fieldwork in the Pabbi Hills, and a man with a genuine and passionate interest in archaeology. He was an excellent host, and the author and his wife remember fondly the time he spared for us. Patrick Wogan, the Deputy High Commissioner in Karachi, is also remembered with great affection for being such pleasant company when I had to visit Karachi. Other staff at the High Commission are thanked for looking after our vehicles in our absence.

The fieldwork in the Pabbi Hills took place under an agreement between the British Archaeological Mission to Pakistan and the Dept. of Archaeology and Museums, Govt. of Pakistan. The Director-Generals of the latter organisation are thanked for allowing this work to proceed, and Dr. S. Ashfaque is thanked for his assistance as government representative on the fieldwork, and for forfeiting the comforts of Karachi for the rigorous and often primitive conditions of our field bases. Mr. Saleem ul-Haq, a young, newly-appointed field officer at the Dept. of Archaeology, was a great asset to the team. The Geological Survey of Pakistan (GSP) was a key participant in this project, one of whose aims was to collect vertebrate fossil material for a reference collection for the GSP. Two of its senior officers are remembered with immense gratitude. These are Dr. Ibrahim Shah, Director of Palaeontology at GSP, Islamabad, and his successor, Dr. Mahmoud Raza. Dr. Shah – or "Shah-Sahib" – was the father figure of the Palaeontology Section at GSP, and adopted us as part of his family, shared our aspirations, and helped in whatever way he could. Mahmoud was an exceptionally fine colleague and friend, and it was a privilege and a pleasure to work with both of them. Dr. Zaki Ahmed, another senior figure at the GSP, was also a steadfast supporter of the project, and amply demonstrated this by helping more than his post required. Dr. Mirza Talib Hasan's support in 1999 is also appreciated.

The field seasons from 1986 to 1990 spanned a total of 11 months. Those taking part could expect bitterly cold nights in January, oven-like days and nights in April, extremes of torrential rain and blistering heat, a six-day working week, long hours and only the most basic of accommodation, frequently lacking running water and a regular electricity supply. The survey itself was very arduous – sometimes a hour's hard walking to reach the starting point, a day of walking across loose and often steep surfaces, and often a rucksack heavy with fossils to carry back at the end of the day. I mention these points to emphasize the amount of stamina and enthusiasm (and good humour) that this type of fieldwork requires. The field teams were magnificent, and easily the best I have ever worked with. My abiding memories of most of them are laughter, as each brought their own brand of humour. To begin with those of the stronger sex: Professor Helen Rendell (then Sussex, now Loughborough), the project's geologist and palaeomagnetic expert, with whom I started working in 1981 in the Soan Valley, and one of the most meticulous and professional colleagues I have met. The survey teams comprised several remarkable figures, many of whom worked for several field seasons in the Pabbi Hills. Dr. Sheila Sutherland (Sheffield) was tall, superbly fit, laconic, and with the nose of a bloodhound when it came to following fossil horizons. The daily entries in her field notes combined accuracy, succinctness, humour, and a continually sharp interest in her both her fieldwork and her food. Hannah Jones (Cambridge) was equally invaluable, irrepressibly enthusiastic, good-natured, hard-working and faultless. Mags Peace (Sheffield) joined us in 1987, and besides being depressingly fit, did a superb job in collating the results of the 1987 survey in Choawala, Bangial and Sohawa Kas, as well as the recording of the material from the excavations of site 73.

Helen Wilson, another Sheffield recruit in 1987, performed brilliantly in making casts of fossil specimens. Dr. Linda Hurcombe (now my wife) was another Sheffield recruit in 1987, and took charge of recording and analysing the lithic material that we found as well as playing a full role on the survey teams. Finally, of the long-serving members of the survey teams, thanks are due to Eddie Moth, another Sheffield graduate, who first worked with me in Pakistan in 1982. Eddie is remembered as unflappable, as having a bizarre and often devastating sense of humour, and for being a very patient and careful field worker and surveyor. Dr. Rogan Jenkinson, then at Creswell Crags Visitor Centre, did a superb job in the preliminary cataloguing of the fossil material, and was another excellent field companion.

The team members mentioned above were the backbone of the field surveys the Pabbi Hills. Others came for shorter periods, but made their own valuable contributions. Dr. Pat Carter (Cambridge) – sadly now deceased - gave us his own insights from his long experience of African prehistory when he visited the project one field season. Dr. Sue Stallibrass (Sheffield, now Liverpool) did a fine job in 1987 in helping to excavate locality 73. Andy Palmer was another that year who did excellent work. Peter Whybrow, Natural History Museum, London, joined us in the Pabbi Hills on two occasions, did sterling work at site 73, helped on the survey, and contributed his own brand of humour; sadly, he died at the time of writing. Gillian Comerford, also from the Natural History Museum helped in 1990 in the excavation of 642, and did so as professionally as one would expect of that organisation. In 1989 and 1990 we were joined by Dr. Mark Beech (Sheffield, now Abu Dhabi), who did a superb job excavating localities 362 and 642 with M. Anwar. Two other excellent Sheffield graduates, Graham McElearney and Ian Smith, and Mike James (Southampton) were other new members in 1990, and each did a great job under difficult conditions. Dr. Ros Coard (Sheffield, now Lampeter) came out in 1990 and continued her involvement with the project by her Ph.D. study of the bovid material from 642, and the taphonomic aspects of the three fossil localities that were excavated. Dr. Andrew Chamberlain (Sheffield) is thanked for advice on programming dBase III, and for helping identify some of the fossil material in Islamabad in 1992. Finally, Dr. Jim Williams (Sheffield), who joined myself and my wife on our last field season in the Pabbi Hills in 1999, is thanked for being as excellent as previous members of the field team.

During the fieldwork, various junior officers from the GSP were seconded to the project. The first was Asif Jah, who later went to Cambridge to study for a Ph.D. on the fluvial sedimentology of the Pabbi Hills. Asif had a mischievous sense of humour, was a good companion in the field, and often very perceptive in his field observations. Tragically, he died in Cambridge of an asthsma attack whilst completing his Ph.D. After Asif departed for Cambridge, we were joined by M. Anwar. He reminded us of a large, protective bear, and he looked after us well. He was also a very competent excavator, and proved an excellent partner with Mark Beech in their investigations of sites 362 and 642. Malik Sikander and Bahadur Khan, two other GSP field officers who helped us in the field between 1986 and 1990 are also thanked. In 1998, in a smaller and final season of fieldwork, by Dr. Imran Khan who proved to be a good field companion as well as an excellent fluvial sedimentologist.

All of us who took part in the fieldwork look back with many happy memories to the services given by the cooks and drivers of the GSP. Most respected of these in the GSP was the late Munsif Khan, or "Chacha" (or "grandfather") Khan. He was inseparable from his beloved and battered Land-Rover, unimpeachably honest, and a very fine man. Another much-liked driver was Javed (Archaeology Dept.), who drove his jeep on tarmac like a fighter pilot. Field teams, like armies, march on their stomachs, and we were always fortunate with the GSP cooks. Amin was larger than life, a wonderful cook, camp manager, negotiator, comedian and handyman; Ulum-Din, Seftar and Chacha Sherif were also brilliant in making camp life as pleasant as possible, and, when required, for producing exquisite and elaborate dinners when we had to entertain visitors.

One of the nicest aspects of conducting fieldwork in the Pabbi Hills was the way we were able to interact as part of a small but very active Siwalik research group. Our American colleagues were always very generous hosts, and we greatly enjoyed the mid-season break that some of us were able to spend at their field camp at Chinji. Catherine Badgley, John Barry, Kay Behrensmeyer, and Jay Quade were fine field companions, and we were delighted that CB, AB and JQ were able to visit us in the field. The Dutch group at Mangla-Samwal were also valued as friends and colleagues, especially Paul Sondaar, Jon de Vos and Gert van den Bergh. Dr. Peter Friend (and Asif Jah's supervisor) from Cambridge fully lived up to his name whenever we met in Pakistan. Julie Hutt, one of Peter's students, is remembered and thanked for her comments on the sedimentology around locality 642.

The completion of this volume would not have been possible without the contribution of a number of people. The obvious ones are those who have contributed text to it. Derek Roe (Oxford) is thanked for paying such critical attention to the first two chapters, which did much to clarify my thinking and writing about the wider context of this work. Andy Currant (Natural History Museum, London) was always pleased to help and advise on some of the more perplexing identifications. I owe a great deal to Mike Rouillard (Dept. of Archaeology, Exeter), who demystified for me the workings of Adobe Photoshop Illustrator, and encouraged me to learn the techniques of computerised illustrations. Without this stimulus, the illustrations for this publication would have taken far longer. Thanks also to Rob Craigie and Shane Earles (Sheffield) for technical help . Finally, there are various academic friends who have been a constant source of encouragement and healthy debate over the years, and they too are thanked, notably Clive Gamble (Southampton), Naama Goren-Inbar, Jerusalem), Wil Roebroeks (Leiden), and Nic Rolland (Victoria, Canada).

Introduction

This volume is the second on the fieldwork conducted by the British Archaeological Mission to Pakistan as part of its Palaeolithic and Pleistocene research programme in the period 1986-1990 in collaboration with the Dept. of Archaeology, Govt. of Pakistan, and the Geological Survey of Pakistan (GSP). The first (Rendell, Dennell and Halim 1989) focused on the Soan Valley, where de Terra and Paterson (1939) had conducted their fieldwork in 1935. They had concluded that the Soan Culture - a simple flake tradition analogous to the Clactonian of southern England - constituted the earliest palaeolithic of the region from the Middle Pleistocene onwards. Our first volume substantially re-evaluated their conclusions on both the geological and archaeological sequence of that area, and also presented the results of our own field observations, mainly compiled in 1981-2. These included a study of the post-Siwalik, Upper Pleistocene loess deposits of the Potwar Plateau (of which the Soan Valley is a part); the excavation of an open-air site near Riwat dated by TL to 45 kya; and the dating and description of a set of artefacts from a conglomerate horizon near Riwat that was dated to a minimum of 1.9 Mya. The latter, not surprisingly, proved highly controversial when first announced. Rendell's investigations of the Soan Syncline continued through the mid-1980's, and the previous volume included the results of a detailed survey of this important artefact-bearing horizon in 1988.

This volume focuses on the second area where the British Mission undertook fieldwork between 1986-1990 and 1999 - the Pabbi Hills, between the river Jhelum and the small town of Kharian. The broad aims here were the same as in the Soan Valley: to find evidence for early hominid occupation and to place this in its environmental and chronological context. Aspects of this work have been published in several interim publications (see e.g. Ashfaque and ul-Haq 1988, 1991; Dennell and Allchin 1988; Dennell 1989, 1991a, 1991b, 1993, 1995, 1999; Dennell et al. 1990, 1992, Dennell and Hurcombe 1999, Hurcombe and Dennell 1992, Jenkinson et al. 1989), but here, the detailed results of the field surveys and excavations are presented for the first time.

The decision to concentrate on the Jhelum basin after 1983 was logical for several reasons. Middle and Lower Pleistocene exposures in the Soan Valley are not only very limited, but also very poor in fossil remains: there was little further prospect of sustained, large-scale investigations of early Pleistocene strata, and little prospect of finding numerous fossil vertebrate remains or archaeological sites in that area. The Jhelum area looked much more promising, as a wide range of vertebrate fossil remains had been collected from often long sedimentary sequences that collectively spanned the Miocene to Mid-Pleistocene, and these had not long before been dated palaeomagnetically by joint Pakistani - American teams in the 1970's. Their work is of inestimable importance, first in making the Upper Siwaliks of the Jhelum basin one of the best-calibrated terrestrial sequences of its age range in the world, and secondly, in providing us with a starting point for our own fieldwork.

In 1983, four of us, accompanied by M. Halim (Dept. of Archaeology and Museums) as Field Officer, conducted a brief 10-day reconnaissance of the Jhelum basin, based in the town of Jhelum. The main results were the discovery of a hand-axe in a conglomerate exposure at Dinar, and of two more in a sandstone exposure at Jalalpur (Rendell and Dennell 1985). Both could be dated to the early Middle Pleistocene; i.e. ca. 500-700 kya. (These are still among the earliest well-dated bifaces in the Indian subcontinent; in a wider continental context, they also indicate that hand-axes were being used in the subcontinent at around the same time they first appeared in western Europe). In addition, that field season also resulted in the discovery of cruder-looking artefacts in the lower conglomerate horizon at Riwat. As related in the previous volume, the age of this horizon was unknown until the necessary palaeomagnetic work was undertaken by Rendell, but was later dated to a minimum of 1.9 Mya. Finally, some vertebrate fossil remains were also noted on erosional surfaces at Dina, Kas Guma, Jalalpur and Rohtas.

The reconnaissance of 1983 therefore showed the possibility of finding artefacts in Middle Pleistocene (and possibly earlier) contexts. The presence of fossil remains in the Jhelum area also opened up prospects of conducting bio-stratigraphic investigations, as well as the exciting but remote possibility of finding hominid remains, and possibly also lower palaeolithic sites. However, such objectives would clearly require longer field seasons and larger teams.

A beginning was made in 1985, when a team of six from the British Mission and two Pakistani field officers spent six weeks in late March and April in the areas of Rohtas, Mirpur and Jalalpur. This season was also the first in which the Mission collaborated directly with the Geological Survey of Pakistan (GSP), which provided a Land-Rover, driver, cook and field officer, as well as invaluable local knowledge and access to Government Rest Houses. From the viewpoint of the GSP, our work provided a means whereby they could obtain Upper Siwalik fossil vertebrate remains, and in time, a fossil reference collection. (Pakistan is in the odd position of having the finest Siwalik exposures, and one of the longest-investigated Neogene fossil sequences in the world, yet lacking major fossil collections of its own. Most Siwalik fossils found before Independence in what is now Pakistan were either removed to western museums before Independence, or remained in India, which retained custody of the India Geological Survey's collections after Partition in 1947).

The 1985 field season was a useful introduction to the exigencies of successful fossil collecting in this part of the world. It revealed the logistic problems of finding and maintaining a suitably located field-base, the problems of conducting a field survey in poorly mapped, deeply dissected and often densely vegetated terrain, and the necessity of ending fieldwork before the onset of the hot season in early April, when temperatures routinely reach 40°C. As far as results are concerned, a large amount of vertebrate fossil material was found but most was very fragmented and poorly preserved. Diagnostic elements tended to derive from large animals such as elephant, and/or comprised teeth, the most durable part of the skeleton. The density of fossil remains across erosional surfaces was usually very low, and they were found in significant localised amounts in only a few places, chiefly at Rohtas and Jalalpur. Although the results were disappointing, the experience gained led to the devising of a field recording system was devised in the light of the 1985 survey that took account of the spatial patterning of fossil remains across the landscapes as well as their geological age. This recording system - described later in the volume - worked well, and was used throughout the five field seasons in the Pabbi Hills between 1986 and 1990.

Subsequent reflection after that field season led to a number of important decisions. The first was to spend more time in the field. Given the high cost of air fares to and from Pakistan (relative to the field budget), the comparatively low field costs once there, and the fact that the Mission had two vehicles in Islamabad that would otherwise be idle, it made sense to keep people in the field as long as possible. In any case, field survey of this nature rarely yields quick results, especially if its objectives include the discovery of items as rare as hominid remains and early palaeolithic sites. The second decision was to complete any year's fieldwork by the end of March, i.e. before the onset of the hot season. A third was to locate the team's field-base, however basic, within a maximum of a 30-minute drive from the area under study (previous field bases had often been over an hour's driving away).

The final decision, to work in the Pabbi Hills the following year, was a gamble, as the team had not had time to visit it previously. However, Keller et al. (1977) and Opdyke et al. (1979) had shown that not only did this anticline have a kilometre-thick sequence (!) from the late Pliocene to the Mid-Pleistocene, but it was also by far the most productive of the areas they had visited in terms of fossil vertebrate remains. The value of their work to our project cannot be overstated, as they showed clearly the age of the deposits, and their palaeontological potential.

Consequently, Dennell and Rendell were allowed by their respective universities to take three months leave without pay to undertake a ten-week field survey from January to March in 1986. The risk of undertaking a long field season in an unknown area paid off handsomely, albeit after a nervous opening period in which almost no fossils were found. By the end of that season, however, some 12,000 specimens (including undiagnostics) had been found from over 150 places, and one extremely rich fossil locality (site 73) had also been partially excavated. The success of the 1986 field season led to three further long field surveys in the Pabbi Hills: another three- month one in spring, 1987 (when Dennell had sabbatical leave); four out of eight weeks in 1988 (the remainder being spent in the Soan Valley and consolidating fossil material at Taxila); and three more months in both 1989 and 1990 (whilst Dennell was on leave from his university on a Senior Research Fellowship). Each of these field seasons is described in greater detail in the relevant chapters of this book. A final season was planned for the spring of 1991, but had to be cancelled because of the outbreak of the First Gulf War.

During each field season, systematic attempts were made not only to collect stone artefacts from fossil-bearing erosional surfaces, but also to find examples in secure geological contexts. The results are described by Dr. Linda Hurcombe, later in this volume. The palaeomagnetic investigations are summarised by Professor Helen Rendell, the longest-serving member of the project. During the course of the field surveys, some exceptionally rich fossil localities were found. Three of these were excavated: site 73 in 1986 and 1987; site 362 in 1989; and site 642 (by far the largest) in 1990. The results are presented by three of the team, Drs. Mark Beech and Ros Coard, both then at Sheffield University, and M. Anwar from the GSP.

The fossil remains from the surveys constitute the main body of evidence, and several chapters are devoted to them. A few preliminary points need to be made here about this material. First, it constitutes a very large body of data. By the end of the 1990 field season, over 40,000 specimens (including undiagnostic fragments) had been found in the Pabbi Hills from over 600 places. On survey, every piece of fossil that was found (with the exception of dense concentrations of ostrich shell and badly fragmented elephant tusk) was collected. Back at the field base, the number of fossils from each fossil occurrence was counted; each piece was cleaned; and any specimen considered diagnostic to skeletal part and/or taxon was marked for later cataloguing. The publication of this material has had to attempt a balance between on the one hand presenting as detailed an account as possible of the location and contents of each fossil occurrence, and on the other, not burying the reader under a mountain of data. In particular, the information in the field notes on where material was found, and the computer files on the fossil remains from each fossil occurrence have been appropriately (and laboriously) compressed. Nevertheless, every effort has been made to present as complete a record as possible of what was found where, and with what. Given the lack of detailed "hard" data on the vertebrate fossil content of the Upper Siwaliks (and Early Pleistocene in Asia generally), it is hoped that this report can be used by future scholars interested in the Upper Siwaliks in particular, but also the taphonomic aspects of fluvial assemblages in general.

We had to face two major constraints throughout this fieldwork during the 1980's. The first concerned the recording of where material was found. Except for a brief period in 1987, we were not able to obtain air photographs of the Pabbi Hills because of various security restrictions, nor could we then afford satellite imagery. The only maps we could use in the field were photocopies of the reprints of the 1908 1": 1 mile topographical series of the (then) India Survey. Whilst admirable, they were of little help in delineating the details of the deeply dissected gully formations in which we had mostly to work. In any case, such is the rate of erosion in the Pabbi Hills that many of these have probably changed substantially since the maps were made. During the fieldwork, we relied on our own sketch maps, which were usually sufficient to enable us to re-locate major fossil occurrences. It was not until the last field season, in 1999, that we were able to work in the Pabbi Hills with satellite imagery and a GPS navigation system, and record accurately fossil localities in terms of their precise location.

The second constraint concerned the identification of fossil material. As mentioned earlier, one of the objectives of our work was the creation of a fossil reference collection for the GSP. This meant, of course, there was no reference collection that we ourselves could use in Pakistan. Additionally, the identification of Upper Siwalik vertebrates is not only a highly specialised field, but also one in need of substantial revision and refinement. Consequently, and as will be seen in the tabulated data, the identifications are often at a family level. The main groups affected are the herbivores, particularly the cervids and suids. Exceptions are the carnivore remains, studied by Professor Alan Turner, whose report is included in this volume, and the bovids, thanks to the investigations of Dr. Ros Coard. Now that the best specimens have been formed into a fossil reference collection at the Geological Survey of Pakistan, specialists should be able to examine this material in more detail at a later date.

Even arriving at a catalogue of what was found at each fossil occurrence was a daunting task. Each field season,a preliminary field catalogue was compiled by Dr. Rogan Jenkinson from the Creswell Crags Field Centre near Sheffield. He worked under very difficult conditions, invariably for long hours, trying to keep pace with the rate of discovery, and often in poor light. Because of his responsibilities in the U.K., he could not take part in the entire field season, and so had to process an enormous amount of material very quickly. The rechecking and completion of the fossil catalogue were done by Dennell in greater comfort at GSP, Islamabad, in the autumn of 1991, and during a further six months in 1992. In the process of rechecking all the material, he was assisted by Ros Coard (then a Ph.D. student at Sheffield University) and Dr. Andrew Chamberlain, also from Sheffield. We were also allowed by the GSP to export on a temporary basis the main carnivore specimens and some other material for identification in the U.K. This work resulted in numerous additions and revisions to Jenkinson's preliminary catalogue, as well as the creation of a reference collection of Upper Siwalik material for the Geological Survey of Pakistan. This is now available for study by any *bona fide* scholar, and it is hoped that this collection will be one of the more valuable permanent results of the Mission's collaboration with the Geological Survey of Pakistan. It also goes a small way to returning the enormous goodwill, support and friendliness that all their staff showed over the years.

Future historians of archaeological fieldwork might note the problems encountered when working in a period of rapid technological change. Jenkinson's original catalogue was on Reflex, and stored on 100 kb. 5.25" floppy discs; Dennell converted these to Dbase III prior to rechecking the identifications, and acquired access to computers with (eventually) sufficient hard disc storage. The conversion of the Dbase files into text was initially done in Word (3, then 5), and next Windows 2, 5, and finally 6.

Finally, a short comment is in order as to why this volume has been delayed longer than any of its contributors would have wished. In autumn, 1993, Dennell returned from his Senior Research Fellowship to resume full-time employment at his Department. On his return, he was made Head of the Undergraduate School, and thus responsible for over-seeing some 400 students, as well as fulfilling his own teaching requirements. His return also coincided with the introduction at Sheffield of semesterisation and modularisation, and all its consequences, including a heavy administrative burden. An administrative interlude followed until 1998, when he served as Head of Department until 2002, during which period there was very little time for sustained work on this monograph. Following that, an additional year was required for the illustrations and formatting. It is thus a great relief to realise that the research begun so long ago, and so often interrupted, will finally now be published.

Part 1: Background to the Fieldwork in the Pabbi Hills

Chapter 1: The Palaeoanthropological Background to the Investigations in the Pabbi Hills

Introduction

The fieldwork in the Pabbi Hills set out to test a model of human origins in the Old World that was prevalent in the 1980's and early 1990's. According to this model, hominins in general, and our own genus *Homo* in particular, originated in East Africa. By 2 Mya ago, the earliest forms of *Homo*, such as *Homo habilis*, had learnt how to make simple stone tools, and were scavenging or hunting animals for their meat and bone marrow. By at least 1.6 Mya ago, an early form of *Homo* had evolved into (or been replaced by) *Homo erectus*, which had a larger brain, and was bipedal in much the same way as ourselves. By 1.5 Mya ago, *H. erectus* was using skilfully-made bifacial "hand-axes" (often called Acheulean, after the place where they were first found in France), and soon after, by 1 Mya ago, it was the only type of hominin in Africa. This model further postulated that at some time around a million or so years ago, some populations of *H. erectus* left Africa and entered Asia, taking with them their hand-axes and Acheulean technology (see Figure 1.1). They rapidly spread across Asia, and eventually reached Indonesia, where their remains were first discovered in the 1890's. One puzzling feature of this process of colonisation was that the Acheulean technology was used by *Homo erectus* as far east as India, but not in Central Asia, China, S.E. Asia and Indonesia, where simpler flake and core technologies were used instead. This problem aside, *Homo erectus* was regarded as the first hominin to inhabit two, and possibly three continents. Europe was thought to have been colonised later, between 1 and 0.5 Mya ago, by populations of (depending on one's viewpoint) a late form of *H. erectus*, an early form of *Homo sapiens*, or a distinctive type of hominin called *Homo heidelbergensis* . Whichever, these first Europeans used both an Acheulean, bifacial technology, and simpler, flake and core assemblages.

Figure 1.1: The dispersal of *Homo erectus* out of Africa and into Asia and Europe, as commonly portrayed in the 1980's and early 1990's. (This version is based on Foley 1987, Figure 10.1a). Recent research suggests that *Homo erectus* had dispersed across southern Asia much earlier, by or even before 1.8 Mya. The Movius Line demarcates Acheulean assemblages to the west from simpler flake and core assemblages to the east.

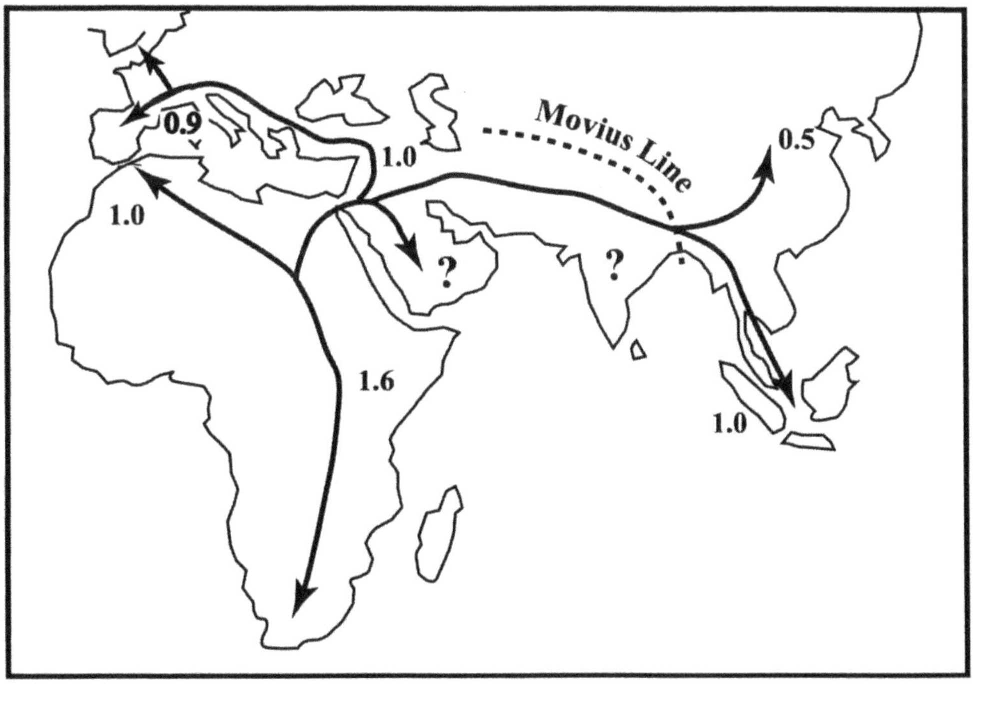

Although plausible, and consistent with the data then available, the model of human origins outlined above left unanswered several important questions. Why did hominins remain in Africa so long? Why did *H. erectus* remain in Africa for so long after it and an Acheulean technology had first appeared? Why did it not leave Africa before the deserts of the Sahara and South West Asia effectively isolated Africa from Asia? Once proto-humans had reached Asia, why were Acheulean bifacial assemblages not used in the Far East? Finally, why was Europe not colonised until much later than eastern Asia?

Since 1994, a new synthesis has emerged which suggests that hominins may have left Africa much earlier, by at least

1.8 Mya ago (see e.g. Etler 1998; Larick and Ciochon 1996, Tattersall 1997). According to this new model, the East African successor of *Homo habilis* was *Homo ergaster* rather than *Homo erectus*, and it was *Homo ergaster* that first left Africa, taking with it an Oldowan-type technology long before the Acheulean was developed. Having rapidly crossed Asia, *H. ergaster* is thought to have reached China and Indonesia ca. 1.8 Mya ago, where it developed into *Homo erectus*. (What the terms *Homo ergaster* and *Homo erectus* signify to different researchers is discussed more fully in section 2.1.1 below). Later, around 1.4 Mya ago, a second wave of *H. ergaster* may have left Africa, this time with an Acheulean technology, which was used in S.W. and S. Asia as far east as India. Europe is still an enigma: whilst some maintain that at least parts of it may have been first occupied as early as 1.8 Mya, the earliest widely-accepted evidence for the first Europeans stems from only ca. 1.0 Mya years ago (e.g. Carbonell et al. 1995a and Oms et al. 2000 for Spain; and Ascenzi et al.1996 and Milliken 1999 for Italy).

Curiously, the earliest Europeans do not seem to have used an Acheulean technology, even though it had been in use for several hundreds of thousands of years in Africa and S.W.Asia; even after 500,000 years ago, simpler flake assemblages were still used over much of Europe alongside the Acheulean.

This model differs profoundly from its predecessor in three crucial respects: first it argues that the genus *Homo* colonised Asia much earlier than first thought; secondly, *H. ergaster* is seen as the first hominin outside Africa; and thirdly, *H. erectus* is regarded as a lineage that is confined to eastern and south-eastern Asia, and which eventually became extinct.

The new model has several advantages over its predecessor. First, it avoids the problem of explaining why the genus *Homo* remained in Africa long after it could have left; and secondly, it neatly explains the absence of the Acheulean in S.E. Asia by arguing that hominins had arrived there long before the Acheulean is supposed to have been developed in Africa. It also accommodates the suspicions of some palaeoanthropologists that the Asian and African specimens of *H. erectus* are in fact two distinct taxa. Europe, however, becomes even more problematic than before, as hominins may have ignored it for a million rather than a few hundred thousand years.

Like its predecessor, the current consensus incorporates evidence that is often complex, ambiguous, and varying in reliability. Before we consider this evidence, we need to examine some of the basic assumptions concerning the origin and early diffusion of hominins across Eurasia (meaning both Europe and Asia, rather than just the vague area where they join). Both the older, "short" chronology for early hominins in Eurasia, and the newer "long" chronology share two key assumptions: first, that hominins, including our own genus *Homo*, originated in East Africa, and secondly, that *Homo erectus* or *H. ergaster* (depending on how one classifies the relevant African specimens) was the first hominin to leave Africa and colonise Eurasia. Each assumption merits careful evaluation.

1) An East African genesis?

The assumption that hominins (including the genus *Homo*) evolved in East Africa is often treated these days as an incontrovertible, self-evident truth: East Africa has by far the longest and most detailed sequence of fossil hominins and stone tools; it contains the grasslands and open woodlands in which hominins evolved; and our closest relatives, the chimpanzee and gorilla, are still found in the region. It is worth remembering, however, that in the first half of this century, it seemed just as certain that hominins originated in Central Asia: what is self-evident for one generation is rarely so for another.

Although currently robust and stoutly defended, the argument that hominins originated in East Africa is vulnerable on three counts.

1.1: How well-documented is the East African fossil hominin record?

Since Leakey's discovery of *Zinjanthropus boisei* and *Homo habilis* at Olduvai in 1959 and 1960, most key discoveries of fossil hominins have occurred in East Africa. Today, East Africa boasts the longest, largest, best-dated and most diverse record of fossil hominins in the world. Faced with this mass of evidence, it is therefore unsurprising that East Africa is widely regarded as "the cradle of mankind".

Nevertheless, the evidence that hominins originated in East Africa remains very meagre: for the 10 million years before 5-6 Mya, there are the remains of perhaps only 11 individuals, and these are derived from a small part of Kenya and Ethiopia, representing ca. 0.1% of the African continent (Hill and Ward 1988: 52). (Recent discoveries will not have greatly changed these statistics). Of the three million years between 1 and 4 Mya ago, "about one-half of that time remains virtually undocumented" (Kimbel 1995: 435) so far as human evolution is concerned. The fact that new types of hominin are still being discovered or recognised in East Africa (such as *Ardipithecus ramidus* (White et al. 1994, 1995), *Australopithecus anamensis* (Leakey et al. 1995), *Australopithecus garhi* (Asfaw et al. 1999), *Homo rudolfensis* (Wood 1999), and *Kenyanthropus platyops* (Leakey et al. 2001) shows that we have not yet reached the position of acquiring only redundant data on human origins from that region. Although the East African fossil hominin record is impressive after 4 Mya, the argument that hominins originated in East Africa "is based entirely on the fact that comparably aged sediments are either absent from or have not yet been located in southern, central and western Africa" (Brunet et al. 1995: 274). When these are found and investigated, we might reasonably expect East Africa to lose some of its current pre-eminence. The recent discovery of *Sahelanthropus tchadensis*, 6-7 Mya-old from Chad, and possibly the oldest hominin yet found (Brunet et al. 2002) is perhaps a further indication that that hominins did not originate in the Rift Valley.

1.2: Where were those grasslands in which hominins evolved?

One of the most persuasive ideas in palaeoanthropology has been that climatic change has driven human evolution by creating the grasslands in which hominins evolved as bipedal, intelligent, tool-making social carnivores. An early expression of this grand theme was Matthew's (1915) *Climate and Evolution*, in which he argued that tectonic uplift in Central Asia during the Neogene created a drier and harsher climate, and this in turn forced the emergence of new species (including hominins), which radiated outwards, displacing earlier and simpler forms. *Contra* Darwin, the existence of gorilla and chimpanzee in Africa indicated the backwardness of Africa, as representing primitive forms displaced by more advanced types that had originated in Central Asia. When *Australopithecus africanus* was later found at the southern end of Africa, it seemed perfectly placed and configured to exemplify a primitive type of hominoid that had been displaced to a peripheral region. Although the location of human origins has been relocated from Central Asia to East Africa since Matthew's time, the arguments have remained much the same. Global cooling and drying of the world's climates around 2.5 Mya ago, it is argued, led to a reduction in forests and woodlands, and a concomitant expansion of grasslands. This in turn caused the extinction of many taxa, and their replacement by ones better adapted to cooler and drier conditions. Vrba in particular has argued in several publications (e.g. Vrba, 1995, 1996) that bovids, micromammals and hominins all share the same macro-evolutionary patterns, with a major "faunal turnover pulse" of grassland-adapted species after 2.5 Mya in response to global cooling. On this model, the emergence of larger-brained, fully bipedal, tool-making hominins that were adapted to the grasslands of East Africa after 2.5 Mya is often seen as the critical "package" of anatomical and behavioural changes that eventually enabled *Homo* to move beyond the East African Rift Valley, and to colonise Asia.

Vrba's papers are rightly recognised as being among the most influential contributions in palaeoanthropology in the 1980's. Whilst no-one doubts as a general premise that early forms of *Homo* emerged against a background of global cooling as well as increased aridity and seasonality, there are substantial doubts as to whether causal links can be established (even leaving aside doubts that 2.5 Mya marks a particularly sharp climatic boundary (see Kimbel 1995: 435)). There are two aspects to these problems. The first is regional, and thus focuses on East Africa as the best data set available; the other is continental, and concerns the antiquity and extent of grasslands before global cooling got under way after 2.5 Mya.

1.2.1. Hominins, grasslands and East Africa

Whilst Vrba's general model is plausible in principle, the East African record does not give it strong support. Cerling (1992: 244), for example, suggests from soil carbonate analyses that before 1.7 Mya ago, the main fossil hominins localities may have been "somewhat open [but] nothing like the grasslands or wooded grasslands of East Africa today". White (1995) argues that there are no clear correlations between global climatic change and the evolution of pigs and hominins. Indeed, most speciations and extinctions occurred between 1.6 and 2.0 Mya, during which time there were no clear global climatic changes. Kimbel (1995) also concludes that whilst hominin evolution may have been forced by climatic change, the fossil data available from Africa are inadequate to confirm or refute this suggestion. In a larger scale study, Behrensmeyer et al. (1997: 1593) argue that "late Pliocene evolution in East Africa was affected by the cumulative ecological consequences of cooler, drier, and more variable climatic conditions rather than by a sudden change toward more open habitats". As East Africa has the best-studied African sequence, this finding "weakens the case for rapid climatic forcing of continent-scale ecological change and faunal turnover" (ibid. 1593). Additionally, the East African hominin localities between 2.5 and 1.6 Mya do not now appear to have been open environments. Various authors (e.g. Kappelman et al. 1997, Reed 1997, Spencer 1997) have shown from large mammal remains that both Australopithecines and *Homo habilis* lived in woodland environments, and Fernandez-Jalvo et al. (1998) have shown from micromammalian evidence that Olduvai in Bed 1 times, ca. 1.8 Mya, was far more densely wooded than today. This is consistent with anatomical evidence that *Homo habilis* retained arboreal capabilities as late as 1.8 Mya (Susman and Stern 1982). Overall, there is little indication that East African hominins were habitually living in grasslands before 1.6 Mya.

1.2.2: Hominins and grasslands before 2.5 Mya ago

Even if the East African evidence indicates that hominins before 1.8 Mya were living in wooded environments, grasslands were nevertheless present in the region, and indeed were very widespread across Africa and Asia. The oldest grasslands are probably those of South Asia, where soil carbonate analyses indicate that they may have originated as early as 7.0 - 7.4 Mya ago (Quade et al. 1989, Quade and Cerling 1995). Grasslands were certainly widespread by 3 Mya. Recent reconstructions of global climate during the Pliocene indicate that they were far more extensive in Africa and Asia 3.0 Mya ago than today (Dowsett et al. 1994, 1999). According to their reconstruction (see Figure 1.2), it would have been possible to remain within a grassland environment almost anywhere within an enormous area between the Atlantic Ocean and North-East China, and between Central Asia and the Cape of Good Hope. Most significantly of all, *the desert barrier between Africa and Asia did not then exist.*

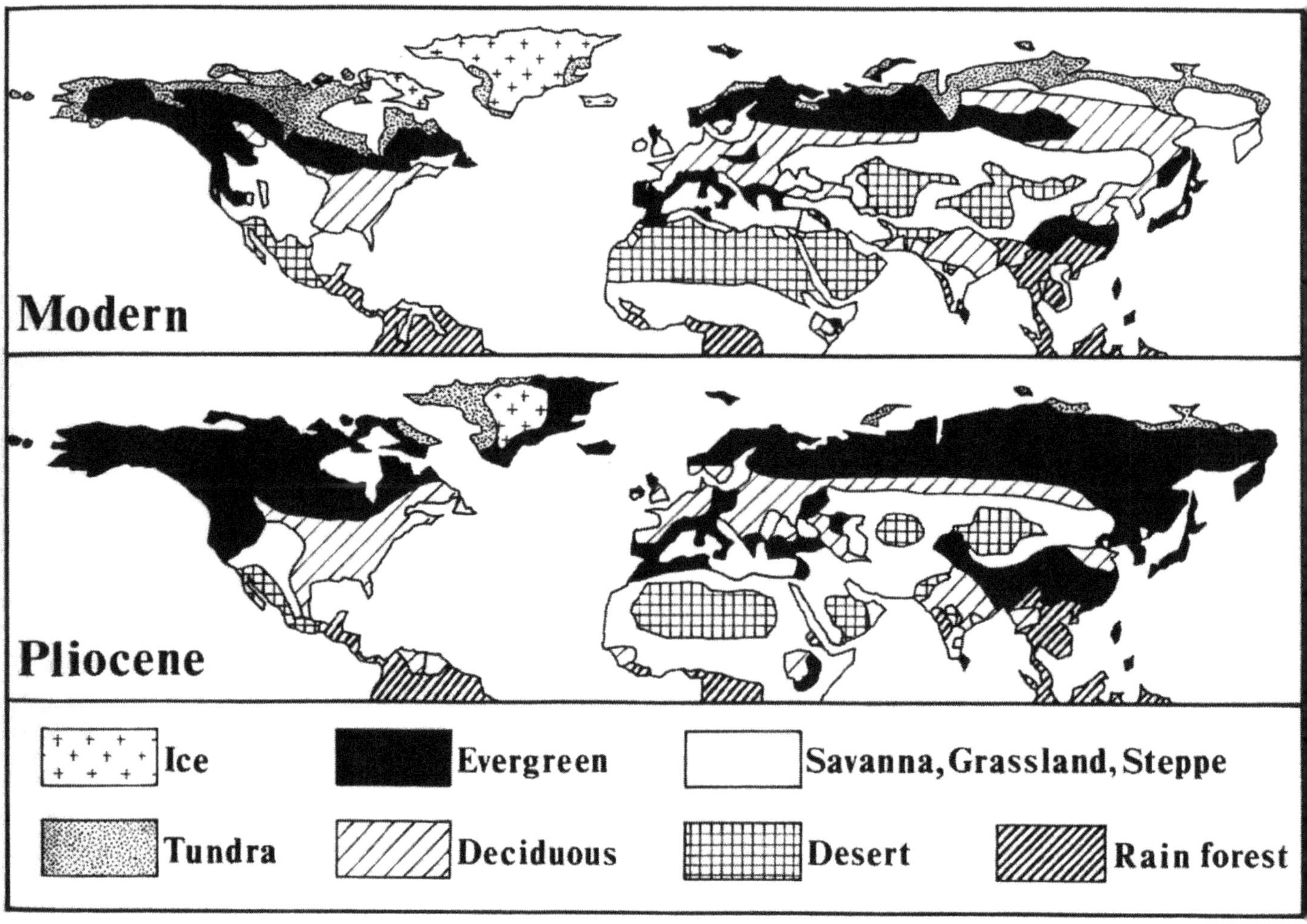

Figure 1.2: Grasslands in the northern hemisphere now and in the Late Pliocene. As indicated, grasslands were much more extensive than now between Africa and Asia, and there was no desert barrier between the two continents that might have impeded hominin dispersals. Early Pleistocene conditions were broadly similar to those of the Late Pliocene.

Source: Dowsett et al. 1994, Figure 11 (Reproduced with kind permission of the publishers).

1.3. When did African hominins first live beyond the East African Rift Valley?

Three recent discoveries indicate that hominins existed well beyond the East African Rift Valley by at least 3 Mya ago. Two are from Chad, 1,500 miles west of the Rift Valley: the 6-7 Mya-old find of *Sahelanthropus tchadensis*, 6-7 Mya-old, and the 3.5 Mya-old specimen of *A. bahrelghazali* (Brunet et al. 1995). As Brunet al. (1995: 274) point out, this particular discovery implies that by 3.5 Mya ago "hominids were distributed throughout the woodland and savannah belt from the Atlantic Ocean across the Sahel through eastern Africa to the Cape of Good Hope". The third recent discovery is the nearly complete, 4 Mya-old *Australopithecus* skeleton (StW 573) from Sterkfontein (Partridge et al. 2003). Another recent discovery, this time of *H. rudolfensis* in Malawi (Schrenk et al. 1993), shows that early *Homo* may have lived along much of the African Rift Valley in the late Pliocene, and additionally demonstrates our uncertainties over how much of Africa was occupied by hominins at this time, and by which types. Again, one should raise the point about distance: if *H. rudolfensis* was living 1,000 miles south of Koobi Fora, where it is better known, it could also have been living the same distance north, in S.W.Asia.

The discovery of *A. bahrelghazali* in particular has profound implications for palaeoanthropology. If hominins were already 1,500 miles west of the Rift Valley 3.5 Mya ago, there is no obvious reason why they were not in Israel, the same distance to the north; and indeed, there was no obvious barrier preventing them from occupying the Asian grasslands as far east as China. To repeat, but to develop further, the argument of Brunet et al. (1995: 274), "if the origins of hominids occurred rapidly, followed by rapid range extension, as seems likely, it may be as futile to seek a specific and localised place of origin for hominids as it is for any other group". What our currently African-dominated fossil hominin record may therefore indicate is merely the fortunate outcome of good preservation, fortuitous re-exposure, intensive fieldwork, and luck.

1.4. What does the distribution of chimpanzees and gorillas tell us?

Many writers have repeated Darwin's (1871) famous suggestion that "it is somewhat more probable that our early progenitors lived on the African continent than elsewhere". As Hill and Ward (1988: 50) point out, fewer have cited the continuation of this quotation: "But it is useless to speculate on this subject.....since so remote a period the earth has certainly undergone many great revolutions, and there has been ample time for migration on the largest scale". Two further points are worth making. First, the gorilla and chimpanzee lack a fossil record, so their previous whereabouts (and anatomy) remain, as in Darwin's day, matters of conjecture. (The only possible exception is a chimpanzee-like tibia from Sterkfontein, South Africa, that is currently attributed to *Australopithecus* (Berger and Tobias 1996). The second point is that the orang-utan - the only extant ape other than ourselves that appears to have a fossil record - should make us wary about assuming too readily that modern apes have remained in the same region and environment as their ancestors. Today, orang-utans live in Borneo and Sarawak, in high canopy rain forest, and spend little time on the ground. They are thought to have descended from the Miocene *Ramapithecus* or *Sivapithecus* clade, which is known from China, India, Pakistan and Turkey (Andrews and Cronin 1982). This hominoid probably inhabited open woodland, and may have been partly terrestrial. (A new discovery of *Khoratpithecus*, a late Miocene specimen from Thailand, may prove to be a more credible ancestor (Yaowalak Chaimanee et al. 2004); interestingly, it lived in dense woodland). During the Middle and Upper Pleistocene, orang-utans lived in South China (de Vos 1984), and possibly in Vietnam until the late Pleistocene (Bacon and Long 2001). The orang-utan's complex evolutionary history might caution us against assuming that the ancestors of gorillas and chimps lived in the same areas or types of environments as now, that they resembled their modern descendants, or that they behaved in the same manner.

2. Out of Africa?

Before examining the assumption that *Homo erectus* or *ergaster* was the first hominin to inhabit Asia, a small digression is necessary on what these terms mean.

2.1: *Homo erectus* and *H. ergaster*: a taxonomic digression

H. erectus - or more properly, *Pithecanthropus erectus* - was first discovered at Trinil in Java by Dubois in 1891. Later discoveries in Indonesia at Sangiran and Modjokerto in the 1930's, and at Chou-kou-tien (now Zhoukoudian) in China were seen as broadly similar, even if initially given their own generic names (such as *Meganthropus* and *Sinanthropus*). In 1950, Ernst Mayr re-classified all this material as *Homo erectus*, with the Trinil specimens as the type fossils. Subsequently, African specimens such as those from Ternifine, Algeria (found in 1954), and at Olduvai (OH9) (found in 1960) were also called *Homo erectus*, as were much later specimens from Europe. *H. erectus* thus became for a while the earliest hominin that was thought to have lived in Asia and Europe, as well as Africa; for authorities such as Rightmire (1991, 1998), it is seen as broadly similar over three continents, and over a million year span.

In the last few years, some specialists have doubted whether the East African specimens should be placed within the same palaeospecies as the Asian ones (see e.g. Qian Wang and Tobias (2001)). In the light of discoveries at Koobi Fora, some have suggested that the earliest African examples should now be called *H. ergaster*, after the name given to KNM-3733, a 1.8 Mya-old specimen from Koobi Fora. This taxon also now includes WT15000, the magnificent 1.6 Mya-old skeleton of a young boy (Walker and Shipman 1996) from Nariokotome that was initially published as *H. erectus* (Brown et al. 1985). Consequently, it is the African *H. ergaster* that is now seen as the hominin that first colonised Asia and formed the founding population of what later became *H. erectus* in China and S.E. Asia (see e.g. Tattersall 1997). Regarding Europe, those specimens once regarded as late examples of *H. erectus* or "archaic *Homo sapiens*" are now increasingly classed under the revived taxon of *H. heidelbergensis*, a term first used to classify the mandible from Mauer, Germany, that was found in 1908. (However, some authorities continue to argue that some European specimens are best classed as *H. erectus*, so the issue is still far from resolved). With this brief history in mind, we can return to the assumption that *H. ergaster* was the first occupant of Asia.

2.2 *Homo ergaster (*or *H. erectus)*: the first hominin outside Africa?

Current controversies over whether *Homo ergaster* (or *Homo erectus*, depending on one's taxonomic choice) left Africa one or two million years ago assume that it was the first hominin to live outside Africa. Most of the arguments about why members of this species were able to colonise Asia invoke their brains or their bodies. If, as now seems likely, they left with a simple Oldowan-type tool-kit and before the development of a bifacial, Acheulean technology, technological proficiency can no longer been seen as the "magic ingredient" (Tattersall 1997: 49) behind the colonisation of Asia. Instead, the "magic ingredient" is supposed to have been *H. ergaster's* ability to walk in a recognisably human form, as well as its thermoregulatory abilities (Wheeler 1991) that enabled it to colonise the grasslands of Asia. According to Tattersall (ibid.), "because the first human diaspora apparently followed hard on the heels of the acquisition of more or less modern body form, it seems reasonable to conclude that the typically human wanderlust emerged in concert with the emancipation of hominids from the forest edges that had been their preferred habitat". Leaving aside the point that equids, elephants, or rodents are not attributed with a "typical wanderlust" in their inter-continental migrations, two major issues are at stake. The first is whether *H. ergaster* was the first hominin to inhabit the grasslands of Africa and Asia; and the second, whether it originated in Africa.

2.2.1: Were the grasslands of Africa and Asia inhabited by hominins before *H. ergaster*?

We have seen above that the late Pliocene grasslands extended from West Africa to north China ca. 3 Mya ago, and those in Africa were occupied by several types of hominin *(e.g. A. afarensis, A. africanus, A, amanensis, A. bahrelghazali* and *Kenyanthropus platyops*). Given the absence of any desert barrier at this time between Africa and Asia, it is hard to see why the Asian grasslands would have remained uninhabited until the appearance of *Homo ergaster* 1.8 Mya ago. Secondly, and as will be seen below, there is evidence from Pakistan that stone tools were being made in Asia *before* the first appearance of *H. ergaster* in Africa; i.e. before 1.8 Mya ago; and also evidence of hominins in Java by 1.8 Mya. Additionally, there is no particular reason why *Homo ergaster* should have been the first tool-using hominin outside Africa. At present, the oldest African stone tools are those from Kadar Gona, Ethiopia, dated to ca. 2.6 Mya (Semaw et al. 1997). As Wood (1997) has suggested, these may have been made by *Paranthropus*, which is regarded by some as a credible maker of the oldest stone tools from Swartkrans, S. Africa (Susman 1988, 1991; but see Marzke (1992) for a more cautious assessment). Stone tools should not therefore be automatically attributed to the genus *Homo* unless there is clear evidence that it was the only resident hominin where they occur. If *Paranthropus* was making stone tools in both East and South Africa, there is no *a priori* reason why it might not have done the same in at least S.W. Asia, especially given the absence of any Saharan barrier at the time between East Africa and S.W. Asia.

2.2.2: Is *Homo erectus/ergaster* East African in origin?

This is probably one of the most controversial issues in palaeoanthropology. If *H. erectus/ergaster* is East African in origin, its most obvious antecedent is *Homo habilis*. The latest appearance of *H. habilis* is probably ca. 1.6 - 1.7 Mya (White 1995: 383); if the earliest specimen of *H. erectus/ergaster* is KNM3733, its first appearance is ca. 1.75 Mya (White ibid.). The origin of *Homo erectus* remains obscure, however. Wood (1984: 99) notes that "the evidence for incipient *H. erectus* traits in *H. habilis* is equivocal". Likewise, Walker and Shipman (1996: 109) see the differences between *H. habilis* and *H. erectus/ergaster* as too great to have developed anagenetically in the time available, and see *H. erectus* as "without a clear ancestor, without a past". In a recent paper, Wood and Collard (1999) suggest that *Homo habilis* and *H. rudolfensis* should be excluded from the genus *Homo*, and that *Homo ergaster* should be regarded as its earliest representative. This scarcely strengthens the case for an East African origin of *Homo ergaster*. White (1995: 383) has been even bolder in suggesting: "It seems more likely that *H. erectus* is an immigrant from Asia to the eastern Africa area than an anagenetic, *in situ* derivative from *Homo habilis*". (A similar suggestion was also made by Dennell, 1995: 24). Much depends upon the relative dating of the earliest specimens of *H. ergaster* in Africa, and *H. erectus* in East Asia. If, as Swisher et al. (1994) argue, the earliest Indonesian specimens of *H. erectus* are actually older than the earliest African specimens of *H. ergaster, H. erectus* in Asia might even have given rise to *H. ergaster* in Africa, and not *vice versa*. If we bear in mind the profound differences between *H. ergaster* and only slightly earlier African forms of *Homo*, such as *Homo habilis*, in terms of brain capacity, body size and proportion, and reduced sexual dimorphism, an origin for *H. ergaster* outside East Africa might not prove so far-fetched as it might now appear. If *H. erectus/ergaster* turned out to be intrusive into East Africa, many of the East African fossil hominins antedating its appearance would of course lose much of their relevance to studies of human evolution.

The future offers therefore a number of interesting possibilities: that hominins may have inhabited the Asian grasslands well before two, or even three, million years ago; that *Homo ergaster/erectus* might not have been the first tool-maker in Asia; and that it might not even have been indigenous to East Africa. Notwithstanding these possibilities, however, the actual situation is that there is no evidence at present of hominins outside Africa before 2.5 Mya. Even now, many palaeoanthropologists would contest at least some of the Asian evidence for hominins before 1.5 Mya ago, and some might even contest all of it. An immediate question to ask is, how genuine is the absence of evidence from Asia?

2.3: Asia, the unknown continent

Both the number and density of well-dated sites in Asia with early hominin remains and/or early, unambiguous stone artefacts remain extremely low, especially when set against the enormity of the Asian landmass. This point is made in Figure 1. 3, which shows the areas over which key sites are distributed in Europe, Africa and Asia. As shown, most European localities are within a circle of 1000-miles diameter, whereas the Asian ones (between, for example, 'Ubeidiya and Sangiran) are scattered over a radius of some 5000 miles. As a further indication of the vastness of Asia: Boxgrove, in southern England is the same distance from Makapansgat in South Africa as 'Ubeidiya in Israel is from Sangiran in Indonesia.

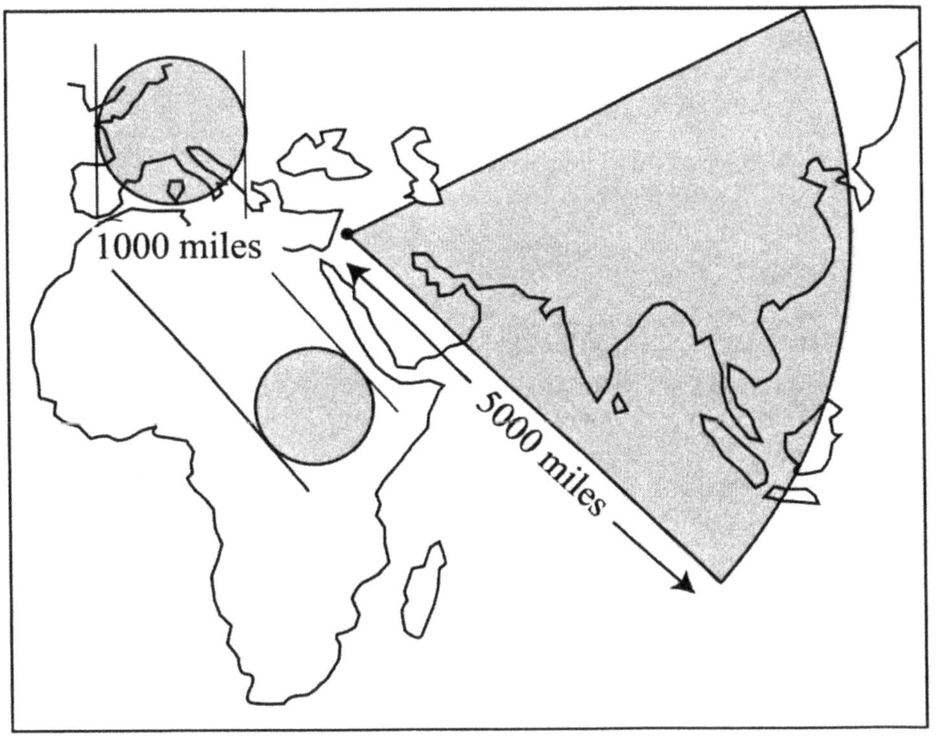

Figure 1.3: The size of areas that need to be sampled for early hominin sites in Europe, Africa and Asia respectively. As indicated, the density of Asian localities is far lower in Asia than in Europe or East Africa.

Vast areas of Asia are still wholly devoid of evidence for or against the presence of early hominins. This is particularly true of South West Asia. As Table 1.1 shows, this area is larger than East Africa (Kenya, Tanzania, Ethiopia) and the countries comprising the European Community in 1998, or larger than the whole of Europe west of the former Soviet Union. South West Asia should be a key link between Africa, Europe, southern and eastern Asia; however, apart from 'Ubeidiya and possibly Yiron in Israel, its earliest occupation is wholly unrecorded. There is therefore almost no information against which individual claims, such as Riwat in Pakistan or Longgupo in China can be checked. Indeed, west of Riwat, the nearest reasonably well-dated early Lower Pleistocene sites are Dmanisi, 1,600 miles to the north-west, and 'Ubeidiya, 2,200 miles west. In contrast, the *maximum* distances between European localities (e.g. Boxgrove (U.K.) or Atapuerca (Spain) to Petralona (Greece) is only 1,300 miles, and in East Africa, all the localities between Laetoli and Olduvai in Tanzania and the Omo Research Area in Ethiopia lie in a strip only 600 miles long.

Table 1.1: Some comparisons of the size (in square miles) of S.W. Asia, East Africa, the European Union (EU), and Europe west of the former Soviet Union.

S.W. Asia	**2,362,400.**
East Africa	977,000
The EU (as in 1998 before enlargement)	1,084,760
The EU and East Africa	2,061,760
Europe west of the former Soviet Union	1,720,760

Notes: S.W. Asia comprises the countries encompassed by the three largest: Saudi Arabia (927,000 sq. mls.), Iran (628,000 sq. mls.) and Turkey (296,000 sq. mls.). East Africa comprises Tanzania (362,000 sq. mls.; included because of Olduvai Gorge on its northern border), Kenya (220,000 sq. mls.) and Ethiopia (395,000 sq. mls.). The four largest European countries west of the former Soviet Union are France (213,000 sq. mls.), Spain (195,000 sq. mls.), Sweden (173,000 sq. mls.) and Germany (138,000 sq. mls.). East Africa and Europe contain abundant evidence for when these areas were first occupied. In contrast, 'Ubeidiya in Israel and Dmanisi in Georgia are the only sites in S.W. Asia that usefully document when that region was first colonised.

Source: Dennell 1998, Table 9.3.

These enormous distances between the few known key early hominin and archaeological localities in Asia are critically important to the level of confidence that can be attached to assessments of when Asia was first colonised. There are simply not enough well-dated finds from Asia to indicate whether a new claim, such as that relating to Riwat, is inherently improbable at a regional level. This point is clearer if set in European terms. Enough is now known for it to be a reasonable prediction that any new discoveries of fossil hominins or archaeological sites in northern Europe will be less than 500,000 years old, given the amount of evidence indicating the absence of hominins before that time as well as their presence thereafter (see Roebroeks 1996). Anomalous claims from any one part of northern Europe can thus be constrained on the basis that "any colonisation of one region of Europe could only be colonised if it was colonised at the same time as most of the others" (Gamble 1995: 290). Even given the imperfections of the East African fossil hominin record, it would seem a reasonable prediction that *H. erectus* will *not* be found there in deposits more than 2.5 or 3 Mya old. These types of constraints do not prevail in Asia, where far fewer sites are known from a much greater land-mass to indicate confidently the last absence of hominins (or any particular palaeospecies of them) as well as their first appearance in any part of it. The essential factor that makes the Asian situation so different from that in Europe or East Africa is that so much of the landmass is simply unexplored. It is not that archaeologists and palaeontologists have looked and found no evidence for early hominins: it is that in many areas, they have not searched at all.

2.4: The earliest Eurasian evidence

Assessments of the very small number of finds that are relevant to an understanding of the earliest hominin colonisation of Eurasia are especially problematic because of major uncertainties over the context, dating, and identification of hominin remains and stone tools. What is absolutely essential is that each case is treated on its own merits, rather than being accepted or rejected simply because it does or does not fit a preferred scenario. With this *caveat* in mind, we can consider the main claims for the initial colonisation of Eurasia. Relevant sites are shown in Figure 1.4.

Figure 1.4: The earliest hominin and archaeological localities in southern Eurasia.

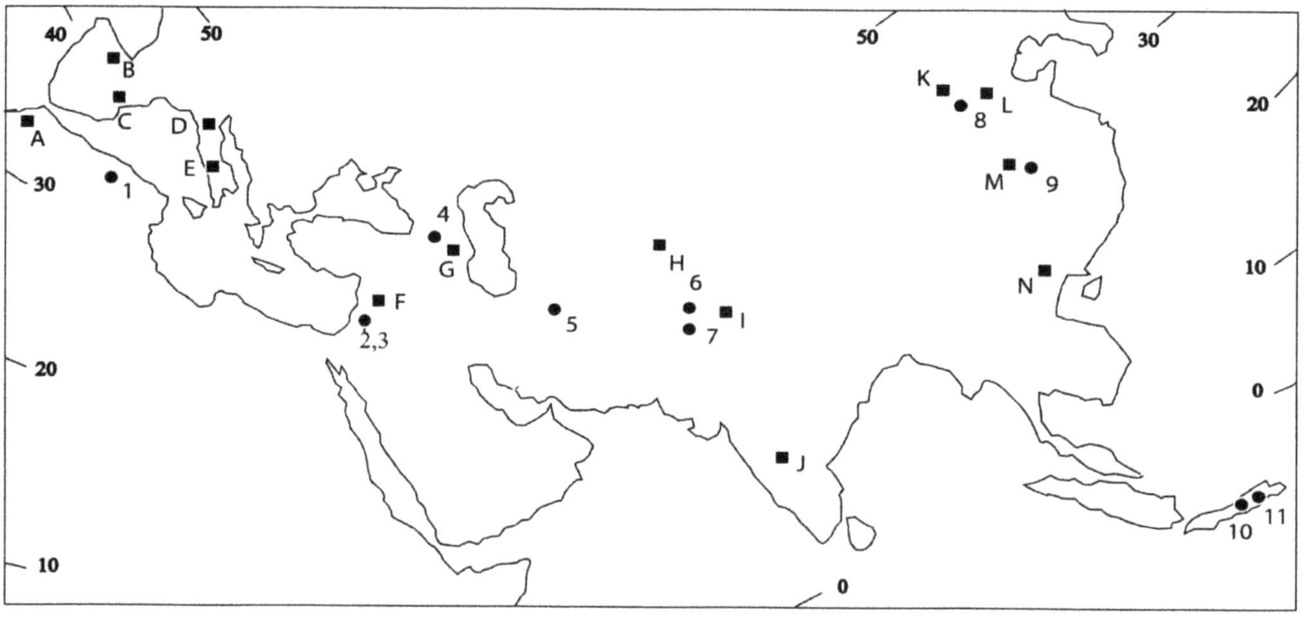

The circles indicate the earliest evidence for hominins from each area considered in the text: 1) Ain Hanech; 2) 'Ubeidiya, 3) Evron; 4) Dmanisi; 5) Kashafrud; 6) Riwat; 7) Pabbi Hills; 8) Xiaochangliang; 9) Gongwangling; 10) Sangiran; 11) Modjokerto.

The squares represent the *next youngest* indication of hominins in each region: A) Casablanca; B) Atapuerca; C) Orce; D) Monte Poggiolo; E) Ceprano; F) Gesher Benot Ya'qov; G) Treugol'naya Gora; H) Kul'dara; I) Dina, Jalalpur; J) Bori; K) Nihewan;L) Zhoukoudian; M) Chenjiawo; N) Bose.

2.4.1: Europe

Opinion is still sharply divided as to when Europe was first colonised. Most authorities believe it was colonised substantially later than Asia, and perhaps as recently as only 0.5 Mya ago. A minority, however, hold the view that parts of southern Europe were occupied by or even before 1.8 Mya ago.

The short chronology has had several advocates. In 1983, Dennell (1983: 39) argued that Europe was the "empty continent" until around a half million years ago, even though it may have been sporadically occupied in earlier times. This case was recently restated in greater detail by Roebroeks and Kolfschoten (1994, 1995), who saw no evidence of hominins in Europe prior to 0.5 Mya. According to them, claims for hominins before 0.5 Mya in Europe are unproven because of uncertainties over the identification of flaked stone as genuinely the work of hominins (or over whether particular fossils are in fact hominin), their context, and/or the dating of the layer from which they came, or were supposed to have come. Roebroeks and Kolfschoten (1994) and Roebroeks (1996) also argued that the absence of clear, incontrovertible evidence for hominins in northern Europe before 0.5 Mya is indeed genuine because of the sustained failure of numerous archaeologists and geologists to find it over the past 150 years. Additionally, it is only after 500,000 years ago that one finds in Europe undisputed hominin remains, artefacts that can be conjoined, and flaked stone assemblages in fine-grained sediments, often associated with bones of large mammals. The 500,000-year-old site of Boxgrove (Roberts and Parfitt 1999) is perhaps the clearest example of a European lower palaeolithic site with all these features, as well as being one of the earliest sites in northern Europe.

Recent research at two Spanish localities have recently added to the debate. Carbonell et al. (1995a) tested the short chronology by further excavation at the Gran Dolina section at Atapuerca. Here, the TD6 horizon contained indisputable hominin remains and stone tools, and was also re-dated palaeomagnetically to just below the Brunhes-Matuyama boundary; i.e. >780,000 b.p. They have thus proposed a "mature" chronology, whereby southern Europe at least was colonised ca. 800,000 years ago (Carbonell et al. 1995b). Subsequently, Dennell and Roebroeks (1996) restated a modified version of the short chronology, in which southern Europe was probably colonised by 800,000 b.p., but northern Europe remained unoccupied until ca. 500,000 years ago.

The most dramatic recent claim for a long chronology in Europe has come from the area of Orce, in southern Spain, where stone tools (Gibert *et al.* 1998, 2001) have been claimed in contexts which they dated palaeomagnetically and faunistically to 1.6 - 1.8 Mya. Evidence that hominins were also present has been challenged (e.g. Moya-Sola and Kohler 1997). At the time of writing, the jury is still undecided (Dennell and Roebroeks 1996:), and the dating in particular will need considerable improvement before one can be certain that Spain was first colonised in the early part of the lower Pleistocene. Emphatically, however, Orce should not be ignored or dismissed out of hand. Whilst the hominin remains are not especially convincing (at least to non-specialists), the artefacts are genuine, from an excavated context, and may well turn out to be older than the TD6 horizon at Atapuerca. (A recent, and credible, dating of oldest finmds from Orce to ca. 1 Mya-old has recently been made by Oms et al. (2000)). Two Italian finds - the hominin from Ceprano (Ascenzi et al. 1996) and the artefacts from Monte Poggiolo (Milliken 1999) - may be of the same age-range.

2.4.2: South West Asia

The two best-known early hominin sites from this enormous region are 'Ubeidiya in Israel, and Dmanisi, Georgia. Two other Israeli sites, Erq el-Ahmar and Yiron also merit attention.

2.4.2.1: Yiron, Israel

At Yiron, near Haifa, Ronen (1991) found several pieces of flaked stone in a limestone gravel in a red clay matrix that extends under a basalt 50-metres away that was dated by K/Ar to 2.47 Mya. (Professor Ronen kindly gave me casts of some of the flaked items, and some seem convincing examples of intentional flaking). There are, however, serious doubts over the dating of this evidence, and it is not currently accepted as indicating that hominins were making stone tools in S.W. Asia before 2.0 Mya.

2.4.2.2: Erq el-Ahmar.

According to recent palaeomagnetic investigations of the artefact-horizons at Erq el-Ahmar, the Oldowan-type artefacts there may date to the Olduvai Subchron (1.77 – 1.95 Mya) (Ron and Levi 2001). If so, this locality is the oldest one with hominin evidence yet found in the eastern Mediterranean.

2.4.2.3: 'Ubeidiya, Israel

'Ubeidiya is not only the best-known early Acheulean site in S.W. Asia, but, until recently, was also the earliest generally accepted archaeological site in the whole of Asia. Because of its location at the northern end of the African Rift Valley and at the "gateway to Asia", its age is a matter of some importance. Estimates have ranged from 0.8 to 2.6 Mya; faunal studies by Tchernov (1989) suggested an age range of 1.0 - 1.4 Mya, and a probable age of 1.4 Mya (see Bar-Yosef 1995: 510). Even if it has lost its place as the oldest hominin site outside Africa, it is still vitally important as the oldest outside Africa with an Acheulean assemblage.

2.4.2.4: Dmanisi, Georgia

During restoration work on the medieval castle at Dmanisi, an indisputable mandible of *Homo sp.* and equally unambiguous stone artefacts were found in water-lain sediments below the medieval layers. The faunal remains in these sediments were initially dated to the early Galerian (Gabunia and Vekua 1990), i.e. ca. 0.9 Mya old. Later, the basalt immediately underlying these sediments was dated by potassium-argon to 1.8 Mya. As this basalt is thought to have formed an outcrop behind which water-lain sediments accumulated almost immediately afterwards, the stone tools, mandible and recently discovered cranial remains are assumed to be of the same age (Gabunia et al. 2000a, Gabunia and Vekua 1995). Recent work has succeeded in finding three superb cranial specimens (Gabunia et al. 2000a, 2000b), dated to or shortly after 1.8 Mya. As such, these finds lend powerful support to the notion that hominins were in S.W. Asia in the Early Pleistocene.

2.4.3: South Asia (India, Pakistan)

The earliest hominin occupation of this enormous area is still largely unknown. Studies of the early palaeolithic in India and Pakistan suffer from a lack of reliable absolute dates, lack of exploration, and perhaps also a concomitant over-reliance on fieldwork conducted before 1939. The foundations were laid by de Terra and Paterson (1939), who conducted fieldwork in 1935 in the Soan Valley in what is now Pakistan, as well as in Kashmir and the Narmada Valley. According to them, the region was first occupied in the Middle Pleistocene, during the 2nd Himalyan Glaciation that they correlated with the Mindel glaciation in Europe, within the old, four-glacial system of the time. Moreover, the earliest inhabitants used a simple flake and core technology that Paterson called the Soan Flake Culture, or Soanian. According to him, this was used throughout the Middle and well into the Upper Pleistocene, and could be divided into four stages, each showing slight improvements in flaking techniques over its predecessor. Additionally, there was also a Middle Pleistocene, Acheulean bifacial component that Paterson saw as intrusive and short-lived.

It is rarely remembered that Paterson's main pre-war research was on the Clactonian flake tradition of southern England, which he examined in exactly the same way as he viewed the Soanian (Dennell and Hurcombe 1992). It is worth noting that few of his conclusions on its internal development are accepted today by British researchers, and some would even doubt the independent existence of the Clactonian (e.g. Ohel 1977, Ashton et al. 1992).

De Terra and Paterson's work survived far longer than any comparable fieldwork of their contemporaries elsewhere. Our previous monograph (Rendell et al. 1989) included a lengthy critique of their work, and concluded that none of their geological and archaeological arguments could be sustained in the light of our own field observations, and in the light of what has been learnt since 1935 about the Pleistocene in general, and the formation of the Himalayan forelands in particular.

2.4.3.1: The earliest current evidence from Pakistan and India

The earliest evidence for hominins in the sub-continent is the small assemblage of flaked stone from Riwat, dated stratigraphically and palaeomagnetically to a minimum of 1.9 Mya (Dennell et al. 1988, Rendell et al. 1987, Rendell et al. 1989). When first published, the age of this material was totally anomalous in Asia, and unsurprisingly was accordingly rejected out of hand or ignored. In the last few years, however, it has become consistent with other finds of comparable age, such as Dmanisi (see above), or Longgupo, and some (e.g. Etler 1996, Larick and Ciochon 1996, Tattersall 1997) have included it within syntheses of the early hominin colonisation of Asia.

In a South Asian context, however, the Riwat discovery is still highly anomalous. This is because the next oldest evidence for hominins in the Indian subcontinent consists of the bifacial, hand-axes from Dina and Jalalpur, Pakistan, that were found in fluvial contexts just above the Brunhes-Matuyama boundary; i.e. slightly less than 780,000 years old (Rendell and Dennell 1985), and the Acheulean material at Bori, India, dated to 670,000 b.p. (Mishra et al. 1995). If hominins were in Pakistan by 1.8 Mya ago as well as large areas of Asia, there is clearly over a million years missing from the Indian palaeolithic sequence. How much of that gap may have been filled by the Acheulean is currently a matter of conjecture that will be returned to later in this volume, in the discussion of the stone tools from the Pabbi Hills.

2.4.4: China and Central Asia

There have been occasional claims for hominins in China substantially earlier than 1 Mya. One such site was Yuanmou, which supposedly contained two hominin incisors and some stone artefacts (Olsen and Miller-Antonio 1992) that have been claimed on palaeomagnetic grounds to be ca.1.6-1.7 Mya-old. The finds are very poorly provenanced, however, and most independent authorities familiar with this material suggest either its age is indeterminate (Schick and Zhuan 1993), or probably less than 730,000 years old (e.g. Olsen and Miller-Antonio 1992; Hyodo et al. 2002: 11).

The oldest site that has recently featured in discussions of the earliest colonisation of Asia is that of Longgupo (Wanpo et al. 1995). This is a cave fissure that was excavated in the 1980's , and then re-dated by ESR and palaeomagnetism. It is reported as containing a hominin mandible and incisor, and two stone artefacts made from water-worn andresitic porphyrite cobbles, all of which are dated to ca. 1.8 Mya ago.

Both the identification of this material and its dating are very problematic. According to Wood and Turner (1995: 240) "the Longgupo hominid may be much more primitive than *H. erectus*". Etler (1998), however, suggests that the "hominid" remains are in fact those of a descendant of the Miocene ape *Lufengpithecus*. Recently, Wu (2000) has also argued that the Longgupo hominid is an ape. Schwartz and Tattersall (1996) also raise doubts as to whether the mandible is in fact a hominin, whilst expressing confidence in the incisor, although "there is nothing about it that aligns it with any particular hominid species" (Tattersall 1997: 50). Schwartz and Tattersall also state (ibid) "in any case the presence of artefacts establishes the presence of hominins". These are problematic, however. The flaked stone "artefacts" lack clear flake scars, ripple scars and bulbs of percussion or other evidence of intentional flaking, and are unconvincing (personal observation). Additionally, there is no evidence of débitage or small utilised flakes. Surprisingly perhaps, the hammer stone –a minimally-modified cobble with some pitting at one end – is much more convincing. *Gigantopithecus* is also evidenced in these deposits. As Etler (1998) comments, there may have been a "dazzling array of hominoids" in Asia at this time, and (as also stated above), there is no reason why *Homo* should have been the only genus capable of using a hammer stone.

According to the investigators, the 20 metre-deep fissure at Longgupo contains sediments that accumulated in a gradual and continuous manner over a million years, consequently showing all the palaeomagnetic changes of the lower pleistocene. Yet the sediments include evidence of higher-energy flow: pebbles and coarsely sorted gravels; one tooth (used for ESR dating) was found embedded within a long bone, which hardly implies a lack of post-depositional disturbance. A more likely depositional history is one of episodic rapid in-filling, washing-out, and reworking of deposits. Unfortunately, the stratigraphic sequence is poorly documented as the site was excavated in one-metre spits, which is not a technique commonly used on cave excavations in Europe during the last 40 years. Overall, one incisor and one flaked piece of stone from Longgupo is not adequate evidence (to this author at least) to demonstrate the unequivocal presence of hominins (still less *H. ergaster*) in eastern Asia as early as 1.9 Mya ago.

At present, the earliest evidence of hominins in China is from the Nihewan basin (latitude 40° N.), where Zhu et al. (2001) have recently dated the hominin occupation at Xiaochangliang to 1.36 Mya, thus making it the oldest so far recognised in East Asia, as well as the earliest occupation this far north. This age-estimate is also similar to a recent re-dating of the Lantian calotte (34° N.) to 1.2 Mya by Hyodo et al. (2002), and the evidence from Gongwangling at 1.15 Mya (Wang et al. 1997). These dates are broadly comparable to the earliest evidence from Central Asia, from sites such as Kul'dara, dated to ca. 900 kya (Ranov et al. 1995). These sites all have simple flake and core assemblages, and lack any component that could be called Acheulean.

2.4.5: South East Asia and Indonesia

There is little well dated material that indicates when hominins first entered mainland S.E. Asia. The oldest stone tools from Thailand are from contexts dated to ca. 0.8 Mya (Pope et al. 1986, Reynolds 1993). The site of Kota Tampan, Malaysia, which was thought to be early Middle Pleistocene and thus one of the oldest in the region, may be as young as 31,000 years old (Reynolds 1993: 10). Sadly, data from the mainland of S.E. Asia do little to illuminate the probable age of the earliest hominins in Indonesia.

Regarding Indonesia, there is a considerable amount of detailed Dutch, Indonesian and Japanese research that dates the first appearance of hominins in Indonesia to ca. 1.3 Mya (Pope 1983, Pope and Cronin, 1984) or perhaps a little later, ca. 1.0 Mya ago (Leinders et al. 1985). However, Swisher et al. (1994) argued on the basis of their argon-argon dates that the hominins from Modjokerto and Sangiran, Java, were around 1.81 and 1.6 Mya-old respectively. (These dates revive an earlier claim by Ninkovich and Burckle (1978) that the earliest Javan hominins dated from 1.9 Mya). Huffman's (2001) detailed re-examination of Perning/Mojokerto supports the original observation that the hominin (found in 1936) was penecontemporaneous with the volcanic unit subsequently dated by Swisher and his team to 1.81 Mya. Swisher et al.'s results are also compatible with other recent Ar^{39}/Ar^{40} dates obtained by Larick et al. (2001). They examined sediments from the Bapang (Kabuh) Formation which has yielded almost 80 *H. erectus* specimens, over 50 of which have known find-spots, and obtained dates for the deposits containing hominin remains ranging from 1.51 ± 0.08 Mya to 1.02 ± 0.06 Mya. The earliest date of 1.51 ± 0.08 Mya lay above an unconformity, and also 10m above the earliest hominin remain (Brn 1996.04) in the sequence, so hominins were clearly there before 1.5 Mya. - assuming of course that the dates are correct.

If Swisher and his team, and other researchers, are correct, Asia was colonised when *Homo habilis* was still scavenging at Olduvai. Their work has had an undoubted important, positive effect in forcing palaeoanthropologists to consider the possibility that Asia was colonised much earlier than 1.0 Mya ago. Yet serious, unresolved issues remain concerning the relationship of these dates to the hominin remains themselves, and the layers from which they are supposed to have come. In addition, they do not tally with the large amount of geological, palaeomagnetic and faunal evidence so far collected (e.g. Leinders et al. 1985). De Vos and Sondaar (1994) argue that the dates may well be "technically correct"; what is at issue is the relationship of those dates to the hominins, particularly those collected (in a generally unsystematic manner) in the 1930's.). More recently, Hyodo et al. (2002) has re-asserted the primacy of palaeomagnetic evidence, and claimed that the dates obtained by Swisher et al. (1994) and Larick et al. (2001) are simply "too old", and the result of reworking. A similar argument has been made by de Vos and Sondaar (1994), and also Langbroek and Roebroeks (2000), who argue for a much younger dating on the basis of the tektite evidence. (According to Antón (2002), however, this particular criticism is based on a single date of a tektite of uncertain provenance. This last point seems unfair – the provenance of the tektite is shown in great detail in a section published by Itahara's team in 1985, and is reproduced in Langbroek (2003: 27)). Overall, although the absolute dates are consistent in indicating that hominins were in Java before 1.5 Mya, doubts still persist from those primarily reliant on biostratigraphic and palaeomagnetic evidence.

3: Discussion

It is evident that views on hominin origins, particularly in Europe and Asia, have changed dramatically in the years since Swisher et al.'s paper of 1994. At present, many favour an early colonisation of parts of Eurasia by or around 1.8 Mya ago. However, the Eurasian evidence for its colonisation by hominins such *Homo ergaster* cannot be treated in isolation from the African evidence for their origin and evolution. As we have seen, the evidence from both Africa and Asia is often ambiguous and of varying reliability, so each piece of evidence has to be treated in its own right, and not accepted or rejected simply because it does or does not agree with a preferred paradigm.

The present evidence for early hominin evolution and dispersal across Eurasia can be summarised in a number of ways. One such way is as follows: the hand-full of fossil hominin remains older than 5 Mya from East Africa may indicate that humans actually originated there, or may merely indicate where fieldwork has been persistent and successful. By 3.0 - 3.5 Mya ago, hominins were certainly living in East Africa, South Africa, and areas such as Chad, and were probably quite widely distributed through the African grasslands and open woodlands. Given the absence of a desert barrier at that time between Africa and Asia, they may even have been in South West Asia, and perhaps further east. Tool-making may well already have been part of their normal behaviour: in any case, by 2.5 Mya, some hominins were making, using and discarding large numbers of stone tools in localised areas. Currently the oldest examples of the multiple discard of flaked stone are from Kada Gona, Ethiopia, and possibly Yiron, Israel, both ca. 2.5 Mya-old. The maker(s) of these is unknown, although *Paranthropus* seems a likely candidate at Kada Gona.

Between 2.5 and 1.8 Mya ago, East African hominins lived in predominantly wooded environments, and probably spent much of their lives in or near trees. The first hominin that habitually lived in East Africa in open grasslands and was an accomplished bipedal walker was *Homo ergaster* (or, if one prefers, *H. erectus*), which supplanted the indigenous *H.* (or perhaps, *Australopithecus) habilis* and *H.* (perhaps, *Australopithecus*) *rudolfensis*, and eventually *Paranthropus* as well. It is not yet clear whether *H. ergaster/erectus* originated in East Africa, another region of Africa, or Asia. It is however reasonably clear from Dmanisi that it was in S.W. Asia, and perhaps also Java by 1.8 Mya.

Between 1.5 and 1.0 Mya ago, descendants of *Homo ergaster* probably expanded their range into northern China, and possibly southern Europe using initially a simple core and flake technology. An Acheulean, bifacial technology was used in sub-Saharan Africa and Israel by at least 1.4 Mya. Its origins may be East African, but it may equally well have originated over a large area encompassing East Africa and South West Asia. Alternatively, it may have originated in either, and then spread rapidly from one to the other.

Major gaps in our current knowledge of hominin origins in the Old World are: the Pliocene record for hominins outside East and South Africa; the hominin record in East Africa between 1.5 and 0.5 Mya ago; in S.W. Asia before 1.0 Mya; in South Asia prior to 400,000 years ago; and China and Indonesia before 1 Mya ago. With these limitations in mind, we can now assess the geological and faunal background to the fieldwork in the Pabbi Hills.

Chapter 2: The Palaeontological Background to the Fieldwork in the Pabbi Hills

The fieldwork described in this volume focused on fossil-bearing deposits in the Pabbi Hills between 2.2 and 0.9 Ma-old. The palaeontological aspects of this fieldwork had three main objectives. One was to improve our knowledge of the biostratigraphy of the late Pliocene and early Pleistocene of South Asia, primarily for the benefit of researchers in Pakistan and India. The second objective, of greater relevance to those interested in human origins in the Old World, was to find hominin remains in or eroding from deposits more than one million years old. The third and most general objective was to understand better the fossil record of the type of riverine palaeolandscape represented by the Pabbi Hills. Each of these objectives has specific problems, which need to be taken in turn.

2.1: Siwalik Biostratigraphy

In regional and chronological terms, vertebrate fossils from areas such as the Pabbi Hills are placed within the Pinjor Faunal Stage of the Upper Siwaliks. Before examining these terms in more detail, it is worth discussing the wider background.

The Siwaliks are named after the Siwalik Hills of northern India, and comprise several thousand metres' depth of predominantly fluvial sediments that have been deposited along the forefront of the Himalayas and Hindu Kush over the last 18 Mya (Figure 2.1). They form one of the longest, most continuous and detailed terrestrial Neogene sequences in the world, even including the better-known East African ones. A good historical overview is provided by Pilbeam (1997).

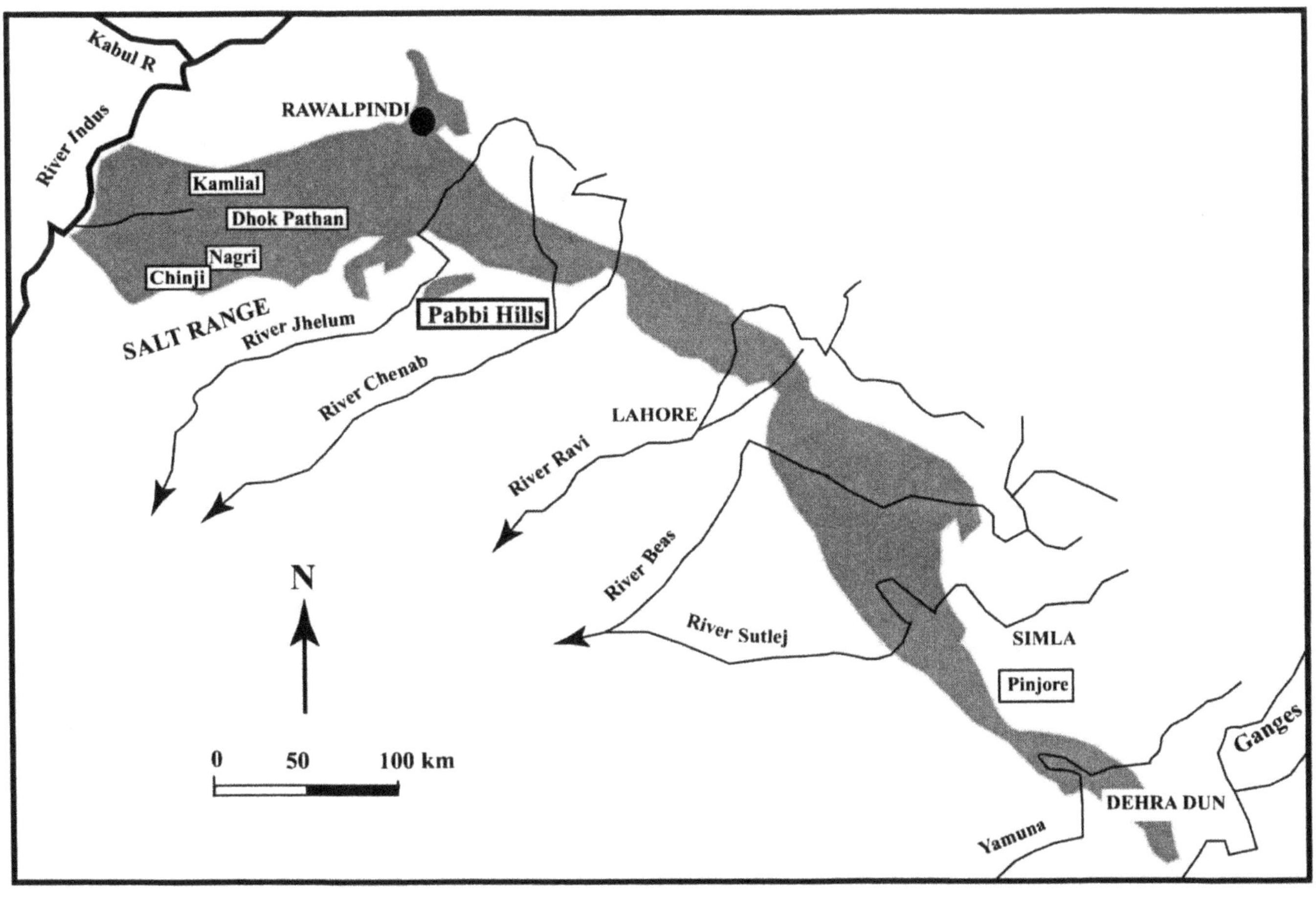

Figure 2.1: Map of Siwalik deposits across northern Pakistan and India, showing location of type sections (in boxes) and the Pabbi Hills. (Based on Nanda 1981, Figure 1).

2.1.1. History of research

The richness of the Siwalik fossil record was appreciated at an extraordinarily early stage, especially when compared with Africa. The first fossils were collected in the Himalayas in 1819 by a Captain Webb, and mentioned by Buckland in his *Reliquiae Diluvianae*, published in 1826. Other fossils were collected by Crawford in 1826 along the Irrawaddy in Burma . The first fossil primates ever discovered were found in the Siwalik Hills of northern India as early as 1836 by Baker and Durand, who were two of many enterprising (and presumably bored) members of the British military who helped record the geology and antiquities of India during the 19th century. Their discoveries prompted Hugh Falconer (1808-1865), then curator of a botanical garden and Proby Thomas Cautley, a military engineer, to collect further specimens; through Falconer's detailed studies, these two pioneers demonstrated the Tertiary age of their material, and published a selection in the first monograph on Siwalik palaeontology (Falconer and Cautley 1845).

Following the replacement of the East India Company by the British Crown as the formal power in India, the Geological Survey of India (GSI) was established in the 1860's, and its officers made several notable contributions to the study of the Siwaliks. Medlicott (1864) demonstrated that the Siwaliks could be divided into a lower (unfossiliferous) and upper (fossiliferous) series, and he and Wynne helped correlate sequences in the Siwalik Hills of northern India with those in the Potwar Plateau, several hundred kilometres west in what is now Pakistan. Others such as Theobald and Fedden organised major collections of fossils in the Potwar Plateau and Sind that were later described in detail by Lydekker, who published numerous lengthy papers on Siwalik fossils between 1876 and 1891.

By far the greatest contribution to Siwalik studies by a GSI officer in the British period was made by Guy Pilgrim (1875-1943), who was attached to the GSI from 1902 to 1930, but continued working on Siwalik material thereafter. His main contribution was to divide the entire Siwaliks into three stratigraphic units, the Lower, Middle and Upper, each of which he later subdivided into faunal zones (see Figure 2: 2) and correlated with those thought to be equivalent elsewhere in Eurasia (Pilgrim 1910, 1913). As an example, the Upper Siwaliks were correlated with the Villafranchian of the Mediterranean basin. Having established the overall regional sequence and placed it in a continental framework, he then proceeded to publish numerous lengthy papers on various taxa represented in Siwalik fossil collections, notably the carnivores (1932), giraffids (1911), tragulids (1915a), antelopes (1937), bovids (1939), and the primates (including hominoids) (1915b, 1927).

Figure 2.2: The lithological and palaeontological zonation of the Siwaliks.

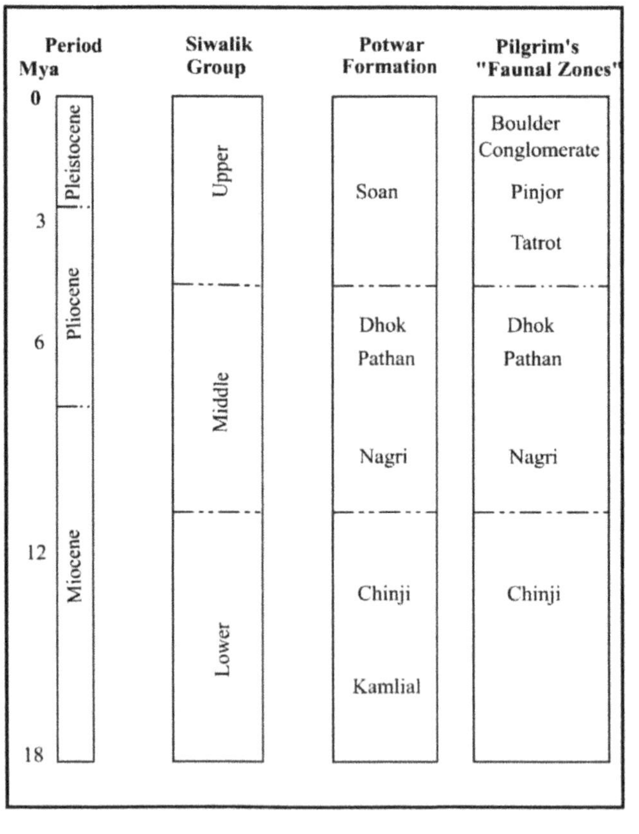

Further major collections of Siwalik fossils were organised in the Potwar Plateau and Chandigarh areas in the 1920's by Barnum Brown for the American Museum of Natural History (AMNH). This material was published in numerous papers and monographs, and often in considerable detail; major examples are Matthew (1929) and Colbert (1935). Matthew and Colbert also studied the collections of Falconer and Cautley in the Natural History Museum, London, and those of Lydekker and Pilgrim in Calcutta; additionally, Pilgrim (1937) studied the bovid and antelope specimens in the AMNH collections. The Yale-Carnegie expedition of 1934-35 to northern India was also important: it included Lewis who found the type specimen of *Ramapithecus brevirostris* at Haritalyangar, northern India, which he claimed was a hominid (Lewis 1934), as well as de Terra and Paterson, whose monograph (Terra and Paterson 1939) on the early palaeolithic sequences of post-Siwalik deposits in northern India is still influential in South Asia, despite numerous short-comings in the light of what is known today (see Rendell et al. 1989). By the outbreak of war in 1939, the Siwaliks were probably the best documented Neogene faunal sequence in the world, with impressive reference material in New York, London and Calcutta that was known to several researchers.

As a result of Partition in 1947, India was left with the fossil collections in Calcutta but few type-sections (most notably Pinjor near Chandigarh), and Pakistan with no fossil collections but most of the key fossil localities and type-sections in the Potwar Plateau. In both countries, there was a lengthy hiatus in Siwalik research, and it was not until the 1970's that it was resumed on a large scale. The main Siwalik research has taken place in Pakistan, and has been undertaken by the Yale Peabody Museum (and later Harvard) and the Geological Survey of Pakistan (GSP); this has focussed since 1973 on the Miocene sequence of the northern Potwar Plateau (see Figure 2.3).

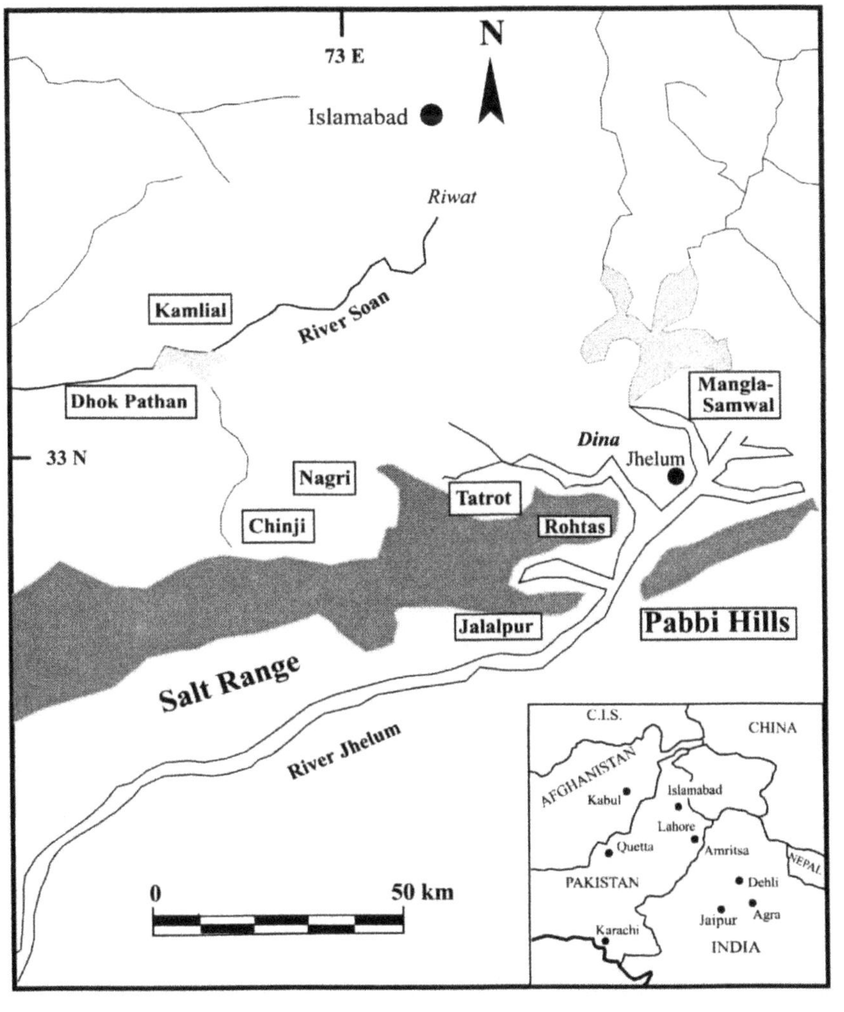

Figure 2.3: Principal fossil localities (i.e. names in boxes) in the Potwar Plateau and adjacent areas. The main Pliocene and Pleistocene localities such as the Pabbi Hills and Mangla-Samwal lie to the east of the main Miocene localities. Sections at Riwat and Dina contain early Palaeolithic artefacts.

As a result of this research programme, the Miocene sequence of this area is once again one of the best-studied Neogene terrestial sequences in the world. As Figure 2.4 shows, it is especially detailed for the period between 18 and 7 Mya. Within this time-period, the sedimentological record is exceptionally complete and thoroughly-studied (see for example Behrensmeyer 1982, Khan 1994, Khan et al. 1997, Willis 1993), and also well-dated by palaeomagnetism (e.g. Behrensmeyer and Tauxe 1982, McCrae 1990). Palaeontological data have been critical in helping to establish a secure biostratigraphic sequence, which will be discussed in the next section. These data can also be used to show the main taphonomic processes affecting the burial and preservation of animal carcasses in this region (Badgeley 1986a, 1986b, Badgeley and Tauxe 1990), the likely composition of the faunal community when alive (Badgley and Behrensmeyer 1980), and the modes of evolutionary change (Flynn et al. 1990). The time-ranges of some genera - such as the hominoid *Sivapithecus* and the rodent *Sayimys* - can also be estimated with confidence limits of 80 and 95% (Flynn et al. op.cit.). Finally, carbonate analyses of palaeosols (Quade et al. 1989) indicate an expansion of grasslands in Pakistan at the expense of woodlands around 7.4 Mya, with concomitant major faunal consequences. This change may indicate a regional shift to a monsoonal climatic regime (Quade et al. 1989), the gradual emergence of grasslands independent of climatic change (Morgan et al. 1994), or a global drop in mean temperatures and precipitation (Cerling et al. 1997). Differences of opinion notwithstanding, the Siwalik data are critical to studies of environmental and climatic change in the Neogene.

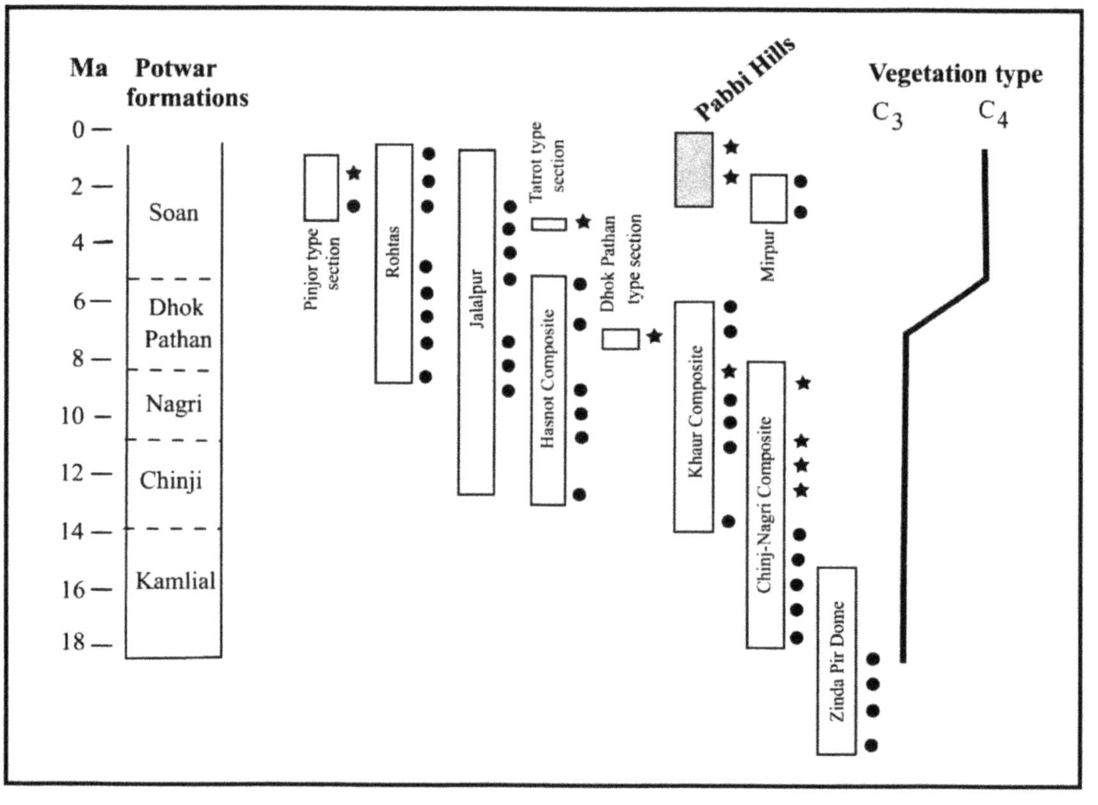

Figure 2.4: Stratigraphic sections, age ranges, and occurrences of fossils in Siwalik sediments. Asterisks indicate poor-to-good preservation; circles, levels with very good representation. Source: Barry 1995, Figure 9.1.

The 1970's also witnessed a major programme of palaeomagnetic dating of Siwalik sequences in Pakistan that was initiated by the Dept. of Geology at Peshawar University and various American universities. A major target of this research was to clarify the mode and rate of uplift of the Karakorum mountains, and the development of the modern drainage system of the Indus and its triburaries. Uplift rates in the Karakorum have been studied by several teams, working in Tibet (e.g. Patriat and Achache 1984), the Nanga Parbat area of the Karakorum (Butler and Prior 1988, Zeitler et al. 1982) and even by those studying detrital material in sediments in the Bay of Bengal (Amano and Taira 1992). Key components of palaeomagnetic research into the Siwaliks of northern Pakistan were lithological descriptions of a series of major stratigraphic sections over 1 or even 2 km-thick (in terms of total sediment accumulation) that were also dated palaeomagnetically (e.g. Keller et al. 1977, Barnt et al. 1978, Johnson, N.M. et al. 1982, Opdyke et al. 1982, Tauxe and Opdyke 1982, Burbank and Johnson 1982, 1983, Johnson N.M. et al. 1985, Raynolds and Johnson 1985). Most of these sections were in Pakistan, but some of the above-mentioned personnel also dated major sections in India (e.g. Johnson and Vondra 1972, Johnson G.D. et al. 1983). Volcanic tuffs that probably originated from the Dasht-i-Nawar area of Afghanistan ca. 600 km east of the Potwar Plateau could also be used to date some of these sections by fission-track techniques (Visser and Johnson 1978, Johnson G.D. et al. 1982). Fortuitously, some of these tuffs bracket the Gauss-Matuyama boundary at ca. 2.5 Mya and thus help anchor this major boundary within the Upper Siwaliks.

2.1.2. Siwalik Zonation

As noted above, it was Pilgrim who first subdivided the Siwaliks into three major sub-groups, and introduced the faunal zones (in order of decreasing age, Chinji, Nagri, Dhok Pathan, Tatrot, Pinjor and Boulder Conglomerate) that are still used today. Research in the last twenty years, particularly by the Yale/Harvard team, have highlighted a number of problems arising from the use (or mis-use) of these terms.

Early studies of the Siwaliks were confronted with the persistent problem that even major sedimentary units are rarely continuous over large distances, and thus regional - or even local - correlations were extremely difficult on lithological grounds alone. For this reason, early researchers such as Pilgrim used vertebrate fossil remains as the main (and often, only) means of relative dating. Vertebrate fossil remains were often localised, however, and could not be used to date those units that were poor or lacking in fossils. Geologists were thus forced into the assumption that unfossiliferous lithological units were contemporaneous with similar units that contained fossils. Consequently, Pilgrim's "Faunal Zones" were initially based on biostratigraphic criteria, but were unfortunately soon used by him and others in a lithological manner as well. As an example, a "Nagri fauna" has meant i) only those animals represented at the Nagri type-site ii) those animals at the Nagri type-site as well as in nearby deposits that can confidently be correlated to that particular locality; iii) or all those animals represented in deposits attributed to the (lithologically-defined) Nagri Formation. In the inter-war years, researchers such as Pilgrim, Colbert and Lewis used the last-listed criterion, and consequently, the "Nagri Fauna" came to contain many older and younger taxa that have never been found at the type site itself (see Barry et al. 1982: 97 for fuller details). As another example of how biostratigraphic and lithological units could be confused, Pilgrim (1913) initially defined his Dhok Pathan Faunal Zone by material from Dhok Pathan itself, plus that from Bhandar, which was assumed to be of the same age because the deposits seemed similar. We now know from recent absolute dating that the Bhandar material is some two million years younger (Johnson, N.M. et al. 1982: 35).

One of the main achievements of the Yale/Harvard-GSP programme has been to maximise the opportunities of palaeomagnetic dating in the construction of a reliable biostratigraphic framework. Fossil vertebrate remains from the erosional surfaces of Siwalik strata have been recorded in the field in terms of their horizontal and vertical occurrence, and tied to long stratigraphic sections that have been mapped and dated palaeomagnetically. Variations in the quality of data have also been taken into account (Barry et al. 1980) by grading fossil finds and/or assemblages on a scale of 0 - 6 in terms of the quality of their stratigraphic and spatial provenance (see chapter 3), with much material from early collections being graded at 5 or 6, or very poorly provenanced. Taphonomic factors, degree of access to fossiliferous exposures, and length of time spent searching for fossils are other criteria built into assessments of the quality of the fossil record.

Fossils found within these sections have been used to construct a biostratigraphic sequence that is unlike Pilgrim's earlier "Faunal Zones" in being wholly independent of lithological criteria (Barry et al. 1982, Barry 1995, Keller et al. 1977, Opdyke et al. 1979). Four "Interval-Zones" (Figure 2.5) have been recognised, defined by "the stratigraphic interval between two distinctive biostratigraphical events" (Barry et al. 1982: 98). It is recognised that the boundaries of these "interval-zones" may change in the light of new discoveries, especially for those stratigraphic intervals that are poorly sampled. These tend to be younger than 5 Mya old: partly because they have been studied much less than the earlier ones; partly because sediments less than 7 Mya old are poor in fossils, and partly because of difficulties in dating sections palaeomagnetically without the help of absolute dates from volcanic ashes (Barry 1995: 130). Both the first and last appearances of many taxa after 5.0 Mya are poorly known, and even major taxa are poorly documented.

What we do know in outline at least is that between 5.3 and 2.9 Mya, (i.e. during Barry's *Hexaprotodon sivalensis* to *Elephas planifrons* Interval-zone) several African taxa including *Elephas, Sivatherium giganteum,* various suids (e.g. *Hexaprotodon)* and possibly also cercopithecoid primates (Barry 1987) imigrated from Africa. Around 2.5 Mya, *Equus* immigrated into northern India and Pakistan from North America (Lindsay et al. 1980), and cervids also arrived from Europe or Central Asia (Barry et al. 1982: 127). The arrival of these animals probably constituted a protracted dispersal event, in contrast with that which witnessed the more sudden arrival of equids ("*Hipparion* s.l."), suids, and large giraffes into the region at ca. 9.5 Mya (Barry et al. 1982: 126). Immigrations and extinctions after *Equus*, cervids and elephants arrived remain unclear. These uncertainties arise from poor calibration of the Upper Siwaliks.

Figure 2.5: The faunal zonation of the Siwalik deposits on the Potwar Plateau, based on "interval zones" that are palaeomagnetically dated. (Source: Barry, 1982 et al., Figure 3).

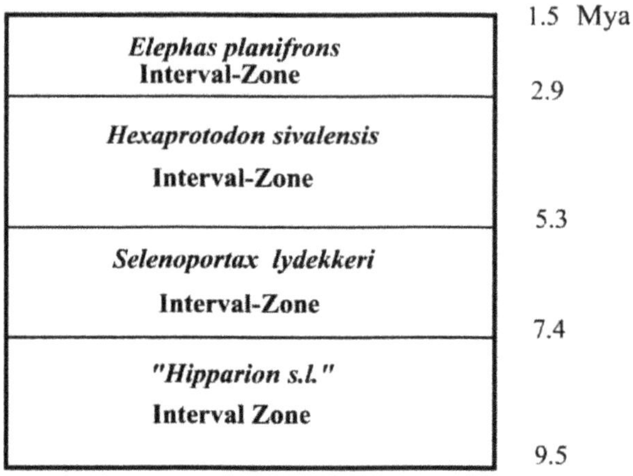

2. 1.3. The Upper Siwaliks: Tatrot, Pinjor, and Boulder Conglomerate "Faunal Stages"

The Upper Siwaliks are the least well-documented part of the entire Siwalik sequence. Its basic subdivisions into the Tatrot, Pinjor and Boulder Conglomerate Faunal Zones have several drawbacks. Tatrot was an unfortunate choice of type locality, in that the type section in the Potwar Plateau (now in Pakistan) is a 60-metre thick section (short by Siwalik standards) that is underlain by an unconformity, and overlain by much later, post-Siwalik silts. It is thus a short "floating" section that cannot be easily dated, or correlated stratigraphically with neighbouring sections, still less with comparable ones in India. Fossils within it are localised, and not particularly numerous. The dating of the Tatrot type-section is especially problematic. It was originally dated to between 2.5 and 2.9 Mya by Barry et al. (1982: 124), but is now placed at ca. 3.3 Mya (Barry 1995: 117)). In India, however, the junction of the Tatrot and overlying Pinjor units has been dated to 2.5 Ma; i.e. at the Gauss-Matuyama boundary (see e.g. Azzaroli and Napoleone 1982, Ranga Rao et al. 1988, 1995, Tandon et al. 1984). This means that the Tatrot type-section in Pakistan probably represents just a short episode within what Pilgrim intended as the Tatrot Faunal Stage. It additionally means that in India, the latest Tatrot deposits (i.e. those immediately below the Gauss-Matuyama boundary at 2.5 Ma) are up to 800,000 years younger than the Tatrot type-section in Pakistan.

Pinjor is the faunal stage that Pilgrim (1913) suggested as the successor to Tatrot. The type locality is near Chandigarh in India, ca. 600 kilometres to the east of Tatrot. Deposits assigned to the Pinjor Stage at Pinjor itself and elsewhere in northern India have yielded large amounts of fossil specimens since the 19th century, and many of these have been described in detail for taxonomic purposes by British-period and Indian researchers. Sadly, earlier workers tended to be interested only in knowing the faunal stage from which a specimen came, and usually paid villagers to collect material for them. Unsurprisingly, the provenance of most specimens obtained before 1947 is unknown beyond stating that they are, for example, Upper Siwalik and came from an area 5 miles west of a particular village or rest house. (Barnum Brown, who collected fossil specimens in the 1920's for the American Museum of Natural History, was exceptional in giving precise map co-ordinates of his principal specimens; see Colbert 1935: 37-55)). As a result of much unsystematic collecting, the "Pinjor Faunal Stage" remains a largely undifferentiated unit that spans some two million years from the late Pliocene to middle Pleistocene. In European terms, this is equivalent to treating the whole of the Villafranchian and Cromerian as one unit; or, in African terms, even worse than treating the entire Olduvai sequence from 1.8 - 0.7 Ma as one unit. Even now, hardly any "Pinjor Stage" fossil material can be precisely dated relative to a local palaeomagnetic stratigraphy, and the timing of most local appearances and extinctions remains unknown. Additionally, "Pinjor" deposits (and fossils) cannot be correlated to the Tatrot type-section in Pakistan in any meaningful geological sense.

2.1.4. Recent research into the Pinjor

Recent research in India and Pakistan has developed independently, and along different lines in the last 30 years. Indian researchers have tended to retain Tatrot and Pinjor as valid stages, albeit ones in need of better dating and definition, instead of replacing them with the type of "interval zones" that are independent of lithological criteria, as advocated by Barry et al. (1982). Nevertheless, there have been some successes in dating Upper Siwalik sequences palaeomagnetically at, for example, Pinjor itself (Azzarroli and Napoleone 1982), the Jammu Hills (Ranga Rao et al. 1988), and Surai Kola (Nepal) (Corvinus and Nanda 1994). Indian palaeontologists, such as Badam (1979), Basu (2004), Ganjoo (1984, 1985), Gaur (1987), Gaur and Chopra (1984), Sahni and Kahn (1988), Vasishat (1985), and especially A.C. Nanda have also been very active in collecting and describing fossil vertebrate specimens (see e.g. Nanda 1976, 1977, 1978a, 1978b, 1978c, 1979, 1988, 1992, 1997, Nanda and Sehgal 1993). Useful work has also been undertaken on the Siwalik sequence in Nepal (e.g. Nanda and Corvinus 1992, Corvinus and Nanda 1994, Corvinus and Rimal 2001). Although Pinjor has been treated as a single unit in India and Nepal, there are at least detailed descriptions of specimens of most of the large mammals.

In Pakistan, the Pinjor has been one of the poorest relations of the Siwalik Group, and is documented much more sparsely than preceding periods. The main palaeontological data from recent research projects has been based on small-scale collecting in the areas of Jalalpur, Rohtas and Dina by the British Archaeological Mission to Pakistan in 1985, work by a Pakistani-Dutch group at Mangla-Samwal (Arif 1985, Arif and de Vos, 1989, Hussain et al. 1992), as well as the research reported here in the Pabbi Hills (Figure 2. 3). In all cases, fossil material could be related to local palaeomagnetic sections (Keller et al. 1977, Opdyke et al. 1979, Lindsay et al. 1980). According to their work, elephants (*E. planifrons)* first appeared in the Indian subcontinent ca. 2.9 Mya b.p. A little later, around the Gauss-Matuyama boundary ca. 2.5 Mya, *Equus* and cervids with antlers immigrated into the sub-continent, and their arrival provides a useful way of defining the base of the Pinjor Faunal Zone (Barry 1995: 118). (Because Pilgrim and earlier workers claimed cervids without antlers in pre-Pinjor contexts, it is sensible to restrict the term "cervid" to species in which the males at least had antler).

It is unclear when *Hipparion* and the giraffids such as *Sivatherium* became extinct: the uppermost, very poorly-defined boundary for the last appearance of *Hipparion* was set at ca. 1.5 Mya from data in the Pabbi Hills (Keller et al. 1977: 199; Opdyke et al. 1979: 27). This would be a very late date for the last appearance of *Hipparion* in South Asia, as it is absent from faunal sequences elsewhere after 2.5 Mya: examples are Rohtas, Jalalpur and Samwal (Hussain et al. 1992) in Pakistan, and in India, the Jammu Hills (Ranga Rao et al. 1988), Khetpurali (Tandon et al. 1984), Pinjor (Azzaroli and Napoleone 1982), and the Jammu, Chandigarh areas and Nepal (Nanda 1997). As will be seen in a later chapter, our own data do not confirm the presence of *Hipparion* in the Pabbi Hills as late as 1.5 Mya.

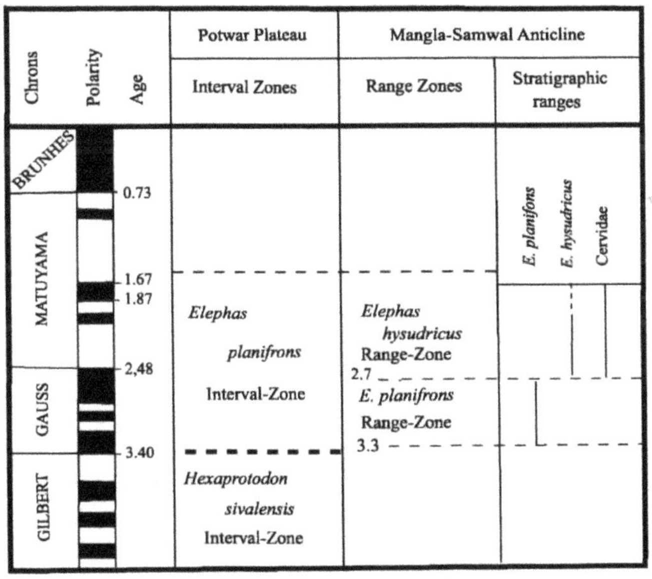

Figure 2. 6: Correlations of the Faunal Interval Zones of Barry et al. (1982) with those proposed by Hussain et al. (1992) for the Mangla-Samwal anticline. Note the replacement of the *E. planifrons* interval zone with a restricted *E. planifrons* zone, followed by an open-ended one based on *E. hysudricus*. Redrawn from Hussain et al. 1992, Figure 6.

In an important recent paper, Hussain et al. (1992) have clarified the faunal changes that occurred between 3.5 and 1.8 Mya. Their data from the Mirpur-Samwal anticline near Mirpur (Figure 2.6) comprise fossil specimens that can be securely linked to a carefully-mapped series of sections that were dated palaeomagnetically to between 3.4 to 1.1 Mya, with additional support from fission-track dating of a volcanic ash at ca. 2.5 Mya. Fossil specimens were found in deposits from 3.4 to ca. 1.8 Mya. Hussain et al. (ibid) propose that the lower limit of Barry et al.'s "*E. planifrons* Interval Zone" should be revised downwards, from 2.9 to 3.3 Mya. They also terminate the upper limit at 2.7 Mya, instead of leaving it open-ended, and propose an additional interval-zone defined by the presence of cervids and *Elephas hysudricus*, which are first evidenced in the Samwal sections at 2.7 Mya (when *E. planifrons* and *E. hysudricus* were briefly contemporary). At around the same time, or perhaps a little later, *Equus* makes its first appearance in the area. As *E. hysudricus* is present throughout the remainder of the fossiliferous part of the Samwal sections (i.e. up to 1.8 Mya), the upper limit of this interval zone is left open-ended pending further research. As will be seen in a later chapter, our own data from the Pabbi Hills confirm Hussain et al.'s conclusions, and also extend the time-span of the *E. hysudricus* faunal interval well beyond 1.8 Mya to less than 1.0 Mya-old.

To summarise, it is difficult to integrate the Upper Siwalik research undertaken in India and Pakistan since Independence for both geological and political reasons. At present, Indian researchers need better calibration of fossil material within deposits assigned to the Pinjor Stage, and researchers in Pakistan need their own sections equivalent to those at Pinjor in India. At present, the exposures at Mangla-Samwal and the Pabbi Hills may provide the best prospects in Pakistan for establishing a well-calibrated biostratigraphy for the late Pliocene to middle Pleistocene. This issue will be raised after the results of the surveys in the Pabbi Hills have been presented.

2.1.5. The Boulder Conglomerate Faunal Stage

Pilgrim (1913) identified the "Boulder Conglomerate" stage at several localities in the Siwalik Hills and Salt Range area as succeeding the Pinjor Faunal Stage. At that time, it was assumed to be extremely rich in fossils - richer indeed than the underlying Pinjor Zone. Pilgrim assigned no less than 34 fossil species as definitely from the Boulder conglomerate, with a further 8 as probably and 22 as possibly from that horizon. Later, almost all this material was re-assigned to the underlying Pinjor Zone, leaving the Boulder Conglomerate as a faunal stage that was virtually fossil-free.

As a faunal stage, it is hard to imagine one that is now less useful than the Boulder Conglomerate Stage. As might be guessed from a term implying the high-energy deposition of stones as large as boulders, fossil preservation varies from extremely poor to non-existent. More importantly, however, the "Boulder Conglomerate Stage" is not a "stage" at all, but "represents individual responses of ancient distributary systems to local changes in hydrologic regimen" (Opdyke et al. 1979: 32). The timing of these responses varies considerably in age from the Pliocene to well into the Middle Pleistocene over even a short distance. Thus at Samwal near Mirpur the oldest conglomerates are 3.3 Mya old; those in the Soan Valley, are 2.1 - 1.9 Mya old, but those only 90 km away in the Pabbi Hills are only ca. 0.4 Mya-old.

The discovery that the age of "boulder conglomerate" horizons varies considerably over short distances has immense repercussions on documenting the first appearance of hominins in this region. This is because de Terra and Paterson (1939) had assumed that the Boulder Conglomerate (as a Formation) was uniformly Middle Pleistocene in age, when noting that it contained the earliest (Soan Culture) stone artefacts in the sub-continent. Clearly, some of these artefacts have the potential to be much older. Although hominins and stone tools have not been reported from the underlying Pinjor contexts in India or Pakistan, there is no reason why some should not eventually be found, if, as seems increasingly likely, hominins had colonised Asia by 1.8 Mya ago (Tattersall 1997).

2.1.6. Summary

The fossil collecting that we undertook in the Pabbi Hills between 1986 and 1990 is part of a long-term attempt to document better the faunal communities of the Upper Siwaliks. Barry et al.'s (1982) use of "faunal interval zones" is a major contribution to the construction of an accurate biostratigraphic sequence for the Siwaliks. More recently, Hussain et al.'s (1992) proposal of two successive zones marked by the appearance of *E. planifrons* at 3.3 Mya and *E. hysudricus* at 2.7 Mya marks a further important step in subdividing the Upper Siwaliks. Hopefully, the material that we collected in the Pabbi Hills between 1986 and 1990 will begin to help extend this more detailed record up to ca. 0.9 Ma. The fact that the faunal subdivisions of the late Pliocene and early Pleistocene in this part of the world are still being defined and clarified at such a basic level is an eloquent statement of the extent to which studies of the Siwaliks have lagged behind those of comparable regions in Europe and Africa.

2. 2. Fossil hominin remains and fluvial landscapes

Since the Pabbi Hills, like other Siwalik sequences across northern Pakistan and India, are predominantly fluvial, it is appropriate to consider the likelihood of finding hominin remains in such fluvial contexts.

On the whole, hominin remains are rare in fluvial (as opposed to stream and lakeside) palaeolandscapes, and most finds have been intermittent and often accidental (see Table 2. 1). Most finds have been of crania or mandibles; post-cranial remains are rare, although one complete and three incomplete femora were found at Trinil, two tibiae at Solo (Ngandong) (Day 1986), and a clavicle at Narmada in India (Sankhyan 1997). Additionally, hominin remains from river deposits are usually "one-offs", in the sense that further searching in the same deposits has rarely led to the discovery of other hominin remains.

Table 2.1: Hominin skeletal remains from fluvial deposits

Site	Date	Element(s)	Age
Trinil, Indonesia	1891-8	Calotte, 5 femora	Early Pleistocene
Mauer, Germany	1907	Mandible	Middle Pleistocene
Ngandong (Solo), Indonesia	1931-3	12 Crania, 2 tibiae	Late Pleistocene
Steinheim, U.K.	1933	Calvaria	Middle Pleistocene
Swanscombe, U.K.	1935-6, 1955	3 conjoining skull fragments	Middle Pleistocene
Olduvai OH9	1960	Calvaria	Lower Pleistocene
Omo SL7A, Ethiopia	1967	Mandible	Late Pliocene
Sambungmachan, Indonesia	1973	Calvaria	?Upper Pleistocene
Dali, China	1978	Cranium	Middle Pleistocene
Awash, Ethiopia	1981	Femur and frontal fragments	Late Pliocene
Narmada, India	1982	Skull fragment, clavicle	Middle Pleistocene

Source: Day, 1986; Sankhyan 1997 and Sonkalia 1985 for Narmada. Notes: the Awash specimens were surface finds; the context is problematic at Sambunmachan.

There are a few exceptions: one partial exception is Swanscombe, where a third piece of cranium was found in 1955 that miraculously joined the other two, conjoining pieces found at the same locality in 1935 and 1936; even so, all three can be counted as part of just one skeletal element. Sangiran, Indonesia, may be another exception as several hominins are known, but these are so poorly provenanced that their original context is often unknown. The most unusual set of hominin remains from an obviously fluvial context are the 12 crania and 2 tibiae from Solo (Ngandong), again in Indonesia. One possibility here is that these represent mass drowning in a catastrophic flood event; another is that they originally derived from a cemetery. This latter possibility is not implausible, given their age may be as recent as 27-54,000 years old (Swisher et al. 1996).

Additionally, the hominin remains so far found in fluvial contexts have come from sediments related to small- to medium-sized river systems, such as the Thames (U.K.), Narmada (India), or Solo (Indonesia); sequences associated with large rivers such as the Rhine, Nile or Indus have not yet yielded any fossil hominin remains. The rarity of hominins in fluvial contexts is perhaps surprising, given the enormous numbers of hand-axes and other types of early palaeolithic stone artefacts that have been found in riverine contexts over the past 150 years, particularly in northern Europe, but also along major and medium-sized rivers elsewhere in the Old World. Early hominins clearly did not avoid rivers; and indeed, sought them out: the principal question is why their skeletal remains are so rarely preserved or found in those deposits where their tools are often so abundant. An additional question is whether the rarity and type of hominin remains in fluvial contexts are substantially different from those of other animals of comparable size.

The poor tally of discoveries to date tells us something about the prospects of finding hominin remains in a fluvial palaeolandscape such as the Pabbi Hills. First, the chances of success are obviously low, and certainly substantially less than in lakeside environments, or from small stream deposits. One has only to think of the hundreds of hominin remains from low-energy lakeside and small stream channels in East Africa to appreciate this point. Secondly, the elements most likely to be found are mandibles and cranial fragments. This is only partly because they are the most easily recognised parts of the human skeleton, and thus the most likely to be noticed during commercial gravel digging or by local inhabitants; after all, Leakey and Marston were on-site at Olduvai, and Swanscombe respectively, and would have checked other material carefully for additional hominin remains. Hominin post-cranial remains are likely to be extremely rare, and probably some distance from any hominin cranium or mandible: as an example, Dubois (1898) found the complete femur at Trinil 15m from the cranial fragment. The two main reasons for the preponderance of cranial and mandibular material in fluvial contexts are likely to be first, the destruction of post-cranial material both before and after it entered a river channel, and secondly the preferential transport of skulls (with mandibles still attached) prior to their incorporation into channel deposits (Boaz and Behrensmeyer 1976).

Although the chances of finding hominin remains in fluvial systems such as the Upper Siwaliks are remote, our surveys in the Pabbi Hills tried to narrow the odds in three ways. One was by extensive survey, in which everything was collected, just in case hominin remains were part of the stream load. A second was to look for minor streams and abandoned channels, as these low-energy environments are usually kinder to the preservation of hominin remains. Thirdly, we worked in the hope that an early Pleistocene carnivore might have done the collecting for us through either predation or scavenging. For that reason, we investigated sites where crocodile remains were common, given the evidence for crocodile damage to the hominin remains at Sangiran 8 mandible (Day 1986: 354), and also Olduvai 7 (Davidson and Solomon 1990)). We also searched actively for early Pleistocene localities of large mammalian carnivores: the *Zinjanthropus* site at Olduvai, Lyanimok (Kenya), Hoojies Poort (South Africa) and possibly also Vente Micena at Orce, Spain, are all carnivore localities that include hominin remains. Such searches obviously need to bear in mind the incompleteness of the fossil record. If, for example, there had been just one pack of hyaena forming a new den each decade in what are now the Pabbi Hills, there should be the remains of somewhere around 180,000 of their dens in the 1.8 million-year-long sequence of the Pabbi Hills. Looking ahead to some of the survey data presented later in this volume, we have so far found around six that might have been, and excavated just three!

2. 3. Fossils, rivers and landscapes

Vertebrate palaeontology has been dominated by biostratigraphic studies: in part, to help date geological sequences, partly to document changes in climate and environment, and partly as a means of studying long-term evolutionary changes in animal lineages or communities. The sub-discipline of hominin palaeontology has been dominated by the searching for and classification of hominin remains, and dating those that have been found. All these are entirely valid and worthwhile goals, but do not easily accommodate other research objectives. These include taphonomic investigations of how the fossil record was formed in the first place, understanding what factors might skew our knowledge of biostratigraphical records, and understanding how different types of palaeolandscapes were used by creatures such as early hominins.

These objectives have generated an enormous amount of research in recent years within palaeoanthropology, particularly in East Africa. One important focus has been on early hominin archaeological sites, notably those at Olduvai Gorge and Koobi Fora. The realisation in the early 1980's (e.g. Binford 1981) that the so-called "living sites" were most unlikely to represent base camps where animal carcasses that males had obtained by hunting were dismembered and shared with females and off-spring opened up alternative interpretations. These included the possibility that hominins had not hunted, but scavenged, that the animals represented at these sites may have been hunted by other predators, and later scavenged by hominins, or had died naturally and then been scavenged by both carnivores and hominins. In addition to much productive research on cut-, tooth- and percussion-marks on animal remains (e.g. Potts and Shipman 1981, Selvaggio 1994, Blumenschine and Selvaggio 1988), fragmentation patterns and skeletal element representation (e.g. Binford 1988, Todd and Rapson 1988), rates at which bone weathered before burial (e.g. Potts 1986), much attention was also paid to the type of debris left by modern humans and other carnivores. Consequently, there have been many excellent recent studies on the feeding habits and debris of leopard (e.g. Cavallo and Blumenschine 1989), hyaena (e.g. Cruz-Uribe 1991, Kerbis-Peterhans and Horwitz 1992), lion (e.g. Schaller and Lowther 1969), chimpanzee (e.g. Plummer and Stanford 2000), as well as contemporary hunter-gatherers and pastoralists (e.g. Gifford and Behrensmeyer 1977), as well as on disarticulation patterns of modern animal carcasses, different types of post-mortem damage to bones, and rates of burial. All these studies, as well as the related explosion of taphonomic studies in the last twenty years, has resulted in a fuller understanding of the myriad of factors that influence the composition of these very ancient archaeological sites. These data, especially when integrated with the stone tools and animal remains from early archaeological sites and insights from hominin skeletal remains, have considerably improved our understanding of early hominins, even if investigators often disagree among themselves.

As a result of recent work, we now know a great deal about the fate of animal remains in modern landscapes; the type and distribution of bone at the feeding places of modern humans, chimpanzees, and different kinds of carnivores; and a fair amount about the type of debris at late Pliocene and early Pleistocene localities where hominins were active. What we know much less about is the fate of animal remains on palaeolandscapes. Whilst it is true that there are no modern direct analogues to environments such as Plio-Pleistocene Olduvai, it is equally true that there are few fossil data sets comparable to those of modern animal remains observed and collected in areas that are broadly applicable to early hominins. Indeed, the best studies of fossils as parts of their palaeolandscapes have concerned much earlier circumstances, notably the dinosaur beds of Wyoming (e.g. Wood et al. 1989, Willis and Behrensmeyer 1995) and the Miocene landscapes of northern Pakistan (see previous section).

There are many sound reasons why the fossil remains of palaeolandscapes cannot be directly compared with the modern remains found in present-day landscapes. Fossil exposures may be limited, selective collection by previous researchers may have seriously affected the type of fossils that are likely to be found; fossil preservation may be poor, and erosion rates may be low. Yet there are instances where the data might have been collected had it not been for an over-riding concern with biostratigraphical goals. Biostratigraphic objectives normally result in very selective fossil collecting, as it is rarely necessary to collect more than a small proportion of the fossils in an area. These tend to be dentitions, as these survive well, and are often the most diagnostic part of the skeleton. Other body parts, such as horn cores and antlers, may be essential for the identification of groups such as bovids and cervids. Isolated post-cranial fragments are rarely of much use in biostratigraphic work, and elements such as ribs and vertebrae are of minimal importance. Abundance of taxa is often given impressionistically: for example by "x" for present, "xx" for common, and "xxx" for very common; and few details are generally available on skeletal part representation, or on the spatial distribution of fossils. When fossil data has been collected in a manner appropriate for testing against the distribution of modern animal remains in comparable landscapes, the assemblages have tended to be small. For example, in her survey of the Okote Tuff FxJj horizon, Koobi Fora, Stern (1993, Table 3) found only 200 specimens in the scatters (or low-density occurrences) of fossils between high-density localities. In another study, Gaur and Chopra (1984) had only 357 mammalian fossils from both the Tatrot and Pinjor zones on which to base their reconstructions of habitat and changes in ecology.

The Pabbi Hills provided excellent opportunities to answer some of these questions. First, it contains several square kilometres of late Pliocene to middle Pleistocene exposures that were often exposed along several kilometres of scarps. Secondly, there was no significant bias caused by previous, large-scale collecting. Pilgrim is known to have visited the area briefly early this century, but commented only that "the Pabbi Hills have yielded no identifiable fossil but *Stegodon*" (Pilgrim 1913: 271). De Terra and de Chardin (1936) also collected some elephant and horse remains "from the crest of the ridge" in 1935. (This was probably around the Banni Rest House, which we used as a field base). The collecting by Opdyke et al. (1982) was again small-scale, and focussed on exposures near their palaeomagnetic sections. Thirdly, much of the Pabbi Hills is sparsely vegetated, thereby enhancing the visibility of fossil remains; fourthly, erosion rates are extremely high as the sediments are soft and rainfall heavy, particularly in the monsoon. Finally, the remote possibility of finding a hominin specimen was not allowed to impinge (for better or worse) on the attention paid to other taxa. As will be argued after the evidence has been presented, the results gained by 12 months of fossil collecting over four major field seasons have many interesting implications on the interpretation of the fossil record elsewhere.

Part 2: Results of the Surveys in the Pabbi Hills

Chapter 3: The Pabbi Hills: Introduction and Overview.

Introduction

Before the outcomes of the surveys and excavations in the Pabbi Hills are presented, it is worth discussing the problems of conducting palaeontological fieldwork in an area such as the Pabbi Hills. This chapter is intended primarily for those readers who are not familiar with this type of landscape, or with working in an area such as Pakistan. Additionally, it details the procedures used for recording the occurrences of fossil and other material that were found.

The Pabbi Hills cover ca. 40,000 hectares (400 sq. km), most of which are uninhabited, badland scenery. They form an anticline ca. 40 km long, trending from north-east to south-west, and ca. 10 km wide. Today, the Pabbi Hills are dissected by the main road from Islamabad to Lahore (still called the G.T. or Grand Trunk Road from the British days), and by the railway line that runs alongside it (see Figure 3.1).

Figure 3.1: The areas surveyed in the Pabbi Hills, 1986-1989.

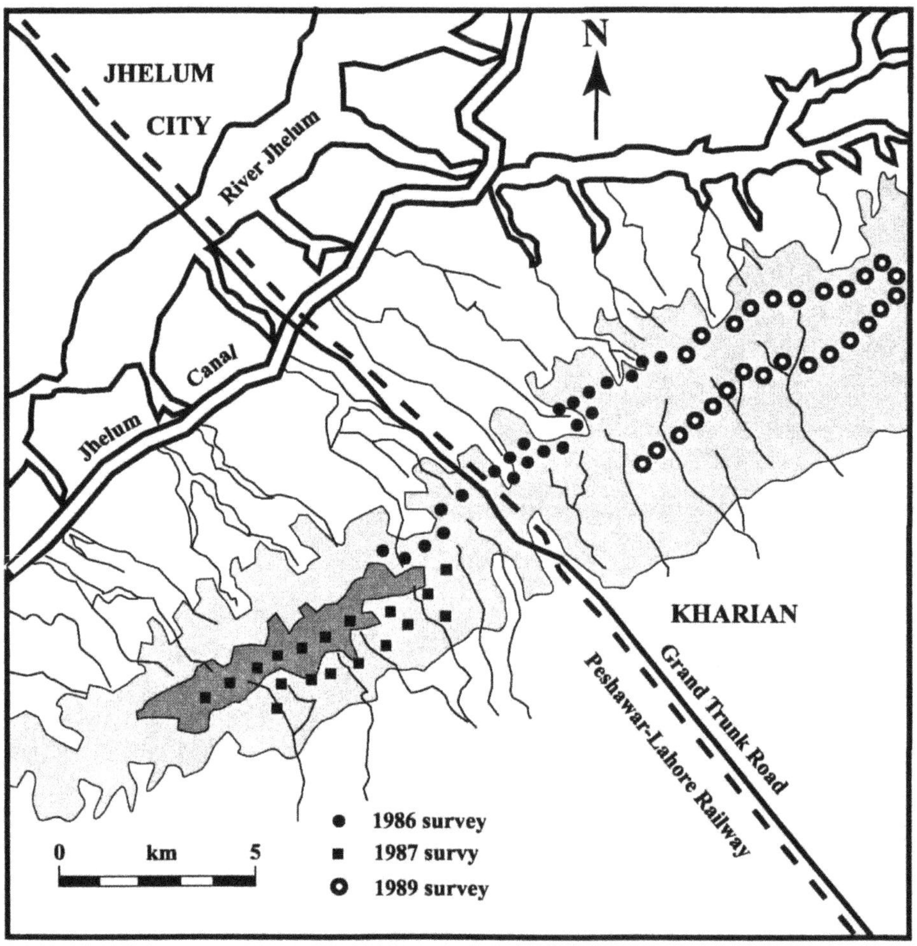

The plains of the Punjab, generally level and heavily farmed, lie on either side of its flanks, between the river Jhelum to the west, and eastwards towards the river Chenab and Lahore. Although densely settled on either side, the Pabbi Hills themselves are largely empty. There is no permanent water, and much of it is too dissected and difficult of access to farm. Vegetation is dense in places, particularly of thorn scrub on scarp faces, and oleander on many expanses of dip slopes, particularly south of the G.T. road. Much of the Pabbi Hills is a designated protected area owned by the Forestry Department, and thus grazing by livestock, and woodcutting by local villagers and migrant groups of *kuchi* nomads, is in theory constrained. In practice, both occur, and many areas away from the G.T. road are only sparsely vegetated. As the sediments are soft and the monsoon rains heavy, these areas have little opportunity to re-establish a protective vegetation cover. Although bad in terms of soil conservation, the sparseness of vegetation was a major benefit to fossil collecting. On the other hand, the Pabbi Hills is a finite, and probably short-term, national asset as a source of Plio-Pleistocene fossils. As the pressure for arable land continues to rise, so inevitably will attempts be made to extend crop farming into those parts of the interior that are flatter, and where a tractor with a bulldozer blade can enter. The Forestry Department has also terraced several areas south of the G.T. road in order to plant trees. As a result, some of the fossil localities discovered in 1987 were bulldozed by 1990. Whilst these inroads are understandable, they will also be at the expense of fossil localities; sadly, some of this work will not be repeatable.

3. 1 Fieldwork problems

The surveys and excavations undertaken in the Pabbi Hills encountered various problems, of which the main ones are outlined below:

3. 1.1. Field base

The first problem was a simple one of finding a field base. Camping under canvas was unwise because of lack of privacy and security. Government rest houses are also unsuitable for lengthy stays, or for use by a large team. Local town houses and hotels can be rented, but tend to be small, cramped and often noisy: however, one rented in Kharian in 1986 enabled the team to be based near the survey areas. The surveys of 1987-89 were largely possible because of Banni Bangla, a derelict Rest House near the Grand Trunk (G.T.) road and on the watershed of the Pabbi Hills. Banni Bangla lacked electricity, water and sanitation, and in wet weather it leaked like a sieve; on the other hand, what it lacked in way of facilities, it amply made up for in being large, relatively private and quiet, and in the centre of the survey area. Its occupation was regularly shared with mice, gekkoes, birds, frogs, and various donkeys, a three-legged water buffalo, snakes, spiders and centipedes. On a more practical note, there was a large verandah around the entire circumference that provided an excellent place for storage and processing fossils. The goodwill of local officials led to the provision of electricity and occasional use of a water tanker. Sadly, it was demolished in 1990 shortly before it would have fallen down. (In insecure times, however, it could never have guaranteed personal security; see below, dacoity).

Village houses can also be rented. For example, when working in the north-western end of the Pabbi Hills in 1989, we rented a house in a small road-side settlement called Bulani (=Place of Ghosts), an hour's drive along the road running alongside the Jhelum canal and north of the G.T. road. This was small and very cramped, and necessitated most of the team sleeping on the roof at nights, which are remarkably cold in January. Bulani is also vulnerable to dacoity (see below), and is best avoided as a field base. In 1990, we had to base ourselves in the village of Doga, on the north eastern side of the Pabbi, in order to gain access to site 642, near the smaller village of Baroth. Doga lay some 1.5 hours by vehicle (in fair weather only) from the G.T. road and some 45 minutes drive (again, fair weather only) from the village of Baroth. The accommodation itself was extremely cramped, totally lacking in privacy, and exceptionally noisy; in wet weather, some team members had to sleep in the vehicles. In complete contrast, the house rented at Khanpur (near Sarai Alamgir) in 1999 was by far the most comfortable and relaxing field base ever used while working in the Pabbi Hills.

3. 1.2. Access

The main access by vehicle to the interior of the Pabbi Hills is along a dirt track cut by the Forestry Department along the watershed both north and south of the G.T. road. The northern track is reached by driving along the G.T. road towards Kharian, and then turning left through some brickearth quarries shortly after the railway line. The southern track is reached by driving along the G.T. road towards Sarai Alamgir; and then turning left a few hundred metres past a petrol station near the Banni Bangla radio mast. Both watershed tracks require a four-wheel-drive vehicle and fair weather; by the time the survey ended in 1990, both tracks had deteriorated considerably during the summer monsoons, and were no longer navigable along their entire length. Access to the north-west part of the Pabbi Hills in 1986 was usually by vehicle along the watershed track to a convenient point at which the team(s) could disembark; usually thereafter the survey areas could be reached within an hour or so of brisk walking. (An alternative route was to walk along the railway line to a convenient point and then head along a path or gully, or along one of the sandstone ridges to a point where one could then reach prospective areas). In 1987, the watershed track on the south side of the G.T. road was the main line of entry to the survey areas; this track was upgraded in 1999 by hiring a tractor with a front-blade. In 1999, a service track for a gas pipeline running parallel to and ca. 1 km south of the G.T. road provided an additional way of reaching fossil exposures. The main landmark on the south side of the Pabbi Hills is "Flag Hill", a low but very conspicuous feature with a shrine at its base, often a flag on top, and by the watershed track ca. 5 km. from the G.T. road.

The northern and southern extremities of the Pabbi Hills are best reached by a service road that runs beside the Jhelum Canal west of Sarai Alamgir and allows access up various tracks and stream beds to the western (Jhelum) side of the anticline. The eastern (Lahore) side is more difficult to reach by vehicle, as tracks and roads between villages are poorly connected, and often some distance from the Pabbi Hills. Additionally, the army base immediately north of Kharian, and the associated firing range immediately south should be avoided.

3. 1.3. Natural hazards

The main natural hazard perceived by local people was snakes, which can be troublesome in the hot season. These were encountered, but only rarely, and less often than spiders, scorpions and hornets. Dogs were a more serious hazard, whether owned by villagers or by the *kuchi*, or nomads, who come down from the Karakorum each winter with their flocks and horses. Although none of the team suffered a dog bite in the Pabbi Hills, most of us walked with some stones handy. Thorn bushes were a constant hazard, and secateurs are advisable items of field equipment. As thorns can be up to 7 cm. long, they can easily puncture vehicle tyres and cause much inconvenience, especially if two tyres are punctured on the same vehicle. The potential for spraining ankles was always high because of the steep and loose nature of many of the escarpments, although luckily no accidents ever occurred. For safety reasons (leaving aside dacoity, see below), people never worked on their own, and a team of three was considered desirable as that would have provided some backup in case of an accident. The army firing range south of the G.T. road and near the Pabbi Hills also caused some nervous moments when working in Kotha Kas, as parts of it are within range.

3. 1.4. Banditry

Areas such as the Pabbi Hills are ideal for *dacoits*, the Urdu term for bandits, as the area is difficult to police, especially away from the main road, and easy to hide in. Their presence prevented us from working north of the G.T. road in 1987, and forced us to terminate the survey of the north-western part in 1989 after an attempted ambush and an unpleasantly tense encounter near Bulani. At the writing of writing, dacoity and terrorism are serious problems in many areas of Pakistan. After a very close shave in 1999, I consider it very unwise to contemplate fieldwork anywhere in the Pabbi Hills without at least one armed body-guard. Additional precautions include not working in any one area for more than a few days at a time; not having more than one team in the field at any time; keeping informed of developments from villagers and local officials, and having a satellite telephone in case of emergencies. Needless to say, vehicles should never be left unattended, and people should always work in groups. The selection of any field base also requires a consideration of personal security, and preferably a house with a secure compound. Even if Banni Bangla had not been demolished, for example, it could not be used if dacoits were operating in the area. Likewise, it is now unwise to drive locally at night, especially along minor rural roads.

3. 2. Field procedures and survey methods

Each team had to be self-sufficient in the field, and take its own food, drink and equipment; and of course, carry all the fossils that had been collected. During the field seasons of the 1980's, two survey teams of 2-4 people plus a field officer from the Geological Survey of Pakistan and/or the Dept. of Archaeology, Govt. of Pakistan worked simultaneously along different transects. Teams of three were found to be the optimum number; although people sometimes worked in pairs, they often found it difficult to cover the ground adequately; conversely, a team of four sometimes got in each others' way.

Navigation within the Pabbi Hills is often difficult because of dense vegetation, the badland topography, and depth of gullies. For day-to-day field operations, teams quickly made their own reference points and mental maps which allowed them to work without (usually) being lost. Transferring that information to maps, along with precise details of where each fossil occurrence was found, remained a major problem. Ideally, we should have had sets of up -to-date maps at 1:50,000, and good air photographs at a smaller scale. Sadly, the only maps available were photocopies of reprints of the 1: 63,360 maps of the India Survey, 1911. Although good for their time, they were out of date, and not of much use within the complex landscape inside the Pabbi Hills. For security reasons, we were not allowed to consult air photographs of the area, although Asif Jah, as field officer of GSP, was able to obtain a set in 1987 whilst undertaking the fieldwork for his Ph.D. We therefore lacked a decent ground map of the surveyed areas, although sketch plans of where sites were found, and descriptions in day-books were usually sufficient to re-locate those sites not subsequently lost through erosion or terracing. (In 1999, for example, we did not encounter any problems in re-locating fossil occurrences that had last been visited a decade or more before). Field recording was transformed in 1999, as we were then able to use a GPS system and SPOT satellite imagery (neither of which were available in the 1980's at an affordable price); as a result, the key fossil localities could now be located to within 50 metres. (At that time, non-military GPS readings were still subject to built-in inaccuracies).

As stated earlier, the surveys in the Pabbi Hills had three main objectives: to find hominin remains, or at least stone tools that might indicate their presence; to help establish a biostratigraphy for the lower Pleistocene of Pakistan; and to improve our understanding of the fossil record from fluvial palaeolandscapes. The methods used in the survey reflected these objectives in three ways.

3.2.1. Collecting strategy

All fossil material was collected, including specimens that were undiagnostic to taxon and/or skeletal element. A few exceptions were made, such as when dense concentrations of fragmented elephant tusk or ostrich eggshell fragments were found. In these instances, a small sample was taken. Another was when the remains of elephant crania were found in large sandstone blocks that often weighed over 200 kg. Items such as these were noted in the field, and left.

3.2.2. Time-banding and stratigraphic zonation

A basic prerequsite for the establishment of a biostratigraphy is that the recording of the probable age of diagnostic specimens is as accurate as possible. The Pabbi Hills have two assets in terms of establishing a lower pleistocene biostratigraphy despite its lack of volcanic ash horizons. First, the area has been well-dated palaeomagnetically by Keller et al. 1977; see Figures 3.2,3.3 and 3.4), and thus the Jaramillo and Olduvai Events of normal polarity provide useful marker horizons at 0.9 and 1.6- 1.8 Ma respectively. In many other areas with deposits spanning this time-range, those between 0.9 and 1.6 Mya would be poorly dated. However, the second major biostratigraphic asset of the Pabbi Hills is Sandstone (SST) 12, as identified by Keller et al. (1977). This unit is especially prominent, as it forms a continuous scarp over several kilometres along both side of the anticline, and is stratigraphically intermediate between the sandstone units assigned to the Jaramillo and Olduvai events. As Keller et al. (1977) estimated its age as ca. 1.2 - 1.4 Mya (see Figure 3.4), it thus forms a very useful way of dating material between 0.9 and 1.8 Mya. Even more fortunately, the deposits capped by Sandstone 12 are very fossiliferous, and produced ca. 50% of the total number of specimens found.

Figure 3. 1 has already shown the areas covered by the three main field surveys of 1986, 1987 and 1989. Our surveys relied on SST12 as a major marker horizon for 1986 and 1989, and most of the material found in those seasons can be estimated as between 1.2 and 1.4 Mya old. The 1987 material was largely derived from the south side of the GT road, and from deposits older than SST12. (See Figure 3.1)

Rendell's palaeomagnetic sections along Choawala and Sohawa Kas (see appendix 1) identified the Olduvai Event as a useful marker horizon in both drainage systems. Fossil occurrences tied to those sections thus enable related sets of occurrences to be dated relative to her palaeomagnetic sections. Consequently, most of the fossil material found between 1986 and 1990 can be assigned to time-bands that are each 200-300,000 years long. In the absence of repetitive volcanic ash horizons, the length of those time-bands is probably as short as can be managed at present.

Table 3:1 provides an overview of the fossil totals from each season, arranged by time-band. As is evident, there is abundant material from before and during the Olduvai Event, and a great deal more from SST12. The main deficiencies are between 1.4 and 1.6 Mya, i.e.the end of the Olduvai Event and the beginning of the SST12 sequence, and after 1.2 Mya, i.e. younger than SST12. The main reasons are that deposits between 1.4 and 1.6 Mya are not well-exposed on the south side of the GT road, and the youngest strata in the Pabbi Hills are generally poor in fossils. (There is a large area of deeply dissected badlands in the interior of the Pabbi Hills north of the GT road that has not yet been explored, and collections from here might one day help improve our data set of material between 1.4 and 1.6 Mya.) These shortcomings notwithstanding, the best of the 42,000 fossils found to date should considerably help improve our understanding of Upper Siwalik biostratigraphy during the Lower Pleistocene.

Table 3.1: Numbers of fossil specimens collected by field season and time-band in the Pabbi Hills.

	0.9 – 1.2	**1.2 – 1.4**	**1.4 – 1.7**	**1.7 – 1.9**	**1.9 - 2.2**	**Total**
1986	**480**	**10818**	**311**	**0**	**0**	**11169**
1987	**697**	**40**	**148**	**9140**	**9630**	**19655**
1989	**278**	**10984**	**520**	**0**	**0**	**11782**
Total	**1455**	**21842**	**979**	**9140**	**9630**	**42606**

Figure 3.2: Cross-section and palaeomagnetic dating of the Pabbi Hills anticline. (Redrawn from Raynolds and Johnson 1985, Figure 16).

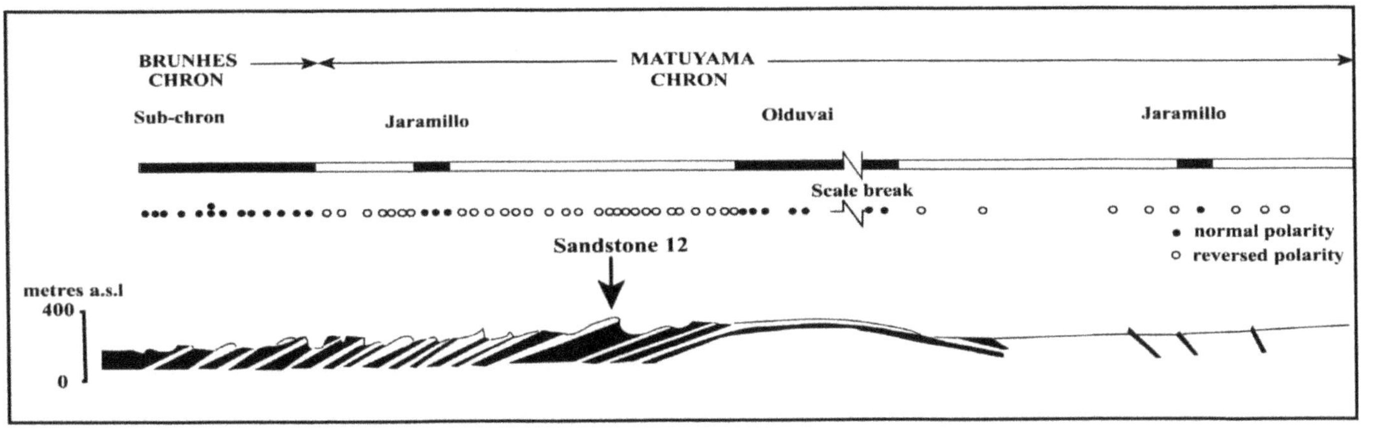

Figure 3.2:Palaeomagnetic cross-section of the Pabbi Hills anticline. (Redrawn and modified from Raynolds and Johnson 1985, Figure 16). Here, the vertical scale (and hence the angle of slope) has been doubled to show the anticline's profile more clearly.

Figure 3.3: Correlation of palaeomagnetic sections in the Pabbi Hills (redrawn from Keller et al. 1977, Figure 12). The palaeomagnetic timescale has been updated from Cande and Kent (1995).

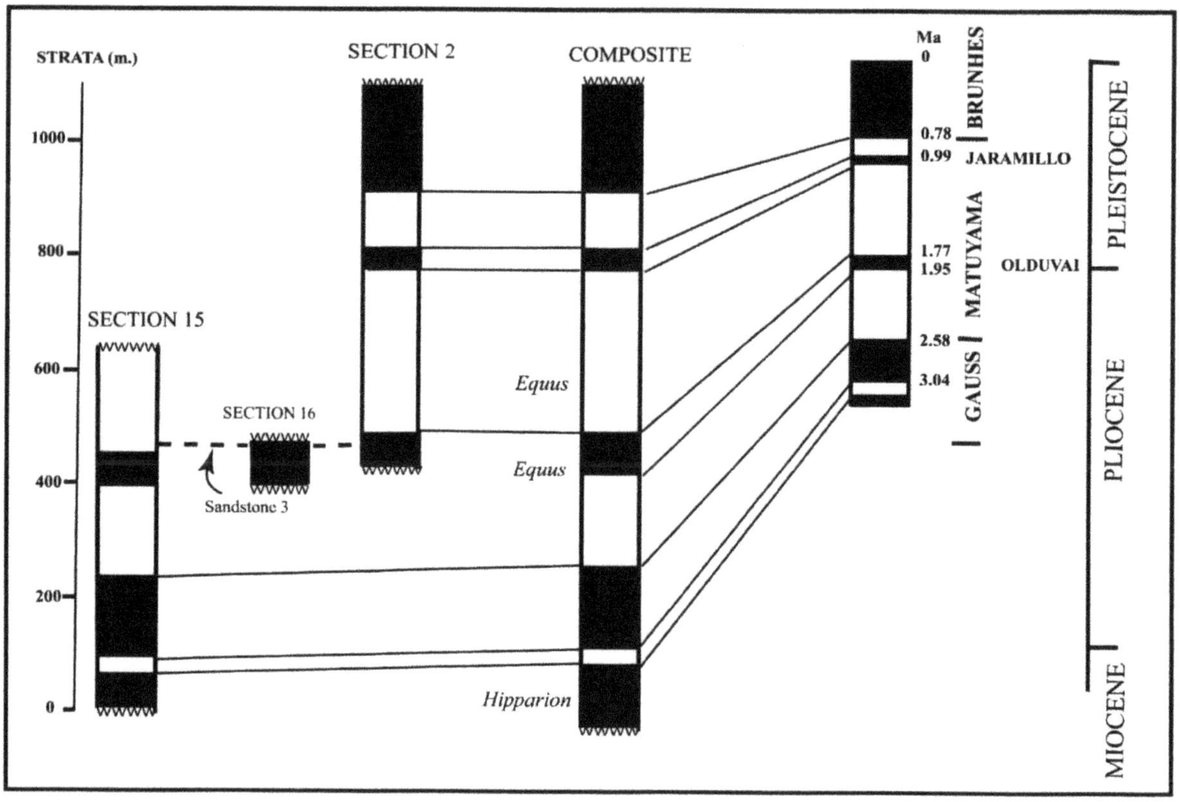

Figure 3.4: Keller et al.'s (1977) palaeomagnetic zonation of the Pabbi Hills. The solid bars show the location of their main sections. (Redrawn from Keller et al. 1977, Figure 14).

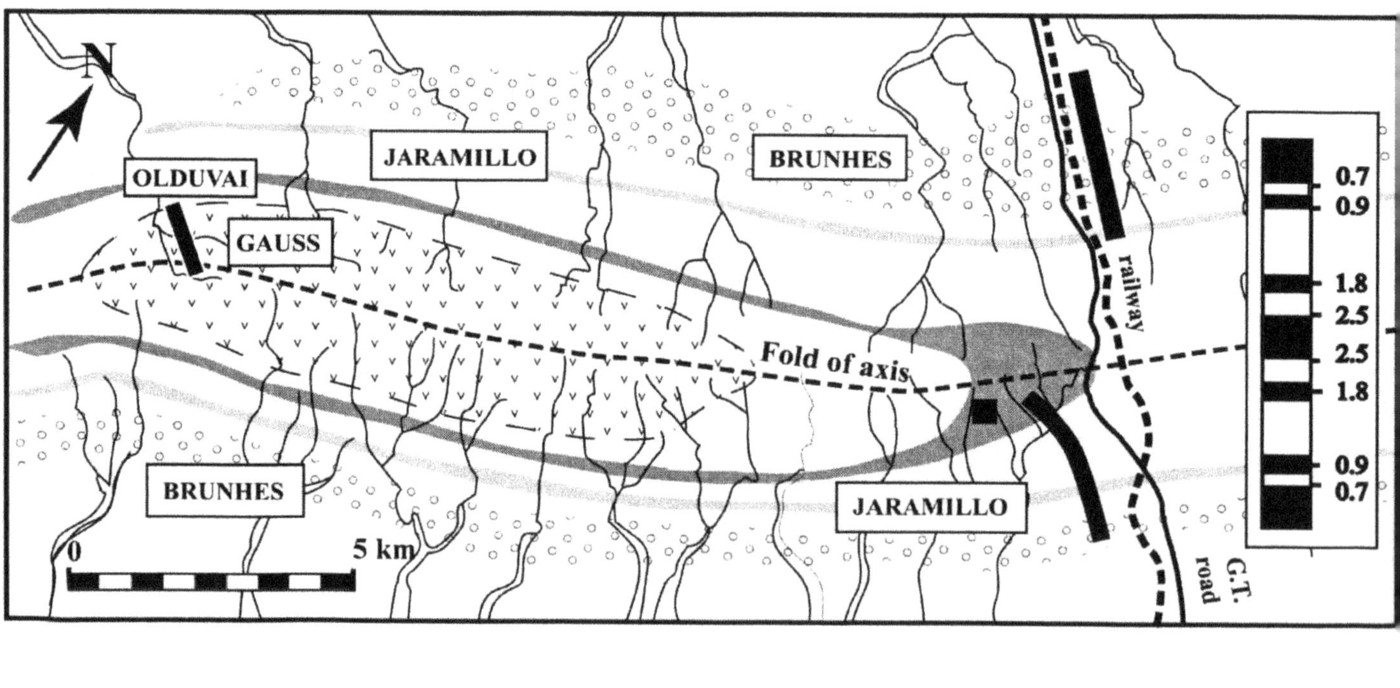

3.3. Recording methods

The spatial distribution of fossil vertebrate remains across the landscape of the Pabbi Hills was recorded in terms of their frequency and spatial density. Three main types of fossil occurrences were noted (see also Figure 3.4):

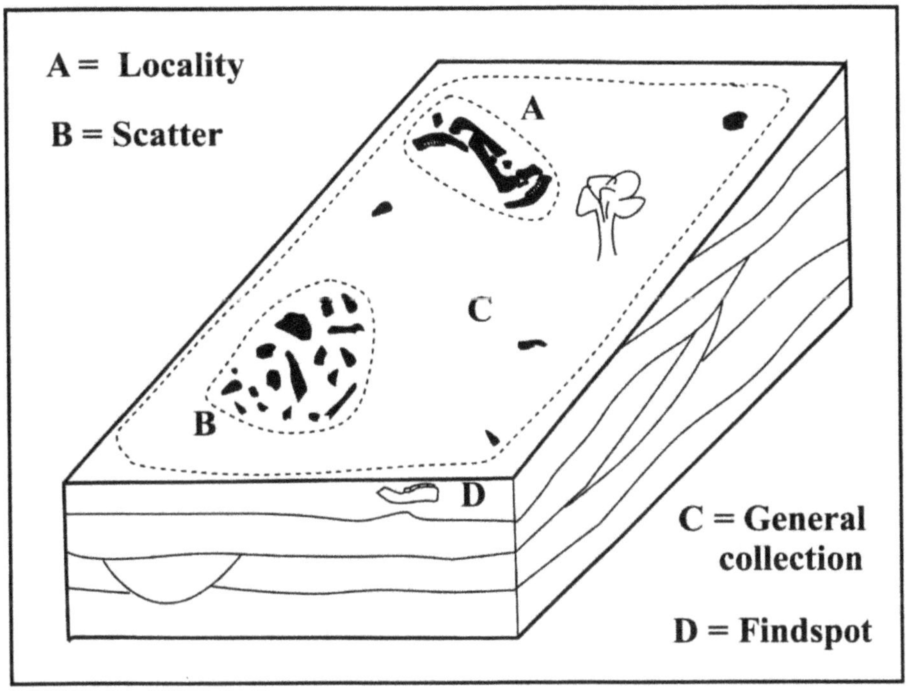

Figure 3.5: The main type of fossil occurrences in the Pabbi Hills. In this schematic representation, a high-density fossil locality (A) and a lower-density scatter (B) are found eroding from an area covered by a general collection (C), which would comprise material from more than one layer. A specimen (D) in its original context is awaiting exposure as a find-spot.

a) Localities. These were large and dense concentrations of fossil bones and teeth, typically between 500-5,000 specimens, from a small area that was as little a few square metres or as large as 100 sq. m. (e.g. locality 642). These have a precise geographic and stratigraphic provenience.

b) Scatters. These were smaller, and usually more diffuse concentrations of fossils. A typical one might yield 50 - 100 specimens from an area of 50 - 100 sq m. Most of these also have a precise geographic and stratigraphic provenience.

Most scatters contained fossilised remains of a small number of animals. Some comprised fossilised fragments of just one animal (e.g. occurrence 636), and thus indicated a single death event. Closely adjacent scatters at roughly the same stratigraphic level might represent discrete clusters from different death-events, or could indicate the edges of a much larger locality. Some scatters in gullies incised into a steep scarp face (as often happened on e.g. Sandstone 12) were probably accumulations through down-slope transport.

c) General collections. These were very diffuse occurrences from areas covering several hundred if not thousands of square metres, and probably more than one sedimentary layer. They were usually defined topographically: for example, as the material from a slope within a particular drainage system, or from an area between two obvious topographic features. They correspond to the "survey blocks" used by Barry et al. (2002: 5) in their discussion of Siwalik Miocene material. Most general collections were from along slopes, and thus tried to follow a defined stratigraphic horizon (e.g. the slopes below a major sandstone unit). As most of these general collections contained scatters within them, their contents can usually be assigned the same stratigraphic position, even if the spatial location of individual finds is less exact.

In addition, a small amount of material fell into one of the following categories:

d) Find-spots: as the term implies, these were individual finds in an undisturbed geological context. An example might be a bone or tooth eroding from a sandstone unit.

e) Parallel gullies. Some fossil material was found in the beds of gullies that flowed along the base of a major sandstone unit. In most cases, it was likely that the fossils were derived from that unit, even though they had clearly been re-deposited.

f) Transverse gullies. These cut through numerous units, and fossils found in them could have been derived from any of several units, and thus have very limited biostratigraphical value.
(Only a small amount of material was collected from gully beds, and usually when walking in or out of an area at the

beginning or end of a day in the field). The methods used here are broadly similar to those from comparable surveys. Barry et al. (1980: 12-13) used a two-part classification scheme to express the degree of confidence over the geographical and stratigraphical provenance of fossil material (see Table 3:2). In their case, they had to take account of fossil material that had been collected decades previously by earlier teams, and it was often difficult, if not impossible, to establish precisely where specimens had been found.

Table 3.2: The Chinji classification scheme for recording fossil provenance. (Source: Barry et al. 1980:12-13).

Geographic provenance:

Class	Criteria
1	Location known to within 10m. Location marked on aerial photographs and maps, and sketches and ground photographs have been made. The exact position can usually be confirmed by the presence of bone scrap.
2	Location known to within 100 or 200m. but either because of the absence or extra abundance of bone scrap it is now impossible to find the exact location. In some cases these are localities which have a very wide collection area, encompassing what are really several different localities. Locations are marked on topographic maps and sometimes aerial photographs. Most Class 2 localities probably cannot be upgraded to Class 1. **Comment**: this type of fossil occurrence relates to collections by earlier workers, and does not apply to the Pabbi Hills data.
3	Location is also known to within 100 or 200m. but not yet confirmed by actually revisiting the site. Locations are marked on topographic maps. In most cases these can be upgraded to Class 1.
4	Location not known to within 200m. Locations marked on topographic maps, but we have not yet attempted to revisit them. Some, but not all, could be upgraded to either Class 1 or 2.
5	Location not known to within 200m. Locations marked on topographic maps, but it is not possible to fix the locality's position more precisely.
6	Location essentially unknown other than to region (i.e. Chinji, Hasnot). Probably cannot be upgraded.

B. Stratigraphic provenance

Class	Criteria
1	Stratigraphic position marked on one or more local measured sections and accurate to within 3m. It should be possible to unambiguously order stratigraphically all the Class 1 localities tied to each local section.
2	Stratigraphic position marked on a local section with a range of precision between 3m and 15m. Some placements could be upgraded to Class 1.
3	Stratigraphic position marked on a local section with a range of precision between 15 and 50m. Composed mainly of Geographic Class 2 localities where the uncertainty about the geographic location makes it difficult to fix the stratigraphic horizon more accurately. Cannot be upgraded.
4	Stratigraphic position marked on a local section with a range of precision between 15 and 50m. Composed mainly of Geographic Class 3 localities. Could probably be upgraded to either Class 1 or 2.
5	Stratigraphic position tentatively marked on a local section with a range of precision greater than 50m. May only be possible to state that a locality is older or younger than some datum. Can be upgraded.
6	Stratigraphic position known only in terms of Pilgrim's faunal zones (i.e. "Chinji") or completely unknown. Not on measured marked sections.

This simple but effective scheme allowed the Yale/Harvard-GSP team to integrate fossil material from previous surveys into their own, high-quality collections. The same type of scheme can be used here as a way of grading uncertainties over the spatial and stratigraphic provenance of material. In general terms, all the localities and find-spots, most of the scatters, and some of the general collections can be graded as Class 1 on the above geographic and stratigraphic criteria. The remaining scatters and general collections are mostly Class 3, and the small amount of material from gully beds that is graded as Class 4 - 6. Further details are found with the relevant tables of contents of each fossil occurrence.

3.4. Relationship between fossil occurrences and palaeolandscapes

When collecting fossils in the field, the discovery and numbering of fossil occurrences is largely dictated by the present-day drainage: for example, a modern drainage basin of several stream systems (as in Kotha Kas), or a smaller gully system that cuts through a fossiliferous horizon (for example, Kanarawala Kas), or a series of adjacent gullies incised into the face of a scarp (as often along Sandstone 12). In order to record the spatial and/or stratigraphic context of fossil material as closely as was feasible, several collecting numbers were often used in the course of a day. The problem thereafter lies in establishing whether each collecting number relates to a specific, discrete event in the past, or whether a series of adjacent collecting numbers are simply points at which one single event was sampled (see Figure 3: 4). A channel bar on which animal remains were fossilised might, for example, have originally extended for several hundred metres, and now be partially exposed over a similar distance along an eroding scarp face. Without excavation and systematic, closely-spaced sedimentary descriptions (both of which are very time-consuming), it is often difficult to establish whether adjacent collecting numbers are sampling one large fossil occurrence, or several smaller ones. In a small number of cases, a fossil occurrence could be accurately delineated by only one collecting number. Examples are localities 68, 73, 362 and 642, all of which were exceptionally rich occurrences of fossil material from small, well-defined areas. These exceptions aside, adjacent occurrences of similar types of fossil material from a spatially-defined area of exposures are best grouped into sets and treated as samples of an original extensive palaeo-surface.

Chapter 4: Magnetic Polarity Stratigraphy of Upper Siwalik Sediments in the Pabbi Hills

Introduction

Magnetic polarity stratigraphy (MPS) provides a time framework for the vertebrate fossils collected in the Pabbi Hills. Recent erosion of the gently folded Siwalik deposits forming the Pabbi Hills anticlinorium has revealed sections through just over a 1 km thickness of these fluvial sediments. These sediments were deposited by a precursor of the present day Jhelum River and comprise a sequence of mostly single storey sands and overbank sandy silts, silts and clays. Some of the sand units are reasonably well cemented and form prominent escarpments, or hog's back ridges, that can be traced over 10's of km along the strike, forming key marker horizons for stratigraphic correlation.

The basis of the method

When sediments are deposited any grains of ferrimagnetic minerals such as magnetite or haematite become orientated in the direction of the Earth's magnetic field and therefore preserve a record of that field. Throughout geological history the Earth's magnetic field has repeatedly changed polarity. At present the Earth's geographical North Pole and magnetic North Pole broadly coincide and the magnetic remanence of samples from sediments deposited under these conditions will have a *declination* of about 360° and a positive *inclination* and therefore *normal* polarity. But for about 50% of geological time the Earth's geographical North Pole has coincided with the magnetic South Pole. Under these conditions sediments will, on deposition, have a *declination* of about 180° and a negative *inclination* and a *reversed* polarity. Magnetic polarity stratigraphy is not a dating technique, instead it can be used to infer ages provided other independent evidence is available to 'fix' one of the observed polarity changes in time. A magnetic polarity timescale has been established using volcanic rocks that not only contain a strong magnetic signal, but also can be dated independently using radiometric techniques such as $^{40}K/^{40}Ar$ or $^{40}Ar/^{39}Ar$. The most recent revision of the magnetic polarity timescale for the last 3 million years is given in Table 4. 1.

Table 4.1: Magnetic polarity timescale for the last 3.22 million years (after Cande and Kent (1995) and Berggren et al., (1995))

Polarity Interval (Ma)	Polarity	Code	CHRON or subchron
0.00-0.78	Normal	C1n	BRUNHES
0.78-0.99	Reversed	C1r.1r	MATUYAMA
0.99-1.07	Normal	C1r.1n	Jaramillo
1.07-1.77	Reversed	C1r.2r	MATUYAMA
1.77-1.95	Normal	C2n	Olduvai
1.95-2.14	Reversed	C2r.2r	MATUYAMA
2.14-2.15	Normal	C2r.1n	Reunion
2.15-2.58	Reversed	C2r.2r	MATUYAMA
2.58-3.04	Normal	C2An.1n	GAUSS
3.04-3.11	Reversed	C2An.1r	Kaena
3.11-3.22	Normal	C2An.2n	GAUSS

One important assumption when employing MPS techniques is that deposition has been more or less continuous and there have been no breaks in sequence e.g. erosional hiatuses. In contrast to sedimentation in the deep oceans, the deposition of terrestrial sediments is notoriously discontinuous. Fortunately, the deposition of the Siwalik sediments, eroded from the rapidly uplifting Himalayan mountain range, in the foredeep area to the south of these mountains provides a more or less continuous sequence and has proved ideal for palaeomagnetic studies.

Siwalik Magnetic Polarity Stratigraphy

In the 1970's and early 1980's a succession of masters and doctoral students from Dartmouth College, USA, sampled Siwalik sections throughout the Potwar area and established the regional pattern of magnetic polarity zonation (Johnson et al., 1986). These workers established a dating framework for these fluvial molasse deposits with critical tie points, in the form of absolute dates, provided by fission track dating of zircon grains in volcanic ash deposits in many, but not all, of the sections sampled (Johnson et al., 1982)

Previous work on Pabbi Hills by Keller (1975) and Keller et al. (1977) established a broad framework for dating these deposits. Samples were taken along the section conveniently provided by the railway cutting that bisects the Pabbi Hills anticline, and, in addition, along the Khoharawala Kas, and the headwaters of the Kotha Kas on the northern limb of the anticline. The results of Keller et al. (1977) demonstrate that the sediments of the Pabbi Hills cover the timespan from Late Pliocene to mid Pleistocene (ca. 3.2 Ma-0.5 Ma). In their schematic magnetic polarity zonation diagram the fold is shown as an anticlinorium – an anticline with its axis orientated ENE-WSW and with the axis plunging both ENE and WSW. The oldest sediments are exposed in the core of the fold structure WSW of the GT road and Railway cutting and were sampled by Keller (1975) in the headwaters of the Khoharawala Kas. On the basis of their three sections, Keller et al. (1977) created a composite section for the Pabbi Hills which they correlated with the magnetic polarity timescale of Opdyke (1972). They also made use of one of the prominent sandstone horizons as a stratigraphic marker, linking their sections together. Unlike sections further west in the Potwar, the Pabbi Hills deposits do not contain any volcanic ashes that could be independently dated. Instead the assignment of ages to the magnetic zones identified within the Pabbi Hills was based on the discovery of fossil remains of *Equus* and *Hipparion* (Keller, 1975; Keller et al, 1977). The dating was based existing evidence of the first appearance of *Equus* in Asia, which was argued to have occurred more recently than 3.5 Ma. (Later, Lindsay et al. (1980) showed that *Equus* first appeared in South Asia ca. 2.5 Ma, most probably from North America).

Sampling Strategy

The main reason for extending the palaeomagnetic sampling within the Pabbi Hills anticline was to confirm the extrapolated and interpolated magnetic polarity zonation and associated stratigraphy derived by Keller (1975) and Keller et al. (1977). The objective was to sample sequences of deposits containing changes, or switches, in magnetic polarity that could be correlated unambiguously over space in order to provide a robust temporal framework for the vertebrate fossil finds.

Choice of sections for the present study

The sampling in the present study makes use of the natural sections exposed along the torrents (nullas) which dissect the Pabbi Hills with main channels running more or less at right angles to the strike of the strata (ie. consequent streams). Three sections were sampled: Choawala Kas and Ratial Kas on the northern and southern limbs of the anticline in the central section of the hills to the southwest of Keller's railway section and Kanarawala Kas to the northwest of the railway section. The aim was to try to sample through the Olduvai Sub-Chron in the Choawala Kas and the Jaramillo Sub-Chron in the Kanarawala Kas section. Both of these sections, together with Keller's (1975) Railway section, contain Sandstone 12 as a prominent marker horizon, a ridge-forming sandstone that outcrops for at least 15km along the strike on the northern limb of the anticline.

Sampling Method

Sampling was limited to the finer grained units comprising silty sands, silts and clays. The sand units are frequently poorly cemented, making recovery of intact samples difficult, and also weakly magnetised. Sections were first cleaned back and then two blocks were excavated at each sampling location with the field orientation i.e. strike and dip direction, measured using a Brunton compass/inclinometer, marked in pencil on each block. Details of sampling for each section are given in Table 4. 2. The large blocks were then wrapped in aluminium foil, labelled and returned to field base. Each block was then cut into smaller blocks, maintaining orientation, and inserted into perspex cubes for palaeomagnetic measurements. Once sampling had taken place, sample locations were painted and numbered. On completion of sampling each section was surveyed using a theodolite and staff. In addition to surveying in all of the sample locations, the location of the tops and bases of all of any sandstone units were fixed. Detailed sections and stratigraphic logs were compiled using the survey data coupled with field measurements of the dip of the sedimentary units.

Table 4. 2: Details of palaeomagnetic sampling in the Pabbi Hills

Section	Number of sample locations	Section thickness (m)
Choawala Kas	39	200
Ratial Kas	9	60
Kanarawala Kas	70	245

Palaeomagnetic Analysis

The intensity of the natural remanent magnetism was measured for all samples. Samples were then incrementally demagnetised in 50°C up to a maximum temperature of 600°C and were re-measured after each step. Previous work involving an intercomparison between thermal and alternating frequency (AF) demagnetisation had established the former as the most appropriate method for Siwalik samples (Rendell et al., 1987). The final values for declination and inclination were obtained using vector end point (VEP) plots. The different components of sample magnetisation were resolved using principal component analysis of the VEP. The lower temperature component is thought to represent a viscous magnetic overprint of the contemporary magnetic field, whereas the high temperature component reflects the characteristic remanent magnetism (ChRM), reflecting the polarity of the Earth's magnetic field at the time of deposition. All samples were corrected for tilt.

Results

The mean high and low blocking temperature (Tb) results are given in Table 4. 3. A (normal polarity) viscous overprint is apparent at low blocking temperatures. The high temperature data show the overall polarity zonation of the samples. Choawala Kas samples show a normal polarity zone between two reversed polarity ones whereas the data sets for both Ratial Kas and Kanarawala Kas show consistent evidence of reversed polarity magnetisation only.

Table 4. 3. Mean blocking temperature (Tb) components for sampled sections.

	Declination	**Inclination**	α_{95}
Choawala Kas			
High Tb: B1-B13	169.9°	-32.7°	9.6°
High Tb: B14-B31	352.9°	30.0°	6.9°
High Tb: B34-B39	183.3°	-26.1°	9.8°
Ratial Kas			
Low Tb	2.1°	51.6°	10.7°
High Tb	176.6°	-41.2°	14.4°
Kanarawala Kas			
Low Tb	345°	17.0°	12.7°
High Tb	181.0°	-19.6°	9.3°

α_{95} radius of the 95% confidence circle about the mean.

Fig. 4.1 shows the magnetic polarity zonation of the sampled sections together with those of Keller (1975) and Keller et al. (1977). Sandstone 12 can be identified unambiguously and, as noted above, can be traced laterally for in excess of 10km along the strike. Given its prominence, Sandstone 12 is used as a stratigraphic marker and links the sections in Choawala Kas, where it is 10m thick, and Kanarawala Kas, where thickness is 15m, with Keller's (1975) long Section 2 (The Railway section) in which Sandstone 12 is 12m thick. Keller (1975) and Keller et al. (1977) also use Sandstone 3 as another stratigraphic marker. It should be noted that the numbering of the sandstone units is somewhat arbitrary since it depends on the starting point used, and is also influenced by the fact that some sandstone units are laterally discontinuous.

The interpretation of the magnetic polarity zonation is that the Choawala Kas section encompasses the Olduvai subchron and ends at Sandstone 12. The much shorter Ratial Kas section is interpreted as pre-Olduvai Matuyama Chron (C2r.2r), while the longer Kanarawala Kas section is post Olduvai, pre-Jaramillo Matuyama Chron (C1r.2r) on the basis of their polarity and relative stratigraphic positions. The lack of the expected normal polarity subchron (Jaramillo) at the top of the sampled section along the Kanarawala Kas raises some issues concerning comparability of measured sections. However a difference of only 30m over a total strata thickness of 1 km would be sufficient to account for 'missing' this sub-chron.

Discussion

The Choawala, Ratial and Kanarawala Kas sections broadly fit into the magneto-stratigraphic framework developed by Keller (1975) and Keller et al. (1977). The new data reinforce the dating framework for the vertebrate fossil finds. They also enable questions about the synchronous versus diachronous nature of the main sand units to be explored in more detail. Within sequences of fluvial sediments, it is not uncommon to have units that are

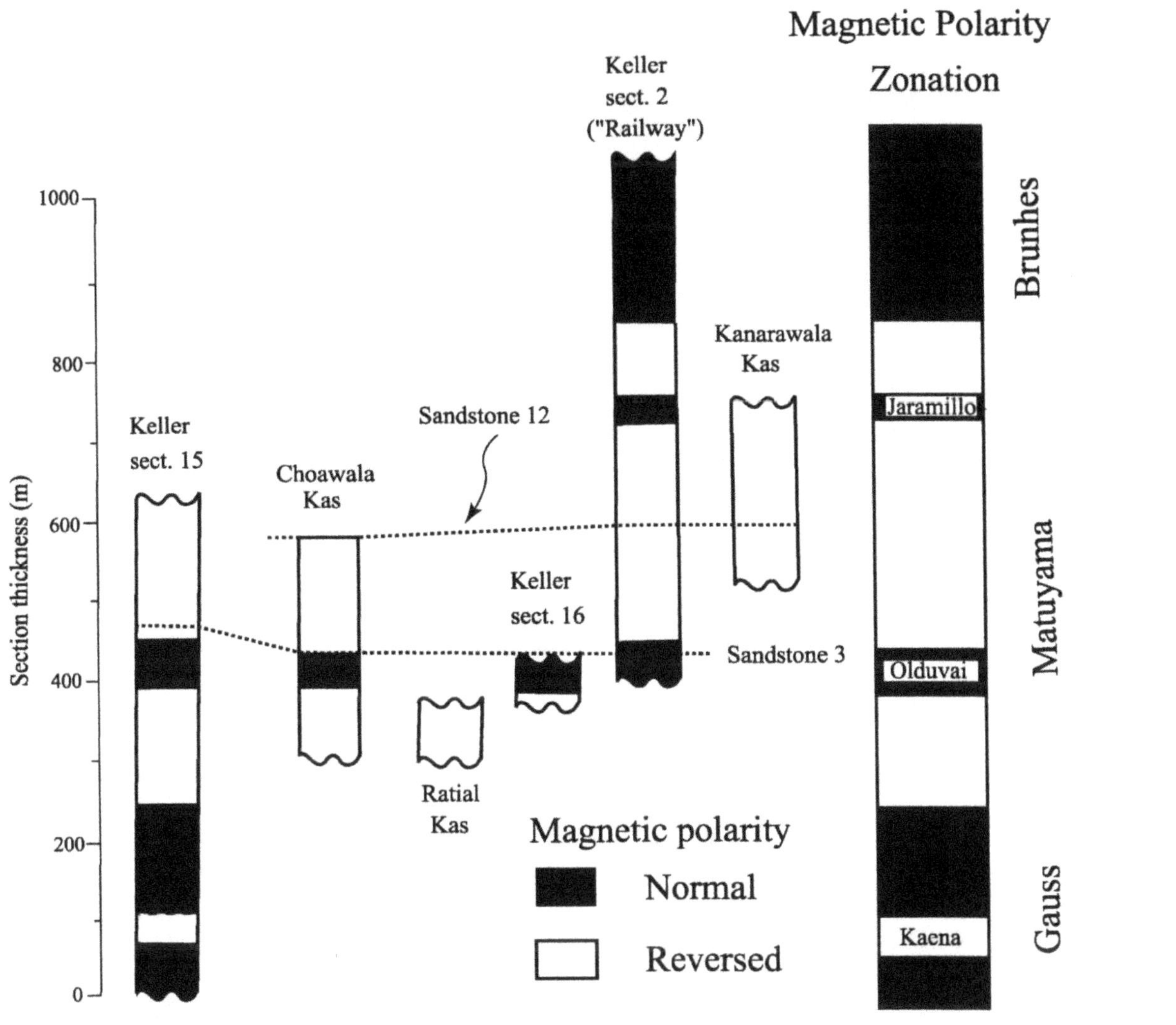

Figure 4.1: Combined palaeomagnetic stratigraphy of the Pabbi Hills

time transgressive (e.g. as an alluvial fan system progrades into a sedimentary basin). Keller (1975) shows that Sandstone 3, in his nomenclature, appears to be slightly diachronous with normal polarity in Sections 2 and 16 but reversed polarity in section 15. He comments that "*the discrepancy can be attributed to error in strike-mapping the ridge on the aerial photo and/or to the time-transgression of the strata*" (Keller, 1975 p.47). Keller's (1975) Sandstone 3 was mapped as Sandstone 5 in the present study. In Choawala Kas Sandstone 5 is a major ridge forming sandstone some 21 m thick with normal polarity silts immediately beneath and with a silty-sand lens with reversed polarity magnetisation 6m from the top of this unit. The unit therefore straddles the magneto-zone boundary between Matuyama (C1r.2r) and Olduvai (C2n). In Keller's (1975) log of the Section 2 (Railway Section) this sandstone is shown within a normal polarity zone some 12m below this C1r.2r/C2n boundary. The Choawala Kas data therefore reinforce the interpretation of Sandstone 3 as time transgressive with respect to the magnetic polarity zones defined here. This is a significant finding since the sandstone units are used both by Keller et al, (1977) and in the present study as though they were actually time-lines. The question then arises as to the uncertainty values which need to be placed around units such as sandstone 3 that would encompass the age range exhibited in the sampled sections. Assuming a mean sedimentation rate of 0.37m/1000yr (Keller, 1975), the age difference of sandstone 3 between Keller's (1975) Section 15 and Section 2 is estimated to be 70,000yr. Thus the age of Sandstone 3 at Section 15 is 1.733Ma and at Section 2 it is 1.802Ma, values within 2% of 1.77Ma.

Correlations of fossil occurrences

The palaeomagnetic zonation established by Keller and this study provides a framework for placing most of the fossil material found in the Pabbi Hills between 1986-1990 within timebands of 200-300 ka. During the field surveys, fossil occurrences were recorded in respect of the major sandstone "marker" horizons described in this study (see

Figure 5.3). Those most of those found in 1987 south of the GT road could thus be dated as earlier than, during, or later than the Olduvai Event but still older than Sandstone 12. Those found along Sandstone 12 can be dated to ca. 1.2 – 1.4 Ma. Given the poor dating of fossil material from many Upper Siwalik sections, this marks a significant advance in establishing the time range of Upper Siwalik vertebrate taxa.

Conclusions

The additional palaeomagnetic sections at Choawala Kas, Ratial Kas and Kanarawala Kas add support to the magnetic polarity zonation of the Upper Siwalik sediments within the Pabbi Hills originally derived by Keller (1975) and Keller et al. (1977). Ridge-forming sandstone units provide prominent stratigraphic markers that can be traced for many km along the limbs of the anticline. However, the position of one of the sandstone units close to the Matuyama (C1r.2r)/ Olduvai (C2n) boundary varies along the strike relative to that boundary. Although the unit is therefore time-transgressive (diachronous) the uncertainty in age that is introduced is only of the order of ±2%. Given the magnitude of these uncertainties the sandstone horizons can still be used as stratigraphic markers, while the magnetic polarity zonation provides a sound framework for the correlation of the vertebrate faunas.

Chapter 5: The 1986 survey (fossil occurrences 14 - 156) of Sandstone 12 and adjacent units

Introduction

Sandstone 12 was an obvious place to begin the surveys in the Pabbi Hills: first, is very easily recognised in the field, and secondly, it can be easily traced over several kilometres on both sides of the anticline, and on both sides of the G.T. road. In addition, its estimated age of 1.2 - 1.4 Mya (Keller et al.,1979) was close to what many then thought was the age of the earliest hominins in S.E. Asia, which Pope and Cronin (1984) had assessed as being no older than ca. 1.3 Mya.

The choice of Sandstone 12 was fully justified by the results. Figure 5.1 shows the main area surveyed in 1986. The first major discovery was on 30/01/86, when site 14 was discovered in the area of Kanarawala Kas. Working thereafter along the scarp-face exposures of Sandstone 12, site 73 was found only 10 days later. Survey was then halted for five days because of heavy rains, but proceeded thereafter with few interruptions. The exposures along Sandstone 12 north of the G.T. road had been surveyed by the end of February, and a week was then spent collecting fossil along Sandstone 14 (occurrences 90 - 108); thereafter, the team split into two groups. One surveyed the badlands in front of Sandstone 12 on the north side of the G.T. road, whilst the other surveyed the area to the south. The survey ended with the discovery of occurrence 156 on 29/03/86: overall, over 140 fossil occurrences and over 12,000 specimens had been discovered in eight weeks. The remainder of the field season was spent conducting a short excavation of locality 73 - the richest found that season - and working at base to clear the backlog of fossil material that had to be cleaned, marked and catalogued.

Figure 5.1: The main areas surveyed along Sandstone 12 in 1986

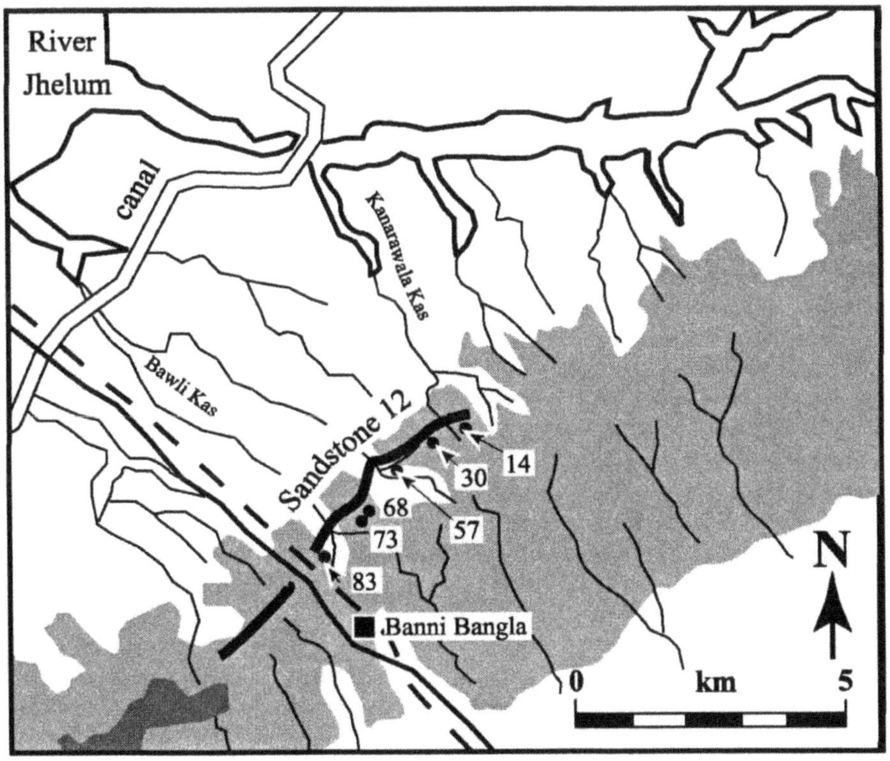

For the most part, the terrain was very dissected, and lacked obvious landmarks. Sandstone 12 itself in the area surveyed in 1986 was easy to recognise in the field, as the ridge-forming sandstone uniformly caps a distinctive pink clay horizon that in turn overlies a thick sequence of predominantly yellow sands and silts. The main reference points used during the surveys along Sandstone 12 were the stream channels (= *nullah* or *kas*) that cut through units such as Sandstone 12; electricity pylons (notably the two on top of Sandstone 12 at no. 33; see below); concrete triangulation points near occurrences 14, 83 and 133; a brick weir in one of the channels in the badlands north of the G.T. road; and the radio mast at Banni Bangla, near the G.T. road and on top of the watershed. As this volume may be used by future research groups in the Pabbi Hills, some information is given on how to locate the main fossil occurrences. Relevant GPS readings from 1999 are also included. Tables on the contents of fossil occurrences are given at the end of this chapter.

Overview

Table 5.1: Fossil occurrences of the 1986 survey grouped according to age, sandstone (SST) unit, and location north or south of the Grand Trunk (G.T.) Road.

Older than Sandstone 12	Sandstone 12 badlands	Sandstone 12 scarp	Sandstone 12 dipslope	Sandstone 14
1.4 - ?1.7 Mya	**1.2 - 1.4 Mya**	**1.2 - 1.4 Mya**	**0.9 - 1.2 Mya**	**0.9 Mya**
		42 - 49		
	24, 25	14 - 23, 109		
	41	26 - 28, 30 - 40	29	
		51 - 54, 56 - 63	55	
	75	64 - 74		
	76 - 80	81 - 85	87	
				90 -108
	110 - 150 even nos.			
North of Grand Trunk (G.T.) Road				
South of Grand Trunk (G.T.) Road				
		111 - 137 odd nos.		
139 - 147 odd nos.				
153 - 156				

Table 5.1 summarises the age and stratigraphic context of fossil material collected in 1986. Most was collected from the scarp face of Sandstone 12 north of the G.T. road (see Figure 5.1). The remainder was derived largely from the "badlands" in front of, but belonging to Sandstone 12; from the scarp face of Sandstone 14 that lies behind Sandstone 12; and from exposures south of the G.T. road.

Table 5.2 (see end of chapter) lists the fossil occurrences from 1986 in terms of the area in which they were found, their age, size and type. As indicated, the fossil occurrences found in 1986 can be grouped into eight main sets, defined primarily in topographic terms by the area in which they were found. The first two (14-25, 43-50 and 26-41) came from over a kilometre of exposures on both sides of Kanarawala Kas, and contained dispersed but abundant remains of turtle and a wide range of herbivores. Fossil occurrences 14, 30-34, 109 and 815 (found in 1999) were particularly productive. The third set (51-63) was meagre. The fourth set (64-75) included two major concentrations – localities 68 and 73 - that probably accumulated on land, did not contain turtle or crocodile remains but had good evidence of mammalian carnivores. Apart from these two major localities, fossil material was scarce in this area. The fifth set (74-85), near the G.T. road, was generally unremarkable apart from locality 83 and an egg-shell occurrence (number 85). The sixth set (110-150, even numbers only) included some large, turtle-rich, fossil localities, notably occurrences 116, 124 and 128 from the badlands immediately in front of locality 83 and in sand units belonging to Sandstone 12. A seventh set (111-137, odd numbers only) came from deposits along and in front of Sandstone 12 on the south side of the G.T. road. The last set was obtained from exposures along Sandstone 14 between Kanarawala Kas and the G.T. road, and which are probably ca. 0.9 Ma old.

A preliminary report on the 1986 field season was published by Jenkinson et al. (1989), which summarised the main features of the sedimentary sequence along Sandstone 12, and the types of fossil occurrences that were found. As indicated in that report, the main fossil horizons occurred in fine sands and silts, representing overbank and abandoned/slow-water channel deposits. In 1986, Rendell obtained a detailed sedimentary profile along Sandstone 12 between locality 14, Kanarawala Kas, and the section by the railway line, near the G.T. road by excavating and recording all sedimentary units at each major locality, and at intervening intervals. This is summarised in the following diagram (Figure 5.2).

A note on the tables of fossil specimens: Fossil specimens were numbered by the occurrence (e.g. the number of the locality, scatter or general collection in which they were found, and then by "GB" (to denote the Geological survey of Pakistan/British Archaeological Mission to Pakistan), and then by their own unique number. An example would be 18GB1. Each year had its own series of numbers: 1-199 for 1986; 200-599 for 1987; 600-699 for 1989 and 700-799 for 1990."W" in the tables of specimens refers to their condition, and is based on Behrensmeyer's (1978) scheme of recording bone weathering. Here, a score of 1 would denote a specimen with perfect surface condition; a score of 5 would denote one in extremely poor condition. The scoring differs from Behrensmeyer's in that it denotes overall condition, and thus includes all changes that were sub-aerial prior to burial, ones that occurred during burial, and those occurring after re-exposure. Measured specimens are indicated by bold text. All indeterminates are lumped, irrespective of skeletal element, to save space. All specimens diagnostic to taxa are stored in GSP, Islamabad.

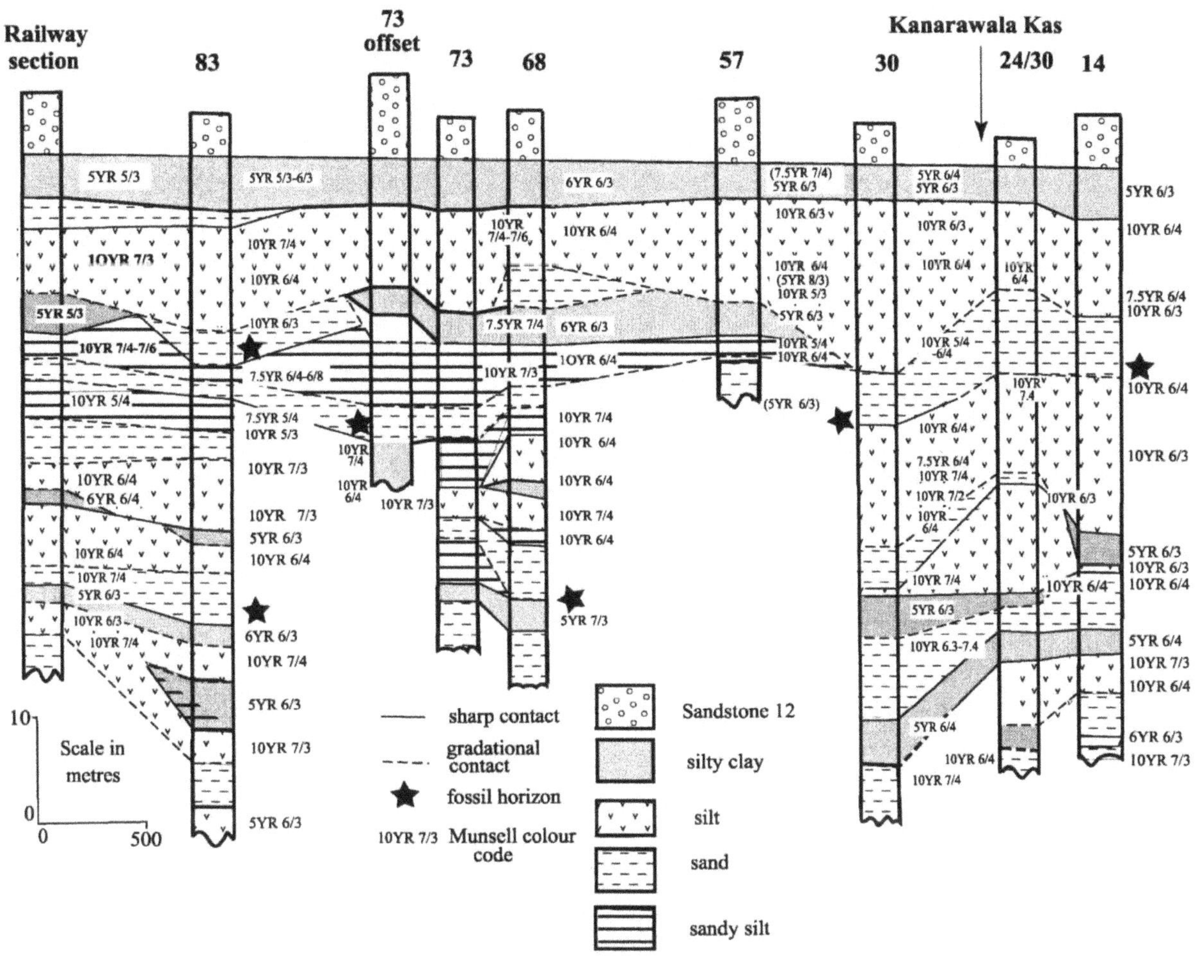

Figure 5.2: Sedimentary section along Sandstone 12 between Kanarawala Kas and the G.T. road

Figure 5.2: Sedimentary section along Sandstone 12 between Kanarawala Kas and the G.T. road. Redrawn from Jenkinson et al. 1989, Figure 2.

Set 1: The North-East Side of Kanarawala Kas: 14-25, 42-50, and 109

To reach locality 14, drive along the watershed track (see chapter 3) to a point directly above a good, well-trodden footpath (see below for the relevant GPS point). Follow this track down into Kanarawala Kas until one reaches a very deep section on the right-side side of the kas. (Occurrence 109 lies above and to the left of the deep section). At this point follow the kas channel running along the base of Sandstone 12 for about 300-500m. Locality 14 is on one of the ridges leading up to the top of Sandstone 12. The top can be reached by a good exit at the left-hand (west) end of the sandstone ridge: there are the remains of a very weathered trigometric point nearby which is roughly above locality 14. Alternatively, one can drive up Kanarawala Kas from the canal road (see chapter 3) and then walk for ca. 1 km to the already mentioned conspicuous section of yellow deposits at the edge of Sandstone 12. (It is prudent to mark the path into and down Kanarawala Kas with small cairns so that the exact route can be followed on the return trip. On no account should any other return route be attempted).

Relevant GPS points:

Watershed track at top of path into Kanarawala Kas:	32° 51' 840"N., 73° 52' 052"E.
Trigometric point above locality 14:	32° 52' 342"N., 73° 51' 362"E.
Base of locality 14:	32° 52' 337"N., 73° 51' 460"E.

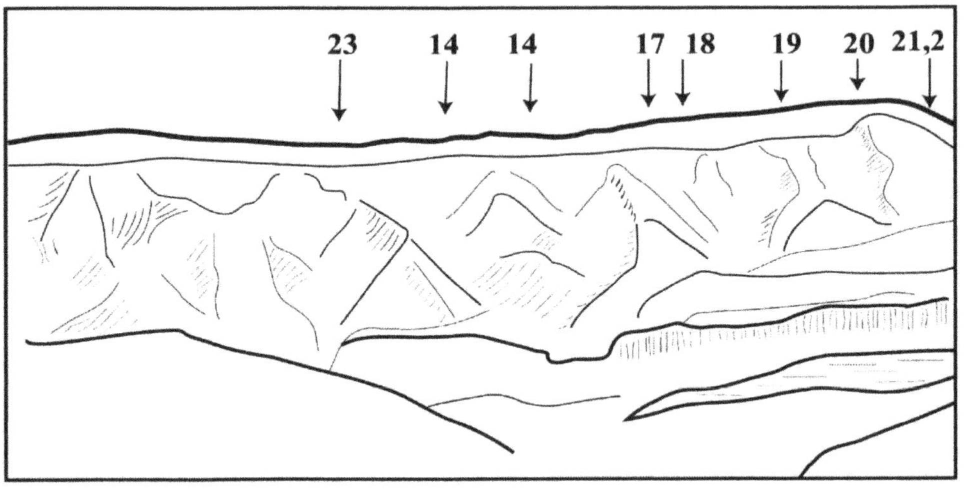

Figure 5.3: Schematic view towards fossil occurrences 14-22, scarp face of Sandstone 12, north bank of Kanarawala Kas. Drawn from a field photograph, taken on the approach to Sandstone 12 from the watershed track. Kanarawala Kas and occurrence 109 are slightly to the left of the figure.

Figure 5.3 shows where fossil occurrences were found on this part of the scarp face of Sandstone 12 north of Kanarawala Kas.

Locality 14: The fossils found at locality 14 were eroding from and down slope of a small hummock near the top of the scarp face. During collection, locality 14 was divided into 14A (upslope) and 14B (down slope). The fossil material is summarised in Table 5.3. Adjacent fossil occurrences 15-25 were generally smaller and more diffuse (see Table 5.4). Number 15 was a general collection from the area around site 14. Fossil occurrence 16 was from the gully bed parallel to the scarp face and below 14 and 15, from which this material was probably derived. Numbers 17 - 21 were scatters from areas defined by gullies cutting into the scarp face to the north of site 14. Number 17 lay ca. 100m north of locality 14; and number 21/22 was at the head of a small saddle which defines the start of the gully that run along the base of this part of Sandstone 12. Interestingly, the area between numbers 14 and 17 produced a considerable amount of new material (fossil occurrences 815 and 816) in 1999 that was almost as good as that found at locality 14 in 1986.

The scarp-face exposures between locality 14 and Kanarawala Kas contained a small scatter (23), and a small but rich concentration of material (occurrence 109) near the top of the scarp face, and overlooking Kanarawala Kas. This scatter was 3m by 5m, eroding from pink silts overlying a sand unit with grit at base, and contained numerous equid and artiodactyl bones as well as some ostrich shell. A considerable amount of new material was found in 1999. This included remains of a gazelle, some suid remains, several equid post-cranial remains, and a superb pair of equid mandibles that were lifted in a plaster jacket.

The exposures north of fossil occurrence no. 22 were surveyed as far as the next drainage system (Bangdwara Kas), but produced little of note. Table 5.5 summarises the contents of fossil occurrences 42-50.

Set 2: The West side of Kanarawala Kas (26 - 41)

This area of exposures can be reached by walking along the top of Sandstone 12 towards a pair of pylons which stand on top of the sandstone directly above occurrence 34. From here, Banni Bangla is 190°, and Flag Hill 220° from the pylon nearer the G.T. road. One can also reach these pylons by driving along the watershed track north of the G.T. road. Shortly before another pylon and nearby trigometric point by the track, there is a smaller track on the left leading downhill. Drive as far as is prudent, and then walk towards the pylons on the top of sandstone 12. These can be reached by an easy ascent up a good goat-path up the gully below the pylons, trending left to an easy exit to the top of Sandstone 12.

GPS point: larger of two pylons above occurrence 34: 32° 57' 982"N., 73° 50' 797"E.

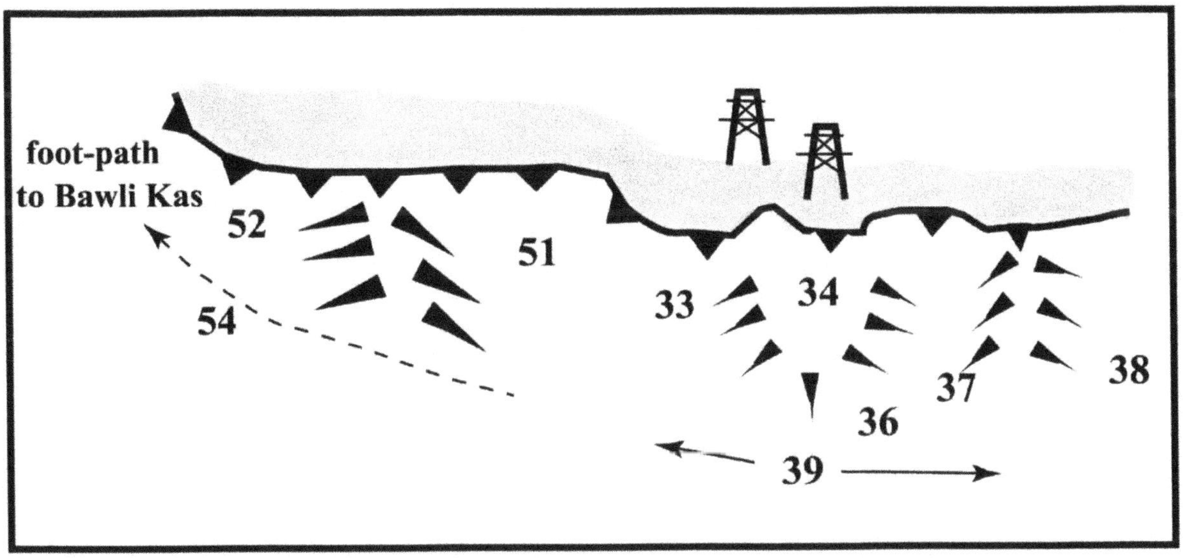

Figure 5.4: Fossil occurrences 33-39, Sandstone 12. Based on a field sketch by Hannah Jones.

Figure 5.4 summarises where most fossils were found in this area, and Table 5.6 summarises the contents of each fossil occurrence. This set included some productive scatters (30-38), some of which may have been eroding areas of a single, larger locality rather than discrete death events. No. 29 was on the dip-slope behind Sandstone 12, and is thus younger than those occurrences on the scarp face of SST12. No. 33 lay slightly left (facing upslope) of the pylons on top of Sandstone 12. In this area, ground visibility was generally good, with reasonably rapid erosion, little vegetation and few weathered sandstone blocks. No. 34 was immediately below the pylons in a series of small gullies in a horseshoe. Bone and carapace fragments were found eroding from a gritstone horizon ca. one third of way down from top of the ridge. Most fragments were found in the central gully. Number 35 was at the bottom of the scarp, where mainly medium sized fragments were eroding from clay layers with some gritty lenses outcropping from a side gully and the main gully bed. Number 36 was located right (looking upslope) of the pylon on top of ridge, east of scatters 30 and 31, and on a SSE-facing slope. Bone was found near/below the junction of yellow sand and red clays, and over the gritstone horizon noted at scatter no. 35. Scatter no. 37 comprised a small number of bone fragments from a small area on a 30° south-east-facing slope adjacent to a steep ridge separating one series of scarp-face gullies from the next. Number 38 was east of no. 36, and found at the junction of a yellow sand unit over pink clay, in turn over a thin gritstone horizon; several pieces of turtle carapace had a gritstone matrix. Scatter no. 39 included some weathered bone found on a red clay surface at the base of the section below nos. 33-38. This material was probably derived from further upslope, as there was no evidence that it was in the red clays themselves.

Figure 5. 5 shows a schematic section of the deposits at fossil occurrence 30 that was sketched in the field on the day that this occurrence was discovered. This should be compared with the more detailed section subsequently recorded by Rendell, and shown in Figure 5.2. As can be seen, the field sketch correctly recorded the sequence of deposits, and identified the source of the fossils. This implies that other sketch sections done in the field during fossil collecting probably provide a rough but accurate indication of local sequences.

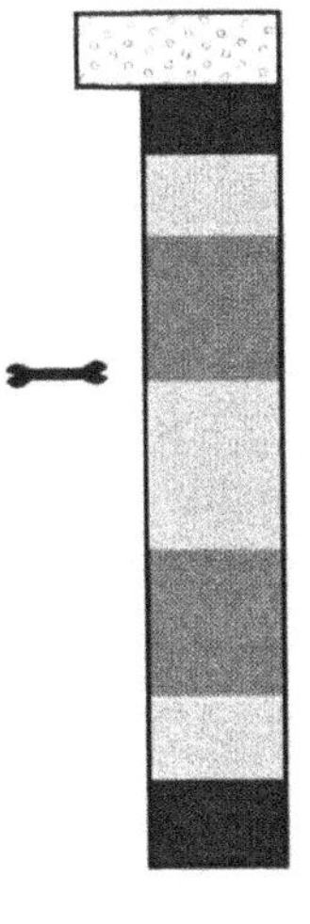

Figure 5.5: Schematic section at fossil occurrence 30. Based on a field sketch by Hannah Jones. Approximate section height is 60m.

Set 3: Fossil occurrences 51-63.

The contents of this set are itemised in Table 5. 7. Apart from one reasonably productive general collection (no. 57), there was little of significance in this area between the area covered by the previous set and Bawali Kas. The main finds were a canid mandible (no. 54) from a gully parallel to the base of Sandstone 12, an elephant cranium in a sandstone block (no. 57, left in situ), and a suid scapula (no. 58) near Bawali Kas. Scatter no. 60 was found in an area in front of Sandstone 12.

Set 4: Fossil occurrences 64 – 75.

The scarp face of Sandstone 12 south of Bawali Kas would also have been rated as very unproductive in terms of fossil material had localities 68 and 73 not been found. General collections no. 65 and 67, for example, produced only 22 fragments each from ca. 300 - 400m of scarp face, despite good ground visibility. (As will be seen later in this volume, other major fossil localities (notably 362 and 642) were also found in areas that were otherwise very poor in fossil yield). The contents of the fossil occurrences in this set, excluding localities 68 and 73, are summarised in Table 5.8. General collection 75 was made in the badlands immediately in front of Sandstone 12.

Localities 68 and 73

Localities 68 and 73 were among the richest fossil localities found in the Pabbi Hills. Both were very localised concentrations in areas that were generally poor in fossil. **Locality 68** was in a gully cutting into the scarp face of Sandstone 12, and eroding from a soft, yellow-buff sand near the base of the section. The area - covering some 5m horizontally and 7-8m vertically - was divided into zones A, B, C and X, and bagged accordingly (see Figure 5. 6). As can be seen from Table 5.9, ungulate remains were very common, and carnivores were also represented. There was no indication of turtle, crocodile, or any other aquatic species. Many of the fossils were very fragmented, light, white and fragile, and needed consolidating with PVA once back at base. Some from a slightly higher level were better preserved, and also had a sandy matrix. It is unclear why fossil preservation here was different from elsewhere. A small scatter (no. 69) immediately adjacent to (and possibly even part of) site 68 contained well-preserved fossil; however, at the top of the dip slope in front of site 68, similarly light fossils were found in a similar yellow sand. Some local factor may have affected the normal course of bone diagenesis.

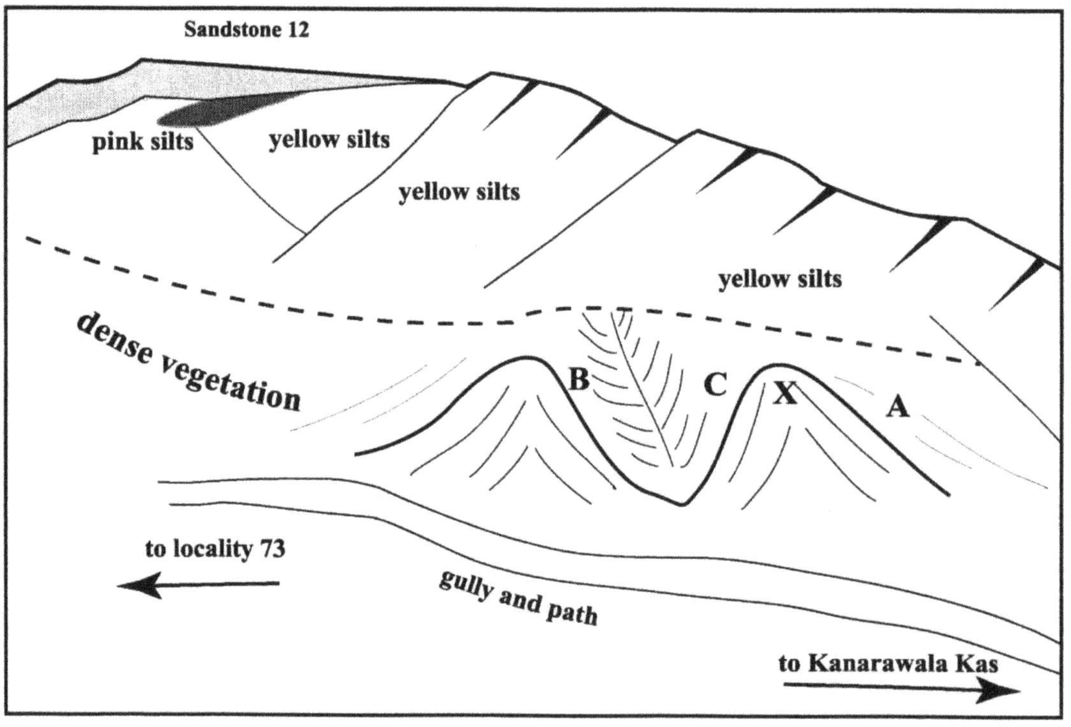

Figure 5. 6: Schematic view of locality 68. (Based on a field sketch by Sheila Sutherland).

Locality 73

To reach the area of locality 73, start at the brick weir mentioned in the description of how to reach locality 83 (see next box). Proceed further downstream to the kuchi camp, and bear rightwards up a tributary. This leads towards Sandstone 12; locality 73 was on a small spur in front.

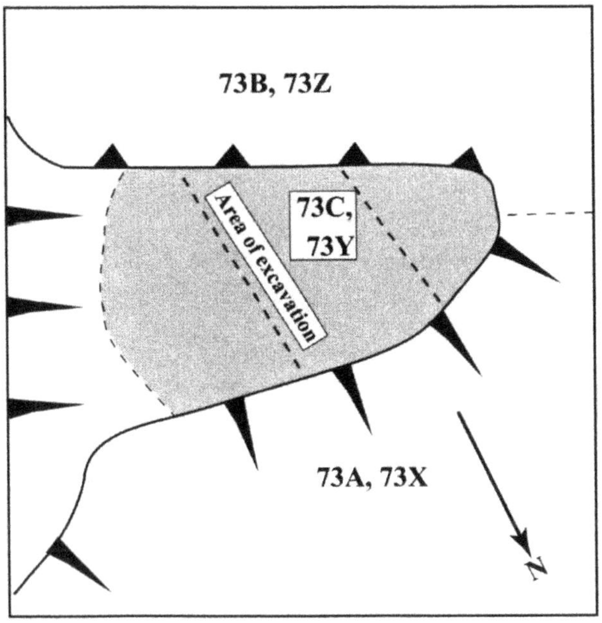

Figure 5. 7: Schematic map of locality 73. Based on field notes by Sheila Sutherland

Locality 73 was a small but very rich fossil locality and was found a few hundred metres west of locality 68 in a deeply dissected area of Sandstone 12, where the scarp face features a series of gullies and spurs. Ground visibility was good, with fairly active erosion, little grass, and few weathered sandstone blocks. The site lay on one of these small spurs, facing ca. 245 - 250° N, whilst the scarp face here faced 160° N. When discovered, the two most obvious features were a heavy concentration of bone fragments on top of and on both sides of the spur, and a complete *Rhinoceros* scapula eroding from the spur's surface, and overlying an almost complete but very fragile carnivore skull. At the time, it was late in the afternoon, and heavy rain was also imminent. The scapula, cranium and a few other smaller elements were therefore removed, and carried back to base, and the top of the spur was protected temporarily with sand and brushwood. When fieldwork resumed after five days of rain, the site was divided into three main zones, A, B and C (see Figure 5. 7 and Tables 5.10-12); the subsurface of each of these was also searched with trowels and some small-scale sieving as zones X, Y and Z respectively (Tables 5.13-15). Like site 68, site 73 was very localised, as all the material was eroding from a small area of little more than 1.5 sq. m. of light-brown silts on top of the spur. In general, large but fragmented bones were found on the top of the spur (zone C), with a great variety of material spilling down either side into zones A and B. The *in situ* deposits were excavated; the results are presented later in this volume. As can be seen from Tables 5.10-15, bovids, cervids and equids were the commonest remains; there were also remains of several carnivores, but no evidence of turtle, crocodile, or elephant.

Set 5: Fossil occurrences 74 - 85 (Site 73 to the G.T road), Sandstone 12

As with locality 68, there was little other fossil material between locality 73 and the next major occurrence, in this case locality 83. The material found is listed in Table 5.16. The adjacent general collection (no. 74) covered a large area of the slopes and gullies in the area below 73 and towards the bed of the main *kas* in this area, but it was wholly unrelated to locality 73, and there were no obvious fossil concentrations, or increase in frequency towards site 73. Ground visibility was also good, with only sparse grass and bushes. General collections 76-78 and 80 were unremarkable, as was scatter no. 79 that had probably eroded from buff sand/silts at the base of the local sequence. Nos. 76 - 80 were east of the area between the weir and the *kuchi* camp (see box below), and west of site 73 . The terrain in this area is heavily dissected, and ground visibility was often poor because of dense vegetation (although less so on south facing slopes). There were also two find-spots: no. 81, an artiodactyl scapula in a block of grey sandstone that had fallen from the ridge, and no. 82, large amounts of fragmented elephant tusk.

Site 83

There are two main ways of reaching locality 83. The easier is from the G.T. road. Start in front of Sandstone 12 from a culvert that runs under both the dual carriage-way and the railway line. Once past or through the culvert, follow a good footpath that runs roughly parallel to the scarp face of Sandstone 12 for about 0.5 km. The path then traverses a low watershed ridge that leads into the next local drainage system. Once over this watershed ridge, the scarp face of Sandstone 12 can be seen. Here, it stretches for about 300m, with a blunt promontory at the right hand end. Locality 83 lies along the ridges and gullies at this end. The top of the sandstone can be reached via a short climb at the northern end. The top is heavily vegetated, but the G.T. road can be clearly seen. In 1986, a trigometric point on the edge of sandstone and directly over locality 83 was a useful marker, but this had fallen downslope after part of the scarp face collapsed in the 1990's.

The other way of reaching occurrence 83 is from Banni Bangla. Walk down the railway line, and then follow a reasonably good footpath that leads into a small gully. This in turn leads into a larger gully with a brick weir a little downstream of the confluence. (This weir is one of the few reference points in this part of the area). Immediately upstream of the weir, walk towards the G.T. road until the promontory above locality83 can be seen. Fossil occurence 83 lies on the right-hand end of this face. Downstream from this weir, there is sometimes a small *kuchi* camp.

Relevant GPS points:

Brick weir:	32° 5'1 284"N., 73° 49' 655" E.
Sandstone ridge above occurrence 83:	32° 51' 003"N., 73° 49' 609" E.
83E (ostrich egg shell):	32° 51' 013"N., 73° 48' 540" E.

Site 83 was another large but isolated discovery along this part of Sandstone 12. As with sites 14, 68 and 73, the locality was divided into several collecting zones (A - G), (see Figure 5. 8) although these are unlikely to have much significance. Ground visibility was good, with relatively active erosion and little vegetation. The fossil material (summarised in Table 5. 17) was found on and downslope of a series of adjacent ridges and gullies on the scarp face of Sandstone 12. Although some fossils had very little surface damage, others were more abraded and weathered, and this may reflect some degree of reworking and redeposition. Given the high incidence of erosion, and the 40°-slope of the scarp face, it is unlikely that fossil has been extensively weathered and abraded since being exposed.

The uppermost sandstone unit (i.e. capping Sandstone 12) is about 5 - 8 m thick, and the dominant colours of the underlying units are (in descending order) pink, buff, pink and buff (see Figure 5. 8). The upper buff unit has a horizon of small quartzite pebbles, mostly 2-3 cm. in diameter, but up to 8 cm. in diameter. Most of the fossil was found on the lower buff unit (zones A-C), or downslope (zones D and F). A couple of fragments were found in a thin gritstone horizon ca. 0.5 m thick (zone G) above the upper buff unit and on top of a low ridge. This grit horizon is ca. 8m wide, dips at ca. 10°, and shows some cross bedding. There was also a small patch of *in situ* eggshell fragments (zone 83E in the 1986 collection) in a pink mudstone on the right-hand side of a spur ridge (looking downslope) that defines the northern limit of 83.

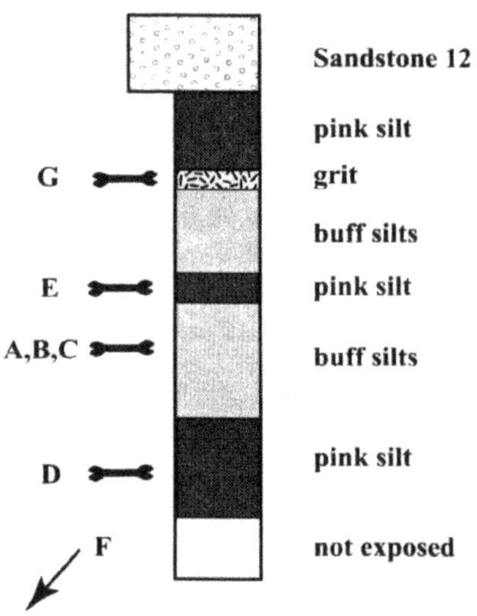

Figure 5. 8: Schematic section at locality 83, based on a field sketch by Sheila Sutherland. The section is ca. 30m high.

The discovery of site 83 was followed by two general collections westwards along the scarp of sandstone 12. (Table 5.16). Despite good ground visibility, no. 84 produced little except a few clusters of scrappy material at the base of the scarp. No. 85 was the last along the scarp of Sandstone 12 before the G.T. road, and largely comprised *Struthio* egg-shell fragments. Near the main road, ground visibility was less good, as many slopes were vegetated, stable and covered with weathered sandstone blocks.

After the survey along Sandstone 14 (see later, this chapter), the team was split into two groups. One, using even-numbers for recording fossil occurrences, worked in the badlands in front of (and still belonging to) Sandstone 12 in the area of locality 83. The other team, using odd-numbers, worked south of the GT road, to follow Sandstone 12 southwards, and to reconnoitre the area for future surveys.

Set 6: The badlands in front of Sandstone 12: (nos. 110 - 150, even numbers only)

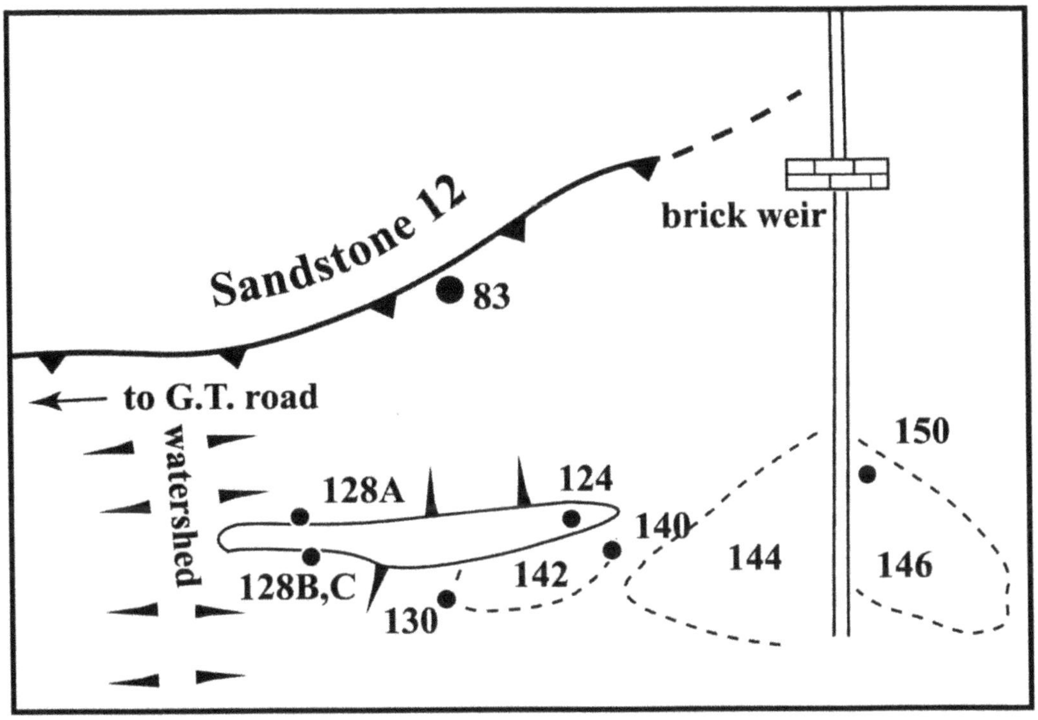

Figure 5. 9: Schematic map of localities 83, 124 and 128. Based on field notes by Hannah Jones.

The fossil material found in this area (Figure 5. 9) is summarised in Tables 5.18-20. The first found was 110, which was below and 232° from site 83, and ca. 100m distant. The fossil from this scatter came from an area of yellow sands 10m x 10m, just above a layer of red clays. No. 110 lay within an area of a general collect (no. 112) in the sparsely vegetated area in front of no. 83, going away from the G.T. road and towards the first main *kas*. No. 114 was a scatter from an area 20 m sq, facing west, and on a gentle slope. It was found 260° and ca. 500m from site 83, east of the large sandstone block above 79 - 80, and near the *kas* with the brick weir (see above, nos. 79 - 80). A larger scatter, no. 116, was found ca. 50m. east of 114, along the same junction of pink and yellow silts, in the area shown in the next chapter (Figure 5.xx). This was sufficiently large to sub-divide into four zones, A - D. Most material appeared to be derived from yellow silts (see next chapter, Figure 5. xx); indeed, a proximal metatarsal (116CGB18) was embedded in these, as was an indeterminate shaft fragment in zone B. However, 116AGB1 had a clay skin, and 116DGB1 had a gritty matrix.

No. 116 lay within a large general collection, no. 118, that lay between the *kas* with the brick weir, and the next one to the north. Within this area, and SSE of site 73 and 330° from the highest point of the nearby ridge of Sandstone 12, there was also a large scatter of ostrich egg-shell (occurrence 120), which covered some 4 sq m, and was eroding from the lowest yellow silt unit at the base of a local sequence of alternating pink and yellow silts. Other bone was found in the same general area, but was more scattered.

Scatter no. 122 was collected the following day near the base of no. 83 when it was approached from a different direction from the usual one. Although it may have been part of no. 83, it will nevertheless be treated as distinct from it.

The area of "badlands" in front of nos. 83 and 122 produced two very notable occurrences of fossil inside a large area of a general collect, no. 126. The first major occurrence was 124 (Table 5.19), which lay at 292° from the triangulation point near no. 83, and was evidenced by a very dense concentration of turtle bone coming from an area of silty sands 30 by 10m near the top of a low ridge parallel to Sandstone 12, with some gritstone and mudballs on the surface.

No. 128, the richest fossil occurrences in this area of badlands, was found ca. 200 east of no. 124, and 325° from the triangulation point near no. 83. The fossil came from both sides of a low ridge of yellow sands, but at a higher level than at locality 124. It was sub-divided into five zones. Zone A lay on the north side of the ridge, zone B on the south, C at the west end, D, within zone C, and E along the base of a small gully (see Table 5.20).

The area of badlands behind nos. 124 and 128, and between the G.T. road and the *kas* with the brick weir, was extensively covered in 1986 (see Figure 5.8). No. 130 was a large scatter of fragments from an area ca. 30m laterally by 6m downslope, lying on silty sands among grits, although some bones were also found at a higher level. A number of bones had a grey, gritty matrix; many were small and very weathered. Ground visibility was poor because of vegetation and weathered sandstone blocks. Nos. 132 and 134 were other general collections in this area, and contained no. 136, a scatter of elephant remains.

Scatter no. 140 was immediately behind 124, and 200m downstream of no.130, on the left hand side of a small gully. The fossil (largely turtle) was in slope wash derived from the erosion of a sand unit. The area between nos. 130 and 140 was covered by a general collection, no. 142. Nos. 144 and 146 were two further general collections in this area, and no. 148 was a small scatter within the area of 146.

The last major scatter found in this part of the survey was no. 150, which lay between nos. 116, and 83, and covered a large area of small ridges and gullies. The fossil was very weathered and fragmented, and some pieces had a sandstone matrix.

The survey south of the G.T. road: Set 7: Nos. 111 - 137 (odd numbers only), Sandstone 12

The scarp face of Sandstone 12 and associated "badlands" were also surveyed south of the G.T. road in 1986, as far as Choawala Kas. These are summarised in Tables 5. 21 and 5.22.

Scatter no. 111 was found in a heavily vegetated area on top of a pink silt, and immediately below a buff sand unit. No. 113 was the general collect of that area, within which there were two small scatters, nos. 115 and 117. No. 119 was a general collection further along the scarp of Sandstone 12, up to where it is close to the watershed track, and before a very vegetated section. No. 121 was a small scatter of bone fragments adjacent to this track, and was found just in front of a hillock of partly case-hardened sand. The fossil remains were found on the surface of a pink sediment, covered with grits and weathered fragments of sandstone.

No. 123 was found in an area of "badlands" in front of, but belonging to, the Sandstone 12 complex. It was a small scatter of fragments, most of which were probably from the same bone, eroding from a pink horizon, in turn overlain by a sandy buff layer. Two other small scatters were also in this area. No. 125 was found on pink silt, below an orange-buff sand horizon. No. 127, 10 - 20m away, was a small cluster of largely amphibian remains from an orange-buff sand horizon. Both were found within a general collection area, no. 129, in the "badlands" between the scarp face of Sandstone 12 and the watershed track. Most of the specimens found in this area were found in small clusters on small humocks of gritty material.

The main reference point for nos. 131, 133 and 135 is a concrete triangulation point on the right-hand side of the watershed track (proceeding from the G.T. road). (The GPS reading is 32° 49' 659"N., 73° 048' 459"E.) .This marker is on a small ridge at right angles to the main ridge of Sandstone 12, and left of a gully. No. 131 was found to its left (if facing the main ridge) at the end of a small ridge that is roughly parallel to the strike of Sandstone 12. It was a scatter of egg shell fragments (very like nos. 85 and 120 on the north side of the G.T. road) found on a gently sloping remnant of pink silt on top of a buff sand. Fossil locality 133 (Table 5. 22) was on the other side of the gully, on the slopes of a ridge at right angles to the main scarp face, and ca. 100° from the trigometric point. There is also a pylon visible at 275° from no. 133. It was a very large, diffuse scatter of bone, much of which was weathered and fragmented. Many fragments were partially embedded in slope wash, and similarly weathered. Ground visibility was only moderate because of thick tufts of grass. The fossils were derived from just below a pink silt which forms a slight bench through case hardening, in the middle of a long sequence of alternating pink silts and buff sands. No. 135 was a find-spot of fragmented elephant tusk eroding out of a greyish sand (also visible in the section containing no. 133), in the middle of a major buff horizon above the thin pink silt containing no. 133. No. 135 was on the second ridge projecting from the main scarp face beyond no. 131. The general collect for the area containing fossil occurrences 131-135 was no. 137.

The survey south of the G.T. road: Set 8: Nos. 139-153 (odd numbers only), 154, 156; older than Sandstone 12

This part of the survey ended with a brief reconnaissance of the area south of the G.T. road and between (and thus older than) Sandstone 12 and the watershed track. Finds were largely unexceptional, and are summarised in Table 5.23.

The main discoveries were a reasonably good scatter, no. 139, that contained a piece of *Stegadon*, and a small locality, no. 156, the last discovery of the field season and located near the watershed track. Details are given in Table 5.24.

Set 9: The survey of Sandstone 14 (fossil occurrences 90 - 108)

In addition to collecting material from Sandstone 12 in 1986, some material was also collected from Sandstone 14. However, this proved to be disappointing, as no major fossil localities or archaeological evidence were found, although fossil preservation was often good. Most of the area surveyed along Sandstone 14 from Kanarawala Kas to the G.T. road was covered by a series of general collections over large areas, with a small number of unexceptional scatters. These are summarised in Table 5:25. As with Sandstone 12, the main areas of survey tended to be defined by sections of scarp face between the stream channels that cut through them, notably Bawali Kas and Kanarawala Kas. The electricity pylons on top of Sandstone 12 and above the area of 33 - 57 were among the few topographic landmarks that could be used. Access to the area was either by a long walk from the watershed track down Kanarawala Kas past Sandstone 12, or via the G.T. road.

Parts of Sandstone 14 are capped by a conglomerate of poorly cemented quartzite clasts. Consequently, many exposures on the scarp face were obscured by stone debris from this conglomerate. As will be detailed later, several pieces of struck stone were also found in this part of the survey, although sadly none was in context (see chapter 8).

The first fossil finds were made when approaching Sandstone 14 from Kanarawala Kas, within the general collect area no. 90. Occurrence no. 91 was a scatter of crocodile and turtle remains from an area 20m by 20m, at the base of the face scarp, and 80° from the twin pylons visible on the top of Sandstone 12. At the time of the survey, there was a *kuchi* camp 500m before this scatter. Bone was heavily mineralised and iron stained, and most was concentrated in a small patch. The fossil was probably derived from yellow silty sands towards the base of the section. Another scatter (no. 92) was found ca. 100m to the east, and ca. 250m before a second *kuchi* camp in Bawali Kas, close to a very conspicuous deep section. (This is easily recognised by the large numbers of parrotts that nest in this section). As at no. 91, the fossil material came from a yellow silty sand in the lower part of the section. No. 93 was a general collect on a north-west facing slope on the way from no. 92 to site 73 (Sandstone 12), east of the *kuchi* village, and 120° from the top of site 73. Number 94 was a general collect of virtually indeterminate material along the base of Sandstone 14, between the two *kuchi* camps; the scarp itself was covered by another general collect (no. 95) involving four people, but produced almost nothing. However, two small scatters (nos. 96 and 97) were found further along from 91. No. 96 was on a stone covered surface overlying yellow silty sands; nearby, no. 97 was a small scatter 10m by 10m, on a hummock of pebble-covered pink sediment protruding through the more recent valley fill. Finally in this series, nos. 98 and 100 were general collects, obtained by contour walking towards Kanarawala Kas and starting ca. 500m north of the second *kuchi* camp. Three small scatters were found near Kanarawala Kas: no. 99 was a small scatter 5m by 10m in an area of yellow silty sands, the surface of which was covered with leaf, sandstone blocks, and some quartzite pebbles. Nos. 101 and 102 were similar scatters of poor material on a pebble-covered surface over yellow silty sands. A small amount of material was also collected between Bawali Kas and the G.T. road. No. 103 was a poor general collection, obtained by contour walking from the *kuchi* camp in Bawli Kas towards the G.T. road. This section contained a few small scatters, such as no. 104 and 105, both on pink silts covered by pebbles, and also nos. 106 and 107. This meagre set of results ended with the final general collection in this area, no. 108, which resulted in a general collection of just a few pieces of elephant tusk fragments from a large area of scarp face.

Table 5.2: Fossil occurrences from the 1986 field survey of the Pabbi Hills

No.	Area	Type	Set	Horizon	Total
14	Kanarawala Kas NE	Locality	Set 1	SST12 scarp, 1.2-1.4 Mya-old	693
15	Kanarawala Kas NE	General collect	Set 1	SST12 scarp, 1.2-1.4 Mya-old	21
16	Kanarawala Kas NE	Parallel gully	Set 1	SST12 scarp, 1.2-1.4 Mya-old	12
17	Kanarawala Kas NE	Scatter	Set 1	SST12 scarp, 1.2-1.4 Mya-old	29
18	Kanarawala Kas NE	Scatter	Set 1	SST12 scarp, 1.2-1.4 Mya-old	31
19	Kanarawala Kas NE	Scatter	Set 1	SST12 scarp, 1.2-1.4 Mya-old	26
20	Kanarawala Kas NE	Scatter	Set 1	SST12 scarp, 1.2-1.4 Mya-old	15
21	Kanarawala Kas NE	Scatter	Set 1	SST12 scarp, 1.2-1.4 Mya-old	5
23	Kanarawala Kas NE	Scatter	Set 1	SST12 scarp, 1.2-1.4 Mya-old	68
24	Kanarawala Kas NE	General collect	Set 1	In front of SST12	7
25	Kanarawala Kas NE	General collect	Set 1	In front of SST12	7
26	Kanarawala Kas West	General collect	Set 2	SST12 scarp, 1.2-1.4 Mya-old	8
27	Kanarawala Kas West	General collect	Set 2	SST12 scarp, 1.2-1.4 Mya-old	8
28	Kanarawala Kas West	General collect	Set 2	SST12 scarp, 1.2-1.4 Mya-old	10
29	Kanarawala Kas West	Scatter	Set 2	Above SST12	24
30	Kanarawala Kas West	Scatter	Set 2	SST12 scarp, 1.2-1.4 Mya-old	116
31	Kanarawala Kas West	Scatter	Set 2	SST12 scarp, 1.2-1.4 Mya-old	23
33	Kanarawala Kas West	Scatter	Set 2	SST12 scarp, 1.2-1.4 Mya-old	43
34	Kanarawala Kas West	Scatter	Set 2	SST12 scarp, 1.2-1.4 Mya-old	51
35	Kanarawala Kas West	Scatter	Set 2	SST12 scarp, 1.2-1.4 Mya-old	32
36	Kanarawala Kas West	Scatter	Set 2	SST12 scarp, 1.2-1.4 Mya-old	68
37	Kanarawala Kas West	Scatter	Set 2	SST12 scarp, 1.2-1.4 Mya-old	20
38	Kanarawala Kas West	Scatter	Set 2	SST12 scarp, 1.2-1.4 Mya-old	117
39	Kanarawala Kas West	Scatter	Set 2	SST12 scarp, 1.2-1.4 Mya-old	10
40	Kanarawala Kas West	Parallel gully	Set 2	SST12 scarp, 1.2-1.4 Mya-old	3
41	Kanarawala Kas West	Scatter	Set 2	In front of SST12	18
42	Kanarawala Kas NE	Scatter		SST12 scarp, 1.2-1.4 Mya-old	11
43	Kanarawala Kas NE	General collect	Set 1	SST12 scarp, 1.2-1.4 Mya-old	11
44	Kanarawala Kas NE	General collect	Set 1	SST12 scarp, 1.2-1.4 Mya-old	10
45	Kanarawala Kas NE	Scatter	Set 1	SST12 scarp, 1.2-1.4 Mya-old	17
46	Kanarawala Kas NE	Scatter	Set 1	SST12 scarp, 1.2-1.4 Mya-old	17
47	Kanarawala Kas NE	Scatter	Set 1	SST12 scarp, 1.2-1.4 Mya-old	63
48	Kanarawala Kas NE	Scatter	Set 1	SST12 scarp, 1.2-1.4 Mya-old	7
49	Kanarawala Kas NE	General collect	Set 1	SST12 scarp, 1.2-1.4 Mya-old	45
50	Kanarawala Kas NE	Find-spot, gully	Set 1	SST12 or older	75
51	Bawali Kas South	General collect	Set 3	SST12 scarp, 1.2-1.4 Mya-old	17
52	Twin pylons to Bawali Kas	General collect	Set 3	SST12 scarp, 1.2-1.4 Mya-old	5
53	Twin pylons to Bawali Kas	Find-spot, gully	Set 3	SST12 scarp, 1.2-1.4 Mya-old	1
54	Twin pylons to Bawali Kas	Find-spot, gully	Set 3	SST12 scarp, 1.2-1.4 Mya-old	1
55	Twin pylons to Bawali Kas	Scatter	Set 3	SST12 dip slope	7
56	Twin pylons to Bawali Kas	Scatter	Set 3	SST12 scarp, 1.2-1.4 Mya-old	18
57	Twin pylons to Bawali Kas	General collect	Set 3	SST12 scarp, 1.2-1.4 Mya-old	152
58	Twin pylons to Bawali Kas	General collect	Set 3	SST12 scarp, 1.2-1.4 Mya-old	12
60	Twin pylons to Bawali Kas	Scatter	Set 3	SST12 scarp, 1.2-1.4 Mya-old	60
61	Twin pylons to Bawali Kas	General collect	Set 3	SST12 scarp, 1.2-1.4 Mya-old	14
62	Twin pylons to Bawali Kas	Find-spot	Set 3	SST12 scarp, 1.2-1.4 Mya-old	1
63	Twin pylons to Bawali Kas	Find-spot	Set 3	SST12 scarp, 1.2-1.4 Mya-old	96
64	Bawali Kas South	Scatter	Set 4	SST12 scarp, 1.2-1.4 Mya-old	53
65	Bawali Kas South	General collect	Set 4	SST12 scarp, 1.2-1.4 Mya-old	22
66	Bawali Kas South	Find-spot	Set 4	SST12 scarp, 1.2-1.4 Mya-old	1

Table 5.2 continued: Fossil occurrences from the 1986 field survey of the Pabbi Hills

67	Bawali Kas South	General collect	Set 4	SST12 scarp, 1.2-1.4 Mya-old	22
68	Bawali Kas South	Locality	Set 4	SST12 scarp, 1.2-1.4 Mya-old	672
69	Bawali Kas South	Scatter	Set 4	SST12 scarp, 1.2-1.4 Mya-old	15
73	Bawali Kas South	Locality	Set 4	SST12 scarp, 1.2-1.4 Mya-old	1673
74	Bawali Kas South	General collect	Set 5	SST12 scarp, 1.2-1.4 Mya-old	326
75	Bawali Kas South	General collect	Set 5	In front of SST12	123
76	Site 73 to GT road	General collect	Set 5	SST12 scarp, 1.2-1.4 Mya-old	3
77	Site 73 to GT road	General collect	Set 5	SST12 scarp, 1.2-1.4 Mya-old	3
78	Site 73 to GT road	General collect	Set 5	SST12 scarp, 1.2-1.4 Mya-old	7
79	Site 73 to GT road	Scatter	Set 5	SST12 scarp, 1.2-1.4 Mya-old	40
80	Site 73 to GT road	General collect	Set 5	SST12 scarp, 1.2-1.4 Mya-old	43
81	Site 73 to GT road	Find-spot	Set 5	SST12 scarp, 1.2-1.4 Mya-old	1
82	Site 73 to GT road	Find-spot	Set 5	SST12 scarp, 1.2-1.4 Mya-old	161
83	Site 73 to GT road	Locality	Set 5	SST12 scarp, 1.2-1.4 Mya-old	1563
84	Site 73 to GT road	General collect	Set 5	SST12 scarp, 1.2-1.4 Mya-old	91
85	Site 73 to GT road	General collect	Set 5	SST12 scarp, 1.2-1.4 Mya-old	153
90	GT road to Kanarawala Kas	General collect	Set 9	Sandstone 14	11
91	GT road to Kanarawala Kas	Scatter	Set 9	Sandstone 14	101
92	GT road to Kanarawala Kas	Scatter	Set 9	Sandstone 14	25
93	GT road to Kanarawala Kas	General collect	Set 9	Sandstone 14	15
94	GT road to Kanarawala Kas	General collect	Set 9	Sandstone 14	17
96	GT road to Kanarawala Kas	Scatter	Set 9	Sandstone 14	32
97	GT road to Kanarawala Kas	Scatter	Set 9	Sandstone 14	5
98	GT road to Kanarawala Kas	General collect	Set 9	Sandstone 14	63
99	GT road to Kanarawala Kas	Scatter	Set 9	Sandstone 14	58
100	GT road to Kanarawala Kas	General collect	Set 9	Sandstone 14	17
101	GT road to Kanarawala Kas	Scatter	Set 9	Sandstone 14	12
103	GT road to Kanarawala Kas	General collect	Set 9	Sandstone 14	13
104	GT road to Kanarawala Kas	Scatter	Set 9	Sandstone 14	7
105	GT road to Kanarawala Kas	Scatter	Set 9	Sandstone 14	3
106	GT road to Kanarawala Kas	Scatter	Set 9	Sandstone 14	23
107	GT road to Kanarawala Kas	Scatter	Set 9	Sandstone 14	4
108	GT road to Kanarawala Kas	General collect	Set 9	Sandstone 14	11
109	Kanarawala Kas NE	Scatter	Set 1	SST12 scarp, 1.2-1.4 Mya-old	304
110	Brick weir area, near 83	Scatter	Set 6	SST12 badlands	45
111*	GT road to Choawala Kas	Scatter	Set 7	SST12 scarp, 1.2-1.4 Mya-old	260
112	Brick weir area, near 83	General collect	Set 6	SST12 badlands	11
113*	GT road to Choawala Kas	General collect	Set 7	SST12 scarp, 1.2-1.4 Mya-old	21
114	Brick weir area, near 83	Scatter	Set 6	SST12 badlands	52
116	Brick weir area, near 83	Scatter	Set 6	SST12 badlands	195
117*	GT road to Choawala Kas	Scatter	Set 7	SST12 scarp, 1.2-1.4 Mya-old	38
118	Brick weir area, near 83	General collect	Set 6	SST12 badlands	38
119*	GT road to Choawala Kas	General collect	Set 7	SST12 scarp, 1.2-1.4 Mya-old	67
120	Brick weir area, near 83	Scatter	Set 6	SST12 badlands	334
121*	GT road to Choawala Kas	Scatter	Set 7	SST12 scarp, 1.2-1.4 Mya-old	22
122	Brick weir area, near 83	Scatter	Set 6	SST12 badlands	29
123*	GT road to Choawala Kas	Scatter	Set 7	SST12 badlands	143
124	Brick weir area, near 83	Locality	Set 6	SST12 badlands	306
125*	GT road to Choawala Kas	Scatter	Set 7	SST12 badlands	33

Table 5.2 continued: Fossil occurrences from the 1986 field survey of the Pabbi Hills

126	Brick weir area, near 83	General collect	Set 6	SST12 badlands	27
127	GT road to Choawala Kas	Scatter	Set 7	SST12 badlands	12
128	Brick weir area, near 83	Locality	Set 6	SST12 badlands	586
129*	GT road to Choawala Kas	General collect	Set 7	SST12 badlands	35
130	Brick weir area, near 83	Scatter	Set 6	SST12 badlands	400
131*	GT road to Choawala Kas	Scatter	Set 7	SST12 badlands	194
132	Brick weir area, near 83	General collect	Set 6	SST12 badlands	11
133*	GT road to Choawala Kas	Scatter	Set 7	SST12 badlands	820
134	Brick weir area, near 83	General collect	Set 6	SST12 badlands	8
135*	GT road to Choawala Kas	Find-spot	Set 7	SST12 badlands	100
136	Brick weir area, near 83	Scatter	Set 6	SST12 badlands	5
137*	GT road to Choawala Kas	General collect	Set 7	SST12 badlands	22
139*	SST12 to watershed track	Scatter	Set 8	older than SST12	206
140	Brick weir area, near 83	Scatter	Set 6	SST12 badlands	169
141*	SST12 to watershed track	Scatter	Set 8	older than SST12	29
142	Brick weir area, near 83	General collect	Set 6	SST12 badlands	4
143*	SST12 to watershed track	General collect	Set 8	older than SST12	9
144	Brick weir area, near 83	General collect	Set 6	SST12 badlands	12
145*	SST12 to watershed track	Find-spot	Set 8	older than SST12	1
146	Brick weir area, near 83	General collect	Set 6	SST12 badlands	10
147*	SST12 to watershed track	General collect	Set 8	older than SST12	6
148	Brick weir area, near 83	Scatter	Set 6	SST12 badlands	20
149*	SST12 to watershed track	Scatter	Set 8	older than SST12	3
150	Brick weir area, near 83	Scatter	Set 6	SST12 badlands	165
153*	SST12 to watershed track	General collect	Set 8	older than SST12	21
154*	SST12 to watershed track	Scatter	Set 8	older than SST12	3
156*	SST12 to watershed track	Locality	Set 8	older than SST12	287
Total					**11710**

Notes: Fossil occurrences 21 and 22, and 113 and 115 were lumped; 59 was a find spot of 2 flakes near 60; 70-72, 86-89 and 95 had no finds recorded; 102, and 138 can also be ignored; and recording numbers 151 and 152 were not used. Occurrences 111-156: those marked with * were south of the G.T. road

Table 5.3. Contents of fossil locality 14, Sandstone 12. Measured specimens are indicated by bold text in this and subsequent tables.

Zone A (upslope):

No.	W	Taxon	Element	Part	Other	N
166		Artiodactyla	Cuboid			1
23		Artiodactyla	Femur			1
126		Artiodactyla	Metapodial			1
0	5	Artiodactyla	Molar	Spall		1
350		Artiodactyla	Patella	Complete		1
370	2b	Artiodactyla	1st Phalange	Proximal	Split	1
69,83	2	Artiodactyla	Rib			2
0	3	Artiodactyla	Scapula	Fragment		1
0	4	Artiodactyla	Tooth	Spall		1
167, 363	2	Artiodactyla	Tooth	Molar	Fragment	2
179,351		Artiodactyla	Vertebra	Complete		2
612		Artiodactyla	Vertbra	Lumbar		1
344, 354	3	Artiodactyla	Vertebra	Flange		2
26		Bovid	1st Phalange	Proximal	Complete	1

Table 5. 3 continued: Contents of fossil locality 14, Sandstone 12, Zone A.

67	2	Bovid	Mandible	P4, left	Fragment	1
193	2	Bovid	Mandible	DM3,M1,M2	Complete	1
199	2	Bovid	Mandible	M1,2,M3	Juvenile	1
190	2	Bovid *(Damalops)*	Maxillary	P2-4, M3		1
191	2	Bovid (*Damalops*)	Maxillary	Premolars		1
513, 601		Bovid	Metapodial	Shaft		2
3		Bovid	Tooth	3rd Molar	Upper	1
28 221	4	Bovid	Tooth	Premolar		2
89, 98		Bovid	Tooth	3rd Molar		2
347		Bovid	Tooth	Molar		1
19,22,41,52		Cervid	Antler			4
2		Bovid (*Damalops*)	Tooth	3rd Molar	Complete	1
37, 13, **87**, 340,342	2	Cervid	Astragalus	Complete		5
35		Cervid	Humerus			1
203, 368, 622		Cervid	Humerus	Distal		3
75		Cervid	Metacarpal	Proximal	Complete	1
112,132, 175		Cervid	Metacarpal	Shaft		3
43, 84, 597		Cervid	Metapodial	Shaft		3
111		Cervid	Metapodial	Distal		1
17,193, 482		Cervid	Metatarsal	Shaft		3
62, 222		Cervid	Pelvis	Acetabulum		2
200,283		Cervid	Radius	Shaft		2
100		Artiodactyla	Tooth	Root		1
25	2	Cervid	Radius	Proximal	Complete	1
38, 20		Cervid	Ramus	Fragment		1
137		Cervid	2nd Phalange			1
49,66,242,289		Cervid	Tibia	Shaft		4
4		Cervid	Molar	Upper	Complete	1
29,33,303		Cervid	Tooth	Molar	Fragments	3
142,339,548	2	Cervid	Tooth	Molar	Split	3
148		Cervid	Tooth	Molar	Deciduous	1
173		Cervid	Tooth	3rd Molar		1
337	2	Cervid	Premolar	Complete	Upper	1
341		Cervid	Tooth	P2		1
various		Chelonia	Scute			81
99		*Crocodylus*	3rd Phalange	Complete		1
99		*Crocodylus*	Tooth			1
345		Proboscidean	Tooth	Fragment		1
278		Proboscidean	Tusk	Fragment		1
213		Equid	Atlas			1
42		Equid	Humerus			1
359	2b	Equid	Tooth	Incisor	Upper	1
150		*Struthio*	Egg shell			1
352, 357		Suid	Metatarsal			2
362	4	Suid	Metatarsal	Distal		1
65		Suid	Tooth	Molar		1
211		Suid?	Metatarsal	Proximal		1
		Indeterminate	Fragments			209
Total						**384**

Note: Nos. 193 and 199 are probably from the same animal, as are 190 and 191.

Zone B:

No.	W	Taxon	Element	Part	Other	N
76, 201		Artiodactyla	Crania			2
122		Artiodactyla	Cuboid			1
377		Artiodactyla	Femur	Distal	Shaft	1
311		Artiodactyla	Navicular			1
35		Artiodactyla	Shaft			1
385		Artiodactyla	Tooth	Molar		1
210		Artiodactyla	Vertebra			1

Table 5. 3 continued: Contents of fossil locality 14, Sandstone 12, Zone B.

29		Bovid	Astragalus	Complete		1
183		Bovid	Humerus	Distal	Condyle	1
389,696		Bovid	Horncore			2
387		Bovid (*Damalops*)	Mandible	M3	Left	1
258	3	Bovid	Metacarpal	Shaft		1
208	2b	Bovid	Metapodial	Distal	Condyle	1
210		Bovid	Metapodial	Distal		1
277		Bovid	Phalange 2	Complete		1
182		Bovid	M3 Upper	Complete	Left	1
227,229,331,346		Bovid	Tooth	Molar	Fragment	4
236,259		Bovid	PM	Upper	Complete	2
331,346		Bovid	Tooth	Molar	Fragment	2
303		Bovid	Tooth	M3 upper	Right, worn	1
184,256,275,388		Cervid	Antler			4
391		Cervid	Antler	Tine		1
185,253		Cervid	Calcaneum	Shaft		2
273	2	Cervid	Calcaneum	Complete		1
783		Cervid	Femur	Distal		1
108, 113	2	Cervid	Metacarpal	Shaft		2
285	2	Cervid	Metacarpal	Proximal		1
322	4	Cervid	Metacarpal	Distal		1
312		Cervid	Metatarsal	Shaft		1
261		Cervid	Metatarsal	Proximal		1
84,181		Cervid	Pelvis	Fragment		2
84		Cervid	Phalanage	Distal		1
204		Cervid	Tibia			1
16,18,23,24,26, 27, 30,273,296, 340,737,738		Cervid	Tooth	Molar		12
213,245,314,318,338		Cervid	Tooth	Premolar	Fragment	5
225		Cervid	Molar 2	Complete	Upper	1
226		Cervid	Premolar	Fragment	Worn	1
0		Chelonia	Scute			59
232	2	*Crocodylus*	Tooth	Complete		1
240,336		*Crocodylus*	Tooth	Fragment		2
287		*Crocodylus*	Vertebra			1
188,207		Equid	Tooth	Molar	Upper	2
228,292		Equid	Tooth	Incisor	Fragment	2
381,734		Equid	Tooth	Molar	Fragment	2
150		*Struthio*	Eggshell			1
72		Suid	Femur	Distal		1
257		Bovid	Horncore (Type C)	Pedicel		1
		Indeterminate	Fragments			163
_**Total**						**300**
Overall Total						**693**

Table 5. 4: Contents of fossil occurrences 15-25 and 109 (Set 1), Sandstone 12, Kanarawala Kas North East, 1986.

Fossil occurrence 15: general collection, scarp face:

Specimen	W	Taxon	Element	Part	Other	N
8	3	Cervid	Molar	Lower		1
3	2a	Equid	Tooth	Incisor	Crownless	1
11	4	Equid	Molar	Crown	Worn	1
0		Indeterminate	Fragments			18
Total						**21**

Fossil occurrence 16: gully collection, parallel to scarp face:

Specimen	W	Taxon	Element	Part	Other	N
3	2a	Bovid	Astragulus	Split		1
9	2a	Bovid	Molar	Complete	Upper	1
0	2a	Bovid	Molar	Split		1
1,2		Proboscidean	Tooth	Tusk	Fragment	2
11	3	Equid	Tibia	Distal	Complete	1
8	3	Suid	Molar 3, lower	Split	V. worn	1
		Indeterminate	Shaft	Fragments		4
Total						**11**

Fossil occurrence 17: scatter, scarp face:

Specimen	W	Taxon	Element	Part	Other	N
0		Artiodactyla	Humerus	Shaft		1
0		Bovid	Horncore			2
4	1	Bovid	Lunar			1
2	3	Bovid	Metatarsal	Proximal	Complete	1
6	3	Bovid Large	Premolar	Fresh		1
5		Cervid	Scapula	Right	Fragment	1
		Chelonia	Scute			5
		Chelonia	Scute	Honeycomb		1
3	2a	Equid	Tibia	Distal	Complete	1
1		Equid	Molar	Complete	Left, upper	1
0		Indeterminate	Fragments			14
Total						**29**

Fossil occurrence 18: scatter, scarp face:

Specimen	W	Taxon	Element	Part	Other	N
5	4	Bovid Large	Tooth	M1/2	Upper	1
8	4	Bovid	Molar	Upper		1
15		Bovid Large	Vertebra	Axis		1
2		Cervid	Metacarpal	Distal	Complete	1
1		Cervid	Pelvis	Acetabulum		1
10		Cervid	Tooth	Fragment		1
		Chelonia	Scute			4
0		Indeterminate	Fragments			22
Total						**31**

Fossil occurrence 19: scatter, scarp face:

Specimen	W	Taxon	Element	Part	Other	N
7		Artiodactyla	Molar	Deciduous	Upper	1
1	1	Cervid	Pelvis			1
		Chelonia	Scute			5
0,3,10,12		Proboscidean	Molar	Plate		5
5	1	Rhinoceritidae	Molar	Split		1
		Indeterminate	Fragments			13
Total						**26**

Table 5. 4 continued: Fossil occurrence 20, scatter, scarp face:

Specimen	W	Taxon	Element	Part	Other	N
4	1	Bovid	Tooth	3rd Molar	Lower	1
6	4	Cervid	Astragulus	Split	Rolled	1
9		Cervid	Metapodial	Shaft		1
5		Cervid	Metatarsal	Shaft		1
2	2	Cervid	Molar	Complete	Upper	1
	1	Chelonia	Scute			1
1	1	Proboscidean	Tooth	Molar		1
	4	Indeterminate	Fragments			8
Total						**15**

Fossil occurrence 21: scatter, scarp face:

Specimen	W	Taxon	Element	Part	Other	N
2	2	Cervid	Femur	Shaft	?Gnawed	1
1	2	Cervid	Molar	M2	Lower	1
3		Cervid	Molar	Complete	Upper	1
0		Indeterminate	Fragments			2
Total						**5**

Fossil occurrence 23: scatter, scarp face:

Specimen	W	Taxon	Element	Part	Other	N
45		Artiodactyla	Metatarsal			1
39		Artiodactyla	Pelvis	Acetabulum		1
40	2b	Artiodactyla	Vertebra	Complete		1
19	3	Artiodactyla	Rib			1
38	4	Bovid	Mandible	Molar 3	Left, worn	1
6	4	Bovid	Tooth	Molar	Worn	1
41	2a	Cervid	2nd Phalange			1
47	2	Cervid	Calcaneum			1
24	5	Cervid	Femur	Condyle		1
23		Cervid	Mandible			1
41		Cervid	Metapodial			1
36	2b	Cervid	1st Phalanage	Proximal		1
3	3	Cervid	PM3	Upper	Little wear	1
46		Chelonia	Maxilla			1
0		Chelonia	Scute			23
		Chelonia	Scute	Smooth		6
2		Chelonia	Scute	Honeycomb		1
0, 8, 22		Proboscidean	Tooth	Tusk		3
0		Proboscidean	Tooth			2
17		Proboscidean	Shaft			1
0		Indeterminate	Fragments			18
Total						**68**

Fossil occurrence 24: general collection; dip-slope opposite scarp face:

Specimen	W	Taxon	Element	Part	Other	N
2	2a	Bovid	Calcaneum	Unfused	Right	1
1	2	Bovid	Radius	Proximal	Complete	1
0		Indeterminate	Fragments			5
Total						**7**

Fossil occurrence 25 continued: general collection; dip-slope opposite scarp face:

Specimen	W	Taxon	Element	Part	Other	N
1,6,9	2	Bovid	Horncore			3
5	3	Bovid	Metacarpal	Shaft		1
2		Cervid	Molar	Split		1
3	2	Cervid	Premolar	Split		1
		Indeterminate	Tibia	Shaft		1
Total						**7**

Table 5.4 continued: Fossil occurrence 109: scatter, scarp face:

Specimen	W	Taxon	Element	Part	Other	N
14		Artiodactyla	1st Phalange			1
8		Artiodactyla	Astragulus			1
3,5,45		Artiodactyla	Femur	Shaft		3
1,12		Artiodactyla	Humerus			2
9,10,13, 15,34,39		Artiodactyla	Metapodial	Shaft		6
6		Artiodactyla	Navic-cuboid			1
39		Artiodactyla	PM3/4	Upper	Right, worn	1
7,28		Artiodactyla	Pelvis			2
39		Artiodactyla	Premolar 3/4	Fragment		1
22		Artiodactyla	Tibia			1
30,32		Bovid	Molar	Lower		2
43	3	Bovid	Premolar 3	Lower	Fragment	1
2,19,32	4	Equid	Molar	Lower	Fragment	3
21		Equid	Premolar	Lower		1
23		Equid	Premolar 3	Split	Upper, worn	1
4		Equid	Tooth	Molar		1
16,42		Equid	Tooth	Premolar	Lower	2
20	5	Equid	Molar	Lower	Fragment	1
0		*Struthio*	Egg-shell			19
29	4	Ungulata	Molar	Spall		6
		Indeterminate	Fragments			148
Total						**204**

Notes: Artiodactyl shaft fragments for occurrence 109: 4,11,18,24,25,31,35,40,44,46,47,52; Ungulata molar spalls: 29, 37, 41,48, 49,50,51. Specimen 16/1 was suncracked, with a coarse sand matrix; 20/1 and 21/1 had a sand matrix; 23/24 was suncracked and crushed; 23/40 had a fine sand matrix.

Table 5.5: Contents of fossil occurrences 42 – 50 (Set 1), Sandstone 12, Kanarawala Kas North East, north of site 14

Fossil occurrence 42: scatter on dip slope in front of Sandstone 12:

Specimen	W	Taxon	Element	Part	N
3		Cervid	Antler	Pedicel	1
4		*Crocodylus*	Tooth		1
1,2		*Struthio*	Egg shell		2
0		Indeterminate	Shaft	Fragments	7
Total					**11**

Fossil occurrence 43: general collection, scarp face of Sandstone 12:

Specimen	W	Taxon	Element	Part	N
1	1	Cervid	Humerus	Shaft	1
		Chelonia	Scute		1
		Indeterminate	Fragments		9
Total					**11**

Fossil occurrence 44: general collection, scarp face of Sandstone 12:

Specimen	W	Taxon	Element	Part	N
7		Artiodactyla	Humerus	Distal	1
9	2	Artiodactyla	Phalange	Distal	1
6	1	Bovid	Molar 3, right	Complete, little wear	1
5	3	Bovid	Premolar 3	Complete	1
4		Cervid	Radius	Proximal, complete	1
8,10	2	*Chelonia*	Scute		2
1	2	*Crocodylus*	Tooth		1
3		Equid	Radioulna	Shaft	1
		Indeterminate	Shaft	Fragment	1
Total					**10**

Table 5.5 continued: Fossil occurrence 45: scatter, scarp face of Sandstone 12:

Specimen	W	Taxon	Element	Part	N
6	2	Artiodactyla	Glenoid		1
3	4	Cervid	Astragulus	Complete	1
10	2b	Cervid	Tooth	Molar	1
17		Cervid	Tooth	Spall	1
	2	Indeterminate	Fragments		13
Total					**17**

Note:.45/:3 and 45/13 are water-worn. Indeterminate shaft numbers are 4,5,7-9,11,12,14-16; 45/1 was in sand

Fossil occurrence 46: scatter, scarp face of Sandstone 12:

Specimen	W	Taxon	Element	Part	N
1		Artiodactyla	Shaft		1
4	3	Bovid	Molar	Complete	1
		Chelonia	Scute		1
		Indeterminate	Fragments		15
Total					**17**

Note: 46/1 has a grit matrix; 46/4 a clay matrix.

Fossil occurrence 47: scatter, scarp face of Sandstone 12:

Specimen	W	Taxon	Element	Part	N
39	3	Artiodactyla	Sesamoid		1
1	2	Artiodactyla	Vertebra	Centrum	1
6	4	Bovid	Acetabulum	Fragment	1
8		Bovid	Astragulus	Complete	1
0		Bovid	Horncore		2
4		Bovid	Molar 3, upper	Complete, little wear	1
5	1	Bovid	Molar	Lower M2	1
23	2	Bovid	Phalange	Proximal	1
40		Bovid	Phalange 3	Complete	1
29	3	Bovid	Tooth	Fragment	1
2,7,10,12, 30	2	Cervid	Antler		5
14	2	Cervid	Antler	Tine	1
21		Cervid	Molar	Spall	2
24	1	Cervid	Molar	Complete	1
27,31,32		Cervid	Molar	Split	2
		Chelonia	Scute		10
		Indeterminate	Fragments		31
Total					**63**

Note: 47/5 has a grit matrix

Fossil occurrence 48: scatter, scarp face of Sandstone 12:

Specimen	W	Taxon	Element	Part	N
6		Bovid	Molar	Fragment	1
		Chelonia	Scute		3
		Indeterminate	Fragments		3
Total					**7**

Fossil occurrence 49: general collection, scarp face of Sandstone 12:

Specimen	W	Taxon	Element	Part	N
24		Bovid	Phalange	Fragment	1
10		Canid	Tooth	Fragment	1
16	4	Cervid	Antler		1
19	3	Cervid	Premolar	Split	1
		Chelonia	Scute		5

Table 5.5 continued: Fossil occurrence 49:

18	4	Equid	Molar	Split		1
15	4	Equid	Premolar	Split		1
17		Rhinoceritidae	Tooth	Molar		1
		Canid	Tooth	Fragment		1
		Indeterminate	Fragments			33
Total						**45**

Find-spot 50 comprised 5 pieces of Proboscidean (probably *Elephas*) molar plate fragments, and 70 fragments of tusk.

Table 5.6: Contents of fossil occurrences 26 – 41 (Set 2), Sandstone 12, Kanarawala Kas West

Fossil occurrence 26: general collect, scarp face of Sandstone 12:

Specimen	W	Taxon	Element	Part	Other	N
1	5	Equid	Mandible	M2,M3	Left	1
2,3	5	Equid	Mandible	P4,M1	Left	2
		Indeterminate	Fragments			5
Total						**8**

Note: 26/1, 276/2 and 26/3 join.

Fossil occurrence 27: general collect, scarp face of Sandstone 12:

Specimen	W	Taxon	Element	Part	Other	N
1	4	Artiodactyla	Metapodial	Shaft		1
6		Bovid	Metapodial	Shaft		1
2,4,8		*Elephas*	Molar	Plate		3
7		*Elephas*	Tooth	Molar		1
3		*Struthio*	Egg-shell			1
		Indeterminate	Shaft			1
Total						**8**

Fossil occurrence 28: general collect, scarp face of Sandstone 12:

Specimen	W	Taxon	Element	Part	Other	N
		Chelonia	Scute			1
3,4,5	3	Equid	Molar	Split		3
	4	Indeterminate	Fragments			6
Total						**10**

Fossil occurrence 29: scatter, dip-slope of Sandstone 12:

Specimen	W	Taxon	Element	Part	Other	N
29GB/8	3	Artiodactyla	1st Phalange	Proximal		1
4	3	Artiodactyla	Tooth	Premolar	Lower	1
11	1	Cervid	1st Phalange			1
1	2	Equid	2nd Phalange	Complete		1
		Indeterminate	Fragments			20
Total						**24**

Fossil occurrence 30: scatter, scarp face of Sandstone 12:

Specimen	W	Taxon	Element	Part	Other	N
30		Artiodactyla	Vertebra	Lumbar		1
42,46	2	Artiodactyla	Rib			2
44	2	Artiodactyla	Frontal			1
31	4	Artiodactyla	Humerus	Distal		1
6,35,58		Bovid	Horncore			3
2		Cervid	1st Phalange	Complete		1
8		Cervid	1st Phalange			1
39	3	Cervid	Calcaneum	Shaft	Left	1
16	1	Cervid	2nd Phalange	Complete		1
		Chelonia	Scute			30

Table 5. 6 continued: Fossil occurrence 30:

		Chelonia	Scute	Smooth		1
		Chelonia	Scute	Honeycomb		3
		Chelonia	Scute	Fishnet		2
		Chelonia	Scute	Dimpled		1
3, 40	2	Suid	Femur	Proximal	Unfused	2
68		Crocodylus	Mandible			1
4,32		Crocodylus	Tooth	Complete		2
64	2b	Equid	1st Phalange	Proximal		1
50, 67		Equid	Accessory carpal	Complete		2
9,65	2b	Equus	Metacarpal	Distal		2
5		Equid	Metapodial			1
28	4	Proboscidean	Molar	Plate		1
		Indeterminate	Fragments			53
Total						**116**

Notes: 30GB1 comprised 6 fused plates; 30/2 and 30/16 (Cervid), and 30/64 and 30/65 (Equid) probably belonged to the same animals. Scute fragments: 12,13,20,21,23,24,27,45,49,56,60,61,64 and 17 unmarked fragments. Rib fragments: 7,22,29,36,47,52,53,55,66,67. Matrix: 30/39 was sandy; 30/11was grit.

Fossil occurrence 31: scatter, scarp face:

Specimen	W	Taxon	Element	Part	Other	N
3,4	2	Artiodactyla	Vertebra			2
5		Cervid	3rd Molar	Lower		1
8		Cervid	Antler			1
13		Cervid	Metapodial	Shaft		1
1	2b	Cervid	Radius	Proximal	Complete	1
9	1	Cervid	Ulna	Proximal	Unfused	1
		Chelonia	Scute	Honeycomb		3
		Indeterminate	Fragments			13
Total						**23**

Note: 31/1 was suncracked; 31/9 was unfused; both joined.

Fossil occurrence 33: scatter, scarp face:

Specimen	W	Taxon	Element	Part	Other	N
		Bovid	Metatarsal	Shaft		1
15,22		Bovid	Molar	Lower	Split	2
16		Bovid	Molar 1/2,	Lower	Complete	1
24		Bovid	Phalange	Fragment		1
10	1	Cervid	2nd Molar	Lower	Split	1
12	1	Cervid	Molar 1/2	Lower,split	Fresh	1
		Chelonia	Scute			6
6-8,11,13		Proboscidean	Tooth	Molar		6
14	3	*Elephas*	Tooth	Tusk		1
1-5		*Struthio*	Egg-shell			5
		Indeterminate	Fragments			18
Total						**43**

Fossil occurrence 34: scatter, scarp face:

Specimen	W	Taxon	Element	Part	Other	N
54	3	Artiodactyla	3rd Molar	Split		1
26	3	Artiodactyla	Metacarpal	Distal	Condyle	1
8, 36	4	Artiodactyla	Metapodial	Shaft		2
15		Artiodactyla	Molar	Fragment	Little wear	1
40,45		Artiodactyla	Premolar	Fragment		2
2,11	2b	Bovid	Horncore			2
12		Bovid	Phalange	Fragment		1
43		Cervid	Antler			1
7	2a	Cervid	Calcaneum	Shaft	Right	1
9	1	Cervid	Vertebra, unfused	Thoracic	Complete	1
		Chelonia	Scute			13
		Chelonia	Scute	Smooth		2

Table 5. 6 continued: Fossil occurrence 34:

14		Equid	Cuneiform	Complete		1
27,36	4	Equid	Molar	Fragment		2
21	1	Equid	Premolar 1	Complete	Fresh	1
1		Equid	Radius	Proximal	Unfused,	1
		Indeterminate	Coprolite			1
		Indeterminate	Fragments			16
Total						**51**

Notes: 34/15 and 34/21 have little wear; Scute fragments:16-20,22,23,28-31,33,34,39,51,52,59.; Matrix: 34: one piece of scute had grit.

Fossil occurrence 35: scatter, scarp face:

Specimen	W	Taxon	Element	Part	Other	N
24		Bovid (Medium)	3rd Molar	Upper, left	Mod wear	1
8		Cervid	Cuboid			1
11	4	Cervid	Metapodial	Shaft		1
3	2	Cervid	Metatarsal	Distal	Fragment	1
4		Cervid	Molar	Lower	Split	1
		Chelonia	Scute	Fishnet		1
		Chelonia	Scute	Smooth		1
		Chelonia	Scute			3
5,6,23		*Elephas*	Molar			3
1		*Elephas*	Ramus			1
		Indeterminate	Fragments			18
Total						**32**

Fossil occurrence 36: scatter, scarp face:

Specimen	W	Taxon	Element	Part	Other	N
2	2	Artio (Small)	Metacarpal	Shaft		1
12		Artiodactyl	Humerus	Fragment		1
0		Cervid	Antler			1
5	3	Cervid	Antler			1
1	3	Cervid	Astragulus	Split		1
15	2	Cervid	Metatarsal	Shaft		1
11	2	Cervid	Molar	Fragment		1
3	4	Cervid	Molar	Split	Upper	1
		Chelonia	Scute			4
23		Crocodylus	Maxilla	(Dubious)		1
7	3	Elephas	Cranium	Fragment		1
20-22		Proboscidean	Tooth	Root		3
24-45		*Struthio*	Egg-shell			17
	2	Indeterminate	Fragments			34
Total						**68**

Notes: 36/21 and 36/22 join

Fossil occurrence 37: scatter, scarp face:

Specimen	W	Taxon	Element	Part	Other	N
1,2,14,15	3	Bovid	Horncore	Fragment		4
3		Cervid	Scapula	Fragment		1
4-13		Elephas	Tusk	V. scrappy		10
		Indeterminate	Fragments			5
Total						**20**

Fossil occurrence 38: scatter, scarp face:

Specimen	W	Taxon	Element	Part	Other	N
5		Artio	Tibia	Fragment		1
34	5	Artio	Molar	Fragment	Rolled	1
44	5	Artio	Molar	Fragment		2

Table 5.6 continued: Fossil occurrence 38:

29	3	Bovid (Large)	Tooth	M1/2;		1
30,31	3	Bovid	Molar	Fragment		2
11		Bovid	Ulna Carpal	Complete		1
35	3	Bovid	Molar	Fragment	Worn	1
22		Canid (Small)	Ulna	Proximal		1
1,10		Cervid	Antler			2
24		Cervid	Femur	Distal		1
23		Cervid	Mandible	M3	Left	1
28	1	Cervid	Metacarpal	Condyle		1
82		Cervid	Orbital			1
6, 17	2	Cervid	Pelvis	Fragment		2
33	3	Cervid	Tooth	Upper M1/2	Fresh	1
4, 25,36		Cervid	Tooth	Molar	Fragment	3
3		Cervid	Ulna	Proximal		1
		Chelonia	Scute			22
27	2	Crocodylus	Vertebrae	Complete	Rolled	1
20		Proboscidean	Vertebrae	Epiphysis		1
15		Suid	Molar	Decid.	Lower	1
		Indeterminate	Fragments			69
Total						**117**

Fossil occurrence 39: scatter, scarp face:

Specimen	W	Taxon	Element	Part	Other	N
4	2	Artiodactyla	Vertebra	Flange		1
5	2	Cervid	Antler	Pedicel		1
	2a	Chelonia	Scute	Smooth		1
7	3	Suid	3rd Phalange	Complete		1
1		Indeterminate	Fragments			6
Total						**10**

Note: 39/1 had two bite marks.

Fossil occurrence 40: gully collection, parallel to scarp face:

Specimen	W	Taxon	Element	Part	Other	N
1		Proboscidean	Femur	Shaft		1
2	5	Proboscidean	Molar	Fragment		1
		Indeterminate	Shaft			1
Total						**3**

Note: 40/1 had a sandy matrix

Fossil occurrence 41: scatter on dip slope in front of Sandstone 12:

Specimen	W	Taxon	Element	Part	Other	N
		Chelonia	Scute	Smooth		3
		Chelonia	Scute	Fine dimpled		7
10-17		*Elephas*	Molar			8
Total						**18**

Table 5.7: Contents of fossil occurrences 51-63 (Set 3), Sandstone 12, south of occurrence 30

Fossil occurrence 51: general collection scarp face of Sandstone 12:

Specimen	W	Taxon	Element	Part	N
2,8		Artiodactyla	Molar, lower	Fragment	2
1		Bovid	Tibia	Distal	1
6		Cervid	Antler		1
3		Cervid	Phalange	Complete	1
		Chelonia	Scute		3
		Indeterminate	Fragments		9
Total					**17**

Table 5. 7 continued: Fossil occurrence 52: general collection, scarp face of Sandstone 12:

Specimen	W	Taxon	Element	Part		N
3	2b	Artiodactyla	Tooth	Spall		1
		Artiodactyla	Molar	V.worn		1
		Indeterminate	Tooth	Spall		3
Total						**5**

Fossil occurrence 56: scatter scarp face of Sandstone 12:

Specimen	W	Taxon	Element	Part		N
11	5	Artiodactyla	Tooth	Fragment		1
7,13,17		Proboscidean	Molar	Plate		3
		Indeterminate	Shaft	Fragments		14
Total						**18**

Fossil occurrence 57: general collection scarp face of Sandstone 12:

Specimen	W	Taxon	Element	Part	Other	N
82	2	Artiodactyla	Tooth	Fragment		2
70	3	Artio/Suid	Incisor	Split		1
10	2	Bovid	M1/2	Complete	Upper	1
62,65	2	Bovid	Molar	Split		2
32	3	Bovid	Molar 1/2	Complete	Lower	1
64	2	Bovid	PM2	Complete		1
28	3	Bovid	Premolar	Split	Lower	1
68	1	Cervid	Humerus	Distal	Left	1
6	4	Cervid	Metatarsal	Proximal		1
5	2a	Bovid	M3 right	Lower	Mod. wear	1
38		Cervid	Pelvis	Fragment		1
24,25	1	Cervid	Premolar	Split	Lower	2
		Chelonia	Scute			1
0*		*Elephas*	Molar	Plate		41
7		*Elephas*	Tooth	Tusk		1
11,	5	Proboscidean	Cranium	+ 2 limb frags		3
53,55		Rhinoceritidae	Molar	Fragments		2
54		Rhinoceritidae	Molar	Fragment	Lower	1
		Indeterminate	Fragments			88
Total						**152**

Fossil occurrence 58: general collection scarp face of Sandstone 12:

Specimen	W	Taxon	Element	Part	Other	N
4		Artiodactyla	Humerus			1
2,3,5	2	Artiodactyla	Molar	Split	Lower,	3
6	2a	Artiodactyla	Molar	Complete	Lower,	1
		Chelonia	Scute			2
12	3	Suid?	Scapula	Complete		1
		Indeterminate	Fragment			4
Total						**12**

Fossil occurrence 60: scatter in front of scarp face of Sandstone 12:

Specimen	W	Taxon	Element	Part	Other	N
1,4-8,38,40		Proboscidean	Rib			8
2,3,30,31,36,39		Proboscidean	Tooth	Tusk		37
	3	Indeterminate	Fragments			15
Total						**60**

Table 5. 7 continued: Fossil occurrence 61: general collection scarp face of Sandstone 12:

Specimen	W	Taxon	Element	Part	Other	N
13	2	Artiodactyla	Femur	Shaft		1
5		Chelonia	Scute			1
3		Proboscidean	Tooth			1
14	2	Proboscidean	Femur	Shaft		1
		Indeterminate	Shaft			10
Total						**14**

Notes: 58/12 was cased in sandstone. **Matrix:** Tusk fragments 60/1,8,30,31 had a quartzite pebbly matrix; Ribs 60/1 is suncracked, and has a gritty matrix that includes some small quartzite pebbles. 60/6 also has a quartzite pebbly matrix. 60/33 is waterworn and rolled; 60/38 has a sandstone matrix. **Other finds**: Find-spot 53, an artiodactyls femur fragment in sandstone; a canid right mandible in a gully parallel to scarp face of Sandstone 12 at occurrence 54; a small scatter (occurrence 55) of 4 Proboscidean molar plate fragments and 3 indeterminate fragment; no. 62, an artiodactyl proximal humerus in a gritstone channel fill; and find-spot 63 on the scarp face of Sandstone 12, 91 Proboscidean tooth fragments and 5 indeterminate fragments.

Table 5. 8: Contents of fossil occurrences 64 – 75 (Set 4), Sandstone 12, excluding localities 68 and site 73

Fossil occurrence 64: scatter, scarp face:

Specimen	W	Taxon	Element	Part	Other	N
8		Artiodactyla	Radius	Proximal		1
50		Artiodactyla	2nd Phalange	Distal		1
7	2	Bovid	Molar	Fragment	Lower	1
13		Bovid	Vertebra	Fragment		1
24	2	Cervid	Molar	Fragment		1
		Chelonia	Scute			4
22		Proboscidean	Molar plate			1
31,35	5	Equid	Molar	Fragment	Lower	2
51,52		*Struthio*	Egg-shell			4
H	2	Indeterminate	Fragments			37
Total						**53**

Fossil occurrence 65: general collection scarp face:

Specimen	W	Taxon	Element	Part	N
3	5	Artiodactyla	Premolar	Fragment	1
1,2		Proboscidean	Molar		2
		Indeterminate	Fragments		19
Total					**22**

Fossil occurrence 67:general collection scarp face:

Specimen	W	Taxon	Element	Part	N
8		Artiodactyla	Molar		1
5-7,9,13,15		Proboscidean	Molar		6
19		Equid	Molar	Worn,split	1
		Indeterminate	Fragments		14
Total					**22**

Fossil occurrence 69: scatter scarp face:

Specimen	W	Taxon	Element	Part	N
9	2b	Artiodactyla	Calcaneum	Complete	1
1, 4	3	Equid	M3 lower	Very worn	2
9,10-13,15		Proboscidean	Tooth	Fragments	6
		Indeterminate	Fragments		6
Total					**15**

Table 5. 8 continued: Fossil occurrence 74: general collection scarp face:

Specimen	W	Taxon	Element	Part	Other	N
19	2	Artiodactyla	Metapodial			1
92		Artiodactyla	Metatarsal	Shaft		1
10	2	Artiodactyla	Metatarsal	Distal	Condyle	1
101		Artiodactyla	Phalange 1/2	Fragment		1
210	2a	Bovid	Cuneiform	Complete		1
4,44,62,7,1 07	3	Bovid	Horncore			5
1	3	Bovid	Molar 1/2	Complete	Upper,fresh	1
105		Bovid	Phalange			1
52		Bovid	Premolar	Split		1
74	2	Cervid	2nd Phalange	Distal		1
24		Cervid	Metapodial			1
3		Cervid	Tooth	Premolar		1
6,75		Chelonia	Scute			2
63,212		*Crocodylus*	Radius			2
33		*Crocodylus*	Tooth			1
3,28,31-2, 64,75		Proboscidean	Molar	Fragments		6
35	2	Rhinocerotidae	Metacarpal	Complete		1
		Indeterminate	Fragments			300
Total						**326**

Fossil occurrence 75: general collection "badlands" in front of Sandstone 12 near no. 60:

Specimen	W	Taxon	Element	Part	Other	N
17	3	Bovid	3rd Molar	Complete, worn	Upper	1
34		Bovid	Horncore			1
21,27	2	Bovid	Premolar	Split		2
48	3	Bovid	Premolar	Split	Lower	1
0		Proboscidean	Tooth	Molar	Plate	18
0		Proboscidean	Tusk			8
28		*Struthio*	Egg-shell			1
		Indeterminate	Fragments			91
Total						**123**

Notes: Fossil occurrence 66 comprised a bovid metapodial fragment. Nos 74/35 and 74/52 have a clay skin.

Table 5. 9: Contents of fossil locality 68 (Set 4), Sandstone 12.

Zone A:

Specimen	W	Taxon	Element	Part	Other	N
8, 71		Artiodctyla	Astragulus			2
222	3	Artiodctyla	Calcaneum			1
13		Artiodctyla	Glenoid			1
5		Artiodctyla	Humerus	Distal		1
100	3	Artiodctyla	Humerus	Proximal	Condyle	1
23		Artiodctyla	Humerus	Shaft		1
75	2	Artiodctyla	Mandible	P2 socket		1
217		Artiodctyla	Metapodial	Shaft		10
83	2b	Artiodctyla	Metatarsal	Proximal	Split	1
218	1	Artiodctyla	Navicular	Complete		1
42		Artiodctyla	Pelvis			1
12, 51,310	2	Artiodctyla	Rib	Fragment		3
14		Artiodctyla	Scapula			1
103		Artiodctyla	Tooth	Split	Premolar	1
257		Artiodctyla	Tooth	Premolar		1
323	3	Artiodctyla	Tooth	Fragment	Worn	1
288		Artiodctyla	Tooth	Premolar	Lower	1
287		Artiodctyla	Tooth	Molar		1
351		Artiodctyla	Tooth	P2	Lower	1
299		Artiodctyla	Tooth	Premolar	Lower	1

Table 5. 9 continued: Contents of fossil locality 68, Sandstone 12.

67, 92		Artiodctyla	Ulna	Proximal		2
78		Aves	Tibia	Proximal		1
2	2	Bovid	Carpal	2nd/3rd.		1
22, 74, 104		Bovid	Horncore	Tine		3
47		Bovid	Tibia			1
70		Bovid	Tooth	Molar	Upper	1
77	2	Bovid	Tooth	M1/2	Upper	1
3	3	Bovid	Tooth	Molar	Lower	1
36		Canidae	Vertebra	Axis		1
10		Cervid	1st Phalange	Fragment		1
32		Cervid	Cuneiform	Complete		1
4, 11		Cervid	Humerus	Distal	Condyle	2
59, 69		Cervid	Mandible	Fragment	Premolar	2
258, 259	2b	Cervid	Metacarpal	Distal		2
16		Cervid	Metapodial	Shaft		1
68		Cervid	Metatarsal	Epiphysis		1
312		Cervid	Tooth	Premolar	Upper	1
356		Cervid	Tooth	Molar		1
15		Cervid	Tooth	Premolar		1
17		Cervid	Tooth	M1/2 upper	Upper	1
60		Cervid	Tooth	Molar	Upper	1
50		Cervid	Tooth	Molar	Lower	1
214		Equid	Metacarpal			1
215		Equid	Scapula			1
308		Suid	3rd Phalange	Complete		1
286		Suid	Tooth	Molar	Upper	1
28,53,55,88, 138, 322,339		Suid	Tooth	Molar		7
		Indeterminate	Fragments			199
Total						**270**

Zone B:						
127, 361		Artiodctyla	3rd Phalange			2
310		Artiodctyla	Atlas			1
326		Artiodctyla	Calcaneum	Left	Shaft	1
136		Artiodctyla	Cranium			1
240		Artiodctyla	Humerus	Distal	Condyle	1
397		Artiodctyl	Humerus	Proximal		1
5, 335		Artiodctyla	Humerus	Shaft		2
15		Artiodctyla	Mandible	Fragment	Right	1
321		Artiodctyla	Metacarpal	Shaft		1
328		Artiodctyla	Metapodial	Distal	Condyle	1
248		Artiodctyla	Metapodial			1
318, 343,393		Artiodctyla	Metatarsal	Shaft	Fragment	3
256		Artiodctyla	Ramus	Fragment		1
406		Artiodctyla	Scapula			1
344, 346		Artiodctyla	Shaft			2
33		Artiodctyla	Tooth	Premolar	Lower	1
8		Artiodctyla	Tooth	Molar		1
25, 36	2b	Artiodctyla	Tooth	Incisor		2
27		Artiodctyla	Tooth	Premolar	Lower	1
293		Artiodctyla	Tooth	Molar/PM3	Upper	1
32, 35		Artiodctyla	Tooth	Molar	Spall	2
284		Artiodctyla	Tooth			1
88		Artiodctyla	Tooth	PM2 lower	Complete	1
200,405, 408		Artiodctyla	Vertebra			3

338		Aves?	Shaft			1
313		Bovid	Calcaneum			1
28	2	Bovid	Cuneiform	Complete		1
322		Bovid	Horn-core	Fragment		14
342		Bovid	Phalange	Distal		1
103	2	Bovid	Tooth	P3 Upper	Mod.wear	1
83		Bovid	Tooth	P4	Upper	1
414		Bovid	Tooth	Molar	Upper, fresh	1
31		Bovid/Artio	Tooth	Upper	Premolar	1
29		Bovid/Artio	Tooth	Premolar		1
44		Bovid/Artio	Tooth	Premolar	Lower	1
314	3	Carnivora	Mandible	Ramus		1
311		Carnivora	Vertebra			1
339	3	Cervid	Antler	Palmate		1
34, 74		Cervid	Antler	Tine		2
105, 302, 331, 333, 337,		Cervid	Antler	Fragment		5
129		Cervid	Mandible	Molar	Left	1
23		Cervid	Mandible	M1,M2, unworn	Left	1
327		Cervid	Mandible	Dm3	Left	1
309		Cervid	Metacarpal	Proximal	Complete	1
325, 329		Cervid	Tooth	Molar		2
92		Cervid	Tooth	Molar	Complete	1
330		Cervid	Tooth	Premolar		1
95	2b	Cervid	Tooth	Premolar	Complete	1
102	2b	Cervid	Tooth	Premolar	Lower	1
106		Cervid	Tooth	P3/4	Lower	1
13		Cervid	Tooth	Molar	Fractured	1
243	3	Cervid/Artio	Metatarsal	Shaft		1
30		Equid	Tooth	Premolar	Lower	1
415		Hippo	Tooth	Incisor		1
41,54		Indet/Suid	Tooth	Molar		2
24	3	Suid	Canine	1		1
75		Suid	Mandible	P1,P2 roots only		1
72,73,81,89		Suid	Tooth	Fragment		4
324		Suid	Tooth	Molar, upper	Worn	1
		Indeterminate	Fragments			230
Total						**321**
Zone C:						
3-7		Bovid	Horncore			4
1		Bovid/Cervid	Metacarpal	Distal		1
36		Canidae	Vertebra			1
22		Cervid	1st Phalange			1
26		Cervid	Mandible	M1/2	Premolar	1
27	2	Cervid	Maxilla	Premolars		1
25		Cervid	Molar	Complete	Upper	1
23		Equid	Metacarpal	Distal		1
2		cf. *Hexaprotodon*	Tooth	Incisor	Upper	1
24		Suid	Tooth , M3	Molar, right	Lower	1
28,29,30		Suid	Tooth	Fragment		3
		Indeterminate	Fragments			16
Total						**32**
Zone X:						
9		Artiodctyla	Metapodial			1
5		Bovid	Premolar	Complete	Upper	1
1, 3		Bovid	Molar	Complete	Upper	2
2		Bovid	Molar	Complete	Lower	1
4		Bovid	Molar			1
8		Carnivora	Metatarsal			1

Table 5. 9 continued: Contents of locality 68, Zone X

Specimen	W	Taxon	Element	Part	Type	N
6		Carnivora	Tooth			1
7		Chelonia	Cranium			1
10		Elephas	Tooth			1
0		Indeterminate	Fragments			28
Total						**38**
Overall total						**672**

Table 5. 10: Contents of locality 73, zone A, Sandstone 12

Specimen	W	Taxon	Element	Part	Type	N
292		Artiodactyla	3rd Phalange			1
213	3	Artiodactyla	Maxilla	Fragment		1
16		Artiodactyla	Scapula	Glenoid		1
159		Artiodactyla	Tooth	Molar		1
137	2b	Artiodactyla	Tooth	Spall		1
149,330		Artiodactyla	Tooth	Molar, lower	Split	2
0		Artiod/Bovid	Calcaneum	Fragment		1
Various		Artiodactyla	Crania			7
Various		Artiodactyla	Metapodial	Shaft		4
288	2	Artiodactyla	Metatarsal	Proximal		1
	2	Artiodactyla	Metatarsal	Shaft		4
284	1	Artiodactyla	Phalange	Complete		1
223, 224		Artiodactyla	Rib			2
273		Artiodactyla	Shaft			42
283		Artiodactyla	Tibia	Shaft		1
18		Artiodactyla	Tibia	Distal		1
181		Artiodactyla	Tooth	Premolar		1
118		Artiodactyla	Ulna	Proximal		1
190	2	Artiodactyla	Vertebra	Lumbar		1
188	2	Bovid	2nd Phalange			1
7	2	Bovid	Astragalus	Complete		1
2	2	Bovid (Large)	Calaneum	Shaft		1
250		Bovid	Femur	Shaft		1
9	2	Bovid	1st Phalange	Complete		1
176	3	Bovid	1st Phalange	Proximal	Complete	1
3	2	Bovid	Humerus	Distal	Gnawed	1
102		Bovid	Maxillary	Complete	M3 erupting	1
20	2	Bovid	Metacarpal	Proximal	Complete	1
12	2b	Bovid	Metacarpal	Distal	Condyle	1
11		Bovid	Parietal			1
39		Bovid	Parietal	+ Horn		1
105	2	Bovid	Radius	Distal	Complete	1
14		Bovid	Tibia	Distal		1
17	2b	Cervid	Calaneum	Fragment		1
148	1	Cervid	Mandible	Right	PM1	1
145		Cervid	Mandible	Premolars	Young	1
100	2	Cervid	Mandible	Left	M2,M3, fresh	1
289	2	Cervid	Mandible	M3, left	Unworn	1
193	2	Cervid	Mandible	PM4, right	Worn	1
146		Cervid	Mandible	Complete		1
147	2b	Cervid	Maxilla	Right	M2,3 worn	1
106	2	Cervid	Metacarpal	Proximal	Complete	2
5	2	Cervid	Metacarpal	Distal		1
21,30,31, 38, 339	2b	Cervid	Metapodial	Shaft		5
10, **107**	2b	Cervid	Metatarsal	Proximal	Complete	2
26, 30, 40		Cervid	Metatarsal	Shaft		3
101	2	Cervid	Maxilla	Right	P4-M2	1
1		Cervid	1st Phalange	Complete		1

Table 5. 10 continued: Contents of locality 73, zone A

Specimen	W	Taxon	Element	Part	Type	N
33	2b	Cervid	Ramus			1
63	3	Cervid	Tooth	Premolar	Split	1
90		Cervid	Tooth	Incisor		1
139	2b	Cervid	Tooth	Molar	Mod. wear	1
157	1	Cervid	Tooth	PM2	Fresh	1
57	2b	Cervid	Tooth	PM2		1
55,70,72, 82, 86, 87		Cervid	Tooth	Premolar	Spall	6
150	1	Cervid	Molar	Split	Upper, fresh	1
154, 158,73		Cervid	Tooth	Premolar	Split	3
51,64,155		Cervid	Tooth	Molar (155=upper)	Spall	3
298	1	Cervid	Molar	Complete	Upper, fresh	1
44		Equid	Cuboid			1
153	2b	Equid	Maxilla	M3	Left, worn	1
203		Equid	Maxilla	PM2	Right	1
77,78		Equid	Maxillary	Incisor		2
104	2	Equid	Metacarpal	Distal		1
22		Equid	Metapodial	Shaft		1
109	2b	Equid	Metapodial	Distal		1
6		Equid	Metapodial			1
108	2b	Equid	Metapodial	Shaft		1
83, 329		Equid	Tooth	Premolar		2
65,74,156		Equid	Tooth	Molar	Fragment	3
151	1	Equid	PM2	Complete	Lower, worn	1
329	3	Equid	Mandible	M3	Fragment	1
76		Ungulata	Tooth	Fragment		1
249		Carnivore	Pelvis	1		1
270	5	Carnivore	Mandible	Crushed	Premolar,	1
191		Carnivore	Maxilla			1
260		Carnivore	Scapula			1
97		Carnivore	Tooth	Incisor		1
93		Carnivore	Tooth	Canine		1
152	3	Carnivore	Tooth	Fragment		1
59	3	Carnivore	Scapula	Large		1
13		Felidae	Metatarsal	Distal		1
1		Hyaenidae	Crania			1
1		Hyaenidae	Maxilla			1
1		Hyaenidae	Occipital			1
1		Hyaenidae	Premaxilla	Complete		1
2		Hyaenidae	Mandible			1
202		Hyaenidae	Mandible	P2-4	Left	1
75		Hyaenidae	Tooth	Fragment	Premolar	1
328	1	Hyaenid	Calcaneum	Complete	Left	1
52	1	Herpestid	Mandible	M1	Fresh, right	1
		Indeterminate	Fragments			97
Total						**264**

Table 5. 11: Contents of fossil locality 73, zone B, Sandstone 12.

Specimen	W	Taxon	Element	Type	Other	N
33		Artiodactyla	Astragalus	Proximal		1
18		Artiodactyla	Astragalus			1
236		Artiodactyla	Femur			1
14	2a	Artiodactyla	Femur	Distal		1
233		Artiodactyla	1st Phalange			1
15, 25		Artiodactyla	1st Phalange	Distal		2
47	3	Artiodactyla	Mandible	Fragment		1
535		Artiodactyla	Mandible	Ramus		1
532		Artiodactyla	Mandible	PM2,	Worn	1
28		Artiodactyla	Metacarpal	Proximal		1

Table 5. 11 continued: Fossil locality 73, zone B, Sandstone 12

19,22,27,241, 307, 342,556		Artiodactyla	Metacarpal	Shaft		8
29		Artiodactyla	Metapodial	Proximal		1
17		Artiodactyla	Radius	Distal		1
319	2a	Artiodactyla	Rib			1
13		Artiodactyla	Scapula	Fragment		1
36		Artiodactyla	2nd Phalange	Shaft		1
128		Artiodactyla	Shaft			1
31		Artiodactyla	Tibia	Proximal		1
50, 57,96, 485	2a	Artiodactyla	Tooth	Spall		4
517,523	1	Artiodactyla	Tooth	Incisor	Complete	2
206	2	Artiodactyla	Tooth	M3, lower	Split, fresh	1
51,54,56		Artiodactyla	Tooth	Molar		3
37		Bovid	Horncore			1
12	1	Bovid	Metacarpal	Distal	Condyle	1
499,512		Bovid	Tooth	Molar		2
108	2	Bovid	Tooth	Molar	Complete	1
81		Bovid	Vertebra			1
79		Cervid	Antler			1
1	1	Cervid	Astragalus	Complete		1
429		Cervid	1st Phalange	Proximal	Epiphysis	1
3		Cervid	Mandible	M2	Split	1
451	4	Cervid	Mandible	Premolar	Split	1
441	2a	Cervid	Mandible	DM3, M1	Fresh	1
11	2a	Cervid	Metatarsal	Shaft		1
10		Cervid	Pelvis			1
85		Cervid	3rd Phalange			1
526	1	Cervid	Tibia	Distal		1
66,77		Cervid	Tooth	Incisor		2
Various	2a	Cervid	Tooth	Molar		9
83,92,94,97, 444, 448, 450	2a	Cervid	Tooth	Molar	Spall	7
72, 452	1	Cervid	Tooth	Molar	Fresh	2
9		Cervid	Tooth	M1/2	Upper, fresh	1
43		Cervid	Tooth	DM3	Lower	1
8	1	Cervid	Tooth	M1/2 upper	Complete	1
449	2a	Cervid	Tooth	M1/2 upper	Fresh	1
4	2a	Cervid	Tooth	M3 upper	Complete	1
		Cervid	Tooth	Premolar	Fragment	2
98,192,206,215, 423,442,516	2	Cervid	Tooth	Premolar		7
453	2a	Cervid	Tooth	PM3	Lower	1
49		Cervid	Tooth	Fragment		1
64		Cervid	Tooth	Root		1
5, 513	2	Equid	Mandible	M3 (5= left)	Worn	2
6,7	1	Equid	Tooth	Incisor		2
55	5	Equid	Tooth	Molar	Lower	1
3		Equid	Tooth	M3		1
213		Equid	Tooth	Premolar	Fragment	1
67		Carnivora	Femur			1
80		Carnivora	Humerus	Shaft		1
78		Carnivora	Metapodial	Shaft		1
63		Carnivora	2nd Phalange			1
75		Carnivora	Tibia	Shaft		1
66,67, 68,90, 95		Carnivora	Tooth	Incisor		5
518		Carnivora	Tooth	Premolar	Cusp	1
408	2	Carnivora	Vertebra	(Small)		1
183	2	Canid	Metatarsal	Proximal		2
258	3	Canid	Scapula			1
409		Canid	2nd Phalange			1
61		Felid	Tooth	DM4		1
62		Felid	Tooth	Incisor	Lower	1
73		Felid	2nd Phalange			1

Table 5. 11 continued: Fossil locality 73, zone B, Sandstone 12

Specimen	W	Taxon	Element	Type	Other	N
86, 407		Felid	Metatarsal	Distal		1
84, 181		Felid	Metatarsal	Proximal		1
505		Felid	Tooth	P3	Split	1
184		Hyaenid	Mandible, right	Canine	Errupting	1
520		Hyaenid	Tooth	Canine		1
515	1	Ursidae	Tooth	M1	Right	1
2,50		Suid	Tooth	Fragment		2
519, 521		Ursid	Tooth	Canine		2
521		Ursid	Tooth	Canine		1
		Indeterminate	Fragments			433
Total						**556**

Table 5. 12: Contents of fossil locality 73, zone C, Sandstone 12

Specimen	W	Taxon	Element	Type	Other	N
717		Artiodactyla	Femur			1
116		Artiodactyla	Metapodial			1
27		Bovid	Horncore			1
91	2a	Bovid	3rd Phalange	Large		1
60-63, 711		Canidae	Metatarsal	Shaft		5
7		Canidae	3rd Metatarsal			1
100, 719		Canidae	3rd Phalange			2
17, 195, 712		Carnivora	Rib			3
134, 722		Carnivora	Tooth	Canine		2
110		Carnivora	Metatarsal	Proximal		1
0		Carnivora	Scapula	Fragment		1
125		Carnivora	Tooth	Canines	Small	2
21		Cervid	Atlas			1
11		Cervid	Cranium			1
710		Cervid	Femur			1
73		Cervid	1st Phalange			1
162		Cervid	1st Phalange	Proximal	Epiphysis	1
76	4	Cervid	1st Phalange	Complete		1
83		Cervid	2nd Phalange			1
93		Cervid	2nd Phalange	Proximal	Split	1
0		Cervid	3rd Phalange			1
42		Cervid	Maxilla	Premolar		1
7		Cervid	Metacarpal			1
39		Cervid	Pelvis			1
142		Cervid	Phalange	Distal	1	1
12		Cervid	Tooth	M1/2	Upper	1
64		Cervid	Tooth	Molar		1
82		Cervid	Tooth	Premolar		1
4	2a	Cervid	Mandible	M3	Split	1
47	2a	Cervid	Maxilla	M2,3	Right	1
77		Equid	Cuneiform			1
1		Equid	Femur	Shaft		1
139		Equid	1st Phalange			1
46		Equid	Humerus			1
2	2b	Equid	Mandible	Premolar M3,right		1
3	2b	Equid	Mandible	Molar	M2,right	1
5		Equid	Mandible	Molar3 M3,Left		1
6	3	Equid	Mandible	Molar T	M1,right	1
69	3	Equid	Mandible	PM2, broken, PM3	Right	1
7		Equid	Maxilla	Molar		1
8		Equid	Maxilla	Premolar4	Left	1
138	2a	Equid	Metacarpal	Shaft		1
11		Hyaenidae	Tooth	Molar		1
133		Hyaenidae	Tooth	Canine	Root	1
164		Hyaenidae	Tooth			1

Table 5. 12 continued: Contents of fossil locality 73, zone C, Sandstone 12

107		Rhinocerotidae	Astragulus			1
9		Rhinocerotidae	Fibula			1
713		Rhinocerotidae	Metatarsal			1
38		Rhinocerotidae	Radius			1
74		Rhinocerotidae	2nd Metatarsal			1
36		Suid	Radius	Proximal		1
65	2a	Suid	3rd Phalange	Lateral		1
114		Suid	3rd Phalange			1
		Indeterminate	Fragments			78
Total						**140**

Table 5. 13: Contents of fossil locality 73, zone X, Sandstone 12

Specimen	W	Taxon	Element	Part	Other	N
21	2a	Artiodactyla	Humerus			1
32	2b	Artiodactyla	Mandible	Fragment		1
17	2a	Artiodactyla	Metacarpal	Shaft		1
2,35, 36,80, 84		Artiodactyla	Metapodial			5
1, 19	2a	Artiodactyla	Metatarsal			2
7		Artiodactyla	Pelvis			1
18	2a	Artiodactyla	Rib			1
14	4	Artiodactyla	Tooth	Split	Molar	1
		Bovid	Horncore			20
13	5	Bovid	Tooth	Molar	Lower	1
28	2b	Bovid	Tooth	M1/2, lower	Complete	1
11	2a	Bovid (Large)	Tooth	Molar		1
36		Canid (Small)	Metatarsal			1
59		Canid (Small)	Calcaneum			1
15		Cervid	Cuboid			1
25	2b	Cervid	1st Phalange			1
8		Cervid	Mandible	M3	Worn	1
9	2a	Cervid	Tooth	dm3		1
50	2a	Cervid	Tooth	Plate		1
16	2b	Cervid	Tooth	Molar, upper	Split	1
0		Cervid	Tooth			2
12		Equid	Calcaneum			1
4		Equid	Pelvis			1
29		Equid	Ramus			1
10	1	Equid	Tooth	Incisor	Worn	1
12		Equid	Tooth	PM2	Split, worn	1
22		Equid	Ulna			1
20		Equid	Metacarpal			1
51		Felid	Tooth	Canine	Fragment	1
58		Hyaenidae	Mandible	M2		1
57		Hyaenidae	Mandible	PM2,3	Left	1
56		Hyaenidae	Mandible	P4, M1	Right	1
60		Hyaenidae	Metatarsal			1
3	3	Hyaenidae	Ramus			1
86		Hyaenidae	Scapula			1
5		Rhinoceritidae	Calcaneum	Shaft		1
		Indeterminate	Fragments			139
Total						**200**

Table 5. 14: Contents of fossil locality 73, zone Y, Sandstone 12.

Specimen	W	Taxon	Element	Part	Other	N
17		Artiodactyla	Metatarsal	Proximal		1
11		Artiodactyla	Metatarsal			1
28		Artiodactyla	Humerus	Distal		1
44	4	Bovidae	Horncore			1

Table 5.14 continued: Contents of fossil locality 73, zone Y, Sandstone 12.

43		Bovidae	Humerus	Distal		1
1		Bovidae	Metatarsal	Proximal	Complete	1
2		Cervid	Humerus	Distal		1
10	3	Equus	Tooth	P3-M2	Lower, worn	1
14		Ungulata	Tibia			1
62		Indeterminate	Fragments			97
Total						**106**

Table 5.15: Contents of fossil locality 73, zone Z, Sandstone 12.

Specimen	**W**	**Taxon**	**Element**	**Part**	**Other**	**N**
1		Artiodactyla	Calcaneum	Shaft		1
25		Artiodactyla	Mandible	Premolars	Fragment	1
66, 190		Artiodactyla	Metacarpal			2
85,86 + 2 unmarked		Artiodactyla	Metatarsal			2
Various	3	Artiodactyla	Metapodial	Shaft		13
60, 72	5	Artiodactyla	Tooth	Fragment		2
107		Bovid	Astragalus			1
370		Bovid	1st Phalange			1
68 + 3 Unmarked		Bovid	Horncore			4
106	2b	Bovid	Shaft	Complete		1
178		Canidae	Radius	Distal		1
82		Carnivora	Cuboid			1
81	2b	Carnivora	Metatarsal	Distal		1
80		Carnivora	Rib			1
191		Carnivora	Scapula			4
44		Cervid	Humerus			1
15		Cervid	Mandible	Premolars		1
4	2b	Cervid	Metatarsal			1
2, 82	2a	Cervid	Metapodial	Shaft		2
85		Cervid	Ramus			1
78		Cervid	2nd Phalange			1
101		Cervid	Tooth	Molar		1
102		Cervid	Tooth	Molar	Spall	1
94	2a	Cervid	Molar	Upper	Fresh	1
64, 66, 92, 96, 172		Cervid	Tooth	Premolar		5
Unmarked		Cervid	Tooth	Premolar		3
3		Equid	Lunar			1
93	4	Equid	Mandible	Molar		1
13		Equid	Mandible	PM2	Root	1
Unmarked		Equid	Metacarpal	Fragment		2
26	2b	Equid	Metacarpal	Shaft		1
891		Equid	Metatarsal	Shaft		1
8, 18		Equid	Metapodial	Shaft		2
64		Equid	3rd Phalange			1
92		Equid	Tooth	Molar	Spall	1
98		Equid	Tooth	Incisor		1
95, 99, 100, 122 + 1 Unmarked		Equid	Tooth	Molar		5
867		Hyaenidae	4th Metatarsal			1
6		Hyaenidae	Metatarsal	Proximal		1
39		Hyaenidae	Scapula			1
42		Hyaenidae	Tooth	Incisor		1
		Indeterminate	Fragments			314
Total						**407**

Notes: 95 equid is worn; 172 cervid is upper. Artiodactyl metapodial fragments: 27,57, 100, 176, 111, 113, 114, 125, 128, 131, 135,141 + 1 unmarked.

Table 5. 16: Contents of fossil occurrences 74 – 85 (Set 5), Sandstone 12, between locality 73 and the GT road

Fossil occurrence 74: general collection scarp face:

Specimen	W	Taxon	Element	Part	Other	N
19	2	Artiodactyla	Metapodial			1
92		Artiodactyla	Metatarsal	Shaft		1
10	2	Artiodactyla	Metatarsal	Distal	Condyle	1
101		Artiodactyla	Phalange 1/2	Fragment		1
210	2a	Bovid	Cuneiform	Complete		1
4,44,62,7,107	3	Bovid	Horncore			5
1	3	Bovid	Molar 1/2	Complete	Upper,fresh	1
105		Bovid	Phalange			1
52*		Bovid	Premolar	Split		1
74	2	Cervid	2nd Phalange	Distal		1
24		Cervid	Metapodial			1
3		Cervid	Tooth	Premolar		1
6,75		Chelonia	Scute			2
63,212		*Crocodylus*	Radius			2
33		*Crocodylus*	Tooth			1
3,28,31,32,64, 75		Proboscidean	Molar	Fragment		6
35	2	Rhinoceritidae	Metacarpal	Complete		1
		Indeterminate	Fragments			300
Total						**326**

Fossil occurrence 75: general collection, "badlands" in front of Sandstone 12, near no. 60:

Specimen	W	Taxon	Element	Part	Other	N
17	3	Bovid	3rd Molar	Complete	Upper,worn	1
34		Bovid	Horncore			1
27	2	Bovid	Premolar	Split		1
48	3	Bovid	Premolar	Split	Lower	1
21	2b	Cervid	Premolar	Split	Little wear	1
0		Proboscidean	Tooth	Molar	Plate	18
0		Proboscidean	Tusk			8
28		*Struthio*	Egg-shell			1
		Indeterminate	Fragments			91
Total						**123**

Note: tusk fragments 11,40,41,43,44,46,54,56.

Fossil occurrence 79: scatter, scarp face of Sandstone 12:

Specimen	W	Taxon	Element	Part	N
10	2	Artiodactyla	Humerus	Proximal	1
17,23,33		Chelonia	Scute		3
26		*Struthio*	Egg-shell		1
10		Indeterminate	Coprolite?		1
0*		Indeterminate	Fragments		34
Total					**40**

Fossil occurrence 80: general collection scarp face of Sandstone 12:

Specimen	W	Taxon	Element	Part	N
2,6,8,15,24		Proboscidean	Molar	Plate	5
0,12,14,16,20		Proboscidean	Tooth	Tusk	5
0*		Indeterminate	Fragments		33
Total					**43**

Fossil occurrence 84: general collection scarp face of Sandstone 12; between locality 83 and the GT road:

Specimen	W	Taxon	Element	Part	N
9,10,12		Artiodactyla	Molar	Fragment	3
8,13		Chelonia	Scute		2

Table 5. 16 continued: Fossil occurrence 84

6		Equid	Premolar	Fragment	1
3	2b	Equid	Pelvis		1
0		Indeterminate	Fragments		84
Total					**91**

Fossil occurrence 85: general collection, scarp face of Sandstone 12; between locality 83 and the GT road:

Specimen	W	Taxon	Element	Part	N
4		Chelonia	Scute		1
8		*Struthio*	Egg-shell		146
1-3,5-7		Indeterminate	Shaft		6
Total					**153**

Other finds: also included in this set is a findspot (no. 81) of an artiodactyl scapula. General collections 76 comprised 1 cervid PM and 2 fragments; no. 77, 2 tusk and 1 other fragment; no. 78, 4 artiodactyl tooth fragments and 3 other fragments. Find-spot 82 comprised 1 piece of tusk, 80 Proboscidean molar plate fragments and 80 other fragments. All these occurrences were on the scarp face of Sandstone 12.

Table 5. 17: Contents of locality 83 (Set 5), Sandstone 12

Specimen	W	Taxa	Element	Part	Other	N
Zone A:						
13		Artio (Medium)	Metapodial	Distal	Condyle	1
30		Artio (Medium)	Molar	Lower	Worn	1
21	3	Artiodactyla	Metapodial	Shaft		1
27	4	Bovid (Large)	Horncore			1
2	2b	Bovidae	Astragalus	Complete		1
31	2b	Bovidae	M3	Upper	Left	1
3	2a	Cervid	Astragalus	Complete		1
25	3	Chelonia	Scapula			1
3,32,36		Chelonia	Scute	Smooth		3
1	3	*Elephas*	Molar	Lower	Complete	1
0	4	Proboscidean	Tooth	Molar	Plate	8
11,100,101,1 07		Proboscidean	Tooth	Tusk		4
22	5	Equid	Astragalus	Complete		1
19	2a	Equid	Tooth	PM3		1
		Indeterminate	Fragments			80
Total						**106**
Zone B:						
32	2b	Artiodactyla	Humerus	Distal		1
11		Artiodactyla	Metapodial	Distal		1
23	2	Artiodactyla	Metatarsus	Proximal		1
30, 31	2b	Cervid	Antler	Tine		2
5	2b	Cervid	Mandible, left	M1, M2,M3 broken	Little wear	1
7	2a	Cervid	Tooth	M1/2	Fresh	1
28	3	Cervid	Tooth	Premolar		1
22,28,29,33		Chelonia	Scute	Smooth		4
24,25,34		Chelonia	Scute	Dimpled		3
		Indeterminate	Fragments			10
Total						**25**
Zone C:						
9	4	Artiodactyla	Vertebrae	Centrum	Complete	1
11	3	Artiodactyla	Vertebrae			1
1,4	3	Cervid	Antler			2
6	3	Cervid	Antler (Palmate)	Pedicel		1
0	3	Chelonia	Scute	Smooth		6
13,14,31	3	Chelonia	Scute	Dimpled		3
3,22	3	Chelonia	Scute			2
31	3	*Crocodylus*	Palate			1
32-34		Proboscidean	Tooth	Molar		7
11	3	Equid	Metatarsal	Proximal	Split	1
0		Indeterminate	Fragments			27
Total						**52**

Table 5. 17 continued: contents of locality 83

Zone D:						
0		Artiodactyla	Tooth	No crowns	Split	7
5		Artiodactyla	Ulna	Proximal		1
39	3	Artio/Cervid	Scapula	Glenoid	Right	1
111	4	Artiodactyla	Metapodial	Shaft		1
11		Artiodactyla	Radius	Shaft		1
25	2b	Bovid (Medium)	Metapodial	Distal	Condyle	1
0	4	Bovid/Cervid	Tooth	Split		3
0		Bovid/Cervid	Tooth	Premolar	Mod. Wear	1
0	3	Chelonia	Scute	Dimpled		6
0	4	Chelonia	Scute	Smooth		15
87	4	Chelonia	Scute	Netted		1
456	3	Chelonia	Scute			1
55,68	4	*Crocodylus*	Humerus	Shaft		2
23	4	*Crocodylus*	Shaft			1
511	3	Equid	M3	Complete	Upper	1
10	5	Proboscidean	Pelvis			1
0	4	Proboscidean	Rib			5
29	4	Proboscidean	Scapula			1
6	4	Proboscidean	Shaft			1
31,454	4	Proboscidean	Tooth	Molar	Plate	2
20	5	Proboscidean	Vertebrae			1
4	4	Proboscidean	Carpal	Split		1
15	5	Proboscidean	Carpal			1
0	5	Proboscidean	Tooth	Fragments		800
8	4	Ungulata	Femur	Proximal epiphysis	Unfused	1
		Indeterminate	Fragments			466
Total						**1324**
Zone E:						
1	3	Artiodactyla	Cuneiform	Complete		1
12	4	Artiodactyla	Tooth	Molar	Lower	1
7	3	Cervid	Antler			1
4	4	Cervid	Metacarpal	Proximal	Complete	1
9	3	Chelonia	Scute	Smooth		1
8	4	Proboscidean	Tooth	Molar	Fragment	1
0		*Struthio*	Eggshell	Fragments		34
		Indeterminate	Fragments			11
Total						**51**
Zone F:						
1-3	2b	Cervid	Antler			3
5	2b	Cervid/Bovid	PM3 upper	Complete	Worn	1
	2b	Indeterminate	Rib	Fragments		1
Total						**5**
Overall total						**1563**

Note: 83EGB4 metacarpal has a possible bite mark.

Table 5. 18: Contents of fossil occurrences in the "badlands" in front of Sandstone 12 north of the G.T. road, 1986 (Set 6).

Fossil occurrence 110: scatter:

Specimen	W	Taxon	Element	Part	Other	N
11, 16	3	Artiodactyla	Premolar	Spall		2
7	5	Bovid	Horncore			1
2,12	3	Bovid (Large)	Molar	Split	Worn	2
4		Felid	Metapodial	Distal		1
		Indeterminate	Fragments			39
Total						**45**

Table 5. 18 continued: Contents of fossil occurrence 112: general collection:

Specimen	W	Taxon	Element	Part	Other	N
3	2b	Artiodactyla	Tooth	Fragment		1
1,4	3,5	Cervid	Antler			2
9	4	Proboscidean	Tooth	Molar		1
		Indeterminate	Fragments			7
Total						**11**

Fossil occurrence 114: scatter:

Specimen	W	Taxon	Element	Part	Other	N
2-6,8		Artiodactyla	Tooth	Molar		6
12		Chelonia	Scute	Rippled		1
1		*Crocodylus*	Tooth			1
7		*Struthio*	Egg-shell			1
		Indeterminate	Fragments			43
Total						**52**

Fossil occurrence 116: scatter:

Zone A:						
Specimen	**W**	**Taxon**	**Element**	**Part**	**Other**	**N**
3	5	Bovid	Horncore			1
126		Cervid	Antler			1
18		Chelonia	Scute			6
19	3	Chelonia	Scute	Smooth		21
10	2a	*Crocodylus*	Tooth	Broken		1
1		Indet/Suid	Humerus	Shaft	Unfused	1
	4	Indeterminate	Fragments			17
Zone B:						
7, 9	5	Bovidae	Horncore			2
1, 39	2a	Cervid	1st Phalange	Complete		2
6	3	Chelonia	Scute	Smooth		3
8	4	Chelonia	Scute	Netted		1
10		Chelonia	Scute			1
		Indeterminate	Fragments			39
Zone C:						
1	2b	Bovid	2nd Phalange	Complete	?Bite-mark	1
18	2a	Cervid	Metatarsal	Proximal	Complete	1
30	2b	Cervid	Tooth	Complete	Incisor	1
		Chelonia	Scute	Smooth		2
		Chelonia	Scute			6
		Indeterminate	Fragments			36
Zone D:						
27		Artio (Small)	Metapodial	Shaft		1
		Chelonia	Scute	Netted		2
		Chelonia	Scute	Smooth		19
2		cf. *Hexaprotodon*	Tooth	Tusk		1
		Indeterminate	Fragments			29
Total						**195**

Fossil occurrence 118: general collection:

Specimen	W	Taxon	Element	Part	Other	N
0	3	Artiodactyla	Molar	Fragment		2
1		*Struthio*	Egg-shell			29
		Indeterminate	Fragments			7
Total						**38**

Table 5. 18 continued: Fossil occurrence 120: scatter:

Specimen	W	Taxon	Element	Part	Other	N
2		Chelonia	Scute			1
1,3		Equid	Tooth	Molar		2
5		*Struthio*	Egg-shell			328
		Indeterminate	Shaft	Fragments		3
Total						**334**

Fossil occurrence 122: scatter:

Specimen	W	Taxon	Element	Part	Other	N
1		Cervid	Metatarsal	Proximal	Complete	1
25		Cervid	Tooth			2
		Chelonia	Scute			11
		Indeterminate	Fragments			15
Total						**29**

Fossil occurrence 126: general collection:

Specimen	W	Taxon	Element	Part	Other	N
14	3	Artio	Tooth	Molar	Spall	1
7	5	Cervid	Tooth	Molar, upper	Complete	1
17		Cervid	Molar	Upper	Fresh	1
		Chelonia	Scute			6
0		Indeterminate	Fragments			12
Total						**27**

Fossil occurrence 130: scatter:

Specimen	W	Taxon	Element	Part	Other	N
203,209		Aves (unconfirmed)	Tibia	Shaft		2
73	4	Artiodactyla	Glenoid	Fragment		1
		Artiodactyla	Metapodial			9
293,320		Artiodactyla	2nd Phalange			2
136,405		Artiodactyla	Tooth	Fragment		2
90		Artiodactyla	Vertebra			1
204		Artiodactyla	Vertebra	Complete		1
406		Bovid	1st Phalange			1
104		Bovid	2nd Phalange	Fragment		1
165		Bovid	Vertebra			1
0		Chelonia	Scute			1
		Proboscidean		Tooth	Plate	12
89,93	4	Proboscidean		Tooth	Tusk	2
80	4	Equid	Molar	Split	Lower	1
210		Struthio	Eggshell			1
						2
Total						**400**

Fossil occurrence 132: general collection:

Specimen	W	Taxon	Element	Part	Other	N
13		Artiodactyla	Metapodial			1
6		*Crocodylus*	Tooth			1
0		Indeterminate	Shaft			9
Total						**11**

Table 5. 18 continued: Fossil occurrence 134: general collection:

Specimen	W	Taxon	Element	Part	Other	N
7		Bovid	Mandible	M3	Left, fresh	1
1		Chelonia	Scute	Smooth		4
8		Proboscidean		Tooth	Root	1
4	3	Indeterminate	Rib			2
Total						**8**

Fossil occurrence 136: scatter:

Specimen	W	Taxon	Element	Part	Other	N
7		Proboscidean	Femur			1
8, 60, 61,59		Proboscidean	Mandible			4
Total						**5**

Fossil occurrence 140: scatter:

Specimen	W	Taxon	Element	Part	Other	N
110,111		Bovid	Horncore			2
96		Chelonia	Scute	Fine mesh		1
0		Chelonia	Scute	Smooth		77
114		Proboscidean	Rib			1
		Indeterminate	Fragments			44
Total						**125**
Zone A						
Specimen	**W**	**Taxon**	**Element**	**Part**	**Other**	**N**
12,57		Artiodactyla	Metapodial			2
60		Artiodactyla	Tooth	Premolar		1
17		Artiodactyla	Tooth, upper	Split	Molar	1
10	2b	Cervid	1st/2nd	Phalange	Distal	1
85		Cervid	Radius			1
15		Chelonia	Scute			1
1	3	Proboscidean	Shaft			1
		Proboscidean	Tooth	Molar		23
33		Equid	Tooth	Molar		1
		Indeterminate	Fragments			7
Total						**39**
Zone B						
Specimen	**W**	**Taxon**	**Element**	**Part**	**Other**	**N**
28		Artiodactyla	Tooth	Molar		1
1		Chelonia	Scute	Fine mesh		1
		Chelonia	Scute	Dimpled		2
23		*Crocodylus*	Tooth			1
Total						**5**
Overall total						**169**

Note: the indeterminates listed for occurrence 130 include 1 record of fish scales and 1 reptile limb shaft fragment, both unconfirmed.

Fossil occurrence 142: general collection:

Specimen	W	Taxon	Element	Part	Other	N
1	4	Bovid	Horncore			1
		Indeterminate	Shaft			2
		Struthio	Eggshell			1
Total						**4**

Table 5. 18 continued: Fossil occurrence 144: general collection:

Specimen	W	Taxon	Element	Part	Other	N
3,6	4	Artiodactyla	Metapodial			2
4	4	Bovid	Horncore			1
1,2		Chelonia	Scute			6
		Indeterminate	Fragments			3
Total						**12**

Fossil occurrence 146: general collection:

Specimen	W	Taxon	Element	Part	Other	N
1		Bovid	Horncore			1
5	5	Proboscidean	Tooth	Molar		1
6		*Struthio*	Eggshell			4
		Indeterminate	Tibia			4
Total						**10**

Fossil occurrence 148: scatter:

Specimen	W	Taxon	Element	Part	Other	N
1	1	Cervid	Tooth, upper	Complete	Molar	1
3		Proboscidean	Tooth	Tusk		3
4		*Struthio*	Eggshell			1
		Indeterminate	Shaft			15
Total						**20**

Fossil occurrence 150: scatter

Specimen	W	Taxon	Element	Part	Other	N
65		Artiodactyla	Metapodial			1
119		Bovid	3rd Phalange			1
54,83		Bovid	Horncore			2
59		Cervid	Antler			1
6		Cervid	Humerus			1
17	2b	Cervid	Tooth, upper	Complete	Molar	1
78		Cervid	Tooth	Molar		1
		Chelonia	Scute			4
53,64		Proboscidean	Rib			2
1,2,43		Proboscidean	Shaft			3
		Proboscidean	Tooth	Molar		78
21		Proboscidean	Tooth	Tusk		1
53		Indeterminate	Fragments			69
Total						**165**

Note: 110/1 has possible signs of rodent-gnawing. :114GB 4-6 was probably part of one tooth, and 2, 3, 8 of another.

Table 5. 19: Contents of fossil locality 124 (Set 6), badlands in front of Sandstone 12

Specimen	W	Taxon	Element	Part	Other	N
519		Artiodactyla	1st Phalange			1
502		Artiodactyla	Metacarpal			1
511,518		Artiodactyla	Metapodial			2
154		Artiodactyla	Molar	Lower	Split	1
500,501		Artiodactyla	Vertebrae			2
51	2a	Bovid (Large)	PM4	Upper	Fresh	1
116		Bovid (Large)	1st Phalange	Proximal	Complete	1
0,25,191		Bovid	Horncore			4
52,180,		Bovid	Tooth	Lower	Split	3
999		Bovid	Tooth	M1/2	Lower	1

Table 5. 19 continued: contents of fossil locality 124

Specimen	W	Taxon	Element	Part	Other	N
140		Bovid	Tooth	M1/2	Lower	1
521		Carnivora	Crania			1
39		Cervid	Antler	Palmate		1
144		Cervid	Antler	Rolled		1
53		Cervid	Tooth	Incisor		1
64	2a	Cervid	Tooth	Molar 3	Upper	1
122		Cervid	Tooth	Molar		1
51	2a	Cervid/Bovid	Tooth	P4, fresh	Right, upper	1
0		Chelonia	Scute	Smooth		136
0		Chelonia	Scute	Dimpled		17
0		Chelonia	Scute	Netted		10
		Chelonia	Scute			2
11		*Crocodylus*	Mandible	Unf. symphysis		1
33	2b	*Crocodylus*	Premaxilla	Left	Sand matrix	1
56		*Crocodylus*	Crania			1
	4	*Struthio*	Eggshell			1
		Indet/*Crocodylus*	Femur	Proximal		1
		Indeterminate	Fragments			110
Total						**306**

Note: Two Chelonia scute had a fine sand matrix, as did *Crocodylus* specimens 124/11 and 124/31.

Table 5. 20. Contents of fossil locality 128 (Set 6), badlands in front of Sandstone 12

Specimen	W	Taxon	Element	Part	Type	N
Zone A:						
22		Artiodactyla	Vertebra			1
100		Bovid	Tooth	Premolar		1
35	2b	Cervid	2nd Phalange	Complete		1
8	2	Cervid	Ulna	Proximal		1
		Chelonia	Scute	Smooth		42
		Chelonia	Scute	Dimpled		1
		Chelonia	Scute	Rippled		1
		Indeterminate	Fragments			28
Total						**76**
Zone B:						
13		Artiodactyla	Cuneiform			1
200		Artiodactyla	Cuboid	Complete		1
3		Artiodactyla	Humerus	Proximal		1
133.		Artiodactyla	Phalange			1
16		Artiodactyla	Ramus	Condyle		1
224		Artiodactyla	Ramus			1
221		Artiodactyla	Sesamoid			1
214		Artiodactyla	Tooth	Fragment		1
37.	2b	Artiodactyla	Ulna	Proximal		1
82	3	Artiodactyla (Large)	Vertebra			1
50, 56		Artiodactyla	Vertebra	Fragment		2
110		Artiodactyla	Vertebra	Centrum		1
7	2a	Artio/Cervid	Femur	Distal		1
5, 6, 57, 133, 141	2b	Bovid	Horncore			5
137	1	Bovid	Mandible	M2,M3	Fresh	1
53		Cervid	Calcaneum			1
62	2	Cervid	Humerus	Distal	Split	1
102		Cervid	Metacarpal	Distal	Unfused	1
39	2a	Cervid	Pelvis			1
128		Cervid	Ramus			1
35	2a	Cervid	2nd Phalange			1
44		Cervid	Tibia	Proximal	Epiphysis	1
		Chelonia	Scute	Smooth		87

Table 5. 20 continued: Contents of fossil locality 128, badlands in front of Sandstone 12

Specimen	W	Taxon	Element	Part	Type	N
		Chelonia	Scute	Netted		1
0		Chelonia	Scute	Dimpled		1
36, 58		*Crocodylus*	Rib			2
181		*Crocodylus*	Tooth			1
22, 105	2b	Carnivore (Small)	Scapula			2
126	2b	Cervid	Fibula	Distal		1
234		Chelonia	Scute	Smooth		1
93	4	Equid	Tooth	Molar		1
106		*Struthio*	Eggshell			2
222		Indeterminate	Fragments			148
Total						273
Zone C:						
2, 23		Artiodactyla	Metapodial	Shaft		2
98	2b	Artio (Medium)	Metapodial	Proximal	Split	1
14	2b	Artiodactyla	Metatarsal	Shaft		1
4	2b	Artiodactyla	Tooth	Incisor		1
30		Artiodactyla	Tooth	Premolar		1
126		Artiodactyla	Fibula	Distal		1
16	3	Bovid (Large)	Metacarpal	Proximal	Split	1
24	3	Bovid (Medium)	Calcaneum	Split		1
18		Bovid (Medium)	1st Phalange	Distal		1
15.	2a	Bovid (Medium)	1st Phalange	Complete		1
30		Bovid (Medium)	1st Phalange	Proximal	Complete	1
249		Bovid	Tooth	Premolar		1
250		Carnivora	1st Phalange			1
253		Carnivora	Pelvis			1
7		Carnivora	3rd Phalange	Small		1
252		Carnivora	Ulna			1
		Chelonia	Scute	Smooth		54
		Chelonia	Scute	Netted		3
		Chelonia	Scute	Dimpled		1
6, 8		*Crocodylus*	Tooth			2
		Indeterminate	Fragments			104
Total						**181**
Zone D:						
39		Artiodactyla	Axis	Fragment		1
11	3	Artiodactyla	Humerus			1
		Chelonia	Scute	Smooth		22
		Chelonia	Scute	Netted		1
		Chelonia	Scute	Rippled		1
8		Indeterminate	Fragments			28
Total						**54**
Zone E:						
		Chelonia	Scute			11
		Indeterminate	Fragments			6
Total						**17**
Overall total						**586**

Notes: Zone A 128GB102 Cervid has a clay skin; 44 is gritty; one Chelonia scute had a silt matrix; 128GB108 is gritty; Crocodylus B58 is suncracked.

Table 5. 21: Contents of fossil occurrences along Sandstone 12 on the south side of the G.T. road, 1986 (excluding locality 133), Set 7

Fossil occurrence 111: scatter, Sandstone 12 scarp:

Specimen	W	Taxon	Element	Part	Other	N
4, 55,94,101		Artiodactyla	Metapodial			4
95		Artiodactyla	Tooth	Molar, lower	Fragment	2
88		Chelonia	Scute			1
3		*Crocodylus*	Tooth	Complete		1
1, 4, 92	3	Proboscidean	Rib			3
69,91		Proboscidean	Shaft			2
3, 9		Proboscidean	Tusk			4
		Proboscidean	Tooth	Molar	Plate	22
		Indeterminate	Fragments			221
Total						**260**

Fossil occurrence 113: general collection, Sandstone 12 scarp:

Specimen	W	Taxon	Element	Part	Other	N
2	3	Bovid (Large)	Molar	Fragment		1
0		Bovid (Small)	Premolar	Upper		1
20		Chelonia	Scute			1
19		Chelonia	Skull			1
18	2a	Artiodactyl (Small)	Molar	Complete		1
		Indeterminate	Fragments			16
Total						**21**

Note: This table includes 2 indeterminate fragments from occurrence 115.

Fossil occurrence 117: scatter, Sandstone 12 scarp:

Specimen	W	Taxon	Element	Part	Other	N
2	2	Cervid	Cuneiform	Magnum		1
5,6		Chelonia	Scute	Smooth		2
7		Chelonia	Scute	Rippled		1
9,10,13		Chelonia	Scute			3
0		Indeterminate	Fragments			31
Total						**38**

Fossil occurrence 119: general collection, Sandstone 12 scarp:

Specimen	W	Taxon	Element	Part	Other	N
34		Cervid	Tooth	Molar		1
2		Cervid	Metacarpal	Proximal	Complete	1
4	2b	Cervid	Tooth	M1/2,, fresh		1
49	5	Cervid	Tooth	Premolar	Complete, fresh	1
13,46,51		Chelonia	Scute			3
15,23,28		Chelonia	Scute	Dimpled		3
9,58		Chelonia	Scute	Rippled		2
3,21		Chelonia	Scute	Smooth		2
8		Ungulate	Tooth	Molar		1
0		Indeterminate	Fragments			52
Total						**67**

Fossil occurrence 121: scatter, Sandstone 12 scarp:

Specimen	W	Taxon	Element	Part	Other	N
2,5	3	Bovidae	Horncore			2
6		Chelonia	Scute			1
0		Indeterminate	Fragments			19
Total						**22**

Note: 121GB5 has a gritty matrix.

Table 5. 21 continued: fossil occurrence 125: scatter, Sandstone 12 badlands:

Specimen	W	Taxon	Element	Part	Other	N
11		Bovid	Scaphoid			1
2, 3		Bovid (Large)	Mandible	Right, M3		2
1	4	Cervid	Antler			1
		Cervid	Cuneiform			1
13, 22	3	Cervid	Tooth	Premolar		2
4		Chelonia	Scute	Fishnet		1
15		Chelonia	Scute	Pitted		1
5		Chelonia	Scute	Smooth		1
0		Chelonia	Scute			9
0		Indeterminate	Fragments			14
Total						**33**

Fossil occurrence 127: scatter, Sandstone 12 badlands:

Specimen	W	Taxon	Element	Part	Other	N
		Artiodactyla	Shaft			1
3		*Crocodylus*	Humerus			1
4		*Crocodylus*	Scapula			1
8	2a	*Crocodylus*	Tooth	Complete		1
1		Rhinocerotidae	Patella			1
2		Indeterminate	Fragments			7
Total						**12**

Fossil occurrence 129: general collection, Sandstone 12 badlands:

Specimen	W	Taxon	Element	Part	Other	N
7	5	Artiodactyla	Molar	Split		1
1		Chelonia	Scute	Pitted		8
6		Chelonia	Scute	Smooth		13
2	4	Equid	Molar, Worn	Split, Lower		1
19		Equid	Tooth	Fragment		1
0		Indeterminate	Fragments			11
Total						**35**

Fossil occurrence 131: scatter, Sandstone 12 badlands:

Specimen	W	Taxon	Element	Part	Other	N
0		*Struthio*	Eggshell			194

Fossil occurrence 137: general collection, Sandstone 12 badlands:

Specimen	W	Taxon	Element	Part	Other	N
22	5	Artiodactyla	Tooth	Spall		1
4, 6	5	Equid	Molar	Split	Worn	2
13,14		*Struthio*	Egg Shell			2
11	5	Ungulata	Tooth	Fragment	Molar	1
0	4	Indeterminate	Fragments			16
Total						**22**

Other finds: Occurrence 123, a small scatter in the badlands in front of Sandstone 12, contained 1 bovid incisor, 1 Proboscidean limb shaft fragment, and 141 other fragments. Occurrence 135 was a patch of 100 Proboscidean tusk fragments.

Table 5. 22: Fossil material from locality 133, Sandstone 12 (Set 7)

Zone A:

Specimen	W	Taxon	Element	Part	Other	N
220	5	Artiodactyla	Cuneiform	Complete		1
423	3	Artiodactyla	Mandible	Molar	Fresh	1
365	4	Artiodactyla	Pelvis	Fragment		1
374,450	4	Artiodactyla	Vertebra	Fragment		2
375	3	Artio/Bovid	3rd Phalange	Fragment		1
107		Artio/Cervid	Metacarpal	Distal epiphysis	Unfused	1
159,222	4	Artio/Cervid	3rd Phalange	Complete		2
102	3	Artio/Cervid	Humerus	Distal		1
360,426	5	Artiodactyla	Metapodial	Distal	Condyle	2
458	5	Artiodactyla	Metapodial	Distal		1
0		Artiodactyla	Tooth	Molar		13
386	3	Artiodactyla	Ulna			1
0	4	Bovid	Horncore			9
17,104,351		Bovid	Tooth	Split		3
409	4	Bovid	M1	Lower	Complete	1
0, 415		Bovid	Tooth	Molar	Fragment	13
0		Bovid	Tooth	Premolar	Fragment	11
380,393	4	Bovid	M3	Lower,worn	Complete	2
312	3	Cervid	Mandible	Fragment P3	Worn	1
29,66	3	Cervid	Metapodial	Shaft		2
14	2b	Equid	M3	Complete	Upper, worn	1
2	2b	Equid	P2	Complete	Upper	1
4		Equid	Tooth (Complete)	Premolar	Upper	1
9	2b	Equid	P3/M2	Complete	Upper, worn	1
36, 67	4	Equid	PM2	Complete	Upper, worn	2
42	2b	Equid	PM3, right	Complete	Upper, worn	1
73,363	4	Equid	Molar	Split	Worn	2
376,389		Equid	Molar	Complete	Upper, worn	2
368		Equid	Molar	Complete	Lower, worn	1
0		Equid	Tooth			4
417	3	Equid	Tooth	Incisor	Fragment	1
		Struthio	Egg shell			15
111,113,205	5	Ungulata	Tooth	Molar	Fragments	3
		Indeterminate	Fragments			389
Total						**493**
Zone B:						
15,77,121,236,241	5	Artiodactyla	Tooth	Premolar	Fragment	9
0		Artiodactyla	Tooth	Molar	Fragment	22
7		Bovidae	3rd Phalange			1
0		Bovidae	Horncore			7
23	3	Bovidae	Maxilla	+ P4, M1	Worn	1
55		Bovidae	Tibia	Shaft		1
19	3	Bovidae	M3	Lower	Fragment,	1
26		Bovidae	Tooth	Molar		1
21,43,46,132		Bovidae	Tooth	Premolar	Fragments	4
128,172,200	4	Bovidae	Molar	Split	Fragments	8
various	4	Bovidae	Premolar	Split		14
115	3	Bovidae	Premolar	Upper	Complete	1
312	3	Cervid	Mandible	Premolar3		1
89	4	Cervid	Metapodial	Shaft		1
37	4	Cervid	Metatarsal	Shaft		1
197	3	Cervid	2nd Phalange	Complete	Unfused	1
33	4	Cervid	Tibia	Distal	Complete	1
81,92	4	Cervid	Premolar	Split		12
100	4	Cervid	M3	Split	Worn	1
130,284		Cervid	Premolar			2
326		Cervid	Tooth	Molar		1
265		*Crocodylus*	Tooth	Fragment		1

Table 5. 22 continued: Fossil material from locality 133, Sandstone 12

31	3	Equid	Metacarpal	Proximal	Complete	1
25		Equid	Radius	Proximal		1
24	5	Equid	3rd Phalange	Complete		1
68		Equid	3rd Phalange			1
72	3	Equid	M3	Fragment	Upper	1
14	3	Equid	M3	Right	Upper	1
12,38,109,306	3	Equid	Tooth	Molar	Split	4
144		Hyaenidae	Tooth	Broken	Molar	1
various	5	Ungulata	Tooth	Molar	Fragments	8
112,164, 289	5	Ungulata	Tooth	Premolar	Fragments	3
		Indeterminate	Fragments			214
Total						**327**
Overall total						**820**

Note: Equid tooth 133A/67 and Cervid tibia 133B/33 have a clay skin. Ones in **bold** are shown in Figure 10.5

Table 5. 23: Contents of fossil occurrences south of the G.T. Road and in front of Sandstone 12 (Set 8).

Fossil occurrence 139: scatter:

Specimen	W	Taxon	Element	Part	Other	N
40	2b	Artiodactyla	Tibia			1
64	5	Artio	Tooth	Molar	Very worn	1
3	2a	Bovid	1st Phalange	Complete		1
5	3	Bovid	Tooth	Molar, upper	Split, worn	1
21		Chelonia	Scute	Fishnet		1
28		Chelonia	Scute			2
69		*Elephas*	Humerus			1
68		*Elephas*	Mandible			1
0		*Elephas*	Tooth	Molar	Plate	7
67		*Elephas*	Tooth	Tusk		8
1	3	Equid	Metatarsal	Shaft		1
7	5	Equid	Tooth	Premolar	Split, lower, worn	1
60	3	Proboscidean	Cranium	Fragment		1
57	3	Proboscidean	Pelvis	Fragment		1
58	3	Proboscidean	Shaft	Fragment		5
56		*Stegodon*	Tooth	Fragment		1
11		*Struthio*	Eggshell			1
0		Indeterminate	Fragments			171
Total						**206**

Fossil occurrence 141: scatter:

Specimen	W	Taxon	Element	Part	Other	N
11	5	Artio	Tooth	Premolar	Complete	1
0	5	Artio	Tooth	Fragment		10
7	4	Bovid	Metapodial	Distal		1
2	2b	Bovid	Phalange	Distal		1
16	5	Bovid	Tooth	Fragment		2
0		Indeterminate	Shaft			14
Total						**29**

Fossil occurrence 143: general collection:

Specimen	W	Taxon	Element	Part	Other	N
3	4	Bovid	Horncore			1
7	4	Bovid	Premolar	Complete	Lower	1
6	5	Bovid	Tooth	Premolar	Split	1
	4	Indeterminate	Shaft	Fragments		6
Total						**9**

Notes: Number 145 was not used. Scatter 149 contained 1 fragment of equid right scapula, an equid complete upper molar, and an equid upper molar fragment. General collection 153 comprised only 20 indeterminate fragments and 1 piece of ostrich egg-shell. Scatter 154 contained only 3 pieces - 1 piece of scute, a piece of proboscidean molar plate, and an indeterminate fragment. 139GB40 had a sand matrix; and 139GB57 had a possible crocodile puncture mark. The identification of *Stegadon* specimen 139/56 was confirmed by A.C. Nanda.

Table 5. 24: Fossil material from locality 156 (Set 8).

Specimen	W	Taxon	Element	Part	Other	N
Zone A:						
52		Artiodactyla	Metapodial			1
29		Artiodactyla	Pelvis			3
36	2b	Artio	Ulna	Proximal		1
10		Bovid	Femur	Shaft		1
38	5	Bovid	Horncore	Pedicel		1
51	3	Bovid	Horncore			2
5	3	Bovid	Humerus	Distal		1
34	5	Bovid	3rd Phalange			1
45	3	Bovid	Tooth	P4, left, upper	Fragment	1
9		Bovid	Tooth, worn	Molar, left, lower	Complete	1
39		Bovid	Cuneiform	External	Complete	1
31		Cervid	Antler	Tine		1
43		Cervid	Metapodial			2
69	2a	Cervid	Tooth, little wear	Molar, left, upper	Complete	1
44	3	Cervid	Tooth, worn	Premolar, upper	Complete	1
30		Chelonia	Scute			1
23		Chelonia	Scute	Smooth		1
20		Chelonia	Scute	Fishnet		1
49		Proboscidean	Tooth	Molar	Plate	3
4		Rhinocerotidae	Astragulus	Split		1
1		Rhinocerotidae	Vertebra	Centrum		1
3	3	Rhinocerotidae	Vertebra	Atlas		1
		Indeterminate	Fragments			214
Zone B:						
191		Artiodactyla	Mandible			1
185		Artiodactyla	Metapodial			19
124	5	Artiodactyla	Tooth	Premolar, lower	Fragment	1
91	5	Bovid	1st Phalange	Proximal		1
217		Bovid	Cranium			1
216		Bovid	Horncore			1
195		Bovid	Tooth	Molar		1
187,192,193		Bovid	Tooth	Premolar		3
114	4	Cervid	Tooth	Premolar, lower	Complete	1
151		Cervid	Tooth	Molar		1
98,109		Chelonia	Scute			2
98.10		Chelonia	Scute	Dimpled		1
112		Chelonia	Scute	Fishnet		1
116		*Crocodylus*	Tooth	Complete		1
117		Proboscidean	Tooth	Molar		1
110		Equid	Navicular			1
188		Rhinocerotidae	Metacarpal			1
104,194		Rhinocerotidae	Tooth	Molar		2
		Indeterminate	Fragments			5
Total						**287**

Table 5. 25: Contents of fossil occurrences from Sandstone 14, 1986 (Set 9)

Fossil occurrence 90, general collection:

Specimen	W	Taxon	Element	Part	Other	N
1,2 (joins)		Rhino/Bovid	Vertebra	Cervical		1
4		*Struthio*	Eggshell			4
		Indeterminate	Fragments			6
Total						**11**

Table 5. 25 continued: Fossil occurrence 91:

Specimen	W	Taxon	Element	Part	Other	N
		Artiodactyla	Astragalus	Split		1
		Artiodactyla	Fibula	Distal		1
73	3	Artiodactyla	Tooth	M3	Lower	1
8, 68		Cervid	Antler			2
64	2a	Cervid	Tibia	Distal		1
78		Cervid	Astragalus			1
0		Chelonia	Scute	Smooth		36
63		*Crocodylus*	Femur			1
66		*Crocodylus*	Humerus			1
5		*Crocodylus*	Tooth			1
74		*Crocodylus*	Ulna			1
		Indeterminate	Fragments			54
Total						**101**

Fossil occurrence 92, scatter:

Specimen	W	Taxon	Element	Part	Other	N
3		Artiodactyla	Tooth	Fragment		1
1		Proboscidean	Tooth	Molar	Plate	1
		Indeterminate	Fragments			23
Total						**25**

Fossil occurrence 93, general collection:

Specimen	W	Taxon	Element	Part	Other	N
5	3	Bovid (Large)	Tooth, upper	M1/2, right	Complete	1
8,10,63,73,113		Proboscidean	Tooth	Molar		5
1,3,13,93		Proboscidean	Tooth	Tusk		4
		Indeterminate	Fragments			5
Total						**15**

Fossil occurrence 94, general collection:

Specimen	W	Taxon	Element	Part	Other	N
1		Artiodactyla	Metatarsal	Shaft		1
8	4	Cervid	Tooth	Spall		1
		Indeterminate	Fragments			15
Total						**17**

Fossil occurrence 96, general collection:

Specimen	W	Taxon	Element	Part	Other	N
4,6,7,9,20,22	4	Artiodactyla	Tooth	Molar		6
13	3	Suidae	Tooth	Molar		1
		Indeterminate	Fragments			25
Total						**32**

Fossil occurrence 97, scatter:

Specimen	W	Taxon	Element	Part	Other	N
3	4	Artiodactyla	Ulna carpal	Complete		1
7	4	Artiodactyla	Tooth	Molar	Upper	1
4	4	Cervid	Tooth	Fragment	Lower	1
1,2	4	Proboscidean	Tooth	Molar	Root	2
Total						**5**

Fossil occurrence 98, general collection:

Specimen	W	Taxon	Element	Part	Other	N
48		Artiodactyla	Femur	Shaft		1
9,36		Artiodactyla	1st Phalange	Distal		2
37		Artiodactyla	1st Phalange	Shaft		1
27		Artiodactyla	Phalange	Proximal		1

Table 5. 25 continued: Fossil occurrence 98, general collection:

53		Artiodactyla	Tooth	M1/2	Complete	1
85		Artiodactyla	Tooth	Fragment		1
38-43		Artiodactyla	Tooth	Molar		6
1		Bovid	Horncore			1
44		Bovid	Ulna	Fragment		1
47		Cervid	Radius	Distal		1
6		Chelonia	Scute			1
		Indeterminate	Fragments			46
Total						**63**

Fossil occurrence 99, scatter:

Specimen	W	Taxon	Element	Part	Other	N
4	3	Artiodactyla	Astragalus	Split		1
2		Artiodactyla	Calcaneum	Shaft		1
15		Artiodactyla	Cuneiform	Fragment		1
6,9, 44	3	Artiodactyla	Humerus	Distal		3
30,39		Artiodactyla	1st Phalange	Distal		2
13		Bovidae	Vertebrae	Lumbar		1
8		Cervid	Metatarsal	Shaft		1
23		Cervid	Scapula	Glenoid		1
33	4	Bovidae	Tooth	Molar		1
35		Bovidae	Vertebrae	Fragment		1
		Chelonia	Scute			7
		Indeterminate	Fragments			38
Total						**58**

Fossil occurrence 100, general collection:

Specimen	W	Taxon	Element	Part	Other	N
8	4	Artiodactyla	Tooth	Molar	Spall	1
1	3	Bovid (Large)	Tooth	M3, Lower	Split	1
4,10		Chelonia	Scute			2
5	5	Equid	Molar, lower	Split	Worn	1
		Indeterminate	Fragments			12
Total						**17**

Fossil occurrence 101, scatter:

Specimen	W	Taxon	Element	Part	Other	N
1	4	Bovid	Tibia	Distal		1
2	3	Bovid (Large)	M2, Upper	Med. wear	Complete	1
	3	Indeterminate	Fragments			11
Total						**13**

Fossil occurrence 102, scatter:

Specimen	W	Taxon	Element	Part	Other	N
		Struthio	Eggshell			2
		Indeterminate	Shaft	Fragments		14
Total						**16**

Fossil occurrence 103, general collection:

Specimen	W	Taxon	Element	Part	Other	N
12		Artiodactyla	Tibia	Shaft		1
4,8,10	5	Bovid	Tooth	Molar		3
5	4	Bovid (Large)	Tooth	Molar	Split, worn	1
2	2	Chelonia	Scute	Smooth		1
		Indeterminate	Fragments			6
Total						**12**

Table 5. 25 continued: Fossil occurrence 106, scatter:

12,22	5	Artiodactyla	Tooth	Molar		2
7	3	Bovidae	Tooth	P2, upper	Complete	1
10	3	Bovidae	Molar	Lower	Worn	1
9	4	Bovidae	Molar	Upper	Worn	1
4,5	3	Bovidae	M1/2 complete	Upper	Med. wear	2
18,19	5	Bovidae	Tooth	Premolar	Spall	2
3		Chelonia	Scute	Dimpled		1
8,11		Chelonia	Scute	Smooth		2
6,13,14		Chelonia	Scute			3
2	4	Proboscidean	Tooth	Molar	Plate	1
		Indeterminate	Fragments			7
Total						**23**

Fossil occurrence 107, scatter:

Specimen	W	Taxon	Element	Part	Other	N
1		Bovid (Large)	M1/2, upper	Complete	Worn	1
3	4	Equid	Tooth lower	M2, right	Worn	1
4		Bovid (Large/*Hemibos*)	M3, right	Complete	Lower	1
		Indeterminate	Rib			1
Total						**4**

Notes: 93GB1 is suncracked; the Artiodactyl teeth fragments from 96 are probably from the same specimen. 98GB38-43 probably from same tooth, and 103GB4,8,10 are probably from the same tooth.

Other finds: General collection 95 contained 2 Proboscidean tusk and 3 tooth fragments. Scatter 104 comprised a piece of Proboscidean molar plate and 6 shaft fragments. Scatter 105 contained 1 cervid upper M3 and 2 other fragments. Occurrence 108 comprised 11 pieces of Proboscidean tusk.

Chapter 6: The 1987 survey of the Pabbi Hills

Introduction

The 1987 survey focused on the south side of the G.T. Road, with the exception of some fossil collecting in the badlands on the north side, and produced some 20,000 fossil specimens from over 350 places. This was not only the largest amount collected in one field season, but most were also the oldest collected in the Pabbi Hills. Most specimens were from deposits relating to or earlier than the Olduvai Event, and are thus between 1.6 and 2.2. Mya old. These provide much-needed information on the late Pliocene and early Pleistocene of the Upper Siwaliks, and provide a useful contrast with the predominantly younger, 1.2 - 1.4-Ma.-old, material collected from sandstone 12 in 1986 and 1989. Other material from Sandstone 14 compliments the otherwise disappointing material collected from this horizon in 1986.

The survey on the south side of the G.T. road was more complicated than in 1986, when most material was collected from along the long and easily recognised scarp face of Sandstone 12. Most of the deposits surveyed in 1987 were located in the head-waters of drainage basins that cut deep into the centre of the anticline. The main ones on the eastern (Jhelum) side of the anticline were Choawala Kas, Bangial Kas, and Sohawa Kas, and Kotha Kas and Ratial Kas on the western (Kharian) side. Figure 6.1 provides an overview of the survey on this side of the G.T. Road. Three figures (Figures 6. 2, 6. 9 and 6. 16) based on the 1933 reprint of the 1911 1" = 1 mile topographic map provide more detailed information on the areas surveyed in Choawala, Sohawa and Kotha Kas. Each of these is complemented by small-scale sketch maps made in the field during the survey that show where material was found.

Local sequences tended to be more complex than in 1986, and often difficult to link with their neighbours. Correlations were accomplished in the field by identifying marker horizons (usually prominent sandstone bodies) above and below major fossil localities, and then tracing them on foot between different areas. The main stratigraphic position and ages of principal fossil sites are summarised in Figure 6. 3. As can be seen, the key marker horizons on the northern limb of the anticline in both Choawala Kas and Sohawa Kas are Sandstones 2 and 5, both of which can be assigned to the Olduvai Event, i.e. ca. 1.6 - 1.9 Mya old. A sandstone unit between Sandstones 2 and 5 should correspond to Sandstone 3 or 4. These are laterally discontinuous, however, and are approximated here by the term "Intermediate Sandstone". On the southern side of the anticline in Ratial Kas and Kotha Kas, Sandstones 1 and 2 are the main marker horizons, and almost all fossil material was found on exposures between these two units. These are therefore older than the Olduvai Event, and probably between 1.9 and 2.2 Mya old. The dating and correlation of deposits are discussed more fully by Rendell elsewhere in this volume.

The main access to the southern part of the Pabbi Hills was along the watershed track, which joins the G.T. road near the crest of the anticline. The main landmark is Flag Hill, the highest part of the Pabbi Hills at 1302' (ca. 400m.) OSL. Flag Hill was so-named because of the flag pole and flag on its summit. At its foot and by the watershed track, there is a small shrine marked by pieces of cloth suspended in a tree, and usually a container of water. For the survey team, Flag Hill was one of the most conspicuous landmarks in the Pabbi Hills, as it can be seen even as one approaches the River Jhelum from Rawalpindi. The climb to its top is rewarded by superlative views over most of the Pabbi, westwards across the Jhelum Valley, eastwards across the plains past Kharian and towards Lahore, and northwards as far as the mango tree at Ban Barilla near Baroth.

The overall results of the 1987 survey are shown in Table 6.1, listing sequentially the 350 fossil occurences found that year as well as the area in which each was collected, the type of fossil occurrence, probable age, and numbers of fossil specimens. As shown in Figure 6.1, fossil occurrences are also grouped into 21 sets of adjacent or related collecting numbers. Table 6.2 summarises the total number of fossil specimens, and the average number per site, from each set. Each can be taken in turn, beginning with Choawala Kas (sets 1-5). Figure 6. 2 shows the collecting areas covered in this area.

Set 1: Choawala Kas headwaters (1.7 - 1.9 Mya-old): 200-218 (even nos. only), 365, 367, 388.

To reach this area, drive along the watershed track until a few hundred metres before it crosses a concrete-lined embankment (also known by the survey teams as "the viaduct"). A small pond by the side of the watershed track was a good landmark in 1987, but later fell into disuse. From here, a good foot-path on the NW side leads into a deep valley that defines the headwaters of the northern part of Choawala Kas. Three sandstone units are exposed in the sides of this valley, of which the lowest is incised by a deep gully. Most of the fossil material was derived from yellow sands between the lowest and middle sandstone bodies, but no. 218 was located between the middle and topmost ones.

GPS points:

start of path at pond:	N32° 49' 591"	E43° 48' 239"
locality 214:	N32° 49" 787	E73° 48'033"
site 218:	N32° 49' 798"	E73°48' 035"
Flag Hill to 214	1.02km	

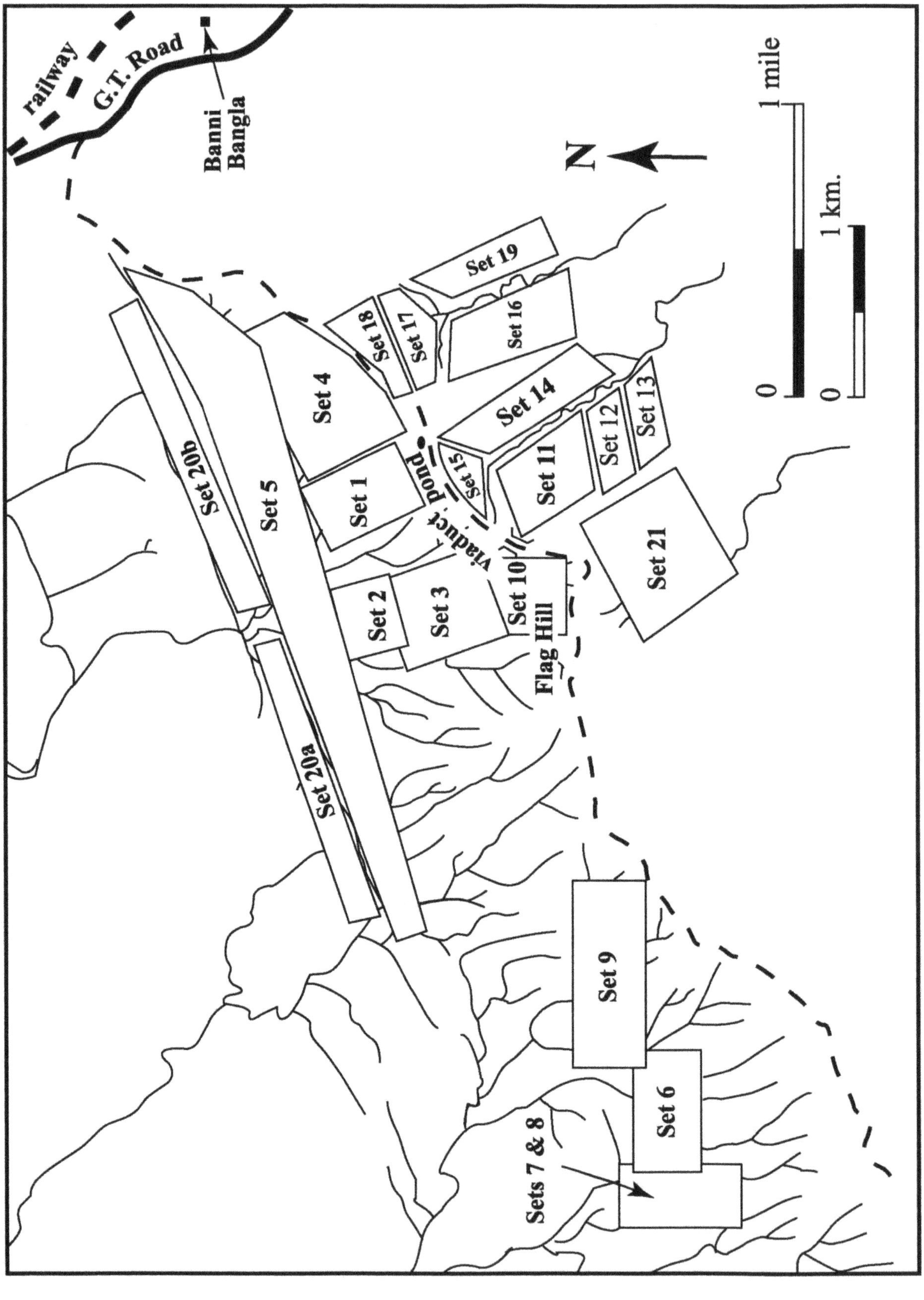

Figure 6. 1: Area surveyed in 1987 on the south side of the G.T. road.

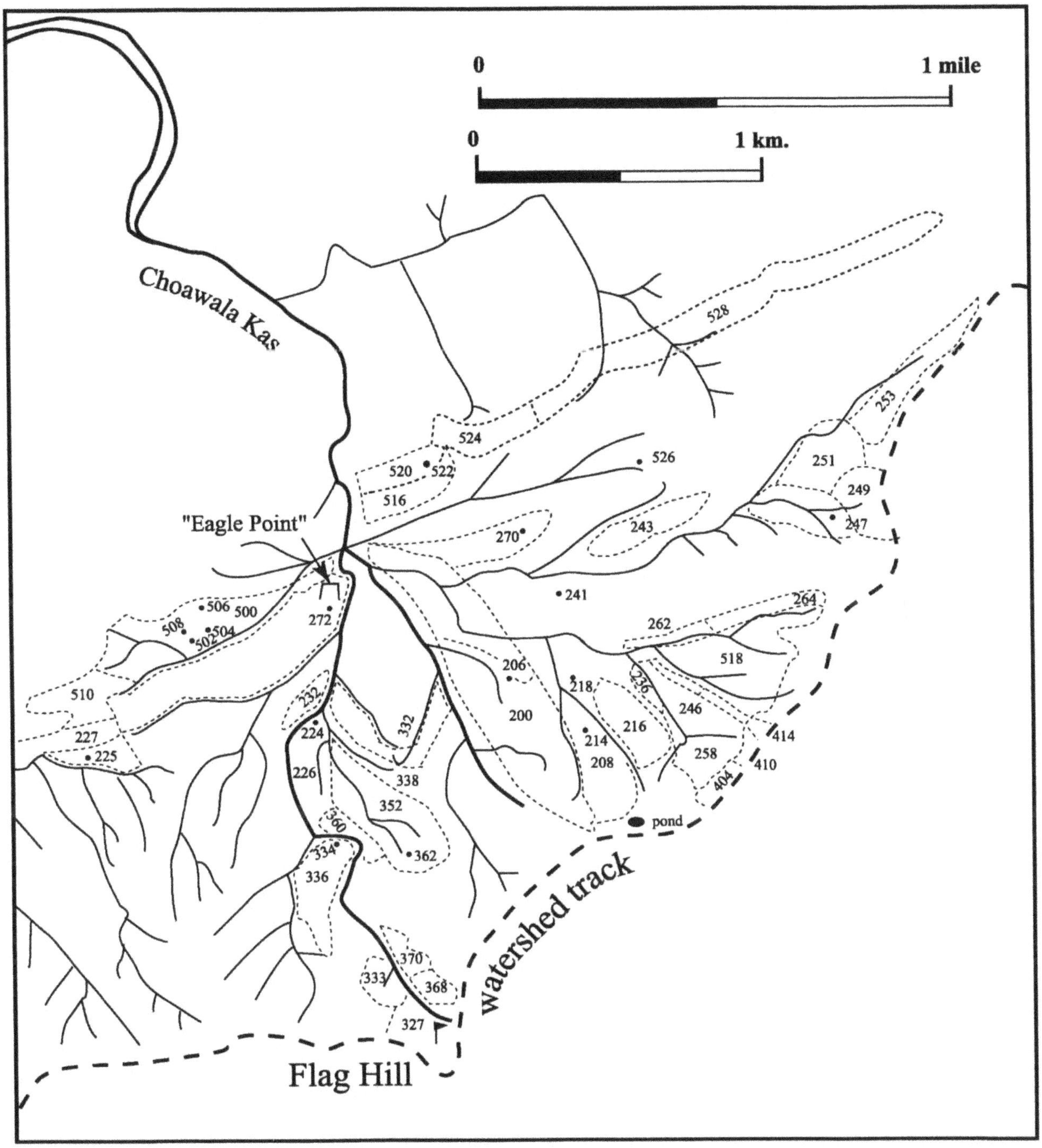

Figure 6. 2: Principal collecting areas in Choawala Kas (based on records by Mags Peace).

Figure 6. 3: The stratigraphic position and age of the principal fossil occurrences found in 1987 (based on data from Helen Rendell). **Note that Sandstone 5 in this study corresponds to Sandstone 3 in Keller (1975)**; see chapter 4 for details.

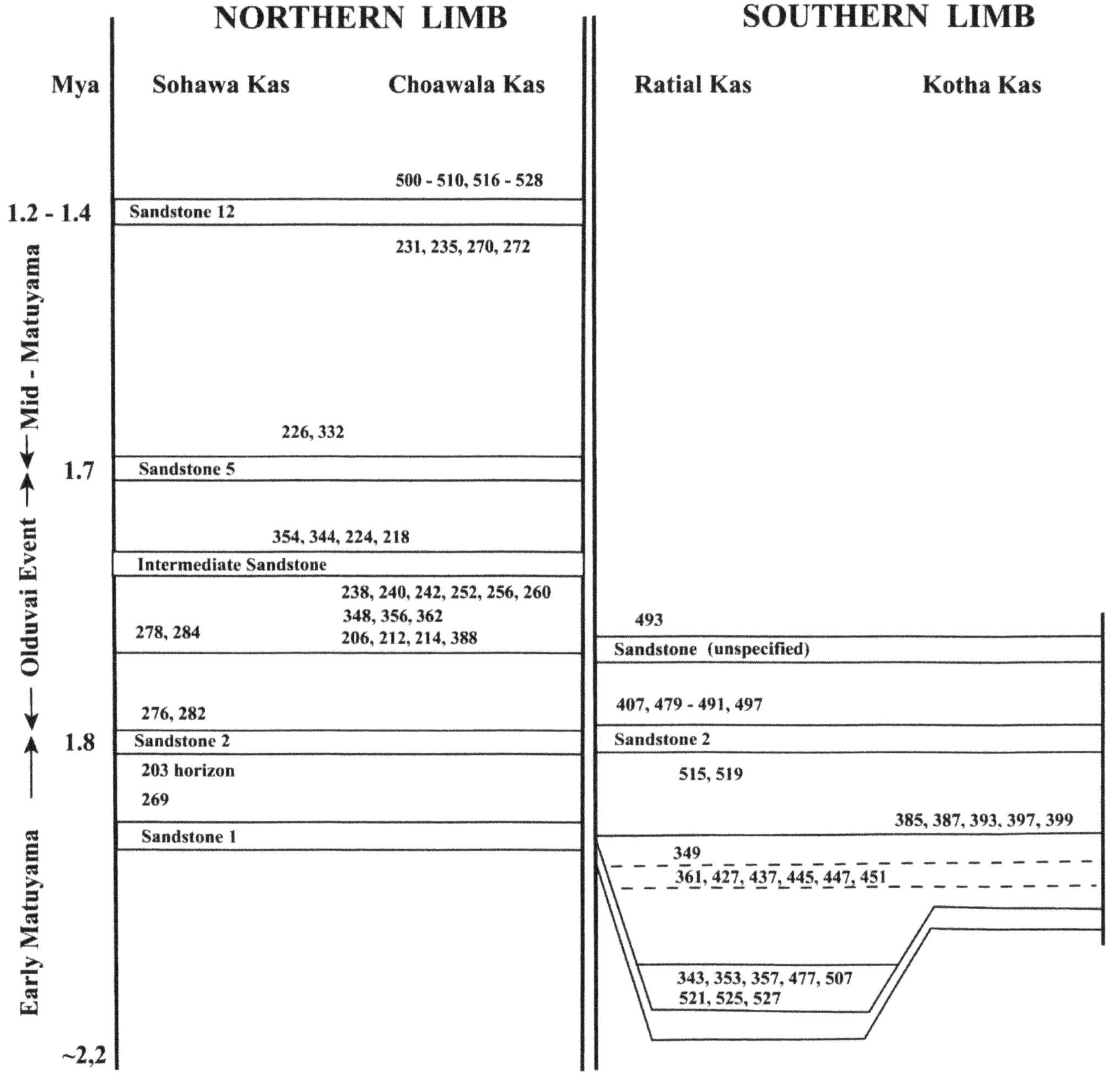

Set 1: This set marked the beginning of the 1987 survey of the areas south of the G.T. road. Table 6.3 summarises the fossil material from this area, and the location of each collecting area in this set is shown in Figure 6.4. The survey in this area began with the discovery of a bovid distal radius (200GB1) ca. two-thirds of the way down the foot-path into the valley. The resulting general collections 200, 202 and 204 covered the western side of the valley. Nos. 202 and 204 were used to denote sub-areas that were on 30-40°, WSW-facing slopes of buff sands that had a slightly greater density of fossils, but are best lumped with general collection 200. Fossil occurrence no. 206 was a scatter ca. 20 x 20m on a 30° slope within general collection 200; most of the fossils were collected from the surface of a pink clay unit, but some fragments were found eroding from the overlying yellow/buff silts. General collection 208 covered the eastern side of the valley, and is divided by a deep *kas* that cuts into the lowest sandstone in this area. It contained two small scatters: no. 210, on top of pink silts, and no. 212, on yellow silts and sands. The latter lay near and below locality 214, the first major discovery of the field season. It comprised a large number of fossils (including ones of a large hyaena) which had originated from a small area of soft yellow silts immediately below an outcrop of the sandstone unit that the caps the sediments in this part of the valley. As noted already, this should be the Intermediate Sandstone. As at 109 (chapter 5) and 362 (see below), most fossils were found downslope. General collection 216 covered the area north of and along the same level as locality 214, but produced very little material, thus confirming the localised nature of locality 214. A subsequent find of a piece of bovid cranium (220GB1) can be subsumed within 216.

Figure 6.4: Location of fossil sites in Choawala Kas, Set 1 (200-220, even nos. only; 365, 367, 368). Numbers in rectangles denote general collections; those surrounded by broken lines were scatters. (Based on a field sketch by Linda Hurcombe). The area covered is ca. 1 km across.

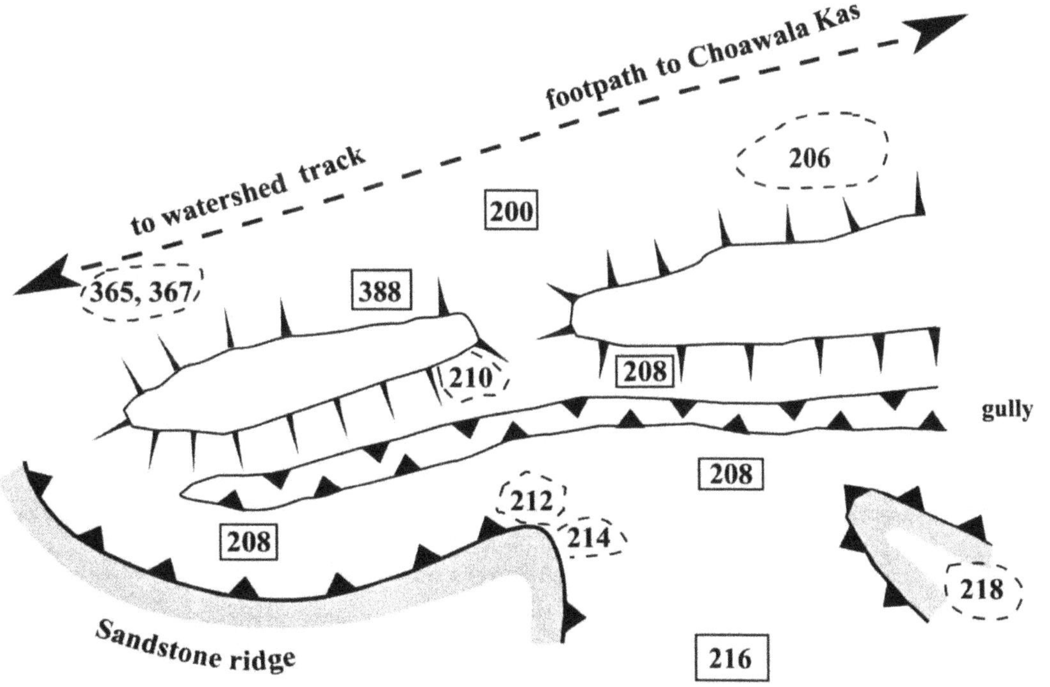

Prospection of the area above the Intermediate Sandstone capping the area covered by 200-216 resulted in the discovery of a scatter (no. 218) of some fossil material and flaked stone on a west-facing, 30-35° slope below Sandstone 5. The latter are discussed in the chapter by Hurcombe. (This area is best reached by walking through a small gorge beyond 206 and then a steep clamber to the top of the middle sandstone. Locality 214 is visible from the edge of the nearby cliff).

This series ends with three later discoveries. Two small scatters of fossils (nos. 365 and 367) were found eroding from yellow silts and sands ca. 100m. north of the watershed track at the head of this drainage basin. General collection 388 was found after a period of heavy rain on the low ridge near scatter 210. Interestingly, this area had previously been surveyed three weeks earlier by four people, but nothing had then been found.

Set 2: Choawala West (1.4 - 1.7 Mya-old): 222 – 232 (even numbers); also 332, 334, 336

This small group of unremarkable finds lies in the area between the headwaters of Choawala Kas and Bangial Kas. Figure 6.5 shows a sketch map of this area, and Table 6.4. summarises the fossils that were found. The first numbers in this series are reached by following the main foot path into Choawala Kas mentioned in the previous section, and branching off left (facing downhill). There is a prominent sandstone-capped bluff that faces the watershed track, and is a convenient point from which to view the area covered in this series, and another flat-topped scarp a little further west with three small trees on top. This was "Three-Tree Point", a convenient land mark for the area. Two pylons are also visible from it, at 009° and 020°, on what is probably Sandstone 12, and Flag Hill is at 152°. (On the 1911 survey 1:63,360 map, this point is also on the "S" of "Reserved" in the text "Reserved Forest Area"). The end point in this series leads one into the eastern part of Bangial Kas, from which one can exit to the watershed path a little west of Flag Hill.

Figure 6. 5: Location of fossil sites in Choawala Kas West, Set 2 (222-236, even nos. only, 332,334 and 336). (Based on field records by Mags Peace).

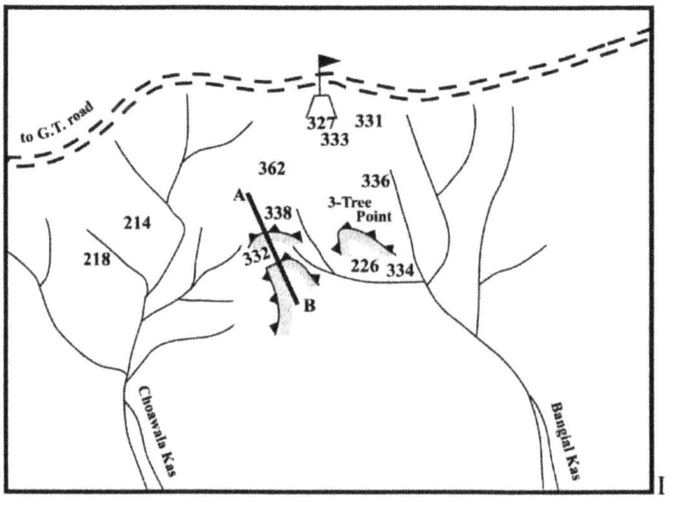

The stratigraphic context of these fossil occurrences is shown in Figure 6.6. As can be seen, most fossil occurrences were from the same horizon as locality 218, (i.e., between the Intermediate Sandstone and Sandstone 5), whilst the remainder (332, 334 and 336) were younger, and above Sandstone 5. In most cases, fossils were found in or on the surfaces of yellow sands, with occasional pebble inclusions (noted at 222, 226, 232). Scatters 222 and 224 are best subsumed within general collection 226, and 332 and 334 can be lumped within 336. Table 6.2 shows how poor this area was in fossil material: only 5.9 fossil specimens per collecting area, compared with 40.5 in the previous set.

Figure 6.6: Stratigraphic context of fossil occurrences in Sets 2 and 3, Choawala Kas West. (Based on a field sketch by Mags Peace).

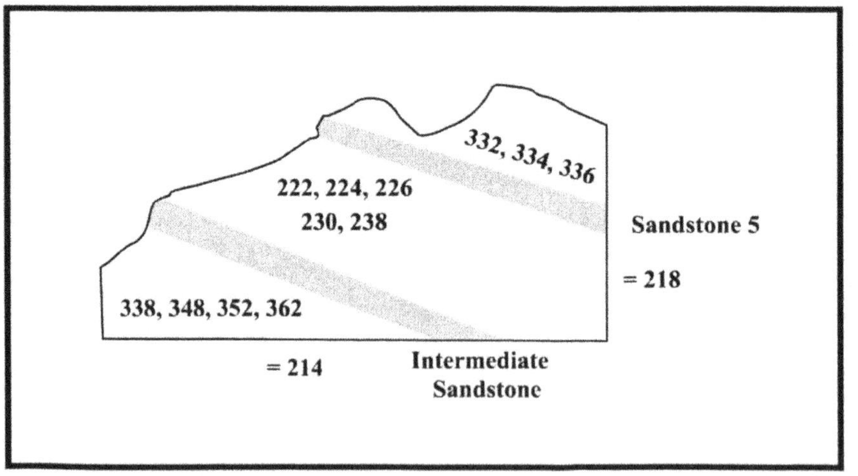

Set 3: Choawala West, 362 area (1.7 - 1.9 Mya-old): 338 - 366.

Figure 6. 7 shows the location of this suite of fossil occurrences from the western part of the Choawala drainage system, and Table 6. 5 summarises their contents. Most of the fossil occurrences in this area lie below the Intermediate Sandstone, and are thus within the Olduvai Event and similar in age to locality 214. The results were generally unremarkable except for the discovery of locality 362, which was small but one of the best found in 1987. It compares in size with locality 214, but is otherwise distinguished by post-cranial remains of *Sivatherium giganteum*. The excavation of locality 362 by Beech and Anwar in 1989 is described in a later chapter.

To reach locality 362: the easiest way to find it from the watershed track is to start at a path near a bend shortly after the viaduct but before Flag Hill. Follow the path upwards over the slight incline, and then skirt around the head of a gully. From there, head across generally level ground to a sandstone bluff. Locality 362 was on the other side, at the top of a steep scarp face.

GPS readings: the viaduct:	N32° 49' 390";	E73° 48' 062"
Flag Hill:	N32° 49' 239";	E73° 47' 927"
Locality 362:	N32° 49' 577";	E73° 47' 850"
Flag Hill to locality 362:	0.63km	

Figure 6. 7: Location of fossil sites in Choawala West (362 series), Set 3 (338-366). (Based on field records by Mags Peace).

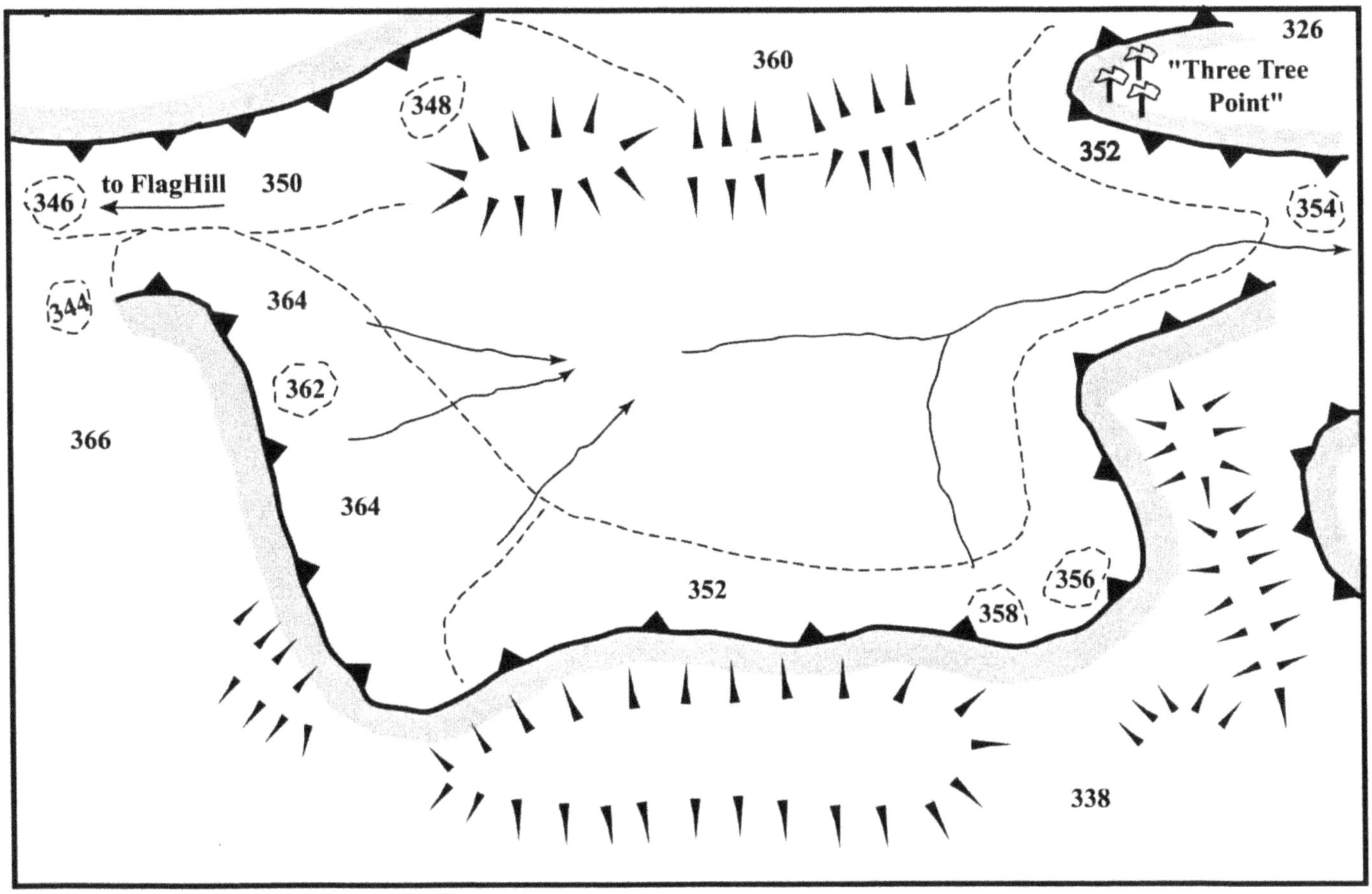

The series began with general collection 338 along a horizon of yellow sands overlying pink silts that was equivalent to those in the 214 horizon. Within it, there were two small, 10m sq scatters. Fossil occurrence 340 was on a 25° degree, west-facing slope, and no. 344 was on a 25° degree, south-east-facing slope, partly obscured by sandstone boulders. The fossils here were probably derived from yellow/buff silts immediately below a sandstone unit. Scatter 346 was found along the same horizon as 338, 340 and 344. However, it contained pottery, a piece of polished stone, other pieces of flaked stone as well as fossil remains, and later inspection showed that it was a palimpsest of material resulting from the erosion of the in-fill of a shallow and probably recent (i.e. holocene) gully that had cut into earlier,

Siwalik deposits. It is therefore excluded from the main results of the survey. No. 348 was a general collection in the area north of 346 along an area of east-facing slopes of the same horizon as 338. General collection 352 was on the other side of the valley, along 60° degree slopes of yellow-brown silty sands, above a pink clay unit, and below the ridge-forming Intermediate Sandstone. It was depressingly unrewarding, as it produced only five pieces from an area some 750 by 30m, or one find per 4,500 sq.m. However, it did include three small scatters, nos. 354, 356 and 358, each in small gullies on the scarp face. General collection 360 produced all of three specimens from a large area of yellowish sands and silts exposed along the other side of the valley, north of 348. At that point, the team moved back to the area of general collection 352, and shortly afterwards found locality 362. Locality 362 comprised a small pocket of fossils eroding from yellow silty sands immediately below the Intermediate Sandstone, and spilling 100 metres down a steep gully. Fossils were collected in eight units, with A at the point of origin, and H at the base of the gully. Those from units A-E were fresh, and those further downslope more weathered. Given the steppness of slope and softness of deposits, the erosion of these fossils was probably very recent. The specimens are summarised in Table 6. 6. As with locality 214, which also lay in yellow silty-sands immediately below a sandstone, 362 was an isolated instance of a dense fossil concentration in an area otherwise poor in fossil. The subsequent general collections (no. 364) along the same level as locality 362 actually produced nothing at all, and no. 366, nothing of interest.

Set 4: Choawala East (1.7 - 1.9 Mya-old). (Even numbers 236-260 for Happy Valley, and 262, 264, 416, 418, 434 and 518, Death Valley).

The eastern drainage area of Choawala Kas produced two very different sets of results. The first (nos. 236-260) was a good haul of fossils in what became "Happy Valley": some 530 fossils from 14 occurrences, or an average of 38 per occurrence (roughly the same as in the area of 214). Although there were no major fossil concentrations, fossils were commonly found over a large area. In contrast, the adjacent valley and the ridge in between yielded an average of 9 specimens per occurrence, from what became known as Death Valley because of both its poor fossil yield, and a dangerously steep exit. Figure 6. 8 shows the collection units in this area, and Table 6. 7 summarises the fossils from both valleys.

Figure 6.8: Location of fossil sites in Choawala East, Set 4 (Happy Valley, Death Valley). (Based on field records by Mags Peace).

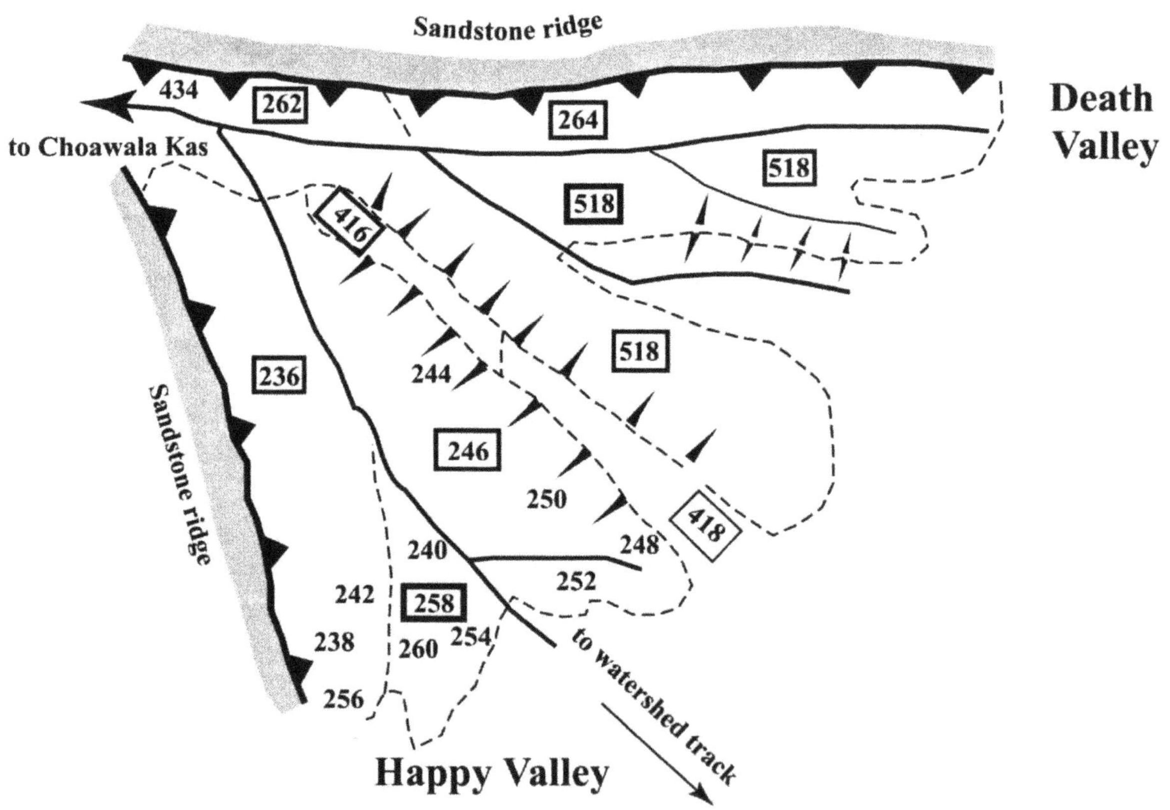

The starting point for this part of the survey was the same pond as already mentioned in the survey of the area 200-214. This series started in the second valley north-east of the pond, and most of the fossils came from the head of the valley, from yellow sands and silts capped by the same sandstone as that over 214. The survey of this valley entailed three general collections (236, 246, 258). No. 236 contained scatters 238, 242, and 256; general collection 246 contained scatters 244, 248, 250, 252; and general collection 256 contained scatters 240, 254 and 260. The Death Valley series entailed general collections 262, 264, 418, 434 and 518, and general collection 416 on the ridge between Happy Valley and Death Valley. Most of the finds in this set are from the same horizons as 200-216 (Set 1).

236 was the general collection along the southern side of the valley, and 238 was the first scatter. It was found on both sides (zones A, B) and at the base (zone C) of a small gully that cut into the hill-side; the fossil was eroding from buff sandy-silts, but some in zone C were on and in the surface of pink silts. The specimens in scatter 240 were found eroding downslope and collecting on the flat bank of the major *kas*. Scatter 242 was in a small gully on a 40° slope. Scatter 244 was a dense concentration of fossils from an area 2 by 2m on a south-west facing, 30° slope, and probably derived from buff sand/silts. 246 was the general collection on the north side of the valley, and included various scatters (248, 250, 252) of fossil material from exposures of yellow silts and sands near the base of the modern valley. Those in scatter 250 were found on top of pink silts but had probably eroded from the overlying yellow silts. Scatter 252 was in a small gully, and was subdivided into two units, A at the top, and B at the base. General collection 258 and 260 covered the head of the valley, and contained scatters 240 and 254. A quartzite core was also found between 252 and 254 and at the same level of sands and silts as the fossils; there was no obvious source of derivation from later deposits. The fossils in scatter 256 were found on pink clay at the base of a yellow silt unit on both sides of a small gully.

Death Valley

This area was covered by four general collections (262, 264, 418, 518) and contained two small scatters (416, 434). The derisory amount of material that was found was collected from yellow silts and sands. The two sides of Death Valley were covered by general collections 264 and 518, and the ridge between Happy Valley and Death Valley by general collection 418.

Set 5: Bangial Kas and Choawala Kas (mostly 1.4 - 1. 7 Mya-old): 225-255, 268, 270, 272, 532.

Table 6. 8 summarises the material in this heterogenous set from an extensive area in front of and along the scarp slopes of Sandstone 12 between Choawala and Bangial Kas. A prominent pinnacle of Sandstone 12 in front of the main scarp face provides a good reference point for this area. Ground visibility was generally good. General collections 227 and 231, and scatter 235 came from the base of Sandstone 12 and should thus be ca. 1.2 - 1.4 Mya-old. General collection 225 was slightly older, and came from the ridge in front; it included a highly fragmented elephant tusk, a few pieces of which were collected as a sample, and the rest left in the field. (Nos. 229, 234, 268 and 532 were gully collections and can be ignored).

The rest of the material in this set came from an area between Sandstone 12 and the watershed track, and the G.T. road and a prominent set of pylons that cross the Pabbi Hills in the eastern part of Choawala Kas. (The GPS reading for the pylon is 32° 50' 138"N., 73° 48' 876"E.). Scatter 241 and general collection 243 were from the next *kas* east of Death Valley, and came from a sandstone-capped unit immediately in front of (i.e. older than) Sandstone 12, hence the estimated age of 1.4 - 1.7 Mya. No. 241 comprised a small number of heavily eroded pieces lying on a pink silt horizon immediately above the sandstone ridge older than Sandstone 12 and which immediately precedes it as this point. Exposures were good, with very little vegetation. General collection 243 was made along the south face of a small ridge in front of and parallel to Sandstone 12. Fossil was scanty and generally heavily eroded. Most came from the lower horizons, as the top buff sand/silts yielded virtually nothing, despite good ground visibility.

Most of the fossil occurrences 245-253 came from an area between the pylons, the watershed track and Sandstone 12; a trigometric point on a small knoll near the north side of the watershed track is within the area of 253. All are general collections apart from scatter 247; ground visibility was generally very good, apart from 251, where there was a moderate amount of grass. Scatter 247 comprised fossils spilling into and down a small gully. Many were probably missed because several of those that were found were half-buried. Many had a red clay matrix. General collects 249, 251, 253 and 255 were very poor: field notes refer to the density of fossil in 249 as "extremely low"; in 251 as 1 per 400 sq.m; 253 as 1 per 500-600 sq m.; and 255 as "a waste of time" despite the sighting of some wild pig.

The final two occurrences (270 and 272) were from between Sandstone 12 and the watershed ridge, on the south side of the G.T. road and from the eastern branch of the Choawala Kas drainage area. Both lay west of the already-mentioned pylons. Find-spot 270 was a bovid tooth in yellow sand, in a small tributary of Choawala Kas at the base of the Sandstone 12 section. No. 272 was a small scatter in the western part of the Choawala Kas drainage and below Sandstone 12, and contained some elephant bone from yellow silty sands. (It is possible that these finds were in areas surveyed briefly in 1986).

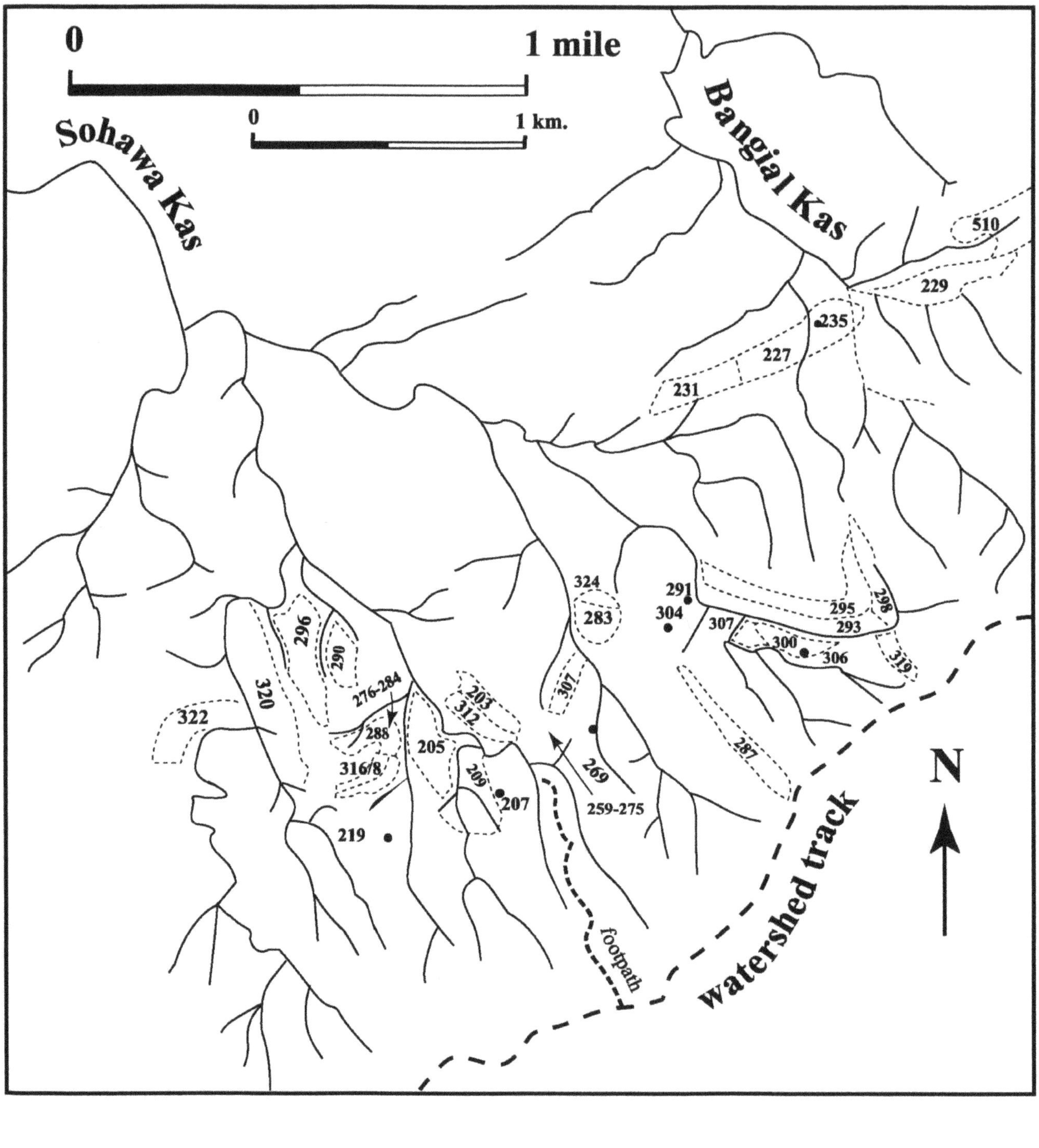

Figure 6.9: Principal collecting areas in Sohawa Kas. (Based on field records by Mags Peace).

The headwaters of Sohawa Kas form a complex and deeply-dissected set of gullies, many of which have cut deep into the earlier deposits in the Pabbi Hills. There were four sets of material from this area, most of which are from deposits spanning the Olduvai Event.

Set 6: Sohawa Kas: the 203 horizon (1.9 - 2.2 Mya-old).

The odd-numbered team began in the Sohawa Kas area, further along the watershed track from Choawala Kas. The first major discovery was what came to be known as the 203-horizon. The location of the main fossil occurrences is shown in Figure 6. 10, and Table 6. 9 summarises the fossils found in this area.

To find the 203 horizon, start from Sohawa village and drive up the main stream bed of Sohawa Kas, avoiding soft sand. When well inside the anticline, a tributary joins from the south-west, and the main course veers east along the foot of a very prominent scarp face, capped by a sandstone unit, and with a thick sandstone at its base. (A useful landmark is a piece of antler embedded in sandstone in the bed of a tributary of Sohawa Kas (= find-spot 207 on Figure 6.9); this was well-preserved in 1987, and still visible in 1999). This scarp is the 203 horizon, and the survey started at a convenient point of entry from the kas bed. The 203 horizon can also be reached via a good foot-path from the watershed track. This track lies ca. 4.5 miles beyond Flag Hill. The junction of this path with the watershed track is hard to detect; however; it is thus easier either to ask a local, or to proceed from Sohawa village. The path itself requires care when descending into Sohawa Kas, particularly after heavy rain, as there are a few places where a fall could be fatal.
GPS readings:
Junction of path to Sohawa Kas with watershed track: N32° 48' 600"; E73° 46' 397"
site 269: N 32° 49' 034"; E73° 46' 228"
Distance along watershed track from Flag Hill to foot-path junction: 4.5 miles
Flag Hill to 269: 2.69 km

Figure 6.10: Location of fossil sites in Sohawa Kas (in and below the 203 horizon). (Based on field records by Hannah Jones).

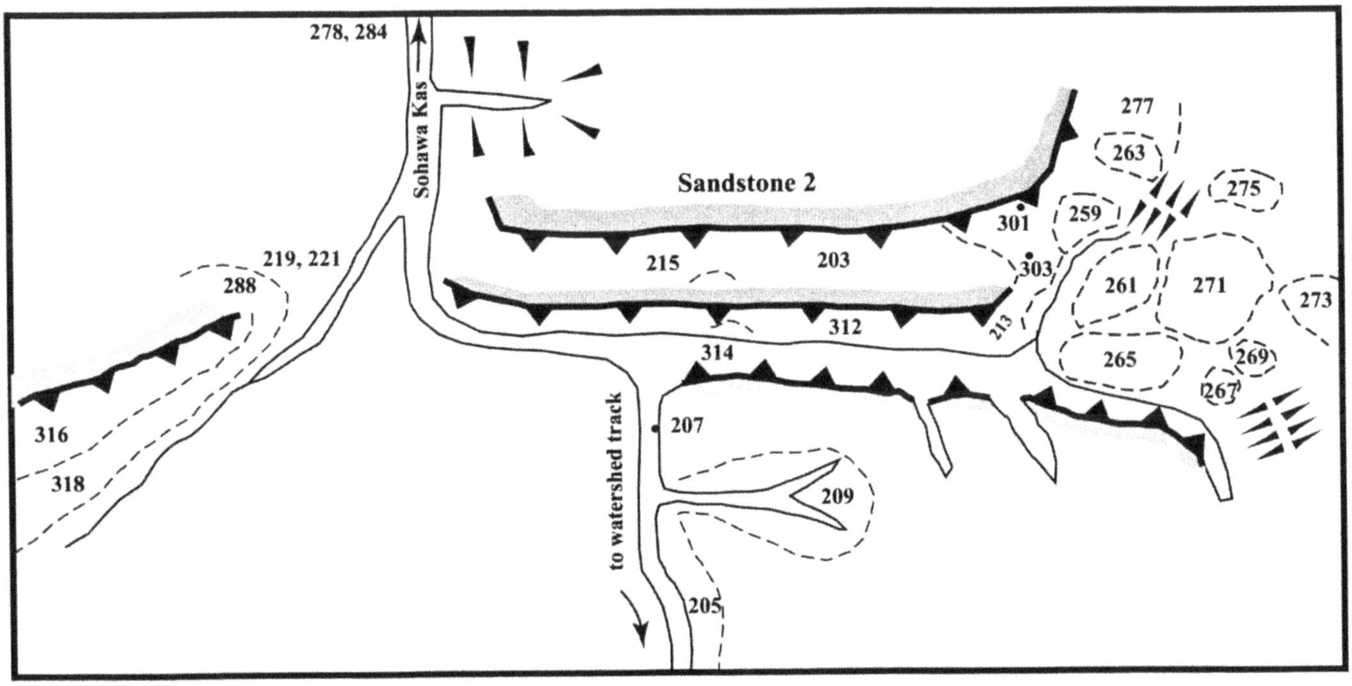

The scarp face of the 203 horizon has three prominent sandstone units. The uppermost, ridge-forming one is Sandstone 2, and thus the fossil material below it antedates the Olduvai Event. Fossil occurrence 203 was a general collection along the scarp face between the middle and uppermost sandstone. Fossils were found regularly and commonly, and ground visibility was excellent. As can be seen in Figure 6.11, most fossils were found in or eroding from sands and silts immediately above the middle of three sandstone units. Slightly higher in the profile, there is a thin gritstone horizon: this is discontinuous, and re-appears at 269 and 271, and possibly (if the same grit unit) at 285 and 291. Two small scatters were found in the 203 collecting area: no. 213, in a small gully at the eastern end of the 203 area, and no. 215, mid-way along and near the base of the 203 horizon. At the eastern end of the 203 horizon (i.e. around 213), the scarp face becomes less steep, with various areas of comparatively level ground separated by low ridges and shallow gullies. These were covered by several collecting numbers (259-273 (odd numbers only)). Ground and fossil visibility was generally very good, with little vegetation.

Figure 6.11. Stratigraphic provenance of fossils from the 203 horizon, Sohawa Kas. (Based on field notes by Hannah Jones). The section is ca. 80m high; see also Figure 11.5.

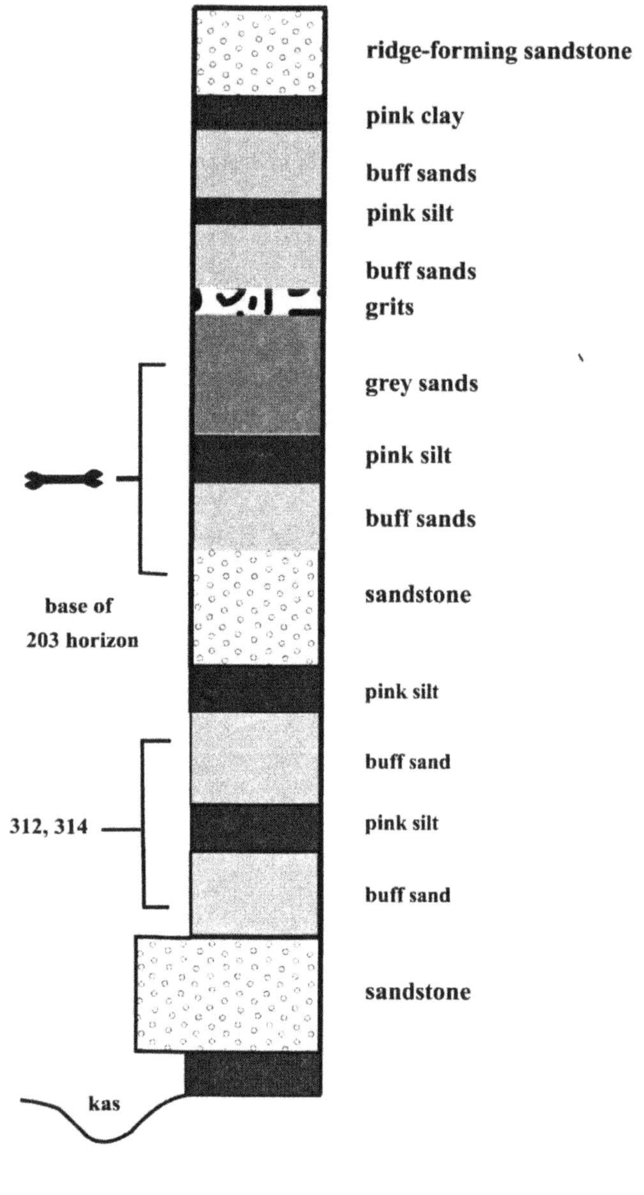

The most interesting discovery in this area, from an archaeological viewpoint, was scatter no. 269, which produced four conjoining pieces of elephant scapula and a quartzite artefact from the base of an eroding sand unit. The artefact was found immediately beside the piece of scapula that lay at the base of the section. Two other conjoining pieces were found slightly further upslope, and the fourth piece was balanced precariously in a small thorn bush growing out of the side of the section. It is thus safe to assume that the pieces of elephant scapula had eroded very recently from the overlying sand unit. It also seemed reasonable to assume that the artefact was derived from the same source. This area was later explored in detail, and the sand unit was also excavated; details of this work are presented in chapter 8 by Hurcombe.

The deposits upslope of (and thus slightly younger than) the numbers considered so far were covered by collecting numbers 297, 299, 301 and 303, which covered the deposits immediately below the ridge-forming sandstone. The main discovery was of a complete bovid metacarpal (301GB1 and 301GB2) that had recently eroded from buff sands immediately below a red mudstone, in turn capped by the ridge-forming sandstone. The exposures on the lower slopes of the 203 horizon (i.e. below the middle sandstone, and thus slightly older than the 203 horizon) were also briefly explored, but produced little (see general collection 312 and scatter 314) despite good ground visibility. Details are given in Table 6. 9.

Set 7: Sohawa Kas headwaters (205, 207, 209) and Sohawa Kas West (217-223 (odd nos. only), 288, 316, and 318): 1.8 - 2.2 Mya-old.

This small group from deeply-dissected terrain in the headwaters of Sohawa Kas is shown in Figure 6.12, and summarised in Table 6.10. Most are stratigraphically older than those from the 203 horizon, and were found between the base of the modern stream beds and the sandstone unit that defines the base of the 203 horizon. No. 205 was a gully collection from the base and sides of an adjacent *kas*,and included the already-mentioned piece of antler (207) embedded in sandstone at the base of the gully. Other material included in this collecting number was found within a 10m radius on the dip-slope of the adjacent sandstone. No. 209 was a general collection in a nearby small side valley on the left-hand side of Sohawa Kas (facing upstream).

A related group of fossil occurrences resulted from an attempt to follow the 203 horizon westwards. The area surveyed was reached by proceeding downstream from the antler embedded in the gully base (207), past the end of the face containing 203; and then up a small tributary of Sohawa Kas. This contained some unremarkable scatters, nos. 217, 219, 221 and 223 (Figure 6. 9). At a later date, there were three further general collections, nos. 288, over an area ca. 500 x 50m. of a generally south-facing 60° slope equivalent to the 203 horizon and below that of 278; no. 316, along the upper slopes (roughly equivalent to the 203 level); and no. 318 on the lower slopes, and at an equivalent level to 312. Quartzite pebbles were noticed in the lower yellow-brown silty-sands. Fossils were found in both, but at low densities although ground visibility was good. Both these collects overlap in part with the area containing scatters 219, 221, 223. No further survey was undertaken in this area.

Figure 6. 12: Location of fossil sites in Sohawa Kas West (based on field records by Mags Peace).

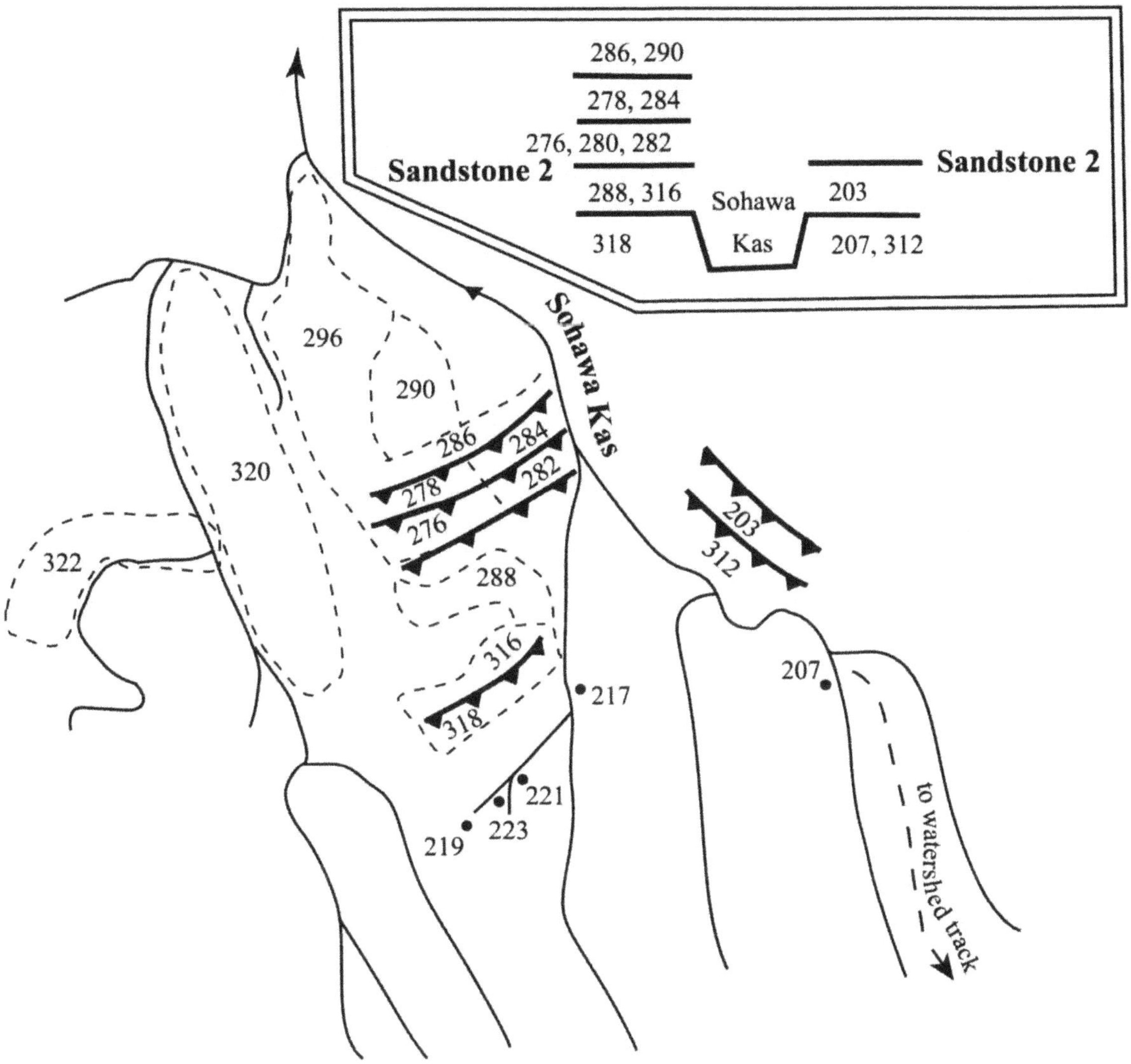

Set 8: Sohawa Kas West (1.7 - 1.9 Mya-old): 276-286 (even nos. only), 290.

This set came from the western side of the Sohawa Kas area, and further downstream from the 203 horizon (Figure 6. 12). The fossil yield was wholly unremarkable (see Table 6.11), but it was intriguing on account of the stone artefacts that were discovered in the course of fossil collecting, particularly at occurrences 276, 278 and 284. These are discussed by Hurcombe in chapter 8. This particular part of the Pabbi Hills is stratigraphically complex, as there are several sandstone units that dip into and are exposed along Sohawa Kas. Three are relevant in this area (see Figure 6. 11). Fossil occurrences 276, 280 and 282 lie on top of the lowest sandstone unit; occurrences 278 and 284 were above the middle sandstone unit; and 286 and 290 were from the top of the highest sandstone unit. Through field observations and by walking along each of these units, 276, 280 and 282 are from the sandstone unit above the 203 horizon (Sandstone 2); 278 and 284 are younger, and from the sandstone that may be equivalent to the Intermediate Sandstone in Choawala Kas above locality 214. General collection 286 covered the top of the sandstone unit above 278.

Set 9: Sohawa East, 291 area (1.9 - 2.2 Mya-old): 237, 257, 263, 273-295 (odd. nos.), 296-308 (even nos.), 320-330 (even nos.), 315-319 (odd nos.), 376; 305, 309, 311, 319.

This area lies in the eastern part of the Sohawa Kas drainage area, and almost extends into the western part of Bangial Kas. The approximate location of the main collecting areas is shown in Figure 6. 13, and the results are summarised in Table 6. 12. The best access is to this area is from the watershed track, and the best local landmark is a concrete trigometric point on a small bluff. In stratigraphic terms, the deposits in this area represent the continuation of the 203 horizon, and are thus assumed to be the same age, i.e. 1.7-1.9 Mya. old. Exact correlations with the sandstones over and under the 203 horizon were not always obvious in this area, but any differences in age are unlikely to be significant. This was not a very productive area, and there were many instances when very little was found over a large area. General collection 287, for example, produced only 36 fossils from an area ca. 200m x 200m of gentle, 10° slopes, or an average of less than 1 per 1000 sq.m. The most significant palaeontological finds were of a large felid at scatter 291.

Figure 6. 13: Location of fossil sites in Sohawa East, 291 area. (Based on field records by Hannah Jones and Sheilagh Sutherland).

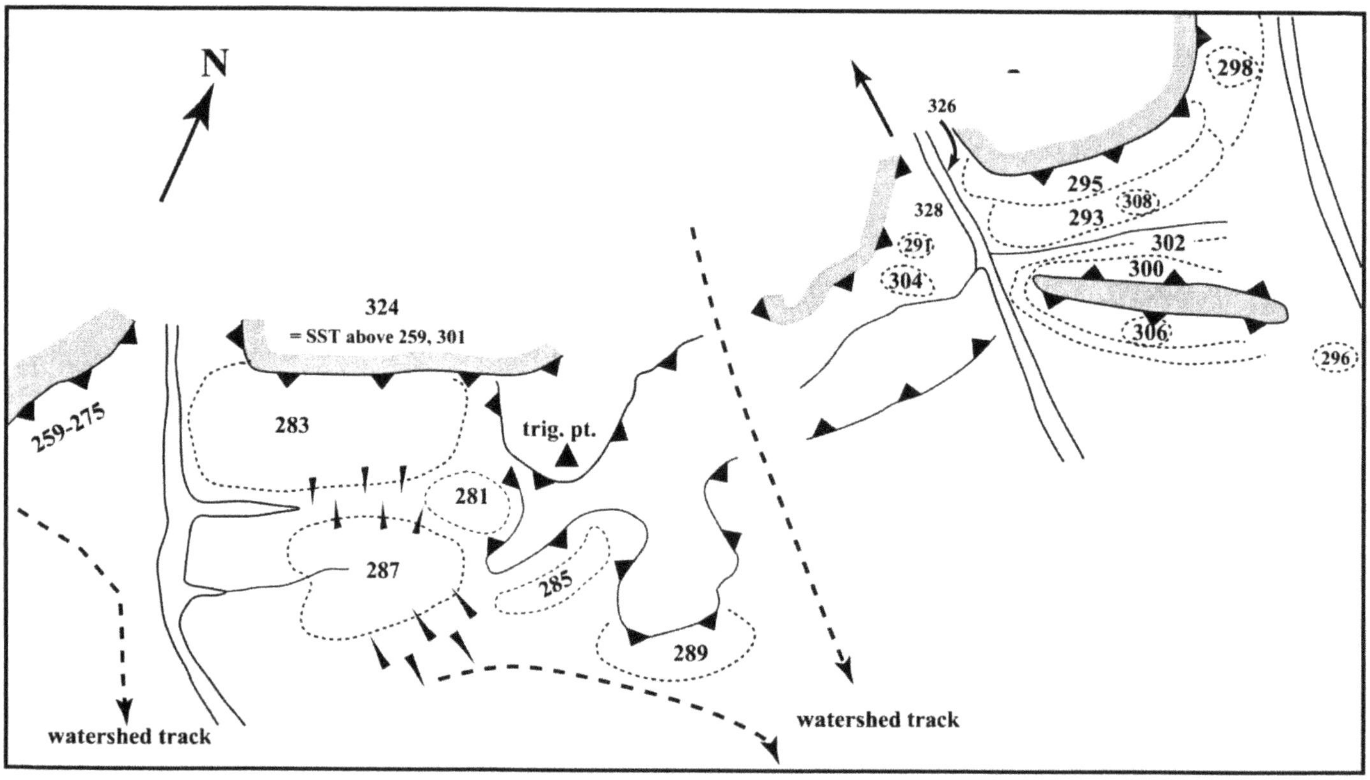

General collection 281 produced a proboscidean mandible fragment, scapula and almost complete tibia. General collection 283 came from south-facing slopes below a sandstone; on a later occasion, the slopes above this sandstone ridge were covered by general collection 324, which produced only one specimen. Fossil occurrence 285 yielded a large number of fossils, mostly fragmented and from an area only 15 by 5m. The material probably derived from a gritty horizon immediately below a greyish case-hardened sand, a remnant of which forms a spur and separates this scatter from fossil occurrences 283/281 etc. Fossil occurrence 289 was another small scatter, 3m x 3m on a 30 ° slope near a path leading to the watershed track.

Scatter 291 - the main discovery in this area - lay near a track, on a 30 ° slope, and on the same horizon as 285/289. It was found as a result of following the greyish sand and buff silt/gritty horizon seen at occurrence 285 north- eastwards. Fossil occurrence 291 was moderately rich in an area that was otherwise very poor in fossil material, with ca. 120 specimens from an area only ca. 5m by 4m. There were several vertebrae, and other fragments of a large felid.

East of 291/328, there is a prominent *kas*, beyond which there is a prominent south-facing ridge capped by a sandstone before another *kas* is met. Fossil occurrences 293 and 295 were two general collections on the slopes below this prominent sandstone ridge (see Figure 6. 14). 293 contained two small scatters, nos. 298 and 308. Ground visibility was good in this area, with little vegetation, but fossil density was very low, especially for 295, which produced only two fossils over a large area. Occurrences 293, 298 and 308 are probably the same geological age as occurrences 281, 285 and 291; occurrence 295 is slightly younger as it lies on the slopes above 293. However, the sandstones is this area appeared laterally discontinuous from those in the area between 285 and 291, and further investigations are necessary to clarify this point

Figure 6. 14: Schematic section of fossil occurrences 293 and 295 (based on field notes by Hannah Jones). No. 293 is the same age as 298 and 308, and probably 281, 285 and 291.

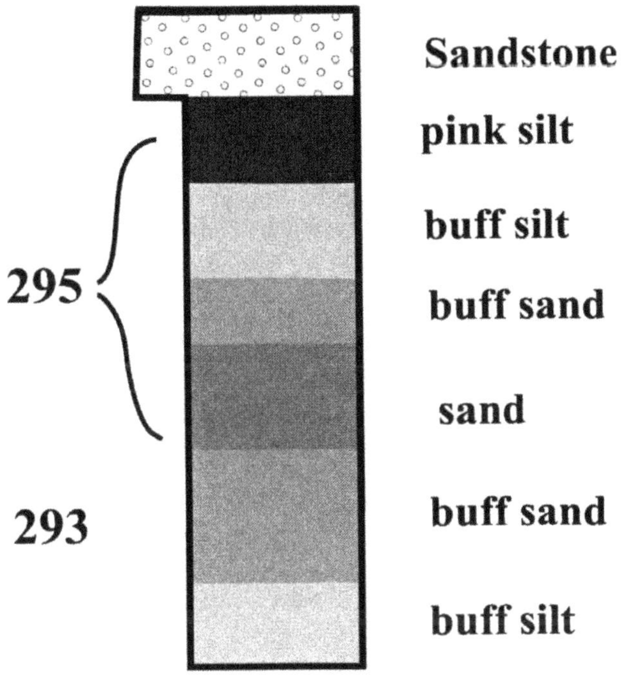

There is a stream bed at the base of the 293/295 scarp face, and another sandstone-capped ridge between it and the watershed track. Fossil occurrences 300 and 302 were two poor general collections on the sides of this ridge; scatter (296) was found at the same level and a little beyond general collection 300.

There are also the following finds from the Sohawa basin that resulted from several visits from both teams. Some of these overlap with the area covered by Set 6. Scatter 237 was found on a west-facing 20° slope when walking up a branch of Sohawa Kas towards the watershed ridge, ca. 1 km or so east of old Sohawa village (map reference, 73°, 46'E., 32°, 49'W.). On the east bank of this gully, a small number of remains were found, including a large heavily-stained astragalus (326GB1). A thorough search was not undertaken, and more material may have been present, but the density appeared very low. Scatter 257 faced south, and was 40 m sq; it may have been part of site 156, found in 1986.

Material was also collected between the 203 horizon and 291 areas. No. 305 was a 5 by 5m scatter on a south facing, 10° slope near a much-used goat track. It was a small dense patch of heavily eroded fossil, most of which were small fragments of large bones/tusk of elephant. Only 25% was therefore collected. The area was heavily vegetated and the fossil very weathered - therefore both ground and fossil visibility were poor. A stone artefact was also found. No. 307 was a very poor general collection in the gullies between the 305 and 277 collecting areas. A stone piece was found on a north facing slope, in a heavily vegetated slope wash below the main sand unit. Nos. 309-17 (odd numbers) were found on the eastern side of Sohawa Kas (see map on 309, 315). No. 315 was from the sides of a gully, and no. 317 was on the dipslope between 257, 291 and 300. 317 was also north of 257 and 300, and above them on a gently dipping NNE-facing slope between Bangial and Sohawa Kas.

At this stage, we can consider one set of material from around Flag Hill before moving on to the large and complex set of information from Kotha Kas.

Set 10: the Flag Hill area (1.7 - 1.9 Mya-old): 327-335 (odd nos.), 368-372 (even nos.).

The survey around Flag Hill produced a surprisingly large amount of material from general collections, although no scatters or localities of any note were found. The deposits in this area are all considerably older than Sandstone 12 (1.2 – 1.4 Mya-old), and are probably part of the same horizon as locality 362 in Choawala Kas, i.e. 1.7 – 1.9 Mya-old. Table 6. 13 summarises the fossil material found in this area, and Figure 6. 15 shows the area around Flag Hill. As shown, this area was covered by a series of general collections (nos. 327, 331, 333, 368, 370), with one very poor scatter of indeterminate fragments (no. 329) and a find spot (no. 335) of a crocodilian mandible within the area of no. 333. No. 372 was a gully collection.

General collect 327 around the western side of Flag Hill produced much material from slopes averaging 40°; densities were often as high as 6 per sq.m. Most of the fossils were found in gullies, but several fragments were found in and on the top of pink silts. One was found on top of the shrine in a sand unit, and another, in a sandstone unit. Because 327 covered a large area, it was arbitrarily divided into zones A - D. No. 329 was a scatter 3m by 5m of small and badly eroded fossils on level ground north of Flag Hill. General collections 331 and 333 were northwest of Flag Hill; slopes were generally east- and west-facing at 30°; fossils were usually fragmented, and often several small pieces were found close together that had belonged to the same bones. General collections 368 and 370 were near Flag Hill and immediately north of the watershed track over exposures of yellow silts. No. 372 was a small collection of fragments that had been washed down a small gully within the collecting area of 370.

Figure 6. 15: Location of sites in the Flag Hill area (based on field records by Linda Hurcombe).

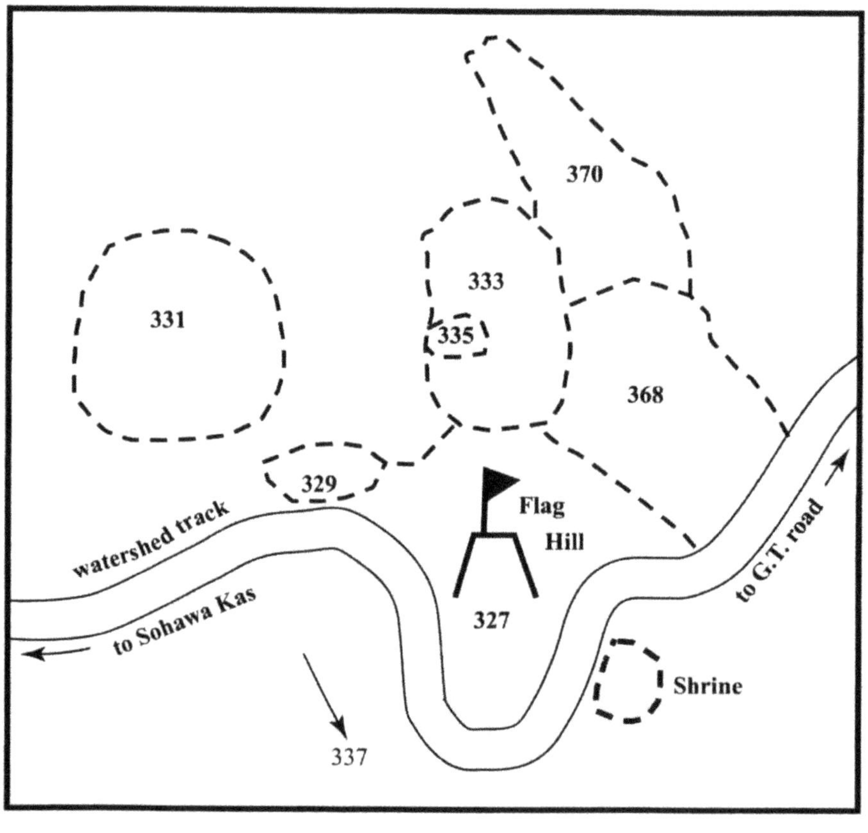

Figure 6. 16: Map of discoveries in Kotha Kas (based on a field map by Linda Hurcombe). The scale and orientation can be inferred from Figure 6.1. Sandstone 1 (the base of the local sequence) is exposed along the main drainage channels.

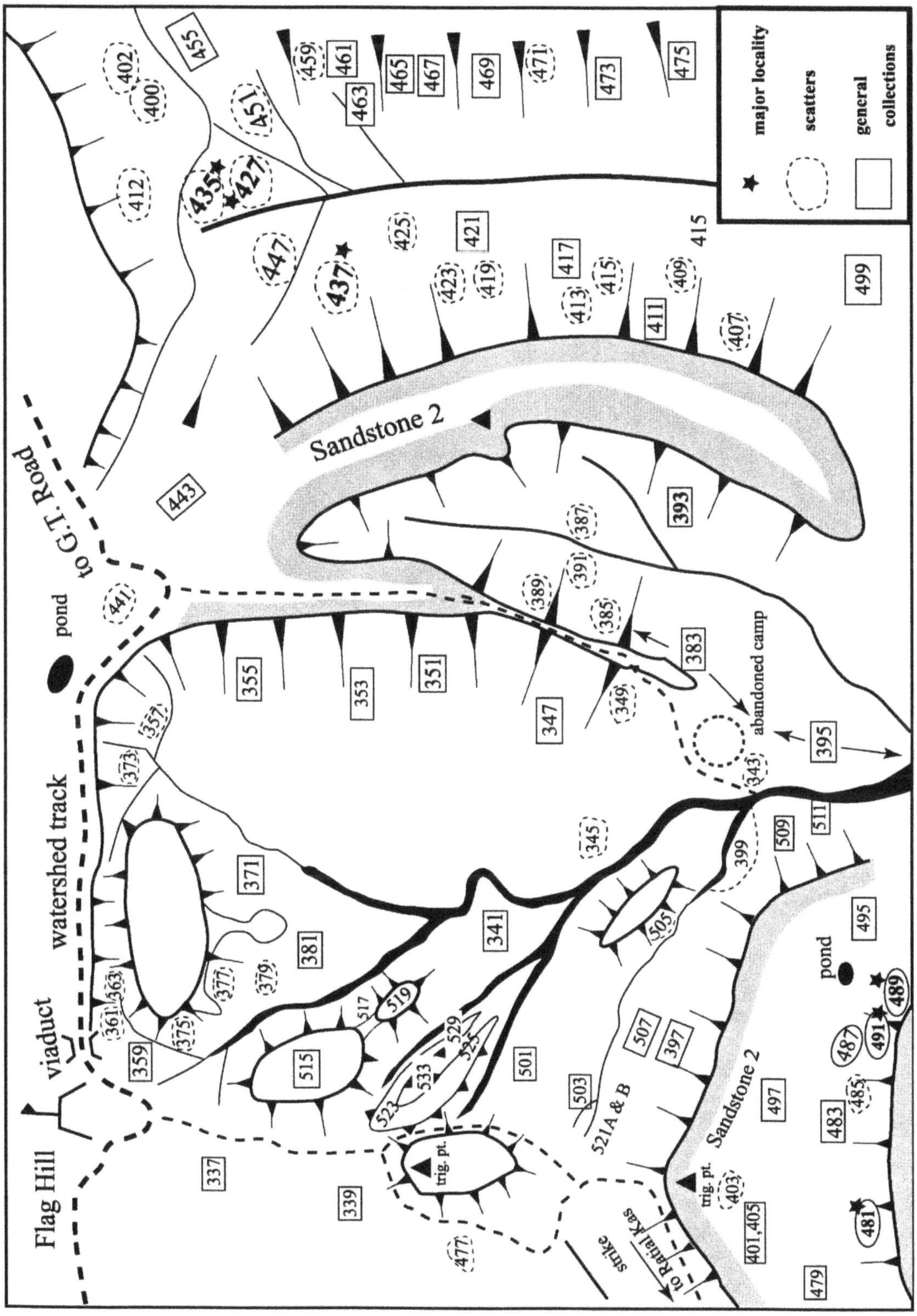

Kotha Kas (Sets 11-19)

The largest set of data from the 1987 survey came from the drainage system of Kotha Kas. This is a large drainage basin with two main branches (east and west) and several minor ones. An enormous amount of material was discovered in this area, including many of the largest sites and best specimens. There is a large amount of complex information from this area, which will be presented as a series of subsets, roughly in the order in which they were discovered, but also taking into account their geological age.

Figure 6. 16 shows the drainage and main collecting areas in the Kotha Kas area; the stratigraphic position of material has been summarised already (see Figure 6.2 and Rendell's section). Figure 6.17 provides a schematic cross-section of the west branch of Kotha Kas, showing the main sandstone units and fossil occurrences. The sets of data within Kotha Kas are shown in Figure 6. 1.

The stratigraphic sequence in this area is as follows: the lowest unit is a thick sandstone unit (Sandstone 1), which is exposed along the sides of the main gullies. The deposits adjacent to these gullies are overlain by Sandstone 2, which is clearly exposed in the southern part of the drainage system, and is in turn overlain by a third sandstone unit, which has no formal designation. Most material could thus be identified as lying between Sandstone 1 and 2, or between Sandstone 2 and the younger, unnamed sandstone unit. As indicated in Figure 6. 3, Sandstone 1 precedes the Olduvai Event, and is thus >1.9 Mya-old. The material overlying Sandstone 2 lies within the Olduvai Event, i.e. between 1.7 and 1.9 Mya-old.

Figure 6. 17: Schematic cross-section down the West branch of Kotha Kas, showing the sandstone unit underlying the main fossil occurrences, and the relevant sets of material. The sandstone unit under 493 (and thus above 481-491) has no formal number. Based on a field sketch of Sheila Sutherland.

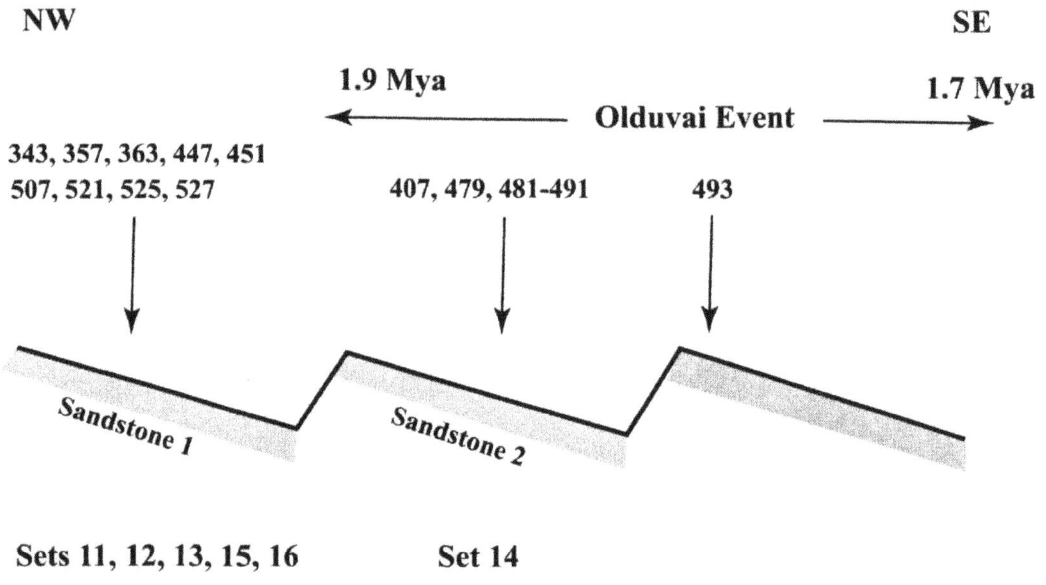

With this background information, we can now consider the first set from Kotha Kas.

Set 11: West Branch of Kotha Kas (various numbers from 337 to 535)

This set began with a reconnaissance that produced three general collections, nos. 337, 339 and 341 (see Figure 6.16). A subsequent more thorough survey generated collecting numbers 515 – 535, odd nos. only). As shown in Figure 6.15, this set of material came from deposits between Sandstone 1 and Sandstone 2, and is thus between 1.9 – 2.2 Mya-old. Table 6. 14 summarises the evidence from this area. Although no major localities were found, there were some excellent finds: a complete rhinoceros mandible from 341, a complete pig skull at 525, and a superb elephant maxilla at 535. As elsewhere in Kotha Kas, the yellow-buff sands in the main part of the exposed section (and which usually produced most of the fossils) often overlay a thin gritty/swampy horizon that in turn lay above a thin pink precipitate layer. (At occurrence 515, the fossil came from this grit horizon). At 525, the suid skull and other bones were found in a pocket of pink sandy silt within the yellow-buff sands: this may have been a burrow, possibly dug into an exposed channel bank. An alternative explanation (Imran Khan, pers. comm.) is that 529 (the matrix as well as the bones in it) represents a large mud-ball that was incorporated into the surrounding deposits.

Set 12: West Branch of Kotha Kas (397, 399, 477, 501- 511, odd. nos.)

This area lies on the other side of a gully from the previous set to the edge of Sandstone 2 (see Figure 6. 15), and was surveyed twice: first, a reconnaissance (nos. 397, 399) and then at greater length (nos. 501-511 and 521). Scatter 477 was found during the walk-in from Flag Hill to the area described in the next section. The material in this set is shown in Table 6. 15. All this material predates the Olduvai Event (see Figure 6. 3)

Set 13: West Branch of Kotha Kas (the 481 and 489 areas)

This material in this extremely rich set came from deposits immediately above Sandstone 2 (and thus lying within the Olduvai Event), but below the sandstone unit that caps the local sequence at localities 481 and 489. There is a thin but extensive grit horizon at these localities. In cross-section along the scarp face behind 489-491, this grit horizon shows a clear channel that thins at either end; the fossil was most abundant where the channel began to shallow. These localities indicate gently-flowing water, as evidenced by several pieces of conjoining turtle scute and other bones. Most of this set (485-495) are probably arbitrary sampling areas of a continuous palaeo-surface.

The easiest way to reach 489/491 is by a good path from Flag Hill that leads to a *kuchi* camp, usually abandoned. Two sandstones before this camp, turn left and head up another track leading gently upslope. The ground flattens near a small pond; this is immediately in front of 489. Alternatively, one can drive through the army firing range (when quiet!) and take the gas-pipeline service track until Kotha Kas is crossed. At this point, follow the stream and then climb to the ridge above.

GPS points: top of 491:	N32° 48' 899"; E73° 48' 625"
end of track for car coming along track from Kharian:	N32° 49' 199"; E 73° 48' 934"

The material from this small area is summarised in Tables 6. 16 – 19. The three main localities (481, 489 and 491) are summarised in Tables 6. 17-19, and the remaining material in Table 6. 16. This comprises some poor general collections (479, 483, 495 and 497) and a scatter (485) collected on the surface of the same gritstone that appears at 481- 489. No. 479 was a collection of material derived from further upslope; 483 was probably either derived downslope from 481 and 495, or was deflated from the same surface. 487 is classed as a small locality derived from the same horizon as 481, 489 and 491, and is probably part of the same surface. Scatter 493 came from the dip slope on top of the sandstone above 481-489, and is thus younger: no other material was collected from this horizon.

Localities 481, 489 and 491 were all found along the scarp slope of deposits capped by the youngest of the three sandstones in this area. Figure 6. 18 shows a schematic plan of locality 481; the local section is shown in Figure 6. 19. A section and plan of localities of 489 and 491 are shown in Figures 5. 20 and 5. 21.

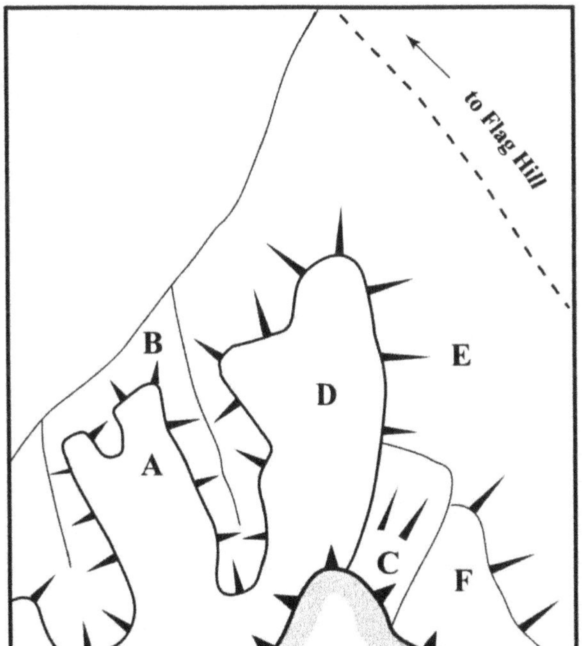

Figure 6. 18: Plan of locality 481 (based on records by Sheila Sutherland).

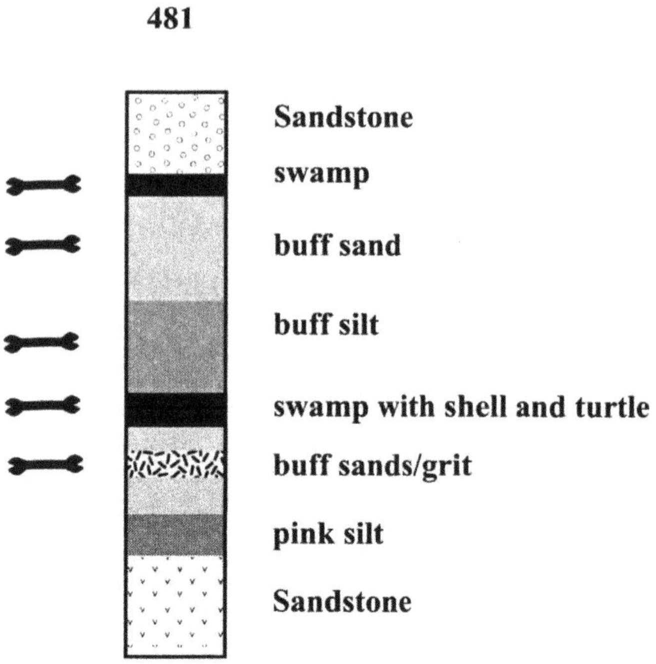

Figure 6. 19: Schematic section at locality 481 (based on records by Sheila Sutherland). The section is ca. 10m high.

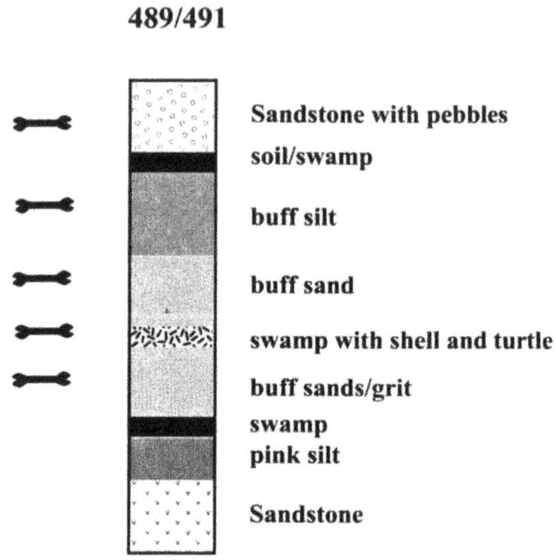

Figure 6. 20: Schematic section at localities 489 and 491 (based on records by Sheila Sutherland). The section is ca. 10m high.

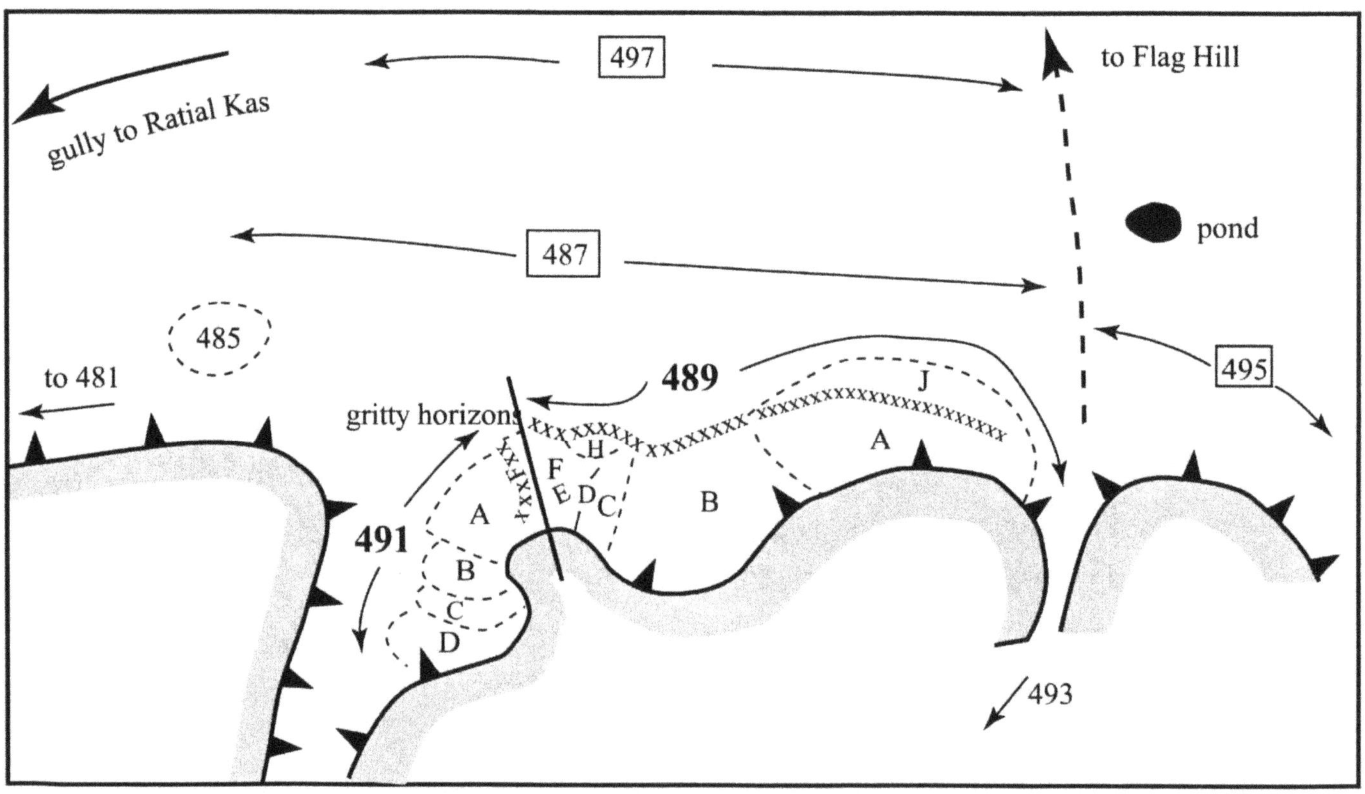

Figure 6. 21: Plan of the area around localities 489 and 491 (based on records by Sheila Sutherland). The area shown is ca. 500m. across.

As Tables 5. 16 – 5. 18 indicate, these localities produced abundant remains of artiodactyls and bovids (most of the artiodactyl remains were probably bovid, given that few specimens were identified as cervid or giraffid). Dentitions and hind limb elements dominated. Cervids, equids, giraffids, Proboscideans and rhinoceros were also present in small numbers. Carnivores included crocodile, hyaena, a large cat, and a canid. Turtle remains were very common. Preservation was excellent, and excavation would almost certainly have produced complete, and even articulated, remains. A mandible found in 1999 – some 12 years after the initial discovery of this locality – has been provisionally identified as a small primate.

Overall, these localities were similar to, but much more extensive than locality 83 from Sandstone 12 (see chapter 4), and similar to other localities elsewhere in Kotha Kas such as 427, 435 and 437 which will be described shortly.

Set 14: West Branch of Kotha Kas (343-357, 383-395)

This area is shown in Figure 6.15, and the fossils summarised in Table 6. 20. This set has two components: the first (occurrences 343-357) came from deposits on the other side of the drainage channel from Sets 11-13, and the second (383-395) came from a tributary side valley. This was named "Pelvis Valley", following the discovery and excavation of a complete bovid/cervid pelvis at 387. All material in this set came from deposits between Sandstones 1 and 2. Fossils were generally well-preserved except in scatter 343, which were white, fragile, and like those from locality 68. (A pig skull was recovered from this bone pocket in 1999). At 343, the fossil was derived from yellow sands above a gritstone horizon (possibly the same as at 349 (see below) that directly overlay the thick basal sandstone. Occurrence 349 was a find spot of fragmented elephant bones, of which only a small sample was taken. These came buff silts immediately below a thin gritstone horizon (see Figure 6. 22 in the 347 collecting area. The survey then proceeded towards the watershed track, following the grit horizon along the lower slopes of the scarp that defines the eastern side of this small valley. This resulted in general collections 351, 353 and 355, and a small scatter (357) in the area of 355.

As Figure 6. 3 shows, all this material predates the Olduvai Event.

Figure 6. 22: Schematic section of fossil occurrence 349 (based on field notes by Sheila Sutherland).

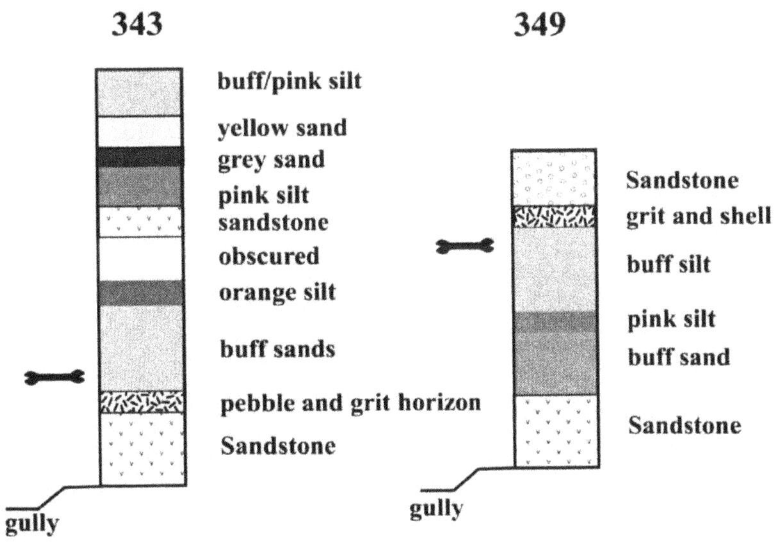

Pelvis Valley (383-395)

This part of Kotha Kas is best reached by a good path leading along a ridge from the pond near the viaduct, and trending gently down slope towards an abandoned *kuchi* camp. Occurrence 383 was a general collection from this camp to the eastern side of a small ridge. The main discovery was a large scatter (or possibly, a small locality), no. 385, which was subdivided into four sub-areas. The deposits here were (from base to top) a pink sandy-silt, then a thin dark clay horizon under grey-yellow sands. One piece from 385A had a grey sand matrix; fragments from 385C lay on top of the dark clay. Nearby, and on the right-hand side of a path (looking towards the watershed track), the top of what turned out to be a complete artiodactyls pelvis was found (occurrence 387) and later excavated. The pelvis lay in fine yellow sand, fining upwards into a silty sand that was capped by a thick (20 cm) layer of hard pink silt under another sand layer. The remainder of fossil collected in this area (collections 389-395) was unremarkable.

Set 15: The Viaduct area (1.9 - 2.2 Mya-old): 359-363, 369-381 (odd nos. only).

This set was derived from close to the Kharian side of the watershed track, between the viaduct and the shrine at Flag Hill. Figure 6. 23 shows where material was found, and Table 6. 21 summarises the fossils from this area. General collection 359 began 50-100m from the shrine by Flag Hill, following a conspicuous yellow sand horizon above Sandstone 1. This general collection contained one large scatter (361), 150m from the viaduct, and three poor ones (363, 365, 367). General collection 371 contained scatter 373; and scatters 375, 377, and 379 were within general collection area 381. The main stratigraphic feature of this area was a black clay/swamp horizon within a sand horizon that was case-hardened and almost ridge forming. At scatter 361, the fossils were derived from and below this black clay "swampy" layer that also contained shells and turtle carapace (see Figure 6. 23); and at 375, material probably eroded from a thin grit layer (with one piece found *in situ*). This scatter lay on both sides of a small spur, and was collected in three units, A-C.

Figure 6. 23: Location of fossil sites in the Viaduct area (359-379, odd. nos.), and section at 361. Based on field records by Sheila Sutherland.

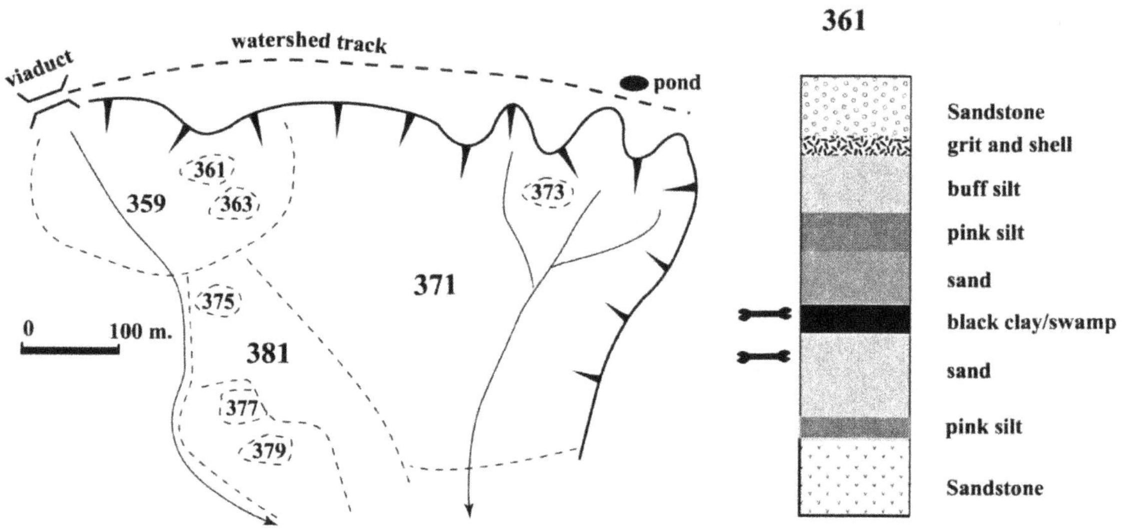

The East Branch of Kotha Kas

The eastern branch of Kotha Kas had very little of interest in the southern part, but the northern end, near the watershed track, contained some of the best localities found in this part of the Pabbi Hills. Localities 427, 435 and 437 were especially noteworthy. As with the previous two sets (of material from the western branch (Sets 14-15), the deposits here are sandwiched between Sandstones 1 and 2, and thus predate the Olduvai Event. We can begin first with the southern end of this part of Kotha Kas.

Set 16: East Branch of Kotha Kas (1.9-2.2 Mya-old): 407-425 (odd nos. only) and 499.

Figure 6. 16 shows the area covered by this set. Nos. 411, 417, 421, 425 and 499 were (poor) general collections along the western side of the valley; 407, 409, 413, 415, 419, 423 were scatters. These are summarised in Table 6. 22. The general collections were along the side slopes and bottom, and the scatters (notably 415, 419 and 423) occurred near the base of small side gullies – presumably because they had collected there after being exposed further upslope. The main discovery was a moderate-sized scatter, no. 413, over an area 20 by 15m. The stratigraphic sequence in this part of the eastern branch is not unlike that seen previously: i.e., a multi-layered sandwich of fine sands and silts, interspersed with thin grit and clay horizons, all indicating a low-energy depositional environment. In the area of 407, some turtle pieces and bivalves were found in the upper part of a grit layer, immediately under a yellow-brown to grey-brown swampy clay. The grit layer was above a thin pink silt, which was in turn over a thin grit and then another pink silt layer.

Set 17: Kotha Kas (1.9-2.2 Mya-old): 427-461 (odd nos.)

This small area produced some of the richest localities in Kotha Kas, most notably localities 427 with 382 fragments; 435 with almost 700, 437 with over 1,300 pieces, and 451 with over 450 fragments, and 445 with over 300. It lies close to the watershed track, shortly after a service track constructed shortly before 1999 for a gas pipe-line that leads towards Kharian. Stratigraphic sections of 427, 435 and 437 are shown in Figure 6. 24, and Figures 6. 25 and 6. 26 show the area covered by this set. The fossils found are summarised in Tables 6. 23 and 6. 24. As can be seen from Figure 6. 24, both 427 and 437 are derived from a grit horizon; indeed, both localities could represent material derived from a former continuous surface. The same grit horizon also outcrops at 445, 447 and 451. As Figure 6. 27 shows, material from 445C and D came from the grit. Locality 435 is slightly younger, being derived from the overlying yellow sand unit. The local general collections 429, 439 and 443 produced surprisingly little material considering the richness of the localities nearby. (Sadly, at the time of writing, 435 has been largely destroyed by bull-dozing).

Figure 6. 24: Schematic section of localities 427, 435 and 437. (Based on field records by Sheila Sutherland).

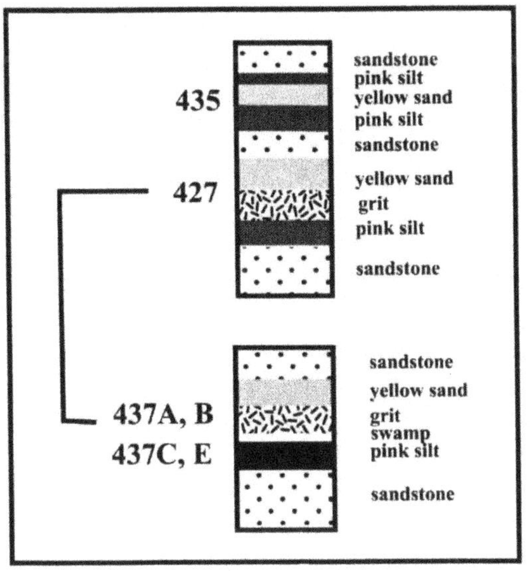

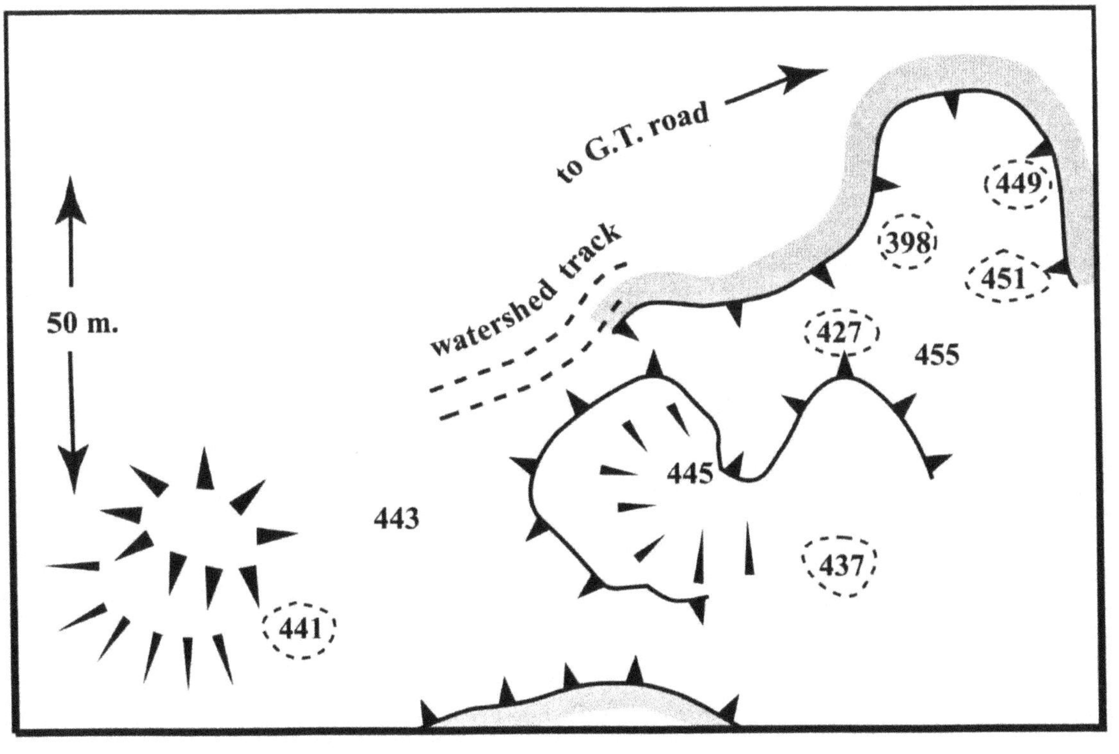

Figure 6. 25: Principal occurrences in the area of localities 427 and 437. (Based on field records by Sheila Sutherland).

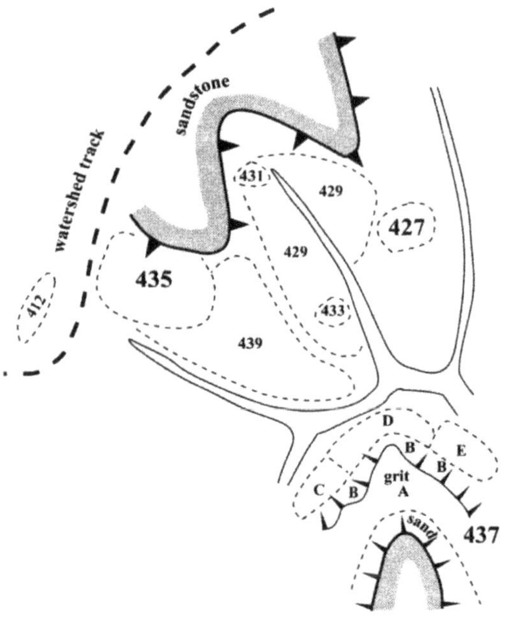

Figure 6. 26: Field plan of the 437 area. (Based on field records by Sheila Sutherland). (This shows the area covered in Figure 6. 24 at a smaller scale; both are sketch plans, drawn on different days, so the match between them is not perfect).

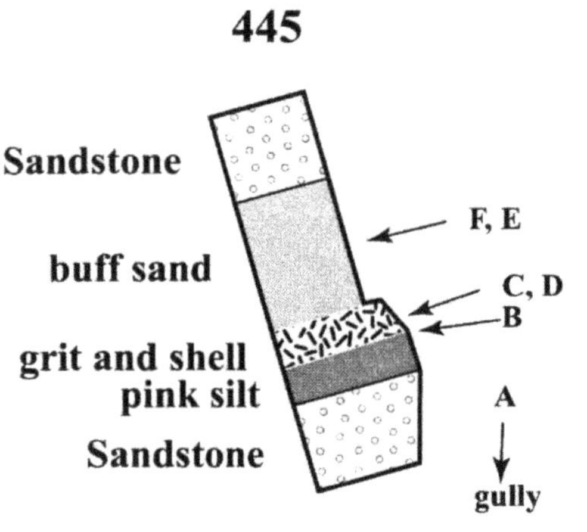

Figure 6. 27: Schematic section of locality 445. (Based on field records by Sheila Sutherland).

Set 18: Head of Kotha Kas, between Happy and Death Valleys (?1.7 - 1.9 Mya-old): 398 - 414 (even nos. only).

The fossil occurrences in this small set of general collections are located close to the watershed track, and between the heads of Happy Valley and Death Valley. (The pond mentioned in the previous section lies a short distance beyond 404 and 406). In this area, the watershed track is below the crest of the anticline, and part of the track cuts through sections that contain an extensive grit horizon. The age of this unit is probably between 1.7 - 1.9 Ma.-old; i.e. within the Olduvai Event. Figure 6. 28 shows the local section and collecting units, and Table 6. 25 summarises their contents.

General collections 404 and 414 were undifferentiated; 404 because the local sequence had not been recognised, and 414, because exposures were poor and little was found. The material in scatter 412 was very fragmented, and derived from the grit unit and overlying fine sands. Although the contents of this set are unremarkable, the survey in this area served as a useful link between the headwaters of Kotha and Choawala Kas, and alerted the team to the fossiliferous grit horizon that was later traced downstream into the east branch of Kotha Kas.

Figure 6.28: Schematic section of occurrences 400-412 (based on records by Mags Peace and Helen Wilson).

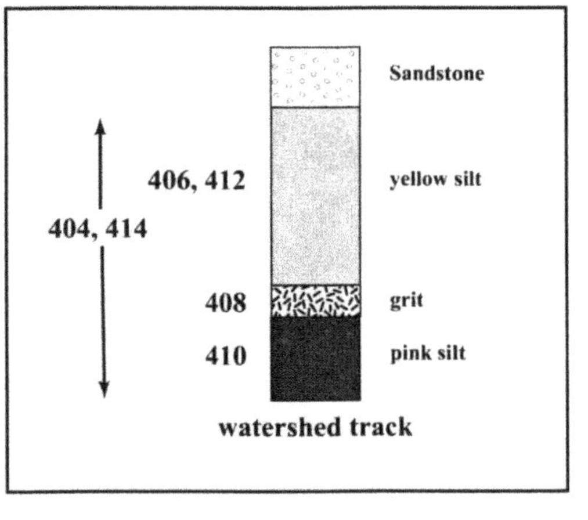

An additional reason why the survey in this area was useful was that it helped resolve one problem that arose when following the grit horizon shown above. It soon became apparent that this was at a higher level than a short distance away in the area of localities 427 and 435. This was because of a fault between these localities and the watershed track.

As Figure 6. 29 shows, this could be either normal or reversed; given the asymmetric folding of the anticline, the latter is more likely. Either way, the grit horizon identified at the section by the track is likely to be the same as traced further along the east branch of Kotha Kas.

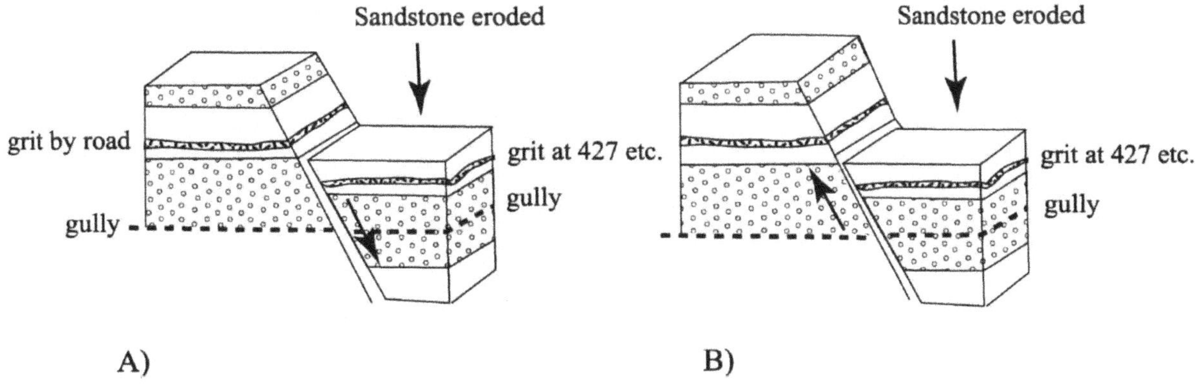

Figure 6. 29: A schematic view of the fault between localities 427 and 435, and the watershed track (based on field notes by Sheila Sutherland).

Set 19: East Branch of Kotha Kas (453-475, odd. nos. only)

The last set from Kotha Kas was probably the most disappointing, and comprised a series of meagre general collections and scatters along the scarp slopes and side gullies along the eastern bank of the eastern branch. The approximate location of these is shown in Figure 6. 15, and their contents in Table 6. 26. The most useful outcome of this part of the survey was further stratigraphic information on the context of fossil material. As shown in Figure 6. 29, the material from collection 465 (and probably the others in this series) came from deposits below the major sandstone unit that underlies 427 and 437.

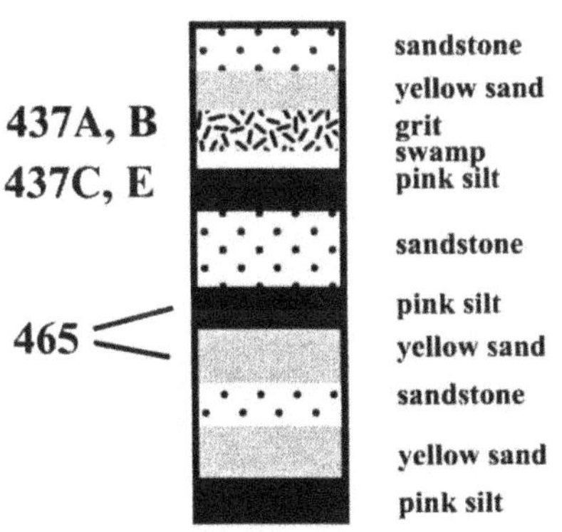

Figure 6. 30: Schematic section of stratigraphic relationships of fossil material at occurrences 427 and 465, East Branch of Kotha Kas. Based on field notes by Hannah Jones.

Set 20: Sandstones 14 - 16, Choawala Kas (0.9 - 1.2 Mya-old): 500-510, 516, 520-528 (even nos. only).

The penultimate set in the area surveyed in 1987 on this side of the G.T. road covered the exposures behind (and thus younger than) Sandstone 12 in the Choawala Kas drainage basin. The area covered is shown in Figures 5. 1 and 5. 2. The main landmark in this area is a large, sandstone-capped mesa on the left-hand bank (facing downstream) of Choawala Kas. Occurrences 500-510 (Set 20A) were on the left-hand-side of the kas, and 516-528 (Set 20B) on the other. The scarp faces behind Sandstone 12 looked promising because they were extensive and generally clear of vegetation, and offered good prospects of supplementing the meagre amount of material collected from Sandstone 14 north of the G.T. Road in 1986.

The results are summarised in Table 6.27. Stratigraphic sections of occurrences 504 and 516 are shown in Figure 6. 31.

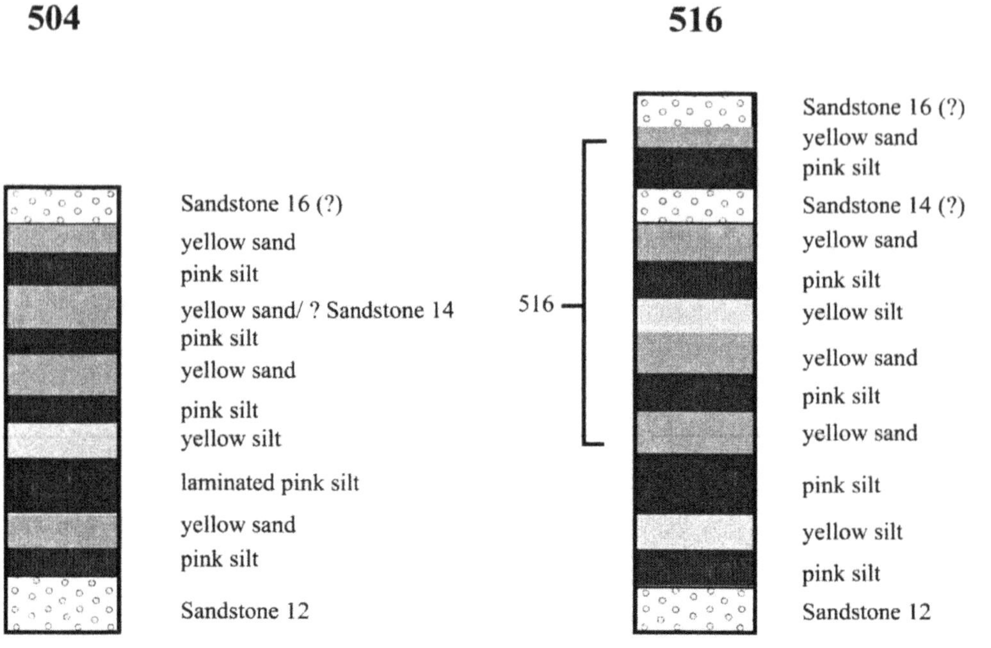

Figure 6. 31. Schematic sections of occurrences 504 and 516 (based on field records by Hannah Jones).

As Figure 6. 30 shows, the exposures were multi-storied, with at least two major sandstone units overlying Sandstone 12. Following Keller et al.'s (1979) convention of assigning major sandstone units even numbers, it was assumed that these were equivalent to their Sandstones 14 and 16. If so, the deposits in these sections should lie between the Jaramillo Event (ca. 0.9 Mya) and 1.2 Mya, the estimated age of the top of the Sandstone 12 sequence. However, the identification of Sandstones 14 and 16 was tentative, as it was not possible to link these sections to their roadside section.

Although a large area was covered, the results were meagre. General collections 500, 510, 516, 520, 524, 528 provided good exercise, but were scarcely worth the effort. General collections 500, 510 and 516 covered scarp slopes that probably contained Sandstone 14 and 16. Scatters 504 and 506 were described as immediately above Sandstone 12; and scatter 522, as younger than Sanstone 12 but below Sandstone 18. This scatter (subdivided into a number of sub-areas during collection) was reasonably good, and probably came from a thin, discontinuous grit horizon.

Set 21: Ratial and Dhorian Kas (382-386, 420-432)

The last set from the this side of the G.T. road came from Dhorian and Ratial Kas. Although a large area was surveyed, including extensive slopes that were free of vegetation, the results were very meagre, and are included only for the sake of completeness. The results are summarised in Table 6. 28.

Set 22: The badlands north of the G.T. Road (436-528 even nos.)

One area that had not been extensively covered north of the G.T. road in 1986 was a large area of "badlands" in front of Sandstone 12 between the G.T. road and Kanarawala Kas. As noted in the previous chapter, this was partially explored in the area of locality 83, and resulted in some useful discoveries such as localities 124 and 128 (see Figure 5. 9). Further investigation thus seemed appropriate. This part of the anticline is a broad valley, as most of the deposits originally capped by Sandstone 12 have been eroded, leaving only a few sandstone-capped inselbergs that show the original profile of the anticline. This area was surveyed in 1987; the fossils found are summarised in Table 6. 29, and details of the main parts of this extensive area in Figures 5. 32 and 5. 33; stratigraphic sections are shown in Figures 5. 34 - 5. 36. Sadly, the survey in this area was not especially productive, although it did produce one of the finest specimens found in the Pabbi Hills, namely a remarkably complete turtle cranium from pink silts at occurrence 492 (see Figure 6. 35).

According to Keller et al. (1977; Figure14), these deposits should belong to the interval between the Olduvai Event and Jaramillo. As they are in front of and below the Sandstone 12 sequence, and under an earlier sandstone (possibly Sandstone 10), they should therefore be between 1.4 and 1.7 Mya old. If so, they are a useful addition to the scant amount of material from the Pabbi Hills that can be assigned to this time-range.

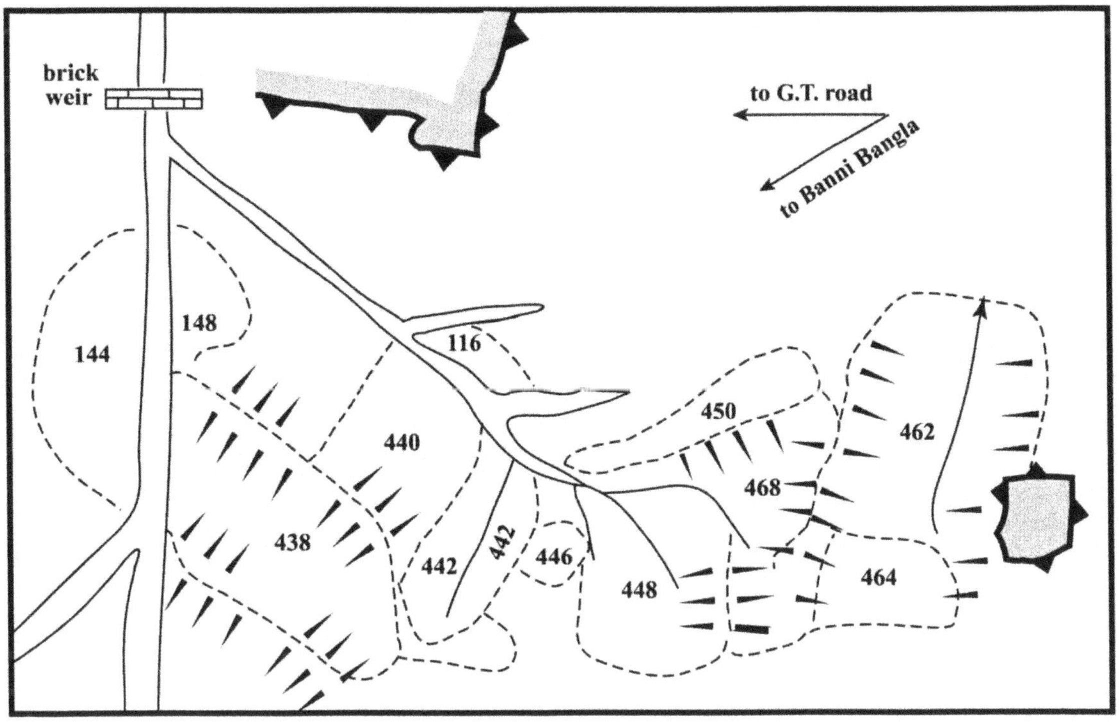

Figure 6. 32: Principal collecting areas in the badlands near the brick weir north of the G.T. road (based on field records by Hannah Jones). The brick weir is the same one as shown in Figure 5. 9.

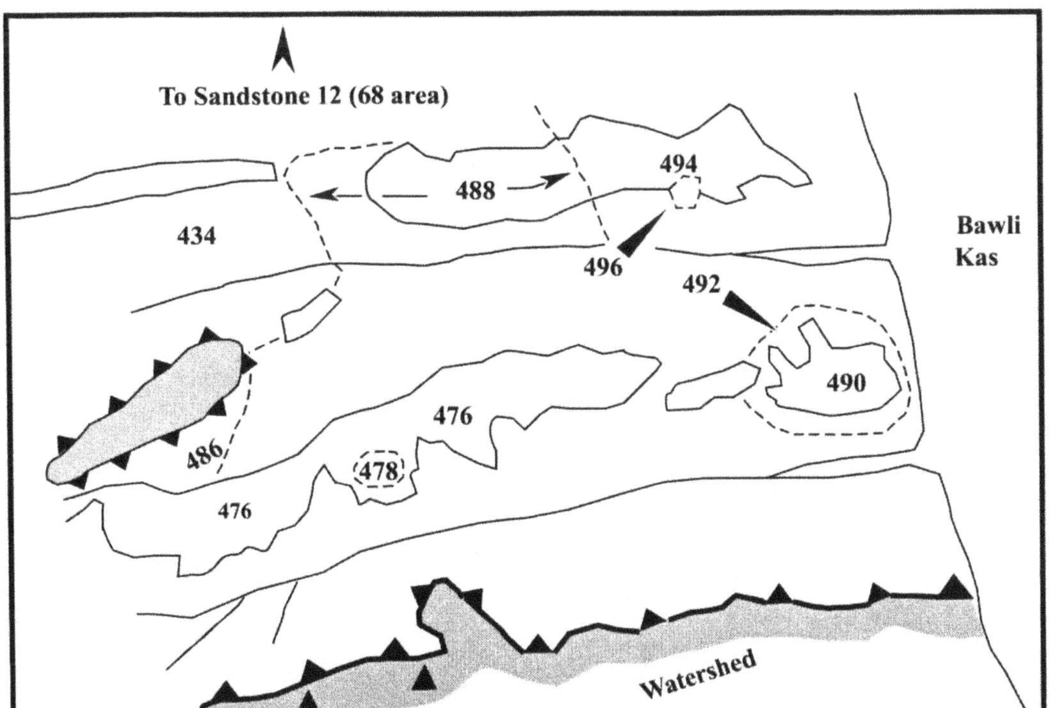

Figure 6. 33: Principal collecting areas in the badlands near Bawli Kas, north of the G.T. road (based on field records by Hannah Jones). This area lies to the east of that shown in the previous figure.

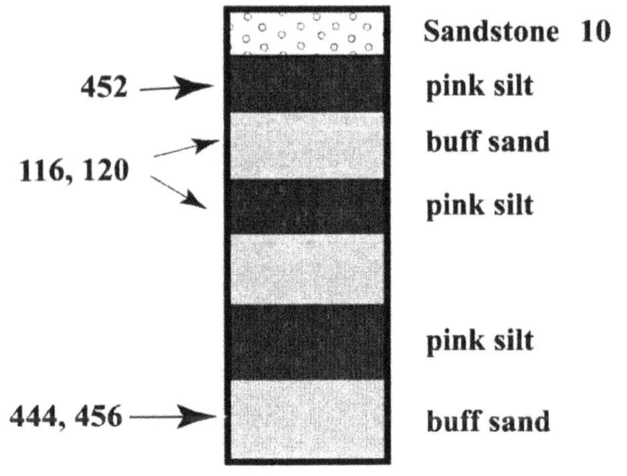

Figure 6. 34: Schematic section of occurrences 116, 120, 444, 452 and 456. General collection 450 covered the entire profile.The sandstone capping this profile is probably Sandstone 10. (Based on field records by Hannah Jones).

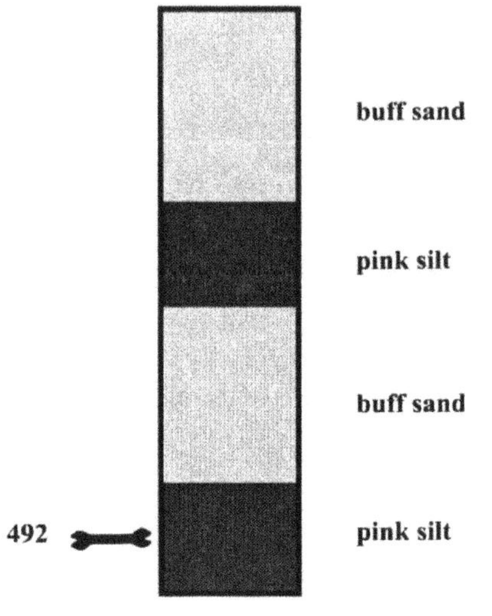

Figure 6. 35: Schematic section of occurrence 492. The turtle cranium came from the pink silts at the base. General collection 490 covered the entire profile. (Based on field records by Hannah Jones).

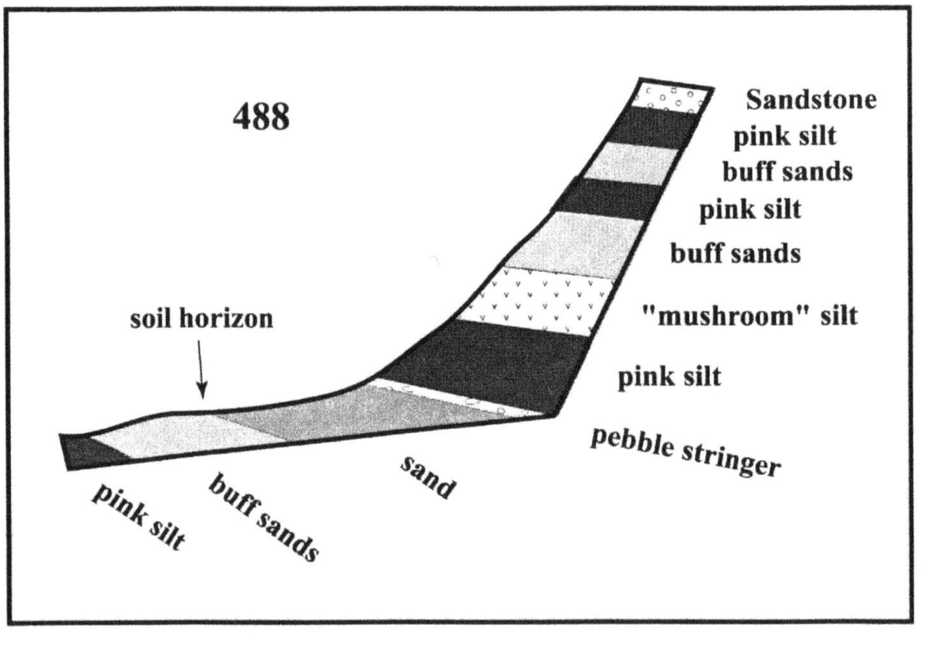

Figure 6. 36: Schematic section of occurrence 488. Here, it was not possible to identify the precise origin of fossil material, but most were found near the base. The sandstone at the top of this sequence (ca. 10-15 m high) is probably Sandstone 10. (Based on field records by Hannah Jones).

Table 6:1. Fossil occurrences found in 1987, showing each in terms of collecting area, set, type of occurrence, age and total.

Time-bands	Age (Mya)
Younger than SST 12:	0.9 - 1.2
Sandstone 12:	1.2 - 1.4
Top of the Olduvai Event to the base of Sandstone 12:	1.4 - 1.7
Olduvai Event:	1.7 - 1.9
Pre-Olduvai Event:	1.9 - 2.2

No.	Area	Type	Set	Age	Total
200	**Choawala**	**General collection**	**Set 1**	**1.7 - 1.9***	**30**
202	**Choawala**	**General collection**	**Set 1**	**1.7 - 1.9***	**24**
204	Choawala	General collection	Set 1	1.7 - 1.9	10
206	**Choawala**	**Scatter**	**Set 1**	**1.7 - 1.9***	**53**
208	Choawala	General collection	Set 1	1.7 - 1.9	72
210	Choawala	Scatter	Set 1	1.7 - 1.9	21
212	**Choawala**	**Scatter**	**Set 1**	**1.7 - 1.9***	**35**
214	**Choawala**	**Locality**	**Set 1**	**1.7 - 1.9***	**234**
216	Choawala	General collection	Set 1	1.7 - 1.9	10
218	**Choawala**	**Scatter**	**Set 1**	**1.7 - 1.9***	**25**
220	Choawala	Findspot	Set 1	1.7 - 1.9	1
365	**Choawala**	**Scatter**	**Set 1**	**1.7 - 1.9***	**7**
367	Choawala	Scatter	Set 1	1.7 - 1.9	5
388	**Choawala**	**General collection**	**Set 1**	**1.7 - 1.9***	**40**
222	Choawala West	Scatter	Set 2	1.4 - 1.7	5
224	Choawala West	Scatter	Set 2	1.4 - 1.7	26
226	**Choawala West**	**General collection**	**Set 2**	**1.4 - 1.7***	**6**
230	Choawala West	General collection	Set 2	1.4 - 1.7	5
232	Choawala West	General collection	Set 2	1.4 - 1.7	12
332	Choawala West	Scatter	Set 2	1.4 - 1.7	1
334	Choawala West	Scatter	Set 2	1.4 - 1.7	1
336	Choawala West	General collection	Set 2	1.4 - 1.7	3
338	Choawala West	General collection	Set 3	1.7 - 1.9	25
340	Choawala West	Scatter	Set 3	1.7 - 1.9	17
342	Choawala West	General collection	Set 3	1.7 - 1.9	3
344	Choawala West	Scatter	Set 3	1.7 - 1.9	30
348	**Choawala West**	**Scatter**	**Set 3**	**1.7 - 1.9***	**99**
350	Choawala West	General collection	Set 3	1.7 - 1.9	1
352	Choawala West	General collection	Set 3	1.7 - 1.9	5
354	Choawala West	Scatter	Set 3	1.7 - 1.9	14
356	**Choawala West**	**Scatter**	**Set 3**	**1.7 - 1.9***	**28**
358	Choawala West	Scatter	Set 3	1.7 - 1.9	25
360	Choawala West	General collection	Set 3	1.7 - 1.9	3
362	**Choawala West**	**Locality**	**Set 3**	**1.7 - 1.9***	**217**
366	Choawala West	General collection	Set 3	1.7 - 1.9	32
236	Choawala, Happy Valley	General collection	Set 4	1.7 - 1.9	29
238	**Choawala, Happy Valley**	**Scatter**	**Set 4**	**1.7 - 1.9***	**60**
240	**Choawala, Happy Valley**	**Scatter**	**Set 4**	**1.7 - 1.9***	**23**
242	**Choawala, Happy Valley**	**Scatter**	**Set 4**	**1.7 - 1.9***	**8**
244	**Choawala, Happy Valley**	**Scatter**	**Set 4**	**1.7 - 1.9***	**25**
246	Choawala, Happy Valley	General collection	Set 4	1.7 - 1.9	13
248	**Choawala, Happy Valley**	**Scatter**	**Set 4**	**1.7 - 1.9***	**108**
250	Choawala, Happy Valley	Scatter	Set 4	1.7 - 1.9	81
252	**Choawala, Happy Valley**	**Scatter**	**Set 4**	**1.7 - 1.9***	**66**

254	Choawala, Happy Valley	Scatter	Set 4	1.7 - 1.9	61
256	**Choawala, Happy Valley**	**Scatter**	**Set 4**	**1.7 - 1.9***	**5**
258	Choawala, Happy Valley	General collection	Set 4	1.7 - 1.9	3
260	**Choawala, Happy Valley**	**Scatter**	**Set 4**	**1.7 - 1.9***	**31**
262	Choawala, Death Valley	General collection	Set 4	1.7 - 1.9	18
264	Choawala, Death Valley	General collection	Set 4	1.7 - 1.9	3
416	Choawala, Death Valley	Scatter	Set 4	1.7 - 1.9	3
418	Choawala, Death Valley	General collection	Set 4	1.7 - 1.9	3
434	Choawala, Death Valley	Scatter	Set 4	1.7 - 1.9	3
518	Choawala, Death Valley	General collection	Set 4	1.7 - 1.9	35
225	Bangial, Choawala	General collection	Set 5	1.4 - 1.7	16
227	Bangial, Choawala	General collection	Set 5	1.2 - 1.4	12
229	Bangial, Choawala	Gully bed	Set 5	>0.5	1
231	Bangial, Choawala	General collection	Set 5	1.2 - 1.4	14
234	Bangial, Choawala	Gully bed	Set 5	>0.5	4
235	Bangial, Choawala	Scatter	Set 5	1.2 - 1.4	47
241	Bangial, Choawala	Scatter	Set 5	1.4 - ?1.7	14
243	Bangial, Choawala	General collection	Set 5	1.4 - ?1.7	56
245	Bangial, Choawala	General collection	Set 5	1.4 - ?1.7	31
247	Bangial, Choawala	Scatter	Set 5	1.4 - ?1.7	49
249	Bangial, Choawala	General collection	Set 5	1.4 - ?1.7	43
251	Bangial, Choawala	General collection	Set 5	1.4 - ?1.7	2
253	Bangial, Choawala	General collection	Set 5	1.4 - ?1.7	7
255	Bangial, Choawala	General collection	Set 5	1.4 - ?1.7	4
268	Bangial, Choawala	Gully bed	Set 5	>0.5	3
270	Bangial, Choawala	Find spot	Set 5	1.2 - 1.4	1
272	Bangial, Choawala	Scatter	Set 5	1.2 - 1.4	39
314	Bangial, Choawala	Scatter	Set 5	1.9 - 2.2	15
532	Bangial, Choawala	Gully bed	Set 5	>0.5	1
203	**Sohawa Kas**	**General collection**	**Set 6**	**1.9 - 2.2***	**120**
213	Sohawa Kas	Scatter	Set 6	1.9 - 2.2	11
215	Sohawa Kas	Scatter	Set 6	1.9 - 2.2	21
259	Sohawa Kas	General collection	Set 6	1.9 - 2.2	94
261	Sohawa Kas	Scatter	Set 6	1.9 - 2.2	59
263	Sohawa Kas	General collection	Set 6	1.9 - 2.2	33
265	Sohawa Kas	Scatter	Set 6	1.9 - 2.2	82
267	Sohawa Kas	Scatter	Set 6	1.9 - 2.2	25
269	Sohawa Kas	**Scatter**	**Set 6**	**1.9 - 2.2***	**4**
271	Sohawa Kas	Scatter	Set 6	1.9 - 2.2	195
273	Sohawa Kas	Scatter	Set 6	1.9 - 2.2	71
275	Sohawa Kas	General collection	Set 6	1.9 - 2.2	54
277	Sohawa Kas	General collection	Set 6	1.9 - 2.2	13
297	Sohawa Kas	General collection	Set 6	1.9 - 2.2	15
299	Sohawa Kas	General collection	Set 6	1.9 - 2.2	8
301	Sohawa Kas	Scatter	Set 6	1.9 - 2.2	4
303	Sohawa Kas	Scatter	Set 6	1.9 - 2.2	17
312	Sohawa Kas	General collection	Set 6	1.9 - 2.2	33
205	Sohawa, Headwaters and West	Gully	Set 7	1.9 - 2.2	8
207	Sohawa, Headwaters and West	Scatter	Set 7	1.9 - 2.2	12
209	Sohawa, Headwaters and West	General collection	Set 7	1.9 - 2.2	36
217	Sohawa, Headwaters and West	Scatter	Set 7	1.9 - 2.2	23
219	Sohawa, Headwaters and West	Scatter	Set 7	1.9 - 2.2	9
221	Sohawa, Headwaters and West	Scatter	Set 7	1.9 - 2.2	33
223	Sohawa, Headwaters and West	Scatter	Set 7	1.9 - 2.2	25

288	Sohawa, Headwaters and West	General collection	Set 7	1.9 - 2.2	14
316	Sohawa, Headwaters and West	General collection	Set 7	1.9 - 2.2	14
318	Sohawa, Headwaters and West	General collection	Set 7	1.9 - 2.2	42
276	**Sohawa Kas, West**	**General collection**	**Set 8**	**1.7 - 1.9***	**99**
278	**Sohawa Kas, West**	**Scatter**	**Set 8**	**1.7 - 1.9***	**5**
280	Sohawa Kas, West	General collection	Set 8	1.7 - 1.9	65
282	**Sohawa Kas, West**	**Scatter**	**Set 8**	**1.7 - 1.9***	**1**
284	**Sohawa Kas, West**	**Scatter**	**Set 8**	**1.7 - 1.9***	**48**
286	Sohawa Kas, West	General collection	Set 8	1.7 - 1.9	4
290	Sohawa Kas, West	General collection	Set 8	1.7 - 1.9	2
296	Sohawa Kas, West	Scatter	Set 8	1.9 - 2.2	81
320	Sohawa Kas, West	General collection	Set 8	1.7 - 1.9	10
378	Sohawa Kas, West	Find spot	Set 8	1.7 - 1.9	1
237	Sohawa Kas, East	Scatter	Set 9	?1.7 - 1.9	9
257	Sohawa Kas, East	Scatter	Set 9	?1.7 - 1.9	120
279	Sohawa Kas, East	General collection	Set 9	1.9 - 2.2	11
281	Sohawa Kas, East	General collection	Set 9	1.9 - 2.2	38
283	Sohawa Kas, East	General collection	Set 9	1.9 - 2.2	56
285	Sohawa Kas, East	Scatter	Set 9	1.9 - 2.2	194
287	Sohawa Kas, East	General collection	Set 9	1.9 - 2.2	37
289	Sohawa Kas, East	Scatter	Set 9	1.9 - 2.2	11
291	Sohawa Kas, East	Scatter	Set 9	1.9 - 2.2	118
293	Sohawa Kas, East	General collection	Set 9	1.9 - 2.2	35
295	Sohawa Kas, East	General collection	Set 9	1.9 - 2.2	2
298	Sohawa Kas, East	Scatter	Set 9	1.9 - 2.2	13
300	Sohawa Kas, East	General collection	Set 9	1.9 - 2.2	8
302	Sohawa Kas, East	General collection	Set 9	1.9 - 2.2	6
304	Sohawa Kas, East	Scatter	Set 9	1.9 - 2.2	42
305	Sohawa Kas, East	Scatter	Set 9	?1.7 - 1.9	17
306	Sohawa Kas, East	Scatter	Set 9	1.9 - 2.2	7
308	Sohawa Kas, East	Scatter	Set 9	1.9 - 2.2	9
309	Sohawa Kas, East	General collection	Set 9	?1.7 - 1.9	16
311	Sohawa N.E.	Find-spot	Set 9	?1.7 - 1.9	7
315	Sohawa N.E.	Gully	Set 9	?1.7 - 1.9	5
317	Sohawa N.E.	General collection	Set 9	?1.7 - 1.9	3
319	Sohawa Kas, East	General collection	Set 9	1.9 - 2.2	12
322	Sohawa Kas, East	General collection	Set 9	1.7 - 1.9	5
324	Sohawa Kas, East	General collection	Set 9	1.9 - 2.2	1
326	Sohawa Kas, East	Transverse gully	Set 9	>0.5	1
328	Sohawa Kas, East	General collection	Set 9	1.9 - 2.2	3
330	Sohawa Kas, East	Scatter	Set 9	1.9 - 2.2	3
376	Sohawa Kas, East	Scatter	Set 9	1.9 - 2.2	2
327	Flag Hill	General collection	Set 10	?1.7 - 1.9	249
329	Flag Hill	Scatter	Set 10	?1.7 - 1.9	41
331	Flag Hill	General collection	Set 10	?1.7 - 1.9	170
333	Flag Hill	General collection	Set 10	?1.7 - 1.9	111
335	Flag Hill	Find-spot	Set 10	?1.7 - 1.9	1
368	Flag Hill	General collection	Set 10	?1.7 - 1.9	73
370	Flag Hill	General collection	Set 10	?1.7 - 1.9	5
372	Flag Hill	Gully	Set 10	?1.7 - 1.9	20
337	Kotha Kas, West Branch	General collection	Set 11	1.9 - 2.2	16
339	Kotha Kas, West Branch	General collection	Set 11	1.9 - 2.2	181
341	Kotha Kas, West Branch	General collection	Set 11	1.9 - 2.2	270
515	**Kotha Kas, West Branch**	**General collection**	**Set 11**	**1.9 – 2.2***	**84**

517	Kotha Kas, West Branch	General collection	**Set 11**	1.9 - 2.2	12
519	**Kotha Kas, West Branch**	**Scatter**	**Set 11**	**1.9 - 2.2***	**22**
521	**Kotha Kas, West Branch**	**Find-spot**	**Set 11**	**1.9 - 2.2***	**4**
523	Kotha Kas, West Branch	General collection	**Set 11**	1.9 - 2.2	341
525	**Kotha Kas, West Branch**	**Scatter**	**Set 11**	**1.9 - 2.2***	**34**
527	**Kotha Kas, West Branch**	**Scatter**	**Set 11**	**1.9 - 2.2***	**43**
529	Kotha Kas, West Branch	Scatter	**Set 11**	1.9 - 2.2	21
531	Kotha Kas, West Branch	General collection	**Set 11**	1.9 - 2.2	21
535	Kotha Kas, West Branch	Find-spot	**Set 11**	1.9 - 2.2	1
399	**Kotha Kas, West**	**General collection**	**Set 12**	**1.9 - 2.2***	**80**
507	**Kotha Kas, West**	**General collection**	**Set 12**	**1.9 - 2.2***	**241**
477	**Kotha Kas, West**	**Scatter**	**Set 12**	**1.9 - 2.2***	**60**
501	**Kotha Kas, West**	General collection	**Set 12**	1.7 - 1.9	208
503	**Kotha Kas, West**	General collection	**Set 12**	1.7 - 1.9	131
505	**Kotha Kas, West**	Scatter	**Set 12**	1.7 - 1.9	50
509	**Kotha Kas, West**	General collection	**Set 12**	1.9 - 2.2	16
511	**Kotha Kas, West**	**General collection**	**Set 12**	**1.9 - 2.2***	**16**
401	Kotha Kas, West	General collection	Set 13	1.9 - 2.2	5
403	Kotha Kas, West	Scatter	Set 13	?1.7 - 1.9	4
405	Kotha Kas, West	General collection	Set 13	?1.7 - 1.9	14
479	**Kotha Kas, West**	**General collection**	**Set 13**	**1.7 - 1.9***	**26**
481	**Kotha Kas, West**	**Locality**	**Set 13**	**1.7 - 1.9***	**1287**
483	**Kotha Kas, West**	**General collection**	**Set 13**	**1.7 - 1.9***	**210**
485	**Kotha Kas, West**	**Scatter**	**Set 13**	**1.7 - 1.9***	**219**
487	**Kotha Kas, West**	**Locality**	**Set 13**	**1.7 - 1.9***	**345**
489	**Kotha Kas, West**	Locality	**Set 13**	1.7 - 1.9	1330
491	**Kotha Kas, West**	**Locality**	**Set 13**	**1.7 - 1.9***	**630**
493	**Kotha Kas, West**	**Scatter**	**Set 13**	**1.7 - 1.9***	**25**
495	Kotha Kas, West	General collection	**Set 13**	1.7 - 1.9	111
497	**Kotha Kas, West**	**General collection**	**Set 13**	**1.7 - 1.9***	**71**
397	**Kotha West**	**General collection**	**Set 14**	**1.9 - 2.2***	**168**
343	**Kotha Kas, West**	**Scatter**	Set 14	**1.9 - 2.2***	**130**
345	Kotha Kas, West	Scatter	Set 14	1.9 - 2.2	127
347	Kotha Kas, West	General collection	Set 14	1.9 - 2.2	13
349	**Kotha Kas, West**	**Find-spot**	Set 14	**1.9 - 2.2***	**4**
351	Kotha Kas, West	General collection	Set 14	1.9 - 2.2	62
353	**Kotha Kas, West**	**General collection**	Set 14	**1.9 - 2.2***	**25**
355	Kotha Kas, West	Scatter	Set 14	1.9 - 2.2	84
357	**Kotha Kas, West**	**General collection**	Set 14	**1.9 - 2.2***	**75**
383	Kotha Kas, West	General collection	Set 14	1.9 - 2.2	45
385	**Kotha Kas, West**	**Scatter**	Set 14	**1.9 - 2.2***	**289**
387	**Kotha Kas, West**	**Scatter**	Set 14	**1.9 - 2.2***	**45**
389	Kotha Kas, West	Scatter	Set 14	1.9 - 2.2	4
391	Kotha Kas, West	Scatter	Set 14	1.9 - 2.2	15
393	**Kotha Kas, West**	**General collection**	Set 14	**1.9 - 2.2***	**162**
359	Viaduct	General collection	Set 15	1.9 - 2.2	26
361	**Viaduct**	**Scatter**	**Set 15**	**1.9 - 2.2***	**175**
363	Viaduct	Scatter	Set 15	1.9 - 2.2	38
371	Viaduct	General collection	Set 15	1.9 - 2.2	93
373	Viaduct	Scatter	Set 15	1.9 - 2.2	100
375	Viaduct	Scatter	Set 15	1.9 - 2.2	421
377	Viaduct	Scatter	Set 15	1.9 - 2.2	81
379	Viaduct	Scatter	Set 15	1.9 - 2.2	46
381	Viaduct	General collection	Set 15	1.9 - 2.2	47

407	**Kotha Kas, East Branch**	**General collection**	**Set 16**	**1.7 - 1.9***	**130**
409	Kotha Kas, East Branch	Scatter	Set 16	1.7 - 1.9	62
411	Kotha Kas, East Branch	General collection	Set 16	1.7 - 1.9	13
413	Kotha Kas, East Branch	Scatter	Set 16	1.7 - 1.9	105
415	Kotha Kas, East Branch	Scatter	Set 16	1.7 - 1.9	61
417	Kotha Kas, East Branch	General collection	Set 16	1.7 - 1.9	24
419	Kotha Kas, East Branch	Scatter	Set 16	1.7 - 1.9	125
421	Kotha Kas, East Branch	General collection	Set 16	1.7 - 1.9	7
423	Kotha Kas, East Branch	Scatter	Set 16	1.7 - 1.9	39
425	Kotha Kas, East Branch	General collection	Set 16	1.7 - 1.9	13
499	Kotha Kas, East Branch	General collection	Set 16	1.7 - 1.9	90
427	**Kotha Kas, East Branch**	**Locality**	**Set 17**	**1.9 - 2.2***	**382**
429	Kotha Kas, East Branch	General collection	Set 17	1.9 - 2.2	55
431	Kotha Kas, East Branch	Scatter	Set 17	1.9 - 2.2	10
433	Kotha Kas, East Branch	Scatter	Set 17	1.9 - 2.2	26
435	Kotha Kas, East Branch	Locality	Set 17	1.9 - 2.2	695
437	**Kotha Kas, East Branch**	**Locality**	**Set 17**	**1.9 - 2.2***	**1346**
439	Kotha Kas, East Branch	General collection	Set 17	1.9 - 2.2	82
441	Kotha Kas, East Branch	Scatter	Set 17	1.9 - 2.2	70
443	Kotha Kas, East Branch	General collection	Set 17	1.9 - 2.2	126
445	**Kotha Kas, East Branch**	**Scatter/Locality?**	**Set 17**	**1.9 - 2.2***	**337**
447	**Kotha Kas, East Branch**	**Scatter**	**Set 17**	**1.9 - 2.2***	**48**
449	Kotha Kas, East Branch	Gully	Set 17	1.9 - 2.2	16
451	**Kotha Kas, East Branch**	**Scatter**	**Set 17**	**1.9 - 2.2***	**466**
461	Kotha Kas, East Branch	General collection	Set 17	1.9 - 2.2	17
398	427 area	Find-spot	Set 18	1.9 - 2.2	1
400	427 area	Scatter	Set 18	1.9 - 2.2	29
402	427 area	Scatter	Set 18	1.9 - 2.2	34
404	Kotha Kas, Headwaters	General collection	Set 18	?1.7 - 1.9	32
406	Kotha Kas, Headwaters	General collection	Set 18	?1.7 - 1.9	100
408	Kotha Kas, Headwaters	General collection	Set 18	?1.7 - 1.9	4
410	Kotha Kas, Headwaters	General collection	Set 18	?1.7 - 1.9	12
412	Kotha Kas, Headwaters	Scatter	Set 18	?1.7 - 1.9	124
414	Kotha Kas Headwaters	General collection	Set 18	?1.7 - 1.9	13
453	Kotha Kas, East Branch	General collection	Set 19	1.9 - 2.2	8
455	Kotha Kas, East Branch	General collection	Set 19	1.9 - 2.2	16
457	Kotha Kas, East Branch	Scatter	Set 19	1.9 - 2.2	23
459	Kotha Kas, East Branch	Scatter	Set 19	1.9 - 2.2	77
465	Kotha Kas, East Branch	General collection	Set 19	1.9 - 2.2	20
467	Kotha Kas, East Branch	General collection	Set 19	1.9 - 2.2	28
469	Kotha Kas, East Branch	General collection	Set 19	1.9 - 2.2	14
471	Kotha Kas, East Branch	Scatter	Set 19	1.9 - 2.2	75
473	Kotha Kas, East Branch	General collection	Set 19	1.9 - 2.2	17
475	Kotha Kas, East Branch	General collection	Set 19	1.9 - 2.2	54
463	Kotha Kas, East Branch	General collection	Set 19	>1.7	46
500	Choawala, Sandstones 14-16	General collection	Set 20	0.9 - 1.2	42
502	Choawala, Sandstones 14-16	Scatter	Set 20	0.9 - 1.2	31
504	Choawala, Sandstones 14-16	Scatter	Set 20	0.9 - 1.2	77
506	Choawala, Sandstones 14-16	Scatter	Set 20	0.9 - 1.2	26
508	Choawala, Sandstones 14-16	Scatter	Set 20	0.9 - 1.2	137
510	Choawala, Sandstones 14-16	General collection	Set 20	0.9 - 1.2	10
516	Choawala, Sandstones 14-16	General collection	Set 20	0.9 - 1.2	4
520	Choawala, Sandstones 14-16	General collection	Set 20	0.9 - 1.2	23
522	Choawala, Sandstones 14-16	Scatter	Set 20	0.9 - 1.2	286

524	Choawala, Sandstones 14-16	General collection	Set 20	0.9 - 1.2	24
526	Choawala, Sandstones 14-16	Scatter	Set 20	0.9 - 1.2	31
528	Choawala, Sandstones 14-16	General collection	Set 20	0.9 - 1.2	5
382	Ratial	General collection	Set 21	?1.7 - 1.9	41
384	Ratial	General collection	Set 21	?1.7 - 1.9	4
386	Ratial	General collection	Set 21	?1.7 - 1.9	12
420	Dhorian	Scatter	Set 21	>1.7	19
422	Dhorian	General collection	Set 21	>1.7	31
424	Dhorian	General collection	Set 21	>1.7	27
426	Dhorian	General collection	Set 21	>1.7	56
428	Dhorian	General collection	Set 21	>1.7	57
430	Dhorian	General collection	Set 21	>1.7	2
432	Dhorian	General collection	Set 21	>1.7	10
436	Badlands, N. of G.T. Road	General collection	Set 22	?1.7 - 1.9	5
438	Badlands, N. of G.T. Road	General collection	Set 22	?1.7 - 1.9	10
440	Badlands, N. of G.T. Road	General collection	Set 22	?1.7 - 1.9	19
442	Badlands, N. of G.T. Road	General collection	Set 22	?1.7 - 1.9	1
444	Badlands, N. of G.T. Road	Scatter	Set 22	?1.7 - 1.9	1
446	Badlands, N. of G.T. Road	Scatter	Set 22	?1.7 - 1.9	14
448	Badlands, N. of G.T. Road	General collection	Set 22	?1.7 - 1.9	15
450	Badlands, N. of G.T. Road	General collection	Set 22	?1.7 - 1.9	53
452	Badlands, N. of G.T. Road	Scatter	Set 22	?1.7 - 1.9	86
454	Badlands, N. of G.T. Road	Scatter	Set 22	?1.7 - 1.9	14
456	Badlands, N. of G.T. Road	Scatter	Set 22	?1.7 - 1.9	90
458	Badlands, N. of G.T. Road	Scatter	Set 22	?1.7 - 1.9	11
460	Badlands, N. of G.T. Road	Scatter	Set 22	?1.7 - 1.9	23
462	Badlands, N. of G.T. Road	General collection	Set 22	?1.7 - 1.9	9
464	Badlands, N. of G.T. Road	General collection	Set 22	?1.7 - 1.9	8
466	Badlands, N. of G.T. Road	Scatter	Set 22	?1.7 - 1.9	63
468	Badlands, N. of G.T. Road	General collection	Set 22	?1.7 - 1.9	14
470	Badlands, N. of G.T. Road	General collection	Set 22	?1.7 - 1.9	21
472	Badlands, N. of G.T. Road	Scatter	Set 22	?1.7 - 1.9	65
474	Badlands, N. of G.T. Road	Scatter	Set 22	?1.7 - 1.9	30
476	Badlands, N. of G.T. Road	General collection	Set 22	?1.7 - 1.9	14
478	Badlands, N. of G.T. Road	Scatter	Set 22	?1.7 - 1.9	18
480	Badlands, N. of G.T. Road	General collection	Set 22	?1.7 - 1.9	1
482	Badlands, N. of G.T. Road	Scatter	Set 22	?1.7 - 1.9	1
484	Badlands, N. of G.T. Road	Scatter	Set 22	?1.7 - 1.9	7
486	Badlands, N. of G.T. Road	General collection	Set 22	?1.7 - 1.9	15
488	Badlands, N. of G.T. Road	General collection	Set 22	?1.7 - 1.9	27
490	Badlands, N. of G.T. Road	General collection	Set 22	?1.7 - 1.9	10
492	Badlands, N. of G.T. Road	Scatter	Set 22	?1.7 - 1.9	35
494	Badlands, N. of G.T. Road	General collection	Set 22	?1.7 - 1.9	18
496	Badlands, N. of G.T. Road	Scatter	Set 22	?1.7 - 1.9	10
Total					**20053**

Those in **bold** and with * are those occurrences specifically tied in to the geological sequence by Rendell. Related ones from the same area are entered in normal fonts. Best estimates for others are given as e.g. "?1.7 - 1.9".

Notes: The following numbers can be ignored: 201 (1 piece from the Banni Bangla car-park); and 346 (Set 3), recent. Nos. 211, 228, 364, 380, 498, 512, 514, 530, and 534 had no finds. Nos. 374, 395, 513, and 533 were used to record stone. Nos. 233, 239, 274, 292, 294 and 310 were gully collections along Sohawa Kas (60 finds in all, but none significant, let alone dateable). 321, 323 and 325 (35 specimens in total) were used to record some inconsequential scatters near the watershed track in the area of 296, Sohawa Kas area, but cannot be dated or tied to other sets. No. 307 was a general collection in the Sohawa Kas area, but had only 20 indeterminates, as was 313, with only 4 indeterminates.

Table 6. 2: Total numbers of specimens per set and the average number per fossil occurrence from the 1987 survey.

Set	Area	Age (Mya)	Occurrences	Total	Average	Key numbers
Set 1	Choawala Kas	1.7 – 1.9	14	567	40.5	200-228; 214 area
Set 2	Choawala Kas, West	1.4 – 1.7	8	59	7.4	222-232; 332-336
Set 3	Choawala Kas, West	1.7 – 1.9	13	499	31.6	338-366; 362 area
Set 4a	Choawala East/Happy Valley	1.7 – 1.9	13	513	39.4	236-260
Set 4b	Choawala East/Death Valley	1.7 – 1.9	6	65	10.8	262 + misc.
Set 5	Bangial, Choawala	1.4 – 1.7	18	344	19.1	225-255
Set 6	Sohawa Kas	1.9 – 2.2	30	874	29.1	203 area, 269
Set 7	Sohawa, Headwaters and West	1.9 – 2.2	10	216	23.1	205-209
Set 8	Sohawa Kas, West	1.7 – 1.9	10	316	35.0	276-286, 290
Set 9	Sohawa Kas, East	1.9 – 2.2	29	791	29.1	273-295, 296-308
Set10	Flag Hill	1.7 – 1.9	8	670	92.9	327-335, 368-372
Set 11	Kotha Kas, West Branch	1.9 – 2.2	13	1050	80.8	337-341, 515-535
Set 12	Kotha Kas, West Branch	1.9 – 2.2	8	802	100.25	397,399,501-511
Set 13	Kotha Kas, West Branch	1.7 – 1.9	13	4277	329.0	481, 489, 491
Set 14	Kotha Kas, West Branch	1.9 – 2.2	15	1248	83.2	343-357, 383-395
Set 15	Viaduct area	1.9 – 2.2	9	1027	114.1	359-363, 369-381
Set 16	Kotha Kas, East Branch	1.9 – 2.2	11	669	60.8	407-425, 499
Set 17	Kotha Kas	1.9 – 2.2	14	3676	81.5	427-461
Set 18	Kotha Kas Headwaters	1.7 – 1.9	9	349	38.8	398-414
Set 19	Kotha Kas, East Branch	1.9 – 2.2	11	378	34.4	453-475
Set 20	Choawala, Sandstones 14-16	0.9 – 1.2	12	696	58.0	500-510,520-528
Set 21	Dhorian and Ratial Kas	>1.7	10	259	25.9	382-386, 420-432
Set 22	Badlands, N. of G.T. Road	1.4 – 1.7	31	708	22.8	436-528
Total			**315**	**20053**	**63.3**	

Notes: These averages are given solely to indicate the relative abundance of evidence from different parts of the Pabbi Hills – from a miserly average of only 7.4 for Set 2, to over 100 in Sets 12 and 15, and over 300 in Set 13. The averages given are not significantly affected by a small number of very poorly provenanced finds from gully beds: Set 5, 9 finds from 4 gullies; Set 7, 8 finds from 205; Set 9, 6 finds from 315 and 326; Set 10; 20 finds from 372; and Set 17, 16 from 449.

Table 6. 3. Fossil material from Set 1, 1987, headwaters of Choawala Kas (200-220, even nos. only; 365,367, 388).

Fossil occurrence 200: general collection:

Specimen	W	Taxon	Element	Part	Other	N
3,4	2b	Artiodactyla	Radioulna	Complete		1
1		Bovid	Radius	Distal	Complete	1
2	4	Bovid	Tibia	Proximal	Complete	1
12	2b	Bovid	Molar	Lower	Split	1
2		Bovid	Molar	Upper	Complete	1
5		Bovid	Tooth	Molar		1
6	3	Cervid	Antler			1
11		Chelonia	Scute	Smooth		2
7	3	Proboscidean	Shaft			1
8,9,13	5	Proboscidean	Tooth	Molar		3
		Indeterminate	Fragments			17
Total						**30**

Fossil occurrence 202: general collection:

Specimen	W	Taxon	Element	Part	Other	N
7	3	Bovid	Molar	Upper	Mod.wear	1
6	3	Bovid	Premolar	Split	Little wear	1
5	3	Bovid	Tooth	Molar	Mod. wear	1
3		Chelonia	Scute	Netted		1
		Indeterminate	Fragments			20
Total						**24**

Table 6.3 continued: Fossil occurrence 204: general collection:

Specimen	W	Taxon	Element	Part	Other	N
8		Bovid (?*Hemibos*)	Mandible	P4,M1-3	Right	1
		Chelonia	Scute	Netted		1
		Indeterminate	Shaft	Fragments		8
Total						**10**

Fossil occurrence 206: scatter:

Specimen	W	Taxon	Element	Part	Other	N
11		Artiodactyla	1st Phalange	Distal	Split	1
10		Artiodactyla	Carpal			1
7		Artiodactyla	Cuneiform			1
9		Artiodactyla	Metapodial	Distal	Split	1
3	4	Bovid	Horncore			1
12	3	Bovid	M1/2	Complete	Little wear	1
2		Bovid	Tooth	Premolar		1
4		Bovid	Tooth	Premolar	Upper, split	1
5		Bovid	Tooth	Premolar	Upper,split	1
1		Bovid	Tooth	Fragment		1
		Chelonia	Scute	Smooth		1
		Indeterminate	Fragments			42
Total						**53**

Fossil occurrence 208: general collection:

Specimen	W	Taxon	Element	Part	Other	N
16	5	Artiodactyla	Tooth	Fragment		1
5	2b	Artiodactyla	Vertebra	Lumbar	Unfused	1
13	2a	Artiodactyla	Vertebra	Caudal		1
11	4	Bovid	Horncore			1
3	2b	Bovid	Premolar	Split	Mod. wear	1
4	2b	Bovid	Premolar	Upper	Split, mod. wear	1
12	2b	Bovid	Molar	Upper	Split, mod. wear	1
1		Cervid	Antler			1
6	3	*Elephas*	Rib	37x25		1
10	5	*Elephas*	Tooth	Tusk		1
9	2b	Equid	Scaphoid	Complete		1
0	5	Proboscidean	Tooth	Fragment		1
0		Indeterminate	Fragments			60
Total						**72**

Fossil occurrence 210: scatter:

Specimen	W	Taxon	Element	Part	Other	N
2		Artiodactyla	Carpal	Complete		1
1	3	Suid?	Humerus	Shaft		1
		Indeterminate	Fragments			19
Total						**21**

Fossil occurrence 212:scatter:

Specimen	W	Taxon	Element	Part	Other	N
4	3	Artiodactyla	Shaft			1
8		Bovid	Mandible			1
17		Bovid	Tooth	Premolar	Split	1
14	3	Bovid	$M^{1/2,}$ right	Complete	Little wear	1

Table 6.3 continued: Fossil occurrence 212

13	4	Bovid	Tooth	Molar	Split	1
11	5	Bovid	Tooth	Molar	Split	1
1-3, 5-9	3	Proboscidean	Rib			5
1	2	Proboscidean	Tooth	Tusk		1
7		Proboscidean	Tooth	Molar		1
10	3	Equid	Molar, upper	Complete	Mod. wear	1
		Indeterminate	Fragments			21
Total						**35**

Fossil occurrence 214: locality:

Zone	**Specimen**	**W**	**Taxon**	**Element**	**Part**	**Other**	**N**
214	13	3	Carnivora	Scapula	Fragment		1
214	30		Chelonia	Scute	Pitted		1
214	19	3	Hyaenid	Tooth	P4	Lower, split	1
214A	1		Hyaenid	Maxillary	Tooth	Carnassial	1
214B	3	2b	Artiodactyla	Metatarsal			1
214B	6		Artiodactyla	Metatarsal	Shaft		1
214B	18	3	Artiodactyla	Tibia	Shaft		1
214B	14	4	Bovid	Horncore			1
214B	94		Bovid	Horncore			1
214B		12	Bovid	Nav.-cuboid			1
214B	1	2b	Bovid (Large)	Humerus	Shaft		1
214B	15	2b	Bovid (Large)	M1/2 lower	Complete	Mod. wear	1
214B	8		Cervid	Mandible	DM3	Fresh	1
214B			Indeterminate	Fragments			57
214B	20	4	Proboscidean	Tooth	Tusk		1
214R	24	5	Artiodactyla	Tooth	Fragment		1
214R	53,63		Artiodactyla	Tooth	Fragment		2
214R	**2**	2b	Bovid	Metacarpal	Complete	Proximal	1
214R	0		Bovid (Large)	Tooth	Fragments		2
214R	13		Bovid (Large)	Tooth	PM^3	Complete	1
214R	2	3	Cervid	Antler			1
214R	0		Chelonia	Scute			1
214R	19		Hyaenid	Tooth	Carnassial	Lower	1
214R	0		Pisces?	Vertebra			1
214R		4	Proboscidean	Tooth	Tusk		1
214R			Indeterminate	Fragments			151
Total							**234**

Fossil occurrence 216: general collection and find-spot 220:

Specimen	**W**	**Taxon**	**Element**	**Part**	**Other**	**N**
220GB1		Bovid (cf *Hemibos*)	Cranium	+ horncore		1
8		*Crocodylus*	Tooth	Complete		1
0, 1, 2	3	Proboscidean	Rib			2
4, 7	4	Proboscidean	Tooth	Molar	Plate	2
		Indeterminate	Fragments			4
Total						**10**

Fossil occurrence 218: scatter:

Specimen	**W**	**Taxon**	**Element**	**Part**	**Other**	**N**
1-3, 5	2a	Cervid	Antler	Palmate		4
		Proboscidean	Tooth	Fragments		9
		Indeterminate	Shaft			12
Total						**25**

Table 6.3 continued: Fossil occurrence 365: scatter:

Specimen	W	Taxon	Element	Part	Other	N
7		Artiodactyla	2nd Phalange			1
4	4	Bovid	2nd Phalange	Complete		1
6	5	Bovid	Horncore			1
		Indeterminate	Fragments			4
Total						**7**

Fossil occurrence 367: scatter:

Specimen	W	Taxon	Element	Part	Other	N
4	4	Bovid	Tooth	Premolar	No crowns	1
1	2b	Cervid	Antler			1
3	5	Proboscidean	Tooth	Molar	Plate	1
		Indeterminate	Fragments			2
Total						**5**

Fossil occurrence 388: general collection:

Specimen	W	Taxon	Element	Part	Other	N
10		Artiodactyla	2nd	Phalange		1
15		Artiodactyla	Metapodial			1
1		Bovid	Horncore			1
11	5	Bovid	Metacarpal	Complete	Distal	1
10		Bovid	Scaphoid	Complete		1
8		Bovid	Tooth	M3, lower	Complete	1
12	4	Bovid	Molar, upper	Complete	Mod. wear	1
9		Bovid	Tooth	Molar, lower	Complete	1
10		Bovid	Trapezoid/Magnum			1
9		Carnivora	Radius	Distal		1
3		Equid	Tooth	Premolar		1
6		Proboscidean	Tooth	Crown		1
13	3	Proboscidean	Tooth	Tusk		1
7		Suid	Tooth	Premolar		1
		Indeterminate	Fragments			26
Total						**40**

Notes: General collections 202 and 204 are best subsumed within 200; 220GB1 is subsumed with general collection 216.
Matrix: 204GB8 had an orange sand matrix; 214GB3 and 214GB6 a clay skin and were probably part of the same bone but did not join. 367GB1 and 388GB13 had a sandy matrix. One of the indeterminate fragments from 210 was rolled.

Table 6 .4: Fossil material from Set 2, Choawala West (222-236, even nos. only, 332,334 and 336).

Fossil occurrence 224: scatter:

Specimen	W	Taxon	Element	Part	Other	N
224A GB3	1	Proboscidean	Shaft	Suncracked		1
224B GB5	1	Artiodactyla	Humerus	Shaft		1
224B GB5	5	Artiodactyla	Metacarpal	Distal		1
224B GB2	5	Artiodactyla	Metatarsal	Proximal		1
224B GB4	4	Bovid	Tooth	Premolar	Mod.wear	1
224B GB4	4	Bovid	Tooth	Premolar	Mod. wear	1
224B GB5	2	Proboscidean	Shaft			1
224B GB4		Indeterminate	Fragments			19
Total						**26**

Fossil occurrence 226: general collection:

Specimen	W	Taxon	Element	Part	Other	N
5	6	Proboscidean	Tooth	Tusk		1
3,4	3	Suid	Tooth	Incisor		2
	0	Indeterminate	Fragments			3
Total						**6**

Table 6.4 continued: Fossil occurrence 230: general collection:

Specimen	W	Taxon	Element	Part	Other	N
5	1	Bovid	Premolar	Split	Little wear	1
5	0	Indeterminate	Fragments			4
Total						**5**

Fossil occurrence 232: general collection:

Specimen	W	Taxon	Element	Part	Other	N
3	1	Bovid	Vertebra		Cervical	1
	0	Chelonia	Scute		Pitted	1
5	0	Indeterminate	Fragments			10
Total						**12**

Fossil occurrence 336: general collection (and subsuming 332 and 334):

Specimen	W	Taxon	Element	Part	Other	N
332GB1	1	Artiodactyla	Metacarpal	Distal		1
334GB1	2b	Proboscidean	Tooth	Tusk		1
2	5	Carnivora	Humerus	Distal		1
3	5	Equid	Tooth	Premolar	Split,worn	1
	5	Indeterminate	Rib		35x14	1
Total						**5**

Notes: general collection 226 includes 5 pieces of indeterminate shaft from scatter 222. No fossil was found in general collection 228. General collection 336 includes two specimens from 332 and 334. 224GB25 had a clay skin; 334GB1 had a grey sand matrix. Gully collection 234 comprised 2 Artiodactyl tooth fragments, a rhino distal metapodial fragment, an equid distal metacarpal fragment and 1 indeterminate fragment.

Table 6. 5: Fossil material from Set 3, Choawala West (338-366).

Fossil occurrence 338: general collection:

Specimen	W	Taxon	Element	Part	Other	N
2	3	Bovid	Tooth	M1/2	Upper	1
6,7	3	Bovid	Tooth	Molar	Split,worn	2
8		Bovid	Tooth	Premolar	Split	1
1	2b	Cervid	Antler			1
5	5	Proboscidean	Tooth	Tusk		1
0		Indeterminate	Fragments			19
Total						**25**

Fossil occurrence 344: scatter:

Specimen	W	Taxon	Element	Part	Other	N
0, 6, 7		Cervid	Antler			5
1	5	Proboscidean	Tooth	Tusk		6
3		Proboscidean	Tooth	Molar	Plate	9
0		Indeterminate	Fragments			10
Total						**30**

Fossil occurrence 348: scatter:

Specimen	W	Taxon	Element	Part	Other	N
3, 12		Bovid	Horncore			2
10	2b	Bovid	Radius	Distal		1
14		Bovid	Scapula			1
5		Bovid	Tooth	Molar		1
6	4	Bovid	Tooth	Molar	Split	1
17		Chelonia	Scute	Netted		1
11		Proboscidean	Shaft			1
0, 1,4		Proboscidean	Tooth	Tusk		10
8		*Rhinoceros*	Tibia	Distal		1
0		Indeterminate	Fragments			79
Total						**99**

Table 6.5 continued: Fossil occurrence 356: general collection:

Specimen	W	Taxon	Element	Part	Other	N
10	5	Artiodactyla	Humerus	Distal		1
2	2b	Bovid	Metatarsal	Proximal		1
1		Bovid	Navicular	Complete		1
9		Proboscidean	Tooth	Molar		1
0		Indeterminate	Fragments			24
Total						**28**

Fossil occurrence 358: general collection:

Specimen	W	Taxon	Element	Part	Other	N
13		Artiodactyla	Femur	Condyle	Proximal	1
10	4	Bovid	Tooth	M1/2	Worn	1
11	2b	Bovid	Tooth	M1/2 upper	Complete	1
4		Chelonia	Scute	Netted		1
1,2,6,8,9		Proboscidean	Tooth	Molar	Plate	5
0		Indeterminate	Fragments			16
Total						**25**

Notes: Scatter 340 contained only 7 pieces of Proboscidean shaft fragments and 10 Proboscidean molar plate fragments. General collection 342 contained 1 bovid split upper molar and 2 indeterminate fragments. Scatter 346 contained 6 miscellaneous ungulate fossil fragments, but also pottery and can be disregarded. General collection 350 produced only 1 piece, an artiodactyl astragalus fragment. General collection 352 contained a bovid split premolar, a *Sivatherium* proximal 1st phalange and 3 indeterminate fragments. General collection 354 produced 2 bovid horncore fragments (1 attached to the cranium) and 12 indeterminate fragments. General collection 360 produced only 1 piece of Proboscidean tusk and 2 indeterminate fragments. There was no material from 364. General collection 366 contained a bovid lower premolar, split, a Proboscidean molar fragment and 30 indeterminates.

Table 6. 6. Fossil material from locality 362, Set 3, Choawala West.

Zone	Specimen	W	Taxon	Element	Part	Other	N
	100		*Sivatherium*	Metacarpal			1
	2		Indeterminate	Femur	Distal	Diaphysis	1
	0		Indeterminate	Shaft	Fragments		63
A	24		Artiodactyla	Metapodial	Shaft		2
A	2		Artiodactyla	Ulna			1
A	19		Bovid	3rd Phalange			1
A	**4**		Bovid	Astragulus	Split		1
A	3		Bovid	Horncore			1
A	**15**		Bovid	Humerus	Proximal	Complete	1
A	**8**, 9		Bovid	Metacarpal	Proximal		2
A		1	Bovid	Metatarsal	Proximal		1
A	7		Bovid	Radius	Distal		1
A	6		Rhinocerotidae	Astragulus	Complete		1
A		1	Rhinocerotidae	Unciform			1
A			Indeterminate	Fragments			12
B	0		Bovid	Humerus	Distal		1
B	**1**		*Sivatherium*	Metacarpal	Distal		1
B	2		*Sivatherium*	Metacarpal			1
B	13		*Sivatherium*	Metacarpal	Shaft		1
B	18		*Sivatherium*	Metacarpal	Shaft		1
B	17		*Sivatherium*	Unciform	Complete		1
B		1	Bovid	Crania +	Horncores		1
B			Indeterminate	Shaft	Fragments		11
C	7		Artiodactyla	Mandible			1
C	0		Bovid	Horncore			1
C	2		Bovid	Humerus			1
C	6		Bovid	Humerus	Distal	Unfused	1
C	3		Bovid	Pelvis			1
C	**8**		Bovid	Radius	Distal		1
C	11		Proboscidean	Humerus	Distal		1
C	0		*Sivatherium*	Metapodial	Split		1
C	**0**		*Sivatherium*	Radius	Distal		1
C	10		Indeterminate	Rib			1

Table 6.6 continued: Fossil material from locality 362, Set 3

D	11		Artiodactyla	Calcaneum	Shaft		1
D	7		Artiodactyla	Femur	Shaft		1
D	16		Artiodactyla	Mandible			1
D		1	Artiodactyla	Metapodial	Shaft		1
D	**9, 12**		Bovid	Astragulus	Complete		1
D	3		Bovid	Horncore			1
D	10		Bovid	Humerus			1
D	**14**	4	Bovid	Humerus	Distal		1
D	4		Bovid	Metacarpal	Proximal		1
D	**2**		Bovid (Large)	Radius	Distal		1
D	10, 15		Bovid (Large)	Scapula	Glenoid		2
D			Indeterminate	Fragments			4
E	10		Bovid	Scapula	Proximal		1
E	**2**		Bovid (Medium)	Mandible	DM3,M1	Erupting	1
F	48		Artio (Medium)	Vertebra	Caudal		1
F	12		Artiodactyla	Astragulus			1
F	6		Artiodactyla	Calcaneum	Proximal		1
F	16		Artiodactyla	Humerus			1
F	17,19,23		Artiodactyla	Metacarpal			3
F	15,41,43		Artiodactyla	Metapodial			3
F	22		Artiodactyla	Metatarsal			1
F	3		Artiodactyla	Radius	Distal		1
F	45		Artiodactyla	Tooth	Plate		1
F	5		Bovid	Horncore			1
F	**51**		Bovid	Mandible + Tooth	dm3		1
F	40		Bovid	Metatarsal	Proximal		1
F	20		Bovid	Radius	Distal		1
F	7		Bovid (Large)	Horncore			1
F	10		Bovid (Large)	Radius	Distal		1
F	13		Cervid	Antler			1
F	31,56,70		Rhino	Tooth	Fragment		3
F	2		Rhinocerotidae	Astragulus			1
F	1,8	3	Rhinocerotidae	Radius	Distal		2
F	11		Rhinocerotidae	Vertebra			1
F			Indeterminate	Fragments			10
G	0,4		Artiodactyla	Cranium			2
G	0		Artiodactyla	Humerus			1
G	19		Artiodactyla	Mandible	Fragment		1
G	0		Artiodactyla	Metapodial			1
G	0		Artiodactyla	Rib			2
G	0		Artiodactyla	Shaft			4
G	18		Artiodactyla	Tibia			1
G	0		Bovid	1st Phalange			1
G	2		Bovid	Astragulus	Complete		1
G	15		Bovid	Astragulus			1
G	0		Bovid	Horncore			3
G	5,8		Bovid	Humerus	Shaft		2
G	0		Bovid	Radius			1
G	0,7,10		Bovid	Tibia			3
G	**9**		Bovid (Large)	Metapodial	Distal	Complete	1
G	6		Bovid (Medium)	Humerus	Distal		1
G	0		Proboscidean	Tooth	Molar		1
G	3		Rhinocerotidae	Astragulus	Complete		1
G	0		Rhinocerotidae	Phalange			2
G	0		Rhinocerotidae	Tibia			1
G			Indeterminate	Fragments			7
H	2,3		Cervid	Antler			2
H	0		Proboscidean	Tooth	Tusk		1
H		1	Indeterminate	Shaft			2
H		1	Carnivora	Radius			1
Total							**217**

Table 6. 7: Fossil material from Set 4, Choawala East (Happy Valley 236-260 even nos. only; Death Valley 262, 264, 416, 418, 434 and 518).

Fossil occurrence 236: general collection:

Specimen	W	Taxon	Element	Part	Other	N
6	3	Artiodactyla	Humerus	Shaft		1
4,8	3	Artiodactyla	Tooth	Molar	Split, worn	2
7		Artiodactyla	Tooth	M1/2	Little wear	1
0		Chelonia	Scute	Smooth		1
7		*Crocodylus*	Tooth			1
2	3	Proboscidean	Rib	51x32		1
0,1,9	5	Proboscidean	Tooth	Molar	Plate	13
3	4	Proboscidean?	Shaft/pelvis			1
	5	Indeterminate	Shaft			8
Total						**29**

Fossil occurrence 238: scatter:

Specimen	W	Taxon	Element	Part	Other	N
6	3	Bovid (Large)	Tooth	M1/2	Upper	1
7		Bovid/Giraffid	Tooth	M1/2	Erupting.	1
8		Bovid	Tooth	Premolar		1
238A GB		Bovid	Horncore			8
238A GB1	4	Bovid	Metacarpal	Distal	Barrel	1
238A GB2	3	Bovid	Metacarpal	Shaft		1
238A GB3	5	Proboscidean	Tooth	Molar		8
238A		Indeterminate	Shaft			5
238B		Indeterminate	Fragments			8
238C GB3		Artiodactyla	Shaft			1
238C GB6		Chelonia	Scute	Dimpled		1
238C GB0		*Elephas*	Tooth	Molar	Fragments	18
238C GB4		Proboscidean	Tooth	Molar	Worn	1
238C		Indeterminate	Shaft			5
Total						**60**

Fossil occurrence 240: scatter:

Specimen	W	Taxon	Element	Part	Other	N
3	3	Artiodactyla	Pelvis			1
8		Bovid	Cuneiform			1
0,2	3	Bovid	Tooth	Molar	Split,worn	2
6		Bovid	Unciform			1
4,9		Cervid	Antler	Palmate		2
7	3	Cervid	Antler	Pedicel		1
1	3	Cervid	Tibia	Proximal		1
0		Chelonia	Scute	Smooth		1
5		Indeterminate	Fragments			13
Total						**23**

Fossil occurrence 242: scatter:

Specimen	W	Taxon	Element	Part	Other	N
2		Artiodactyla	Tooth			1
8	4	Bovid (Large)	M3	Upper	Erupting	1
3		Cervid	Antler	Pedicel	Branched	1
1,7		Proboscidean	Shaft			2
4		Indeterminate	Shaft			3
Total						**8**

Fossil occurrence 244: scatter:

Specimen	W	Taxon	Element	Part	Other	N
8	2b	Artiodactyla	Metacarpal	Proximal		1
4		*Crocodylus*	Ischium			1

Table 6.7 continued: Fossil occurrence 244: scatter:

0,5,10		*Crocodylus*	Pelvis			4
2		*Crocodylus*	Sacrum			1
1,3,9		Hippopotamid	Mandible	Roots only	No teeth	3
7		Hippopotamid	Tooth	Root		1
0		Indeterminate	Shaft			14
Total						**25**

Fossil occurrence 248: scatter:

Specimen	W	Taxon	Element	Part	Other	N
10	3	Artiodactyla	Calcaneum	Fragment		1
14	3	Artiodactyla	Humerus	Proximal		1
12	3	Artiodactyla	Vertebra	Centrum		1
2	4	Bovid	Tooth	Molar	Lower	1
3	3	Bovid	Tooth	Molar	Large,worn	1
22	3	Bovid	Tooth	Premolar	Lower	1
15	3	Cervid	Antler	Pedicel		1
0		Chelonia	Scute	Dimpled		14
0		Chelonia	Scute	Smooth		6
4		*Crocodylus*	Tooth			1
16,23,24		Proboscidean	Shaft			3
0		Proboscidean	Tooth	Tusk	Unmarked	1
8	3	*Hippotamus* sp.	Tooth	Tusk		1
0		Indeterminate	Fragments			75
Total						**108**

Fossil occurrence 250: scatter:

Specimen	W	Taxon	Element	Part	Other	N
11	3	Bovid	1st Phalange	Distal		1
8		Bovid	Astragulus	Complete		1
7		Bovid	Cuneiform	Fragment		1
0		Bovid	Horncore			2
1		Bovid	Magnum	Proxima		1
3	2b	Bovid	Radius	Distal	Complete	1
2	3	Bovid	Tooth	P2		1
4		Bovid	Tooth	Spall		1
8	3	Bovid	Tooth	Premolar	Split	1
9		Chelonia	Scute	Smooth		3
10		Chelonia	Scute	Dimpled		1
0		Proboscidean	Tooth	Tusk		13
0		Indeterminate	Shaft			42
0		Indeterminate	Shaft	?Proboscidean		12
Total						**81**

Fossil occurrence 252: scatter:

Specimen	W	Taxon	Element	Part	Other	N
252A GB11		Bovid	3rd Phalange			1
252A GB2	3	Bovid	Tibia	Proximal		1
252A GB8	3	Bovid	Cranium			1
252A GB6	3	Bovid	Horncore			1
252A GB13	3	Indeterminate	Fragments			21
252A GB12		Proboscidean	Indet	Flake		1
252B GB 6	5	Artiodactyla	Tooth	Fragment		1
252B GB 4	3	Bovid	Premolar	Upper	Mod.wear	1
252B GB 11		Bovid	Tooth	Premolar		1
252B GB 19	3	Bovid	Tooth	Molar	Split	1
252B GB 5	3	Cervid	3rd Phalange	Broken		1
252B GB1,2,7,9,34		Cervid	Antler			5
		Indeterminate	Fragments			29
252B GB3		Proboscidean?	Condyle	Fragment		1
Total						**66**

Table 6. 7 continued: Fossil occurrence 254: scatter:

Specimen	W	Taxon	Element	Part	Other	N
7	3	Artiodactyla (Small)	Vertebra	Centrum		1
8	5	Bovid	Horncore			1
4	5	Bovid	Molar	Upper	Split,worn	1
6	5	Bovid	Tooth	Molar, lower	Split	1
9	5	Bovid	Tooth	Premolar	Split	1
15	5	Bovid	Tooth	Molar	Large	1
1		Bovid	Ulna	Proximal		1
3	3	Bovid (Large)	Metacarpal	Proximal	Complete	1
5		Chelonia	Scute	Smooth		7
12		Chelonia	Scute	Dimpled		5
13		Indeterminate	Condyle			1
0	4	Indeterminate	Shaft			40
Total						**61**

Fossil occurrence 256: scatter:

Specimen	W	Taxon	Element	Part	Other	N
2	4	Bovid	Metatarsal	Distal	Barrel	1
1	5	Proboscidean	Tooth	Root		1
4	4	Proboscidean	Tooth	Tusk	Fragment	1
0		Indeterminate	Fragments			2
Total						**5**

Fossil occurrence 260: scatter:

Specimen	W	Taxon	Element	Part	Other	N
3	5	Artiodactyla	Calcaneum			1
1	4	Bovid	M1/2	Complete	Upper	1
1		*Crocodylus*	Vertebra			1
2	4	Equid	P3-4/M1-2	Lower	Split,worn	1
0		Indeterminate	Shaft			27
Total						**31**

Fossil occurrence 262: general collection:

Specimen	W	Taxon	Element	Part	Other	N
8	4	Artiodactyla	Horncore			1
3	3	Bovid	3rd Phalange	Fragment		1
2,5	4	Bovid	Horncore			2
0		Bovid	Tooth	Premolar		1
1		Cervid	Antler			1
3		Chelonia	Scute	Smooth		1
		Indeterminate	Fragments			11
Total						**18**

Fossil occurrence 518: general collection:

Specimen	W	Taxon	Element	Part	Other	N
4	2b	Artiodactyla	Femur	Proximal		1
0		Chelonia	Scute	Pitted		2
1		Proboscidean	Tooth	Tusk	Rolled	1
0		Indeterminate	Fragments			31
Total						**35**

Notes: Cervid antler 240GB4 and 240GB7 join. The following had a sandy matrix: 236GB2 (Elephant rib); 236GB3 (Proboscidean shaft/pelvis); 240GB1 (Cervid tibia); 434GB1 (Equid tooth); 434GB0 (Indeterminate femur); 518GB4 (Artiodactyla femur). 252GB10 (Indeterminate femur) had a silt matrix.

Other finds: General collection 246 – 5 pieces of bovid horncore, 1 Proboscidea molar plate fragment and 7 indeterminate fragments. General collection 258 had only 1 piece of cervid antler (palmate) and 2 indeterminate shaft fragments. General collection 264 contained 1 piece of bovid horncore, 1 Giraffid premolar (probably *Sivatherium)* and an indeterminate shaft fragment. Scatter

416 comprised an artiodactyls ramus fragment and 2 indeterminate fragments. General collection 418 contained only three indeterminate fragments. Scatter 434 consisted of an equid upper molar/premolar and 2 indeterminate shaft fragments.

Table 6.8: Fossil material from Set 5 (Choawala and Bangial Kas): 225-255 (odd nos.), 270-272, 314, 532.

Fossil occurrence 225: general collection:

Specimen	W	Taxon	Element	Part	Other	N
6		Chelonia	Scute	Smooth		1
1,5	5	*Elephas*	Tooth	Molar	Plate	2
2,6,10	5	*Elephas*	Tooth	Tusk		3
8	4	Equid	Tooth	Premolar	Split	1
7	3	Equid	Tibia	Distal	Complete	1
0		Indeterminate	Shaft	Fragments		8
Total						**16**

Fossil occurrence 227: general collection:

Specimen	W	Taxon	Element	Part	Other	N
5	5	*Elephas*	Mandible	Fragment		1
1,2,11		*Elephas*	Tooth	Molar	Crown	3
4	5	*Elephas*	Tooth	Molar	Root	1
8-10	5	*Elephas*	Tooth	Molar	Plate	3
6		Bovid	Mandible+tooth	Premolar		1
7		Bovid	Tooth	Molar		1
0		Indeterminate	Shaft			2
Total						**12**

Fossil occurrence 231: general collection:

Specimen	W	Taxon	Element	Part	Other	N
6		Chelonia	Scute	Smooth		2
1		Equid	P2-M2, lower	Complete	Worn	1
2	5	Equid	Premolar	Split,	No crowns	1
0	5	Ungulata	Tooth	Fragments		8
0		Indeterminate	Shaft			2
Total						**14**

Fossil occurrence 235: scatter:

Specimen	W	Taxon	Element	Part	Other	N
2	5	Artiodactyla	Scapula	Glenoid		1
10		Artiodactyla	Ulna			1
11,26	5	Bovid	Horncore			2
0,18	2a	Cervid	Antler			2
3	5	Cervid	Astragulus	Fragment		1
13	3	Cervid	Astragulus	Rolled		1
0		Chelonia	Scute	Smooth		20
21	3	Equid	Tooth	Incisor	Worn	1
25	5	Equid	Tooth	Premolar	Split, worn	1
1		Suid (small)	Femur	Distal	Condyle	1
0		Indeterminate	Fragments			16
Total						**47**

Fossil occurrence 241: scatter:

Specimen	W	Taxon	Element	Part	Other	N
7		Artiodactyla	Femur	Proximal		1
6	5	Bovid	Tooth	Molar	Split	1
2		Chelonia	Scute	Pitted		1
1		Hippopotamus	Tooth	Incisor		1
0		Indeterminate	Fragments			10
Total						**14**

Table 6.8 continued: Fossil occurrence 243: general collection:

Specimen	W	Taxon	Element	Part	Other	N
14	3	Artiodactyla	Tooth	Premolar	Split	1
11		Artiodactyla	Unciform			1
5	2a	Bovid (cf. *Hemibos*)	Frontal+Horns			1
6	3	Bovid	Horncore			1
1	3	Bovid	Metapodial	Distal	Barrel	1
2	3	Bovid	Nav.-cuboid			1
17		Chelonia	Scute	Smooth		1
3		Elephas	Scapula			1
16		*Elephas*	Tooth			1
19		*Elephas*	Tooth	Molar	Root	1
9	4	Equid	M2	Complete	Upper	1
12	2a	Equid	M3 upper	Complete	Worn	1
21	4	Equid	PM2/M3	Lower	Worn	1
10	3	Equid	Premolar	Lower	Worn	1
0	3	Indeterminate	Fragments			42
Total						**56**

Fossil occurrence 245: general collection:

Specimen	W	Taxon	Element	Part	Other	N
245-2		Artiodactyla	Metacarpal	Distal	Barrel	1
1		Bovid	Astragulus	Fragment		1
6	2a	Bovid	Calcaneum	Proximal		1
12	4	Bovid	Horncore			1
18		Bovid	Mid-cuneiform			1
10		Bovid	Tooth	Premolar		1
5	3	Bovid	Radius	Distal		1
22	4	Cervid	Antler			1
3		Chelonia	Scute	Smooth		5
9		Chelonia	Scute	Netted		2
14	5	*Elephas*	Tooth	Molar	Plate	1
13	5	Equid	PM2 upper	Right	Worn	1
11		Suid	Metapodial	Lateral		1
0		Indeterminate	Shaft			13
Total						**31**

Fossil occurrence 247: scatter:

Specimen	W	Taxon	Element	Part	Other	N
247-3	2a	Artiodactyla	Humerus	Distal		1
4	2a	Artiodactyla	Humerus	Shaft		1
15	2b	Artiodactyla	Metapodial	Shaft	Proximal	1
24		Artiodactyla	Metapodial	Shaft		1
19	4	Bovid	Calcaneum	Shaft		1
13	5	Bovid	Cranium			1
10,12,22	5	Bovid	Horncore			3
1	2b	Bovid	Humerus	Distal		1
14	3	Bovid	Mandible	M3 split	Little wear	1
8		Bovid	Metacarpal	Proximal		1
15	2a	Bovid	Metatarsal	Shaft	Proximal	1
20	2a	Cervid	Antler	Palmate		1
6	2a	Cervid	Mandible	Molar	Erupting	1
16	2a	Cervid	Metatarsal			1
11	3	Cervid/Bovid	Tooth	Molar	V.worn	1
17	5	*Elephas*	Mandible	No teeth		1
5	2a	Equid	Radius	Distal		1
	5	Indeterminate	Fragments			30
Total						**49**

Table 6. 8 continued: Fossil occurrence 249: general collection:

Specimen	W	Taxon	Element	Part	Other	N
249-4,7,9	4	Bovid	Horn-core			3
6		Chelonia	Scute	Smooth		3
5	4	*Elephas*	Shaft			1
0,1,2, 6		*Elephas*	Tooth	Molar Plate		7
3	4	Equid	Tooth	Molar	Worn	1
0		Proboscidean	Tooth	Fragments		3
0		Indeterminate	Fragments			25
Total						**43**

Fossil occurrence 272: scatter:

Specimen	W	Taxon	Element	Part	Other	N
272-5,6		Chelonia	Scute	Netted		2
1	5	*Elephas*	Tooth			1
8,9	4	*Elephas*	Tooth			2
2-4	3	Indet/?Proboscidean	Shaft			3
7	3	Indet/ProboscidShaft				1
0		Indeterminate	Fragments			30
Total						**39**

Other finds: gully collections: **229**: one piece of Proboscidean molar plate; **234**, an artiodactyl upper molar fragment, 1 equid distal metatarsal and 1 rhino distal metapodial fragment, and one other fragment. **268**, 1 Chelonia scute, 1 Proboscidean rib fragment, and 1 other fragment; **532**, a large artiodactyl proximal unfused humerus fragment. General collection **251** yielded only 1 bovid lower molar fragment and 1 piece of Chelonia scute; general collection **253**, 1 piece of cervid antler and 6 fragments; general collection **255**, 4 indeteminate fragments. **Notes**: Bovid specimens 227GB6 and 227GB7 join, as do 235GB13 and 235GB18 (Cervid antler) and 249GB4 and 249GB7 (Bovid horn core). **Matrix**: 225GB8 (Equid tooth) and 231GB3 (Ungulate tooth) had a clay skin; 225GB7 (Equid tibia) had a sand matrix, as did 243GB2 (Bovid navicular-cuboid). All specimens from 247 were encrusted or encased in red clay, as was 249GB2 (Elephant tooth) and 272GB8 (Elephant tooth).

Table 6. 9. Fossil material from Set 6, the 203 horizon, Sohawa Kas (1.9 - 2.2 Mya-old).

Fossil occurrence 203: general collection:

Specimen	W	Taxon	Element	Part	Other	No
51	4	Artiodactyla	Mandible	+Tooth root		1
9	3	Artiodactyla	Metatarsal	Shaft		1
42	3	Artiodactyla	Pelvis			1
33	2b	Artiodactyla	Phalange	Distal		1
38		Artiodactyla	1st Phalange	Distal		1
50		Artiodactyla	Radius	Shaft		1
49		Artiodactyla	Tibia			1
47		Artiodactyla	Tooth			1
2		Bovid	Calcaneum	Shaft		1
5	3	Bovid	Horn-core			1
3	3	Bovid	Magnum			1
34, 37	3	Bovid	Nav-Cuboid			2
4	3	Bovid	Tooth	Premolar	Mod. Wear	1
6	2b	Bovid/*Hemibos*	M3, upper, right	Mod. Wear	Split	1
10, 15, 22	3	Bovid	Tooth	Premolar	Split	3
20		Bovid	Tooth	Molar	Split	1
39	2b	Bovid (Large)	M3	Lower	Right	1
16-18	2b	Cervid	Antler			3
31		Chelonia	Scute	Dimpled		1
37,41, 48		Chelonia	Scute	Netted		3
40		Chelonia	Scute	Rippled		1
43, 44		Chelonia	Scute	Smooth		2
45	3	Proboscidean	2nd Phalange	Complete		1

Table 6.9 continued: Fossil occurrence 203:

19,30	5	Proboscidean	Shaft			2
21,26,27,32,35, 36	4	Proboscidean	Tooth	Molar	Plate	6
1	4	Proboscidean	Vertebra	Centrum	Rolled	1
0		Indeterminate	Fragments			80
Total						**120**

Fossil occurrence 213: scatter:

Specimen	W	Taxon	Element	Part	Other	No
1	2a	Bovid	Radius	Proximal		1
0		Chelonia	Scute	Smooth		1
5	3	Proboscidean	Cranium	Occipital		1
6	4	Proboscidean	Tooth	Tusk		1
0		Indeterminate	Fragments			7
Total						**11**

Fossil occurrence 259: general collection:

Specimen	W	Taxon	Element	Part	Other	No
7	3	Artiodactyla	Femur	Shaft		1
6		Artiodactyla	Metatarsal	Shaft		1
0		Artiodactyla	Tooth	Molar		2
0	2b	Artiodactyla	Ulna	Fragment		1
3		Bovid	Humerus	Shaft		1
2, 5	2b	Bovid	Radius	Proximal		2
35		Bovid	Scapula	Glenoid		1
0	4	Bovid	Tooth	Fragment		1
20		Bovid	Tooth	Premolar		1
37	2b	Bovid	M1/2, upper	Complete	Med. wear	1
1		Bovid	Vertebra	Cervical		
24	3	Bovid/Cervid	Molar	Upper	Little wear	1
9,28		Cervid	Antler			2
0		Chelonia	Scute			2
18		Chelonia	Scute	Smooth		9
29		Chelonia	Scute	Netted		2
31		Crocodylus	Tooth			1
1		Proboscidean	Mandible			1
11	4	Proboscidean	Rib	>40x24 mm.		1
0		Proboscidean	Tooth	Plate		
0		Proboscidean	Tooth	Tusk		1
29	3	Indeterminate	Fragments			61
Total						**94**

Fossil occurrence 261: scatter:

Specimen	W	Taxon	Element	Part	Other	No
9	3	Artiodactyla	Scapula			1
16	4	Artiodactyla	Tooth	Premolar	Fragment	1
0,12		Bovid	Horn-core			2
11,13		Bovid	Tooth	Premolar		2
15	2b	Bovid	Premolar	Upper	No crowns	1
3	3	Bovid	Ulna	Proximal		1
6		Bovid (Small)	Horncore			1
7	4	Cervid	Antler	Pedicel	Branched	1

Table 6.9 continued: Fossil occurrence 261:

2		Chelonia	Scute	Smooth		1
10		Chelonia	Scute	Pitted		1
15		Chelonia	Scute	Netted		1
5		Proboscidean	Mandible	None	No teeth	1
14	4	Proboscidean	Rib	Suncracked		1
8		Proboscidean	Tooth	Molar	Has crowns	1
5		Proboscidean	Tooth	Molar	Plate	5
0		Proboscidean	Tooth	Tusk		11
1	3	Suid	Humerus	Shaft		1
0		Indeterminate	Fragments			26
Total						**59**

Fossil occurrence 263: general collection:

Specimen	W	Taxon	Element	Part	Other	No
0		Bovid	Tooth	Premolar	Fragments	2
4	4	Bovid	Tooth	Molar	Split,worn	1
3	3	Cervid	Antler			1
0		Chelonia	Scute	Smooth		8
0		Chelonia	Scute	Netted		12
0	5	Proboscidean	Tooth	Plate		1
0	5	Indeterminate	Fragments			8
Total						**33**

Fossil occurrence 265: scatter:

Specimen	W	Taxon	Element	Part	Other	No
2	3	Artiodactyla	Vertebra	Thoracic		1
6	3	Artiodactyla	Vertebra	Centrum		1
5	3	Bovid	Calcaneum	Shaft		1
64,65		Bovid	Horncore			2
11	3	Bovid	Premolar	Complete	Upper, worn	1
14	2a	Cervid	Antler			1
8,10		Chelonia	Scute	Pitted		2
12		Chelonia	Scute	Smooth		1
61		Giraffidae	Tooth	Molar		1
0		Proboscidean?	Carpal			1
4		Proboscidean	Femur	Shaft		1
1,2,8,9,10		Proboscidean	Humerus	Shaft		5
12		Proboscidean	Humerus	Distal	Shaft	1
7,63		Proboscidean	Tooth	Molar		2
58-60		Proboscidean	Tooth	Tusk		3
0		Proboscidean	Tooth	Fragments		3
		Struthio	Eggshell			1
0		Indeterminate	Fragments			54
Total						**82**

Fossil occurrence 267: scatter:

Specimen	W	Taxon	Element	Part	Other	No
1	3	Artiodactyla (Large)	Calcaneum	Shaft		1
3		Artiodactyla	Metapodial	Shaft		1
4,6	4	Artiodactyla	Tooth	Spall		2
5		Bovid	M3	Lower	Left; worn	1

Table 6.9 continued: Fossil occurrence 267:

2		Chelonia	Scute	Smooth		4
7	3	Proboscidean	Tooth	Molar	Crown	1
0		Indeterminate	Fragments			15
Total						**25**

Fossil occurrence 271: scatter:

Specimen	W	Taxon	Element	Part	Other	No
1	4	Artiodactyla	Calcaneum	Shaft		1
33		Artiodactyla	Cranium			1
39		Artiodactyla	Pelvis			1
25		Artiodactyla	Scapula			1
40		Artiodactyla	Tooth	Molar	Fragment	1
44		Artiodactyla	Tooth	Spall		1
6		Artiodactyla	Vertebra	Caudal		1
5	2b	Bovid	1st Phalange	Complete		1
20		Bovid	PM2	Upper	Worn	1
17		Chelonia	Scute			1
18		Chelonia	Scute	Smooth		1
42		Chelonia	Scute	Netted		1
4		Croc/Chelonia	Humerus	Proximal		1
23		Proboscidean	Crania			1
28		Proboscidean	Phalange	Complete		1
2	3	Proboscidean	Rib	38x19		6
13		Bovid	M3 upper	No wear		1
4	4	Proboscidean	Tooth	Molar	Root	1
0		Proboscidean	Tooth	Fragments		16
11		Ungulata	(LargeTooth	Molar	Split	1
0		Indeterminate	Fragments			156
Total						**195**

Fossil occurrence 273: scatter:

Specimen	W	Taxon	Element	Part	Other	No
3		Artiodactyla	Calcaneum	Complete		1
2	4	Artiodactyla	Humerus	Proximal	Condyle	1
5,13		Artiodactyla	Metapodial	Shaft		2
10	3	Artiodactyla	Pelvis			1
0	5	Artiodactyla	Tooth	Spall		1
1	5	Artiodactyla	Vertebra	Complete		1
6		Artiodactyla	Vertebra	Flange		1
0	5	Bovid	Horncore			1
9		Bovid	Nav-cuboid	Split		1
14	5	Bovid	Premolar	Upper	Worn	1
0,16,17		Chelonia	Scute	Smooth		3
18		Chelonia	Scute	Pitted		1
9	5	Proboscidean	Tooth	Molar	Plate	1
7	3	Rhinocerotidae	Metapodial	Distal		1
0		Indeterminate	Fragments			54
Total						**71**

Table 6.9 continued: Fossil occurrence 275: general collection:

Specimen	W	Taxon	Element	Part	Other	No
9,10,15		Artiodactyla	Cranium			3
4	3	Artiodactyla	Metatarsal	Shaft		1
3	4	Bovid	M1/2	Lower	Complete	1
8	4	Bovid	Premolar	Upper	Worn	1
17		Bovid	Tooth	Molar		1
14	3	Cervid	Antler			1
0		Chelonia	Scute	Smooth		2
7,19		Chelonia	Scute	Netted		2
17		Crocodylus?	Pelvis	Fragment		1
20		Proboscidean	Carpal			1
1	3	Proboscidean	Carpal?	Complete		1
18		Proboscidean	Cuboid			1
2	4	Proboscidean	Pelvis/scapula			1
0		Indeterminate	Fragments			37
Total						**54**

Fossil occurrence 277: general collection:

Specimen	W	Taxon	Element	Part	Other	No
1	4	Artiodactyla	Humerus			1
3	4	Cervid	Antler			1
2		Chelonia	Scute	Netted		1
6-8		Chelonia	Scute	Smooth		3
0		Indeterminate	Fragments			7
Total						**13**

Fossil occurrence 299: general collection:

Specimen	W	Taxon	Element	Part	Other	No
2	4	Artiodactyla	Rib	25x13		1
3	2b	Artiodactyla	Rib	33x11		1
1		Chelonia	Scute	Smooth		1
1	2b	Giraffidae	Unciform	Complete		1
4	3	Proboscidean	Rib	35x23 mm.		1
0		Indeterminate	Fragments			3
Total						**8**

Fossil occurrence 301: scatter:

Specimen	W	Taxon	Element	Part	Other	No
1	2b	Bovid	Metacarpal	Proximal	Complete	1
2	2b	Bovid	Metacarpal	Distal	Complete	1
4	3	Bovid	Ulna	Shaft		1
3	2b	Indeterminate	Shaft	Fragment		1
Total						**4**

Fossil occurrence 312: general collection:

Specimen	W	Taxon	Element	Part	Other	No
2,6,11,18, 20	3	Bovid	Horncore			5
7	3	Bovid	Tibia		Split	1
8	3	Bovid	Molar	Upper	Complete	1
13,17	3	Bovid	Tooth	Molar	Split	2
21,22		Bovid	Tooth	Premolar	Split	2

Table 6.9 continued: Fossil occurrence 3125:

2	2a	Cervid	1st Phalange	Complete		1
4		Cervid	Antler			1
0		Proboscidean	Tooth	Molar	Fragments	3
0		Indeterminate	Fragments			17
Total						**33**

Notes: 203GB16-18 are from the same individual animal; 303GB1 has cut-marks (WAC) or porcupine bite marks (RWD). 269GB1- all 4 pieces joined. 259GB6, 263GB3 and 275GB10 had a sandy matrix; 297GB1 had a gritty-sand matrix. 301GB1 and 2 joined; i.e. the metacarpal was complete (length = 257mm).

Other finds: Scatter 215: 2 artiodactyl metapodial fragments and 19 indeterminates; 269 (scatter) – 4 conjoining pieces of Proboscidean scapula; 297 (general collection) – 2 pieces of chelonia scute and 13 other fragments; 303 (scatter) – 1 piece of cervid antler and 16 Probosicdean tibia shaft fragments. 314 contained 4 pieces of Proboscidean scapula and 11 indeterminate fragments.

Table 6. 10. Fossil material from Set 7, Sohawa Kas Headwaters (205, 207, 209) and Sohawa Kas West (217-223 (odd nos. only), 288, 316, and 318).

Fossil occurrence 205: gully:

No.	W	Taxon	Element	Part	Other	N
2	3	Bovid	Horncore	Pedicel		1
7		Bovid	Radius	Shaft		1
1	2b	Proboscidean	Mandible	Molar	No teeth	1
5	3	Proboscidean	Rib			1
6	2b	Proboscidean	Pelvis			1
8		*Struthio*	Egg-shell			1
3,4		Indeterminate	Shaft			2
Total						**8**

Fossil occurrence 207: scatter:

No.	W	Taxon	Element	Part	Other	N
2	4	Bovid	Horncore			1
3	2b	Bovid	Tibia	Distal	Split	1
0		Chelonia	Scute	Smooth		1
1	3	Proboscidean	Shaft			1
		Indeterminate	Fragments			8
Total						**12**

Fossil occurrence 209: gully:

No.	W	Taxon	Element	Part	Other	N
1	2b	Artiodactyla	Shaft			1
5		Artiodactyla	Tooth			1
7	2b	Artiodactyla	Tibia	Distal		1
10	3	Artiodactyla	Vertebra	Cervical		1
17		Artiodactyla	Horncore			1
8		Bovid	Mandible	Premolar		1
16		Bovid	Patella			1
23	3	Bovid	P2 lower	Complete	Worn	1
0		Cervid	Antler	Rolled		1
6	2a	Cervid	Antler	Pedicel	Branched	1
9	3	Proboscidean	Femur	Proximal	Condyle	1
11,13	3	Proboscidean	Rib			2
12	3	Proboscidean	Tibia	Shaft		1
		Indeterminate	Fragments			22
Total						**36**

Table 6. 10 continued: Fossil occurrence 217: scatter:

No.	W	Taxon	Element	Part	Other	N
8		Artiodactyla	Humerus	Proximal	Split	1
1	3	Bovid	Humerus		Shaft	1
2	3	Bovid	Metacarpal	Distal	Barrel	1
6	5	Bovid	Molar, lower	Split	Unworn	1
7	4	Bovid	Metatarsal	Proximal	Split	1
3,4,9	3	Cervid	Antler			3
5		Cervid	Radius	Proximal	Split	1
0		Chelonia	Scute	Fishnet		1
0		Indeterminate	Shaft			13
Total						**23**

Fossil occurrence 223: scatter:

No.	W	Taxon	Element	Part	Other	N
5	3	Artiodactyla	Rib			1
1		Bovid	Radius	Proximal	Complete	1
2,4	3	Bovid	Ulna	Proximal		2
0	5	Indeterminate	Fragments			21
Total						**25**

Fossil occurrence 288: general collection:

No.	W	Taxon	Element	Part	Other	N
2	3	Bovid	Radius	Proximal	Complete	1
1,3	2b	Cervid	Antler			2
6		Chelonia	Scute	Smooth		4
2	3	Proboscidean	Shaft			1
4	3	Proboscidean	Tooth	Tusk	42x24 mm.	1
0		Indeterminate	Fragments			5
Total						**14**

Fossil occurrence 316: general collection:

No.	W	Taxon	Element	Part	Type	Other	N
5	3	Artiodactyla	Ramus				1
1	3	Bovid	Femur	Distal			1
6	5	Proboscidean	Tooth	Molar	Plate	e	1
0,8,10	4	Proboscidean	Tooth	Tusk			5
0	5	Indeterminate	Fragments				6
Total							14

Fossil occurrence 318: general collection:

No.	W	Taxon	Element	Part	Type	Other	N
8	2b	Artiodactyla	Metapodial	Proximal	Split		1
11	3	Bovid	Tooth	Molar	Spall		1
3	5	Proboscidean	Tooth	Molar			3
0		Indeterminate	Fragments				37
Total							42

Notes: 1 indeterminate shaft joins 207GB1; 209GB1 and 209GB2 join 209GB7; 209GB2, joins 209GB3; 217GB3 and 4 (antler) join. 209GB10 has a coarse sand matrix. The list for 207 excludes the antler fragment embedded in sandstone in the bed of Sohawa Kas.

Other finds: Scatter 219 contained 1 piece of cervid antler, 4 pieces of smooth Chelonia scute and 4 indeterminate fragments. Scatter 221 contained part of a bovid left calcaneum, 1 piece of smooth Chelonia scute, and 31 fragments.

Table 6. 11: Fossil material from Set 8, Sohawa West (276-290 even nos. only), 296, 320, 322, 378.

Fossil occurrence 276: general collection:

Specimen	W	Taxon	Element	Part	Other	N
1	2b	Bovid	Tibia	Distal	Split	1
11	3	Bovid	Molar	Lower	Split	1
13	5	Bovid	Tooth	Premolar	Split	1
14	3	Equid	1st Phalange	Complete		1
		Chelonia	Scute	Smooth		16
		Chelonia	Scute	Netted		4
12,26	5	Equid	Tooth	Premolar	Split	2
10		Equid	Tibia	Proximal		1
		Proboscidean	Tooth	Plate		3
		Indeterminate	Fragments			69
Total						**99**

Fossil occurrence 280: general collection:

Specimen	W	Taxon	Element	Part	Other	N
8		Artiodactyla	Astragulus	Proximal	Split	1
20	2a	Artio (Small)	Astragulus			1
10	3	Artiodactyla	Pelvis			1
3	5	Bovid	Tooth	Premolar	Split	1
1,13	5	Cervid	Antler	Palmate		2
9	4	Cervid	Antler	Palmate		1
2		Chelonia	Scute	Plain		2
7		Chelonia	Scute	Netted		4
12		*Crocodylus*	Tooth			1
5,19		Proboscidean	Tooth	Fragment		2
	4	Indeterminate	Fragments			48
Total						**64**

Fossil occurrence 284: scatter:

Specimen	W	Taxon	Element	Part	Other	N
284A GB4	5	Artiodactyla	Tooth	Fragment		1
284A		Chelonia	Scute	Pitted		1
284A GB2		Chelonia	Scute	Netted		11
284A GB6		Chelonia	Scute	Smooth		2
284A GB5,10,11,13-15	5	Proboscidean	Tooth	Molar	Plate	6
284A GB8	3	Equid	Tooth	Incisor	Worn	1
284B GB3	5	Artiodactyla	Tooth	Fragment		1
284B	5	Proboscidean	Tooth	Molar	Plate	12
284A		Indeterminate	Fragments			13
Total						**48**

Fossil occurrence 296: scatter:

Specimen	W	Taxon	Element	Part	Other	N
8		Artiodactyla	Tooth	Split		1
14		Chelonia	Scute	Netted		1
1		Proboscidean	Mandible			1
2		Proboscidean	Mandible	Ramus		1
		Indeterminate	Fragments			77
Total						**81**

Fossil occurrence 320: general collection:

Specimen	W	Taxon	Element	Part	Other	N
2		Artiodactyla	Femur	Distal		1
4	2b	Artiodactyla	Scapula	Glenoid		1
1	2	Cervid	Antler			1
0		Indeterminate	Shaft			7
Total						**10**

Notes: 280GB12 (*Crocodylus* tooth) has a grit matrix; 280GB12 was a mollusc shell. 320GB2 had a sandstone matrix. **Other finds**: Scatter 278 comprised 1 piece of artiodactyl pelvis and 4 indeterminate fragments; general collection 286 had only 4 indeterminate fragments; general collection 290 contained 1 equid premolar and 1 piece of *Struthio* eggshell; general collection 322 contained 1 fragment of Artiodactyl distal metapodial, 1 piece of bovid horncore and 3 indeterminate shaft fragments. 378GB1 was a find-spot of a *Sivatherium* upper M2 (51 x 49 mm.), found near 296.

Table 6. 12. Fossil material from Set 9, Sohawa Kas East (237, 257,279-295 (odd nos.), 298-308 (even nos.), 305, 309-319 (odd nos.), 322-330 (even nos.), 376.

Fossil occurrence 237: scatter:

Specimen	W	Taxon	Element	Part	Other	N
1	2a	Proboscidean	Astragalus	Split		1
2,3	3	Proboscidean	Rib			2
5	3	Proboscidean	Shaft			1
	3	Indeterminate	Fragments			5
Total						**9**

Fossil occurrence 257: scatter (this may also be locality 156 from 1986):

Specimen	W	Taxon	Element	Part	Other	N
2	3	Artio (Large)	Calcaneum			1
3	3	Artio (Large)	Tooth	M1/2	Lower	1
4	3	Artio (Large)	Tooth	M1/2 erupting	Upper	1
11	2b	Artiodactyla	Fibula	Distal		1
19	4	Artiodactyla	Metapodial	Distal	Barrel	1
6,14		Artiodactyla	Radius	Proximal		2
7		Artiodactyla	Ulna	Proximal		1
29	3	Bovid	Horn-core			1
22		Bovid	Tooth			1
20,22,34		Chelonia	Scute	Netted		3
		Indeterminate	Fragments			106
Total						**120**

Fossil occurrence 279: general collection:

Specimen	W	Taxon	Element	Part	Other	N
2	3	Bovid	Calcaneum	Shaft		1
3,7	3	Bovid	Horncore			2
1	2b	Bovid	Tibia	Distal	Complete	1
4	3	Cervid	Antler			1
5	3	Proboscidean	Tooth	Molar	Plate	1
		Indeterminate	Fragments			5
Total						**11**

Fossil occurrence 281: general collection:

Specimen	W	Taxon	Element	Part	Other	N
6	4	Bovid	Premolar	Worn	Split	1
4		Cervid	Antler	Branched		1
7		Chelonia	Scute	Smooth		2
8		*Crocodylus*/Indet	Shaft			1
2	3	Proboscidean	Mandible	No teeth	Fragment	1
11		Proboscidean	Scapula	Complete		1
12		Proboscidean	Tibia			1
0		Indeterminate	Fragments			30
Total						**38**

Fossil occurrence 283: general collection:

Specimen	W	Taxon	Element	Part	Other	N
8	3	Artiodactyla	Calcaneum	Shaft		1
11		Artiodactyla	Pelvis			1
2		Artiodactyla	Radius			1
10	3	Artiodactyla	Shaft			1
3	2b	Bovid	Calcaneum	Proximal		1

Table 6. 12 continued: Fossil occurrence 283:

4		Cervid	Antler	Tine		1
5	3	Cervid	Antler	Pedicel		1
0,6,7,14-16,21	3	Cervid	Antler			10
17		Chelonia	Scute	Smooth		1
18		Chelonia	Scute	Netted		4
9	3	Proboscidean	Rib	38x19 mm.		1
1		Suidae?	Tibia	Proximal		1
0		Indeterminate	Fragments			32
Total						**56**

Fossil occurrence 285: scatter:

Specimen	W	Taxon	Element	Part	Other	N
11	4	Artiodactyla	Orbital			1
39	3	Bovid (Small)	Tooth	M1/2, lower	Worn	1
8	3	Bovid	Cuneiform			1
1	4	Bovid	Femur	Proximal		1
2	4	Bovid	M1/2, upper	Little wear		1
7		Bovid	PM4	Lower	Left	1
10	3	Bovid	M1/2, upper	Mod. wear	Split	1
12		Bovid	Premolar	Upper	Med.wear	1
24	5	Bovid	Tooth	Worn	Split	1
0	4	Bovid (Large)	PM	Upper	Worn	1
30	3	Bovid (Large)	Tooth	Fragment		1
36	4	Bovid (Large)	Tooth	PM	Split	1
4		Cervid	Antler			1
26		Crocodylus?	Patella			1
26		Proboscidean	2nd Phalange			1
0		Proboscidean	Tooth	Plate		5
25	5	Proboscidean	Tooth	Fragmen		1
		Indeterminate	Fragments			173
Total						**194**

Fossil occurrence 287: general collection:

Specimen	W	Taxon	Element	Part	Other	N
3		Artiodactyla	Metatarsal	Shaft		1
4	3	Bovid	Horncore			1
0		Chelonia	Scute	Smooth		3
2		Proboscidean	Femur	Shaft		1
0		Proboscidean	Tooth	Molar	Plate	4
8	5	Proboscidean	Tooth	Tusk		1
		Indeterminate	Fragments			25
Total						**36**

Fossil occurrence 291: scatter:

Specimen	W	Taxon	Element	Part	Other	N
26	4	Artiodactyla	Cranium	+ Horncore	Fragment	1
11		Artiodactyla	Metatarsal			1
31		Artiodactyla	Rib			1
3, 27	2b	Artiodactyla	Vertebra	Lumbar	Complete	2
4,15	3	Artiodactyla	Vertebra	Cervical	Complete	2
13	2a	Artiodactyla	Vertebra	Thoracic	Complete	1
21		Artiodactyla	Vertebra	Axis		1
30,47		Artiodactyla	Vertebra	Centrum		2
45, 51		Artiodactyla	Vertebra	Fragment		2
5, 20	4	Bovid	Horncore			2
23	4	Bovid	PM	Upper	Worn	1
54		Carnivora	Fibula			1
7	2a	Carnivora	Tibia	Shaft		1
10,12	2b	Cervid	Antler			2
11		Chelonia	Scute	Netted		1

Table 6. 12 continued: Fossil occurrence 291:

43		Chelonia	Scute	Dimpled		2
16		Felid	Cuneiform			1
8,9		Felid	Metapodial	Lateral	Distal	2
37		Felid	Pisiform			1
17,29, 55,59, 62		Proboscidean	Tooth	Fragments		5
1,6,18	3	Felid	Pelvis	Acetabulum		3
		Indeterminate	Fragments			83
Total						**118**

Fossil occurrence 293: general collection:

Specimen	W	Taxon	Element	Part	Other	N
8		Artiodactyla	1st Phalange	Complete		1
3	4	Bovid	M1/2	Complete	Upper	1
8	3	Bovid	M1/2, upper	Little wear	Split	1
5		Chelonia	Scute	Dimpled		2
13		Chelonia	Scute	Netted		1
18	2b	Proboscidean	Rib			1
1	3	Proboscidean	Shaft			1
2		**Elephas**	Tooth	Molar		1
11	4	Proboscidean	Tooth	Tusk	Spall	1
0	3	Indeterminate	Fragments			25
Total						**35**

Fossil occurrence 298: scatter:

Specimen	W	Taxon	Element	Part	Other	N
1	2b	Artiodactyla	Vertebra	Cervical		1
3	3	Bovid	Humerus	Distal	Complete	1
2	2a	Equid	PM/Molar	Complete	Lower	1
0	3	Giraffidae?	Tooth	Premolar	Upper	1
		Indeterminate	Fragments			9
Total						**13**

Fossil occurrence 302: general collection:

Specimen	W	Taxon	Element	Part	Other	N
3	2b	Artiodactyla	Vertebra	Neural		1
2	3	Bovid	Horncore			1
6	4	**Elephas**	Tooth	Plate	Crowns	1
0	4	Indeterminate	Fragments			3
Total						**6**

Fossil occurrence 304: scatter:

Specimen	W	Taxon	Element	Part	Other	N
5		Bovid	Tooth	M1/2	Upper	1
7	2b	Bovid	Unciform			1
3	3	Bovid	Vertebra	Lumbar		1
0		Proboscidean	Tooth	Molar	Fragments	35
1	3	Proboscidean	Tooth	Tusk		1
0	5	Indeterminate	Shaft	Fragments		3
Total						**42**

Fossil occurrence 305: scatter:

Specimen	W	Taxon	Element	Part	Other	N
10		Artiodactyla	Tooth			1
3	3	Bovid (Large)	Tooth	Molar	Split	1
2		Cervid	Antler			1
1	3	Proboscidean	Tooth	Tusk		12
0	3	Indeterminate	Shaft			2
Total						17

Table 6. 12 continued: Fossil occurrence 308: general collection:

Specimen	W	Taxon	Element	Part	Other	N
5	3	Proboscidean	Rib	40x27		1
0	5	Proboscidean	Tooth	Molar	Plate	2
4		Indeterminate	Fragments			6
Total						**9**

Fossil occurrence 309: general collection:

Specimen	W	Taxon	Element	Part	Other	N
3,4	3	Bovid	Calcaneum	Complete		1
9	4	Proboscidean	Tooth	Molar	Root	3
1	4	Proboscidean	Tooth	Tusk		3
0		Indeterminate	Fragments			8
Total						**15**

Fossil occurrence 311: find-spot:

Specimen	W	Taxon	Element	Part	Other	N
1	2a	Bovid	Astragulus	Complete		1
9		Proboscidean	Tooth	Molar		1
2		Artiodactyla	Humerus	Shaft		1
0		Indeterminate	Fragments			8
Total						**11**

Fossil occurrence 315: gully collection:

Specimen	W	Taxon	Element	Part	Other	N
4		Artiodactyla	Tooth	Split		1
3		Proboscidean	Tooth	Molar		1
1		Hyaenidae	Mandible	P2-P4		2
3	4	Indeterminate	Shaft			1
Total						**5**

Fossil occurrence 320: general collection:

Specimen	W	Taxon	Element	Part	Other	N
2		Artiodactyla	Femur	Distal		1
4	2b	Artiodactyla	Scapula	Glenoid		1
1	2	Cervid	Antler			1
0		Indeterminate	Shaft	Fragments		7
Total						**10**

Fossil occurrence 322: general collection:

Specimen	W	Taxon	Element	Part	Other	N
1	3	Artiodactyla	Metapodial	Distal	Barrel	1
2	5	Bovid	Horncore			1
0	4	Indeterminate	Shaft			3
Total						**5**

Notes: 285GB11 (Artiodactyl orbital) and 296GB2 (Proboscidean mandible) had a grit/sand matrix; Elephant tooth fragments 302GB6 and304GB1 had a sandstone matrix. The three pieces of Chelonia scute from 287 joined. The 34 elephant tooth fragments from 304 were probably part of the same tooth. 309GB3 and 4 join and so are counted as one specimen. 263GB3 and 275GB10 has a sand matrix; 320GB2 was in a sandstone.

Other finds: Scatter 289 consisted of 2 pieces of Proboscidean rib fragments and 9 indeterminate fragments. General collection 295 (1 cervid antler and 1 indeterminate fragment); general collection 300 (1 probosidean and 7 indeterminate shaft fragments); scatter 306 (7 indeterminate fragments); general collection 307 comprised only 20 indeterminate fragments; general collection 317 comprised only 3 indeterminate shaft fragments. There were also general collection 319 (1 piece of Proboscidean molar plate and 11 indeterminate shaft fragments); general collection 324 (1 piece of bovid horncore); gully find 326, a bovid astragalus; general collection 328 (3 indeterminate fragments); scatter 330 (3 indeterminate fragments); scatter 376 (1 artiodactyl tooth spall and 1 indeterminate shaft fragment).

Table 6. 13. Fossil material from Set 10, Flag Hill area (327-335 (odd nos.), 368-372 even nos.).

Fossil occurrence 327: general collection:

Specimen	W	Taxon	Element	Part	Other	N
327AGB11	5	Bovid (Small)	Molar	Split	Worn	1
327A GB1		Proboscidean	Tooth	Fragment		5
327A GB2	5	Proboscidean	Tooth	Fragment	Tusk	2
327A		Indeterminate	Fragments			10
327B GB4	4	Bovid	Horncore			2
327B		Indeterminate	Fragments			50
327B GB7, 11	3	Proboscidean	Mandible	Fragment		2
327B GB5, 10	4	Proboscidean	Mandible+	Root	Tooth	2
327B GB6	4	Proboscidean	Tooth	Fragment		1
327B GB12	4	Proboscidean	Tooth	Tusk		1
327C GB20	5	Artiodactyla	Tooth	Fragment		1
327C GB8		Cervid	Antler			1
327C GB14	4	Proboscidean	Mandible	Fragment		1
327C GB6	3	Proboscidean	Pelvis/Scapula			1
327C GB5,10	5	Proboscidean	Tooth	Fragment		2
1		Equid	Tibia	Distal		1
7	4	Equid	PM3/4 or M1/2	Lower	Complete	1
327C		Indeterminate	Fragments			55
327C GB3,4, 11	5	Proboscidean	Mandible	Fragment		3
327C GB15		Proboscidean	Tooth	Fragments		15
327C GB5		Artiodactyla	1st Phalange	Split		1
327D		Artiodactyla	Tooth	Spalls		2
327D GB11		Bovid	Tooth	Split		1
327D GB4, 9	2b	Bovid (Large)	Molar, lower	Split	Little wear	2
327D		Indeterminate	Fragments			76
327D GB8,13		Proboscidean	Phalange	Fragment		2
327D		Proboscidean	Tooth	Fragment	Plate	8
Total						**249**

Fossil occurrence 331: general collection:

Specimen	W	Taxon	Element	Part	Other	N
9	4	Artiodactyla	Metapodial	Distal	Barrel	1
2		Proboscidean	1st Phalange			1
0		Indeterminate	Fragments			168
Total						**170**

Fossil occurrence 333: general collection:

Specimen	W	Taxon	Element	Part	Other	N
3	3	Artiodactyla	Cranium			1
0		Bovid	Horncore			5
16		Bovid	Tooth	Premolar	Spall	1
6		Cervid	2nd Phalange	Complete		1
1		Cervid	Antler	Pedicel	Branched	1
7		Proboscidean	Tooth	Tusk	Plate	54
0		Indeterminate	Fragments			48
Total						**111**

Fossil occurrence 368: general collection:

Specimen	W	Taxon	Element	Part	Other	N
2	3	Artiodactyla	Scapula	Fragment		1
34, 36	4	Artiodactyla	Tooth	Spall		2
4		Bovid	Horncore	+base (type C)		1
37	5	Bovid	Humerus	Distal	Complete	1
39	4	Bovid	Humerus	Shaft		1

Table 6. 13 continued : Fossil occurrence 368:

10	4	Bovid	Tooth	Premolar	Worn, split	1
25	5	Bovid	Tooth	Molar	Split, worn	1
38	3	Bovid (Large)	M1/2	Mod.wear	Complete	1
1, 17	3	Cervid	Antler	Palmate		2
12		Equid	Tooth	Molar, lower	Complete	1
5,7		Proboscidean	Tooth	Molar		2
14,16		Rhinocerotidae	Tooth	Molar	V.worn	2
		Indeterminate	Fragments			57
Total						**73**

Fossil occurrence 370: general collection:

Specimen	W	Taxon	Element	Part	Other	N
4	3	Proboscidean	Cranium			1
5	3	Proboscidean	Ramus	Unfused		1
1		Proboscidean	Rib			1
2	4	Proboscidean	Tooth	Tusk		1
3	4	Indeterminate	Shaft	Rolled		1
Total						**5**

Notes: 329 was a scatter of 41 indeterminate fragments; 335 was a findspot of a crocodilian mandible. Gully collection 372 produced 20 specimens, including a rhino distal ulna shaft.

Table 6. 14: Fossil material from Set 11, West Branch of Kotha Kas (337-341 (odd nos. only), 515-535 (odd nos. only).#

Fossil occurrence 337: general collection:

Specimen	W	Taxon	Element	Part	Other	N
1	3	Artiodactyla	Metacarpal	Distal Barrel		1
3	3	Bovid	Metatarsal	Proximal	Complete	1
2,5,7,8		Cervid	Antler	Tine		4
6		Proboscidean	Tooth	Tusk		1
4		Equid	Ulna	Proximal		1
0		Indeterminate	Fragments			8
Total						**16**

Fossil occurrence 339: general collection:

Specimen	W	Taxon	Element	Part	Other	N
17	3	Artiodactyla	Phalange	Distal	Condyle	1
5	2b	Bovid	2nd Phalange	Complete		1
19	2b	Bovid	Horncore			1
2,3,11	2b	Cervid	Antler			3
12, 15		Proboscidean	Tooth	Molar	Plate	2
		Proboscidean	Tooth	Tusk		28
20		Rhinocerotidae	Metatarsal			1
14		Rhinocerotidae	Tooth	Molar	Worn	1
0		Indeterminate	Fragments			143
Total						**181**

Fossil occurrence 341: general collection:

Specimen	W	Taxon	Element	Part	Other	N
8		Artiodactyla	1st Phalange			1
23		Artiodactyla	2nd Phalange			1
24		Artiodactyla	2nd Phalange	Proximal		1
20	2b	Artiodactyla	Calcaneum	Shaft		1
15		Artiodactyla	Metacarpal	Proximal		1
13,26		Artiodactyla	Metapodial			2
32	4	Artiodactyla	Orbital			1
12		Artiodactyla	Phalange	Proximal		1
4		Artiodactyla	Tibia	Distal		1
31	2b	Artiodactyla	Tooth	Premolar	Fragment	1

Table 6. 14 continued: Fossil occurrence 341:

1	3	Bovid	2nd Phalange	Complete	Unfused	1
18		Bovid	3rd Phalange	Distal		1
28	3	Bovid	Horncore			1
11		Bovid	Lunar			1
16	4	Bovid	Tooth	Molar	Split	1
22	2b	Bovid	Tooth	Molar, Complete	Fresh	1
3	3	Cervid	Antler			1
0	3	Equid	M3	Lower, Right	Worn	1
29		Proboscidean	Tooth	Tusk	Flake	1
30	3	Proboscidean	Tooth	Molar		1
33		Rhinocerotidae	Mandible	Complete		1
5		Tragulidae	Horncore			1
0	3	Indeterminate	Fragments			247
Total						**270**

Fossil occurrence 515: general collection:

Specimen	W	Taxon	Element	Part	Other	N
3		Artiodactyla	Tooth	Premolar	Fragment	1
		Artiodactyla	Metapodial	Distal	Split	1
1	3	Bovid	Metacarpal	Proximal	Complete	1
0		Chelonia	Scute	Smooth		19
0		Chelonia	Scute	Netted		15
0		Indeterminate	Fragments			47
Total						**84**

Fossil occurrence 517: general collection:

Specimen	W	Taxon	Element	Part	Other	N
0		Artiodactyla	Tooth	Molar	Spall	1
1	1	Rhinocerotidae	Tooth	Molar	Split, worn	1
0		Indeterminate	Fragments			10
Total						**12**

Fossil occurrence 519: scatter:

Specimen	W	Taxon	Element	Part	Other	N
		Cervid	Antler	Rolled		1
2	5	Equid	Tooth	Fragment		1
1	1	Ungulata (Large)	Magnum			1
4	4	Ungulata (Large)	Phalange	Distal		1
0	0	Indeterminate	Fragments			18
Total						**22**

Fossil occurrence 521: find-spot:

Specimen	W	Taxon	Element	Part	Other	N
521 GB2	2	Artiodactyla	Metapodial	Shaft		1
521A		Indet/Suid	Scapula			1
521A		Suid	Scapula	Fragment		1
521B GB1		Bovid	Vertebra	Atlas		1
Total						**4**

Fossil occurrence 523: general collection:

Specimen	W	Taxon	Element	Part	Other	N
4	4	Artiodactyla	Astragulus	Split		1
13	4	Bovid	Tooth	Molar		1
0		Chelonia	Scute	Netted		1
7	5	Proboscid/Suid	Tooth	Molar	**check**	1
1		Tragulidae	Horncore	and Base		1
0		Indeterminate	Fragments			336
Total						**341**

Table 6. 14 continued: Fossil occurrence 525: scatter:

Specimen	W	Taxon	Element	Part	Other	N
5	5	Artiodactyla	Calcaneum	Shaft		1
12	3	Artiodactyla	Radius	Distal		1
3		Artiodactyla	Rib			1
1		Bovid	Tooth	Premolar		1
0		Chelonia	Scute	Smooth		2
0		Suid	Cranium	Complete		1
0		Indeterminate	Fragments			27
Total						**34**

Fossil occurrence 527: scatter:

Specimen	W	Taxon	Element	Part	Other	N
4	3	Artiodactyla	Humerus	Shaft		1
8		Artiodactyla	Tooth			1
7,10,11		Bovid	Horncore			3
3		Bovid	Humerus	Distal	Split	1
9		Bovid	Mandible	Premolar	Left, worn	1
0		Indeterminate	Fragments			36
Total						**43**

Fossil occurrence 529: scatter:

Specimen	W	Taxon	Element	Part	Other	N
3		Anthrocorcere	Metacarpal	Proximal		1
16		Anthrococere	Ulna	Distal		1
2, 7		Artiodactyla	Metacarpal			2
4		Artiodactyla	Metatarsal			1
21/1		Bovid	Mandible	Left + M3	Unworn	1
12	2b	Bovid	Astragalus	Complete		1
5		Bovid	Astragulus			1
9	3	Bovid	Mandible	Premolar	Left	1
1		Bovid	Tibia	Proximal		1
6		Suid	Humerus			1
0		Indeterminate	Fragments			10
Total						**21**

Fossil occurrence 531: scatter:

12		Artiodactyla	Tibia	Shaft		1
10,11,18	4	Artiodactyla	Tooth	Molar	Fragments	3
	4	Bovid	Astragulus	Complete		1
7		Bovid	Naviculo-cuboid			1
3	3	Bovid	M1/2	Lower	No wear	1
6	3	Bovid	M1/2	Upper	Worn	1
26	2b	Bovid	Mandible	M2,M3	Left	1
2,8,20		Rhinocerotidae	Vertebra	Cervical		3
16		Rhinocerotidae	Tooth	Molar	Plate	1
19		Indet (large)	Cranium			1
1		Indet (large)	Shaft			1
		Indeterminate	Fragments			6
Total						**21**

Notes: There was sandstone casing on several pieces from 525. Indeterminate rib 523GB3 had a clay skin. 535GB1 was a find-spot of an *Elephas* maxilla.

Table 6. 15: Fossil material from Set 12, west branch of Kotha Kas (397-399, 477, 501-511)

Fossil occurrence 397: general collection:

Specimen	W	Taxon	Element	Type	Other	N
21		Artiodactyla	Metapodial			1
50	5	Artiodactyla	Radius	Proximal		1
9,12		Bovid	Calcaneum	Complete		2
18		Bovid	Horncore			1
4	4	Bovid	Metacarpal	Distal	Barrel	1
3	3	Bovid	Radius	Shaft		1
13	4	Bovid	Tibia	Distal	Complete	1
16	2a	Bovid	M3	Upper	Left, Fresh	1
20		Bovid	Tooth	Premolar		1
23	2b	Bovid	Premolar	Upper	Worn	1
2		Bovid	Ulna	Proximal		1
7,22,25,31, 35,41,46	2a	Cervid	Antler			7
5,6	3	Proboscidean	Carpal/tarsal			2
45	4	Proboscidean	Humerus	Proximal	Condyle	1
8		Proboscidean	Rib			1
0		Indeterminate	Fragments			145
Total						**168**

Fossil occurrence 399: general collection:

Specimen	W	Taxon	Element	Type	Other	N
2		Bovid	1st Phalange			1
11		Bovid	Astragulus	Complete		1
24	4	Cervid	Antler			6
25		Equid	Astragulus	Complete		1
22	3	Indeterminate	Fragments			70
0		Ungulata	Tooth	Incisor		1
Total						**80**

Fossil locality 477: scatter:

Specimen	W	Taxon	Element	Part	Other		N
1	3	Artiodactyla	Vertebra	Centrum			1
18,19		Bovid	Pelvis	(joins)			2
16	3	Bovid	Tooth	Spall			1
18	3	Bovid	Tooth	Molar	Worn	Split	1
2		Equid	Radius	Distal			1
19		Giraffid	Tooth	Molar	Spall		1
20		Rhinocerotid	Tooth	Molar			1
0		Struthio	Eggshell				1
		Indeterminate	Fragments				51
Total							**60**

Fossil occurrence 501: general collection:

Specimen	W	Taxon	Element	Part	Other	N
2,6		Artiodactyla	Femur	Proximal		2
1	3	Artiodactyla	Metapodial	Distal	Epiphysis	1
0		Artiodactyla	Tooth	Molar		2
0	3	Cervid	Antler	Tine		1
0		Chelonia	Scute	Netted		1
0		Chelonia	Scute	Smooth		1
0		Proboscidean	Tooth	Molar	Plate	4
7		Proboscidean	Tooth	Molar	Plate	2
0		Indeterminate	Fragments			194
Total						**208**

Table 6. 15 continued: Fossil occurrence 503: general collection:

Specimen	W	Taxon	Element	Part	Other	N
8	3	Bovid	M3	Lower	Mod. wear	1
6,7,9		Cervid	Antler			3
0		Chelonia	Scute	Smooth		3
0		Chelonia	Scute	Netted		4
2	3	Proboscidean	Femur	Proximal		1
1,4,5	3	Proboscidean	Pelvis			3
0		Proboscidean	Tooth			5
0		*Struthio*	Eggshell			1
0		Indeterminate	Fragments			110
Total						**131**

Fossil occurrence 505: scatter:

Specimen	W	Taxon	Element	Part	Other	N
17	3	Artiodactyla	Astragalus	Split		1
8	3	Bovid	Humerus	Shaft		1
21	3	Bovid	Molar	Upper	Split	1
11		Bovid (Large)	Humerus			1
1		Proboscidean	Femur	Distal		1
3,14	2b	Proboscidean	Rib			2
6,7,9,20		Proboscidean	Shaft			4
2	2b	Proboscidean	Vertebra	Lumbar		1
4		Proboscidean	Vertebra			1
		Indeterminate	Fragments			37
Total						**50**

Fossil occurrence 507: general collection:

Specimen	W	Taxon	Element	Type	Other	N
0	4	Artiodactyla	Cranium			2
17		Artiodactyla	Metapodial	1		1
32	3	Artiodactyla	Pelvis	Acetabulum		1
8	3	Artiodactyla	Radius	Proximal		1
26		Artiodactyla	Tooth	Molar		1
2,3,15		Bovid	Horncore			3
18		Bovid	Radius			1
31	2a	Bovid	Radius	Distal		1
29	2a	Bovid	Ramus			1
27		Bovid	Tooth	Premolar	Fresh	1
30	3	Bovid	Molar	Mod. wear	Split	1
38		Bovid	Trapezoid			1
5,7,10-12, 14,22,28,35		Cervid	Antler			9
0		Chelonia	Scute	Netted		8
36		Proboscidean	Astragulus	Complete		1
37	3	Proboscidean	Femur	Proximal		1
40		Proboscidean	Mandible			1
999		Proboscidean	Pelvis			1
34		Proboscidean	Tooth	Molar	Plate	2
1	3	Rhino/Proboscid	Metapodial	Distal		1
23	3	Rhinocerotidae	Scapula			1
0		Rhinocerotidae	Tooth	Molar	Fragment	1
0		Indeterminate	Fragments			200
Total						**251**

Note also 507X, a *Stegodon* mandible

Table 6. 15 continued: Fossil occurrence 509: general collection:

Specimen	W	Taxon	Element	Part	Other	N
0		Artiodactyla	Cuneiform			1
0	3	Artiodactyla	1st Phalange	Distal		1
2,3	5	Cervid	Antler			2
1		Chelonia	Scute	Smooth		1
0		Suid/Bovid	Femur	Distal	Neonatal?	1
		Indeterminate	Fragments			11
Total						**17**

Fossil occurrence 511: general collection:

Specimen	W	Taxon	Element	Part	Other	N
1	3	Artiodactyla	Metacarpal	Distal	Shaft	2
3	4	Artiodactyla	Tooth	Fragment		1
2		Artiodactyla	Vertebra	Centrum		1
0	3	Artio/Indet	Humerus	Distal		1
0		Cervid	Antler	Tine		1
0		Indeterminate	Fragments			11
Total						**16**

Table 6. 16: Fossil material from Set 13, West Branch of Kotha Kas, 401-405, 479-497 (odd nos. only) and excluding localities 481, 489 and 491.

Fossil locality 479: general collection:

Specimen	W	Taxon	Element	Part	Other	N
1		Artiodactyla	Calcaneum	Shaft		1
11	3	Artiodactyla	Pelvis			1
7	2b	Artiodactyla	Tibia	Shaft		1
14	5	Artiodactyla	Tooth	Molar	Worn, split	1
5,6	3	Artiodactyla	Ulna	Proximal		2
3		Bovid	Calcaneum	Proximal		1
16		Bovid	Tooth	Molar		1
27	2b	Bovid	Premolar, complete	Upper	Mod.wear	1
4,8,9,18	4	Cervid	Antler			4
0		Chelonia	Scute	Dimpled		1
0		Chelonia	Scute	Smooth		6
17	5	Proboscidean	Tooth	Molar	Plate	1
0	3	Indeterminate	Fragments			5
						26

Fossil occurrence 483: general collection:

Specimen	W	Taxon	Element	Part	Other	N
43		Artiodactyla	Carpal	Complete		1
35		Artiodactyla	Cuboid	Complete		1
26	3	Artiodactyla	Metapodial	Distal condyle	Unfused	1
2	2b	Artiodactyla	Scapula			1
8		Bovid	Horncore			1
23,27	3	Bovid	Nav.-cuboid	Complete		2
10	2b	Bovid	Radius	Proximal		1
5		Bovid	Molar, lower	Complete	Fresh	1
53	3	Bovid	DM3, lower	Complete	Fresh	1
54		Bovid	Tooth			1
56		Bovid	Tooth	Molar		3
1	3	Cervid	Antler	Tine		1
0		Chelonia	Scute	Dimpled		2

Table 6. 16 continued: Fossil occurrence 483: general collection:

Specimen	W	Taxon	Element	Part	Other	N
0		Chelonia	Scute	Netted		8
0		Chelonia	Scute	Smooth		9
3,4	3	Proboscidean	Shaft			2
0		*Struthio*	Eggshell			1
0		Indeterminate	Fragments			173
Total						**210**

Fossil occurrence 485: scatter:

Specimen	W	Taxon	Element	Part	Other	N
9	3	Artiodactyla	Ulnar			1
4		Bovid	2nd Phalange	Proximal		1
8	2a	Bovid	Astragalus	Complete		1
18	5	Bovid	Astragulus	Rolled	Complete	1
17	3	Bovid	Humerus	Distal		1
8		Bovid	Mandible	Ramus		1
10	3	Bovid	Ulna	Proximal		1
7	2b	Cervid	3rd Phalange	Complete		1
0	2b	Cervid	Mandible	P2-M3	Left, worn	1
0		Chelonia	Scute	Smooth		8
12	5	Equid	Tooth	Premolar	Split	1
11	4	Proboscidean	Phalange	Proximal		1
15	5	Proboscidean	Tooth	Molar	Root	1
0		Indeterminate	Fragments			200
						219

Fossil occurrence 487: locality:

Specimen	W	Taxon	Element	Part	Other	N
4	3	Artio (Large)	Humerus	Proximal	Condyle	1
38	4	Artiodactyla	Femur	Shaft		1
25		Artiodactyla	Metapodial	Distal	Condyle	1
70		Artiodactyla	Metapodial	Proximal		2
75		Artiodactyla	Metapodial			1
12	4	Artiodactyla	Pelvis			1
21	2b	Artiodactyla	Pelvis	Acetabulum		1
28		Artiodactyla	Ramus			1
6,37	3	Artiodactyla	Scapula	Glenoid	Split	2
30	3	Artiodactyla	Tooth			1
15,23		Artiodactyla	Ulna	Proximal		2
7	5	Artiodactyla	Vertebra	Centrum		1
44	2b	Bovid	Astragulus	Split		1
1	4	Bovid	Femur	Distal		1
13		Bovid	Femur	Proximal		1
46		Bovid	Fibula	Distal		1
48		Bovid	Horncore			1
67		Bovid	Magnum	Complete		1
5	4	Bovid	Metapodial	Distal	Condyle	1
19	3	Bovid	Metapodial	Distal epiphysis	Unfused	1
9		Bovid	Radius	Distal		1
53		Bovid	Scaphoid	Complete		1
10	3	Bovid	Molar	Upper	Split, worn	1
11	3	Bovid	M1/2, lower	Complete	Mod.wear	1
22		Bovid	Tooth	Premolar	Upper	1
43, 71		Bovid	Tooth	Molar	Split	2
65	4	Bovid	Tooth	Premolar	Split	1
73		Bovid	Tooth			1

Table 6. 16 continued: Fossil occurrence 487:

Specimen	W	Taxon	Element	Part	Other	N
78	4	Bovid	Tooth	Premolar	Split	1
81		Bovid	Unciform	Complete		1
22		Bovid (Large)	PM upper	Complete	Worn	1
57,72	4	Bovid (Large)	Molar	Split	Mod.wear	2
59,63	3	Bovid (Large)	Tooth	Molar	Split	2
11		Carnivora (Small)	Ulna	Proximal		1
8,16,54		Cervid	Antler			3
58	4	Cervid	Antler	Rolled		1
0		Chelonia	Scute	Smooth		61
21		Proboscidean	Humerus	Distal		1
64		Indet/Proboscid	Tooth			1
14		Indet/Proboscid	Carpal			1
31	4	Indet/Proboscid	Carpal	Complete		2
33	5	Proboscid/Suid	Tooth			1
80		Proboscidean	Tooth			1
41		Suid	Ulna	Distal		1
0		Indeterminate	Fragments			231
						345

Fossil occurrence 493: scatter (on top of the Sandstone unit capping 481, 489 etc.):

Specimen	W	Taxon	Element	Part	Other	N
5		Bovid	Horncore		1	1
8	2b	Bovid	Tooth	M1/2, upper	Complete	1
9	3	Bovid	Tooth	PM3, lower	Complete	1
4		Equid	Podial	Complete		1
1,6,12	2b	Proboscidean	Tooth	Molar		3
		Indeterminate	Fragments			18
						25

Fossil occurrence 495: general collection:

Specimen	W	Taxon	Element	Part	Other	N
4	4	Artiodactyla	Calcaneum	Proximal		1
8		Bovid	Mandible + M3 worn	Molar	Right	1
6	3	Bovid	Tooth	Molar	Mod.wear	1
12	3	Bovid/Cervid	PM/M	Split	Mod.wear	1
0		Cervid	Antler			7
7	3	Suid	Mandible, left + M3	Mod. wear	Split	1
0		Indeterminate	Fragments			99
Total						**111**

Fossil occurrence 497: general collection:

Specimen	W	Taxon	Element	Part	Other	N
8	3	Artiodactyla	1st Phalange	Proximal	Split	1
4	3	Artiodactyla	Calacaneum	Shaft	Unfused	1
2	2b	Bovid	Humerus	Proximal	Unfused	1
9	2b	Bovid	Metacarpal	Distal	Complete	1
3	5	Bovid	Tibia	Proximal		1
5	3	Bovid	Molar	Upper	Split	1
7		Bovid	Unciform	Complete		1
0		Chelonia	Scute	Smooth		2
0		Chelonia	Scute			3
0		Indeterminate	Fragments			59
Total						**71**

Notes: There were just 5 fragments from general collection 401. Scatter 403 comprised 1 bovid astragulus (complete) and 3 indeterminate fragments, 20m from an old trig. point. There were also some large Proboscidean pieces encased in sandstone that were left in the field. General collection 405 comprised 3 pieces of antler, 1 piece of elephant tooth and 10 fragments. 4 pieces of Chelonia scute from 485 joined. 485GB17 had a fine sand matrix; 487GB12 had a silt matrix.

Table 6. 17: Fossil material from locality 481, West Branch of Kotha Kas

Zone	Specimen	W	Taxon	Element	Part	Other	N
	43		Artiodactyla	Carpal			1
	35		Artiodactyla	Cuboid			1
	4		Artiodactyla	Femur	Distal		1
	5	2a	Artiodactyla	Mandible	P2-4,M1	Root	1
	3	2b	Bovid	Calcaneum	Shaft		1
	7,40	5	Bovid	Horncore			2
	2		Bovid	Mandible, left	+M3	Fresh	1
	8	2b	Bovid	Metatarsal	Proximal	Complete	1
	0		Bovid	Tooth			10
	43		Bovid	Tooth	Premolar		1
	1		Pisces	Vertebra			1
	0		Indeterminate	Fragments			101
A	5	2b	Bovid	1st Phalange	Proximal		1
A	0		Chelonia	Scute			7
A	0		Chelonia	Scute	Netted		10
A	54		Chelonia	Scute	Netted	(joins)	3
A	90		Chelonia	Scute	Netted	(joins)	14
A	21		Chelonia	Scute	Rippled	(joins)	3
A	50		Chelonia	Scute	Smooth		3
A	36		Chelonia	Spike			1
A	0		Indeterminate	Fragments			156
B	24		Artio (Small)	Vertebra, unfused	Lumbar	Complete	1
B	12	2b	Artio (Small)	Metatarsal	Shaft		1
B	42		Artiodactyla	Calcaneum	Shaft		1
B	32	3	Artiodactyla	Femur	Distal		1
B	34		Artiodactyla	Femur	Proximal		1
B	4	2b	Artiodactyla	Humerus	Proximal	Condyle	1
B	16	2b	Artiodactyla	Metapodial			1
B	10	3	Artiodactyla	Pelvis			1
B	10,23		Artiodactyla	Phalange			2
B	51	3	Artiodactyla	Radius	Distal	Split	1
B	0,22,31		Artiodactyla	Vertebra			3
B	11	3	Artiodactyla	Vertebra	Flange		1
B	33	3	Bovid	1st Phalange	Proximal	Complete	1
B	45		Bovid	Horncore			1
B	36	4	Bovid	Humerus	Distal		1
B	9		Bovid	Metatarsal	Proximal		1
B	**1**	2b	Bovid	Radius	Proximal	Complete	1
B	**2**	2a	Bovid	Tibia	Distal	Complete	1
B	**6**	2b	Bovid	Tibia	Distal	Complete	1
B	10		Bovid	Ulna	Proximal		1
B	41		Carnivora	Tibia			1
B	3		Carnivora	Vertebra	Atlas		1
B	39		Carnivore (Med)	2nd Phalange	Complete		1
B	21	3	Cervid	Antler			1
B	5	2b	Cervid	Metacarpal	Distal	Complete	1
B	49	3	Giraffid/Cervid	Mandible	P2,P3,P4, M1 root		1
B	37		Tragulidae	Mandible	Molar		1
B	38		Ungulata	Tooth	Incisor	Worn	1
B		3	Indeterminate	Fragments			118
C	2		Artiodactyla	Pisiform	Complete		1
C	13	2b	Artiodactyla	Molar	Upper	Worn	1
C	3		Hyaenidae	Tooth	Molar		1
C	0		Indeterminate	Fragments			53
D	8	2b	Artiodactyla	Humerus	Complete		1
D	9	2b	Artiodactyla	Metapodial	Shaft		1
D	3		Bovid	Navicular	Complete		1
D	1		Bovid	Radioulna	Complete		1
D	0		Indeterminate	Fragments			92

Table 6. 17: Fossil material from locality 481, West Branch of Kotha Kas

Zone	Specimen	W	Taxon	Element	Part	Other	N
E	26		Artiodactyla	1st Phalange	Distal		1
E	21	2b	Artiodactyla	Calcaneum	Shaft		1
E	28		Artiodactyla	Calcaneum			1
E	2	2b	Artiodactyla	Humerus	Distal	Complete	1
E	31	3	Artiodactyla	Humerus	Distal	Unfused	1
E	10	2a	Artiodactyla	Metapodial	Distal condyle	Unfused	1
E	1		Artiodactyla	Pisiform			1
E	1		Artiodactyla	Phalange	Distal		1
E	20	3	Artiodactyla	Radius	Proximal		1
E	25	3	Artiodactyla	Premolar, upper	Complete	Mod.wear	1
E	27	3	Artiodactyla	Tooth	Molar	Mod. Wear, split	1
E	7,19		Artiodactyla	Vertebra	Centrum		2
E	**41**	2b	Bovid	Astragulus	Complete		1
E	3,22		Bovid	Calcaneum	(joins)		2
E	8		Bovid	Calcaneum	Split		1
E	38	3	Bovid	Calcaneum	Distal		1
E	42	3	Bovid	Femur	Distal		1
E	10		Bovid	Lunar			1
E	6	2a	Bovid	Mandible	M2, unworn	Right	1
E	9		Bovid	Navicular			1
E	30		Bovid	Radius	Proximal		1
E	**18**		Bovid	Radius/Ulna	Complete	Distal	1
E	11	3	Bovid	Scapula	Glenoid		1
E	28		Bovid	Tooth	Molar		1
E	46	3	Bovid	Tooth	Molar	Upper, worn	1
E	7,11,40		Bovid	Tooth	Premolar		3
E	5		Carnivora	Mandible			1
E	1		Carnivora	Premaxilla	Canine		1
E	1,5,15,33		Cervid	Antler			4
E	35		Cervid	Antler	Pedicel		1
E	4		Proboscidean	Phalange			1
E	4		Proboscidean	Tooth	Tusk		1
E	23,34	5	Proboscidean	Tooth	Molar		2
E	19		Proboscidean	Ulna			1
E	39		Bovid (Large)	Tooth	Molar 3	Lower, complete	1
E	0		Hyaenid	Maxillary	Fragment		1
E	0		Proboscidean	Tooth	Tusk		1
E	**2**		*Sivatherium*	Humerus	Distal		1
E	14		Indet/Carnivore	Maxilla	Canine	V.worn	1
E	0		Indeterminate	Fragments			136
F	1		Bovid	Magnum			1
F	6		Bovid	Mandible	Molar		1
F	0		Indeterminate	Fragments			99
G	25	3	Artiodactyla	Pelvis			1
G	**12**	2b	Bovid	Radius	Distal		1
G	0		Indeterminate	Fragments			9
H	0		Indeterminate	Fragments			9
J	35		Artiodactyla	Mandible			1
J	40	3	Artiodactyla	Ulna			1
J	5,6	3	Bovid (Large)	Molar, upper	Complete	Mod.wear	2
J	0		Indeterminate	Fragments			347
Total							**1287**

Notes: 481B31 and 32 and 481G GB25 had a sand matrix; 481B28 (indet. shaft) a grit matrix; fine sand was adhering to 481B47 (indet. vertebra), and 481G25. A silt matrix was on: 481BGB47. Several pieces of scute joined from 481 within zones A, B and E; Bovid calcaneum pieces 481EGB8 and 481EGB22 joined.

Table 6. 18: Fossil material from locality 489, west Branch of Kotha Kas.

Zone	Specimen	W	Taxon	Element	Part	Type	N
A	5		Bovid	Radius	Distal		1
A	6		Bovid	Tibia			1
A	1	2a	Cervid	Antler	Palmate	Pedicel	1
A	0		Indeterminate	Rib			1
A	0		Indeterminate	Shaft			1
A	2	3	Indet (Large)	Vertebra	Cervical	Centrum	1
A	3		Indet/*Crocodylus*	Crania			1
B	22	2b	Artio/Cervid	1st Phalange	Distal		1
B	24	4	Artiodactyla	Phalange	Distal		1
B	10	3	Artiodactyla	Radius	Proximal	Unfused	1
B	2,14,32		Artiodactyla	Ramus			3
B	15	3	Artiodactyla	Tibia	Proximal	Unfused	1
B	3	3	Artiodactyla	Vertebra	Caudal	Complete	1
B	18		Artiodactyla	Vertebra	Complete		1
B	21,27	2b	Artiodactyla	Vertebra	Centrum		2
B	**9,13**		Bovid	Astragulus	Complete		2
B	8	2b	Bovid	Calcaneum	Shaft		1
B	37		Bovid	Fibula	Distal		1
B	34		Bovid	Metacarpal	Distal		1
B	31		Bovid	Pisiform	Complete		1
B	9		Bovid	Pelvis			1
B	**6**		Bovid	Tibia	Distal	Complete	1
B	19	3	Bovid	Tooth	M1/2, upper	Worn	1
B	35	3	Bovid	Tooth	Premolar		1
B	33		Canidae	Calcaneum	Proximal		1
B	7		Cervid	Antler			1
B	0		Chelonia	Scute	Smooth		6
B	0		Chelonia	Scute	Netted		1
B	30		*Crocodylus*	Tooth			1
B	25		Equid	Metacarpal	Distal		1
B	23	2a	Equid	Premolar	Complete	Worn	1
B	5		Felidae	Femur	Proximal		1
B	2		Felidae	Scapula			1
B	46	4	Hippopotamid	Femur	Distal		1
B	12	5	Rhino	Metacarpal	Distal		1
B	0		Indeterminate	Fragments			59
C	9	3	Artiodactyla	1st Phalange	Distal		1
C	15		Artiodactyla	1st Phalange			1
C	2	3	Artiodactyla	Femur	Proximal		1
C	15	3	Artiodactyla	Humerus	Proximal	Unfused	1
C	8		Artiodactyla	Mandible			1
C	7		Artiodactyla	Metapodial	Shaft		1
C	25		Artiodactyla	Pelvis			1
C	22		Artiodactyla	Scapula			1
C	34,35		Artiodactyla	Tibia	Proximal	Split	2
C	3,14	3	Artiodactyla	Vertebra			2
C	1	3	Artiodactyla	Vertebra	Atlas		1
C	21		Artiodactyla	Vertebra	Axis		1
C	8,13	3	Artiodactyla	Vertebra	Centrum		2
C	20		Artiodactyla	Vertebra	Lumbar		1
C	26,41		Artiodactyla	Vertebra	Complete		2
C	31	3	Artiodactyla	Vertebra	Cervical		1
C	32		Artiodactyla	Vertebra	Sacrum	Complete	1
C	30		Artiodactyla	Vertebra	Thoracic		1
C	28		Bovid	1st Phalange	Proximal		1

Table 6. 18 continued: Fossil material from locality 489, west Branch of Kotha Kas.

Zone	Specimen	W	Taxon	Element	Part	Type	N
C	**26**		Bovid	Astragulus	Complete		1
C	5	2b	Bovid	Femur	Distal		1
C	22	3	Bovid	Femur	Proximal	Condyle	1
C	30		Bovid	Fibula	Distal	Complete	1
C	25	5	Bovid	Navicular	Complete		1
C	2		Bovid	Pisiform	Complete		1
C	**12**		Bovid	Radius	Distal	Complete	1
C	19		Bovid	Scaphoid	Neural		1
C	4	2a	Bovid	Scapula	Glenoid		1
C	18		Bovid	Tibia	Proximal		1
C	7		Bovid	Tooth	Premolar		1
C	16	3	Bovid	Tooth	PM, upper	Complete	1
C	30		Bovid	Ulna-carpal	Complete		1
C	11		Cervid	Antler			1
C	13		Cervid	Antler	Pedicel		1
C	8		Chelonia	Scute	Smooth		13
C	8	2a	*Crocodylus*	Tooth			1
C	21		Equid	Vertebra	Neural		1
C	0		Indeterminate	Fragments			125
D	62		Artiodactyla	1st Phalange			1
D	2,17	3	Artiodactyla	Femur	Proximal		2
D	37	2b	Artiodactyla	Ulna	Proximal		1
D	0		Artiodactyla	Vertebra			7
D	72	3	Artiodactyla	Vertebra	Complete		1
D	52		Bovid	1st Phalange	Proximal		1
D	1		Bovid	Astragulus	Complete		1
D	4,5		Bovid	Femur	Distal		2
D	14		Bovid	Horncore			1
D	27		Bovid	Navicular			1
D	35		Bovid	Pisiform	Complete		1
D	41		Bovid	Unciform	Complete		1
D	70	3	Bovid (Small)	Tooth	M1/2, lower	Mod.wear	1
D	21		Cervid	Antler			1
D	46		Chelonia	Scute			2
D	3	3	Equid	Tibia	Proximal		1
D	0		Indeterminate	Fragments			109
E	12	4	Artiodactyla	3rd Phalange	Proximal		1
E	13		Artiodactyla	Mandible			3
E	6,16	2b	Artiodactyla	Phalange	Distal		2
E	13		Artiodactyla	Tooth	Molar		1
E	9		Artiodactyla	Vertebra	Complete		1
E	187		Bovid (Small)	Calcaneum	Fragment		1
E	**2**	3	Bovid	1st Phalange	Distal		1
E	3,5	2a	Bovid	2nd Phalange	Complete		2
E	**1**	2b	Bovid	2nd Phalange	Distal		1
E	**10**	3	Bovid	2nd Phalange	Proximal	Complete	1
E	7		Bovid	Pisiform	Complete		1
E	10		Bovid	Phalange			1
E	11	2b	Bovid	Tooth	M1/2, worn	Complete	1
E	13		Bovid	Premolar	Upper	Complete	1
E			Canid (Small)	2nd Phalange	Proximal		1
E	4	2a	Cervid	3rd Phalange	Complete		1
E	0		Chelonia	Scute	Dimpled		60
E	0		Chelonia	Scute	Smooth		100
E	0		Chelonia	Scute	Fishnet		1

Table 6. 18 continued: Fossil material from locality 489, west Branch of Kotha Kas.

Zone	Specimen	W	Taxon	Element	Part	Type	N
E	0		Chelonia	Scute	Netted		1
E	0		Chelonia/Croc.	Scute	Pitted		1
E	0		Croc/Chelonia	Mandible	Fragment		1
E	0		*Crocodylus*	Tooth	Fragment		1
E	0	4	*Crocodylus*	Vertebra	Fragment		1
E	4,10		Giraffidae	Tooth	Molar		2
E	0		Indeterminate	Fragments			17
F	11		Artiodactyla	1st Phalange			1
F	13,29	3	Artiodactyla	1st Phalange	Proximal		2
F	45	4	Artiodactyla	1st Phalange	Distal	Unfused	1
F	28		Artiodactyla	Mandible	Roots		1
F	35		Artiodactyla	Mandible			1
F	43		Artiodactyla	Mandible	PM		1
F	4	2b	Artiodactyla	Metapodial	Distal, unfused	Epiphysis	1
F	32,34		Artiodactyla	Pisiform	Complete		2
F	1	3	Artiodactyla	Tibia	Distal	Unfused	1
F	2		Artiodactyla	Tibia	Proximal		1
F	14		Artiodactyla	Tooth	Premolar		1
F	33,35		Artiodactyla	Tooth	Molar		2
F	48,49		Artiodactyla	Tooth			2
F	8	4	Artiodactyla	Ulna	Proximal		1
F	5		Artiodactyla	Vertebra			1
F	6,7		Artiodactyla	Vertebra	Unfused		2
F	16	3	Artiodactyla	Vertebra	Epiphysis		1
F	21		Artiodactyla	Vertebra	Flange		1
F	39		Bovid	Cuneiform	Complete		1
F	38		Bovid	Fibula	Distal	Complete	1
F	38		Bovid	Fibula	Distal		1
F	10		Bovid	Unciform	Complete		1
F	18	2a	Cervid	3rd Phalange			1
F	0		Chelonia	Scute	Netted		12
F	0		Chelonia	Scute	Dimpled		5
F	0		Chelonia	Scute	Smooth		40
F	0		Chelonia/*Crocodylus*	Ribs			5
F	0		Indeterminate	Fragments			151
G	21	3	Artiodactyla	Axis			1
G	27	3	Artiodactyla	Femur	Distal	Unfused	1
G	9	3	Artiodactyla	Metapodial	Proximal		1
G	18	3	Artiodactyla	Pelvis			1
G	22	2b	Artiodactyla	Pelvis	Acetabulum		1
G	32	2b	Artiodactyla	Sacrum			1
G	10	3	Artiodactyla	Scapula			1
G	15,17,30	3	Artiodactyla	Vertebra			3
G	20,31	4	Artiodactyla	Vertebra	Complete		2
G	26	3	Bovid	1st Phalange	Proximal	Complete	1
G	12		Bovid	Radius	Distal		1
G	1	2a	Bovid	Premolar	Upper	Complete	1
G	5		Bovid	Tooth	Molar		1
G	6		Bovid	Tooth	Premolar		1
G	0		Chelonia	Scute	Dimpled		1
G	2		Chelonia	Scute	Smooth		1
G	29		Indeterminate	Fragments			17
H	2		Bovid	Cuneiform			1
H	1		Bovid	Unciform	Complete		1
H	0		Chelonia	Scute	Smooth		1
H	0		Chelonia	Scute	Netted		1
H	0		*Crocodylus*	Vertebra	Complete		1
H	0		Indeterminate	Shaft			18
J	17	3	Artio (Large)	Vertebra	Axis		1
J	40		Artio (Medium)	Ulna	Fragment		1

Table 6. 18 continued: Locality 489:

J	14	3	Artiodactyla	Humerus	Distal		1
J	35		Artiodactyla (Small)	Mandible	Fragment		1
J			Artiodactyla (Small)	Mandible + M2	Right		1
J	14		Artiodactyla	Metapodial			1
J	20		Artiodactyla	Metapodial	Proximal		1
J	34		Artiodactyla	Ramus			1
J	49	3	Artiodactyla	Tooth	Molar		1
J	11	3	Bovid	Astragulus			1
J	22	3	Bovid	Ext. cuneiform			1
J	1		Bovid	Horncore			1
J	19	3	Bovid	Scapula	Glenoid		1
J	16,22,47	3	Bovid	Tooth	Premolar		3
J	23	4	Bovid	Molar	Complete	Upper	1
J	8	3	Bovid	Ulna	Proximal		1
J	10		Bovid (Large)	2nd Phalange	Complete		1
J	5, 6		Bovid (Large)	Molar, upper	Complete	Mod.wear	2
J	18	3	Bovid (Large)	Ulna	Proximal		1
J	0		Chelonia	Scute	Dimpled		2
J	0		Chelonia	Scute	Netted		3
J	0		Chelonia	Scute	Pitted		1
J	0		Chelonia	Scute	Smooth		19
J	0		Chelonia/Croc.	Rib			2
J	0	5	*Crocodylus*	Vertebra			1
J	7		Indet/Proboscidean	Carpal/tarsal	Complete		1
J	0		Indet	Condyle			1
J	0		Indeterminate	Fragments			373
Overall total							**1330**

Notes: 14 pieces had a silt matrix, 9 had a sand matrix, and one (zone E, indet rib), a clay one. In Zone E, many pieces of turtle scute joined.

Table 6. 19: Fossil material from locality 491, West Branch of Kotha Kas.

Zone	**Specimen**	**W**	**Taxon**	**Element**	**Part**	**Other**	**N**
A	15	3	Artiodactyla	2nd Phalange	Proximal		1
A	24		Artiodactyla	Metacarpal	Distal	Condyle	1
A	25		Artiodactyla	Metapodial			1
A	8	2a	Artiodactyla	Unciform			1
A	5		Bovid	2nd Phalange	Complete		1
A	**0**	3	Bovid	Astragalus	Complete		1
A	11	5	Bovid	Horncore			1
A	2,14	3	Bovid	Humerus	Distal	Split	2
A	1		Bovid	Scapula			1
A	6		Bovid	Tibia	Distal		1
A	21,22	3	Bovid	Tooth	Premolar	Split, lower	2
A	28		Bovid	Tooth	Premolar		1
A	3		Bovid (Large)	P2, lower	Complete	Little wear	1
A	22	3	Cervid	Antler			1
A	28		Proboscidean	Tooth	Molar		1
A	23		Indet/Proboscid	Tooth	Molar	Plate	1
A			Indeterminate	Fragments			77
B	48	2b	Artiodactyla	1st Phalange	Proximal	Split	1
B	1	3	Artiodactyla	Calcaneum	Shaft		1
B	17		Artiodactyla	Calcaneum	Proximal		1
B	1		Artiodactyla	Femur	Distal		1
B	53	4	Artiodactyla	Humerus	Proximal		1
B	51	3	Artiodactyla	Magnum	Complete		1
B	8		Artiodactyla	Pelvis			1
B	0		Artiodactyla	Scapula			1
B	46		Bovid	1st Phalange			1
B	20		Bovid	Horncore			1
B	10	2b	Bovid	Metacarpal	Distal		1
B	47		Bovid	Pisiform	Complete		1

Table 6. 19 continued: Locality 491:

B	2		Bovid	Tibia	Distal		1
B	39	2b	Bovid	M1/2, lower	Complete	Little wear	1
B	11,12		Cervid	Antler			2
B	0		Chelonia	Scute			6
B	0		Chelonia	Scute	Netted		1
B	0		Chelonia	Scute	Fine netted		1
B	0		Chelonia	Scute	Smooth		1
B	15		*Crocodylus*	Humerus	Distal		1
B	38		Giraffidae	Tooth			1
B	9		Indet/Artio	Humerus	Distal	Unfused	1
B			Indeterminate	Fragments			142
C	**7**	2a	Bovid	Astragulus	Complete		1
C	33		Bovid	Calcaneum	Shaft		1
C	6	3	Bovid	Femur	Distal	Unfused	1
	43		Bovid	Lunar			1
C	**11**	2a	Bovid	Metatarsal	Proximal	Complete	1
C	58	3	Bovid	Phalange	Proximal		1
C	**5**	3	Bovid	Radius	Proximal	Complete	1
C	53		Bovid	Premolar	Complete	Upper	1
C	8	3	Bovid (Large)	Tibia			1
C	25	3	Bovid/Cervid	Metapodial	Distal	Condyle	1
C	1-3, 10, 22, 26,32		Cervid	Antler			7
C	0		Chelonia	Scute			40
C	51,56		Proboscidean	Tooth	Molar		2
C	59		Rhinocerotidae	Tooth	Molar	Lower	1
C	0		Indeterminate	Fragments			118
D	13		Artiodactyla	2nd Phalange	Complete		1
D	19		Artiodactyla	Calcaneum	Shaft		1
D	4		Artiodactyla	Humerus			1
D	8		Artiodactyla	Humerus	Distal	Diaphysis	1
D	15		Artiodactyla	Metatarsal	Shaft		1
D	18	2b	Bovid (Large)	Premolar	Complete	Worn	1
D	26	3	Bovid (Large)	Tooth	Molar	Worn	1
D	13		Bovid	2nd Phalange			1
D	17		Bovid	Cuneiform	Complete		1
D	5	4	Bovid	Horncore			1
D	29		Bovid	Mandible	Premolar		3
D	29		Bovid	Mandible	Molar		2
D	**1**		Bovid	Metacarpal	Distal	Unfused	1
D	22	4	Bovid	Metacarpal	Distal		1
D	21		Bovid	Scaphoid	Complete		1
D	11	2b	Bovid	Molar, upper	Complete	V. worn	1
D	12	3	Bovid	Molar	Complete	Mod.wear	1
D	29		Bovid	Tooth	Premolar		1
D	29,33		Bovid	Tooth	Premolar		2
D	5	3	Bovid (Large)	M1/2, upper	Complete	Mod.wear	1
D	38		Carnivora	Tooth	Incisor		1
D	7,9,31,32		Cervid	Antler			4
D	16		Cervid	Antler	Tine		1
D	0		Cervid (Large)	Mandible	P2,3,DM3 worn, M1	Left	1
D	0		Chelonia	Scute			10
D	0		Proboscidean	Tooth	Molar		1
D	0		Proboscidean	Tooth	Tusk		4
D			Indeterminate	Fragments			103
E	0		Chelonia	Scute			10
Total							**630**

Notes: 491A 10 (Bovid astragalus), 491GB39, 491C5, and 491D22 had a fine sand matrix. There was a silt matrix on 491AGB10, 491CGB5, and 491DGB22.

Table 6. 20: Fossil material from Set 14, West Branch of Kotha Kas (343-357, odd nos., 383-395)

Fossil occurrence 343: scatter:

Specimen	W	Taxon	Element	Part	Other	N
7	4	Artiodactyla	Astragalus	Complete		1
0		Artiodactyla	Calcaneum	Shaft	Suncracked end	1
2		Artiodactyla	Humerus	Distal		1
6	4	Artiodactyla	Metapodial	Unfused	Distal Barrel	1
0		Artiodactyla	Scapula	Blade		1
8,10	4	Bovid	Horncore			2
12		Bovid	Metatarsal	Complete	Suncracked	1
1	3	Bovid	Radius	Proximal	Split	1
13-15		Bovid	Radius	Shaft		3
3		Bovid	Tooth			1
5	4	Bovid	M1/2	Complete	Upper	1
11		Cervid/Bovid	Metacarpal	Complete	Distal	1
8		Rhinocerotidae	Tibia	Distal		1
0		Indeterminate	Fragments			114
Total						**130**

Fossil occurrence 345: scatter:

Specimen	W	Taxon	Element	Part	Other	N
7		Artiodactyla	Tooth	Fragment		1
1	2b	Bovid	Astragulus	Complete		1
6	3	Bovid	Tooth	PM3	Lower	1
4		Cervid	Antler	Palmate		1
		Indeterminate	Fragments			123
Total						**127**

Fossil occurrence 347: general collection:

Specimen	W	Taxon	Element	Part	Other	N
16		Artiodactyla	Mandible			1
10		Canidae	Humerus			1
11	3	Proboscidean	Shaft			1
7		Equid	Tooth	Split,worn		1
15	4	*Hippopotamus*	Mandible	Premolar		1
		Indeterminate	Fragments			8
Total						**13**

Fossil occurrence 349: find-spot:

Specimen	W	Taxon	Element	Part	Other	N
4	5	Proboscidean	Mandible		No roots/crown	1
1	3	Proboscidean	Tooth	Molar	Root	1
2	3	Indeterminate	Femur	Proximal		1
3	5	Indeterminate	Vertebra	Centrum		1
Total						**4**

Fossil occurrence 351: general collection:

Specimen	W	Taxon	Element	Part	Other	N
0	5	Artiodactyla	Tooth	PM	Fragment	1
2	2b	Cervid	Antler			1
6	4	Proboscidean	Tooth	Molar	Plate	1
0		Proboscidean	Tooth	Tusk		2
5		Indeterminate	Fragments			57
Total						**62**

Table 6. 20 continued: Fossil occurrence 353: general collection:

Specimen	W	Taxon	Element	Part	Other	N
8	3	Equid	M3	Lower	Right, Worn	1
2		*Sivatherium*	1st Phalange	Proximal	Complete	1
0		*Struthio*	Eggshell			4
0		Indeterminate	Fragments			19
Total						**25**

Fossil occurrence 355: scatter:

Specimen	W	Taxon	Element	Part	Other	N
17	5	Artio (Large)	Tooth	Premolar		1
19,23	3	Artiodactyla	Cranium			2
8		Artiodactyla	Humerus	Distal	Fragment	1
27		Artiodactyla	Metatarsal			1
21	3	Artiodactyla	Shaft			1
0		Artiodactyla	Tooth	Spall		4
14	3	Artiodactyla	Molar	Lower	Split, worn	1
18		Artiodactyla	Tooth	Premolar		4
2,3, 20	3	Bovid	Horncore			3
4-6,10	3	Cervid	Antler			4
25		Cervid	Antler			1
15	3	Giraffid	Premolar	Upper, Split	Worn	1
0, 9, 11,12		Proboscidean	Tooth	Tusk		5
		Indeterminate	Fragments			55
Total						**84**

Fossil occurrence 357: scatter:

Specimen	W	Taxon	Element	Part	Other	N
1	2a	Bovid	Fibula	Distal		1
8		Carnivora	Metatarsal	Proximal		1
5,9		Giraffid	Tooth			2
2,3		Proboscidean	Tooth	Molar		2
0		Indeterminate	Fragments			69
Total						**75**

Notes: Matrix details: 341GB1 had an orange sand matrix; 343GB2 and 343GB7 one of fine sand; 347GB16, 355GB4 and 355GB6 had a sandstone matrix. 343GB13-15 are listed as three specimens but are from the same bone.

Set 14 (Pelvis Valley): 383-395

Fossil occurrence 383: general collection:

Specimen	W	Taxon	Element	Type	Other	N
1,3,5		Bovid	Horncore			3
0		Bovid	Tooth	Premolar		1
4		Cervid	Antler			1
0		Indeterminate	Fragments			39
Total						**44**

Fossil occurrence 385: scatter:

Specimen	W	Taxon	Element	Type	Other	N
385A GB20	3	Artiodactyla	Tooth	Molar	Spall	1
385A GB1		Bovid	Humerus			1
385A GB2, 3,5-7,9-11	2b	Cervid	Antler			8
385A		Indeterminate	Fragments			34
385B GB1		Proboscidean	Scapula			1
385B		Indeterminate	Fragments			52
385C GB10		Artiodactyla	Radius			1
385C GB3		*Hippopotamus*	Radius			1

Table 6.20 continued: fossil occurrence 385:

385C		Indeterminate	Fragments			163
385DGB1, 6,7		Cervid	Antler			3
385D		Indeterminate	Fragments			24
Total						**289**

Fossil occurrence 387: scatter:

Specimen	W	Taxon	Element	Type	Other	N
1	4	Bovid	Calcaneum	Shaft		1
13	3	Bovid/Cervid	Tooth	Molar		1
0	3	Bovid/Cervid	Pelvis	Complete		1
9,12		Cervid	Antler	Palmate		2
3,5	5	Proboscidean	Tooth	Molar	Plate	2
17		Suid	M3	Upper	V.worn	1
0		Indeterminate	Fragments			37
Total						**45**

Fossil occurrence 391: scatter:

Specimen	W	Taxon	Element	Type	Other	N
2		Artiodactyla	Mandible			1
1	2b	Bovid	Mandible	M3, right	Worn	1
3	5	Bovid	Tooth	Premolar	Lower, Split	1
0		Indeterminate	Fragments			12
Total						**15**

Fossil occurrence 393: general collection:

Specimen	W	Taxon	Element	Type	Other	N
7	3	Artiodactyla	2nd Phalange	Proximal		1
8	4	Artiodactyla	Carpal	Complete		1
9	4	Artiodactyla	Fibula	Distal	Complete	1
2		Artiodactyla	Metapodial			3
4	2b	Bovid	Astragulus	Complete		1
21	2a	Bovid	Cranium			1
12	4	Bovid	Lunar	Complete		1
1		Bovid	Tooth	Premolar		1
3	3	Bovid (Large)	M3	Complete	Lower, worn	1
5		Bovid	M3	Lower	Left, Split	1
10	3	Bovid	Premolar	Complete	Upper, worn	1
0	2b	Bovid/Cervid	PM	Upper	Complete	1
6		Cervid	Astragulus			1
20	2a	Cervid	Femur	Proximal	Right	1
15	3	Proboscidean	1st	Phalange	Proximal	1
5		*Hippopotamus*	Mandible			1
0		Indeterminate	Fragments			144
Total						**162**

Fossil occurrence 397: general collection:

Specimen	W	Taxon	Element	Type	Other	N
21		Artiodactyla	Metapodial			1
50	5	Artiodactyla	Radius	Proximal		1
9,12		Bovid	Calcaneum	Complete		2
18		Bovid	Horncore			1
4	4	Bovid	Metacarpal	Distal	Barrel	1
3	3	Bovid	Radius	Shaft		1
13	4	Bovid	Tibia	Distal	Complete	1
16	2a	Bovid	M3	Upper	Left, Fresh	1
20		Bovid	Tooth	Premolar		1
23	2b	Bovid	Premolar	Upper	Worn	1
2		Bovid	Ulna	Proximal		1

Table 6. 20 continued: fossil occurrence 397:

7,22,25,31,35, 41,46	2a	Cervid	Antler			7
5,6	3	Proboscidean	Carpal/tarsal			2
45	4	Proboscidean	Humerus	Proximal	Condyle	1
8		Proboscidean	Rib			1
0		Indeterminate	Fragments			145
Total						**168**

Note: Scatter 389 comprised only 4 indeterminate fragments within general collection 393. Occurrence 395 comprised some flake stone near 387.

Table 6. 21: Fossil material from Set 15, the Viaduct area (359-381, odd. nos. only).

Fossil occurrence 359: general collection:

Specimen	W	Taxon	Element	Part	Other	N
2	3	Cervid	Antler	Pedicel	Palmate	1
3,5,8	2b	Cervid	Antler	Palmate		3
1		Proboscidean	Mandible	No teeth		1
19	5	Proboscidean	Tooth	Molar		1
0		Indeterminate	Fragments			17
Total						**23**

Fossil occurrence 361: scatter:

Specimen	W	Taxon	Element	Part	Other	N
361A GB1		Artiodactyla	Tibia			1
361A GB15		Artiodactyla	Vertebra	Condyle		1
361A GB51		Bovid	Tooth	Molar		1
361A GB12		Bovid (Large)	Tooth	M3 Upper	Mod.wear	1
0		Indeterminate	Fragments			114
10,24		Proboscidean	Tooth			2
361B GB4	4	Artiodactyla	Mandible	P2 socket	No teeth	1
361B GB3		Bovid	Ext.cuneiform	Complete		1
361B GB9	3	Bovid	Horncore			1
361B GB14		Bovid	Radius	Distal		1
361B GB1		Bovid	Tooth	Molar		1
361B GB15	2b	Bovid (Large)	M3 lower	Complete, right	Worn	1
361B		Cervid	Antler			12
361B GB19		Chelonia	Scute	Pitted		1
361B		Indeterminate	Fragments			40
Total						**179**

Fossil occurrence 371: general collection:

Specimen	W	Taxon	Element	Part	Other	N
33		Artiodactyla	Phalange	Distal		1
9	5	Artiodactyla	Tooth			1
2		Bovid	Cuneiform	Cuneiform		1
39	4	Bovid	Horncore			1
40	4	Bovid	Horncore	+Cranium		1
7,10	2a	Cervid	Antler	Palmate		2
18	4	Equid	Tooth, upper	Molar, Split	Mod. wear	1
4		Felid (Large)	Scapula	Glenoid		1
	5	Proboscidean	Tooth	Molar		18
35		Indeterminate	Fragments			66
Total						**93**

Table 6. 21 continued: Fossil occurrence 373: scatter:

Specimen	W	Taxon	Element	Part	Other	N
27		Artiodactyla	Atlas			1
17		Artiodactyla	Radius			1
25		Artiodactyla	Scaphoid			1
14		Bovid	1st Phalange			1
36		Bovid	Radius	Distal		1
3,4, 22		Bovid	Tooth	Fragment		3
11		Bovid	Tooth	Premolar		1
12		Bovid/*Hemibos)*	M3, upper			1
28	5	Bovid	Tooth	Premolar,	Split	1
31	3	Bovid	M3	Lower	Mod. wear	1
19		Bovid	Trapezoid/Mag num			1
25		Bovid (Large)	Lunatc			1
5		Equid	M1/2 upper	Complete	Mod. wear	1
19		Equid	Navicular	Complete		1
32		*Struthio*	Eggshell			1
0		Indeterminate	Fragments			83
Total						**100**

Fossil occurrence 375 scatter:

Specimen	W	Taxon	Element	Part	Other	N
375A GB1		Bovid	Metatarsal			1
375A GB3	3	*Elephas*	Molar			1
375A GB10	3	Proboscidean	Tooth	Tusk		4
375A GB2	3	Rhinocerotid	Scapula	Glenoid		1
375A GB8		Rhinocerotid	Tooth	Premolar	Upper	1
375A	3	Indeterminate	Fragments			224
375B GB8	3	Bovid	2nd Phalange	Complete		1
375B GB3	3	Proboscidean	Femur	Condyle		1
375B2,5,6	2b	Proboscidean	Rib			3
375B GB12, 13	3	Proboscidean	Tooth	Molar		2
375B		Indeterminate	Fragments			102
375C GB11	3	Artiodactyla	Scapula			1
375CGB14, 16	3	Cervid	Antler			2
375C GB4,19	3	Proboscidean	Rib			2
375C GB3	2b	Proboscidean	Molar	Deciduous		1
375C GB9		Proboscidean	Vertebra	Caudal		1
375C		Indeterminate	Fragments			73
Total						**421**

Fossil occurrence 377 scatter:

377A GB14	4	Artiodactyla	Molar	No crowns		1
377A GB19	5	Artiodactyla	Tooth	Molar		1
377A GB4	5	Bovid	Calcaneum	Shaft		1
377A GB5		Bovid	Scaphoid	Complete		1
377A		Indeterminate	Fragments			36
377B GB3		Artiodactyla	Parietal			1
377B GB7		Bovid	Cuneiform	Complete		1
377B GB9,22	5	Bovid	Horncore			2
377B GB14		Bovid	Magnum	Complete		1
377B GB19		Bovid	Pisiform			1
377B GB20	3	Bovid (Large)	M1/2 Lower	Complete,	Erupting	1
377B GB2,18		Cervid	Antler			2
377GB		Indeterminate	Fragments			27
Total						**81**

Table 6. 21 continued: Fossil occurrence 379: scatter:

Specimen	W	Taxon	Element	Part	Other	N
6		Artiodactyla	Metapodial			1
4		Artiodactyla	Phalange			1
8	4	Bovid	Horncore			1
1		Bovid/*Hemibos*	Mandible	Right, +M3		1
10	3	Suid	M3 lower	No crowns	Broken	1
5	4	Indeterminate	Fragments			41
Total						**46**

Fossil occurrence 381: general collection:

Specimen	W	Taxon	Element	Part	Other	N
0	5	Artiodactyla	Tooth			2
3		Bovid	2nd Phalange			1
9		Bovid	Horncore	Tip		1
1	5	Proboscidean	Tooth	Fragment		7
0		Indeterminate	Fragments			36
Total						**47**

Notes: 371GB4 has a sand matrix. Fossil occurrence 363 was a scatter with 38 indeterminate fragments. Scatter 365 contained 2 artiodactyl 2nd phalanges, 1 piece of bovid horn core and 4 fragments; scatter 367 contained 1 bovid tooth fragment, 1 piece of antler, 1 Proboscidean tooth plate and 2 indeterminates. There was no fossil from 369.

Table 6. 22: Fossil material from Set 16, the East Branch of Kotha Kas (407-425 odd nos. only) and 499.

Fossil occurrence 407: scatter:

Specimen	W	Taxon	Element	Part	Other	N
2	5	Artiodactyla	Premolar	Worn	Split	1
10	4	Bovid	Calcaneum			1
26		Bovid	Cuneiform			1
20		Bovid	Mandible	Premolar		1
4		Bovid	Scaphoid			1
4		Bovid	Tooth	Molar		1
15	3	Bovid	Premolar	Worn	Split	1
22,23		Bovid	Premolar	Upper	Fresh	2
24,25		Bovid	Tooth	Premolar		2
18	3	Bovid	Ulna			1
16		Bovid	Unciform			1
21		Chelonia	Scute	Smooth		1
9	3	Equid	Radius	Shaft		1
1		Equid	M3 lower	Complete	Slight wear	1
3		Equid	Unciform			1
6		*Struthio*	Eggshell			1
0		Indeterminate	Fragments			112
Total						**130**

Fossil occurrence 409 scatter:

Specimen	W	Taxon	Element	Part	Other	N
0		Artiodactyla	Rib			2
3	3	Artiodactyla	Scapula	Glenoid		1
0		Artiodactyla	Shaft			46
0		Bovid	Horncore			1
0		Bovid	Tooth	Molar		1
1		Proboscidean	Humerus	Distal	Condyle	1
0		Proboscidean	Shaft			7
0		Proboscidean	Tooth	Tusk		1
2		Proboscidean	Condyle			1
0	3	Proboscidean	Vertebra	Centrum		1
Total						**62**

Table 6. 22 continued: Fossil occurrence 413: scatter:

Specimen	W	Taxon	Element	Part	Other	N
413A GB4	5	Artiodactyla	Pelvis			1
413A GB16	3	Artiodactyla	Scapula	Fragment		1
413A GB5	2b	Bovid	M1/2	Complete	Lower	1
413A GB14	5	Bovid	Premolar	Split	Worn	1
413A GB0,2,7,11		Cervid	Antler			5
413A GB11	3	Cervid	Antler	Tine		1
413A GB3, 12,13		Chelonia	Scute	Netted		3
413A GB10		*Crocodylus*	Tooth			1
413A GB17		Equid	Tooth	Premolar	Split	1
413A GB1	2b	*Hippopotamus*	Tooth	Tusk		1
413A	3	Indeterminate	Fragments			5
413B GB3		Bovid	Astragulus			1
413B GB4, 7,13		Cervid	Antler			3
413B GB8		Cervid	Antler	Tine		1
413B GB3		Proboscidean	1st Phalange	Complete		1
413B GB26		Rhinocerotidae	Humerus			1
413B		Rhinocerotidae	Tooth	Molar		5
413B GB5,6		Rhinocerotidae	Tooth	Molar		2
413B GB9-11		Rhinocerotidae	Tooth	Molar		3
413B GB20-23		Rhinocerotidae	Tooth	Molar		4
413B		Indeterminate	Fragments			63
413C GB3	2a	Bovid	Mandible	M1/2 + Tooth	Right	1
Total						**105**

Fossil occurrence 415: scatter:

Specimen	W	Taxon	Element	Part	Other	N
11		Artiodactyla	Metapodial			1
9	4	Bovid	Tooth	Molar	Split	1
1,2,10	2b	Cervid	Antler			3
		Chelonia	Scute			10
0		Indeterminate	Fragments			46
Total						**61**

Fossil occurrence 417: general collection:

Specimen	W	Taxon	Element	Part	Other	N
2	2b	Bovid	Cuneiform			1
1,3-5	3	Proboscidean	Molar	Plate		4
0		Indeterminate	Fragments			18
417A GB1	4	Bovid	Scapula	Glenoid		1
Total						**24**

Fossil occurrence 419: scatter:

Specimen	W	Taxon	Element	Part	Other	N
22,23		Artiodactyla	Metapodial			2
2	3	Cervid	Antler	Branched		1
		Chelonia	Scute			2
20		Proboscidean	Rib			1
	3	Proboscidean	Tooth	Fragments		7
0		Indeterminate	Shaft			112
Total						**125**

Fossil occurrence 423: scatter:

Specimen	W	Taxon	Element	Part	Other	N
6,7	3	Artiodactyla	Tooth	Fragment		2
3	3	Artiodactyla	Vertebra	Fragment		1
5		Proboscidean	Tooth	Fragment		1
		Indeterminate	Fragments			35
Total						**39**

Table 6. 22 continued: Fossil occurrence 425: general collection:

Specimen	W	Taxon	Element	Part	Other	N
5,6		Bovid	Horncore			2
		Chelonia	Scute			2
0		Proboscidean	Tooth	Spall		1
3		*Struthio*	Eggshell			1
0		Indeterminate	Fragments			7
Total						**13**

Fossil occurrence 499: general collection:

Specimen	W	Taxon	Element	Part	Other	N
6	3	Artiodactyla	Metapodial	Distal	Epiphysis	1
19	2b	Artiodactyla	Nav.-cuboid	Split		1
1	2a	Bovid	Calcaneum	Complete		1
8,10,13	4	Bovid	Tooth	Molar	Fragments	3
20	5	Bovid	Tooth	Molar	Split	1
3	2b	Cervid	Antler			1
0		Chelonia	Scute	Pitted		1
0		Chelonia	Scute	Smooth		6
0		Chelonia	Scute	Netted		5
0		*Struthio*	Eggshell			1
0		Indeterminate	Fragments			69
Total						**90**

Notes: 407GB10 had a grey sand matrix; 407GB17 one of grit and pebbles; 409GB2, coarse sand; 419GB17, sand; 423GB4, fine sand; and 425GB1, cemented sandstone. 411 (general collection) contained a bovid mandible w3ith the right M3, and 11 other fragments; 421 was a general collection with only 7 indeterminate fragments.

Table 6. 23: Fossil material from Set 17, East Branch of Kotha Kas (427-441, odd nos. only)

Fossil occurrence 427: locality:

Specimen	W	Taxon	Element	Part	Other	N
0		Chelonia	Scute	Smooth		8
0		Indeterminate	Fragments			130
1A		Bovid (Large)	Mandible+	M2,M3, left		1
0A		Bovid	Tooth	Fragments		4
0A		Indeterminate	Fragments			50
8B		Chelonia	Scute	Netted		1
45D		Bovid (Small)	M1/2, lower	Complete	Little wear	1
0D		Chelonia	Scute	Smooth		23
4E		Bovid	PM	Upper	Worn	1
0E		Chelonia	Scute			1
0E		Chelonia	Scute	Netted		17
0E		Chelonia	Scute	Smooth		2
0E		Indeterminate	Fragments			66
1F		Artiodactyla	2nd Phalange	Complete		1
0F		Chelonia	Scute			38
0F		Chelonia	Scute	Netted		12
0F		Chelonia	Scute	Fine netted		4
2F		Chelonia	Scute	Dimpled		4
4F		Chelonia	Scute	Smooth		14
7F		*Crocodylus*	Tooth			1
0F		Indeterminate	Fragments			3
Total						**382**

Table 6. 23 continued: Fossil occurrence 429: general collection:

Specimen	W	Taxon	Element	Part	Other	N
3		Artiodactyla	1st Phalange	Proximal		1
11		Artiodactyla	Carpal	Accessory		1
4		Artiodactyla	Humerus			1
1		Cervid	Antler	Pedicel	Branched	1
2		Cervid	Antler			2
10		Chelonia	Scute	Smooth		1
6		Rhino/Hippo	Maxilla	Fragment		1
0		Indeterminate	Fragments			47
Total						**55**

Fossil occurrence 433: scatter:

Specimen	W	Taxon	Element	Part	Other	N
3		Artiodactyla	Ramus			1
9		Bovid (Small)	Molar	Split, mod. wear		1
0		Chelonia	Scute	Dimpled		1
		Chelonia	Scute	Netted		12
0		Chelonia	Scute	Smooth		7
5		Equid	Scaphoid			1
		Indeterminate	Fragments			3
Total						**26**

Fossil occurrence 435: locality:

Specimen	W	Taxon	Element	Part	Other	N
28		Artiodactyla	Humerus	Barrel	Distal	1
3		Artiodactyla	Mandible			1
15,45		Artiodactyla	Metapodial	Barrel	Distal	2
39		Artiodactyla	Pelvis			1
21,26		Artiodactyla	Radius	Proximal	Split	2
64		Artiodactyla	Radius	Distal		1
4		Artiodactyla	Molar	Complete	Erupting	1
5		Artiodactyla	Tooth	Spall		1
105		Artiodactyla	Tooth	Molar		1
6		Artiodactyla	Vertebra	Centrum		1
16		Artiodactyla	Vertebra			1
36		Artiodactyla	Vertebra	Centrum	Split	1
90		Bovid	1st Phalange			1
117		Bovid	1st Phalange	Distal		1
22		Bovid	Ext.cuneiform	Complete		1
14, 37		Bovid	Horncore			2
30		Bovid	Mandible, right	M3+ Tooth	Split	1
1		Bovid	Scaphoid			1
3		Bovid	Premolar	Complete	Worn	1
54		Bovid	Molar	Complete	Worn	1
55		Bovid	Molar	Lower, split	Mod.wear	1
61		Bovid	Tooth	Premolar		1
71		Bovid	M1/2	Lower	Worn	1
74		Bovid	Premolar	Complete	Worn	1
96		Bovid	Molar	Lower, split	Mod.wear	1
119		Bovid	Tooth	Premolar, split	Worn	1
121		Bovid	Molar	Lower	Split	1
130		Bovid	Premolar	Split		1
6		Bovid (Large)	M3, right	Complete, worn	Lower	1
112		Bovid (Small)	Premolar	Complete	Worn	1
10,42,62, 104		Cervid	Antler			4
2		Cervid	Scapula	Glenoid		1
0		Chelonia	Scute	Plain		1
0		Chelonia	Scute	Smooth		21
94		Chelonia	Scute	Netted		94

Table 6. 23 continued: Fossil occurrence 435: locality:

11, 101		*Crocodylus*	Tooth	Complete		2
9		*Crocodylus*	Vertebra			1
5,10,27		*Crocodylus*	Vertebra	Complete		3
27		*Crocodylus*	Vertebra	Complete		1
20		Proboscidean	Tooth	Tusk		2
0		Proboscidean	Tooth	Molar		16
31		Proboscidean?	Rib			1
8		Suid	Mandible	DM3, left	Little wear	1
0		Indeterminate	Fragments			514
Total						**695**

Fossil occurrence 437: locality:

No.	Specimen	W	Taxon	Element	Part	Other	N
437	1		Bovid (Large)	PM	Complete	Little wear	1
437	4		Bovid (Small)	M1/2, upper	Complete	Little wear	1
437	19		Cervid	1st Phalange	Complete		1
437	2		Chelonia	Scute	Smooth		1
437	5		Chelonia	Scute	Smooth		1
437A	0		Artiodactyla	Tooth	Incisor		1
437A	7		Artiodactyla	Molar	Upper, split	Worn	1
437A	0		Chelonia	Scute	Smooth		2
437A	0		Chelonia	Scute	Netted		27
437A	0		Chelonia	Scute	Dimpled		6
437A	0		Indeterminate	Fragments			45
437B	3		Artiodactyla	Femur	Proximal	Condyle	1
437B	2		Bovid	Horncore			1
437B	1		Bovid	Scapula	Glenoid		1
437B	16		Bovid	M1/2, upper	Complete	Little wear	1
437B	4		Chelonia	Scute	Smooth		1
437B	7		Chelonia	Scute	Netted		15
437B	19		Chelonia	Scute	Dimpled		7
437B	2		Chelonia/Croc.	Vertebra	Complete		3
437B	6		Equid	Femur	Proximal		1
437B	13		Giraffidae	Tooth	Premolar		2
437B	7		*Hippopotamus*	2nd Phalange	Complete	Lateral	1
437B	7		Proboscidean	5th Prox.Phal.	(Forelimb)		1
437B	0		Indeterminate	Fragments			142
437C	12		Artiodactyla	Atlas	Fragment		1
437C	34,85		Artiodactyla	Ext. Cuneiform			2
437C	127		Artiodactyla	Femur	Distal		1
437C	23		Artiodactyla	Humerus	Distal		1
437C	32		Artiodactyla	Humerus	Proximal	Condyle	1
437C	16, 33		Artiodactyla	Metatarsal	Proximal		2
437C	36		Artiodactyla	Radius	Distal		1
437C	102		Artiodactyla	Scapula	Fragment		1
437C	39		Artiodactyla	Ulna	Proximal		1
437C	38		Artiodactyla	Vertebra			1
437C	29		Bovid	Astragalus	Split		1
437C	98		Bovid	Lunar			1
437C	10		Bovid	Metacarpal	Distal		1
437C	58,99		Bovid	Metapodial	Distal, split	Unfused	1
437C	13		Bovid	Pelvis			1
437C	132		Bovid	Sacrum			1
437C	20		Bovid	Scaphoid			1

Table 6. 23 continued: locality 437:

437C	121	Bovid	PM, upper	Complete	Worn	1
437C	7	Bovid (Large)	P3, upper	Complete	Mod.wear	1
437C	66	Bovid (Large)	P4, lower	Complete		1
437C	66	Bovid (Large)	P3, upper	Split	Mod..wear	1
437C	30	Bovid (Large)	Trapezoid	+ Magnum		1
437C	47	Bovid (Small)	M3, upper right	Complete	Little wear	1
437C	103	Cervid	M1/2, upper	Split	Mod. wear	1
437C	0	Chelonia	Scute	Smooth		49
437C	61	Chelonia	Scute	Dimpled		1
437C	76	Chelonia	Scute	Netted		8
437C	119	Chelonia	Scute	Pitted		1
437C	66	*Crocodylus*	Tooth			1
437C	0	Indeterminate	Fragments			165
437D	59	Artiodactyla	1st Phalange	Proximal		1
437D	17, 50	Artiodactyla	Calcaneum			2
437D	56	Artiodactyla	Cuneiform	Distal		1
437D	10, 43	Artiodactyla	Humerus			2
437D	85, 129	Artiodactyla	Mandible+ tooth	Molar		2
437D	71	Artiodactyla	Metapodial	Distal	Barrel	1
437D	4,99,132	Artiodactyla	Phalange			3
437D	65	Artiodactyla	Radius	Distal	Split	1
437D	79	Artiodactyla	Tibia	Proximal	Split	1
437D	101	Artiodactyla	Tooth	Fragments		3
437D	2,18,33,72	Artiodactyla	Vertebra			4
437D	55,70,143	Bovid	Ext. cuneiform			3
437D	8,14	Bovid	Horncore			2
437D	62	Bovid	Scaphoid			1
437D	1,5	Bovid	Scapula	Glenoid		2
437D	35	Bovid	M1/2, upper	Complete	Little wear	1
437D	39	Bovid	Tooth	Molar	Split	1
437D	41,45,47,5 1	Bovid	Tooth	Fragment		4
437D	56	Bovid	Tooth	Premolar	Worn	1
437D	69	Bovid	Molar, upper	Split	Little wear	1
437D	76	Bovid	Tooth	Premolar		1
437D	111	Bovid	Tooth	Fragment		1
437D	32,37,40	Bovid	Ulna carpal			3
437D	0	Bovid/Cervid	Molar	Upper	Fragment	1
437D	12	Cervid	Antler			1
437D	0	Chelonia	Scute	Netted		2
437D	28	Chelonia	Scute	Large		1
437D	0	Chelonia	Scute	Smooth		80
437D		Chelonia	Scute	Dimpled		9
437D	30	Equid	Femur	Proximal		1
437D	26	Hyaenidae	Mandible			1
437D	11	*Hippopotamus*	Mandible	No teeth		1
437D	6	*Hippopotamus*	Ramus			1
437D	23	*Hippopotamus*	Unciform			1
437D	0	Indeterminate	Fragments			522
437D	164	Proboscid/Suid	Mandible	Molar		1
437D	45	cf. *Rhinoceros*	Nav.-cuboid	Complete		1
437E	4	Artiodactyla	Tooth	Fragment		1
437E	9	Bovid	1st Phalange	Distal		1
437E	2	Bovid	Calcaneum	Shaft		1
437E	6	Bovid	Lunar	Complete		1
437E	7	Bovid	Tooth			1

Table 6.23 continued: locality 437:

437E	8	Bovid	Premolar	Upper	Worn	1
437E	5	Cervid	Antler	+Pedicel	Branched	1
437E	0	Chelonia	Scute	Smooth		1
437E	1	Chelonia	Scute	Plain		1
437E	10	Chelonia	Scute	Smooth		1
437E	0	Indeterminate	Fragments			38
437F	1	Artiodactyla	1st Phalange	Distal		1
437F	**2**	Bovid	1st Phalange	Proximal	Complete	1
437F	0	Chelonia	Scute	Smooth		2
437F	0	Indeterminate	Fragments			85
437G	1	Bovid (Large)	Calcaneum	Shaft		1
437G	5	Chelonia	Scute	Smooth		1
437G	0	Indeterminate	Fragments			24
Total						**1347**

Fossil occurrence 439: general collection:

Specimen	W	Taxon	Element	Part	Other	N
22		2a	Artiodactyla	2nd Phalange	Complete	1
7		5	Artiodactyla	Molar, worn	Split	1
3			Bovid	Horncore		1
8			Proboscidean	Vertebra		1
5			Proboscidean	Phalange		1
0		Indeterminate	Fragments			77
Total						**82**

Fossil occurrence 441: scatter:

Specimen	W	Taxon	Element	Part	Other	N
8	4	Bovid (Large)	3rd Phalange	Complete		1
2		Proboscidean	Mandible	Ramus		1
4,6		Proboscidean	Tooth	Plate		2
1		Indeterminate/Proboscid.	Mandible	Ramus		1
0		Indeterminate	Fragments			65
Total						**70**

Fossil occurrence 443: general collection:

Specimen	W	Taxon	Element	Part	Other	N
3	3	Artiodactyla	Calcaneum			1
11		Artiodactyla	Calcaneum	Shaft		1
8		Bovid	Tooth	Spall		1
7	2a	Bovid (Large)	M1/2 Upper	Complete	Mod.wear	1
6		Cervid	Antler			1
14		Chelonia	Scute	Dimpled		1
15		Chelonia	Scute	Smooth		1
1,4	4	Hippopotamid	Mandible	Fragment		2
2	4	Hippopotamid	Mandible+	+1 broken tooth		1
10		Hippopotamid	Tooth	Fragment		1
0		Indeterminate	Fragments			115
Total						**126**

Notes: 427GBF3 has an orange sand matrix; there was a sand matrix on 435GB74,435GB29,and 437CGB20; and a fine sand matrix on 435GB51. 427AGB1 has a very gritty matrix; 437BGB7, 437BGB18 and 439GB3 had a mudskin matrix. 435GB60 and 76 joined; 435GB84 joined 102; numerous pieces of Chelonia scute joined, especially from 435. Scatter 431 comprised3 pieces of scute and 7 fragments of a Proboscidean pelvis.

Table 6. 24: Fossil material from Set 17, Kotha Kas (445-461, odd. nos. only).

Fossil occurrence 445: scatter/locality:

Zone	**Specimen**	**W**	**Taxon**	**Element**	**Part**	**Other**	**N**
	2		Artiodactyla	Astragulus			1
	5	3	Bovid	2nd Phalange	Complete		1
	1		Bovid	Femur	Distal		1
	4,18,27		Bovid	Tooth	Molar		3
	9		Bovid	Tooth	Premolar		1
A	1	3	Indeterminate (Large)	Femur	Distal		1
B	21	3	Artiodactyla	3rd Phalange			1
B	4		Artiodactyla	Cuneiform			1
B	3	4	Artiodactyla	Femur	Shaft	Unfused	1
B	4		Artiodactyla	Radius	Distal		1
B	4		Artiodactyla	Shaft			1
B	16,30	4	Artiodactyla	Tooth	Molar		2
B	0	3	Artiodactyla	Vertebra	Centrum		2
B	32		Bovid	1st Phalange	Distal		1
B	32		Bovid	3rd Phalange			1
B	7,33	3	Bovid	Horncore			2
B	23		Bovid	Horncore	Tine		1
B	12		Bovid	Naviculocuboid			1
B	38		Bovid	Naviculocuboid	Complete		1
B	22,46	3	Bovid	Tibia	Distal		2
B	28	2b	Bovid	Molar	Little wear	Split	1
B	3		Carnivora	Maxilla			1
B	various		Cervid	Antler			6
B	27		Cervid	Molar			1
B	9		Chelonia	Scute	Netted		7
B	10		Chelonia	Scute	Smooth		2
B	48		Chelonia	Scute	Dimpled		2
B	5		Proboscidean	Femur	Proximal		1
B	23		Proboscidean	Femur	Distal		1
B	3		Proboscidean	Tooth	Molar		4
B	8		Equid	Femur			1
B	46		Hyaenidae	2nd Phalange	Distal		1
B	2		Proboscidean	Ramus			1
B	0		Indeterminate	Fragments			164
C	2		Artiodactyla	Metapodial	Shaft		1
C	1		Giraff/Camelid	Femur	Distal	Condyle	1
D	3		Bovid	Tooth	Premolar		1
D	9		Bovid (Large)	Tooth	P3 Upper	Complete	1
D	0		Chelonia	Vertebra			1
D	0		Indeterminate	Fragments			28
E	0	3	Artiodactyla	Humerus	Proximal	Epiphysys	1
E	0		Artiodactyla	Phalange	Distal		1
E	16		Artiodactyla	Tibia	Proximal		1
E	38		Bovid	2nd Phalange			1
E	**13**		Bovid	Radius	Proximal	Complete	1
E	18		Bovid	Radius	Distal		1
E	7	3	Bovid	Scapula	Scapula		1
E	5		Carnivora	Astragulus			1
E	1		Chelonia	Scute	Honeycomb		1
E	2		Chelonia	Scute	Smooth		10
E	3		Chelonia	Scute	Netted		3
E	19		Proboscidean	Tooth	Molar		1
E	0		Indeterminate	Fragments			28
F	2	3	Artiodactyla	Tibia	Distal	Unfused	1
F	5		Bovid	1st Phalange			1

Table 6. 24 continued: Fossil occurrence 445:

F	9		Bovid	1st Phalange	Distal		1
F	2		Bovid	Mandible	Premolar		1
F	10	3	Bovid	Metapodial	Distal	Barrel	1
F	19	2a	Bovid	Pisiform	Complete		1
F	6	3	Bovid	Radius	Distal	Complete	1
F	1		Bovid	Ulna			1
F	0		Chelonia	Scute	Dimpled		3
F	0		Chelonia	Scute	Smooth		1
F	8		Chelonia	Scute	Netted		8
F	21-24		Proboscidean	Tooth	Molar		4
F	0		Proboscidean	Tooth	Plate		6
F			Indeterminate	Fragments			4
Total							**337**

Fossil occurrence 447: scatter/locality:

Zone	Specimen	W	Taxon	Element	Part	Other	N
A	0		Indeterminate	Fragments			10
B	4		Chelonia	Scute	Smooth		1
B	1	3	Indeterminate	Shaft			7
B	5		Proboscidean	Tooth	Plate		10
C	1,3	3	Artiodactyla	Pelvis			2
C	18	3	Bovid	PM	Upper	Worn	1
C	19	2b	Bovid	Tooth	PM	Lower	1
C	8		Bovid (Large)	Tooth	PM	Upper	1
C	29		Bovid (Small)	Lunate	Complete		1
C	11		Chelonia	Scute	Smooth		2
C			Indeterminate	Fragments			9
C	17		Proboscidean	Tooth	Plate		3
Total							**48**

Fossil occurrence 451: scatter/locality:

Zone	Specimen	W	Taxon	Element	Part	Other	N
			Crocodylus	Tooth			1
	0		Pisces	Fragment			1
A	3	3	Artiodactyla	Metacarpal	Shaft		1
A	2	2a	Artiodactyla	Metapodial			1
A	9,19		Artiodactyla	Metapodial	Distal		2
A	21	2b	Artiodactyla	Metapodial	Distal barrel	Unfused	1
A	30		Artiodactyla	Phalange			1
A	23		Bovid	1st Phalange			1
A	31	4	Bovid	Horncore			1
A	35	2b	Bovid	Mandible	M3 right	Worn	1
A	3		Bovid	Phalange			8
A	34,37	2b	Bovid	Tooth	Premolar	Upper	2
A	35	2b	Bovid	Tooth	Premolar	Worn	1
A	12, 36,40		Bovid	Tooth			3
A	11, 12	2b	Bovid (Large)	M3, Complete	Lower	Mod.wear	1
A	39	2a	Bovid (Large)	M1/2, upper	Little wear	Complete	1
A	15		Bovid/*Hemibos*	Mandible + M3	Left, worn		1
A	0		Chelonia	Scute	Dimpled		1
A	4		Chelonia	Scute	Smooth		103
A	29		Chelonia	Shaft			1
A	0		Indeterminate	Fragments			261
B	4	2a	Bovid (Large)	PM2, lower	Mod. wear		1
B	0		Chelonia	Scute	Sooth		50

Table 6.24 continued: Fossil occurrence 451:

B	15		Chelonia	Scute	Netted		1
B	16		Chelonia	Scute	Pitted		1
B	2		Chelonia/Croc	Shaft			1
B	0		Indeterminate	Fragments			17
C	1		Bovid	Horncore			1
Total							**466**

Fossil occurrence 461: general collection:

Specimen	W	Taxon	Element	Part	Other	N
4		Bovid	Tooth	Premolar		1
3	3	Bovid (Large)	Tooth	Molar	Split	1
4	3	Cervid	Antler			1
1,2		Proboscidean	Tooth	Molar		2
5		*Struthio*	Eggshell			1
0		Indeterminate	Fragments			11
Total						**17**

Notes: The following had a grit matrix: 445CGB1, 447GB8 and a piece of Chelonia scute from 445B. 451AGB31 had a silt matrix, as did two indeterminate fragments from 451B. One indeterminate fragment from 451A had a fine sand matrix. 445GB3 (Indeterminate) had a possible bite mark. Gully collect 449 contained 1 piece of cervid antler, 1 Proboscidean scapula fragment and 16 other fragments.

Table 6. 25: Fossil material from Set 18, the area between Happy and Death Valleys (398-414 even nos.).

Fossil occurrence 404: general collection:

Specimen	W	Taxon	Element	Part	Other	N
8	3	Artiodactyla	Metapodial	Shaft		1
4	2b	Bovid	Astragulus	Split		1
1,2	4	Bovid	Horncore			2
12		Bovid	Tibia	Proximal		1
7		Chelonia	Scute	Smooth		1
0		Indeterminate	Fragments			26
Total						**32**

Fossil occurrence 406: general collection:

Specimen	W	Taxon	Element	Part	Other	N
23		Artiodactyla	Astragulus			1
1		Artiodactyla	Humerus			1
26,29		Artiodactyla	Pelvis			2
14		Artiodactyla	Shaft			1
3		Bovid	2nd Phalange	Complete		1
7	3	Bovid	Metacarpal	Complete	Distal	1
11		Bovid	Metacarpal			1
8		Cervid	Antler			1
0		Chelonia	Scute	Smooth		1
0		Chelonia	Scute	Dimpled		2
12		Chelonia	Scute	Netted		7
9		Proboscidean	Cuneiform	Complete		1
31		Indeterminate	Fragments			81
Total						**101**

Table 6. 25 continued: Fossil occurrence 408: general collection:

Specimen	W	Taxon	Element	Part	Other	N
3		Artiodactyla	Radius	Distal		1
1		Artiodactyla	Vertebra	Axis		1
4	2b	Bovid	Radius	Complete	Distal	1
2	3	Indeterminate	Rib	Condyle		1
Total						**4**

Table 6. 25 continued: Fossil occurrence 412: scatter:

Specimen	W	Taxon	Element	Part	Other	N
11	4	Artiodactyla	1st Phalange	Distal	Split	1
4	3	Artiodactyla	Tibia	Shaft		1
1		Artiodactyla	Premolar	Lower	Split	1
7		Artiodactyla	Ulna carpal			1
5	3	Bovid	Astragulus	Split		1
10	4	Bovid	Tooth	Molar	Split, worn	1
16	3	Bovid	Tooth	M3	Split, mod.wear	1
13		Chelonia	Scute			3
2		Giraffidae	Tooth	Premolar		1
0		Indeterminate	Fragments			113
Total						**124**

Fossil occurrence 414: general collection:

Specimen	W	Taxon	Element	Part	Other	N
3	3	Bovid	Tooth	Premolar	Worn, split	1
1		Bovid (Large)	P2	Upper	Worn, complete	1
2	2a	Bovid (Small)	PM2/3	Upper	Fresh, complete	1
0		Indeterminate	Fragments			10
Total						**13**

Notes: 398 was a find-spot of a bovid (Hemibos) right mandible with the M1-M3. Scatter 400 contained an artiodactyl tooth spall, an artiodactyl thoracic vertebra fragment, 1 piece each of eggshell and turtle scute, and 25 indeterminate fragments. Scatter 402 comprised a bovid distal metapodial fragment, a bovid M3 (split) and 32 other fragments. General collection 410 contained 1 bovid premolar fragment, 1 piece of egg shell and 11 indeterminate fragments.

Table 6.26: Fossil material from Set 19, East Branch of Kotha Kas (453-475 odd nos. only)

Fossil occurrence 455: general collection:

2	3	Bovid (Large)	Tooth	Premolar	Upper	1
3	2b	Bovid (Large)	Tooth	M3		1
1		Cervid	Antler			1
3	3	Equid	Tooth	Premolar	Upper	1
		Indeterminate	Fragments			12
Total						**16**

Fossil occurrence 457: scatter:

Specimen	W	Taxon	Element	Part	Other	N
6	5	Bovid	Horncore			4
1	3	Rhinocerotidae	Tooth	Molar		1
0		Indeterminate	Fragments			18
Total						**23**

Fossil occurrence 459: scatter:

Specimen	W	Taxon	Element	Part	Other	N
2	5	Bovid	Horncore			1
8		Bovid	Horncore			1
5	3	Bovid (Large)	Tooth	Molar	Mod.wear	1
7	3	Bovid (Large)	Tooth	Premolar		1
0		Chelonia	Scute			3
6		Chelonia?	Shaft			1
0, 3,11, 12		Proboscidean	Tooth	Molar	Plate	4
9,10		Giraffidae	Tooth	Premolar		2
0		Indeterminate	Fragments			63
Total						**77**

Table 6. 26 continued: Fossil occurrence 465: general collection:

Specimen	W	Taxon	Element	Part	Other	N
0		Bovid	Horncore			1
1	3	Bovid	Horncore			1
2	3	Bovid	Tooth	Premolar	Upper	1
0		*Struthio*	Eggshell	3		3
0		Indeterminate	Fragments			14
Total						**20**

Fossil occurrence 467: general collection:

Specimen	W	Taxon	Element	Part	Other	N
1	2b	Bovid	Metacarpal	Proximal	Complete	1
0		Bovid	Tooth			11
2	2b	Bovid (Large)	Tooth	Premolar	Upper	1
4		Chelonia	Scute			1
3		Artio (Medium)	Tibia	Distal	Rolled	1
	5	Indeterminate	Fragments			13
Total						**28**

Fossil occurrence 471: scatter:

Specimen	W	Taxon	Element	Part	Other	N
17		Artiodactyla	Femur	Proximal		1
19		Artiodactyla	Metapodial			1
22		Bovid	3rd Phalange			1
2	2b	Bovid	Femur	Distal		1
7,28,40		Bovid	Horncore			3
42		Bovid	Mandible	Premolar		1
3		Bovid	Tooth	M1/2, lower	Mod.wear	1
		Indeterminate	Fragments			66
Total						**75**

Fossil occurrence 475: general collection:

Specimen	W	Taxon	Element	Part	Other	N
8	4	Artiodactyla	Horncore			1
4	3	Bovid	2nd Phalange	Distal		1
1		Bovid	Horncore			1
3	4	Bovid	Tooth	Molar	Mod.wear	1
9		Bovid	Tooth	Molar		1
7	2b	Proboscidean	Cuneiform			1
0		Indeterminate	Fragments			48
Total						**54**

Note: 459 also had small bivalve mollusca. General collection 453 comprised 1 piece of scute, 1 piece of tusk and 6 other fragments; general collection 463 contained 1 piece of bovid horn core, 1 cervid proximal radius, 1 worn pig tooth (463GB1) and 43 indeterminate fragments. General collection 469 contained 1 piece of Proboscidean molar plate, 1 equid premolar fragment, and 12 other fragments; general collection 473 contained 1 bovid M1/2 (upper) and 16 fragments

Table 6. 27: Fossil material from Set 20, Sandstone 14, Choawala Kas (500-510, 516, 520-528).

Fossil occurrence 500: general collection:

Specimen	W	Taxon	Element	Part	Other	N
2	3	Artiodactyla	Metapodial	Proximal		1
4	5	Artiodactyla	Tooth	Fragment		1
1	3	Bovid (Large)	Molar	Mod. wear		1
6	5	Bovid	Tooth	Molar	Fragment	1
8		Cervid	Antler	Tine		1
0		Chelonia	Scute	Smooth		2
0		Chelonia	Scute	Dimpled		3
7		Proboscidean	Tooth	Molar	Fragment	1
3	3	Equid	Molar	Lower	Complete	1
0		Indeterminate	Fragments			30
Total						**42**

Fossil occurrence 502: scatter:

Specimen	W	Taxon	Element	Part	Other	N
3		Bovid	Tooth	M1/2	Lower	1
4	3	Bovid (Large)	Premolar	Upper	Some wear	1
1		Bovid	PM3	Upper	Complete	1
0		Chelonia	Scute	Fine netted		1
0		Chelonia	Scute	Dimpled		4
0		Chelonia	Scute	Smooth		7
0		Indeterminate	Fragments			16
Total						**31**

Fossil occurrence 504: scatter:

Specimen	W	Taxon	Element	Part	Other	N
1		Bovid	Tooth			1
2		Cervid	Antler			1
0		Chelonia	Scute	Fishnet		1
0		Chelonia	Scute	Smooth		5
0		Chelonia	Scute	Dimpled		5
0		Chelonia	Scute	Pitted		1
3	2b	*Crocodylus*	Tooth	Broken		1
0		Proboscidean	Tooth	Molar	Plate	11
0		Indeterminate	Fragments			51
Total						**77**

Fossil occurrence 506: scatter:

Specimen	W	Taxon	Element	Part	Other	N
0		Chelonia	Scute	Smooth		7
0		Chelonia	Scute	Dimpled		1
2	5	Proboscidean	Tooth	Molar	Root	1
0		Indeterminate	Fragments			17
Total						**26**

Fossil occurrence 508: scatter:

Zone	Specimen	W	Taxon	Element	Part	Other	N
508	5		Artiodactyla	Tooth	Incisor		1
508	5		Cervid	Antler			1
508	0		*Struthio*	Eggshell			4
508A	1		Canidae	1st Phalange			1
508A	3	2b	Cervid	1st Phalnge	Complete		1
508A	2	2b	Cervid	1st Phalange	Proximal	Complete	1

Table 6. 27 continued: Fossil occurrence 508:

508A	0		Chelonia	Scute	Dimpled		3
508A	0		Chelonia	Scute	Smooth		20
508A	0		Indeterminate	Fragments			21
508A	4		Suid	1st Phalange	Distal		1
508B	2	5	Artiodactyla	Tooth	Molar	Fragment	1
508B	3	4	Artio (Large)	Tooth	Molar	Split	1
508B	1	5	Cervid	Antler			1
508B	0		Indeterminate	Fragments			47
508C	1	2a	Cervid	2nd Phalange	Complete		1
508C	0		Chelonia	Scute	Smooth		7
508C	0		Indeterminate	Fragments			14
508D	1		Bovid	Cuneiform			1
508D	0		Chelonia	Scute	Smooth		7
508D	0		Chelonia	Scute	Dimpled		1
508D	0		Chelonia	Scute	Netted		1
508D	0		Indeterminate	Fragment			1
Total							**137**

Fossil occurrence 510: general collection:

Specimen	W	Taxon	Element	Part	Other	N
2	5	Cervid	Antler			1
5	3	Cervid	Antler	Split		1
0		Chelonia	Scute	Smooth		2
1	5	Proboscidean	Tooth	Tusk		1
4	4	Proboscidean	Tooth	Molar	Plate	1
0		Indeterminate	Shaft			4
Total						**10**

Fossil occurrence 522: scatter:

Zone	Specimen	W	Taxon	Element	Part	Other	N
522A	1	2a	Artiodactyla	Vertebra	Axis		1
522A	6		Artiodactyla	Cuboid	Split		1
522A	4	2b	Bovid (Small)	Tooth	Molar	Split	1
522A	2	2a	Bovid (Small)	Tooth	M1/2	Complete	
522A	0		Chelonia	Scute	Smooth		23
522A	0		Indeterminate	Fragments			21
522B	3	2a	Artiodactyla	Metacarpal	Shaft		1
522B	5,8,14		Artiodactyla	Tooth	Molar		3
522B	15	3	Artiodactyla	Tooth			1
522B	4	3	Artio/Suid	Scapula	Glenoid		1
522B	12	3	Bovid	Radius	Distal		1
522B	9	5	Bovid	Tooth	Premolar	Mod.wear	1
522B	7	3	Bovid (Large)	Tooth	Premolar	Split	1
522B	6	3	Bovid (Small)	Tooth	Molar		1
522B	0		Chelonia	Scute	Pitted		1
522B	0		Chelonia	Scute	Smooth		4
522B	2		*Crocodylus*	Metapodial			1
522B	0		Indeterminate	Fragments			34
522B	11	3	Bovid (Small)	Horncore			1
522C	3	4	Artiodactyla	Tooth	Molar	Fragment	1
522C	4		Bovid	Tooth	Premolar		1
522C	5		Bovid	Cuneiform	4th carpal		1
522C	0		Chelonia	Fine netted			1
522C	0		Chelonia	Scute	Smooth		1
522C		0	Indeterminate	Fragments			54

Table 6. 27 continued: Scatter 522:

522D	1	3	Artio (Large)	Metacarpal	Distal	Condyle	1
522D	2	3	Bovid	Horncore			1
522D		0	Indeterminate	Fragments			125
522E	4	3	Bovid (Medium)	Tooth	Premolar		1
Total							**286**

Fossil occurrence 524: general collection:

Specimen	W	Taxon	Element	Part	Other	N
0		Artiodactyla	Tooth	Molar		1
2		Bovid	1st Phalange	Distal		1
6		Bovid	Unciform			1
5	3	Bovid (Large)	Tooth	M1/2	Upper	1
1	4	Bovid	Mandible	M3 worn	Right	1
4		Bovid	Tooth	Premolar		1
0		Proboscidean	Tooth	Molar	Plate	6
0		Indeterminate	Fragments			8
0		*Struthio*	Eggshell			5
Total						**25**

Fossil occurrence 526: scatter:

Specimen	W	Taxon	Element	Part	Other	N
10		Bovid	Tooth	Premolar		1
8,11,12		Proboscidean	Rib			3
9		Proboscidean	Tooth	Molar	Plate	1
1,4	4	Proboscidean	Vertebra			2
0		Indeterminate	Fragments			24
Total						**31**

Notes: There was a sand matrix on 522AGB1; a silt matrix on 508AGB3; and grit on some of the turtle scutes from 500 and 522. General collection 516 contained only 4 indeterminate shaft fragments; general collection 520 contained 1 Proboscidean limb-shaft fragment, 1 piece of Proboscidean molar plate and 21 indeterminate fragments; and general collection 528 contained 1 small bovid upper M1/2, a piece of turtle scute and 3 indeterminate fragments.

Table 6. 28: Fossil material from Set 21, Dhorian Kas and Ratial Kas

Fossil occurrence 382: general collection:

Specimen	W	Taxon	Element	Part	Other	N
5		Artiodactyla	Pelvis			1
11, 12	3	Artiodactyla	Tooth	Molar	Lower	2
4	5	Bovid	Horncore			1
1	3	Bovid	Radius	Proximal		1
8	5	Bovid (Large)	Premolar	Lower		1
2,3,7	2b	Cervid	Antler			3
0		Indeterminate	Fragments			32
Total						**41**

Fossil occurrence 386: general collection:

Specimen	W	Taxon	Element	Part	Other	N
1		Artiodactyla	Cranium			1
2	3	Bovid	Metacarpal	Distal	Complete	1
4	2b	Bovid	M1/2, upper	Complete	Fresh	1
10, 11	5	*Elephas*	Tooth	Molar	Plate	2
3,5,6	5	Equid	Molar	Lower	Split	3
7		*Struthio*	Eggshell			1
0	5	Indeterminate	Shaft			3
Total						**12**

Fossil occurrence 420: scatter:

Specimen	W	Taxon	Element	Part	Other	N
10		Bovid (Small)	M3, lower	Complete	Mod.wear	1
0		Cervid	Antler			1
0		Cervid	Vertebra	Centrum		1
1		*Struthio*	Eggshell			1
0		Indeterminate	Fragments			15
Total						**19**

Fossil occurrence 422: general collection:

Specimen	W	Taxon	Element	Part	Other	N
8	4	Artiodactyla	Humerus	Distal	Fragment	1
7		Artiodactyla	Mandible			1
3	3	Artiodactyla	Tooth	Split, no crowns		1
1		Bovid	M1/2, upper	Complete	Worn	1
4, 5		Bovid	Tooth	Premolar	Upper	2
2,6	3	Cervid	Antler			2
0		Indeterminate	Fragments			22
Total						**30**

Fossil occurrence 428: general collection:

Specimen	W	Taxon	Element	Part	Other	N
7	3	Artiodactyla	Metacarpal	Proximal	Split	1
1,2,4,5	3	Cervid	Antler			4
3	3	Proboscidean	Tooth	Fragment		1
0		*Struthio*	Eggshell			1
0		Indeterminate	Fragments			50
Total						**57**

Notes: Four specimens had a sand matrix: 386GB10, 428GB0 (Struthio shell); 430GB1 and 463GB2. There were the following general collections: 266, 1 piece of turtle scute (smooth) and 5 indeterminate fragments; 384, 1 bovid distal tibia fragment and 3 indeterminates; 424, 3 bits of bovid horncore, 2 molar fragments, 2 pieces of scute and 20 indeterminates; 426, 1 bovid (large) upper M1/2, 1 piece of cervid antler, 1 equid proximal radius fragment and 50 indeterminates; 430, 1 felid distal radius and 1 piece of antler; and 432, a bovid metapodial, astragalus, radius and tooth fragment and 16 indeterminates.

Table 6. 29: Fossil material from Set 22, the badlands north of the G.T. road, 1987 (436-528 even nos.).

Fossil occurrence 440: general collection:

Specimen	W	Taxon	Element	Part	Other	N
2	5	Bovid (Small)	PM	Upper	V.worn	1
4	4	Equid	Tooth	Complete		1
1		Giraffid	Tooth	Molar	Fragment	1
0		Indeterminate	Fragments			16
Total						**19**

Fossil occurrence 446: scatter:

Specimen	W	Taxon	Element	Part	Other	N
7	3	Artiodactyla	Pelvis			1
3	5	Artiodactyla	Tooth			1
8		Bovid	Horncore			1
1,2,10	3	Bovid (Large)	Tooth	Lower	Split	3
5		Chelonia	Scute	Pitted		2
9	5	Equid	Molar	Lower	Split	1
0		Indeterminate	Fragments			5
Total						**14**

Fossil occurrence 448: general collection:

Specimen	W	Taxon	Element	Part	Other	N
5		Bovid	M1/2	Complete	Upper, worn	1
7		Chelonia	Scute	Smooth		1
1		*Elephas*	Tooth			1
0		Indeterminate	Fragments			12
Total						**15**

Table 6. 29 continued: Fossil occurrence 450: general collection:

Specimen	W	Taxon	Element	Part	Other	N
6	3	Artiodactyla	Metatarsal	Shaft		1
10		Chelonia	Scute	Smooth		1
7-9		Proboscidean	Tooth	Plate		3
0		*Struthio*	Eggshell			1
0		Indeterminate	Fragments			47
Total						**53**

Fossil occurrence 452: scatter:

Specimen	W	Taxon	Element	Part	Other	N
9		Artiodactyla	Femur			1
10		Artiodactyla	Femur	Proximal		1
7		Artiodactyla	Metacarpal			1
1		Bovid	2nd Phalange			2
11		Bovid	Femur	Proximal		1
13	3	Bovid	Mandible	M3 right	Mod.wear	1
20		Bovid	Tooth	Premolar		1
21		Canidae	Humerus	Distal		1
22		Canidae	Humerus			1
18		Carnivora	Coprolite?			1
12		Cervid	Antler			1
0		Chelonia	Scute	Netted		2

Fossil occurrence 452: scatter:

0		Chelonia	Scute	Smooth		13
6		Equid	1st Phalange	Distal		1
8		Equid	Tooth	Molar		1
20,25		Equid	Tooth	Premolar		2
27		Equid	Tooth	Incisor		1
15		Indeterminate	Fragments			54
Total						**86**

Fossil occurrence 454: scatter:

Specimen	W	Taxon	Element	Part	Other	N
0	3	Artiodactyla	Calcaneum	Shaft		1
2		Bovid	1st Phalange	Proximal		1
1		Cervid	Antler			1
3		Cervid	Antler	Tine		1
0		Chelonia	Scute	Smooth		6
0		Indeterminate	Fragments			2
Total						**12**

Fossil occurrence 456: scatter:

Specimen	W	Taxon	Element	Part	Other	N
6		Artiodactyla	Femur	Distal	Condyle	1
1		Artiodactyla	Humerus	Proximal		1
7		Artiodactyla	Vertebra			1
5,9	3	Bovid	Femur	Shaft		2
4	3	Bovid	Radius	Shaft		1
0		Indeterminate	Fragments			84
Total						**90**

Fossil occurrence 458: scatter:

Specimen	W	Taxon	Element	Part	Other	N
2	3	Bovid (Large)	Molar	Upper	Mod. wear	1
0		Cervid	Antler			1
1		Cervid	Antler	Pedicel		1
4	3	Equid	Metapodial	Distal	Epiphysis	1
5	5	Equid	Tooth	Molar	Split	1
0		Indeterminate	Fragments			6
Total						**11**

Table 6. 29 cntinued: Fossil occurrence 460: scatter:

Specimen	W	Taxon	Element	Part	Other	N
1	3	Bovid	M3	Upper	Mod. wear	1
2,32	2b	Cervid	Antler			2
4,5,6		Proboscidean	Tooth	Tusk		6
0		Indeterminate	Fragments			14
Total						**23**

Fossil occurrence 464: general collection:

Specimen	W	Taxon	Element	Part	Other	N
1	5	Artiodactyla	Humerus	Distal		1
2	2a	Cervid	Antler			1
3	3	Proboscidean	Tooth	Molar	Root	1
0		*Struthio*	Eggshell			2
0		Indeterminate	Fragments			3
Total						**8**

Fossil occurrence 466: scatter:

Specimen	W	Taxon	Element	Part	Other	N
4	3	Artiodactyla	Pelvis			1
5		Artiodactyla	Vertebra	Caudal		1
7	4	Artiodactyla	Vertebra			1
2	3	Bovid	Vertebra	Axis		1
3		Bovid	Vertebra			1
1	3	Cervid	Antler			1
0		Chelonia	Scute	Netted		1
0		Chelonia	Scute	Dimpled		1
0		Chelonia	Scute	Smooth		6
0		*Struthio*	Eggshell			20
0		Indeterminate	Fragments			29
Total						**63**

Fossil occurrence 468: general collection:

Specimen	W	Taxon	Element	Part	Other	N
1		Equid	Calcaneum	Complete		1
2	3	Equid	Metapodial	Shaft		1
3-5		Proboscidean	Tooth	Molar		3
0		Indeterminate	Fragments			9
Total						**14**

Fossil occurrence 470: general collection:

Specimen	W	Taxon	Element	Part	Other	N
5	3	Artiodactyla	Mandible	None		1
6	5	Artiodactyla	Mandible	+tooth fragment		1
4		Artiodactyla	Metatarsal	Shaft		1
8		Artiodactyla	Vertebra			1
7	3	Bovid	Horncore			1
1	4	Bovid	Metacarpal	Shaft		1
0		Bovid	Tooth	Molar		1
2,3	2b	Proboscidean	Tooth	Molar		2
0		Indeterminate	Fragments			12
Total						**21**

Fossil occurrence 472: scatter:

Specimen	W	Taxon	Element	Part	Other	N
9	4	Artiodactyla	Metapodial	Distal	Condyle	1
1	3	Bovid	Premolar	Upper	Worn	1
2	3	Bovid	M1/2	Complete	Upper, worn	1
6	4	Bovid	Premolar	Complete	Worn	1
7,8	4	Bovid	Tooth	Molar	Split	2
3	5	Equid	Tooth	Molar		1
0		Indeterminate	Fragments			57
Total						**65**

Table 6. 29 continued: Fossil occurrence 474: scatter:

Specimen	W	Taxon	Element	Part	Other	N
1		Cervid	Antler			1
2,3,5-9	5	Equid	Tooth			8
0		Indeterminate	Fragments			21
Total						**30**

Fossil occurrence 476: general collection:

Specimen	**W**	**Taxon**	**Element**	**Part**	**Other**	**N**
2		Bovid	Calcaneum			1
0		Chelonia	Scute	Smooth		1
1	5	Equid	Tooth	Molar	Split ,worn	1
0		*Struthio*	Eggshell			5
0	5	Indeterminate	Fragments			6
Total						**14**

Tale 6. 29 continued: Fossil occurrence 484: scatter:

Specimen	W	Taxon	Element	Part	Other	N
1	3	Equid	3rd Phalange			1
2	5	Equid	Tooth	Premolar		1
0		Indeterminate	Fragments			5
Total						7

Fossil occurrence 486: general collection:

Specimen	**W**	**Taxon**	**Element**	**Part**	**Other**	**N**
1	5	Bovid	Tooth	Premolar	Upper	1
2	5	Proboscidean	Tooth	Molar	Plate	1
3		Suid	Ulnar	Carpal		1
0		Indeterminate	Fragments			12
Total						**15**

Fossil occurrence 488: general collection:

Specimen	**W**	**Taxon**	**Element**	**Part**	**Other**	**N**
3		Artio (Large)	Carpal	Complete		1
5	4	Artiodactyla	Scapula			1
2	3	Bovid	Tooth	Premolar		1
4	4	Bovid	M1/2	Complete	Upper	1
0		Chelonia	Scute	Pitted		1
0		Chelonia	Scute	Netted		1
0		Chelonia	Scute	Smooth		1
1	3	Equid	2nd Phalange	Complete		1
0		Indeterminate	Fragments			19
Total						**27**

Fossil occurrence 490: general collection:

Specimen	**W**	**Taxon**	**Element**	**Part**	**Other**	**N**
0	5	Bovid	Horncore			3
7		Canidae	Tooth	Premolar		1
0		Chelonia	Scute	Netted		1
1		Equid	Tooth	Premolar		1
3		Proboscidean	Tooth	Molar	Plate	1
0	3	Indeterminate	Fragments			3
Total						**10**

Fossil occurrence 492: scatter:

Specimen	**W**	**Taxon**	**Element**	**Part**	**Other**	**N**
1		Bovid	Premolar	Complete	Upper	1
2		Chelonia	Cranium	Complete		1
0		Indeterminate	Fragments			33
Total						**35**

Table 6. 29 continued: Fossil occurrence 494: general collection:

Specimen	W	Taxon	Element	Part	Other	N
0		Chelonia	Scute	Smooth		1
1		Giraffidae	Tooth	Premolar	Upper	1
0		Proboscidean	Tooth	Molar		2
3		Proboscidean	Tooth	Molar		1
0		*Struthio*	Eggshell			1
0		Indeterminate	Fragments			12
Total						**18**

Fossil occurrence 496: scatter:

Specimen	W	Taxon	Element	Part	Other	N
0		Bovid	Tooth	Premolar		1
1	3	Bovid	Premolar	Complete	Upper, Worn	1
4	3	Bovid	M1/2	Complete	Lower, Worn	1
3	3	Cervid	Antler			1
0		Chelonia	Scute	Smooth		1
0		Indeterminate	Fragments			5
Total						**10**

Notes: There was a fine sand matrix on 436GB1, 448GB5, and 456GB1; and coarser sand matrix on 450GB1 and 3, 456GB3, 470GB8 and 490GB0 (horn-core). 466GB2,4 and 7, and 476GB0 (Chelonia) had a clay skin. Antler pieces 460GB2 and 3 joined. The canid and carnivore specimens in 452 and 490 are unconfirmed.

There were also the following minor fossil occurrences: General collection 436 contained 1 bovid complete 2nd phalange, 1 piece of Proboscidean molar plate and 3 indeterminate fragments. 438 (general collection) comprised 3 proboscidean tooth fragments and 7 indeterminate fragments; 442 (general collection); 1 large indeterminate rib fragment; 444 (scatter); 1 bovid distal femur fragment; general collection 462 comprised 1 piece of scute, 1 large (Proboscidean) shaft fragment and 7 other fragments.478 (scatter); at least 18 Struthio egg-shell fragments (large quantities were left in place); 480 (general collection); 1 bovid upper M1/2; and 482 (scatter), 1 bovid complete 1st phalange.

Chapter 7: The 1989 and 1990 surveys of the Pabbi Hills

The survey in 1989 focussed on the exposures along and sometimes in front of Sandstone 12 in the northern end of the Pabbi Hills, from Bangdwara Kas (the first drainage system north of Kanarawala Kas), around the head of the anticline in the area of the famous local shrine of Pir Jaffa to the village of Baroth, and from there, past Ban Barilla, a conspicuous mango tree, and southwards along the eastern side of the anticline (see Figure 3.1). Initially the team was based in the small road-side settlement of Bulani, but the survey in that area was abruptly terminated after an attempted ambush by dacoits. Consequently, there was insufficient time to undertake the usual detail of recording. Thereafter, the team had to operate from the safer but more distant base of Banni Bangla, by the G.T. road. The area of Baroth was reached by either the watershed track followed by a lengthy walk, or by an extremely long and uncomfortable detour by vehicle around the Kharian cantonment and across fields and minor tracks, followed by a shorter walk.

In 1989, the survey produced roughly the same number of fossil specimens as in 1986; that is to say, around 12,000 specimens, including undiagnostics. The principal difference, however, was that over 70% of those specimens were found at only two locations - 610, and particularly 642, which was easily the largest and most impressive fossil site that the team found in the Pabbi Hills.

Locality 642 was partially excavated in 1990, and the results are described in the next chapter. The area was also mapped, and further collecting took place. Although little fossil was collected in 1990, several pieces of flaked stone were found associated with a conglomerate unit: this work is described in chapter 9.

The results of the 1989 survey are summarised in Table 7.1. The survey can be divided into three main parts: first, the material found in the Bulani area between Bangdwara Kas and the area around Pir Jaffa, including locality 610; secondly, that in the area of Baroth village and Ban Barilla, particularly locality 642; and thirdly, the material found between Ban Barilla and the G.T. road.

Area 1: The Bulani area (Bandgwara Kas: 600-607).

The exposures of Sandstone 12 in this area are often steep, impressive but also often obscured by dense thorn scrub. The next drainage system north of Kanarawala Kas is Kinaliwala Kas, which has two small branches, followed by the larger Bangdwara Kas. General collection 601 came from between the two branches of Kinaliwala Kas, and general collection 600 from between Kinaliwala and Bangdwara Kas.The area between Bangdwara Kas and the northernmost limit of the 1986 survey were covered by general collections 602 and 606. No. 604 was a scatter of a highly fragmented long bone that can be included within general collect 606. As Table 7.2 shows, very little was found in this area. This was partly because of dense thorn-scrub, but fossil was scarce even when exposures were good. The other team had the same experience when collecting between Bangdwara Kas and Besa Kas, the next drainage system to the north; as shown by Table 7.2, general collections 601, 603, 607 and a small scatter 605 were very meagre.

Area 2: Besa Kas (608-615); Pir Jaffa (616-625)

Collecting numbers 608-615 covered the area around Besa Kas, a little to the north. Two butes, or small inselbergs of Sandstone 12, are prominent landmarks. This area was more productive than those already surveyed (see Table 7.2). The material from scatter 608 was found in deposits on top of Sandstone 12 a little to the west of Besa Kas, and is thus <1.2 Mya-old. This scatter was fairly large, although the material was very fragmented and had probably been exposed on the surface for several years. It was one of the few finds in this area showing a fluvial assemblage, with remains of turtle, crocodile, and mollusca. Locality 610 lay at the base of Sandstone 12 near the base of one of the inselbergs, and was the second largest find of that season. As Figure 7. 1 shows, the fossil material came from a small gully and surrounding slopes. As with other localities, 610 was divided into several sub-areas during collection. The fossil material had probably been exposed for a long time (perhaps several decades), and the local slopes were much obscured by sandstone blocks and slope wash. As Table 7.3 shows, the fossil material was extremely fragmented: the overwhelming majority of the identifiable pieces were fragmented bovid teeth, with the post-cranial elements mostly reduced to hundreds of small shaft fragments. As shown in a later chapter, the bovid remains may be important in showing a rare type of alcephaline bovid, despite the high rate of fragmentation. In sharp contrast, the surrounding area of adjacent deposits (i.e. general collections 612 and 614) contained almost nothing in the way of fossil material. Material was likewise very poor further along Sandstone 12 in the Pir Jaffa area, which was covered in a series of very poor general collections (616, 618, and scatter 620). The odd-numbered team worked in the same area, but on the dip-slope behind sandstone 12 (i.e. on deposits < 1.2 Mya-old), and also found little: general collections 611, 615, 617 (no finds), 619, 621, 623, one unremarkable find-spot, 613, and a discovery of stone tools, 625 (see also the chapter by Hurcombe).

Results improved after one of the teams made a lengthy reconnaissance into the area around Baroth village, and noted that fossil was more in evidence in that area. In this area, the terrain is gentler, and Sandstone 12 exposed in a series of low escarpments and inselbergs, with much less thorn scrub. The survey in this area fell into two parts: before, and after, the discovery of locality 642, by far the most and most remarkable fossil locality found in the Pabbi Hills.

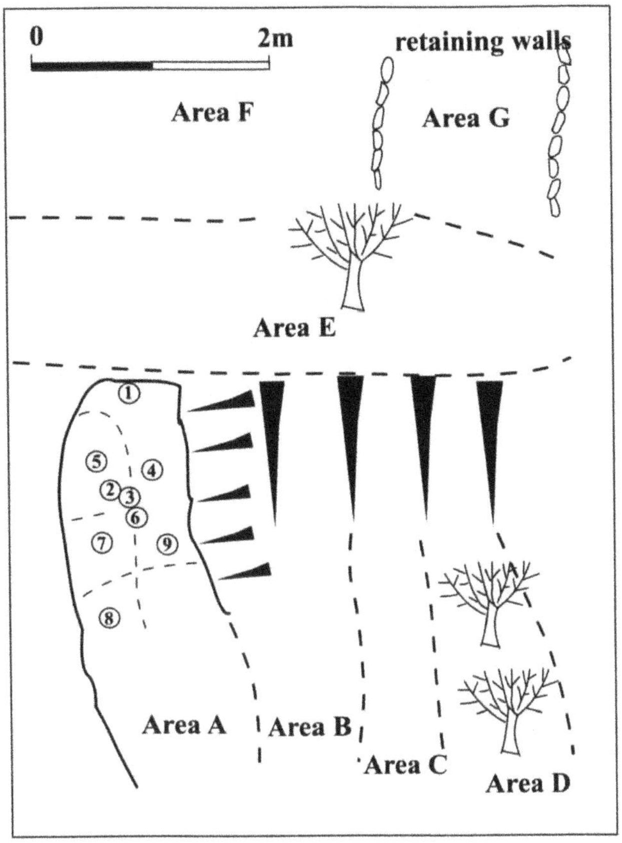

Figure 7.1: Plan of locality 610 (based on field records by Sheila Sutherland), showing the four main collecting areas. The numbers in circles denote smaller clusters of material.

Area 3. The Baroth area (622-640, odd and even numbers)

The Baroth area is shown in Figure 7.2, and was initially surveyed as a series of general collections (622, 627, 630,631, 632, 633, 635, 640, and scatters (624, 628, 636; 638). The more significant discoveries included two find-spots of proboscidean crania (left in place), numbers 626 and 629, and scatter 636, the almost complete skeleton of a medium-sized bovid, found eroding on a foot-path from fine sands below Sandstone 12. This material is summarised in Table 7.4.

Locality 642 is located on a small, flat-topped hill immediately below a small promontory of Sandstone 12 to the north-west of Baroth village (see Figure 7.3). When discovered, the surface was almost completely covered in fossil fragments, including numerous mandibles, maxillae, post-cranial elements, some of which were articulated. As with other major fossil localities, the fossils were collected in a series of zones (1-3), each containing various clusters of specimens (see Figure 7. 4). As Figure 7. 5 shows, the fossils are concentrated in the upper part of a yellow/brown silt deposit. The material collected in 1989 is summarised in Tables 7.5 - 7.10. Most of the material was collected as zones 1 – 3 (Tables 7.5 – 7.7). "Zone X" (Table 7.8) refers to material collected by a helpful if unsystematic shepherd at the time of discovery. Tables 7. 9 and 7. 10 list specimens collected from two clusters in zone 1, and various obvious concentrations of material across the site before the more systematic collections of zones 1 – 3.

As these tables show, over 95% of the identifiable remains were from medium-sized bovids, and the remainder, from rhino, equids, a large hyaenid, a large and a very small bovid, identified as gazelle and found in a fairly small area of the locality. As can be seen (and as at locality 610), the overwhelming majority of remains are isolated and usually fragmented teeth, followed by the remains of hooves and lower limbs. Ribs and vertebrae were also common, and most of these were probably also derived from medium-sized bovids.

The erosion of this hilltop is still rapid. Numerous pieces were collected from the surface in 1990 prior to the excavation of part of the site that season that had not been exposed the previous year. In 1999, a complete bovid cranium (!) was found in a sandstone block a little below the top of the hill, and several other mandibles, maxillae and post-cranial pieces were newly exposed.

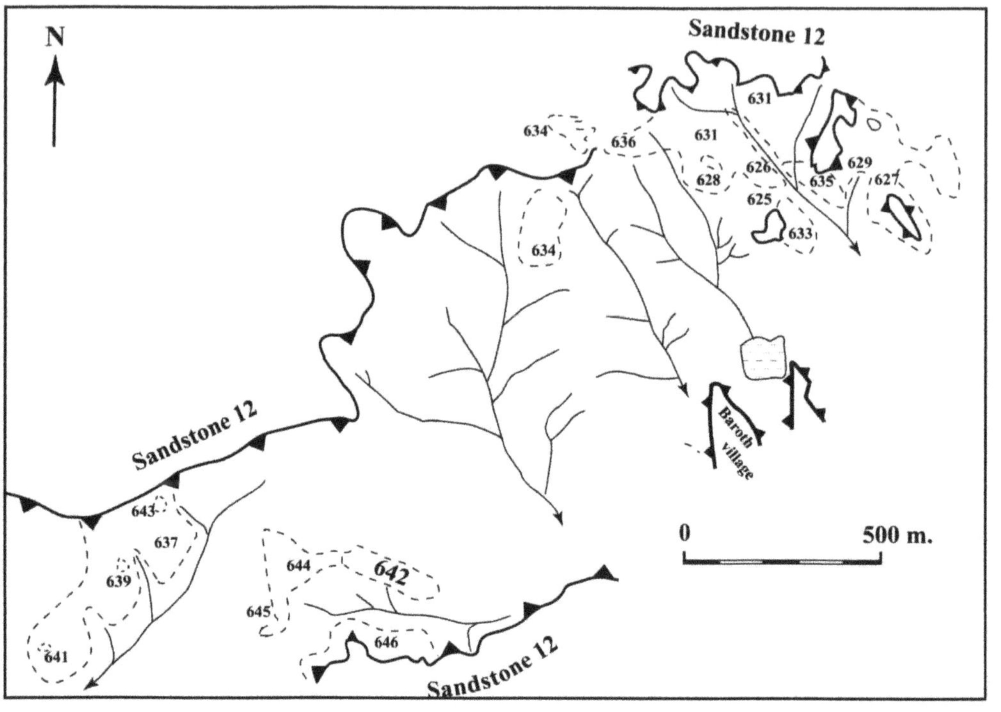

Figure 7.2: Map of the area surveyed in 1989 around Baroth, northern part of the Pabbi Hills (based on a map by Mike James and Saleem ul-Haq).

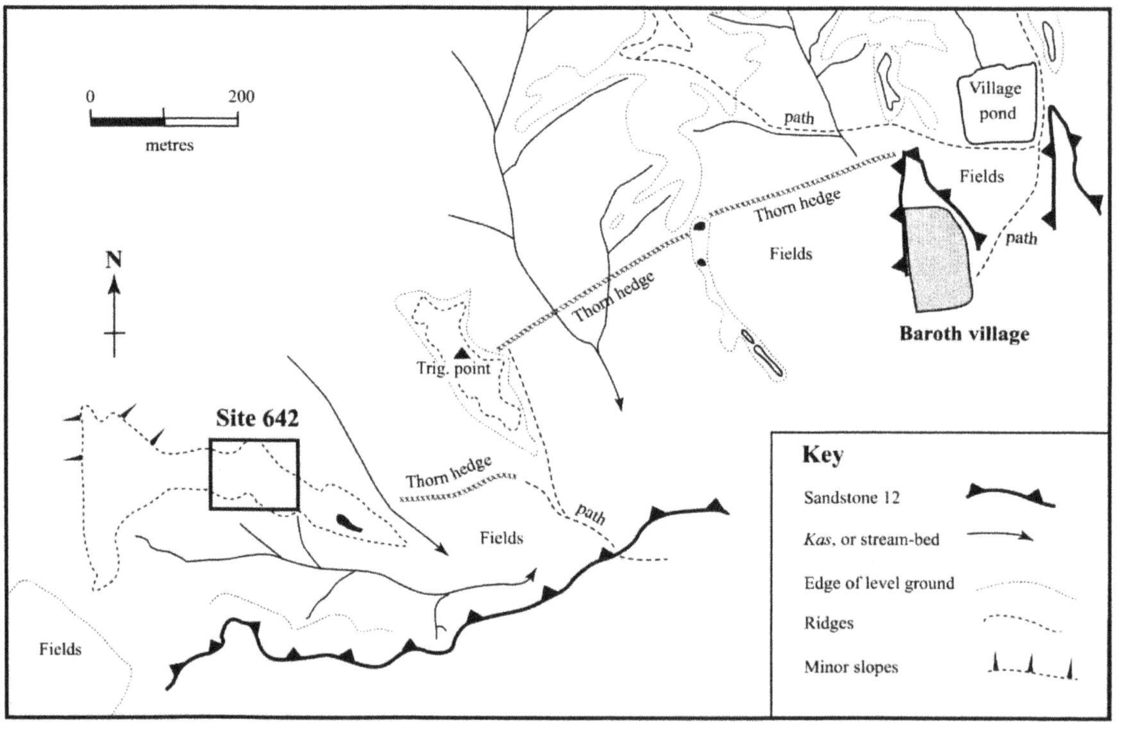

Figure 7.3: Map of the area of locality 642, near Baroth, northern part of the Pabbi Hills (based on a map by Mike James and Saleem ul-Haq).

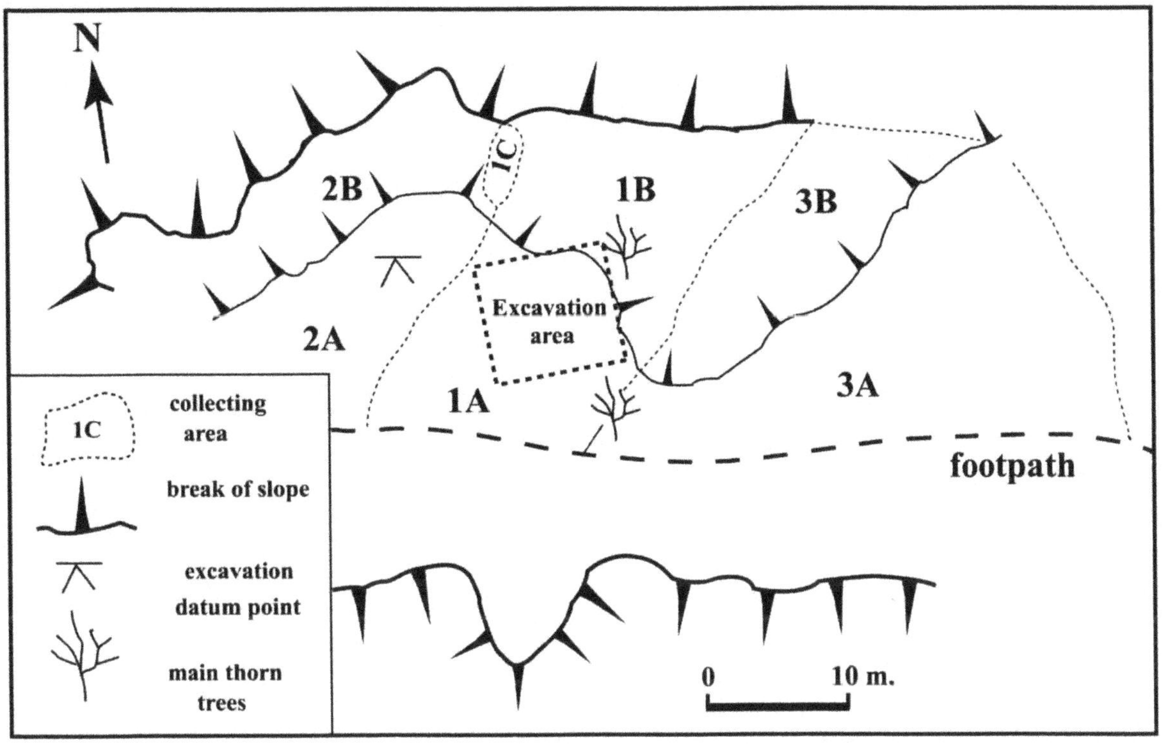

Figure 7.4: Plan of locality 642 ((based on a map by Mike James and Saleem ul-Haq).

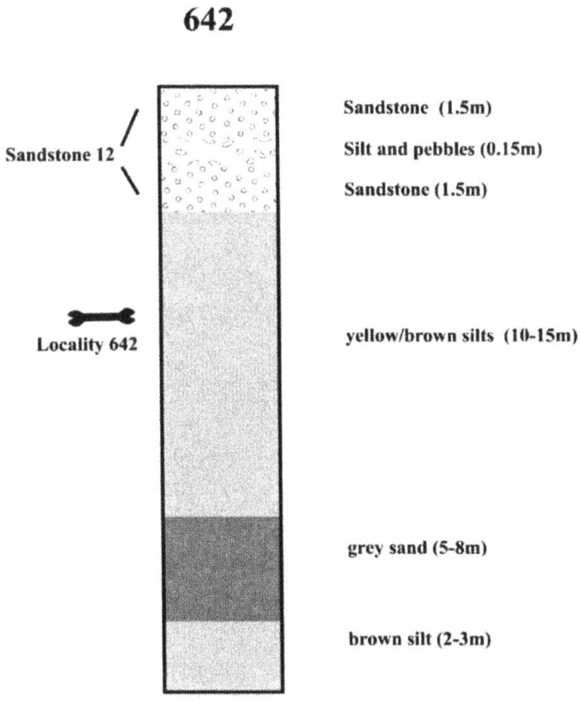

Figure 7.5: Schematic section of profile near locality 642 (based on notes by the author). See Figure 9.20 for a more detailed section of the excavation.

Area 4: Locality 642 to Ban Barilla (637-648)

As with localities 68, 73, 362 and 610, locality 642 was an osteological oasis in a palaeontological wilderness. Very little comparable material was found along the exposures of Sandstone 12 between locality 642 and Ban Barilla, a very prominent mango tree by a small pond about 500m beyond locality 642. (On a clear day, this tree can even be seen from the top of Flag Hill, over 12 kilometres to the south). General collections 643, 644, 646-648, were very poor, as were smaller concentrations, scatters 637, 639, 641. These are summarised in Table 7.11

Area 5: Ban Barilla - Doga (649-674)

The 1989 survey continued along Sandstone 12 towards the G.T. road, but also searched other potentially informative exposures. Coverage was not complete: partly because of difficulty of access to some areas; partly because the quality of the exposures deteriorated; and partly because of agricultural activities near villages such as Doga. There were a few significant discoveries, principally, several proboscidean bones, including a complete femur that were found in a silty deposit within the area of scatter 664; one good locality (666); and an isolated find of a large felid mandible (no. 674) that was probably *Megantereon* (see Turner appendix). A schematic plan of the Proboboscidean remains at 664 is shown in Figure 7. 6.

The results are summarised in Table 7.12. General collections 647, 649 and 653 were made along Sandstone 12 between Ban Barilla and the watershed track, and contained two small scatters, nos. 651 and 655. General collections 652 and 654 were made by the other team between Ban Barilla and the pylons near Doga village, and included two small scatters, 650 and 656. No. 664 comprises specimens found on the surface of a red clay unit associated with Sandstone 12; however the main find was a complete, but very suncracked and fragile, proboscidean femur ca. 2m long, that was excavated, conserved and then successfully transported kback to the G.S.P. facilities in Islamabad. This find was near locality 666, the last major fossil occurrence of that season. As can be seen, crocodile and turtle remains were common, but a mammalian carnivore was also present. A suid maxilla was probably the most useful specimen, given the overall rarity of suids in the Upper Siwaliks. This part of the survey ended with a very extensive but poor general collection (678) along the slopes of Sandstone 12 from locality 666 to the G.T. road.

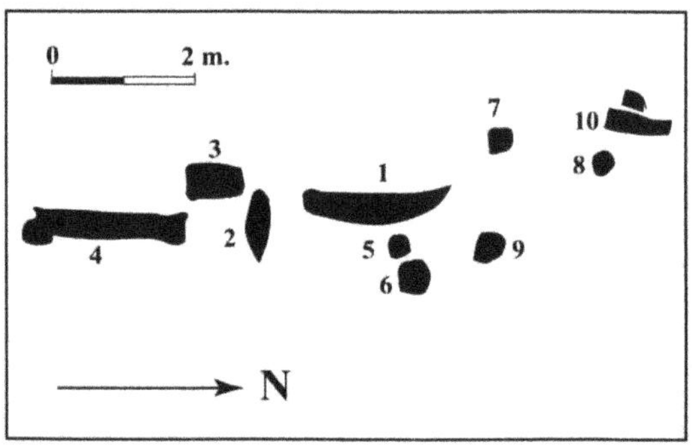

Figure 7. 6: Plan of Proboscidean remains at occurrence 664 (based on field records by Mark Beech)

Key: 1 = Proboscidean ramus; 2 = Proboscidean radius/ulna; 3 = indeterminate; 4 = Proboscidean femur; 5 = Proboscidean astragalus; 6 = Proboscidean femur fragments; 7 = large vertebra fragment; 8 = Proboscidean 2nd phalange; 9 = Proboscidean distal tibia; 10 = Proboscidean tibia. These all lay in a brown-grey clay, which was at the base of a local sequence of (from top to bottom) of sandstone, pink silts, and orange-brown sand.

Area 6: Hairpin bend (657-685)

Finally, Table 7.13 summarises a series of collections on both sides of the watershed track in the vicinity of an extremely sharp, steep (and unmistakable) hairpin bend on the watershed track. Nos. 657, 663, 665, 667, 669, 671, 677, 681, 685 were all general collections; 659 was a find-spot of equid remains, and 661, 675 and 683 were scatters.

Table 7.1: Fossil occurrences found in 1989

Number	Area	Type	Age	Table	Total
600	Bangdwara Kas	General collection	1.2 - 1.4	Table 7.2	5
601	Bangdwara Kas	General collection	1.2 - 1.4	Table 7.2	10
602	Bangdwara Kas	General collection	1.2 - 1.4	Table 7.2	16
604	Bangdwara Kas	Scatter	1.2 - 1.4	Table 7.2	130
605	Bangdwara Kas	Scatter	1.2 - 1.4	Table 7.2	29
606	Bangdwara Kas	General collection	1.2 - 1.4	Table 7.2	2
607	Bangdwara Kas	General collection	0.9 - 1.2	Table 7.2	5
608	Besa Kas	Scatter	1.2 - 1.4	Table 7.2	234
610	Besa Kas	Locality	1.2 - 1.4	Table 7.3	3229
611	Besa Kas	General collection	0.9 - 1.2	Table 7.2	32
612	Besa Kas	General collection	1.2 -.1.4	Table 7.2	7
613	Besa Kas	Find-spot	0.9 - 1.2	Table 7.2	4
614	Besa Kas	General collection	1.2 - 1.4	Table 7.2	35
615	Besa Kas	General collection	0.9 - 1.2	Table 7.2	2
616	Pir Jaffa area	General collection	0.9 - 1.2	Table 7.2	1
618	Pir Jaffa area	General collection	1.2 - 1.4	Table 7.2	39
619	Pir Jaffa area	General collection	1.2 - 1.4	Table 7.2	13
620	Pir Jaffa area	General collection	1.2 - 1.4	Table 7.2	19
621	Pir Jaffa area	General collection	1.2 - 1.4	Table 7.2	41
622	Baroth/642 area	General collection	1.2 - 1.4	Table 7.4	11
623	Baroth/642 area?	General collection	1.2 - 1.4	Table 7.4	6
624	Baroth/642 area	Scatter	1.2 - 1.4	Table 7.4	270
625	Baroth/642 area	Stone only	1.2 - 1.4	Table 7.4	0
626	Baroth/642 area	Scatter/find-spot	1.2 - 1.4	Table 7.4	29
627	Baroth/642 area	General collection	1.2 - 1.4	Table 7.4	17
628	Baroth/642 area	Scatter	1.2 - 1.4	Table 7.4	187
629	Baroth/642 area	Find-spot	1.2 - 1.4	Table 7.4	0
630	Baroth/642 area	General collection	1.2 - 1.4	Table 7.4	60
631	Baroth/642 area	General collection	1.2 - 1.4	Table 7.4	34
632	Baroth/642 area	General collection	1.4 - 1.7	Table 7.4	17
633	Baroth/642 area	General collection	1.2 - 1.4	Table 7.4	0
634	Baroth/642 area	General collection	1.4 - 1.7	Table 7.4	13
635	Baroth/642 area	General collection	1.2 - 1.4	Table 7.4	7
636	Baroth/642 area	Scatter	1.2 - 1.4	Table 7.4	296
637	642 to Ban Barilla	Scatter	1.2 - 1.4	Table 7.11	13
638	Baroth/642 area	Scatter	1.4-1.7	Table 7.4	7
639	642 to Ban Barilla	Scatter	1.2 - 1.4	Table 7.11	60
640	Baroth/642 area	General collection	1.2 - 1.4	Table 7.4	31
641	642 to Ban Barilla	Scatter	1.2 - 1.4	Table 7.11	11
642	Baroth area	Locality	1.2 - 1.4	Tables 7.5-7.10	4219
643	Baroth/642 area	General collection	1.2 - 1.4	Table 7.4	9
644	642 to Ban Barilla	General collection	1.2 - 1.4	Table 7.11	37
645	Baroth/642 area	Scatter	1.2 - 1.4	Table 7.4	70
646	642 to Ban Barilla	General collection	1.2 - 1.4	Table 7.11	10
647	Ban Barilla-Doga	General collection	1.2 - 1.4	Table 7.11	4
648	642 to Ban Barilla	General collection	1.2 - 1.4	Table 7.11	20
649	Ban Barilla-Doga	General collection	1.2 - 1.4	Table 7.12	24
650	Ban Barilla-Doga	Scatter	1.2 - 1.4	Table 7.12	188
651	Ban Barilla-Doga	Scatter	1.2 - 1.4	Table 7.12	21
652	Ban Barilla-Doga	General collection	1.2 - 1.4	Table 7.12	6
653	Ban Barilla-Doga	General collection	1.2 - 1.4	Table 7.12	14

654	Ban Barilla-Doga	General collection	1.2 - 1.4	Table 7.12	17
655	Ban Barilla-Doga	Scatter	1.2 - 1.4	Table 7.12	65
656	Ban Barilla-Doga	Scatter	1.2 - 1.4	Table 7.12	64
657	Hairpin area	General collection	1.4 - 1.7	Table 7.13	22
659	Hairpin area	Find-spot		Table 7.13	8
661	Hairpin area	Scatter	1.4 - 1.7	Table 7.13	40
662	Ban Barilla-Doga	General collection	1.2 - 1.4	Table 7.12	18
663	Hairpin area	General collection	1.4 - 1.7	Table 7.13	20
664	Ban Barilla-Doga	Scatter	1.2 - 1.4	Table 7.12	629
665	Hairpin area	General collection	1.4 - 1.7	Table 7.13	15
666	Ban Barilla-Doga	Locality	1.2 - 1.4	Table 7.12	293
667	Hairpin area	General collection	1.4 - 1.7	Table 7.13	2
668	Ban Barilla-Doga	Scatter	1.2 - 1.4	Table 7.12	4
669	Hairpin area	General collection	1.4 - 1.7	Table 7.13	198
670	Ban Barilla-Doga	Scatter	1.2 - 1.4	Table 7.12	167
671	Hairpin area	General collection	1.4 - 1.7	Table 7.13	3
672	Ban Barilla-Doga	General collection	1.2 - 1.4	Table 7.12	90
674	Ban Barilla-Doga	General collection	1.2 - 1.4	Table 7.13	105
675	Ban Barilla-Doga	Scatter	1.2 - 1.4	Table 7.13	184
676	Ban Barilla-Doga	Scatter	1.2 - 1.4	Table 7.11???	45
677	Hairpin area	General collection	1.4 – 1.7	Table 7.13	57
678	666 to GT road	General collection	1.2 - 1.4	Table 7.13	41
679	Hairpin area	Find-spot	1.4 - 1.7	Table 7.13	1
680	Hairpin area	General collection	1.4 - 1.7	Table 7.13	19
681	Hairpin area	General collection	1.4 - 1.7	Table 7.13	65
682	Hairpin area	General collection			33
683	Hairpin area	Find-spot	1.4 - 1.7	Table 7.13	19
685	Hairpin area	General collection	1.4 - 1.7	Table 7.13	20
Total					**11734**

Notes: Nos. 673 (2 indeterminates only) and 603, 609, 617, 658, 660, and 684 (no finds) can be ignored.

Table 7:2. The Bulani area - Bandgwara Kas (600-607), Besa Kas (608-615) and Pir Jaffa (616-621)

Fossil specimens from Bandgwara Kas area, 600-607: Fossil occurrence 600: general collection:

Specimen	W	Taxon	Element	Part	Other	N
4	4	Artio (Large)	Vertebra	Cervical	Complete	1
1	4	Bovid	Horncore	Tine		1
2	5	Bovid	Metapodial	Proximal	Split	1
5	3	Bovid/Artio	Femur	Distal	Split	1
3	2b	*Struthio*	Eggshell			1
Total						**5**

Fossil occurrence 602: general collection:

Specimen	W	Taxon	Element	Part	Other	N
1	4	Bovid (Large)	Metatarsal	Proximal	Complete	1
4	4	Bovid (Large)	Molar	Lower	Split	1
2	4	Bovid (Medium)	M3 lower	Complete	Worn	1
6	2b	Cervid	Molar	Upper	Fresh	1
0		Chelonia	Scute	Fishnet		1
0	4	Indeterminate	Fragments			11
Total						**15**

Table 7. 2 continued: Fossil occurrence 605: scatter:

Specimen	W	Taxon	Element	Part	Other	N
13	4	Bovid (Medium)	PM	Upper	Complete	1
14	5	Bovid (Medium)	Tooth	Split	Worn	1
7	5	Proboscidean	Tooth	Plate		3
R10	4	Indeterminate	Fragments			24
Total						**29**

Besa Kas (608-615): Fossil occurrence 608: scatter:

Specimen		Taxon	Element	Part	Other	Total
130	5	Artiodactyla	Tooth	Split	No crown	4
25	4	Artio (Medium)	Astragalus	Split		1
112	3	Artio (Medium)	Fibula	Distal	Complete	1
43	4	Artio (Medium)	Humerus	Distal	Split	1
5	3	Bovid	Horncore			5
32	3	Bovid	Horncore	Tine		2
15	4	Bovid (Large)	M1/2, upper	Complete	Little wear	1
20	5	Bovid (Medium)	M3, lower	Complete	Worn	1
110	5	Bovid (Medium)	Molar, upper	Mod.wear	Split	1
110	2b	Carnivora	Scapula	Proximal	Complete	1
154	4	Cervid	Mandible, left	+PM2	Little wear	1
4, 30	4	Cervid	Antler	Palmate		2
156	3	Cervid	Tooth, upper	Complete	PM2	1
82	3	Cervid (Large)	PM, upper	Worn	Complete	1
135		Chelonia	Scute	Dimpled		2
88		Chelonia	Scute	Fishnet		1
7		Chelonia	Scute	Smooth		17
27	5	*Crocodylus*	Vertebra	Complete		1
101,130	5	Equid	Tooth	Fragment		2
81	5	Proboscidean	Tooth	Plate		1
143	4	Snail	Walnut-whip	Spiral		3
		Indeterminate	Fragments			183
Total						**234**

Fossil occurrence 611: general collection:

Specimen		Taxon	Element	Part	Other	Total
8	3	Artiodactyla	Scapula	Glenoid		1
20	5	Artiodactyla	Tooth	Fragment		1
6		Artio (Small)	Mandible	M1,M2 split		1
5		Bovid	Horncore			3
5		Cervid	Antler			1
3	2b	Cervid	PM, lower	Complete	Little wear	1
	3	Indeterminate	Fragments			22
14	5	Indeterminate	Mandible	+ Teeth	Sockets	1
16	5	Probosicdean	Tooth	Plate		1
Total						**32**

Fossil occurrence 612: general collection:

Specimen		Taxon	Element	Part	Other	Total
6	2b	Bovid (Medium)	Metapodial	D.barrel		1
1	4	Cervid	Antler	Fragment		1
2	2b	Chelonia	Scute	Dimpled		1
7	3	Indet (Large)	Facet	Proximal		1
3	5	Indet/Proboscidean	Tooth	Fragment		3
Total						**7**

Table 7.2 continued: Fossil occurrence 614: general collection:

Specimen		Taxon	Element	Part	Other	Total
4	5	Artio (Large)	Tooth	Split	No crown	1
8	4	Bovid (Large)	Metacarpal	Distal	Complete	1
7	3	Bovid (Medium)	Tooth	Split	Med. wear	1
11	3	Cervid	Antler	Palmate		1
2	2b	Chelonia	Scute	Honeycomb		2
6	5	Equid	Tooth	Complete		1
5		Indeterminate	Fragments			22
9	5	Indet/Proboscidean	Shaft			4
13	3	Indet/Proboscidean	Vertebra	Neural	Spine	1
3	5	Proboscidean	Tooth	Fragment		1
Total						35

Pir Jaffa area (616-621): Fossil occurrence 618: general collection:

Specimen		Taxon	Element	Part	Other	Total
2	4	Bovid (Medium)	Molar,	Split, lower	Little wear	1
6	3	Bovid (Small)	Molar	Split, lower	Little wear	1
4	5	Proboscidean	Tooth	Molar	Plate	1
0	5	Proboscidean	Tooth	Tusk	Rolled	1
0	4	Indeterminate	Fragments			35
Total						**39**

Fossil occurrence 620: general collection:

Specimen		Taxon	Element	Part	Other	Total
9	4	Bovid (Large)	Mandible	Ramus		1
1	4	Bovid (Large)	Cranium	+ horns		1
4	4	Bovid (Medium)	M1/2, lower	Complete	Mod. Wear	1
5	2b	Bovid (Medium)	M1/2	Complete	Little wear	1
2	5	Bovid (Medium)	Metacarpal	Distal	Epiphysis	1
0	4	Indeterminate	Fragments			12
7	5	Indet (Large)	Mandible	Fragment		1
3	4	Indet/Proboscidean	Tooth	Tusk		1
Total						**19**

Fossil occurrence 621: general collection:

Specimen		Taxon	Element	Part	Other	Total
8	3	Bovid (Medium)	M1/2, lower	Complete	No wear	1
12	4	Bovid (Medium)	Horncore	Pedicel		1
10	4	Chelonia	Scute	Dimpled		1
9	5	Probosicidean	Tooth	Plate		2
0	5	Indeterminate	Fragments			32
	5	Indet (Large)	Fragments			4
Total						**41**

Notes: 601 has 1 bovid tooth, 5 Proboscidean molar plate fragments and 4 indeterminates. 603 had nothing; general collection 606 had only 2 pieces only; there were 130 indeterminate fragments from scatter 604 and 5 from general collection 607. Fossil occurrences 608, 611-625 (odd numbers) were on top of or behind Sandstone 12, and are thus probably < 1.2 Mya-old. There were also the following general collections, all from large areas: 613 (1 proboscidean tooth root and 3 indeterminate fragments); 615 (1 equid tooth fragment and 1 indeterminate fragment); 616 (1small artiodactyl mandible fragment); 619 (1 bovid worn, split tooth fragment and 12 indeterminate fragments); and 623 (6 indeterminate fragments). There are no fossil finds from 609 and 617.

Table 7.3. Fossil material from locality 610.

Zone	Specimen	W	Taxon	Element	Part	Type	Other	N
	5	5	Artiodactyla	Mandible	+ tooth	No crowns	Split	1
	64	3	Artiodactyla	Metatarsal	Shaft			1
	65	4	Artiodactyla	PM	Upper	Complete	Worn	1
	52	2b	Artiodactyla	Tooth	Incisor	Complete	Worn	1
	39,56,57	4	Artiodactyla	Tooth	Spall			3
	53,54	5	Artiodactyla	Tooth	Split	PM/M		2
	41	5	Artiodactyla	Tooth	Split	No crown		1
	69	4	Bovid	Horncore				1
	63	3	Bovid (Large)	Maxilla	M1	+ Tooth	No wear	1
	24	2a	Bovid (Large)	Tooth	P2	Complete	Upper	1
	37	4	Bovid (Large)	Tooth	P3	Complete	Upper	1
	46	3	Bovid (Medium)	Mandible	PM	Split		1
	47	3	Bovid (Medium)	Maxilla	2 PMs	Complete	Fresh	1
	11	3	Bovid (Medium)	M3, left	Lower	Complete	Mod.wear	1
	12	3	Bovid (Medium)	M3, right	Upper	Complete	Mod.wear	1
	13,50	4	Bovid (Medium)	Molar	Lower	Complete	Little wear	2
	14	4	Bovid (Medium)	Molar	Lower	Split	Little wear	1
	4	4	Bovid (Medium)	Molar	Lower	Split	Mod.wear	1
	7	3	Bovid (Medium)	M1/2	Upper	Complete	Fresh	1
	32	3	Bovid (Medium)	M3	Lower	Split	Mod.wear	1
	78	4	Bovid (Medium)	M3, left	Lower	Split	Mod.wear	1
	15	3	Bovid (Medium)	M3, right	Lower	Split	Mod.wear	1
	17	3	Bovid (Medium)	M3, right	Lower	Complete	Erupting	1
	66	4	Bovid (Medium)	PM	Lower	Complete	Mod.wear	1
	19,51	3	Bovid (Medium)	PM	Lower	Complete	Fresh	2
	60	3	Bovid (Medium)	PM	Lower	Complete	Worn	1
	16,22,28, 29,30,36, 44,49	3	Bovid (Medium)	PM	Upper	Complete	Mod.wear	8
	43	3	Bovid (Medium)	PM	Upper	Split	Worn	1
	61	4	Bovid (Medium)	PM/M	Lower	Split	Worn	1
	31	3	Bovid (Medium)	PM/M	Upper	Split	Fresh	1
	23,34,35	4	Bovid (Medium)	Molar	Lower	Split	Little wear	3
	21,40	3	Bovid (Medium)	Molar	Lower	Split	Mod.wear	2
	25,33,77	3	Bovid (Medium)	Molar	Lower	Split	No crown	3
	38	3	Bovid (Medium)	Molar	Lower	Split	Worn	1
	45	5	Bovid (Medium)	Molar	Upper	Split	Worn	1
	10	3	Bovid (Medium)	Molar	Upper	Split	Mod.wear	1
	42	3	Bovid (Medium)	M1/2	Lower	Complete	Worn	1
	48	4	Bovid (Medium)	M3	Split	No crowns		1
	27	2b	Carnivora	Canine	Broken			1
	3	5	Equid	Molar	Lower	Complete	Mod.wear	1
	2,6	4	Equid	Molar	Upper	Complete	Mod.wear	2
	9	5	Equid	PM2	Lower	Complete	Mod.wear	1
	18,26	4	Equid	PM3/Molar	Lower	Complete	Mod.wear	2
	20	4	Equid	Tooth	Molar	Split	Mod.wear	1
	8	3	Equid	Tooth	P2	Complete	Mod.wear	1
	0		Indeterminate	Fragments				32
A	114	3	Artiodactyla	Carpal	Complete			1
A	11	3	Artiodactyla	Humerus	Shaft			1
A	16	3	Artiodactyla	Metatarsal	Shaft			1
A	0	4	Artiodactyla	Tooth	Fragments			61
A	9	3	Artiodactyla	Tooth	Incisor	Complete		1
A	3	3	Artio (Medium)	Humerus	Shaft			1
A	14	3	Artio (Medium)	Metatarsal	Shaft			1
A	4	4	Artio (Medium)	Radius	Shaft			1
A	30	3	Artio (Medium)	2nd Phalange	Distal			1
A	49	4	Artio (Medium)	Tibia	Distal	Split		1
A	20	4	Artio (Medium)	Tooth	PM	Split	Mod.wear	1
A	79	3	Artio (Medium)	Tooth	PM	Complete	Upper	1

Table 7. 3 continued: locality 610

Zone	Specimen	W	Taxon	Element	Part	Type	Other	N
A	24,47	5	Bovid	Horncore				2
A	3	3	Bovid (Large)	Tooth	M3	Complete	Erupting	1
A	**9**	3	Bovid (Medium)	Astragalus	Split			1
A	27	3	Bovid (Medium)	Metapodial	Proximal	Split		1
A	15	3	Bovid (Medium)	Metatarsal	Proximal	Split		1
A	2	3	Bovid (Medium)	Metatarsal	Shaft			1
A	6	3	Bovid (Medium)	Metatarsal	Proximal	Complete		1
A	8	4	Bovid (Medium)	Molar	Upper	Split	Mod.wear	1
A	2,6	3	Bovid (Medium)	M1/2	Upper	Complete	Mod.wear	2
A	41	3	Bovid (Small)	M3	Right	Complete	No wear	1
A	0	3	Bovid (Medium)	PM	Lower	Complete	Worn	2
A	21,23,58,7 3,87,124	4	Bovid (Medium)	Tooth	Molar	Split	Mod.wear	6
A	22	3	Bovid (Medium)	Tooth	M1/2	Complete	Mod.wear	1
A	10	3	Bovid (Medium)	Tooth, lower	M3,right	Complete	Mod.wear	1
A	11	5	Bovid (Medium)	Tooth	M3	Split	No crown	1
A	0	3	Bovid (Medium)	Tooth	PM	Complete	Mod.wear	8
A	96	4	Bovid (Medium)	Tooth	PM	Complete	Worn	1
A	16,46,120	4	Bovid (Medium)	Tooth	PM	Split	Mod.wear	3
A	34,91,99	4	Bovid (Medium)	Tooth	PM/M	Split	Mod.wear	3
A	0	4	Bovid (Medium)	Tooth	Split	No crown		12
A	0	3	Bovid (Medium)	Tooth	Worn	Split		2
A	93	2b	Bovid (Small)	Tooth	Incisor	Complete	Lower	1
A	60,64	3	Bovid (Small)	Tooth	M1/2	Complete	Little wear	2
A	36	3	Bovid (Small)	Tooth	M3, left	Split	No wear	1
A	117118	2b	Bovid (Small)	Tooth	P2	Complete	Mod.wear	2
A	83,121	3	Bovid (Small)	Tooth	P4	Complete	Mod.wear	2
A	1	3	Carnivora	Tooth	Canine	Broken		1
A	**7**	3	Equid	Tooth	Molar	Complete	Mod.wear	2
A	31,110	3	Equid	Tooth	Molar	Split	Mod.wear	2
A	53	3	Equid	Tooth	Incisor	Complete		1
A	**4**	3	Equid	Molar	Lower	Complete	Mod.wear	1
A	**5**	3	Equid	M3	Lower	Complete	Mod.wear	1
A	123	3	Hyaenid\Felid	Tooth	Canine	Broken		1
A	0		Indeterminate	Fragments				697
A	11	2b	Suid	Tooth	Molar	Fragment		1
A	12	3	Suid	Tooth	PM	Fragment		1
A	86	3	Suid	Tooth	Split			2
B	21	2b	Artiodactyla	Tooth	Incisor	Complete		1
B	18	5	Artiodactyla	Tooth	Root			1
B	12,13,20,	4	Artiodactyla	Tooth	Split	No crown		3
B	4	4	Artiodactyla	Tooth	Split	Mod.wear		1
B	19	3	Artiodactyla	Tooth	Spall			1
B	27	3	Artio (Medium)	Radius	Shaft			1
B	0,25		Artio (Medium)	Tooth	Incisors	Complete		3
B	32,43,46	3	Bovid	Horncore				3
B	8	2a	Bovid (Large)	PM	Upper	Complete	Little wear	1
B	0		Bovid (Medium)	M1/2	Lower	Complete	Little wear	1
B	11,15	3	Bovid (Medium)	M1/2	Lower	Complete	Mod.wear	2
B	3	3	Bovid (Medium)	M3	Upper	Complete	Mod.wear	1
B	9	3	Bovid (Medium)	M3	Upper	Split	Mod.wear	1
B	1	4	Bovid (Medium)	M3, left	Lower	Split	Mod.wear	1
B	0		Bovid (Medium)	PM	Lower	Complete	Little wear	1
B	5	4	Bovid (Medium)	PM	Lower	Complete	Mod.wear	1
B	17	3	Bovid (Medium)	PM/M	Upper	Complete	Mod.wear	1
B	26	5	Bovid (Medium)	Tooth	M1/2	Complete	Lower	1
B	0		Bovid (Medium)	Tooth	Split	Mod.wear		2
B	2,6,7	4	Bovid (Medium)	Tooth	Molar	Split		3
B	14,22	4	Bovid (Medium)	Molar	Lower	Split	Mod.wear	2
B	10	3	Cervid?	PM	Lower	Complete	Little wear	1
B	23	2a	Cervid?	Tooth	PM	Complete	Lower	1

Table 7.3 continued: locality 610

Zone	Specimen	W	Taxon	Element	Part	Type	Other	N
B	30	4	Equid	Metapodial	Split			1
B	29	3	Equid	Metatarsal	Proximal	Complete		1
B	0		Equid	Tooth	Incisor	Complete		1
B	24	2a	Hyaenid	Tooth	M1,	Split	Worn	1
B			Indeterminate	Fragments				679
C	0	3	Artiodactyla	Tooth	Incisor	Fragment	Lower	1
C	0	5	Artiodactyla	Tooth	Split	Mod.wear		1
C	23,38	3	Artiodactyla	Metapodial	Shaft			2
C	26	3	Artio (Medium)	Metacarpal	Shaft			1
C	43	3	Artio (Medium)	Metapodial	D.barrel			1
C	20	3	Artio (Medium)	Radius	Shaft			1
C	7	4	Artio (Medium)	Tooth	PM	Split	Lower	1
C	15	3	Artio (Medium)	Tooth	Incisor	Complete	Mod.wear	1
C	49,99,100	3	Artio (Medium)	Tooth	Incisor	Complete	Lower	3
C	16	3	Artio (Medium)	Tooth	Incisor	Complete	Upper	1
C	47,101	4	Artio (Medium)	Tooth	PM	Complete	Lower	2
C	103	3	Artio (Small)	Tooth	PM/M	Split		1
C	32	3	Bovid (Medium)	1st Phalange	Shaft			1
C	22,27,31, 34, 35,40	4	Bovid	Horncore				6
C	75,76,84	3	Bovid (Large)	Molar	Lower	Split	Erupting	3
C	14,83	3	Bovid (Large)	Tooth	PM	Complete	Mod.wear	2
C	87	3	Bovid (Large)	Tooth	PM/M	Split	Little wear	1
C	66	3	Bovid (Medium)	Mandible	PM2 socket, PM3	Left	Mod.wear	1
C	6,10-12, 48	3	Bovid (Medium)	Tooth	M1/2	Complete	Mod.wear	4
C	60,70	3	Bovid (Medium)	Tooth	M1/2, right	Complete	Mod.wear	2
C	71	4	Bovid (Medium)	Tooth	M1/2	Split	Mod.wear	1
C	17	4	Bovid (Medium)	Tooth	M1/2	Split	Worn	1
C	5	3	Bovid (Medium)	Tooth	M3, left	Complete	Mod.wear	1
C	2	4	Bovid (Medium)	Tooth	M3, left	Split	Mod.wear	1
C	1,9	3	Bovid (Medium)	Tooth	M3, right	Complete	Mod.wear	2
C	72	3	Bovid (Medium)	Tooth	M3, right	Split	Worn	1
C	67	3	Bovid (Medium)	Tooth	M3, right	Split	Mod.wear	1
C	51	3	Bovid (Medium)	Tooth	M3	Complete	Mod.wear	1
C	3,59	3	Bovid (Medium)	Tooth	M3, right	Split	Mod.wear	2
C	13,98	3	Bovid (Medium)	Tooth	Molar	Complete	Fresh	1
C	54,62,68, 69, 77, 78, 79, 81	4	Bovid (Medium)	Tooth	Molar	Split	Mod.wear	8
C	74,104	4	Bovid (Medium)	Tooth	Molar	Split	Worn	2
C	64,65	3	Bovid (Medium)	Tooth	PM	Complete	Mod.wear	2
C	52,53	3	Bovid (Medium)	Tooth	PM	Split	Fresh	2
C	56	3	Bovid (Medium)	Tooth	PM/M	Split	Mod.wear	1
C	61,63	3	Bovid (Medium)	Tooth	PM2	Complete	Mod.wear	2
C	58	3	Bovid (Medium)	Tooth	PM3/4, right	Complete	Worn	1
C	73	3	Bovid (Medium)	Tooth	PM	Complete	Worn	1
C	80,82, 105, 106	3	Bovid (Medium)	Tooth	PM	Complete	Mod.wear	4
C	85	3	Bovid (Medium)	Tooth	PM	Split	Worn	1
C	86	3	Bovid (Medium)	Tooth	PM2	Complete	Mod.wear	1
C	88	3	Bovid (Medium)	Tooth	PM/M	Complete	Worn	1
C	89,90,91, 93, 97	3	Bovid (Medium)	Tooth	PM	Complete	Worn	5
C	92	4	Bovid (Medium)	Tooth	PM/M	Split	Worn	1
C	94	3	Bovid (Medium)	Tooth	PM	Split	Lower	1
C	96	4	Bovid (Medium)	Tooth	PM	Split	Worn	1
C	30	3	Bovid (Small)	Metatarsal	Shaft			1

Table 7.3 continued: locality 610:

Zone	Specimen	W	Taxon	Element	Part	Type	Other	N
C	0	3	Carnivora	Tooth	Fragment			1
C	19	3	Carnivora	Tooth	Canine	Complete		1
C	18	3	Carnivore	Tooth	M1	Complete	Lower	1
C	102	3	Carnivora	Tooth	PM	Complete	Lower	1
C	0	5	Equid	Tooth	Fragment			1
C	8	4	Equid	Tooth	Molar	Complete	Lower	1
C	50	4	Equid	Tooth	Molar	Split	Mod.wear	1
C	57	3	Equid	Tooth	PM2	Split	Worn	1
C	95	4	Equid	Tooth	Split	Mod.wear		1
C	0	3	Indeterminate	Fragments				958
D	5	3	Artio (Medium)	1st Phalange	Distal			1
D	6	3	Bovid (Medium)	Mandible	PM	Mod.wear	Lower	1
D	2	3	Bovid (Small)	Mandible	M2,M3	Left	Erupting	1
D	1	3	Equid	Tooth	Molar	Complete	Lower	1
D	0		Indeterminate	Fragments				149
D	3	3	Indet/Suid	Mandible	PM	+ Teeth	Lower	1
E	1	2b	Artio (Medium)	Tooth	Incisor	Complete		
E	8	4	Artio (Medium)	Tooth	Molar	Split	Mod.wear	1
E	0		Artio (Medium)	Tooth	Fragments			7
E	0	4	Indeterminate	Fragments				109
F	0	5	Artiodactyla	Tooth	Fragments			10
F	0	4	Bovid	Tooth	PM/M	Split	Mod.wear	2
F	0	4	Equid	Tooth	Fragment			1
F	0	4	Indeterminate	Fragments				49
G	0	5	Artio (Medium)	Tooth	Fragment			1
G	4,5	3	Artio (Medium)	Tooth	Incisor	Complete	Worn	2
G	2	5	Artio (Medium)	Tooth	Molar	Split	Mod. wear	1
G	1	3	Bovid (Large)	Tooth	M1/2	Complete	Erupting	1
G	0	4	Bovid (Medium)	Tooth	PM/M	Split	Mod. wear	1
G	3	3	Equid	Tooth	Molar	Split	Mod. wear	1
G	0		Equid	Tooth	Fragment			1
G	0		Indeterminate	Fragments				3
	1	2b	Bovid (Medium)	Mandible	M3	No wear	Right	1
	1	3	Bovid (Medium)	Mandible	P3,4			1
1	1	3	Suid (Medium)	Mandible	M3	Complete+	Left	1
2	1	5	Artio	Tooth	Fragment			1
3	1	3	Suid	Tooth	PM	Complete	Lower	1
4	1	2b	Bovid (Medium)	Mandible	M1/2	Little wear	Right	1
4	2	2b	Bovid (Medium)	Mandible	M3	Complete	Right	1
4	5	2b	Bovid (Medium)	Tooth	M1/2	Complete	Little wear	1
4	0	3	Indeterminate	Fragments				6
4	1	3	Artio (Medium)	Humerus	Distal	Shaft		1
4	1	3	Equid	Tibia	Distal	Split		1
4	1	4	Indet/Artio	Tibia	Distal			1
4	0	3	Indeterminate	Fragments				20
4	1	3	Indet (Medium)	Shaft				3
4	1	3	Suid (Medium)	Maxilla	M2	Erupting		1
5	0	5	Indeterminate	Fragments				46
6	1	3	Indet (Medium)	Shaft				1
6	0	3	Indeterminate	Fragments				17
7	0		Indeterminate	Fragments				15
7	0	3	Struthio	Eggshell				1
8	0	3	Indeterminate	Fragments				9
9	1	4	Bovid (Large)	M3, right		Complete	Mod.wear	1
9	2	4	Equid	Tooth	M3/P3	Split	Mod.wear	1
9	0	4	Indeterminate	Fragments				28

Notes: The overall total was 3,229 specimens. There was a silt matrix on G1, G7; the small bovid may be *Gazella sp.*; the small carnivore may be a canid.

Table 7: 4. Fossil specimens from Baroth area (622-640).

Fossil occurrence 622: general collection:

Specimen	W	Taxon	Element	Part	Other	N
8	4	Bovid	Horncore			1
9	4	Cervid	Antler			2
2	4	Indeterminate	Fragments			7
4	4	Indet/Proboscidean	Tooth	Fragment		1
Total						**11**

Fossil occurrence 624:scatter:

Specimen	W	Taxon	Element	Part	Other	N
48	5	Artiodactyla	Tooth	Split	No crown	1
41	3	Artio (Medium)	Tibia	Shaft	1	1
24	5	Bovid	Horncore			6
27	5	Bovid	Horncore	Tine		4
16	3	Cervid	Antler	Tine		3
31	3	Cervid	Antler			10
30	3	Cervid	Antler	Palmate	+ Pedicel	1
45	4	Cervid	Antler	Palmate		1
0	4	Chelonia	Scute	Dimpled		1
0		Chelonia	Scute	Smooth		72
0		*Struthio*	Eggshell			1
2	4	Proboscidean	Carpal/tarsal			1
		Indeterminate	Fragments			168
Total						**270**

Fossil occurrence 626:find-spot:

Specimen	W	Taxon	Element	Part	Other	N
5	4	Artio (Medium)	Radius/ulna	Proximal	Fused	1
8	4	Artio (Medium)	Calcaneum	Distal		1
2	4	Proboscidean	Cranial/	Fragment		5
0		Indeterminate	Fragments			22
Total						**29**

Fossil occurrence 627: general collection:

Specimen	W	Taxon	Element	Part	Other	N
8	3	Artiodactyla	Tooth	Spall		1
3	3	Bovid (Large)	Tooth, upper	Split	M1/2	1
4	4	Bovid (Large)	Tooth	Split	No crown	1
2	4	Bovid (Medium)	M1/2, upper	Complete	Mod wear	1
5	3	Bovid (Medium)	Tooth	Split	Little wear	1
1	2a	*Struthio*	Eggshell			1
0		Indeterminate	Fragments			11
Total						**17**

Fossil occurrence 628:scatter:

Specimen	W	Taxon	Element	Part	Other	N
6	3	Artio (Medium)	Phalange	Distal		1
1	3	Bovid (*Hermibos*)	Maxilla	+ P2-4	Left, mod. wear	1
3	3	Cervid	Antler	Palmate		4
1	3	Cervid	Antler	+ branch & pedicel	Shed, palmate	1
0	3	Chelonia	Scute	Rippled		39
0	3	Chelonia	Scute	Smooth		10
24	5	Chelonia	Scute/spike			2
0		Chelonia?	Fragments	Non-scute		10
20	3	*Crocodylus*	Tooth	Broken		1
25	2b	Pisces	Vertebra	Complete		1
10	3	Reptillia	Cranial			1
18	3	Reptillia	Shaft			10
17	5	Ungulata	Tooth	Fragment		1
	4	Indeterminate	Fragments			105
Total						**187**

Table 7.4 continued: Fossil occurrence 630: general collection:

Specimen	W	Taxon	Element	Part	Other	N
30	4	Bovid	Horncore			1
39	5	Bovid (Large)	Ulna	Proximal		1
37	3	Bovid (Large)	Metacarpal	Proximal	Complete	1
38	3	Bovid (Large)	Radius	Proximal	Split	1
23	2b	Bovid (Large)	M1/2	Complete	Mod. Wear	1
33	3	Bovid (Large)	Mandible+ M3	Right	Little wear	1
29	3	Bovid (Large)	Radius	Proximal	Split	1
34	3	Bovid (Large)	Femur	Proximal	Complete	1
21	3	Bovid (Large)	Metacarpal	Shaft		1
18	3	Bovid (Medium)	Metacarpal	Distal	Split	1
26	3	Bovid (Medium)	Metapodial	Shaft		1
14	5	Bovid (Medium)	Tooth, lower	Split	Mod.wear	1
3	5	Bovid (Medium)	Tooth, lower	No crowns	split	1
11	4	Bovid (Medium)	Tooth	Crowns broken	Split	1
0	3	Bovid (Small)	Molar	Split	Little wear	1
41	3	Chelonia	Scute	Netted		1
12	3	Chelonia	Scute	Smooth		1
27	3	Indet (Large)	Fragments			4
19	3	Proboscidean	Tooth	Plate		1
15	4	Ungulata	Tooth	Plate		1
0	3	Indeterminate	Fragments			37
Total						**60**

Fossil occurrence 631: general collection:

Specimen	W	Taxon	Element	Part	Other	N
5	4	Bovid	Horncore			1
0	5	Equid	Tooth	Lower	Fragment	1
	5	Indeterminate	Fragments			30
2	5	Proboscidean	Tusk			1
4	5	Proboscidean	Tooth	Plate		1
Total						**34**

Fossil occurrence 632: general collection:

Specimen	W	Taxon	Element	Part	Other	N
11	4	Artiodactyla	Tooth	Root	No crowns	1
2	3	Artio (Medium)	Calcaneum	Proximal		1
9	3	Bovid (Medium)	M3, upper	Complete	No wear	1
5	4	Equid	M3, lower	Worn	Complete	1
10		*Struthio*	Eggshell			1
0	5	Indeterminate	Fragments			12
Total						**17**

Fossil occurrence 634: general collection:

Specimen	W	Taxon	Element	Part	Other	N
0	3	Bovid (Medium)	Tooth	Split	Little wear	1
4	4	Bovid (Medium)	Mandible, left	+ M3	Mod. wear	1
6	4	Bovid (Medium)	Tooth	M1/2 upper	Worn	1
		Chelonia	Scute	Fishnet		3
	5	Indeterminate	Fragments			5
1	3	Proboscidean	Tooth	Fragment		1
0	3	Proboscidean	Tooth	Tusk		1
Total						**13**

Fossil occurrence 635: general collection:

Specimen	W	Taxon	Element	Part	Other	N
1	5	Bovid	Horncore			1

Table 7.4 continued: general collection 635:

7	3	Bovid	Tooth	No wear	Split	1
2	3	Bovid (Large)	Metacarpal	Shaft		1
6	3	Bovid (Large)	M1/2, upper	No wear	Complete	1
5	3	Bovid (Medium)	Molar, lower	Split	Fresh	1
3	3	Indeterminate	Fragments			2
Total						**7**

Fossil occurrence 636: scatter:

Specimen	**W**	**Taxon**	**Element**	**Part**	**Other**	**N**
0,135,136,145,146	4	Artiodactyla	Tooth	Spall		5
0	4	Artiodactyla	Tooth	No crowns	Split	1
116	3	Artiodactyla	Vertebra	Centrum	Split	1
86	3	Artio (Medium)	Cranial			1
52	2b	Artio (Medium)	Femur	Distal	Condyle	1
0	2b	Artio (Medium)	Humerus	Proximal		1
21	3	Artio (Medium)	Humerus	Proximal	Condyle	1
143	2b	Artio (Medium)	Rib	10x9 mm.		1
130	4	Bovid (Medium)	2nd Phalange	Complete		1
0	3	Bovid (Medium)	Astragalus	Complete		1
109	3	Bovid (Medium)	Axis	Fragment		1
81	3	Bovid (Medium)	Axis	+ Vertebra	Joined	1
97	2b	Bovid (Medium)	Calcaneum	+ Astragalus	Joined	1
1	3	Bovid (Medium)	Cranium	Fragment		1
137	2b	Bovid (Medium)	Cuboid	Complete		1
138	2b	Bovid (Medium)	Femur	Distal,	Complete	1
0,85,91,105	4	Bovid (Medium)	Horncore			4
27	3	Bovid (Medium)	Humerus	Complete	Distal	1
106	3	Bovid (Medium)	Humerus	Distal		1
92	5	Bovid (Medium)	Mandible, right	+ Teeth		1
158	3	Bovid (Medium)	Mandible, left	+ Teeth		1
0	4	Bovid (Medium)	Mandible, left	+ Tooth	Mod. wear	1
0	2b	Bovid (Medium)	Mandible	+ Tooth	Split	1
63	4	Bovid (Medium)	Mandible	+ Tooth	Complete	1
87	3	Bovid (Medium)	Maxilla, right	+ 2 Teeth		1
87	3	Bovid (Medium)	Maxilla	+ 4 Teeth		1
93	2b	Bovid (Medium)	Maxilla, left	+ 6 Teeth	Unworn	1
114, 104	2b	Bovid (Medium)	Metacarpal	Distal	Complete	1
26,124	3	Bovid (Medium)	Metacarpal	Proximal	Complete	2
99,111,115	3	Bovid (Medium)	Metapodial	Distal	Complete	3
118,142	2b	Bovid (Medium)	Metatarsal	Proximal	Complete	2
92	3	Bovid (Medium)	Pelvis	Acetabulum		1
37	3	Bovid (Medium)	Radius	Distal	Complete	1
82	3	Bovid (Medium)	Radius	+ 3 Carpals	Distal	1
22	3	Bovid (Medium)	Radius	Proximal	Split	1
36, 45	4	Bovid (Medium)	Radius	Proximal	Complete	2
0	3	Bovid (Medium)	Tibia	Proximal	Split	1
84	3	Bovid (Medium)	Tibia	Proximal	Complete	1
34,**44**	4	Bovid (Medium)	Tibia	Distal	Complete	2
110	3	Bovid (Medium)	Tibia (distal)	+Astragralus	Joined	1
0	4	Bovid (Medium)	Tooth, upper			1
60	3	Bovid (Medium)	Tooth, upper	Complete	Fresh	1
134	3	Bovid (Medium)	Tooth, lower	Complete	Fresh	1
0	4	Bovid (Medium)	Tooth	No crowns	Split	1
32,104	2b	Bovid (Medium)	Ulna	Proximal		2
0*	3	Bovid (Medium)	Vertebra	Centrum		6
0*	4	Indeterminate	Fragments			229
Total						**296**

Table 7.4 continued: scatter 636: Bovid (Medium) vertebra centrums: 83,90,107,108,121,140. There were also 7 indeterminate fragments, and 51 shaft fragments. Marked Indet fragments: 68,70,101,119,153,156,157 (7 in total). Marked indet erminate shaft fragments: 23, 28, 29, 32, 34, 35, 38, 39, 40, 42, 43, 46, 47, 47, 48, 49, 50, 51, 54, 56, 57, 58, 59, 61, 62, 64, 65, 66, 67, 69, 73, 77, 94, 100, 102, 102, 112, 120, 122, 123, 124, 125, 126, 127, 128, 131, 148, 149, 149, 150, 151, 152. (51 in total).

Fossil occurrence 640: general collection:

Specimen	W	Taxon	Element	Part	Other	N
11	4	Bovid (Large)	Horncore			1
12	4	Bovid (Large)	Tooth, upper	PM	No crown	1
2	4	Cervid	Antler			1
4	5	Indet/Artio	Tooth	No crowns	Split	1
3,8	5	Rhino	Tooth	Fragment		2
0	4	Indeterminate	Fragments			25
Total						**31**

Notes: Scatter 637 had one equid lower molar/premolar and 12 fragments, and scatter 638 had only 3 proboscidean/suid tusk fragments, and 4 indeterminate fragments.

Table 7.5: Fossil specimens from locality 642, area 1, 1989

Specimen	Zone	W	Taxon	Element	Part	Other	N
249	A	5	Artio	Astragalus	Split		1
180	C	4	Artio	Atlas	Fragment		1
323	A	5	Artio	Cranial			1
166	B	5	Artio	Mandible	+ Tooth	No crowns	1
321	A	5	Artio	Mandible	+ teeth	2 roots	1
63	B	5	Artio	Metapodial	D.barrel		1
various	C	4	Artio	Tooth	Fragment		7
64	B	3	Artio (Large)	Vertebra	Flange		1
370	A	4	Artio (Medium)	Mandible	+ 2 Molars		1
371	A	3	Artio (Medium)	1st Phalange	Distal		1
346	A	3	Artio (Medium)	1st Phalange	Proximal		1
103	B	4	Artio (Medium)	Axis	Fragment		1
132	B	4	Artio (Medium)	Femur	Distal	Condyle	1
164	B	5	Artio (Medium)	Mandible + tooth	Split		1
293	A	4	Artio (Medium)	Mandible	P2,P3 sockets, P4 root		1
354		5	Artio (Medium)	Mandible	Teeth	No crowns	1
302	A	5	Artio (Medium)	Mandible	Fragment		1
253	C	4	Artio (Medium)	Mandible	Ramus		1
212	C	5	Artio (Medium)	Maxilla	+Tooth socket		1
265,315, 343,392,	A	4	Artio (Medium)	Metapodial	Proximal	Split	4
82,269, 285, 290,300	A	5	Artio (Medium)	Metapodial	Shaft		5
35,130, 197		4	Artio (Medium)	Metatarsal	Shaft		3
414	A	4	Artio (Medium)	Radius	Proximal	Complete	1
109	B	5	Artio (Medium)	Tibia	Distal	Complete	1
312	A	4	Artio (Medium)	Tibia	Distal	Shaft	1
320,395	A	3	Artio (Medium)	Tooth	Incisor	Complete	2
237	C	4	Artio (Medium)	Tooth	PM, lower, worn	Complete	1
79	B	5	Artio (Medium)	Tooth	Split		16
241,345	A	3	Artio (Medium)	Vertebra	Thoracic	Complete	2
96	B	5	Artio (Medium)	Metapodial	Distal	Condyle	1
372	A	5	Artio (Small)	Maxilla	+ Incisors 1,2		1
311,393	A	2b	Artio (Small)	Metatarsal	Shaft	(joins)	2
262,303	A	4	Artio (Small)	Radius	Distal	Complete	2
385	A	4	Artio/Bovid	Tooth	PM, lower	Worn	1
390	A	3	Artio/Bovid	Tooth	PM, upper	Worn	1
227	C	4	Artio/Suid	Tooth	Fragment		1
various		5	Bovid	Horncore			23
110	B	3	Bovid	Horncore	Base		1
143	B	5	Bovid	Horncore	Tine		3
122	B	5	Bovid	Mandible	M2,M3 complete	Left, Mod.wear	1

Table 7.5 continued: locality 642, area 1, 1989

Specimen	Zone	W	Taxon	Element	Part	Other	N
182	C	5	Bovid	Metapodial	Distal	Epiphysis	1
18	B	5	Bovid	Tooth	M3 right, mod. wear	Complete,	1
33	B	5	Bovid (Large)	1st Phalange	Complete		1
43	B	5	Bovid (Large)	2nd Phalange	Complete		1
64	B	5	Bovid (Large)	Astragalus	Complete		1
191,226	C	4	Bovid (Large)	Calcaneum	Facet	(joins)	2
1,178	B,C	5	Bovid (Large)	Calcaneum	Shaft		2
67	B	4	Bovid (Large)	Femur	Distal	Complete	1
46	B	5	Bovid (Large)	Horncore			1
272	A	4	Bovid (Large)	Mandible	1 broken, 1 erupting+ 2 teeth	Left	1
245	A	5	Bovid (Large)	Mandible	+ tooth root	Split	1
273	A	5	Bovid (Large)	Mandible	Split,crowns		1
261	A	3	Bovid (Large)	Mandible	Erupting tooth	Right	1
386	A	5	Bovid (Large)	Maxilla	+ tooth		1
34	B	3	Bovid (Large)	Metatarsal	Shaft		1
120	B	4	Bovid (Large)	Tibia	Proximal	Split	1
247,379	A	5	Bovid (Large)	Tooth	No crowns	Split	2
284	A	3	Bovid (Large)	Tooth	PM worn	Split,	1
316	A	4	Bovid (Large)	Tooth	PM2, upper	Worn	1
45	B	5	Bovid (Large)	Horncore			1
48	B	3	Bovid (Medium)	Astragalus	Complete		1
244,366	A	4	Bovid (Medium)	Horncore			2
243,279, 363	A	4	Bovid (Medium)	Horncore	Tine		3
415,417	A	4	Bovid (Medium)	Humerus	Distal	Split	2
410	A	5	Bovid (Medium)	Mandible	M3 complete, right	Mod.wear	1
427	A	3	Bovid (Medium)	Mandible	M3, right	Mod. wear	1
58	B	5	Bovid (Medium)	Mandible	+ 2 Teeth	Left	1
3	B	4	Bovid (Medium)	Mandible	M and PM errupting+ M split		1
144	B	4	Bovid (Medium)	Mandible	M1+M2 split	Right	1
26	B	4	Bovid (Medium)	Mandible	M1, M2	Split	1
210	C	4	Bovid (Medium)	Mandible	M1/2 broken	+ Teeth	1
123	B	3	Bovid (Medium)	Mandible	M1, M2 left	Split	1
31	B	3	Bovid (Medium)	Mandible	M2 split, M3	Worn, left	1
73	B	3	Bovid (Medium)	Mandible	3 PM sockets + M1		1
93	B	4	Bovid (Medium)	Mandible	M split,	Erupting	1
13,104	B	4	Bovid (Medium)	Mandible	+ M3	+ Tooth	2
8, 57	B	3	Bovid (Medium)	Mandible	+ M3	+ Tooth	2
65	B	5	Bovid (Medium)	Mandible	+ M3?	+ Tooth	1
121	B	4	Bovid (Medium)	Mandible	+ Tooth	No crowns, split	1
9	B	5	Bovid (Medium)	Mandible	+ PM2-M1	Right	1
147	B	4	Bovid (Medium)	Mandible	2 Teeth, 1 split	Worn	1
12	B	4	Bovid (Medium)	Mandible	Molar1/2+socket	Mod.wear	1
318	A	5	Bovid (Medium)	Mandible	1 split,1 root		1
257	A	4	Bovid (Medium)	Mandible	M1;M2 erupting	Left; joins 274	1
274	A	4	Bovid (Medium)	Mandible	M3	Complete; joins 257	1
347	A	4	Bovid (Medium)	Mandible	P2 broken,P3socket	Left	1
283	A	4	Bovid (Medium)	Mandible	+ tooth		1
374	A	5	Bovid (Medium)	Mandible	Tooth broken	+ tooth	1
192	C	3	Bovid (Medium)	Mandible	+ M2 broken + M3	Left	1
344	A	4	Bovid (Medium)	Maxilla	P2,P3	+ teeth	1
181,260	C, A	3	Bovid (Medium)	Metacarpal	Distal	Complete	2
107	B	4	Bovid (Medium)	Metacarpal	Distal	Epiphysis	1
124	B	3	Bovid (Medium)	Metacarpal	Proximal	Complete	1
137,201	B, C	5	Bovid (Medium)	Metacarpal	Shaft		2
101	B	3	Bovid (Medium)	Metapodial	Proximal	Complete	1
138,139	B	4	Bovid (Medium)	Metapodial	Proximal	Split	2
203,248, 251, 277	C	3	Bovid (Medium)	Metatarsal	Proximal	Complete	4

Table 7.5 continued: locality 642, zone 1, 1989

Specimen	Zone	W	Taxon	Element	Part	Other	N
184	C	3	Bovid (Medium)	Metatarsal	Shaft		1
185,266	C	3	Bovid (Medium)	Radius	Proximal	Complete	2
175	C	4	Bovid (Medium)	Scapula	Proximal	Glenoid	1
70,208	B,C	4	Bovid (Medium)	Tibia	Distal	Complete	2
223	C	4	Bovid (Medium)	Tooth	Molar, lower	Broken	1
42	B	3	Bovid (Medium)	M1/2, upper	Complete	Mod. wear	1
148	B	4	Bovid (Medium)	Tooth	M1/2, lower	Complete Mod.wear	1
76	B	4	Bovid (Medium)	M1/2, upper	Complete	Mod.wear	1
224	C	4	Bovid (Medium)	Tooth	M1/2, lower	Complete	1
190	C	3	Bovid (Medium)	Tooth	M3 left, upper	Complete	1
217	C	4	Bovid (Medium)	Tooth	PM, worn	Complete	1
32,125	B	4	Bovid (Medium)	Tooth	PM, lower	Complete Worn	2
87,170	B	4	Bovid (Medium)	Tooth	PM, upper	Complete	2
19	B	3	Bovid (Medium)	Tooth	PM, upper	Complete	1
126,154	B	4	Bovid (Medium)	Tooth	PM, lower	Complete	2
4,90,149,160,171,173	B	3	Bovid (Medium)	Tooth	PM, upper	Complete	6
158	B	5	Bovid (Medium)	Tooth	Spall		1
151,159	B	4	Bovid (Medium)	Tooth	Split	Worn	2
256,313	A	5	Bovid (Medium)	Tooth	Split		2
232	C	5	Bovid (Medium)	Tooth	M/PM	Split	1
75	B	3	Bovid (Medium)	Tooth	M1/2 lower	Split	1
113	B	4	Bovid (Medium)	Tooth	Molar, worn	Split	1
61,80,140, 150,162	B	4	Bovid (Medium)	Tooth	Molar, mod. wear	Split	5
289,314	A	5	Bovid (Medium)	Tooth	No crowns	Split	2
324	A	4	Bovid (Medium)	Tooth	Molar	Split,worn	1
350	A	4	Bovid (Medium)	Tooth	PM, upper	Worn	1
377	A	3	Bovid (Medium)	Tooth	PM	Worn	1
351	A	4	Bovid (Medium)	Tooth	PM2, lower	Worn	1
275	A	3	Bovid (Small)	Humerus	Distal	Complete	1
199,276	A	3	Bovid (Small)	Metacarpal	Distal	Complete	2
364	A	3	Bovid (Small)	Metatarsal	Proximal	Complete	1
169	B	4	Bovid (Small)	Tooth	Molar, worn	Split	1
100	B	3	Bovid/Cervid	Mandible	M1/2, complete	+ Tooth	1
253	A	3	Bovid/Cervid	Maxilla	P3-M3, mod. wear	Left	1
53	B	3	Cervid	Antler	Pedicel	Palmate	1
196	C	4	Equid	2nd Phalange	Complete		1
59	B	5	Equid	Maxilla	Crowns damaged	+ Tooth	1
54	B	5	Equid	Maxilla	M3,poor crowns	Tooth split	1
400	A	3	Equid	Metacarpal	Proximal	Complete	1
183	C	4	Equid	Metatarsal	Proximal	Complete	1
55	B	3	Equid	Tibia	Distal	Complete	1
136	B	2a	Equid	Tooth	M3, lower, left	Complete	1
179	C	4	Equid	Tooth	Molar, upper	Complete	1
1	A	3	Hyaenid	Mandible	P3, P4, right	+ Teeth	1
30	B	3	Rhinoceritidae	Astragalus	Complete	Left	1
15	B	4	Rhinoceritidae	Tooth	Fragment	Upper	1
198	C	4	Ungulata (Med)	Vertebra	Fused		3
			Indeterminate	Fragments			911
Total							**1150**

Table 7. 6: Fossil specimens from locality 642, zone 2, 1989

Specimen	Zone		Taxon	Element	Part	Other	N
290	B	4	Artiodactyl	Ramus	Fragment		1
130	B	4	Artiodactyl	Tibia	Shaft		1
various		4	Artiodactyl	Tooth	Fragment		5
159	B	4	Artiodactyl	Ulna	Proximal		1
8	A	4	Artio (Large)	Humerus	Shaft	Distal	1
2	A	4	Artio (Large)	Humerus	Shaft		1

Table 7. 6 continued: locality 642, area 2, 1989

Specimen	Zone		Taxon	Element	Part	Other	N
233	B		Artio (Large)	Maxilla	+Tooth fragment		1
165	B	5	Artio (Large)	Vertebra	Centrum		1
161	B	4	Artio (Medium)	1st Phalange	Proximal		1
192, 291	B	5	Artio (Medium)	Mandible	Tooth broken		2
188	B	4	Artio (Medium)	Mandible	Ramus frag		1
237	B	5	Artio (Medium)	Maxilla	PM	Worn	1
178	B	5	Artio (Medium)	Pelvis	Acetabulum		1
173	B	5	Artio (Medium)	Radius	Proximal		1
103	A	5	Artio (Medium)	Tooth	Split		1
107	A	5	Artio (Medium)	Tooth	Split		1
198	B	3	Artio (Medium)	Vertebrae	Complete		3
255	B	4	Artio (Small)	1st phalange	Distal	Complete	1
273	B	5	Artio (Small)	2nd phalange	Proximal	Complete	1
190	B	4	Artio/Bovid	Vertebra	Flange		1
various		4	Bovid	Horncore			17
22	A	4	Bovid	Horncore	Base		1
98	A	4	Bovid	Horncore	Curved		1
34	A	4	Bovid	Horncore	Tine		7
189	B	4	Bovid	Metacarpal	Distal	Complete	1
145	B	5	Bovid	Tooth	M1/2		1
0		4	Bovid (Large)	Horncore	Fragments		2
118	B	4	Bovid (Large)	Metapodial	D.barrel		1
26,43	A,B	4	Bovid (Large)	Radius	Proximal	Split	2
205	B	4	Bovid (Large)	Tooth	M1/2	Errupting, Upper	1
210,246	B	5	Bovid (Large)	Tooth	Fragment		2
18	A	4	Bovid (Medium)	Atlas	Complete		1
294	B	3	Bovid (Medium)	Fibula	Distal	Complete	1
85	A	3	Bovid (Medium)	Frontal+	2 horncore		1
134	B	4	Bovid (Medium)	Mandible	M3	Right	1
3	A	4	Bovid (Medium)	Mandible	M1,M2 broken, M3	Left	1
160	B	4	Bovid (Medium)	Mandible	P2 broken, P3,P4, M1	Mod.wear, Left	1
134	B		Bovid (Medium)	Mandible	M3		1
139	B	5	Bovid (Medium)	Mandible	Molar obscured		1
137	B	4	Bovid (Medium)	Mandible	M3 split		1
122,158	B	4	Bovid (Medium)	Tooth	M1/2		2
124,234	B	4	Bovid (Medium)	Tooth	M3 lower	Worn	2
235	B	5	Bovid (Medium)	Tooth	M3 broken		1
72,100	A	2b	Bovid (Medium)	PM Upper	Complete	Mod. wear	2
102	A	4	Bovid (Medium)	Tooth	PM Upper	Complete,worn	1
301	B	4	Bovid (Medium)	Tooth	Fragment		1
238	B	5	Bovid (Medium)	Tooth	Molar	Fragment	1
101,169	A, B	4	Bovid (Medium)	Tooth	Molar	Split, Worn	2
77, 172,208	A,B,B	5	Bovid (Medium)	Tooth	Molar	Split	3
259	B	4	Bovid (Medium)	Tooth	PM	Upper	1
175	B		Bovid (Medium)	Tooth	M1/2	Worn	1
104,226	A, B	4	Bovid (Medium)	Tooth	PM lower	Worn	2
211	B	4	Bovid (Medium)	Tooth	PM	Worn	1
291	B	4	Bovid (Small)	1st Phalange	Proximal	Complete	1
224	B	5	Bovid (Small)	Mandible	+ 1 tooth		1
197	B	3	Bovid (Small)	Metacarpal	Complete	Proximal	1
166,212	B	3	Bovid (Small)	Metapodial	Distal	Complete	2
135	B	3	Bovid (Small)	Metapodial	Distal		1
19	A	4	Bovid (Small)	Nav.-cuboid			1
187	B	3	Bovid (Small)	Tibia	Distal	Complete	1
119	B	4	Bovid/Cervid	Mandible, R.	M2 split, M3 complete	Mod.wear,	1
154	B	3	Bovid/Cervid	Mandible	M2,M3	Right	1
67	A	4	Bovid/Cervid	Mandible+	M2 split, M3 split	V. worn	1
138	B	5	Bovid/Equid	Mandible	+ 2 teeth		1
120	B	5	Equid	Mandible	1 tooth,split		1

Table 7. 6 continued: locality 642, area 2, 1989

Specimen	Zone	W	Taxon	Element	Part	Other	N
33	A	3	Equid	Mandible	P2-4,M1-2	Left	1
68	A	4	Equid	Mandible	Incisor		1
64	A	5	Equid	Mandible	M3, complete	Right	1
66	A	4	Equid	Mandible	Molar		1
131, 132,142	B	5	Equid	Maxilla	Tooth		3
129	B	4	Equid	Metacarpal	Distal	Complete	1
127	B	4	Equid	Metacarpal	Proximal	Complete	1
117	B	3	Rhinoceritidae	Metapodial	Proximal		1
17	A	4	Rhinoceritidae	Carpal	Complete		1
1	A	4	Rhinoceritidae	Femur	Distal		1
various		5	Ungulata	Tooth	Fragment		5
			Indeterminate	Fragments			454
Total							**576**

Notes: specimen 135 (Bovid (Small) distal metapodial) was still attached to its sesamoid and both 1st phalanges. No. 197 (Bovid (Small) proximal metacarpal was attached to its carpals.

Table 7. 7: Fossil specimens from locality 642, zone 3, 1989

Specimen	Zone	W	Taxon	Element	Part	Other	N
345	B	5	Artiodactyl	Calcaneum	Distal	Shaft	1
86	A	4	Artiodactyl	Humerus	Distal	Split	1
293	B	3	Artiodactyl	Humerus	Distal		1
562	B		Artiodactyl	Mandible	P2 broken, P3		1
353	B	4	Artiodactyl	Mandible	P3 socket at back		1
382	B	4	Artiodactyl	Mandible	Ramus		1
319	B	4	Artiodactyl	Pelvis	Acetabulum	Split	1
900	B	4	Artiodactyl	Tooth	PM	Lower	1
501	B	5	Artiodactyl	Tooth	PM, worn	Complete	1
151	A	5	Artiodactyl	Tooth	Fragment		8
356	B	4	Artiodactyl	Tooth	PM	Fragment	1
465	B	5	Artiodactyl	Tooth	PM2, lower	Fragment	1
440	B	5	Artiodactyl	Tooth	Molar, upper	Worn	1
343	B	5	Artiodactyl	Tooth	Split		1
391	B	5	Artiodactyl	Tooth	Split	Fragment	1
13	A	4	Artiodactyl	Ulna	Proximal		1
340	B	5	Artiodactyl (Large)	Mandible	+Tooth, split	Sheared off	1
257	B	5	Artiodactyl (Large)	Mandible	Tooth root	Fragment	1
122	A	5	Artiodactyl (Large)	Mandible	Tooth root	Fragment	1
104	A	5	Artiodactyl (Large)	Tooth	Molar, upper	Little wear	1
184	A	5	Artiodactyl (Large)	Tooth	Split	Fragment	1
886	B	3	Artiodactyl (Medium)	2nd Phalange	Proximal		1
420	B	4	Artiodactyl (Medium)	3rd Phalange	Proximal		1
322,385	B	4	Artiodactyl (Medium)	Atlas	Fragment		2
376	B	5	Artiodactyl (Medium)	Carpal	Complete		1
882	B	4	Artiodactyl (Medium)	Cranial			1
417	B	4	Artiodactyl (Medium)	Humerus	D. barrel		1
39	A	5	Artiodactyl (Medium)	Humerus	Distal	Split	1
324	B	4	Artiodactyl (Medium)	Humerus	Distal	Shaft	1
315	B	5	Artiodactyl (Medium)	Mandible	+ 2Teeth, broken		1
727	B	4	Artiodactyl (Medium)	Mandible	+ 3 sheared off teeth		1
429	B	5	Artiodactyl (Medium)	Mandible	+ Tooth	Root	1
672	B	5	Artiodactyl (Medium)	Mandible	Tooth socket	Split	1
874	B	5	Artiodactyl (Medium)	Mandible	Tooth socket	Fragment	1
133	A	4	Artiodactyl (Medium) Mandible	Ramus			1
38,244	A	3	Artiodactyl (Medium)	Metapodial	Proximal	Split	2
871,880	B	5	Artiodactyl (Medium)	Metatarsal	Shaft		2
631,866	B	5	Artiodactyl (Medium)	Radius	Distal	Complete	2
647	B	5	Artiodactyl (Medium)	Radius	Proximal	Split	1
266	B	4	Artiodactyl (Medium)	Ramus			1
57	A	5	Artiodactyl (Medium)	Tibia	Distal	Complete	1

Table 7. 7: Fossil specimens from locality 642, zone 3, 1989

Specimen	Zone	W	Taxon	Element	Part	Other	N
639	B	4	Artiodactyl (Medium)	Tibia	Distal	Split	1
442	B	3	Artiodactyl (Medium)	Tooth	PM	Worn	1
283	B	4	Artiodactyl (Medium)	Vertebra	Caudal		1
108, 162,250, 271	A	5	Artiodactyl (Medium)	Vertebra	Complete		4
438	B	5	Artiodactyl (Medium)	Vertebra	Flange		1
297	B	3	Artiodactyl (Small)	Metapodial	Shaft		1
379	B	4	Artiodactyl/Bovid	Mandible	Root only	+ Tooth	1
311	B	3	Artiodactyl/Bovid	Metacarpal	Shaft		1
309	B	4	Artiodactyl/Bovid	Tooth	PM, lower	Complete	1
466	B	4	Artiodactyl/Bovid	Tooth	PM	Complete	1
368	B	5	Artiodactyl/Bovid	Tooth	Molar	Split	1
314	B	4	Artiodactyl/Bovid	Ulna	Proximal	Complete	1
various		4	Bovid	Horncore			141
247	B	4	Bovid	Horncore	Tine		14
146	A	5	Bovid	Mandible	M1/2 worn	+ Tooth	1
233	B	4	Bovid	Mandible	P3,P4 right	Mod.wear	1
135	A	4	Bovid	Mandible	M1/2 mod. wear	Lower	1
138	A	5	Bovid	Mandible	M1/2	Mod. wear	1
853	B	5	Bovid	Maxilla	PM, worn		1
561	B		Bovid	Metacarpal	Shaft		1
697	B	5	Bovid	Metapodial	D.barrel		1
15	A	4	Bovid	Metatarsal	Proximal	Complete	1
230	B	5	Bovid	Tooth	Molar, worn	Fragment	1
828	B	5	Bovid	Tooth	Fragment		15
172	A	5	Bovid	Tooth	Molar, worn	Fragment	1
362,461	B	4	Bovid	Tooth	PM	Root	2
449	B	5	Bovid	Tooth	Split	Fragment	15
163	A	5	Bovid	Tooth	M3 right	Split	1
355	B	5	Bovid	Tooth	Molar, worn	Split	1
364	B	5	Bovid	Tooth	PM	Split	1
390	B	5	Bovid	Tooth	M1/2 upper	Worn	1
272	B	5	Bovid	Tooth	PM	Worn	1
64	A	4	Bovid (Large)	1st Phalanage	Proximal	Complete	1
67	A	4	Bovid (Large)	Calcaneum+	Astragalus		1
815	B	5	Bovid (Large)	Horncore			1
5	A	4	Bovid (Large)	Humerus	Distal	Shaft	1
222	B	4	Bovid (Large)	Humerus	Distal	Split	1
802	B	5	Bovid (Large)	Mandible	+ Tooth	Split	1
860	B	5	Bovid (Large)	Mandible	M1/2 split	+ Tooth, left	1
741	B	5	Bovid (Large)	Mandible	PM2	+ Tooth	1
224	B	5	Bovid (Large)	Metacarpal	D. barrel		1
410	B	5	Bovid (Large)	Metapodial	D. barrel		1
854	B	5	Bovid (Large)	Metapodial	D.barrel		1
845	B	4	Bovid (Large)	Pelvis	Acetabulum		1
613	B	4	Bovid (Large)	Phalange	Proximal		1
862	B	5	Bovid (Large)	Tibia	Distal	Split	1
128	A	5	Bovid (Large)	Tooth	Fragment		1
711	B	5	Bovid (Large)	Tooth	Fragment		1
749	B	4	Bovid (Large)	Tooth	PM	Mod.wear	1
305	B	5	Bovid (Large)	Tooth	Molar, upper	Split	1
294	B	5	Bovid (Large)	Tooth	No crowns	Split	1
327	B	4	Bovid (Large)	Tooth	PM, upper	Worn	1
816	B	4	Bovid (Large)	Ulna	Proximal		1
600	B	4	Bovid (Medium)	Astragalus	+ distal tibia	Complete	1
648	B	3	Bovid (Medium)	Atlas	+ Axis	Fused	1
206,234	B	4	Bovid (Medium)	Axis			2
161	A	4	Bovid (Medium)	Axis	Complete		1
131,287	A	4	Bovid (Medium)	Axis	Fragment		2
787	B	5	Bovid (Medium)	Metapodial	D.barrel		1
114,216	A	5	Bovid (Medium)	Humerus	Distal	Complete	2
204	B	4	Bovid (Medium)	Humerus	Distal	Fragment	1
190	A	3	Bovid (Medium)	Humerus, distal	+Ulna fused	+Radius Right	1
772	B	5	Bovid (Medium)	Mandible	Molar, broken		1
645	B	4	Bovid (Medium)	Mandible	2 molars	Split	1
213	B	4	Bovid (Medium)	Mandible	DM3,M1 right	Mod.wear	1
248	B	2b	Bovid (Medium)	Mandible	M2,3 right	Erupting	1

Table 7. 7: Fossil specimens from locality 642, zone 3, 1989

Specimen	Zone	W	Taxon	Element	Part	Other	N
237	B	3	Bovid (Medium)	Mandible	P2,P3 right	Fresh	1
217	B	3	Bovid (Medium)	Mandible	P4,M1 right	Mod.wear	1
341	B	4	Bovid (Medium)	Mandible	P4 socket + M1-2	+M3 socket	1
423,677, 718,740,850,858,	B	5	Bovid (Medium)	Mandible	+ Tooth	Fragment	6
610	B	4	Bovid (Medium)	Mandible	+ 2 Molars	Split	1
748	B	4	Bovid (Medium)	Mandible	M errupting?		1
893	B	4	Bovid (Medium)	Mandible	M1/2 left	Complete	1
236	B	5	Bovid (Medium)	Mandible	M3	Left	1
303	B	4	Bovid (Medium)	Mandible	M3 left	Split	1
296	B	5	Bovid (Medium)	Mandible	M3right	Worn	1
658	B	5	Bovid (Medium)	Mandible	M3 right	Worn	1
124	A	4	Bovid (Medium)	Mandible	M3 complete	V. worn Left	1
839	B	5	Bovid (Medium)	Mandible	M3 complete, left	Mod.wear	1
46	A	5	Bovid (Medium)	Mandible	M3 split	Right	1
704	B	5	Bovid (Medium)	Mandible	M3 split, left	Mod.wear	1
76	A	4	Bovid (Medium)	Mandible	Molar	Split	1
396	B	4	Bovid (Medium)	Mandible	Molar,worn	Split	1
333	B	4	Bovid (Medium)	Mandible	P2 socket,	+ P3,P4	1
766	B	5	Bovid (Medium)	Mandible	PM, worn		1
750	B	5	Bovid (Medium)	Mandible	PM2	Little wear	1
708,717,791	B	5	Bovid (Medium)	Mandible	Root		3
733	B	5	Bovid (Medium)	Mandible	Worn	Split	1
104	A	5	Bovid (Medium)	Mandible	M1/2,worn		1
264	B	3	Bovid (Medium)	Maxilla	M2 split, M3 complete	Worn, right	1
20	A	3	Bovid (Medium)	Metacarpal	Distal	Split	1
9,16,53, 102, 223, 247, 598, 602, 699, **855, 868**	A	5	Bovid (Medium)	Metacarpal	Distal	Complete	11
260,611,745	B	3	Bovid (Medium)	Metacarpal	Proximal	Complete	3
348	B	4	Bovid (Medium)	Metacarpal	Proximal	Split	1
136	A	4	Bovid (Medium)	Metapodial	D.barrel	Split	1
744,874	B	5	Bovid (Medium)	Metapodial	D.barrel		2
210	B	5	Bovid (Medium)	Metatarsal	Proximal	Complete	1
103,256	A	5	Bovid (Medium)	Metatarsal	Proximal	Split	2
622,758,836, 850	B	3	Bovid (Medium) Metatarsal	Proximal	Complete		4
189,199,201	A	4	Bovid (Medium)	Metatarsal	Shaft		3
870, 879	B	4	Bovid (Medium)	Metatarsal	Proximal	Complete	2
10,192,168,768	A	3	Bovid (Medium)	Radius	Distal	Complete	4
262	B	4	Bovid (Medium)	Radius	Distal	Split	1
126,132, 167, 251	A	4	Bovid (Medium)	Radius	Proximal	Complete	4
157,246,346, 623,721	B	4	Bovid (Medium)	Tibia	Distal	Complete	5
3	A	3	Bovid (Medium)	Tibia	Proximal	Complete	1
0	A	5	Bovid (Medium)	Tooth			1
837	B	4	Bovid (Medium)	Tooth	M1/2	Upper	1
737	B	5	Bovid (Medium)	Tooth	Molar		1
780	B	4	Bovid (Medium)	Tooth	PM		1
764	B	4	Bovid (Medium)	PM2/3	Lower		1
479	B	5	Bovid (Medium)	M1/2	Complete, lower	Worn	1
674	B	4	Bovid (Medium)	M1/2	Complete	Upper	1
26	A	3	Bovid (Medium)	Molar	Complete, upper	Mod.wear	1
40	A	4	Bovid (Medium)	P4, M1	Complete	Worn	1
304,365	B	4	Bovid (Medium)	PM	Complete	Upper	2
366	B	4	Bovid (Medium)	PM	Complete, upper	Worn	1
467	B	4	Bovid (Medium)	PM	Complete	Worn	1
332,365	B	4	Bovid (Medium)	PM	Complete	Upper	2
306	B	3	Bovid (Medium)	PM	Complete	Upper	1
663	B	4	Bovid (Medium)	PM	Complete, upper	Mod.wear	1
363	B	3	Bovid (Medium)	PM2	Complete	Lower	1
350	B	4	Bovid (Medium)	M1/2	Mod. wear	Upper	1
415	B	5	Bovid (Medium)	PM	Mod. wear	Lower	1
734	B	5	Bovid (Medium)	M1/2	Molar	Lower	1
736	B	5	Bovid (Medium)	M1/2	Molar, upper	Mod.wear	1
754	B	4	Bovid (Medium)	M1/2	Molar, upper	Mod.wear	1
794	B	5	Bovid (Medium)	M1/2	Molar	Lower	1

Table 7. 7: Fossil specimens from locality 642, zone 3, 1989

Specimen	Zone	W	Taxon	Element	Part	Other	N
803	B	4	Bovid (Medium)	PM	Premolar	Lower	1
763	B	2b	Bovid (Medium)	PM2/3	Premolar	Lower	1
various		5	Bovid (Medium)	Tooth	Split	Worn	7
848	B	5	Bovid (Medium)	M1/2	Split, upper	Worn	1
42	A	4	Bovid (Medium)	Molar	Split, lower	Worn	1
54	A	5	Bovid (Medium)	Molar	Split	Worn	1
300,422	B	3	Bovid (Medium)	Molar	Split		2
682	B	4	Bovid (Medium)	Molar	Split, lower	Worn	1
738	B	5	Bovid (Medium)	Molar	Split	Upper	1
678	B	4	Bovid (Medium)	Molar	Split	Worn	11
876	B		Bovid (Medium)	Molar	Split	Lower	1
388	B	5	Bovid (Medium)	PM	Split		1
830	B	5	Bovid (Medium)	Tooth	PM2/3, upper	Worn	1
807	B	5	Bovid (Medium)	PM	Worn	Lower	1
690	B	5	Bovid (Small)	Metatarsal	Proximal	Complete	1
810	B	3	Bovid (Small)	Tibia	Proximal	Complete	1
89	A	4	Bovid/Cervid	Mandible	M2 split,M3 whole + Teeth	Worn	1
9	A	4	Bovid/Cervid	Maxilla	M3 split, worn	+ Tooth	1
92	A	3	Bovid/Cervid	Metatarsal	Proximal	Complete	1
221	B	3	Carnivora	Radius	Proximal		1
83	A	4	Equid	Astragalus	Complete		1
137	A	4	Equid	Calcaneum	Complete		1
6	A	4	Equid	Mandible	M2,M3 broken	Worn	1
242	B	3	Equid	Mandible	P3,P4,M1	Right, worn	1
83	A	5	Equid	Mandible	Molar	+ Tooth	1
118	A	5	Equid	Mandible	Split tooth, worn	Fragment	1
222	B	5	Equid	Metacarpal	Distal	Complete	1
27	A	4	Equid	Metacarpal	Proximal	Complete	1
80,290,864	B	5	Equid	Tibia	Distal	Complete	3
892	B	3	Equid	Tooth	PM2 erupting, left	Complete	1
641	B	4	Equid	Tooth	M3 upper	Split	1
128	A	4	Equid	Tooth	PM upper, worn	Split	1
50	A	4	Rhinoceritidae	Astragalus	Complete		1
32	A	3	Rhinoceritidae	Metapodial	Proximal	Complete	1
69	A	3	Rhinoceritidae	Astragalus	Complete	Left	1
70	A	4	Rhinoceritidae	Metapodial	Distal	Complete	1
113	A	4	Rhinoceritidae	Nav-cuboid	Complete		1
863	B	4	Rhinoceritidae	Radius	Distal	Complete	1
258	B	3	Suid	Mandible	M2 split, M3 erupting		1
436	B	4	Suid	Mandible	M3	Split	1
310	B	3	Suid	Tooth	Fragment		1
430	B	4	Suid	3rd Phalange	Lateral	Complete	1
301	B	5	Suid (Large)	Mandible	2 Teeth	Broken	1
320	B	4	Suid (Large)	Mandible	Broken	+ Tooth	1
126	A	5	Ungulata	Femur	Distal	Split	1
628	B	4	Ungulata	Femur	Distal	Condyle	1
891	B	5	Ungulata	Mandible	2 tooth roots	Fragment	1
88	A	5	Ungulata	Metapodial	B.barrel	Split	1
			Indeterminate	Fragments			1252
Total							**1724**

Table 7. 8: Fossil specimens from locality 642, zone X, 1989

Specimen	W	Taxon	Element	Part	Other	N
various	5	Artiodactyla	Tooth	Fragments		22
38	5	Artio (Large)	Mandible	+roots	Roots	1
12	5	Artio (Large)	Metacarpal?	Prox.split		1
22	5	Artio (Large)	Rib			1
2-4, 18	5	Artio (Large)	Shaft			4
31	4	Artio (Med)	Atlas	Fragment		1
24	3	Artio (Med)	Humerus	D.shaft		1
55	4	Artio (Med)	Mandible+M3	M3	Left	1
42	4	Artio (Med)	Metacarpal	Shaft		1
74	3	Artio (Med)	Vertebra	Centrum		1

Table 7. 8 continued: Fossil specimens from locality 642, zone X, 1989

Specimen	W	Taxon	Element	Part	Other	N
72	5	Artio (Large)	Tooth	Split fragment		1
14	3	Artio/Bovid	Metacarpal	Shaft		1
27	3	Artio/Bovid	Metacarpal	Proximal		1
64	4	Artio/Bovid	Metatarsal	Proximal		1
88,89,93	5	Artio/Bovid	Tooth	Broken		3
23	5	Artio/Indet	?Radius	?Distal		1
various	5	Bovid	Horncore			19
48	5	Bovid	Mandible	+tooth	M1/2	1
47	3	Bovid	Mandible+Tooth	M3	Left	1
6	3	Bovid	Metatarsal	Proximal		1
77	5	Bovid	Tooth	Broken		1
1	4	Bovid (Large)	Calcaneum	Completish		1
44	5	Bovid (Large)	Mandible	+tooth	M1/2	1
26	4	Bovid (Large)	Mandible	+tooth		1
51	5	Bovid (Large)	Tooth	Split		1
33	5	Bovid (Large)	Tooth	Complete	M1/2	1
29	4	Bovid (Med)	Horncore			1
36	5	Bovid (Med)	Mandible	+tooth		1
35	4	Bovid (Med)	Mandible	+M3 split	M3	1
73	5	Bovid (Med)	Mandible	+tooth	M1/2	1
61	5	Bovid (Med)	Mandible	+Tooth		1
84	5	Bovid (Med)	Metacarpal	Shaft		1
25	5	Bovid (Med)	Metacarpal	Distal		1
30, 62	5	Bovid (Med)	Metatarsal	Proximal		2
43, 67	4	Bovid (Med)	Metatarsal	Shaft		2
60	4	Bovid (Med)	Tooth	M3	Upper	1
82	5	Bovid (Med)	Tooth	Split	M1/2	1
75,79	4	Bovid (Med)	Tooth		M1/2	2
5	4	?Rhino	Metapodial	Shaft		1
17	4	Equid	Tooth	Upper,worn	Molar	1
various	5	Indeterminate	Fragments			146
Total						**233**

Table 7: 9. Fossil material from locality 642, zone 1, clusters 1 and 2, 1989.

Cluster	Specimen	W	Taxon	Element	Part	Other	N
Cluster 1	329	3	Rhino	Femur	Proximal	Condyle	1
Cluster 1	337		Rhino	Carpal	Complete		1
Cluster 1	330-336,338-340	5	Indeterminate	Pelvis	Fragments		9
Cluster 2	449	3	Artiodactyl	Mandible	+ Tooth	Split	1
Cluster 2	450,452-453,456,459	3	Artiodactyl	Tooth	Fragment		5
Cluster 2	446,471, 473-474	3	Artio (Medium)	Metapodial	Proximal	Split	4
Cluster 2	421	4	Artio (Medium)	Axis	Proximal	Fragment	1
Cluster 2	410,413	4	Bovid	Horncore	Tine		2
Cluster 2	417	4	Bovid	Horncore	Base		1
Cluster 2	435	4	Bovid	Horncore			1
Cluster 2	419	3	Bovid (Medium)	M1/2 right upper	Complete	Mod.wear	1
Cluster 2	420	4	Bovid (Medium)	Mandible + Teeth	M1/2 or M2/M3 split	Mod.wear left	1
Cluster 2	403	5	Bovid (Medium)	Mandible + Tooth	P2 socket		1
Cluster 2	406	4	Bovid (Medium)	Mandible + Tooth	M3 complete	Worn	1
Cluster 2	409-410, 427	4	Bovid (Medium)	Mandible + Tooth	M3	Right	3
Cluster 2	464	5	Bovid (Medium)	Mandible+ PM2	Right		1
Cluster 2	404	3	Bovid (Medium)	Maxilla + M1-3	Worn	Left	1
Cluster 2	413	4	Bovid (Medium)	Maxilla + M2,M3	Left	Broken	1
Cluster 2	**439**	3	Bovid (Medium)	Metacarpal	Proximal	Complete	1
Cluster 2	398	3	Bovid (Medium)	Metacarpal	Proximal	+ carpal	1
Cluster 2	399,430, 431	4	Bovid (Medium)	Metacarpal	Distal	Complete	3

Table 7.9 continued: locality 642, zone 1, clusters 1 and 2, 1989

Cluster	Specimen	W	Taxon	Element	Part	Other	N
Cluster 2	411	3	Bovid (Medium)	Metatarsal	Shaft		1
Cluster 2	**400**	3	Bovid (Medium)	Metatarsal	Proximal	Complete	1
Cluster 2	416	2b	Bovid (Medium)	Metatarsal	Shaft		1
Cluster 2	442,450, 457	3	Bovid (Medium)	PM	Upper	Complete	3
Cluster 2	463	4	Bovid (Medium)	Tibia	Distal	Split	1
Cluster 2	451	3	Bovid (Medium)	Tooth	Broken	Worn	1
Cluster 2	454	3	Bovid (Medium)	Tooth	PM or split molar	Mod. wear	1
Cluster 2	455	3	Bovid (Medium)	Tooth	PM	Complete	1
Cluster 2	458	4	Bovid (Medium)	Tooth	Mod.wear	Split	1
Cluster 2	482	3	Bovid (Medium)	Tooth	PM or split molar	Lower	1
Cluster 2	421	5	Indeterminate	Cranial/Maxilla+	Tooth	Socket	1
Cluster 2		4	Indeterminate	Fragments			46
Total							**103**

Table 7. 10: Fossil specimens from find spots on locality 642, 1989

Subarea	Zone	Specimen	W	Taxon	Element	Part	Other	N
IS		1	5	Equid	Mandible	Complete		1
IS1		1	2b	Bovid	1st Phalange	Complete		1
IS1		2	3	Bovid (Medium)	Maxilla	+ M1/2	Mod.wear	1
IS1		3	4	Bovid (Medium)	M1/2 upper	Complete	Mod.wear	1
IS1		4,5	2a	Indet (Small)	Carpal	Complete		2
IS2		1		Bovid (Medium)	Mandible, right	Fragment	P2-4,M2	1
IS3			4	Artiodactyla	Humerus	Distal	Split	1
IS3		40	4	Artiodactyla	Metacarpal	Distal	Shaft	1
IS3		43	3	Artiodactyla	Metapodial	Distal barrel		1
IS3		13,14, 22	4	Artiodactyla	Tooth	Spall		3
IS3		15	4	Artiodactyla	Tooth	Split	No crowns	1
IS3		16	3	Artiodactyla/Bovid	Tooth	Split	Worn	1
IS3			2b	Bovid (Medium)	Astragalus	Complete		1
IS3			5	Bovid (Medium)	Horncore	Complete		1
IS3			4	Bovid (Medium)	Horncore	+ Pedicel +Cranium		1
IS3			3	Bovid (Medium)	Humerus	Distal	Split	1
IS3		11	3	Bovid (Medium)	Mandible	+ ToothM3	Split	1
IS3			4	Bovid (Medium)	Skull	Base& back		1
IS3		38	5	Bovid (Medium)	Tibia	Distal	Complete	1
IS3		17,19, 20	4	Bovid (Medium)	Tooth	Complete	Worn PM	3
IS3		21	4	Bovid (Medium)	Tooth	Split	Worn	1
IS3			5	Indeterminate	Fragments			31
IS3/4		0	5	Indeterminate	Fragments			158
IS5		1-3		Bovid	Horncore			3
IS6		12	4	Artio (Large)	Radius	Distal	Split	1
IS6		5,13	5	Bovid	Horncore			2
IS6		10	4	Bovid	Horncore	Tine		1
IS6		18	4	Bovid	Molar, upper	Split	Worn	1
IS6		1	3	Bovid (Medium)	Horncore			1
IS6		14	4	Bovid (Medium)	Mandible+ Tooth	Molar	Split	1
IS6		17	3	Bovid (Medium)	Metacarpal	Proximal	Complete	1
IS6		3	4	Equid	Metacarpal	Distal	Complete	1
IS6		11	4	Equid	Astragalus	Complete		1
IS6		various	4	Indeterminate	Fragment			7
IS8		1	2a	Artio (Large)	Vertebra	Lumbar	Complete	1

Table 7. 10 continued: Fossil specimens from find spots on locality 642, 1989

Subarea	Zone	Specimen	W	Taxon	Element	Part	Other	N
IS8		2	4	Indeterminate	Shaft			1
IS9			3	Artio (Large)	Humerus	Distal	Split	1
IS9			2b	Bovid (Medium)	Metacarpal	Proximal	Complete	1
IS9		5	5	Equid	Mandible	Complete		1
IS9			4	Indeterminate	Fragments			5
Pt20		5	5	Artio (Medium)	Mandible	+ 2 molars	Split	1
Pt20		6,7	4	Artio (Medium)	Metacarpal	Shaft		2
Pt20		1	4	Bovid (Medium)	M3, right	Fresh	Complete	1
Pt20		9	4	Bovid/Cervid	1st Phalange	Proximal	Complete	1
Pt20		various	5	Indeterminate	Fragments			5
Pt21		2	3	Artio (Medium)	Calcaneum	Shaft		1
Pt21		3		Artio (Medium)	Calcaneum	Proximal		1
Pt21		6	4	Artio (Medium)	Mandible	+ Teeth	1 root, 1 broken	1
Pt21		3,7		Artio (Medium)	Rib			2
Pt21		5	3	Artio (Medium)	PM, upper	Complete	No wear PM	1
Pt21		0	5	Indeterminate	Fragments			34
Pt22	3	40	4	Artiodactyla	Tooth	Fragment		1
Pt22	3	28	5	Artio (Medium)	Tooth	Split	PM	1
Pt22	3	43	3	Artio (Small)	Tooth	Fragment		1
Pt22	3	24	5	Bovid	Horncore			1
Pt22	3	36	4	Bovid (Medium)	M3	Split	V.worn	1
Pt22	3	1	3	Bovid (Medium)	Mandible+ Tooth	Fresh	Molar	1
Pt22	3	35	3	Bovid (Medium)	Mandible+ Tooth	PM	Mod.wear	1
Pt22	3	32	4	Bovid (Medium)	Metapodial	Distal barrel		1
Pt22	3	**12**	3	Bovid (Medium)	Radius	Proximal	Complete	1
Pt22	3	**13**	3	Bovid (Medium)	Tibia	Distal	Complete	1
Pt22	3	26	3	Bovid (Medium)	Tooth,upper	M1/2	Complete	1
Pt22	3	9	4	Bovid (Small)	Metapodial	Distal barrel		1
Pt22	3	34	3	Bovid (Small)	Metatarsal	Shaft		1
Pt22	3	6,7	3	Bovid (Small)	Tooth, lower	Complete	PM	1
Pt22	3	0	5	Indeterminate	Fragments			69
Pt22	3	23	3	Indet (Medium)	Mandible	+ Tooth	Broken	1
Pt22	3	38	5	Indet (Medium)	Metapodial	Distal barrel		1
Pt22	3	20	4	Indeterminate/Artio	Vertebra	Centrum		1
Pt22	3	16	4	Rhino?	Carpal	Complete	1	1
Pt22	3		3	Ungulata	Tooth	Fragment		4
Pt23		1	4	Bovid	Horncore	Base		1
Pt23		6	3	Bovid	Horncore			1
Pt23		3	5	Bovid (Large)	Tooth	Fragment		1
Pt23		7	4	Bovid (Large)	Radius	Distal	Complete	1
Pt23		4,5	5	Indeterminate	Fragment			2
Pt23		2	4	Indet (Large)	Pelvis/Mandible			1
SE1		30, 35	5	Artio (Medium)	Tooth	Fragment		2
SE1		36	4	Artio (Medium)	Complete	PM	Lower	1
SE1		9,15,28	5	Bovid	Horncore			3
SE1		5	4	Bovid	Horncore	Base		1
SE1		**2**	5	Bovid (Large)	Humerus	Distal	Complete	1
SE1		**1**	4	Bovid (Large)	Radius	Distal	Complete	1
SE1		3	4	Bovid (Medium)	Horncore	Twisted		1
SE1		22	4	Bovid (Medium)	Mandible	+ Tooth	Socket	1
SE1		14	4	Bovid (Medium)	Mandible+ Tooth	M3 right	Mod.wear	1

Table 7. 10 continued: Fossil specimens from find spots on locality 642, 1989

Subarea	Zone	Specimen	W	Taxon	Element	Part	Other	N
SE1		21	5	Bovid (Medium)	Mandible+ Tooth	Molar	Split	1
SE1		23	4	Bovid (Medium)	Mandible+ Tooth	Molar	Broken	1
SE1		**8**	5	Bovid (Medium)	Metacarpal	Distal	Complete	1
SE1		25	5	Bovid (Medium)	PM	Split	Worn	1
SE1		19	4	Bovid (Medium)	Tooth	Split	Fresh	1
SE1		24	4	Bovid (Medium)	Tooth	Complete	PM	1
SE1		29,31, 32	5	Bovid (Medium)	Tooth	Split	Molar	3
SE1		13,18	4	Bovid (Medium)	M1/2, lower	Fresh	Complete	2
SE1		17	4	Bovid (Medium)	M1/2, upper	Complete	Worn	1
SE1		16	3	Bovid (Medium)	M1/2 upper	Complete	Mod.wear	1
SE1		12,20	4	Equid	Molar, lower	Complete	Worn	2
SE1		10	5	Indeterminate	Mandible+ Teeth	2 roots, sheared		1
SE1		33	5	Indeterminate	Maxillary	+ Tooth	Socket	1
SE1		7	4	Indeterminate	Sacrum	Fragment		1
SE1		4-6, 26,27	5	Indeterminate	Shaft			5
SE1		11,34	4	Indeterminate	Vertebra	Fragment		2
Total								**434**

Table 7: 11. Fossil specimens from between locality 642 (Baroth area) and Ban Barilla (637-648)

Fossil occurrence 637:scatter:

Specimen	W	Taxon	Element	Part	Other	N
1	5	Equid	Molar, lower	Complete	Mod wear	1
2	5	Ungulata	Tooth	Fragment		1
0	5	Indeterminate	Fragments			11
Total						**13**

Fossil occurrence 639: scatter:

Specimen	W	Taxon	Element	Part	Other	N
9, 10	2b	Artio (Medium)	Metapodial	Distal	Epiphysis	2
4	3	Artio (Medium)	Nav.-Cuboid	Complete		1
11	2b	Bovid	Molar	Fresh	Split	1
15	2b	Bovid (Medium)	Tooth	Spall		1
12	5	Bovid/Cervid	Molar	Little wear	Split	1
1,3,7	3	Cervid	Antler	Split		3
8	2b	Chelonia	Scute	Fishnet		1
0		Indeterminate	Fragments			50
Total						**60**

Fossil occurrence 643: general collection:

Specimen	W	Taxon	Element	Part	Other	N
1	5	Equid	Tooth	PM/M, lower	Complete	1
2	5	Bovid (Large)	PM, upper	Complete	Little wear	1
3	5	Bovid	Horncore			1
4	4	Artio (Medium)	Nav.-cuboid	Complete		1
0	5	Indeterminate	Shaft	Fragments		5
Total						**9**

Fossil occurrence 644: general collection:

Specimen	W	Taxon	Element	Part	Other	N
9	5	Artio	Tooth	Split	No crown	3
30	4	Artio (Large)	Axis	Fragment		1
19	5	Bovid	Horncore			2
34	5	Bovid (Large)	Calcaneum	Distal		1
24	33	Bovid (Large)	M1/2	Complete	Mod. wear	1
8	4	Bovid (Large)	Tooth	Split, lower	Mod. wear	1
14	4	Bovid (Medium)	M1/2, lower	Complete	Mod. wear	1
10,31	4	Bovid (Medium)	Tooth (10=lower)	Fragment	Mod. wear	2
4	5	Proboscidean	Tooth	Tusk	5	5
	5	Indeterminate	Fragments			20
Total						**37**

Table 7: 11 continued: Fossil occurrence 648: general collection:

Specimen	W	Taxon	Element	Part	Other	N
11	4	Artiodactyl	Fibula	Distal		
3	5	Artio (Medium)	Mandible	1 socket	+ 1 broken tooth	1
6	4	Bovid (Medium)	Mandible, left	DM3 (mod wear), M1	Left	1
7	4	Bovid (Medium)	PM	Complete	Worn	1
8	4	Bovid (Medium)	M1/2	Lower	Complete	1
10	5	Bovid (Medium)	Molar	Lower	Split	1
16	5	Indeterminate	Shaft			6
15	5	Indet (Large)	Pelvis/Scapula			8
Total						**20**

Notes: Scatter 641 had 1 artiodactyla carpal, 4 pieces of cervid antler and 6 indetermiinate fragments. Scatter 645 had 1 piece of proboscidean molar plate and 69 indeterminate fragments; general collection 646 had 2 pieces of bovid horncore, 4 pieces of proboscidean molar plate, and 4 indeterminate fragments. General collection 647 had only 4 fragments. Tooth 644GB10 was lower.

Table 7: 12. Fossil specimens from between Ban Barilla and Doga (649 - 674)

Fossil occurrence 649: general collection:

Specimen	W	Taxon	Element	Part	Other	N
1	5	Artio (Medium)	Tooth	Split	No crown	2
4	5	Bovid (Medium)	M3, lower	Complete	Left	1
5	5	Equid	Molar, lower	Split	Little wear	1
2	5	Indeterminate	Fragments			20
Total						**24**

Fossil occurrence 650: scatter:

Specimen	W	Taxon	Element	Part	Other	N
0	5	Artiodactyla	Tooth	Spall		2
8	4	Artio (Medium)	Metapodial	Distal barrel		1
39	4	Bovid (Large)	Tooth, upper	PM	Complete	1
2	4	Bovid (Medium)	PM/M	Split	Little wear	1
18,21	5	Bovid (Medium)	Molar, lower	Split	Mod.wear	2
23	4	Bovid (Medium)	PM/M, lower	Split	Mod.wear	1
0	5	Bovid (Medium)	Split			1
0	2b	Chelonia	Scute	Dimpled		5
0	2b	Chelonia	Scute	Fishnet		6
0	2b	Chelonia	Scute	Ribbed		1
0	2b	Chelonia	Scute	Smooth		55
1	3	*Crocodylus*	Tooth			1
47	5	Proboscid/Rhino	Tooth	Plate		3
40	5	Proboscidean	Tooth	Fragment		1
25	4	Rhinocerotidae	Tooth	Fragment		1
0	5	Indeterminate	Fragments			106
Total						**188**

Fossil occurrence 651: scatter:

Specimen	W	Taxon	Element	Part	Other	N
	3	Artio	Tooth	Spall		4
1	4	Bovid (Large)	M3, lower	Split	Left	1
7	4	Bovid (Large)	Tooth	Fragment		1
3,6	5	Proboscidean	Tooth	Fragment		2
0	4	Indeterminate	Fragments			13
Total						**21**

Fossil occurrence 654: general collection:

Specimen	W	Taxon	Element	Part	Other	N
12	4	Bovid	Horncore			1
13	3	Bovid (Large)	M1/2	Complete	Mod.wear	1
15	5	Bovid (Medium)	Tooth, lower	M3	Mod. wear	1
3	5	Bovid (Medium)	Molar, lower	Split	Mod .wear	1
2	2b	Bovid (Medium)	Molar, upper	Complete	Little wear	1
8	3	Chelonia	Scute	Smooth		1

Table 7. 12 continued: Fossil occurrence 654: general collection:

Specimen	W	Taxon	Element	Part	Other	N
	4	Indeterminate	Fragments			11
Total						**17**

Fossil occurrence 655: scatter:

Specimen	W	Taxon	Element	Part	Other	N
10	4	Artio (Medium)	1st Phalange	Proximal	Complete	1
2	4	Artio (Medium)	Calcaneum	Distal	Shaft	1
14	4	Artio (Medium)	Phalange	Distal	Complete	1
3	3	Bovid (Medium)	Metacarpal	Distal	Complete	1
33	5	Bovid (Medium)	Tooth, lower	Split	Mod. wear	1
31	3	Bovid (Medium)	Radius	Proximal	Complete	1
23	4	Cervid	3rd Phalange	Complete		1
11	3	Chelonia	Scute	Dimpled		2
7	3	Chelonia	Scute	Fishnet		2
18	3	Chelonia	Scute	Smooth		10
0	4	Indeterminate	Fragments			44
Total						**65**

Fossil occurrence 656: general collection:

Specimen	W	Taxon	Element	Part	Other	N
0		Artiodactyla	Tooth	Fragments		3
5	3	Bovid	Tooth	Split	Little wear	1
22	2b	Bovid (Large)	M1/2, upper	Complete	Mod. wear	1
24	2a	Bovid (Medium)	Metatarsal	Proximal	Complete	1
13	3	Cervid	Antler			4
7	3	Cervid/Bovid	PM, upper	Complete	No wear	1
3	3	Chelonia	Limb/rib?			1
2	2b	Chelonia	Scute	Smooth		2
0		Indeterminate	Fragments			50
Total						**64**

Fossil occurrence 662: general collection:

Specimen	W	Taxon	Element	Part	Other	N
0	4	Artiodactyla	Tooth	Fragments		4
13	4	Bovid (Medium)	M3, right	Complete	Mod.wear	1
0	4	Artiodactyl	Tooth	Fragments		1
1,2,3	4	Bovid (Medium)	Tooth	Split	Mod.wear	3
9	5	Indeterminate	Fragments			9
Total						**18**

Fossil occurrence 664: scatter:

Specimen	W	Taxon	Element	Part	Other	N
51	5	Artiodactyla	Tooth	Split	1	
50	4	Artiodactyla	Tooth	Split	Worn	1
27	4	Artio (Large)	Premolar	Complete	Worn	1
54	3	Bovid (Medium)	Metapodial	Proximal	Shaft	1
56	4	Bovid (Medium)	PM, upper	Complete	Worn	1
48	4	Bovid (Medium)	M3, right	Split	Little wear	1
23	3	Bovid/Cervid	2nd Phalange	Complete	Unfused	1
59	3	Bovid/Cervid	2nd Phalange	Distal	Complete	1
31	3	Bovid/Cervid	Metatarsal	Proximal	Split	1
18	3	Bovid/Cervid	Nav.-cuboid	Complete		1

Fossil occurrence 664: scatter:

Specimen	W	Taxon	Element	Part	Other	N
32	3	Bovid/Cervid	Tibia	Distal	Complete	1
65	2a	Bovid/Cervid	PM	Complete	Fresh	1

Table 7.12 continued: Fossil occurrence 664:

20	4	Cervid	Antler			1
14	3	Cervid	Antler	Pedicel		1
0		Chelonia	Scute	Dimpled		17
0		Chelonia	Scute	Fishnet		1
0		Chelonia	Scute	Smooth		149
62		Mollusc	Whirlytop			0
4	3	Proboscidean	Carpal/tarsal			1
1	5	Proboscidean	Cranial			1
11	3	Proboscidean	Phalange	Fragment		1
		Proboscidean	Miscalleneous			10
0		Indeterminate	Fragments			433
Total						**639**

Fossil occurrence 666: locality:

Specimen	W	Taxon	Element	Part	Other	N
29	3	Artio (Medium)	Metapodial	Distal	Epiphysis	1
93	3	Artio (Medium)	Radius	Shaft		1
83	3	Artio (Medium)	Tooth	Incisor	Complete	1
20	3	Bovid (Large)	M1/2, upper	Complete	Little wear	1
88		Bovid (Large)	Mandible	Right	+M2,M3	1
89	3	Bovid (Medium)	Humerus	Distal	Complete	1
87	2b	Bovid (Medium)	Mandible, right	P2-M2	Mod.wear	1
91	3	Bovid (Medium)	Radius	Proximal	Split	1
90	3	Bovid (Medium)	Ulna	Proximal		1
40	3	Bovid (Small)	Radius	Proximal	Complete	1
26	2b	Carnivora	Ulna	Complete		1
19,24,25, 41	3	Cervid	Antler			4
74		Chelonia	Scute	Dimpled		1
75		Chelonia	Scute	Smooth		10
78,80	2b	*Crocodylus*	Tooth	Complete		2.
0		Mollusc	Endocast			1
	3	Pisces	Scale			1
	4	Proboscidean	Tooth	Fragments		18
	4	Proboscidean	Tusk	Fragments		9
1	3	Suid	Maxilla, left	M3	V. worn	1
11	4	Suid	M3, upper	Complete	Right	1
17	3	Suid/crocodile	Humerus	Shaft		1
	5	Indeterminate	Fragments			233
Total						**293**

Fossil occurrence 670: scatter:

Specimen	W	Taxon	Element	Part	Other	N
11	4	Cervid/Bovid	Maxilla, right	P4,M1	Little wear	1
2	4	Indet/Proboscid	Rib			4
34,36	5	Indet/Proboscid	Tooth	Fragment		2
0		Mollusc	Whirly-top			1
37	3	*Struthio*	Eggshell			1
0		Indeterminate	Fragments			158
Total						**167**

Fossil occurrence 672: general collection:

Specimen	W	Taxon	Element	Part	Other	N
4	3	Artiodactyla	Metatarsal	Shaft		1
3	4	Artio (Medium)	Radius	Proximal		1
7	3	Bovid (Large)	M1/2, lower	Complete	V. fresh	1
2,20	4	Bovid (Large)	Molar, upper	Split	Mod. wear	2

Table 7. 12 continued: Fossil occurrence 672:

Specimen	W	Taxon	Element	Part	Other	N
23	2b	Bovid (Medium)	Metapodial	Proximal	Split	1
1,10	2b	Bovid (Medium)	M3, upper	Complete	Fresh	2
15	2b	Bovid (Medium)	M1/2, upper	Complete	Fresh	1
14,31	3	Bovid (Medium)	Tooth	Split	Mod. wear	2
18	5	Bovid (Medium)	Tooth	Split		1
6	3	Cervid	Antler	Pedicel	Split	1
0	3	Chelonia	Scute	Fishnet		1
1	3	Chelonia	Scute	Fishnet		1
11	4	Chelonia	Scute	Smooth		2
16	4	*Crocodylus*	Tooth	Complete		1
0	5	Indeterminate	Fragments			66
	4	Proboscidean	Tooth	Crown		5
29	2b	*Struthio*	Eggshell			1
Total						**90**

Fossil occurrence 674: general collection:

Specimen	W	Taxon	Element	Part	Other	N
23	4	Artio (Large)	Ulna	Blade	1	
4	4	Artio (Medium)	Atlas	Fragment		1
0	5	Bovid	Horncore			1
1	2b	Bovid	Metacarpal	Complete	1	
24	4	Bovid (Large)	Humerus	Shaft	1	
22	5	Bovid (Large)	PM/M, lower	Split	Worn	1
10	4	Cervid/Bovid	M3, lower	Complete	Mod. wear	1
20	4	Cervid/Bovid	Tooth	Spall		1
5,6,11	5	Equid	Molar	Complete	Mod.wear	3
16	5	Equid	Tooth	Split		1
21		*Megantereon*	Tooth	Canine		1
22		*Megantereon*	Mandible	Right	P3,M1	1
0	5	Indeterminate	Fragments			81

Notes: General collection 652 had 3 indeterminate fragments, 1 piece of proboscidean tusk and 1 artiodactyla tooth fragment. General collection 653 contained 1 split artiodactyl tooth, 1 piece of bovid horncore and 12 fragments. Scatter 668 had 1 bovid (large) fibula, 1 bovid (large) right mandible fragment with M3, moderate wear, and 2 indeterminate fragments. There were no fossil specimens from 658, and 660. 662GB2 was lower, and 662GB3, upper. 672GB14 was lower, and 672GB31, upper. Details of the Proboscidean bones found in situ at 664 are shown in Figure 7.6. Teeth 674GB5 and 6 were lower, and 675GB11, upper.

Table 7: 13. Fossil specimens near hairpin bend, watershed track, Pabbi Hills north (657-685).

Fossil occurrence 657: general collection:

Specimen	W	Taxon	Element	Part	Other	N
7	5	Artiodactyla	Tooth	Spall		1
2	5	Bovid (Medium)	Mandible	+2 Teeth	Crowns broken	1
5	5	Bovid (Medium)	Tooth	Split	Little wear	1
3	3	Indeterminate	Fragments			19
Total						**22**

Fossil occurrence 665: general collection:

Specimen	W	Taxon	Element	Part	Other	N
1	3	Artio (Medium)	Scapula	Shaft		1
3	5	Bovid	Horncore	Fragment		1
8	4	Bovid (Medium)	Tooth	Fragment		1
7	3	*Crocodylus*	Tooth	Complete		1
2	4	Equid	Molar, lower	Complete	Little wear	1
4	5	Equid	Tooth	Split	Crowns broken	1
0		Equid	Metacarpal	Complete		1
0	4	Indeterminate	Fragments			9
Total						**16**

Table 7. 13 continued: Fossil occurrence 669: general collection:

Specimen	W	Taxon	Element	Part	Other	N
26	5	Artiodactyla	Metapodial	Distal	Epiphysis	1
0	5	Artiodactyla	Tooth	Fragments		11
4	5	Artiodactyla	Vertebra	Centrum		1
15	2b	Bovid (Medium)	M1/2, upper	Complete	No wear	1
11	3	Bovid/Cervid	Molar, upper	Complete	No wear	1

Table 7.13 continued: Fossil occurrence 669:

20	2b	Bovid/Cervid	PM, upper	Complete	No wear	1
2	3	Cervid	Antler			1
7	5	Equid	Molar	Complete		1
14,25		Mollusc	Bivalve			2
0	5	Indeterminate	Fragments			176
Total						**198**

Fossil occurrence 675: scatter:

Specimen	W	Taxon	Element	Part	Other	N
	4	Artiodactyla	Axis	Fragment	1	1
5,21	5	Artiodactyla	Mandible	+ Tooth	Split	2
8	4	Artiodactyla	Ramus	1		1
9,11,14	5	Artiodactyla	Tooth	Spall	1	3
5	5	Artiodactyla	Tooth	PM	Split	1
3,8,10,12	5	Artiodactyla	Tooth	Split	No crown	4
4,12	5	Bovid	PM/M	Split	Worn	1
6,14	5	Bovid	Tooth	Split	Mod. wear	2
3	4	Bovid (Medium)	Metacarpal	Distal	Complete	1
16	4	Bovid (Medium)	PM	Complete		1
24	4	Bovid (Medium)	PM	Complete	Mod. wear	1
8	5	Bovid (Medium)	Tooth	PM	Split	1
10	5	Bovid (Medium)	Tooth	Split	No crown	1
15	5	Bovid (Medium)	Molar, lower	Split	Mod. wear	1
17	5	Bovid (Medium)	M3, lower	Split	Worn	1
22	4	Bovid (Medium)	PM, lower	Split	Mod. wear	1
1	4	Bovid (Medium)	Mandible, right	M2 split, M3 split	Little wear	1
13	5	Bovid (Medium)	Tooth, upper	PM	Complete Worn	1
11	5	Bovid (Medium)	Tooth, lower	Split	Mod. wear	1
2	4	Equid	Mandible	P2/M3	Worn	1
6	5	Equid	Tooth	Split	No crown	1
26	5	Equid	Tooth	Split	Mod. wear	1
2	5	Equid	Tooth	PM/M	Complete Worn	1
0	4	Indeterminate	Shaft	Fragments		149
Total						**184**

Fossil occurrence 676: scatter:

Specimen	W	Taxon	Element	Part	Other	N
4	3	Artio (Medium)	Metatarsal	Shaft		1
11	5	Artio (Medium)	Tooth	Split		1
13	4	Artio (Medium)	Mandible	+ M1/2	Mod.wear	1
7	5	Bovid/Cervid	Molar, lower	Mod.wear	Split	1
2	3	Cervid	Maxilla, right	M1/2	No wear	1
3	3	Cervid	Antler			1
10	5	Cervid/Bovid	M1/2, lower	Complete	Mod.wear	1
9	2b	Chelonia	Scute	Smooth 1		1
12	3	Chelonia	Scute	Smooth		1
0		Indeterminate	Shafts			36
Total						**45**

Fossil occurrence 677: general collection:

Specimen	W	Taxon	Element	Part	Other	N
42	4	Bovid	Tooth	Spall		1
2	5	Bovid (Large)	Horncore			1
13	4	Bovid (Large)	M3, lower		Complete Right	1
6	5	Bovid (Large)	Tooth	Split	No crown	1
1	3	Cervid	Antler	Palmate		1

21	3	Chelonia	Scute	Smooth		1
2	3	Equid	Tibia	Distal	Complete	1
8	5	Equid	Metacarpal	Distal	Complete	1
11	5	Equid	M3, lower		Complete Worn	1
7	5	Proboscidean	Tooth	Plate		1
0	4	Indeterminate	Fragments			47
Total						**57**

Fossil occurrence 678: general collection:

Specimen	W	Taxon	Element	Part	Other	N
2	4	Artio	Vertebra	Spine	1	1
7	4	Bovid (Medium)	Mandible, left	M3 errupting	+ Tooth	1
8	5	Bovid (Medium)	Tooth	Split	No crown	1
1	4	Cervid/Bovid	Mandible, right	P2-3, P4, M1 split		1
3	5	Equid	Mandible	+ Tooth	Split	1
4,5	5	Proboscidean	Tooth	Fragment		2
0	5	Indet	Fragments			34
Total						**41**

Fossil occurrence 681: general collection:

Specimen	W	Taxon	Element	Part	Other	N
12	5	Artiodactyla	Tooth	Split	No crown	1
3	3	Chelonia	Scute	Fishnet		1
1	5	Equid	Tooth	Split	Mod. weal	1
0, 2,5,9	5	Proboscidean	Tooth	Fragment		4
0	4	Indeterminate	Fragments			58
Total						**65**

Fossil occurrence 682: general collection:

Specimen	W	Taxon	Element	Part	Other	N
3	5	Bovid (Large)	Molar	Complete	Little wear	1
4	5	Bovid (Large)	Tooth	Split	No crown1	1
5	4	Bovid (Medium)	M1/2	Complete	Mod.wear	1
0, 15	3	Cervid	Antler			2
1	5	Equid	PM2	Complete	Right	1
2	5	Equid	Molar, upper	Lower	Complete	1
0	5	Indet	Fragments			26
Total						**33**

Fossil occurrence 683: find-spot:

Specimen	W	Taxon	Element	Part	Other	N
11	4	Bovid (Medium)	Mandible	Molar	Complete	1
12	4	Bovid (Medium)	M3	Lower	Complete	1
1	3	Equid	Mandible, left	M1-3	Little wear	1
2	3	Equid	Mandible, right	M1-2	Little wear	1
4	3	Equid	Mandible, left	I1-3		1
3	2a	Equid	Molar, lower	Complete	No wear	1
5,6	3	Equid	Mandible, right	Molar	No wear	2
0		Indeterminate	Fragments			11
Total						**19**

Fossil occurrence 685: general collection:

Specimen	W	Taxon	Element	Part	Other	N
11	4	Artiodactyla	Tooth	Fragment		1
1	4	Artio (Large)	Vertebra	Centrum	Split	1
2	4	Bovid (Large)	Metacarpal	Distal	Epiphysis	1
7	4	Chelonia	Scute			1
3	3	Crocodylus	Mandible	+ Tooth	Split	1
9	4	Indeterminate	Fragments			15
Total						**20**

Notes: Find-spot 659 recheck; scatter 661 had 2 equid teeth (1 complete molar, 1 fragment) and 33 indeterminate fragments; general collection 663 had 1 artiodactyl tooth fragment, 1 equid distal tibia fragment, and 14 indeterminate fragments; general collection 671 had 1 bovid (medium) molar fragment (mod. wear) and 2 indeterminate fragments; 673 1 indeterminate canine and 1 other fragment; find-spot 679, 1 piece of proboscidean molar plate; and general collection 680, I fragment of bovid horncore.1 equid distal phalange and 14 indeterminate fragments. In scatter 675, teeth 675GB4, 6, 10, and 12 were lower, and 675GB14 was upper.

Chapter 8: The Lithic Evidence from the Pabbi Hills

Linda Hurcombe, School of Geography and Archaeology, University of Exeter, EX4 4QE, U.K.

SECTION A: The context of the survey and the lithic recording system

Introduction: the general evidence for human occupation and use of the Pabbi Hills

Field surveys are most often conducted across ploughed fields, and the lithic material that archaeologists find on the surface can usually be fairly closely related to the point of origin of the archaeological layers from which that material was derived. If the landscape surveyed is hilly, down slope movement of artefacts, whether by natural or human causes, has to be taken into account. It will be clear from the preceding chapters that the Pabbi Hills is a deeply-dissected and actively eroding badlands landscape, exacerbating all of the problems associated with working on slopes. Added to this is the problem that one is not dealing with data from a few thousand or even tens of thousands of years ago, but potentially from geological strata of well over a million years old. Thus at the outset, the lithic recording strategy for the Pabbi Hills material was designed to address issues of where the material was coming from, and under what circumstances it may have been transported through this eroding landscape. This chapter will concentrate on the recording strategy itself and the results derived from it, but clearly all of this is predicated upon a good understanding of the taphonomic history and origins of those stone tools as subject to the processes within the landscape. For this reason, the chapter includes studies of the likely taphonomic processes via experimental and field observations to aid in the interpretation of the survey assemblage. No other large-scale survey has dealt with these specific issues in such detail, and the approach and results have wider implications.

In order to understand the archaeological significance of the lithic finds from the Pabbi Hills, one must be able to "read" the contemporary landscape. The current users of the Pabbi Hills are local shepherds who graze flocks of mostly goats and sheep in the hills, returning to their villages on the level lands at the fringes of the anticline at night, and nomads who make seasonal encampments within the Hills and graze their herds nearby. Small herds of horse and occasionally solitary water buffalo and camels have also been observed in the area. The Forestry Commission has planted saplings on some flatter areas within the landscape, mostly alongside the watershed track and other areas where a road may be bulldozed. Though illegal, local people use the Hills as a source of firewood. The planted areas did not form part of the survey, but the very active grazing and wood-collecting ensure that many slopes are sparsely vegetated, and thus open to erosion and survey. It is these fast-eroding, barren slopes that dominate the landscape of the anticline. At first glance the area appears to be entirely natural, but closer inspection reveals patterns of human alterations to this landscape. This is mostly transient because the structural materials are ephemeral, e.g. plant stem screens, mud or sandstone blocks. The so-called 'sandstones' are very poorly consolidated, and though pieces can be used for structures, once exposed, the surfaces become rounded and gradually the 'stone' disintegrates. Thus evidence for structures is likely only to come from the very recent past. There are occasional concentrated finds of human occupation. Some of these are marked on the maps as the older versions of villages that have recently shifted towards more level ground to take advantage of wheeled transport, e.g. "Old Sohawa". On stable land, most often on the very fringes of the anticline and on the dip slopes of the larger Sandstone units, there is evidence of former settlements via pottery accumulations and occasionally low lines of sandstone blocks. These are typically within a short walk of land currently under the plough. This kind of site is not found in the interior of the anticline. More widespread are the remnants of very recent nomad camps, visible mainly as cleared areas where the temporary huts and enclosures stood. Observations over the years of the survey show that the nomads may use different camps from year to year even if they return to the same general area. The nomad dwellings (plant screen or cloth walls with plant thatching) leave no permanent traces except a few large stones scattered in the vicinity of what would have been the house walls with cleared space in between. The 'wall' delineating the encampment is usually of cut thorns that disappear in a few years. There is evidence that some of the current users of the Pabbi Hills occasionally repair major paths through the landscape by building retaining walls of sandstone blocks. These will weather away over time. Sometimes the same technique has given rise to retaining walls to trap soil. Similar walls, lined with clay form shallow ponds. These gradually silt up unless they are cleared and re-dug. All these relatively recent human features in the landscape can be identified with ease. They are not associated with large quantities of any artefacts, since at most they may have scattered around them a few pot sherds or fire-cracked quartzite stones. Without maintenance none of these features will last. The most long-lived would be the walls retaining soil since these have one side of the sandstone blocks buried. At the very most, these features may be a few hundred years old. Only on the very fringes of the anticline outside the main collecting areas and on more stable surfaces has anything approaching an accumulation of pottery been seen. Thus the Pabbi Hills landscape does have some evidence of human structures and sites from the relatively recent past and contemporary usage, but the landscape is devoid of substantial evidence for more ancient sites and structures.

It is against this background that the lithic artefact distribution and collection must be judged.

One of the most actively eroding scarps in the anticline is the dramatic Sandstone 12 horizon East of the G.T. road and on the northern limb of the anticline, as described in chapter 5. It was in this interior area that the team first found stone tools in 1986. They were not concentrated into discrete lithic scatters and were at first ignored. Over the course of the season there were too many to ignore, and the team began to collect them on an ad hoc basis. Looking at the 1986 assemblage now, it is evident that the collection favoured readily identifiable pieces. In 1987, I became part of the project with special responsibility for the lithic artefacts. One of the first tasks was to set up the collecting and recording strategy. Thus in the 1987 season, the team knew that they were going to be addressing the stone as well as the fossil bone assemblage. The lithic collection and recording strategy devised was specific to the nature of the material. Possible origins are listed below, and it may be a mixture of more than one of the possibilities given.

1: natural agencies causing struck flake features at any time in the last two million years
2: deliberately struck but very recent in origin
3: deliberately struck but derived from reworked residues of post-Siwalik material from ca. 400 kya onwards, i.e. after the anticline was formed and thus Mid-Pleistocene in age onwards
4: deliberately struck and originating from the underlying Upper Siwalik strata dating from about 0.9 to over 2 Mya

The purpose of this chapter is to describe the collected material and analyse it in order to see to what extent each of these possible origins may have contributed to the assemblage, and to make an overall assessment of how much, if any, of the lithic material may be Lower Pleistocene in age.

Stone Collecting and Recording Systems

Collecting surface lithic material in an eroding landscape

The 1986 field season identified the presence of stone tools, and thus the need to incorporate lithic finds in the collecting strategy of future seasons. Both the fossil and lithic finds needed to work within one system, and so the distinctions of gully collect, general collect, locality, site and find spot devised for structuring the fossil evidence (see chapter 3) also served to record the stone tools recovered. The constraints were rather different from those acting on the fossils because the lithic finds were firstly, fewer in number and secondly, were rarely found in even a weak concentration. This pattern is in itself interesting and is discussed later. Here, it is important to outline the lithic issues within the collecting units.

Gully collects would denote where worked stone was found as part of fluvial deposits accumulating at the base of slopes in the seasonal watercourses. In fact, no such accumulations were found.
General collects served to create a unit for widely dispersed lithic material. Although most general collects contained no stone, or only one or two pieces, much of the stone tool assemblage is recorded in the general collects, reflecting a very low level of background artefact presence.
Scatter collects: no lithics were found in these
Locality collects could be initiated by either a relatively dense concentration of fossil or lithic finds. A site collect need not reflect material in primary context, but simply a localised concentration. Thus relatively discrete concentrations of lithic material would be recorded as a 'site'. In the unlikely event that one or more stone tools would be found in primary context, these too would be treated as a 'site', and the primary context described on the survey card. There are very few localities defined on the grounds of lithic concentrations (examples being 346, 625, and 722).

Find spots were reserved for individual items in their original context or of special importance. If a chronologically and typologically distinctive tool such as a hand axe had been found, this too would have been recorded as a separate collecting unit. Only one lithic find spot was noted, which was a stone tool found on a small goat path near the scarp top of Sandstone 14, and which was found to fit back into a socket left in the eroding sandstone bank above this. The socket was *in situ* but in a secondary context within a small channel cut into Sandstone 14.

Even from the short comments on these different kinds of collection units, it is apparent that some would rarely if ever be used for recording the lithic finds. For example, it is difficult to see how, on survey, a stone tool could be found in its primary context, given the high rates of erosion and steepness of slopes in most parts of the Pabbi Hills. It would have to have been found partly eroded from primary sediments, with sufficient of the surface visible for it to attract attention, yet still remain in indisputable context. There is an inherent contradiction here: lithic material found on the surface would, by definition, be derived; material still in its primary context would be all but invisible. Thus the fieldwork from the outset was known to rest mainly upon lithic material found as part of dispersed general collects or within the relative concentrations recorded as localities. Referring back to the known human use of the landscape, there was no evidence that the people who most commonly use the Pabbi today were modifying quartzite cobbles for use. (Stiles (1978: 139) reported that shepherds in the Potwar Plateau flake stones out of boredom, throw stones at other stones (which may induce flaking), and sharpen their wood-cutting axes by pounding the blade on cobbles. From his account, he did not observe these activities first-hand. In the 15 months that we spent in the Pabbi Hills, we never observed shepherds (or villagers) flaking stone, nor did we see any evidence of recent flaking (easily spotted by fresh fractures and fresh, conjoining pieces). Axe-sharpening was usually done on sandstone, and we did not observe blade-pounding). Nor was there any evidence that later prehistoric sites were present in the main survey area though some sites are known on the fringes (e.g. 652 and 672 on the gentle dip slope of Sandstone 12 between Baroth and the GT road, and a very small tell on a gentle dip slope just west of the main kas bed as it leaves the Pabbi Hills in Kotha Kas. The position of 'old' Sohawa and Meira villages are also marked on the 1908 map of the region. It seems likely that the rich fertile soil on either side of the anticline had in the past supported prehistoric populations. The uplift of the anticline sometime after 400 kya created a dissected landscape that was better exploited as pasture and hunting areas than for agricultural settlement. Thus the Pabbi Hills could have few or very small prehistoric sites because it was utilised in a different way. Undoubtedly, the high rate of erosion has increased the loss of any such sites. The most active erosion period is during the monsoon and since this is known as early as the Miocene, even Neolithic sites would have had to survive 4000-5000 monsoon seasons. The most stable areas would be the sandstone dip slopes but even here, material could have been washed into deposits at the base of the dip slopes and become gradually lost in the gullies that form there.

An appreciation of the way stone and other material moves in the landscape is an essential part of the interpretation of the lithic finds and is discussed later. The immediate local situation of the tool may reveal much about its possible origins. The collecting strategy was designed to cover concentrations of derived material via deflation or fluvial accumulation processes, as well as dispersed scatters from disturbed primary contexts. The latter could be from later prehistoric sites had these been superimposed on the much earlier strata, or from primary geological material eroding down slope. The latter possibility was intriguing since the fluvial deposits forming the modern erosional surfaces date from 0.9 to over 2 million years in age. The river that had laid down these deposits was moving laterally across a wide floodplain, making it unlikely that there would be palimpsest sites, as is the case around former lakes in East Africa. Thus the lithic collection and recording had to be prepared to pick up very subtle patterns. We were also interested in trying to establish the absence of artefacts, and more controversially, bear in mind that if there were tools they need not be very sophisticated since they might not even have been made by members of the genus *Homo*. Whilst the last point is perhaps unlikely, it shows the range of options that were being considered. The collection strategy erred on the side of caution, and anything that looked unnatural, even if it did not conform to previously defined types, was collected.

The field personnel were all experienced archaeologists but not necessarily familiar with stone tools. Given the difficulties of recognising flake features on quartzites as mentioned below, all the team were shown what to look for and what to collect. For the first season, the teams were actively encouraged to collect any flaked stone, including even those fractured stones they judged as likely to be natural, in order to ensure a coherent strategy on what would be discarded and what kept. Unworked quartzite pebbles were noted in the field so that the natural occurrence of stone could inform later interpretations. Furthermore, for the 1987 season only, material that was graded as unlikely to be of human origin was also kept and recorded. Overall, the field collection strategy, especially for that first season in 1987, was deliberately over cautious and predicated upon the belief that it was easier to throw away surplus material later than it was to miss elements of the lithic picture inadvertently. The strategy gave confidence that different individuals and different teams collected the lithic material in a uniform way, and thus the lithic assemblage collected from 1987 onwards is a reliable indication of the type of flaked material from a wide area of the total area surveyed. Furthermore, the size ranges (with the smallest item weighing only 1 gm.!), and the subtleties of the pieces collected show the level of care exhibited.

Recording

The recording system was devised for the particular requirements of the Pabbi Hills survey assemblage. In addition to the standard measurements of length, width, breadth and weight, fourteen other variables were noted, and a range of types was identified. The purpose of the recording system was to systematise both objective and subjective information from the lithic finds; to separate the criteria used to decide whether the stone was intentionally or geologically shaped, and above all to record data as continuums to allow broad trends to be identified. Codes were used even for subjective data in order to facilitate computerisation and the listings in Table 1 (end of chapter). Formal tool types were used but this assessment was an additional descriptive tool supplementing the other variables recorded. The typology will be discussed first, followed by sets of the variables according to the purpose behind them.

Conventional Typologies

The stone tools collected in 1986 indicated that there would be few sophisticated retouched types present, and the assemblage would probably be dominated by pebble cores and simple flakes. Artefacts such as fragments of polished stone axes, or scrapers presented no difficulties in identification, but formal unambiguous tool types were a rarity. A number of options were available as typological systems in 1986. For the Indian sub-continent, there was the De Terra and Paterson (1939) terminology of Soan tools, or there was the African Oldowan nomenclature. Each has its own problems. The Soan tool terms have been perpetuated in the region for some time (e.g. Sankalia 1974) but this team in particular has rejected the utility of De Terra and Paterson's scheme (e.g. Rendell et al. 1989, Dennell 1995a, Dennell and Hurcombe 1992) since this was poorly defined, based on a chronological methodology considered problematic today, and was devised from material which is no longer considered reliable fieldwork evidence regarding its context and age.

The classification system for a simple lithic assemblage devised by Mary Leakey has become classic, with the larger categories remaining effective descriptive units (Leakey, M., 1971 and 1994). Her types are certainly the most suited to pebble cores and flake material with two caveats; she paid very little attention to flakes, and many of her core tool types are now seen as overly classified (e.g. the different kinds of choppers and the distinction between discoid, polyhedrons and spheroids which are now considered to be part of a continuum). Furthermore some categories are based on an *assumed* function that presupposes not only that they are tools at all, but also their mode of use (e.g. choppers, scrapers). The Oldowan terms thus include functionally described categories that prejudge their use in a problematic manner. Similarly, on her scheme, flakes tended to be defined on the basis of shape and striking direction. This could be useful, but is really saying more about flaking strategy than end result. A much better system for classifying simple flakes has been put forward by Toth (1985), who has also shown experimentally that simple flakes can be very effective tools (Toth 1987). The dichotomy between flakes and cores is no longer seen as a rigid division, as both could be waste products or used as tools.

By using the terminology associated with the earliest African stone tools, it could also be said that the age and significance of the lithic assemblage was being prejudged. It was not, and it would not help to lay the system open to that criticism. In addition, scholars working in the Indian subcontinent (e.g. Armand 1983) have made a reasoned plea for the archaeology of the region to be defined in its own terms rather than be seen in relation to African or European evidence. In consequence there were powerful reasons for seeing the typology of the Pabbi Hills assemblage in its own terms, but practical issues suggested adherence to at least some of the terms prevalent in work based on the African material, if only to facilitate discussion. The problem of typology was multi-facetted. The technologies observed in the Pabbi Hills assemblage were generally simple but nonetheless contained differences. Both the technological characteristics and the conventional type categories might provide useful information, so two separate but linked systems were devised. Firstly the material was described according to more conventional types, and secondly, the item was given a technological type here referred to as techno-type. Since this refers to the manner of *flaking*, all elements except polished and ground stone tools were classified into this system.

The typological system described below is thus an empirically based system specific to the material studied. It could be applied to other cobble-based simple assemblages. In the discussion section, comparisons are made with other descriptive systems. Most of the types outlined are along standard lines with the exception of ring stones, which are familiar to the region, "straight-edged core tools" and "high action core tools" that form new categories of *tool* recognised in this assemblage. These types together with flakes thought to originate from them are asterisked in the brief characteristics listed as part of Table 8. 1.

Techno-types

In addition to the types outlined above all flakes and cores were placed in a technological category that was useful in defining lithic reduction strategies. Figures 8.1 and 8.2 define these flake and core categories via schematic diagrams. The techno-types illustrated (Figure 8. 1) show that the flake techno-types more or less equate to those of Toth (1985) by distinguishing firstly whether the striking platform is cortex or a fracture surface, then subdividing further on the basis of the amount of cortex present. Flakes are often divided into primary, secondary and tertiary, based on the presence or absence of cortex on their dorsal face. All cores and flakes had the percentage of their entire surface covered by cortex recorded. Thus a decortication flake typically has c. 50% of its whole surface, i.e. its dorsal surface, covered by cortex. The secondary flakes were further subdivided into those that had <25% of their total surface covered in cortex (i.e. those with less than half their dorsal surface as cortex) and those that had > 25% cortex. This system created a neat division of the flakes into eight categories (F1-F8). A few distinctive flakes did not fit neatly into this system. They have been called *salami flakes*, since they resemble salami slices, and labelled F9. They are round, and have a continuous arc or ring of cortex from the platform and extending down around their circumference, with the flake scar from the previous removal forming a neat round dorsal scar. They signify the opening stages of a pebble core reduction strategy by being the second or at most third flake off a cobble, and could either alter the angle of the platform created at the top of a cobble, or widen it to encompass more of the cobble's breadth.

Figure 8.2 shows the different core reduction strategies that seem to have been used. Virtually all the raw material is quartzite, so most material is produced from rounded fluvial cobbles. The cores are more complex but have been divided into evidence for five basic strategies of core reduction. The aim of the basic techno-types was to establish broad reduction strategies and allow comparison of the flakes and cores that would form part of these reduction sequences.

Strategy A, Cobble splitting. The cobble is brought into forceful contact with another hard object with the aim of splitting the cobble up into more workable fragments. This sometimes results in the cobble shattering into several angular pieces (A1), or can create a slightly irregular split (A2) due to a natural weakness in the lithic material, or because unsuccessful blows have created partial Hertzian cones beneath the impact point. The assumed best result of strategy A is a neat single fracture breaking the cobble in two (A3). A3 is the necessary first step in strategies C, D, and possibly E.

Strategy B, Cobble flaking. The ends of rounded cobbles are the easiest places to detach flakes. Some have a few flakes removed from one direction on one end (B1), others have removals from two directions but on different ends (B2) i.e. bi-directional flaking, or exhibit true bifacial flaking (B3) creating an edge with fracture surfaces on the two faces. Some have more complex removals and are termed multi-directional (B4).

Strategy C, Circumference flaking. The flaking may use the cortex as a platform (C1), and take off flakes from the circumference across the main fracture surface (A3) or use the fracture surface as a platform removing flakes circumferentially down the cortical surface (C2). Some cores have both faces flaked (C3). The end process results in a core with very little cortex, and it can look quite angular rather than hemispherical. All of these different types show the same reduction strategy of working around the circumference and have thus been grouped as circumferential flaking. All the types listed here are further subdivided by adding an "A" for simple and a "B" for complex after the category type. This extra division is recognised by how many flakes have been removed, how much cortex remains, and the fineness of execution. Thus a C3B may look like a slightly asymmetrical disc core or tortoise core, whereas a C4B would be so reduced and angular that it would approach a polyhedron or spheroid in shape. This strategy results in cores that can be termed discoids, polyhedrons or even sub-spheroids.

Strategy D, Square platforms. The split fracture (A3) is used as a platform but instead of working around the cortex, the flaking seems to deliberately create sharp angles on the platform face so that the platform has one (D1) or more (D2) squared areas in plan view. The working faces also tend to have high angles approaching 90°. Once this shape has been created and several straight edged flakes have been removed from the faces, the core is capable of producing straight edged flakes or flake blades (D3). The well-produced end result of this strategy is a neat core capable of producing flake blades or short blades (D4).

Strategy E, Oval platforms. A slightly oval rather than round cobble is selected and a fracture platform created (E1). The opening stages of this strategy may use the alternate strategy of taking a flake off one end, followed by a second which uses the round scar left by the first removal, i.e. a salami slice flake is removed. Alternatively, a successful split cobble might form the starting point. However it is produced, the opening stage is flaked along one of the broader faces (E2). Gradually the core becomes slightly tapered and the platform thins down to an elongated ellipse (E3) and in its final stages the core may have little or no cortex remaining on the tapered end, with only the back face retaining any cortex (E4).

It is difficult to place any firm dates on these strategies, but D and E are not thought to be characteristic of the Early or Middle Palaeolithic, but of later periods. The discoidal polyhedrons of strategy C are similar to earlier palaeolithic types but could also be part of later flaking strategies. The tortoise cores of the Middle Palaeolithic could also be part of strategy C but these are rarely so well made as to resemble a true tortoise core. Strategy B could be early but need not be used solely in early periods.

Variables signifying post-depositional histories

Edge rounding, post-depositional damage and patination were all recorded. The absence of these three features need not imply a primary source, but where they are present in quantity they could suggest that material was derived or from mixed sources. Edge rounding was described in five categories based on whether the rounding could be perceived with the naked eye and/or felt. One category was reserved for edges that had no rounding, but a further category defined those whose edges were still so sharp or crisp that they should be regarded with suspicion. The categories used were: *very rounded* (extensive visible rounding, causing a lack of clear scar ridges), *rounded* (visible rounding but retaining clear scar ridges), *slightly rounded* (the rounding is just visible but can also be felt), *fresh* (no visible rounding and none can be felt), *crisp* (edges that are so sharp and fine, that they seem unlikely to be old).

Post-depositional damage was recorded, where present, with comments on its nature, e.g. features such as snap fractures and other irregular scars. Patination was also noted if present. Whilst flint tools are frequently patinated, the majority of the assemblage was of materials that rarely show visible chemical surface alterations. The term patination was used to signify any colour or textural change of a surface that was likely to be a result of post-depositional factors. In practice this was easier to see on those tools which also had some form of post-depositional edge damage so that fresh scar fractures showed the difference between the original and altered surfaces. Patination is far less frequently observed on quartzite artefacts than on finer grained flints. These variables aid the taphonomic interpretation of stone from specific localities and the interpretation of the origin of the Pabbi Hills assemblage in general.

Variables showing the range and quality of the raw materials

These variables were recorded in order to investigate whether the lithic sources changed through time or space. They were devised to pick out any chance finds and as importantly to show that some of the objects in very coarse materials are probably artefacts even though they may not have such clear flake features. Thus these variables record the shades of grey in artefact recognition that are more normally seen either as black or white, or not recorded at all (see also Patterson, 1983). That is to say, most lithic analysts would accept that the Pabbi Hills assemblage contains artefacts. However, artefacts such as handaxes, arrowheads and scrapers are incontrovertibly of human origin in a way that a simple decortication flake might not be. Whilst a handaxe has an infinitesimally small chance of being produced by natural agency, the primary flake has a still small, but nonetheless greater, chance of being so created. Thus the complexity of the lithic assemblage affects the chances of correctly attributing all pieces to natural or deliberate creation. Given that the assemblage might be coming from the underlying geological strata, depending upon which part of the Pabbi Hills was being surveyed, there was a potential time range of between just under one million to older than two million years. It was therefore unwise to make the assumption that the lithic industry would arrive fully fledged with recognisable tools showing several flaking events. Some of the earliest stone tools might prove to be the most difficult to attribute to deliberate agency. Furthermore, at this point in the sequence there was the question of who might be making stone tools. Again, the possibility that the assemblage was generated by a hominin other than *Homo*, or even an ape, was controversial but allowed for. The approach adopted was to try and avoid prejudging the issues, and to see what could be encountered by way of stone tools and any patterning due to increasing complexity. This requirement was compounded by the nature of the dominant raw material, quartzite. This one term masked a wide range of material in terms of both coarseness of the grain and the fracture quality. At one end of the scale the material is so fine grained and homogeneous that *éraillure* scars are visible, yet at the opposite end of the spectrum the grain size can be coarse but homogenous, or coarse with internal preferred fracture planes. These spectra do exhibit conchoidal fracture, but the ability to see fine details of the flaking direction, ripple marks, stress fissures and the like can be lost in the coarse material. This lack of detail might be a problem for artefact recognition - in effect present an archaeological study problem - yet need present no such inhibiting effect in terms of the maker's ability to govern the shape of the tool and its effectiveness. One more factor has been noted for the quartzites; some have the propensity to fracture along natural planes within the raw material. Whilst this tends not to occur in the fine-grained varieties of quartzites, it is not possible to say that this feature always varies with coarseness. In general the rougher surface textures of quartzites in comparison to those of flint tend to show radial stress fissures much better, and ripple marks less clearly, though there are surface undulations that fulfil the same role as ripple marks on flint. The stress fissures on the quartzite flake fractures still occur at right angles to the direction of the force as they do on flint examples, but they tend to cluster much more on the bulbar surface. In contrast the stress fissures on flint are most often observed on the lateral edges of the ventral surface. The quartzite stress fissures show as small reflective fractures whose direction points back to the point of origin of the force that detached the flake. They are thus good indicators of the direction of the blow, but because of their low ability to reflect light they do not show well in photographs. Radial fissures are represented stylistically on the drawings of the quartzite tools.

Thus a description of the raw material consists of three separate pieces of information: the type of rock, its colour, and its conchoidal fracture quality. The first two are recorded to give an indication of the sources of lithic material but they are also used in conjunction with the fracture quality. The latter is described by codes according to both the grain size and the propensity to fracture along planes within the raw materials. *Good*, denotes fine conchoidal fracture properties usually associated with a fine grained flint but just occasionally used to describe exceptionally fine grained quartzites, *rough* where the features of conchoidal fracture are clearly discernible; *very rough*, where these are vaguely present but diluted by the rough natural texture of the surface; the propensity to fracture in planes is recorded by adding a *'p'* to any of the codes for coarseness. Just occasionally there was a granular aspect to the surface texture so this was recorded by adding the symbol *'g'* to the description. The system collectively identifies features that are essentially due to the size of the grain.

Variables of the knapping sequence

This group of four variables was designed to pick out general trends. The surface of the lithic item is perceived as having a total of 100% surface area. The proportion of the surface that is covered by cortex, and negative and positive scars are all estimated as percentages of the total surface area. In this way, general trends of decreasing amount of cortex, i.e. the presence of primary flaking, or extended reduction sequences might be discerned. Where the estimated positive scar surface is zero, the item is a core, whilst those with a substantial proportion of positive scar surface are flakes. Where the raw material's fracture properties make patterns difficult to discern, the system can still be applied in the same way. In addition to the three percentage surface area variables, there was a simple sign system (- or +) denoting whether the piece was largely covered in negative scars, or was a positive scar piece. The advantage of this is that it yet again avoids the use of terms such as flake and core as binary opposites. In some cases there are items that may have been very large flakes that are subsequently used as cores, or there may be cores that are used as tools. These variables were recorded in order to compare percentage cortex profiles for negative pieces and positive pieces from different areas of the Pabbi Hills.

Variables indicating deliberate flaking

As part of the difficulty of dealing with coarser materials small remnants of previous flake scar removals remaining on a tool may not retain the limited sections which clearly show the fracture features, and as mentioned above, some of these will in any event be in materials so coarse that the flaking features are in any case vague or absent. In consequence, it was felt that there should be a measure of how many scar surfaces showed flake features and how many did not. For example a lithic item which mimicked a handaxe, i.e. showed the scars of previous removal all round its edge and over two faces yet did not retain clear flake features, could never the less be regarded as highly unlikely to result from natural agency. In order to record the data by which such degrees of likelihood are estimated three variables were defined and noted. Firstly the straightforward 'number of flake scars showing clear flaking features'; secondly the 'number of scar surfaces without clear flake features'; and thirdly the number of different directions of flaking represented by the nature and location of the scar surfaces. Thus the first two variables would, if added together, give the number of different flaking events to which the piece bore witness, whilst the third would give the number of different directions a blow would have to originate from in order to achieve the shape. Thus to return to the handaxe example, the piece would have been flaked from four different directions on both faces, and so the number of directions would be eight.

The last category of 'credibility' is a subjective assessment of whether the variables recorded suggest the piece is due to natural or human agency. This was thought to be necessary in case there were early, poorly executed flaking events on materials which did not exhibit good flaking features. It was recognised that such pieces would be discriminated against in normal classification and selection processes. The raw material especially could, if poorly chosen in the past, make the flaking difficult to recognise in the present, and the simplest items are more difficult to attribute unequivocally to human agency. Neither factor is normally important to lithic recording strategies because there is usually relatively good quality flaked stone in use and these are mostly used in sophisticated ways. It could not be assumed that flaking events at the very earliest times would show such selection and sophistication. Again the variable was selected precisely for the task in hand.

The scale used was from 0 to 5. A score of 0 denoted a natural piece of the raw material completely covered in unblemished cortex; 1) not at all likely to be humanly formed but one could not be 100% certain it is natural, e.g. there may be a slight bruising of the surface or irregularity caused by usage; 2) may be due to human agency but much more likely to be due to natural causes, e.g. a facet on a cobble or a fracture surface which is very rounded with no clear flake features; 3) may very rarely be due to natural causes but is much more likely to be due to human agency, e.g. a decortication flake with clear flake features; 4) a clear example of an artefact, e.g. with flaking in different directions ; and 5) a textbook example of a complex artefact flaked from many directions and with clear flake features. Thus on this system and way of thinking a decortication flake would be unlikely to score 5 or even 4, and so it would be recorded as a 3. As an example, we applied this system to Kafuan material available in Cambridge that was originally thought to be an early cultural assemblage (Leakey L. 1931), but later shown to be the result of natural agencies (Clark 1958). None of the material examined scored more than a 2 and much was rated 1 or 0. Other material from Olduvai Bed I and Koobi Fora scored 3 or more, i.e. as intentionally-flaked. We thus have every confidence that the material we have examined and classed as artefacts from the Pabbi Hills are in every way comparable with other studies.

It is worth noting that the lithic material in a completely different area of Pakistan, the Riwat assemblage was recorded in the same system as that being used for the Pabbi Hills. None of those who have seen this material have expressed doubt as to its origin from a deliberate intentional and controlled set of flaking activities. However, some (at various conferences) have chosen to see the credibility system as an indication of a problem. It is rather the reverse. The credibility system outlines and reaffirms our certainty that the Riwat assemblage cannot be due to any other cause than 'human' agency in an explicit way that has never been matched by discussions of other simple assemblages made from materials that do not necessarily show flake features well. We have graded our pieces as *artefacts* at grades 3, 4 and 5, recognising the differences explained above, when most other studies would simply call all of these 'artefacts'.

Some comments on the drawings and nomenclature

The drawings reflect the technological classification system and thus flakes are orientated with bulbs at the top. The first angle of rotation has been used rather than the third angle. For quartzite artefacts, stress fissures are more numerous than fine ripple marks so the drawings indicate these lines as a means of showing the directions of striking. Since the material could not be removed from Pakistan, all artefacts graded 4 or 5 and many of the grade 3 items were drawn in the field. They are accurate diagrammatic representations of the technological features rather than refined illustrations showing shade effects. Where a piece has internal flaws or is prone to fracture in planes an attempt has been made to convey this in the drawing. Ripple marks have only been illustrated where they can be seen; often the ripple mark is not a neat arc but a broken line running around the coarse textures so this broken but nonetheless arching feature of fracture has been indicated in the drawing. All the stone pieces collected were numbered as collecting zone number, full stop, stone item number running from 1 onwards for that collecting zone. This numbering system has been retained throughout this report. Not all items collected and labelled were seen as artefacts since the team was encouraged to pick up questionable pieces, so the item number may be higher than the number of artefacts identified from the locality. The 1986 material was originally numbered as individual items. In order to fit this material into the same database and use a naming system that encoded the find area, new numbering systems were adopted using the data in the 1986 field records and the often more detailed recollections of team members.

SECTION B: Results, Analysis and Discussion

The character of the assemblage

The total assemblage consists of 607 artefacts graded 3-5. There are no true lithic 'sites', just a very low density spread of material consisting mostly of locally available materials, and largely composed of simple flake and core types. Table 8. 2 shows that while 211 collecting zones have stone artefacts, 45.5% have only one such find, and 77.7% of the collecting zones have no more than three artefacts. The different areas surveyed can be grouped according to the age of the underlying deposits in a similar way to the fossils, but without assuming that the lithic material *necessarily* comes from the Siwalik deposits. The density of lithic material was so low that only three sets of material can be defined (see Table 8.2). As can be seen, the collecting zones with more than 10 artefacts are all confined to the scarp and areas just below Sandstone 12. This is not just because Sandstone 12 was especially rich in fossil: There were many areas that were poor in fossil but still closely surveyed, and thus if areas had been stone rich but fossil poor, they would have been recorded as such. It might be thought surprising that so few artefacts have been recovered considering the area of ground surveyed, but the assemblage shows the team were picking up the simplest of artefacts (decortication flakes and cobbles with only one flake removed), and the smallest (minimum size 10 x 10 x 2 mm) and lightest (minimum weight 1gm.) of pieces, and therefore the collected assemblage seems to be an accurate reflection of the stone artefacts in the vast area surveyed.

As Table 8. 3 shows, the assemblage is dominated by quartzite artefacts (95.6%), with only 12 flint pieces and 15 items of other rocks. Half the flint pieces are micro-cores (6) while the other, mainly igneous, rocks have mostly been used to make ground or polished stone artefacts (11), though basalt and other fine-grained igneous material has also contributed to the flaked stone assemblage (8). There is unambiguous manufacturing retouch, rather than edge damage from use or incidental damage, in only 17 cases, i.e. 2.8% of the assemblage. Ground or polished stone items form only 1.8 % of the whole assemblage, which is predominantly composed of flakes (57.8 %) and cores (40.7%) of chipped stone. Given the differences between the artefacts of different materials, the description of the assemblage is split into three components based on the lithology, with the non-quartzite minority (4.4%) discussed first.

The non-quartzite artefacts including flint

Flint: The kinds of flintwork known from flint source areas such as the Rohri Hills, southern Pakistan (Biagi and Cremaschi 1988) are entirely absent, but although the survey team has found *natural* flint in the Pabbi Hills it is both rare (only four pieces) and small. Three of these nodules weighed less than 10 gm. and would not have been suitable for tool manufacture, even for micro-cores, but the fourth piece could have been used in this way since it weighed 152 gm. and was 65 x 40 x 40 mm. This could explain why the micro-cores were entirely of flint, and why the flint artefacts found are so small, with an average weight of only 4.4 gm. (range 1 – 14 gm.). The 12 flint pieces comprise 6 micro-cores (numbers 216.2, 339.4, 351.1, 383.10, 722.63, and 840.1), 3 split flakes (411.2, 722.60, 722.72), a simple flake (284.3), an irregular flake (722.97) and a scraper (349.5), although 339.4 could also be termed as such. It is conceivable that the flint used was from relatively local contexts because of the sizes of some of the natural nodules found. Figure 8.3 shows some of the micro-cores and the very small scraper (349.5) with part of the cortex remaining. When compared with the flint micro-cores, some of the flake scars suggest that the piece could also have served as a core, and 339.4 could have served as both core and scraper. The size of these cores makes them consistent with the mesolithic trend towards smaller artefacts, but the pieces could simply reflect the natural nodule size range. For example, the small core 722.63 measures 27 x 18 x 14 mm, and weighs 4 gm., but there is sufficient cortex remaining to show that the original flint nodule was probably a small pebble. It seems most unlikely that these small cores are from the Siwalik deposits, especially as micro-cores are known from both mesolithic/final palaeolithic and chalcolithic contexts. In one case, (840.1), the flint was found near a scatter of undatable pottery, over a rectangular platform in the base of Kohar Kas. This was an isolated survey area visited in 1999, and by far the most western area surveyed.

Polished and ground stone: Polished and ground stone artefacts along with hammerstones or fragments of them total only 15 artefacts. There are four hammerstones (367.1, 409.1, 475.1) or hammerstone fragments (395.1), some of which are illustrated in Figure 8. 4. These show unequivocal damage of a pecked and worn surface consistent with many impacts. Hammerstone 367.1 is a slightly disc-shaped stone with separate areas of wear on the periphery, mostly where it is slightly thinner, while fragment 395.1 shows two distinct wear facets on one area of what would once have been the 'edge' of a rounded quartzite cobble. A bifacial core (393.3) also shows distinct facets of damage on its cortical surface, as though it was a hammerstone that had subsequently been flaked. It is likely that other cobbles were also used as hammerstones, but the wear was not so distinctively patterned for them to be recognised as such. The fact that any were spotted in the field shows the diligence of the teams.

There are six fragments of polished stone axes (224.1, 346.4, 387.25, 389.14, 494.1, 725.6, 799.2) the best of which are illustrated in Figure 8. 4. All except one could be regarded as discarded fragments that occurred in use, though 494.1 could have given useful flakes thereafter. One (346.1) is made from a flat pebble whose end has been shaped. All are recognised as Neolithic or later, and site 346 was recognised at the time of discovery as a very recent deposit (see chapter 6). The axe fragments seem to be isolated finds. None occur at true 'sites' with high numbers of other stone artefacts in the same collecting zone, but 387, 389 and 725 do have 8 or 9 stone artefacts, which is unusually high for the Pabbi Hills, as discussed later.

Three fragments of rubbing stone (426.4, 828.2) were recovered and distinguished by their worn surface. Their profile shapes are not consistent with axe fragments and 389.14 could be a ringstone fragment. Two objects are best described as ringstones (799.1 and 745.1). They have been roughly flaked to a flat disc and then the centres of both faces have been worn and bear the marks of circular grinding. In each case the flattened upper and lower surfaces show grinding in an approximately circular pattern, and the faces are noticeably smoothed. Graziosi (1964: plates 86-9) describes exactly similar objects and the term ringstone, has been retained here. As he found, our examples are broken in half. Other material illustrated and attributed to the same location, 'vicinity of Morgah', included pottery with impressed and applied decoration. Sankalia (1974: 241, 248-249, 254, 302) also describes 'ringstones or maceheads,' giving examples associated with Mesolithic and Neolithic/Chalcolithic material. The illustrations (e.g. Sankalia 1974: 248, Fig. 60, No. 32) and descriptions suggest that these ringstones could have been used as digging stick weights, net sinkers or maceheads since they were shaped like doughnuts and perforated, though they also have been grinding stones. One of the Pabbi Hills 'ringstones' (799.1) is not perforated, and both examples have hour-glass profiles suggesting unsophisticated drilling techniques, and both are more disc-shaped as shown in Figure 8. 4. Such objects seem too large for digging sticks or maceheads but could be grinding stones or door pivot points, judging by their wear and shape. In summary, 'ringstones' as a category cannot be attributed to a specific period but are likely to be Mesolithic or later.

The quartzite and non-flint chipped stone assemblage

The assemblage is grouped into cores and flakes with the most distinctive types discussed first. The pebble cores can be divided into those which are based on a split cobble, giving essentially hemispherical shapes, and those that are unifacially or bifacially flaked pebble cores. Blocky squared cores and flake-blade cores also exist. Some cores show splits, multiple flaking events, or evidence of bi-polar flaking. Four other types are present; choppers, straight-edged core tools (SECTs which may be functionally related to cleavers), possible bifaces and high action core tools (HACTs - see below).

Handaxes: There are no true handaxes in the assemblage, despite handaxes being present at Dina and Jalalpur at ca. 700, 000 b.p. ca. 40 km. away (Rendell and Dennell 1985), although a few pieces had superficial resemblances to them. One piece (830.3, illustrated in Figure 8. 5) could be classed as a "proto-handaxe" on the basis of the intentional creation of a bifacially flaked and roughly symmetrical form, although the edge angle is high and the long flakes removed could make this piece more cleaver-like. Another piece (728.6) was a unifacial core that was handaxe shaped, but the raw material was very rough-textured and prone to planar fractures so that clear intentional flake scars were absent. Two other items (a bi-directional core 196.1, and a complex disc core 97.3) were symmetrical like a handaxe, but had edges with angles of c. 70°. 196.1 also had extensive cortex. None of these is regarded as a true hand-axe.

The absence of hand-axes – the most distinctive type of early Palaeolithic artefact – is regarded as significant. Given the range of other lithic material collected by the team, including pieces as small as 1 gm., it is extremely unlikely that hand-axes were present but missed by the survey teams. There are two likely explanations. The first is that hand-axes were not used in that part of South Asia when the fluvial sediments in the Pabbi Hills were being deposited; and the second is that they were also not used in the later Middle and Upper Pleistocene, when the Pabbi Hills anticline was being eroded and some post-Siwalik deposits were formed. This would imply that the Acheulean was confined to the Levant and S.W. Asia in the Early Pleistocene, and not prevalent in South Asia until after the main part of the Pabbi Hills sequence.

Straight-edged core tools (SECTs): There are five distinctive tools in this category (52.1, 747.3, 812.1, 816.1, 712.1, illustrated in Figure 8. 6). They are mostly made on cores, though in one case with a split surface as the ventral face. They usually have a straight edge in plan view and a robust edge angle, typically 70-85°. They may have functioned as cleaving tools, but they do not conform to the classic definition of 'cleavers' with their gently graded edge angles (although there are simpler versions defined from Africa (Tixier, 1956) and some recent discussion of the tool class (Ranov 2001), and the Pabbi Hills examples seemed too unlike the generally accepted definitions to be classed as cleavers. Nonetheless, such core tools could have been used for similar functions. Some examples could also be classed as 'choppers' (e.g. 816.1 and 712.1) on account of their distinctive working edges. Number 747.3 has a slightly convex working edge formed on one end of an oval cobble. Only one face has fine scarring, which makes it slightly atypical. If it was used in a chopping action it is difficult to explain this asymmetry, but a low working angle more akin to an adzing motion would give this kind of scarring. It is very similar in form to 812.1, with both having >50% cortex. 812.1 has had one main flake removed and used a natural ancient fracture as the opposing face; the working edge is slightly convex in birds' eye view, as illustrated. 816.1 is the only one to be formed from a split cobble, resulting in a smaller edge angle than the others, and perhaps also explaining why the working edge is slightly off the long axis of the piece. 52.1 is unusual as it is formed on a cobble which is square in plan and rhomboid in cross section with two working edges. One main working edge is straight and it has retouch on both surfaces forming the edge. In addition, there is a smaller and less regular section of scarring on the opposing edge. Despite its unusual form the tool's main functional edge is very characteristic of the type. All five SECTs in the assemblage are from erosional surfaces assigned to the geostrata timeband of 1.2-1.4 Mya.

High action core tools (HACTs) with hemispherical and angular worn or battered edges: This type (14 examples) is frequently hemispherical or polyhedral in shape, and is characterised by the pattern of wear, which lies along protruding edges and in some cases forms a worn facet blunting the edge and lying at almost a right angle to it – hence the name 'high action core tool' (HACT). Sometimes the wear is a smoothed surface, with the remnant of small flake scars along the edge (also from use) worn down (10 examples: 97.2, 269.1, 395.11, 403.3, 426.3, 625.5, 722.146, 728.3, 728.8, 731.8), or the wear can be more like a minutely pecked surface with 'bruising' or 'contusions' along the edge in the form of multiple small impact scars at high angles and often with step terminations (four examples; 385.8, 808.1, 828.1, 830.1). Representative pieces are illustrated in Figure 8. 7, although it is difficult to convey the detail of the worn areas on drawings.

The class includes some pieces that have bifacial flaking and a sinuous edge (e.g. 722.146, 269.1, 395.11), but the individual pieces vary in shape. Centripetal flaking achieves the basic circular edge of 426.3. The prominent edges show varying degrees of wear. On some it is only just visible, but 403.3 shows the most extensive wear. Most examples have one or more worn edge that is slightly rounded and flattened. The profile of the edge can be clearly seen on the drawing of 269.1. A few appear to have seen heavier use and have small flake scars along the used zone, e.g. 728.3. The style of wear is best described as smooth battering. The used portions of the surface are limited and the flattening of the edge profile suggests the contact material is relatively hard so that only a small area of the tool edge contacts the worked material. A more detailed wear analysis of these tools would be useful.

This is one of the most distinctive formal tool types in the Pabbi assemblage. It has no obvious parallels in artefacts from the subcontinent from later prehistoric periods. Research on East African discoids, polyhedrons, sub-spheroids and spheroids does suggest a continuum of forms resulting from wear (Willoughby 1987), and these could be related to the HACTs, as defined here. The distinctiveness of this group is even more impressive if a curious group of flakes are considered alongside them. There are five flakes (722.69, 722.70, 722.125, 731.69, 751.2) which could have come from these kinds of tools because they show a distinctive worn area on their proximal surface which has occurred prior to flake detachment, but does not make sense as part of a natural rounding of the core/cobble edge, e.g. 722.69.

Flake-blade cores: There are six flake-blade cores (66.1, 92.4, 96.1, 191.9, 193.2, and 710.1). Different techniques for achieving a core suitable for flake-blade removals are evident in Figure 8. 8. Artefacts and 191.9 and 710.1 are remarkably similar. A cobble has been split to create a broad platform and the flakes have been removed progressively, lengthening the ridges as shown by 191.9. The cores are also achieved by removing the end of an oval cobble to create a shallow platform angle (e.g. 96.1) and removing flakes successively from one face with the other remaining mostly as cortex (e.g. 66.1 and 92.4). A further technique used on a flat oval cobble is the removal of flakes from one end of the oval with some working down of the distal portion, e.g. 193.2. The technique is similar to that shown in the disc cores but the flaking is concentrated mostly on one end. The latter two techniques give the potential for longer flake-blades because they use a greater proportion of the full cobble length. Some cores are very reduced. The cores are found in the later two timebands but 193.2 was found in a deep path and refitted exactly to a socket in pepper-grey sands just off the summit of Sandstone 14. As mentioned below, the section was cleaned back to reveal a later v-shaped cut. These cores could be Late Early/Lower Palaeolithic (e.g. Pappu 1996) or later (see Sankhalia 1974).

Squared cores: There are ten cores that have high angled sides and a platform with some straight edges (65.1, 66.4, 94.3, 389.2, 440.2, 460.2, 504.1, 515.1, 632.1, 731.6 with some illustrated in Figure 8. 9). They are very different from the rounded cobble cores and have been termed squared cores here. In at least a few examples the difference may be due to an angular rather than rounded raw material nodule, e.g. 515.1. However, the group as a whole relies on a different reduction strategy and seems to work towards a preconceived shape. The core is approximately square across its platform and in its cross section. The platform is usually one fracture surface but advantage is also taken of natural planes or fractures, e.g. 389.2. The strategy seems to start with a split pebble, with at least one other side squared to this by means of either another flake removed down the cortical face from the split fracture, or one or more flakes removed across the core to establish the almost right angle planes. The resulting shape of 515.1 would allow straight-edged or flake blades to be removed. The illustrations (Figure 8. 9) show that some flakes did not penetrate the full length of the core, and 440.2 may have been abandoned because several removals ended in step fractures and did not penetrate the length of the core. Overall the group shows some preconception and a pattern of working which has been recognised by the techno-type strategy E (see Figure 8.2).

Split cobbles: Sixteen split cobbles were recovered, and three (399.5, 489.3 and 631.1) are illustrated in Figure 8.10. Whilst there is at least one clear striking point, this is the weakest category since some of the simple splits in this group are those which might be replicated in a natural striking event (see e.g. Jarvis 1988, White 1996) though experimental results show how rare this would be (see below). Flake features show that one striking event sometimes results in two removals (e.g. 489.3), or a very uneven fracture because of internal flaws in the raw material (e.g. 631.1) that may have made further flaking too unpredictable.

Shattered wedge core fragments: It is not surprising that there are 51 pieces that can be described as shattered wedge fragments, which range from pieces with a lot of cortex to angular fragments, as shown in Figure 8.10. It is not easy to split cobbles of tough quartzite and in some cases blows are repeated or the raw material is flawed so that the splitting results in irregular fragments. For example, 722.136 shows *three* distinct Hertzian cones, where the blow which finally resulted in the fragment has split the piece along the lines of previous hard blows to the core. This action resulted in the formation of subsurface cones that did not initially propagate these fractures to give flakes. This is indicative of unsophisticated knapping.

Irregular cores: Irregular cores are also numerous (30 collected, with some illustrated in Figure 8.10) and vary in shape from very angular pieces with a few flake scars, to pieces with more flake scars but a very irregular profile, e.g. 393.1. One, 526.1, shows two impact points for the same flaking event resulting from an irregular surface on the striking implement.

Bi-polar cores: Four pieces show evidence of bi-polar flaking (Figure 8.10). Artefact 651.2 is a fragment with two striking points immediately opposite one another, 653.6 is an angular cobble which has been split by one blow with a minor, shallow scar off the opposite left distal portion, and 639.2 is a small oval cobble with opposing flakes removed from one face. Bipolar flaking using a hard anvil is one way of tackling the problem of flaking rounded cobbles of tough rock.

Unidirectional and unifacial pebble cores: Sixteen unidirectional (and thus also unifacial) pebble cores were found (66.9, 216.1, 284.6, 322.1, 399.4, 411.3, 473.1, 491.2, 605.2, 608.1, 649.2, 725.1, 728.6, 730.12, 734.4, 741.3). The illustrations (Figure 8.11) show that some examples look very simple (e.g. 405.2 and 649.2, whilst others could be used to produce a fine flake (e.g. 216.1), but all have cortex all over the back. This category has been divided from the disc cores because the flake removals are both fewer and cover less of the circumference of the cobble.

Bidirectional and bifacial pebble cores: There were numerous (48) examples of pebble cores with flakes removed from two or more directions (18.1, 39.1, 65.5, 148.1, 196.1, 191.8, 261.2, 348.1, 363.1, 376.5, 389.19, 393.3, 395.4, 457.1, 475.8, 467.1, 473.2, 483.1, 500.8, 502.1, 507.2, 511.1, 516.1, 284.5, 516.2, 531.1, 628.1, 639.1, 719.1, 722.148, 722.154, 722.167, 722.174, 725.7, 728.4, 728.7, 730.9, 741.1, 741.4, 741.8, 741.12, 743.1, 744.1, 745.2, 746.1, 804.1, 830.2, 841.1). A smaller group within this type could be described as choppers with flake fractures on both faces creating an effective edge (21). If Leakey's (1971) categories of 'two-edged', 'side', 'pointed' and 'end' choppers are added, this figure could be increased to 35, and if 'unifacial' choppers on one edge (16 items, as described above) or more than one edge (11) are added the total becomes 62, or 10.2% of the overall assemblage. However, few of these potential 'choppers' have any kind of wear which would justify naming them as tools, though some of the HACTs (3) and SECTs (4) which do show wear as described above could also be termed bi-directional cores and 'choppers'. Some of the discoidal cores described below might also fall into the widest definitions of 'choppers', so it would be fair to say that the Pabbi Hills assemblage has many 'choppers'. However, the classification system used here has distinguished core tools via clear signs of use wear, and treated the remaining objects as cores. A selection of these bi-directional cores is shown in Figure 8.11. A few do not seem to have been intended as tools (e.g. the oddly-shaped 804.1, with a small and inconvenient area of bifacial flaking) because of their size and shape. This group includes ones with pointed or irregular bifacial edges, e.g. 730.9, and 516.1. Some have a split surface on one side (e.g. 389.19, 511.1, and 628.1) similar to some of the disc core types, but they are not described as disc cores because they are larger and heavier than the items placed in that category. 722.154 is made from a quartzite that tends to fracture in planes so that one side has quite complex scarring with one flake removed from the other side. The general shape of this tool suggests that it may have been used as a cleaver, though it was not fashioned finely enough to be included in that category, nor does it have the clear wear to be placed in the SECT category. This piece should also be compared with 65.5. One, 389.19, is made of a quartzite that tends to fracture in planes and has natural flaws, but the shape is effective. Sinuous edges are found on the bi-directional cores 65.5 and 393.3 (Figure 8.11), but also on some simple disc cores (Figure 8.12) and the HACTs 269.1 and 395.11 (Figure 8.7). One piece, 393.3, that has a sinuous bifacially flaked edge also shows clear evidence of wear prior to flaking. There are two facets of worn, slightly pecked, surface on the cortex as shown in the side view in Figure 8.11.

Simple disc cores: There are 23 simple disc cores (11.3, 66.2, 193.1, 278.7, 284.2, 305.1, 328.1, 403.2, 405.3, 405.4, 435.1, 442.2, 491.3, 523.2, 640.1, 703.1, 725.2, 728.1, 731.16, 817.1, 817.2, 835.3). There were probably several ways of flaking them. Five were made by fracturing a cobble and taking off flakes around the circumference using the cortex as the platform (e.g. 193.16, 248.2, 305.1 in Figure 8.12). Twelve used the fracture surface as platform instead (e.g. 731.16, 640.1), and five used both fracture and cortical surfaces as platform (e.g. 435.1) though not many flakes were removed (average numbers of flake scars are 6.6, 6.3, and 6.2 respectively). Artefact 305.1 shows the remnant of a large fracture on one surface. Some artefacts in this category have had so many flakes removed that it is now impossible to tell whether all these originated from a split cobble, but most seem to follow this pattern. If the sequence of flake removals was continued then the less discoidal cores (e.g. 645.1) would have resulted because the flakes form an increasingly abrupt angle to the cortex. It is difficult to date these kinds of disc cores as they could be Lower or Middle Palaeolithic, or much later.

Complex disc cores: There are 14 more complex disc cores (10.5, 91.2, 91.6, 93.1, 97.3, 507.1, 513.1, 645.1, 722.2, 722.100, 722.163, 722.171, 731.5, 744.4.). Four used the cortex as platform (e.g. 722.171 in Figure 8.12), four used the fracture surface (e.g. 507.1, 722.100), and six used both as platforms (e.g. 731.5, 722.2), with average numbers of scars of 9.3, 10.5 and 11.3 respectively. While some of these appear to be just disc cores there is one finely made core (722.2) that could be termed a prepared core. The prepared cores and perhaps others of this category could be seen as part of a Middle Palaeolithic prepared core tradition although disc cores are also known from later periods.

Polyhedral cores and polyhedrons: There are only three polyhedral cores or polyhedrons (91.3 (see Figure 8.12), 105.1 and 713.2). One is a multidirectional pebble core, and the other two are part of the disc core tradition with both fracture and cortex surfaces used as platforms. They average 8.6 flake scars. Artefact 91.3 is the blockiest, measuring 85 x 80 x 70mm. These artefacts could be Lower Palaeolithic but may also be from later periods.

Flake-blades: Fourteen elongated flakes that have comparatively straight and parallel edges, often with parallel dorsal ridges have been placed within this group (10.1, 27.1, 40.1, 91.8, 91.7, 92.3, 105.2, 105.4, 108.2, 196.6, 605.1, 605.4, 619.1, 730.5, 818.3). Some examples are illustrated in Figure 8. 13. There are two trends. Firstly, there is a tendency for the distal dorsal surface to retain a small amount of cortex (e.g. 105.2, 196.6 and 92.3) with often a small striking platform relative to the width of the flake. Secondly, some of the pieces with parallel dorsal scars show retouch or edge damage on one lateral edge (e.g. 605.1, 27.1, 108.2, 91.7, 105.2). For some, the scars form fairly regular trimmed edges, but others (e.g. 730.5) may be edge damage since they are much more irregular.

Rounded simple and complex flakes: There are 76 (12.5%) simple rounded flakes. Some are decortication flakes or show the scars of a few previous flake removals on their dorsal surface (Figure 8.14; 385.10, 376.11 and 722.1, 376.1, 224.2, 730.2 respectively). In contrast, there are only thirteen more complex rounded flakes showing scars from different directions on their dorsal surface (e.g. Figure 8.14; 218.1, 384.1, 515.3, 376.8). These flakes may have come from cobble or disc cores. Three flakes (Figure 8.14; 723.1, 749.1 and 749.2) have a cortex platform that extends down each of the margins, and in the case of 749.1 around almost the entire circumference. These are termed salami slices and believed to be part of a knapping strategy to prepare a platform. No date can be given for these objects.

Straight-edged flakes: There are 37 straight-edged flakes i.e. 6.1% of the overall assemblage. This category covers flakes that almost fit the flake blade category, but were slightly more irregular (see Figure 8.15; 722.173 and 835.1), or flakes whose straight edge seemed more fortuitous but which could make effective tools (e.g. Figure 8.13; 600.1 and 405.8), though the majority might just be knapping debris.

Angular and irregular flakes, split flakes and flakes with double bulbs: The assemblage has examples of longitudinally and laterally split flakes, but the split flakes in this category are both longitudinally split and most importantly, split from the same flaking event as resulted in the fracture. This can clearly be seen in the illustrated examples (e.g. Figure 8.15; 191.1, 405.5, 631.4, 672.2, and 731.9). There are also five flakes with double bulbs where the flaking event had two contact points (e.g. 387.9 and 728.15, illustrated in Figure 8.15). Eleven angular flakes (e.g. 489.2) and 109 irregular flakes were recovered varying considerably in shape and size (see Figure 8.15).

Two are best described as wedge-shaped shattered flakes. Flakes in these categories for the most part indicate knapping problems and strategies rather than a desirable end product. Given the difficulties of the raw material it is not surprising that together they form 30.8% of the whole assemblage.

Retouched flake tools: knives and scrapers: 383.9 (91 x 55 x 32 mm 146 gm; Figure 8.16): this side struck flake has been trimmed by one flake removal, and has fine bifacial retouch on a lateral edge forming a knife. The cortex forms a natural backing and the piece is the finest and largest retouched artefact in the assemblage. The retouch is finely executed. The piece would not be out of keeping in an assemblage from the Mousterian onwards. Two other artefacts with straight edges show distinctive retouch, and one round artefact could be a knife rather than a scraper. 319.1 (65 x 46 x 21 mm; 66 gm.; Figure 8.16) is a broad flake, retouched on the proximal end to remove cortex and create a robust working edge. 353.1 (66 x 55 x 22 mm.; 93 gm.; Figure 8.16) is a straight-edged flake with a small amount of retouch on its lateral edges. The scars are not invasive and might be due to use. Another piece, 446.7, has what appears to be use damage rather than retouch down the fracture surface edge (see Figure 8.13) so it is not included here. Finally, 346.21 (67 x 60 x 23 mm.; 78 gm; Figure 8.16) has been treated as a retouched flake because although it could be a scraper it could also be a round knife since the retouch is very fine and the edge is still only c 30°. This locality was recognised as recent and is associated with pottery and other redeposited material.

There are six scrapers in the overall assemblage. One of flint, 349.5, has been described above. The others are: 377.2 (38 x 31 x 20 mm; 23 gm.), this appears to be a small flake scraper but although no cortex remains, the fracture surfaces do not show clear flake features because the raw material is very coarse; 393.13 (31 x 36 x 13 mm; 16 gm.; Figure 8.16), a small flake with distal cortex, roughly retouched into a scraper; 511.2 (56 x 45 x 34 mm; 80 gm.; Figure 8.16), a crude scraper with a high edge angle. The small retouch has only slightly shaped one portion of the edge. Some cortex remains on the opposing end. This kind of scraper is sometimes attributed to middle palaeolithic periods; 722.165 (51 x 52 x 20 mm; 59 gm.; Figure 8.16), a small and squat flake with rough trimming on the distal end, with an extra flake removed laterally to create a chunky scraper. 731.7 (63 x 63 x 25 mm; 70 gm.; Figure 8.16) is an irregular decortication flake that has been trimmed and retouched to create a concave edge. Artefact 728.13 is quite similar in overall size and shape, but the scars on its concave edge are sporadic and less obviously intentional so it has not been included with the scrapers.

Technological types and knapping strategies

The flake techno-types distinguished the amount of cortex present over the entire surface and also noted whether the striking platform was cortex or fracture surface. Table 8.4 shows the results. Approximately 20% of the flakes were primary decortication flakes with ca. 70% being secondary flakes with some cortex, and ca. 31% showing > 25% cortex, i.e. most of their dorsal surface was cortex. Thus approximately half of the flakes could be said to represent the early stage of core reduction. Only ca. 9% of the flakes were the final stages, i.e. tertiary flakes with no cortex. This pattern need not be explained by a substantial portion of end products being discarded away from the survey area. The rounded cobbles, tough quartzite material, and simple knapping strategies are the prime reasons for the preponderance of cortex flakes and scarcity of tertiary flakes. These same reasons also account for over half the flakes (52.3%) having a cortex platform. Likewise, despite the fact that decortication flakes with a cortex platform might be seen as the most likely outcome of chance flaking events (see below for further discussion), the overall pattern of cortex platforms and large amounts of cortex is entirely consistent with the rest of the assemblage.

Table 8.5 shows how the core and core tool types map across the techno-types. It is evident that there is a large proportion of shattered or awkwardly fractured split cobbles compared to neatly fractured ones. This is to be expected since firstly, neatly fractured cobbles would presumably go on to be used as the basis for disc cores and squared cores rather than remaining to be recovered as such, and secondly shattered cores are the false starts of raw material selection, which was inevitable, given the variable nature of the quartzites. The presence of the A1 and A2s is thus not an indication of natural fracture but rather the reverse . Strategy B is the best-represented strategy for flake production, and interestingly both the proto-biface and the SECTs fall into this category, whereas the HACTs are part of the C disc core strategy. Most strategy C cores use the fracture surface as platform but relatively few are well developed; i.e. the C1B, C2B, C3B and C4B techno-types account for only 15 cores as opposed to 47 represented by the less well-developed strategy C cores. The recognition of a continuum between discoids, polyhedrons and sub-spheroids in the East African assemblages has been shown to be dependent on the propensity of the raw material to fracture. The scarcity of polyhedrons and preponderance of discoidal forms is entirely in keeping with this pattern since the quartzites are such tough raw materials. Strategies D and E that used squared or oval cores to produce longer flakes or flake blades are not prevalent (6.9% of the total cores), which fits with the low numbers of flake blades recovered.

Discussion: the overall nature and affinities of the Pabbi Hills lithic material

A major problem - as well as incentive - of working in an area such as the Pabbi Hills is that there are so few reference points against which results can be judged. The evidence for the Early Palaeolithic in Asia was summarised in chapter 1. As indicated, the Early Pleistocene record (i.e. >800 kya-old) is extremely meagre. Regarding S.W. Asia, there is an early flake and core assemblage from Dmanisi, Georgia, ca. 1.7 Mya, and a well-known sequence of Acheulean assemblages from 'Ubeidiya, Israel, ca. 1.4 – 1.0 Mya. In Central Asia, the earliest evidence is from Kuldara, Tadjikistan, which contained a small flake assemblage (90 pieces) dated to ca. 0.9 Mya (Ranov 1995). In Pakistan, the only archaeological evidence > 1 Mya-old is the small and controversial assemblage from Riwat (Rendell et al. 1987, 1989), with a *minimum* age of 1.9 Mya. In India, there is an unconfirmed, single date of 1.2 Mya for Isampur, which is surprising as the *expected* age of the assemblage was only 500-600 kya (Paddayya et al. 2000). There are no Early Pleistocene assemblages from S.E. Asia and southern China, if the highly dubious material from Longgupo is excluded (see chapter 1, p.11). It is thus difficult to know what an Early Pleistocene assemblage from a riverine environment would look like in southern Asia. Would it, for example, include an Acheulean component after 1.4 Mya, or would it remain based on simple flakes and cores?

Given the possibility that some of the Pabbi Hills assemblage might be residual material from the later Middle Pleistocene – i.e. after the anticline had formed and started to erode – note has to be taken of the much more extensive Middle Pleistocene evidence from southern Asia. In India, the Early Palaeolithic is Middle Pleistocene in age, and comprises both Acheulean and a flake-based "Soanian" industry, all of which is considerably younger than the Upper Siwalik deposits surveyed here (see e.g. Allchin 1995, Allchin and Allchin 1982). The earliest Acheulean dates from ca. 500-700 kya. The earliest is at Bori, recently dated to 670 kya (Mishra et al. 1995), which is similar to the dating of hand-axes at Dina and Jalalpur (Rendell and Dennell 1985) 30 miles north of the Pabbi Hills. In India, hand-axes have a wide distribution on post-Siwalik deposits in northern India (Chauhan 2004, Kumar and Rishi 1986, Mohapatra 1981, Mohapatra and Singh 1979), and elsewhere in the Indian subcontinent (e.g. Gaillard 1996, Gaillard et al 1983, 1985, Mishra 1992, 1995, Mishra et al 1982, 1988, Paddayya 1982, Pappu 2001, Pappu et al. 2003, Petraglia et al. 2003), and also Nepal (Corvinus 1998). Acheulean material has also been reported from Kazakhstan (Moloney et al. 2001), along with simpler pebble-core and flake assemblages, but none is well dated. In Uzbekistan, Sel'oungour (Islamov 1990) contains a "Meridional Acheulean" that is probably Middle Pleistocene and comprises hand-axes, choppers and a wide variety of flakes.

As is well known, there is also a simple, non-hand-axe, flake and core component of the Indian Early Palaeolithic that is known as Soanian, after the Soan Valley, Pakistan, where it was first identified by de Terra and Patterson (1939). As pointed out by Rendell et al. (1989), the assemblages from the Soan Valley are largely undated, lack integrity or clear definition, and thus the Soanian lacks clear meaning as a category of early Palaeolithic lithic assemblage. The term has also been criticised by others: Graziosi (1964: 13-15) was critical of the Soan pebble tools typology when he visited parts of the Soan valley. Although illustrations show some comparable material to the Pabbi Hills assemblage, the finds were from a variety of contexts and there is no clear idea of their relative or absolute chronology. Stiles (1978) also visited the Soan and later offered a very stimulating and critical appraisal of many of the terms used by Terra and Patterson (1939). Additionally, Gaillard (1996) has suggested that in India the Soanian and Acheulian may be contemporaneous and reflect different choices in tool reduction rather than be sequential tool traditions, in the same way that the differences between the Clactonian and Acheulean are now interpreted in southern England (see e.g. Ashton et al. 1992). Further afield, in Tadjikistan, the "loess Palaeolithic", based on very simple tools made from stream pebbles, continues throughout the Middle Pleistocene (Ranov 1995), and comparisons have often been sought with the "Soanian" of India and Pakistan. Even further away, the Chinese early palaeolithic material is problematic due to the way in which layers are combined, and numerous biases over what is kept and how it is reported (see Keates 2000).

The pattern that might exist from these disparate and widely spaced finds from the Middle Pleistocene is the prevalence of discoids, polyhedrons, sub-spheroids and spheroids as part of a continuum and one that spans a wide range of time and space. The key to understanding the assemblages in this part of the world may be to downgrade hand-axes in importance as a relatively late and rare phenomenon, and instead focus on the discoid/core tradition. On this basis, the Early Palaeolithic of much of the Middle Pleistocene in India, Pakistan and Central Asia would comprise a large variety of cores – including discoids, polyhedrons, sub-spheroids and spheroids – and flakes, depending on situation, access to raw material, season etc. and might also, when appropriate, include hand-axes.

With this background in mind, we can consider two aspects of the Pabbi Hills assemblage.

1) The density of material

The most obvious point about the lithic assemblage is that the Pabbi Hills lack major concentrations of lithic material, despite intensive survey of large areas over several field seasons. This is in marked contrast to the best-known early Palaeolithic occurrences in East Africa and S.W. Asia. For example, the 2.5 Mya-old site of Gona, Ethiopia, contained an astounding 2,970 artefacts, of which 1,114 were in situ in a thin, 10 cm layer (Semaw et al. 1997). Similar quantities were found at the 2.34 Mya-old site of Lokalalei 2C (Roche et al. 1999), which also contained extensive (ca. 60) sets of conjoining pieces. Large concentrations are also known from some parts of 'Ubeidiya, and also Dmanisi. The absence of such evidence from the Pabbi Hills could thus be seen as evidence that hominids were absent.

The low density of lithic material from the Pabbi Hills is, however, entirely consistent, however, with the very low density of material found on late Pliocene exposures of the former lake margin at Hata, Ethiopia (Heinzelin et al. 1999). Here, the explanation offered is that hominids were operating in areas where stone was very scarce, in contrast with areas such as Gona, where stone was naturally very abundant. It is worth quoting the discussion of this evidence at some length:

"*Nearly contemporary deposits at Gona, only 96 km to the north, produced abundant surface and in situ 2.6-Ma Oldowan artifacts. In contrast, surveys and excavations of the Hata beds have so far failed to reveal concentrations of stone artifacts. Rare, isolated, widely scattered cores and flakes of Mode I technology appearing to have eroded from the Hata beds have been encountered during our surveys. Most of these surface occurrences are single pieces. Where excavations have been undertaken, no further artifacts have been found...... At the nearby Gona site, abundant Oldowan tools were made and discarded immediately adjacent to cobble conglomerates that offered excellent, easily accessible raw materials for stone-tool manufacture. It has been suggested that the surprisingly advanced character of this earliest Oldowan technology was conditioned by the ease of access to appropriate fine-grained raw materials at Gona . Along the Karari escarpment at Koobi Fora , the basin margin at Fejej , and the lake margin at Olduvai Gorge , hominids also had easy access to nearby outcrops of raw material. In contrast, the diminutive nature of the Oldowan assemblages in the lower Omo [made on tiny quartz pebbles] was apparently conditioned by a lack of available large clasts. The situation on the Hata lake margin was even more difficult for early toolmakers. Here, raw materials were not readily available because of the absence of streams capable of carrying even pebbles. There were no nearby basalt outcrops. The absence of locally available raw material on the flat featureless Hata lake margin may explain the absence of lithic artifact concentrations...... The paucity of evidence for lithic artifact abandonment at these sites suggests that these early hominids may have been curating their tools (cores and flakes) with foresight for subsequent use...... The Bouri discoveries show that the earliest Pliocene archaeological assemblages and their landscape patterning are strongly conditioned by the availability of raw material*" (Heinzelin et al. 1999:628-629).

Likewise, Harris and Cepaldo (1993) adopted a landscape approach to the low-density archaeological record of the Olduvai palaeo-lake basin, and suggest that the earliest phase may be characterised by much more dispersed and less sophisticated, perhaps even seasonal, tool use. The suggested setting for the earlier patterning of low stone tool density is interestingly, riverine fringes with greater cover. The artefacts, as here, are crude choppers and cores plus unmodified flakes.

These two studies, particularly that by Heinzelin et al. (1999) allow one to be much more confident about the Pabbi Hills assemblage, especially as it too would have lacked easy access to large amounts of stone. As in the study of Hata, isolated flakes and cores were the main items recovered.

2) What does an Asian Acheulean and pre-Acheulean look like?

As Roe (1994:309) has pointed out there are glimpses of a tool culture in Europe that is not Acheulean but appears to be perhaps developed Oldowan. Carbonell et al. (1999) take this suggestion further by arguing that the earliest technologies in Europe and Africa were "Mode 1", or simple flake and core assemblages, and successive migrations of hominins later introduced "Mode 2", or Acheulean assemblages. On this model, the non-Acheulean assemblage from Dmanisi represents one colonisation event, and the Acheulean assemblages at 'Ubeidiya (currently the oldest of their kind outside Africa) represent a later series. Given the complete absence of hand-axes, the Pabbi Hills assemblage, whatever it might be, it cannot be explained as Acheulean, which suggests that the period from ca. 600 kya onwards (i.e. post-Siwalik) is not represented to any great extent. (Additionally, none of the fossil material is attributable to the Middle Pleistocene; see chapter 10). Can the lithic assemblage therefore be regarded as pre- Acheulean?

The vast majority of the assemblage represents simple pebble core and flake strategies. A large proportion of the pieces can be regarded as the outcomes of unsuccessful attempts to flake quartzite cobbles, and more successful ones that resulted in multiple flaking. Most of these are simple, but they include straight-edged core tools (SECTs) and artefacts which we have called HACTs, and which have close parallels with the spheroids, polyhedrons and discoids of other Early Palaeolithic assemblages. These are amongst the most numerous identified tool types. Thus in the assemblage as a whole the potentially earlier types are dominant. On typological grounds, by far the greater part of the assemblage is consistent with what would be expected of an Early Paleolithic, non-Acheulean assemblage based on the reduction of quartzite cobbles. Some of the material is consistent with Early Palaeolithic assemblages elsewhere. For example, spheroids, subspheroids and polyhedrons are considered by Willoughby (1987), who concludes that spheroids are a distinct class, but that they also represent one end of a continuum. Under 10% have wear visible to the naked eye as contusions on the surface. Some functions can be eliminated, but it is more difficult to state their specific purpose. Certainly they co-vary with pitted anvils and could be used as hammers in a two-piece kit. There is now a good range of experiments covering the manufacture (techniques and a range of raw materials) and use of the classic early palaeolithic tool types (e.g. Jones 1994, Schick and Toth 1993, Willoughby 1987), suggesting that the raw material may influence the process of manufacture and the way in which the tool will change shape with use overtime. The classic example is the increasing prevalence of 'spheroids' in quartz, a brittle rock, between Olduvai Gorge Beds I and II (Schick and Toth 1993: 129-132). The raw material is thus an important factor. The Pabbi Hills assemblage is dominated by hard quartzites so large numbers of spheroids might not be expected if these objects follow the East African pattern. If Willoughby's continuum is mapped onto the tools which are here termed 'high action core tools' the Pabbi Hills discoids and polyhedrons shapes, some of which have wear visible to the naked eye, there is a similarity in what is described for early African and Near Eastern sources. Material from the site of Sel'oungour, Uzbekistan (kindly shown me by Prof. Islamov) comprised simple flakes and cores originating from a somewhat blocky and tabular raw material; the assemblage had some very large pieces and little evidence of retouch or sophistication. The difference in the character of the starting shapes of the raw material could be expected to result in differently shaped pieces.

We are not claiming that all the lithic material from the Pabbi Hills is Early Palaeolithic; indeed, some of it is demonstrably not. The most obvious is a small but nonetheless interesting component that is probably Neolithic or Chalcolithic.

3) Recent material

In the neolithic and chalcolithic periods the Punjab was, as now, a fertile, relatively well-watered environment with flat lands alongside the river well-suited to agriculture. Modern farmers can grow two crops a year. It could be expected that the flat lands around the Pabbi Hills contained Neolithic settlements but the anticline would not have been such rich agricultural farmland because the monsoon rains would have washed soils on slopes away. The anticline would have been more attractive as an adjacent pastoral area in much the same way as modern flocks from the surrounding villages are taken into the Pabbi Hills to graze today. This kind of pastoral use may have left few stone artefacts since a personal day toolkit need not contain many stone tools, perhaps just an axe and a knife/cutting flake, and these might only be discarded if they broke, became exhausted, or were lost. Nomadic neolithic camps might have had a greater range of objects e.g. scrapers and more artefacts might have been discarded together. There might also have been a range of pottery objects and greater quantities of these, though a nomadic lifestyle may have used containers that were not so susceptible to breakage. All of this is supposition but is not unreasonable and is an essential part of looking for patterns in the lithic assemblage.

The only significant accumulations that include pottery are on the fringes of the anticline and the larger sites were either outside the survey areas or clearly of later origin, e.g. site 652 which showed pottery and iron, and the recognisably recent deposit of 346 which contained a polished stone axe fragment and some pottery. Microliths could be expected from the Mesolithic and possibly Chalcolithic periods, plus a variety of pottery styles from different periods, small 'celts' etc (Agrawal 1982). Wheeler (1950 and 1959) gives classic accounts for the time, of the cultural sequences and pottery styles and (1950:16-17) gives an overview of some of the pottery styles, mentioning that the northern regions seem to have red backgrounds to their painted wares. Sankalia (1974) also illustrates regional pottery sequences. None of the occasional pot sherds found in the interior of the Pabbi Hills were archaeologically datable to a specific prehistoric period, and contemporary local markets often displayed forms and decorations in keeping. The occasional pot sherds were possibly due to the contemporary and very recent usage of the Hills by nomads and shepherds. Thus the dating rests on the lithic pieces themselves, which is notoriously difficult. However, on typological grounds some of the material is consistent with later periods: there is the side struck knife with scalar retouch which could be seen as Neolithic or even mousterian; the micro-cores are consistent with the final Palaeolithic or mesolithic; the stone axes as neolithic or possibly chalcolithic; rubbing stone fragments could be any period from the Neolithic onwards, while the ringstones, if seen as maceheads or weights, could be mesolithic, but could also be much later as some ringstones occur in Iron Age contexts.

Although there are a few pieces which could be regarded as prepared cores amongst the disc core tradition, there are no unequivocal middle palaeolithic/levallois prepared cores. Similarly, the flake blades could be palaeolithic but also possibly Neolithic, but there is no evidence of the type of blade assemblage found at site 55, Riwat, 90 km to the North and dated to ca. 45 kya (Rendell and Dennell 1987, Dennell et al. 1991). Although scrapers can be difficult to assign to a specific period there are relatively few in the assemblage as a whole. Furthermore there is a general paucity of retouched tools (2.8%) in the assemblage; Middle Palaeolithic, Later Stone Age and Neolithic assemblages from the region all show good examples of retouched tools so if these periods were well represented, the percentage of retouched tools should probably be higher. Overall, therefore, if the assemblage was residual Middle and Upper Palaeolithic, it is curiously deficient in its most characteristic types.

In a small number of cases, a post-Siwalik context is likely, or cannot be excluded on geological grounds. One example is the flake blade core from Sandstone 14 that fitted into its socket in a post-Siwalik channel fill in Sandstone 14. Another might be 278/284, where post-Siwalik deposits are nearby. In other cases, there are no obvious nearby later sources from which stone material might have come. A clear example is 269, where the most obvious source was the overlying sand deposit that is dated to 1.9 – 2.2 Mya. As a working hypothesis, we would therefore suggest that the bulk of the flaked stone from the Pabbi Hills is derived from the underlying Upper Siwalik deposits.

4) The overall pattern in relation to the geological strata on which they were found.

If any part of the assemblage is coming from the underlying geological strata then some patterning associated with those strata might be evident. Tables 8.6 and 8.7 show the assemblage split into three different groups based on the age of the geological strata on which the pieces were recovered. The numbers of stone artefacts per group became too small if finer age bands were used (but see below). The irregular pieces dominate all groups, but some types are represented by too few pieces in the assemblage overall for their placement in any group to be significant. However, there seems to be a decrease through time in the unidirectional cores, though the bi-directional ones are more evenly spread across the groups. The simple disc cores are more prevalent in the earlier groups while the more complex ones show the converse. The rounded simple and complex flakes decrease with time. Flake blade cores and flakes both seem to increase in the younger groups. The obviously much later types such as polished and ground stone objects and micro-cores are spread throughout the groups and it is tempting to see this as a pattern of remnants spread across the Hills, but the HACTs also show this pattern. Similar issues arise for the techno-types (Table 8.7). The disc core strategies are complex, but if C1A and C2A are taken to represent very simple disc core forms and all others are seen as more complex, an interesting pattern emerges. In the youngest age band (1), simple disc cores form 11.9% of the assemblages, and complex ones, 21.6% (i.e. double). In the earliest age band (3), the proportions are reversed, with 16.2% simple disc cores and 7.7% complex, whilst in the middle time band (2), the figures are 9.7% and 5.6% respectively. Strategy D (based on square platforms) looks to decrease with time while Strategy E (oval platforms) increases but both of these are represented by very few pieces. The flake techno-types show a clearer overall trend, in that cortex platform flakes with higher percentages of cortex dominate in the earlier groups, while fracture platforms with less cortex are prevalent in the younger ones. A finer breakdown of the percentages of cortex for flakes and cores (Tables 8.8 and 8.9) show that flakes with 10% or less cortex seem to be a feature of the younger age band, despite all groups showing a good percentage of 50% or more cortex, i.e. decortication flakes. The situation for cores is more complex (Table 8.9) probably because of the high proportion of irregular pieces are affecting the overall pattern, but excluding them leaves too few pieces for analysis.

Overall there does seem to be some patterning according to the age of the geological strata on which the tools were found. Taphonomic processes, use of the Pabbi Hills in different periods and discard rates of different tools could affect this pattern (see below). However, if the assemblage was all due to residual material from later periods should there be any patterning at all?

SECTION C: Taphonomic investigations of the distribution, movement and flaking of stone in the Pabbi Hills

Over the course of the project the natural occurrence of stone and the taphomonic processes of erosion and alteration of stone in the landscape were observed and in some cases further investigated by experiments. Some promising areas of the survey were also selected for limited further work, such as cleaning back sections or small excavations. All these have assisted in the interpretation of the stone assemblage, and the main results are documented here.

The occurrence of natural stone in the Pabbi Hills

Keller et al. (1979) divide the Pabbi Hills sequence into two units. They characterised the older and more extensive series as soft sediments, in contrast with the later units that are reported as 'conglomeritic'. There is one major conglomerate deposit in the Pabbi within the area surveyed so far. This is North of the G.T. road, between Kanarawala and Bawali Kas at the top of Sandstone 14. There are more gravel sequences and even clast-supported strata in the younger sequence visible along the drivable portion of Choawala Kas on the other side of the GT road, but these were north of and younger than the area of the survey. However, the new section through the Pabbi Hills anticline created by the widening of the G.T. road since 1990 clearly shows that 'conglomeritic' may be a misleading term. The later sandstone units have more siltstone and mudstones but they are not themselves conglomerates in the true sense of the word. These recent investigations rule out the prospect of an overlying conglomerate (e.g. a "Boulder

Conglomerate") accounting for the presence of 'residual' quartz and quartzite clasts within the sequence. Such a scenario would also have had to suppose that only flaked pieces and a few other clasts had survived, while the vast majority of the clasts (presumably, mostly unflaked) and their matrix had gone. As shown later, our studies of modern erosion rates in the Pabbi Hills make it extremely unlikely that any such material would have survived in the area. Thus the distribution of naturally occurring stone makes an interesting study for two reasons. The fluvial system may have acted both as a lithic resource and as a collecting point for artefacts as it was laid down and, once uplifted and exposed, the natural stone may have been a useful raw material for post-Siwalik populations. Whilst the rate of flow for a river laying down large clasts is considerable, the river would have fluctuated *considerably* in flow according to the monsoon and seasons, exposing beds of larger lithic material in drier seasons in much the same way as can be seen in the modern Soan valley bottom, where it is possible to walk over seasonally exposed pebble beds (see below). The section drawn at Riwat (Rendell et al. 1989, Fig. 7.16) shows how such stringers can occur in a mostly sandy matrix. There are also places within the Pabbi Hills where such pebble stringers are clearly part of the sequence.

An examination of the size range and lithology of natural stone is an important part of understanding the lithic artefact assemblage from the same area. For all these reasons, a number of detailed studies were conducted into the natural lithic distribution.

As stated above, the only distinctive conglomerate in the Pabbi Hills overlies a section of Sandstone 14 between Kanarawala Kas and Bawali Kas. In Bawali Kas the large section (see page 43) shows that sandstone 14 has two storeys, the basal one of which had multiple stringers of extra-basinal clasts of quartzites. At this point the clast size was not suitable for stone tool making though it would encompass the size range of flakes. The most dramatic stone cascades down the slopes are 50-100m from the watershed between the Kanarawala and Bawali drainage systems. Dr. Imran Khan (our field geologist in 1999) thought that since the blocks of sandstone on top of this sequence contain quartzites, at least one source of the quartzite clasts in this area was Sandstone 14. The blocks of capping sandstone show cobbles up to 13 cm in length (Figure 8. 17). The clasts are a mixture of different colours and textures of quartzites and are of a size suitable for making stone tools. Below this are sands with pebble stringers that appear to be part of the sequence (Figure 8. 18). However, in a laterally restricted area, there appears to be a dense lens of larger pebbles forming a slight shoulder to the scarp profile. Dr. Khan was sceptical about the date of this laterally restricted conglomerate, though Rendell had not raised any doubts about its integrity within the sequence. A small (ca. 50 cm) area of the latter was cut back to check its integrity (Figure 8. 19). Here, Khan thought the layer was thinning out, and because the sand around the clasts was loose, coarse and fining upwards, it could be a later deposit filling a cutback into the true Siwalik deposit when the anticline was just beginning to form. There is certainly some Post-Siwalik material at the very top of the section that is much more jumbled and has a slightly darker matrix tone with a greater clay silt component in with the sand. About 50 metres further west the ridge path has cut into the sands just below the summit and clearly shows pebble stringers in the sands appearing to be Siwalik in date, and of a clast size consistent with that of the cores and flakes in the assemblage.

Despite the difficulties of establishing exactly where any tools come from, there are no less than eight sources of pebbles in this landscape. These are: 1) pebbles deflated on the dip slope, 2) pebbles in the consolidated capping sandstone, 3) the soft unconsolidated sands with 4) pebble stringers mostly in the basal section, 5) reworked material at the top of the section where the capping sandstone has gone, 6) a possible post-Siwalik fill, 7) pebbles cascading down the slopes which could have come from any of the above, and 8) the slope wash or fill at the base of the section. These uncertainties are compounded by the refitting of an artefact to a socket near the top of the sequence in the soft sands. When the section was cleaned back, a neat v-shaped cut showed that in at least one place these soft sands have later cuts in-filled by deposits that are at first sight similar to the Siwalik layers. If there are later conglomeritic deposits, these are very restricted since they do not appear on any of the other capping sandstones (12 and 16) that have good exposures nearby. This is very different from the shoal of pebbles seen as a post-Siwalik deposit in Kohar kas (at the far western end of the Pabbi Hills) that went at an obviously different angle that respected none of the sandstone ridges. Thus this area of Sandstone 14 remains a conglomerate in the sense that it has many large pebbles, at least some of which appear to be integral to the Siwalik sequence. It is one of the few places in the Pabbi Hills where a search can be made for in situ artefacts in conglomerate exposures, although the precise deposit would need to be carefully recorded in view of the problems outlined above. It thus remains a mixed set of potential sources for the stone tools found on the scarp. If the clast material is part of a later channel deposit, then it could have acted as a source of stone once exposed; if it is integral to the Siwalik sequence, the clast size suggests rapid flow and thus the pebbles may have been underwater rather than available for tool-making, though the rate of flow in dry periods may have exposed the lithic beds in much the same way as currently happens in the modern Soan river.

Because of the possibility that some of the Pabbi Hills assemblage could have resulted from fluvial fracturing, a modern pebble exposure in a river bed was investigated.

Pebble beds in modern rivers

A pebble bed laid down by the modern Soan river near Islamabad was examined in detail as a point of comparison. The pebbles were of a size suitable for tool making and were exposed in a 30-40 m strip for several hundred metres alongside the river, which in April still had ripples on its surface indicating a fast current. An area ca. 10m long and 9-11m wide was measured for the detailed survey of 100 square metres. Counts of the clasts over 5 cm in diameter in a 10% sample of 1 m squares were used to estimate the total number of clasts surveyed as 10,200. Most of the pebbles were well rounded though occasionally there were fracture surfaces in varying states of being worn back into a rounded profile. Each 1 m square was examined for fractured clasts. Only 91 pieces showed some form of fracture or facet, i.e. 0.89% out of 10200. Of these 91, most showed no flake features and would have scored 1 or 2 on our credibility system for evaluating fractured stone. Only 8 pieces, 0.08%, showed any flake features that could be viewed as perhaps the result of intentional flaking. These were mostly either split cobbles or cobbles with one flake removed, with the newer surface in various stages of rounding. Interestingly, but not surprising in the Soan valley in which de Terra and Patterson (1939) did their original fieldwork, two pieces were assessed as very rolled but real artefacts. One was a blade with very rounded edges but with a worn patch discernible indicating the *éraillure* scar and with worn rings showing the ripple marks on its ventral surface. Overall, therefore, in these 100 square metres of pebble bed only 0.06% of fractures could potentially be *confused* with real artefacts. This fieldwork suggests that most of the fracture patterns in the Pabbi Hills assemblage are unlikely to be caused by fluvial flaking as the stones are much more dispersed, and even where there are dense accumulations of stone such as the conglomerate capping sandstone 14, there would only be in the order of 6-8 'artefacts' per 10,000 clasts.

These results are consistent with those we obtained when investigating the lower conglomerate horizon at Riwat, dated to a minimum age of 1.9 Mya (Rendell et al. 1989). In the conglomerate section in which two intentionally flaked pieces were found 40m apart, 1264 other clasts were recorded. Only one was fractured, and this showed no flake features indicative of hominin modification. Similarly low frequencies were observed along the other 4 km of this conglomerate horizon (see Rendell et al. 1989:121). We thus conclude that fluvial action is most unlikely to account for the Pabbi Hills assemblage.

Walks along the banks of the modern river Jhelum showed that though no conglomerate exposures were observed, there was stone suitable for tool-making. The cobbles existed as odd stringers in sandbanks and overbanks that had been cut back by the river. From these field observations of a modern, much reduced and modified, river, it seems possible that there would have been sources of stone available for tool making at the time the Pabbi Hills deposits were being laid down, though stones may have been available only as rare stringers, or only in dry periods of a year. It is assumed that the monsoon rains and the snow melts might both have affected the ancient river that laid down the Pabbi Hills, creating a complex pattern throughout the year, and from year to year.

Modern stone sources within the Pabbi Hills

Within the Pabbi Hills there are recent construction projects that have brought lithic material into the area. The railway, the GT road, pylon bases, and most recently, the gas pipeline laid within the last few years, have all used hard core. Much of this material is excavated from river gravels derived from the Karakorum, i.e. exactly the same kind of lithology as the clasts incorporated within the ancient river system that laid down the Pabbi sequence. Having seen in 1987 gangs of workmen knapping these pebbles into smaller hardcore to create the G.T. flyover of the widened GT road at the southern edge of the Pabbi, no material was collected anywhere near or down slope from these construction projects. The western watershed track shows concentrations of pebbles in some areas. These have similarly been treated with extreme scepticism. In consequence, all the possible modern imported lithic material has been excluded from both the study of the natural stone accumulations and the artefact assemblage.

Survey data

In keeping with many other surveys, there was limited potential for more detailed recording of small areas within the general survey area, and excavation of promising areas. A handful of examples identify some of the key learning points. **Occurrence 346** was discovered early in 1987. It contained stone and bone and as a relatively rich concentration, was carefully mapped by the field team. A return visit led by Dennell the next day showed that the locality was a recent colluvial assemblage. Elsewhere there were intriguing stone finds, not because of their number but because of the location of the stone finds and the possible choices for where the stone could be originating. A small number were the subject of further study. Their location and general features are described in the relevant parts of chapters 5 –7.

218 area: this was on a small spur very near the top of the deposits as shown in Figure 8. 20. There was a sharp drop of 20 m on one side. The stones, a mixture of artefacts and some natural stones were collected early on in the 1987 season and the position of the stones noted (for 218. 1, see Figure 8.14). Later in the season a small team returned to clean down a small section of hillside immediately above the highest stone find (left of arrow in Figure 8. 20). This showed that there were no residual deposits higher than 218.1. The layers were entirely in keeping with the tilt of the sequence and a layer of ostrocods, indicating shallow water, was identified. Though the section and small step trench were only 1 m by 2m a better understanding of the strata in the vicinity of stone finds was obtained. Although this investigation failed to find in situ lithic material, it showed that there were no later deposits from which the flaked stone could have been derived.

Locality 387 and the Pelvis valley area of Kotha Kas: Occurrence 387 in the East side of 'Pelvis Valley' (see chapter 6) was situated along and at the base of a steep hillside strewn with sandstone blocks and turned into minute steps by myriad animal paths. This kind of hillside was subsequently characterised as 'block-strewn stepped' (see erosion experiments, below). Stone artefacts were found mostly at the base of the slope but a bovid/cervid pelvis was excavated from *in situ* deposits just above the base so trial pits were established there. Excavations in 1989 in the vicinity of the pelvis find and slightly down slope and south of this scraped back to primary tilted geological strata including mud balls in the sands and a gritty layer just below the pelvis strata with clays incorporating shells below this. However the base of the boulder-strewn slope proved to have a mixture of different kinds of deposits and was obviously re-deposited from the scarp above. Whilst the pelvis was in its geological context within the hills, the stone could have come from a variety of deposits which were perhaps originally in the flat area on top of the scarp and in the process of working their way down the 'block-strewn stepped' deposits. This tallies with the kinds of finds, which included a flake from a rubbing stone (389.14 Figure 8. 4), a flake from a polished stone axe (387.25), a retouched side-struck knife (383.9, Figure 8.16), flint micro-core (383.10, Figure 8.3) simple complex and split and double-bulb flakes, and a HACT and squared core. However the excavation showed more natural (unflaked) stones and a further detailed survey of the hillside identified 14 more pieces (maximum size 11 x 8 x 6.5 cm; average 7 x 5.6 x 4.2 cm). Its origin remains unclear although the stones were found in loose sands, on the hillside slopes and at the base of the slopes.

Locality 722 and the Baroth area: The promising area near Baroth was identified in 1989 and intensively surveyed over eight weeks in 1990. The main focus of the 1990 field season was the excavation of locality 642, and the investigation of its environs. The field survey concentrated on finding examples of flaked stone rather than vertebrate fossil remains. Although considerably more examples of flaked stone were found in 1990, the amount (and quality) of fossil found that year was derisory. An account of this fieldwork has been published by Dennell et al. (1993), and also by Ashfaque, the most senior of our Pakistani colleagues from the archaeological service, whose account (Ashfaque and ul-Haq 1991) was derived from the field records we gave him at the end of the field season.

The central part of the Pabbi Hills, north and west of Baroth, is extremely dissected, particularly where the overlying sandstones have been eroded, or had never formed. Erosion around Baroth is much less dramatic, and sandstone 12 is often exposed as a low (25 - 50m) scarp, or as a series of isolated inselbergs. The major archaeological discovery in this area was a gritstone/conglomerate horizon in a complex of sands and silts that was capped by sandstone 12. This conglomerate could be traced over a large area (see Figure 8. 21) as a cemented sandstone, or a conglomerate, between 1 and 2m thick. It is predominantly single-storied, but in the central part of the survey area, was sometimes traced as a multi-storied series of thin sandstones with pebble inclusions. It is often laterally discontinuous, and grades into softer sands and silts. Because it is more resistant to erosion than adjacent softer deposits, much of it is preserved as a series of low ridges. Orientation of ripple marks indicates that the flow was roughly NE - SW, along the main axis of the anticline, and not transverse to it, like the modern and post-Siwalik drainage. The archaeological importance of this conglomerate is that it contained quartzite stones large enough to be used in making tools and because its clast size ranges are comparable with artefacts.

The survey in 1990 covered most of the area shown in Fig. 8. 24, and paid particular attention to exposures of this conglomerate. Good numbers of stone tools were found, and almost all were found on the erosional surfaces of this conglomerate. This is not to say, of course, that they are necessarily of the same age. As we pointed out in our investigations of the alleged "terraces" in the Soan Valley (Rendell et al, 1989), stone tools could easily have been made and discarded at conglomerate exposures long after the conglomerate was deposited. Consequently, it is essential to know not only the age of the conglomerate, but when it was subsequently re-exposed. The Pabbi Hills were uplifted and folded to form an anticline in the last 400,000 years. Therefore, the conglomerate within sandstone 12 has been exposed within at most, the last 400,000 years. The exposure of this conglomerate is likely to be considerably more recent, because it could not have been exposed before the top of SST 12, as well as all the deposits within sandstones 14 and 16, had been eroded. There are circumstantial reasons why the stone tools collected in 1990 were made when the conglomerate was laid down, and not after it was re-exposed over 800,000 years later. First, the tool types found are generally similar to those found elsewhere in the Pabbi Hills on erosional surfaces of comparable age. Secondly, there was no evidence of debitage and in-situ stone-knapping, as is evidenced in, for example, the Soan Valley, where late Upper Pleistocene (<50,000 years old) stone-working at much older conglomerate exposures is often evidenced by conjoining pieces and debitage (see Rendell et al., 1989).

There is also some stratigraphic evidence that suggests that stone artefacts may be integral to the conglomerate horizon within sandstone 12. The first came from a locality known as site 722. Here, several extremely fresh quartzite flakes and cores were found on the surface of silty sands that graded laterally into this conglomerate horizon. Site 722 itself consisted of 51 pieces of flaked quartzite extending upslope from the base of these sands and at the same level as the conglomerate. These included a micro-core, a HACT and other simple cobble and disc cores. The latter show some very neat examples illustrated in Figure 8.12, one of which (722.2) could be regarded as a prepared core. A small (2 by 2m) sounding was dug into this sand unit, and a small disc core (722.100, see Figure 8.12) was found ca. 15 cm below the surface in the sands that formed the matrix within the trench, but because these sands were fairly soft it can be argued that the artefact could have sunk into the sands rather than been a part of them. The soft deposits made excavation physically difficult to control and offered no security of context for any potential finds, so excavations were curtailed. No further specimens of flaked stone were found in shallow step trenches dug further upslope adjacent to the main scatter of flaked stone. There were some pot fragments in this area although these could have been very recent in origin. This locality was revisited briefly in 1999 with Imran Khan, whose assessment was that the section comprising 722 may lie on a fault, and the unit on/in which the flaked stone was found may be equivalent to the uppermost unit of Sandstone 12, rather than one lower in the sequence. Unfortunately, there was not time to investigate this possibility further. The artefacts at 722 are a distinct concentration in the landscape but do contain at least one later type.

In the entire Baroth area, 722 stands out as a distinct concentration but part of a wider area with more stone artefacts than most areas of the Pabbi. Some stone, including a cobble was found on top of Sandstone 12, and there is a spread of excellent pieces from the brown sandy silts and down slope. Three kinds of deposits with stones were observed at 722 and the inselberg of Sandstone 12 immediately above it. The sandstone at the top contained some small stringers c 5 cm and subangular. Just towards the base of this unit there were larger clasts of quartztites and mudballs 15 cm or more in size and more rounded. This two-story effect was visible only on the western exposure of the inseln, but not on the eastern part. Lower down there is a consolidated conglomerate which forms a ledge in places. This contains quartzites of a variety of sizes from 5 - 15 cm that are angular rather than very rounded. There are occasional large pebbles high up on the western slope that could have eroded from the lower story of the sandstone, but most of the pebbles seem to be in the lower consolidated conglomerate.

At another locality, 730, seven artefacts were collected with two pieces of flaked stone found in a section of a small channel that seemed part of a gritstone horizon forming part of this conglomerate. Here, the sandstone is thin (<1m) but multi-storied, and appears to indicate a shallow, braided stream. Finally, at location 732, two artefacts were found at the base of freshly-eroding sections of the same conglomerate, although unfortunately, none was found in section. Further investigations of both localities was planned for 1991, but sadly cancelled because of the Gulf War.

Some of the stone artefacts found in the Baroth area are flakes struck off simple pebble cores and bear clear bulbs of percussion ventrally, with flake scars also on the dorsal surface (e.g. 722.1). The striking platforms for these simple flakes are sometimes a previous flake scar (e.g. 722.164), but many are struck on the cortex surface (e.g. 730.2). Most flakes still have some cortex but occasionally their surface is composed entirely of fracture surfaces (e.g. 722.10). The flakes are thus simple and in keeping with the flaking observed on the cobble cores. For the most part, these cores are rounded cobbles that have been roughly bifacially flaked along part of their circumference. This can create a variety of shapes in edge and plan view but this variation masks the similarity of the technique. Sometimes a slightly sinuous edge profile is obtained, e.g. 722.146 (Figure 8.7). This tool also shows evidence of use-damage in distinct areas of this sinuous edge. Other techniques for the production of cores include splitting the rounded cobbles and then taking flakes off the rounded cortex surface, using the fractured surface as the striking platform (e.g. 722.2 and 731.16). It is clear that despite the more sophisticated shape of the former, the two are produced by the same simple sequence. A variation sometimes occurs where the original split surface of the cobble core is also used as the surface from which flakes are removed. 731.5 and 722.141 both show this sequence.

The lithic assemblage collected from the intensively surveyed Baroth area shows little evidence of retouched or standardised formal tool types and is entirely in keeping with the Pabbi Hills assemblage as a whole. Figure 8. 21 maps the exposures of the intermediate conglomerate and deflated areas with pebbles against groups of survey collecting zones and the artefacts found in them, demonstrating that stone was present naturally, but the artefacts did not co-vary with these exposures but had their own patterning. The comparative study of the pebble beds in the modern Soan suggest that the artefacts recovered from the Baroth area cannot be explained by fluvial fracturing, and there is no suggestion that the concentrations reflect more recent knapping areas.

Locality 625: In 1989 Dennell noted stone eroding from a small area that formed part of the slightly dissected flatlands just east of Pir Jaffa and just below the level of SST 12. Stones were deflated onto a small collecting zone that was visited the next working day by Hurcombe and Moth. While there were many natural pebbles and some natural splits, six pieces (three irregular flakes, a split flake, a straight-edged flake and a HACT) on the surface were graded 3-5. There were many deflated pebbles, some of a size comparable with stone artefacts and suitable for artefact manufacture (maximum size 20 x 15 x 10 cm, but many are 1 cm round or less); colours are typical of the area, (white, brown, maroon, grey, green, honey) and with varying coarseness for knapping (some medium grained, at least one very coarse, and at least one with tendency to fracture in planes). The area has layers of sandstone, silty sand, pebbles, precipitates and dark soil, and the smaller pebbles lie in one zone with larger ones slightly further away but still in a confined zone. The hillside was cleaned back to show the pebble layer but no artefacts were found firmly within the deposit. Further excavation was planned. However, on the way back from fieldwork at 625 that day the vehicle was shot at, and we reluctantly (in academic respects but not otherwise) decided not to return to the area. It has never been possible to return and investigate this locality further, but the later survey of the Baroth area (nearer the village and approached from a safer direction) suggests that 625 might be a further outcrop of the intermediate conglomerate and deflated pebbles since it is in the right position in the sequence and lies a short distance to the northeast of the Baroth area.

Investigations at 269 and surrounding area (GPS : N 32° 49' 034"; E73° 46' 228")

The most intriguing, and frustrating discovery was at fossil occurrence 269, on the 203 horizon. As explained by Rendell in chapter 4, this horizon is ca. 1.9 – 2.2 Mya-old; the survey details are given in chapter 6. In 1987 the survey team had come up a scarp slope in Sohawa Kas, and very near the top, found a core (HACT 269.1; Figure 8. 7) next to a large piece of Proboscidean fossil on ground that was mostly bare earth with odd clumps of grass with their roots exposed (Figure 8.22). The fossil fragment was part of a scapula, and the team found three more conjoining fragments slightly up the slope, with the last piece still in a thorn tree, just below a path (see Table 6. 9). Just above the path was a low sandy bank that was the top of the section at that point in the hillside (see Figures 8. 23 – 8. 25). The team assumed that the fossil had eroded out of the bank, fallen onto the path and then been kicked by animals or people to fall into the thorn bush and on down the slope into an area which came to be known as the little 'horseshoe'. The fossil had a very circumscribed point of possible origin, but of course the stone tool beside it was intriguing not because of a causal relationship but because it was one of the few occasions when a stone tool could, by analogy with the fossil be traced back to a limited section of hillside as a possible point of origin. Figure 8. 23 looks North showing the major sandstone scarp in the background and the position of the little horseshoe, as shown in the sketch plan Figure 8. 24. In 1989, the area was investigated by excavation trenches (Figure 8.26), a cleaned back section, and a detailed survey of the area for stone sources. Trench A was positioned to investigate the sands immediately beside the path where the topmost fossil fragment was recovered in 1987 (see also Figure 8.25) and was later extended as trench B. Other trenches investigated the opposite side of the horseshoe.

Trenches A and B yielded 119 pieces of fossil of varying sizes including both lag and transport elements. Ostrich shell and molluscs, teeth and other skeletal elements were present, some of which were rolled while others were quite fresh. Most came from the lowest layer, consisting in both trenches of a consolidated crust sand layer with coarse grey sands, and mud balls beneath with increasing amounts of grit. The fossil came from the base of this layer, which was much thicker in the South and the plots of finds showed that the fossil was more prevalent in the southwest. The most significant fossils in terms of size and propensity to roll along or sink into the bed were a horse tibia, complete except for the proximal end (350 x 84 x 54 mm), the glenoid (heavier) part of a bovid scapula (160 x 79 x 49 mm), a complete piece of turtle scute (111 x 53 x 10 mm) (though this piece could have floated), all found in trench A, and a Proboscidean tibia fragment (130 x 62 x 58 mm), a ?rhinocerous calcaneum shaft with both ends sharply truncated (83 x 63 x 57 mm) both from the lowest layer of trench B. There were also four quartzite pebbles of varying colour. Whilst these were small (26 x 19 x 12mm, 17 x 8 x 7mm, 21 x 14 x 8mm, 27 x 21 x 18mm) they show the natural presence of quartzites in the bed load. Furthermore both the size of the fossils and the size of the mud balls (see the sections, Figures 8. 27 and 8. 28) show that the clast sizes encompass the size of stone tools and also the size of raw material suited to tool manufacture. The presence of the fossil also shows that objects from land are getting into the fluvial sediments at this point, and that the skeletal elements include items which would be rolled or lag in the bed load. This is as close as we have come to showing that stone tools could be coming from Siwalik strata. If the layers which we found to have clasts of varying sizes and mostly comprising mud balls, occasionally had quartzite clasts this would explain the source of the natural and worked stone in this confined area, because there are no other obvious sources.

The section down the little horseshoe between the place where the thorn bush with the fossil was located in 1987 (and then the 1989 trench) and the location of the stone tool and lower fossil fragment was cleaned down to show that the section was integral to the tilted geological sequence and did not contain slumped deposits. Material had been washed down and caught by the thorn bush roots and so there was one point where a stone tool could have lain concealed in the section. However, the accumulating material still had no source above it other than layers integral to the Siwalik sequence. The erosion experiments (see below) showed how quickly the sediments were eroding even in flat areas, so that any tool on the horseshoe surface could not be expected to remain there for long.

The area can be seen in the detailed plan in Figure 8. 24, which also indicates the positions of natural pebbles mapped in 1989, plus new ones noted in 1999. The photograph (Figure 8. 25) of the horseshoe area looking towards the excavations shows how near the top of the hillside the finds were, and the circumscribed possible source area.

Table 8. 10 indicates the results of the 1989 detailed survey of the area for stone, and shows the range of artefacts (graded (3-5), and the pebbles (graded 0-2) collected in 1987, and those found and left in place in 1989 and 1999. The natural stones were labelled with indelible ink overlain by PVA and left in place. In 1999 a further detailed survey showed a new stone tool in the 261 area (bi-directional core 261.2, Figure 8.11) and other natural pebbles not apparent in 1989.

The area in which 261.2 was found in 1999 is a little shute which turns into a small gully. There are sandstone blocks, but these are sporadic, blackened and rounded with weathering and now stand on plinths of soil as they have shielded the softer deposits on which they rest. There are also pieces of mudstone and caliche. The vegetation is sparse tufts of grass with their roots exposed. A short distance away from the new artefact 261.2 there were other natural pebbles, as noted above. There was also a trickle of fossil down the same gully area including several fragments of a large animal. There are no obvious cuts or redeposited layers in the area to account for the stone. Since none of the pieces picked up in 1999 was marked with a letter, nor did the descriptions match, none of the 1999 pieces seem to be from the set of material left in the field in 1989. This is not surprising given the field observations and experimental data on stone tool movement and sediment erosion rates (see below), but it does show that over a 12-year period, 24 stones of 'stone tool size' were noted. There is clearly a source of quartzite cobbles within this area, but the excavation data suggests that these could be as isolated pieces dispersed within the sands and grits strata. Despite the fact that we had a confined area of hillside to search and that we excavated some 100 cubic metres of the relevant deposit, the odds of finding a stone in situ, let alone a stone tool, were very poor. However, the 269 area is the one where we can argue the case for a low density dispersal of artefacts within Siwalik deposits most cogently.

Places with pottery: 346, 652, 672

At its southern fringe southwest of Baroth and nearing the GT road SST 12 still yields some fossil material, but the rate of erosion is far slower and the height of the scarp is much reduced, and managed land is very close. Two localities along this stretch of SST 12 have pottery and are clearly relatively recent sites. No. 652 has iron slag, molluscs and unfossilised bone and 672 also has pottery. The stone finds from both these sites do not contain distinctive pieces and at least for 652 the site must be metal-using and thus very recent.

Experiments monitoring falling fractures, erosion and tool movement on hillsides and gullies

Experiments with the fracture patterns of falling stones

It had been suggested that some of the Pabbi Hills assemblage could result from cobbles falling down the erosional surfaces of the Pabbi Hills and causing natural fractures to occur or from shepherds throwing stones as part of herding activities. Stiles (1978: 139-140), for example, suggested that stones can be flaked by falling down steep slopes, or when (bored) shepherds throw stones at other stones. Neither seemed likely, but experiments were devised to investigate whether stone can be flaked by these methods. The aim of two sets of experiments were to allow comparable clasts to fall down a slope to land on a hard surface some 20 metres below, thus copying an extreme example of a stone falling down slope onto a hard surface, and to drop clasts vertically 5 metres down onto a densely packed cobble surface.

One hundred cobbles from modern exposures (outside the Pabbi Hills) comparable in lithology and size with those suitable for making stone tools were collected and tossed from the watershed track to land on the concrete encased slope just East of Flag Hill. Not one of the pebbles broke, though they did bounce dangerously near the person waiting at the base of the slope to collect any fractured pieces, and (if possible) to retrieve unbroken ones. The cobbles were then taken back to the top of the slope and thrown as hard as possible onto the concrete; again none broke. It was concluded that pebbles falling down erosional slopes to land on sediments that were much softer than concrete were extremely unlikely to fracture and could not be used to account for any of the recovered assemblage (see also Dennell and Hurcombe 1995).

However, other experiments have shown that rocks falling onto other rocks would be more likely to cause such fractures. The most common problem is the natural chipping of thin edges. Such damage is sometimes observed to be more recent because of differential patination (e.g. Green 1988: 223). Wallis (1988) observed natural fractures in heaps of mostly quartztite cobbles rejected from commercial gravel extraction in the Abingdon area, Upper Thames Valley, England. He found that there were some split cobbles, but that the most frequent form of damage was a flake removed from one edge of the ovoid cobbles, leaving a shallow, often step-terminated scar. This description is similar to other instances where a case has been made for falling fractures being mistaken for intentional flaking (see also White's (1995) comments on the assemblage from Vallonet, France, and also Nash 1993). In order to understand better the kinds of fractures that could be caused in this way a further experiment was conducted in 1999.

Rounded cobbles are more difficult to knap than angular pieces and the material is tougher than flint. The latter quality makes it both harder to shape, but once formed the edges withstand robust activities better than other materials such as flint. However, even from one area, there is a great range of raw material within the 'quartzite' label. One hundred cobbles ultimately from the same source as the archaeological lithic materials, exposed river pebbles from the Punjab derived from water running off the Karakorum range, were selected to show a range of different kinds of quartzites, including some which it was thought would fracture more easily, or ones which had some preferred fracture lines. Rounded cobbles were chosen since the aim of the experiment was to test whether natural cobbles would fracture on falling vertically onto other stones to create objects that could potentially be confused with intentionally flaked artefacts. The chosen cobbles were all of a size suitable for making stone tools (average weight 491.1 gm) and though all were well rounded, some were flat and oval, while others were thicker and more elongated. Forty similar cobbles were placed three cobbles deep within a strong metal box. A thin cotton cloth was placed over these cobbles so that any fractures resulting from the dropped cobbles could be separated from those that might be caused in the cobbles lining the box. In this way the experimental fractures could be limited to the results of one hundred falling cobbles. The cobbles were dropped vertically 5 metres from a roof onto the cobbles in the box. In the eroding landscapes the only *vertical* drops greater than this are down into the bottoms of the gully systems; however, despite searching, we have never recovered artefacts from such gully contexts.

At the end of the experiment all the dropped cobbles were carefully examined and the flakes and fragments greater than 5 mm were collected. The cobbles forming the lined box were also checked but none of these had fractured. The dropped material was separated into two groups A, the fractured material, and B the cobbles which had no fractures. The group A material was refitted so that the sequence of flaking events could be reconstructed (Figure 8. 29). Eight cobbles either split in half on impact or had one flake removed (e.g. nos. 1-3, Figure 8.29). Such flakes frequently broke as other cobbles were dropped on them. Six cobbles became flaked from more than one direction (e.g. nos. 5-7, Figure 8. 29), with five achieving some form of bifacially flaked edge, though these had a very limited edge or poor edge angle, and would not readily be seen as chopper-like tools (e.g. nos. 8, 9, Figure 8. 29). One further nodule had one flake removed and then the whole nodule split in half along with the flake. Lastly, one nodule with a propensity to fracture in planes achieved a bifacially flaked edge that could be termed a 'chopper'. Most of the fragments did not rate highly in the credibility system (see Table 8. 11). However, in an assemblage where large numbers of stones are falling vertically onto other stones it would seem possible that occasional complex flaking patterns could occur. These would still be a rarity and there would be much larger numbers of cobbles with one flake removed, or decortication flakes, many of which would be split.

Whilst fractures do occur from natural falls, in areas where stone is rare and there are few if any vertical drops, the chance fractures from falling will be negligible. Those that do occur are most likely to be first order cobble 'cores', with only one flake removed and 'decortication' flakes with cortex platform. The incidence will be at its highest where there are vertical drops within the erosional slopes, and where there are large numbers of stones so that falling cobbles will strike stone rather than fine sediments. The only area of the Pabbi Hills where cobbles are likely to fall onto other stones is in the area of the Sandstone 14 conglomerate and even here the drop is more often down a slope onto sediment, rather than vertical onto stone. A reassessment of the artefacts from the sandstone 14 conglomerate area (i.e. 91, 191 and 192) shows that the artefacts are not dominated by the kinds of pieces recorded in the falling experiment, but contain a flake blade and flake blade core, complex disc cores and a bi-directional core. There are no uni-directional cores though there are some irregular pieces. It is concluded that natural fracturing from stone falls is not significant in causing the flaking of the Pabbi Hills assemblage.

Rates of erosion and stone artefact movement: experiments and observations

Anecdotal evidence and field observations suggested rapid erosion rates in at least some areas of the Pabbi Hills landscape. Erosion caused by rainwater and small runoff streams on the hillside were seen to be the most relevant erosional forces for stone tools and fossils. As an example, in 1987 locality 362 (see chapters 6 and 9) was protected with cloth and a layer of earth to retard further erosion. Despite this, a large amount of material had been eroded by 1989, with bone fragments several metres down slope (see Table 9. 6). The erosion of turtle scute at 481 in an exposure of bare earth had yielded many fragments on initial discovery in 1987, but more fragments were found in 1988 and in 1989, despite collecting all visible pieces each year. The turtle scute is typically less than a 1 cm thick, and thus comparable with some stone flakes, so the annual finding of further pieces of scute from the same gently sloping area suggests that here more than a centimetre of soil and most likely several centimetres, is being lost each year. There are also locally catastrophic erosion events. Various teams have heard the distinctive 'crump' of a large section of undercut sandstone cliff falling down (usually after heavy rain; see also Figure 11.6), and Kotha Kas West had a very heavy such fall resulting in a dam holding back water in the kas. The following year, most likely with the monsoon rains, the dam had burst and cut back the jumble of sandstone blocks. The resulting stepped bank in the kas base was soon colonised by tall grasses, creating a series of steps and clumps up the hillside. Looking back up the kas, it was obvious that this form of erosion has occurred several times even in this one section of Kotha Kas. It was also evident that animal feet could collapse small sections and contribute to the movement of soil and stones. Watching a horse roll in mud or sand deposits, and seeing the same hillside grazed by two different herds in one day suggested to us that animals could be both dynamic and quantitative factors in erosion patterns. In most areas of the landscape surveyed, goats and sheep predominate with some cattle, but horses are also herded by the nomads, and camels, water buffalo and wild boar are occasionally seen. Field observations suggest that small bluffs collapse under animal feet especially when there has been recent rain which softens the sediments, and animal feet occasionally kick over weathering sandstone blocks which may roll or fragment. Most of the stone artefacts found lie on relatively soft sediments but there is likely to be some trampling damage that could cause split fractures and edge fractures in the thinner pieces (Knudson 1979)

Stone movement on hillsides and in gullies

Taphonomic studies of site formation processes have been an excellent development over the last 30 years and there are good studies of taphonomic processes affecting bones and also some studies on the effects of animal tramping and stream action on stones (e.g. Behrensmeyer et al. 1986, Gifford-Gonzalez et al. 1985, Goldberg et al. 1993, Knudson 1979, McBrearty 1998, Paddayya and Petraglia 1993, Petraglia and Nash 1987, Petraglia and Potts 1994, Schick 1986), but none adequately investigated the likely movement of stone artefacts on the kinds of erosional surfaces seen in the Pabbi Hills. Experiments were set up in 1989 to investigate these issues. How quickly would stones move down slopes into, and then along the gullies? The 1989 experiments on gullies gave clear results within one year, but the down slope movement was much more complex, and data has now been collected over a 10 year period.

Gully experiments

Sets of pebbles were grouped together and left in the minor hillside stream bed on the East slope facing the erosional experiments and in the narrow sandy bedded main kas immediately below the erosional squares. Their starting position was marked by carving crosses on the sandstones forming the side walls of the gully. The stones were all quartzites taken from river gravels and all were maroon coloured in order to facilitate their relocation in the field. The stones were graded into sizes and 148 were used in the gully experiments. The stones were labelled by stating the size letters (A=3 cm, B=5 cm, C=7cm, D=9 cm and E=11 cm or more; thus CB signified that the pebble would fit into a rectangle of 7 x 5 cm.

The gully experiments show that once in stream or kas beds stone would quickly be moved downstream. As shown in Table 8. 12, only three pebbles from 70 were recovered from the stream gully, and only one from 78 in the main kas bed. The entire stream gully surface was searched for the maroon pebbles but it is possible that they were covered by debris, clay and other sediment, and so were not visible rather than not present. The few stones that were recovered seemed to be lodged in irregularities in the stream bed.

The main gully was carefully searched and there was some concern that the stones might have been buried in the sandy kas base, rather than missing because they had been washed down the gully. Careful searching paid off as 94.5 m downstream there were a series of much larger sandstone and mudstone blocks and one EE size pebble weighing 1.183 kg was found caught in this larger material. Since this is the largest size grouping and it had travelled so far in one year it was assumed that the stones had indeed been transported by water. These results also suggest that the stream gully pebbles may have also been transported rather than buried. It is worth noting that the monsoon in the Pabbi Hills must result in torrential rain swelling even the hillside streams. The teams have occasionally been caught in the Hills during intense rain bursts and seen flash floods. In Islamabad one spring, a newspaper reported that 23 mm of rain had fallen in one afternoon, and this was not during the main rainy season. Although it might be academically useful to undertake a study of the Pabbi Hills in the midst of the heaviest rainfall it would be practically impossible even on foot from the GT road.

Slope experiments

Experiments to investigate down slope transport proved more difficult than the gully experiments. In 1987 we tried to set up some six inch nails to act as markers on one area of hillside, and to take polaroids to monitor the erosion of sandstone blocks and areas of hillside. This did not work because the nails were taken overnight (they were seen as valuable finds to shepherds) and a comparison of the photos the following year showed how hard it was to compare patches of hillside from year to year with no reference points when some blocks had disintegrated or rolled and when thorn bushes were being constantly altered by shepherds taking off branches or removing whole bushes as a source of firewood. In 1989 a more systematic set of experiments were created in order to investigate the rates of movement of flat and rounded stones on a variety of sediments and slopes in the Pabbi Hills to aid interpretation of the artefacts found on such erosional surfaces.

A hillside on the west side of the west fork of Kotha Kas near a path leading south from Flag Hill was selected because it had a variety of different kinds of deposits and gradients. It was also fairly easy to reach from the watershed track, but not on a major route through the landscape. Figure 8. 30 shows the hillside with the positions of the experiment stations marked (and see Figures 8. 31- 8.33 for details of some areas). These were chosen to investigate assumed low, medium and high erosion rates as an aid to understanding the movement of stones in different situations. The scale is given by the people standing on stations 1 and 8 and near the skyline in the centre. Eight one metre squares were laid out on the hillside as indicated. Six of the squares were marked by a 2" nail (much less conspicuous than a 6" one) in the top right and bottom left corner to act as reference points for planning. Stations 6 and 7 were so steep that the nails were placed in the top left and right corners to form a horizontal start line. No paint or other marker was used and the squares were laid out only when we thought there were no local shepherds observing us. In each square 13 items were laid out, i.e. 12 rounded pebbles (reflecting the type of natural clasts and rounded cores) with no fracture surfaces and one flake (representing smaller more angular pieces); this piece was drawn so there could be no confusion later. As with the gully experiments, the stones were quartzites from river gravels and all were maroon coloured in order to facilitate their relocation. Stones were labelled with a black marker pen, and the number was protected by two coats of PVA. To hide the fact that they were labelled, all the stones were placed with the number facing down. The stones were graded into sizes using marked out squares or rectangles on graph paper to process the pieces quickly. Ninety-six pebbles and 8 flakes were laid out in the squares. The stones were labelled by stating the size letters (A=3 cm, B=5 cm, C=7cms, D=9 cm and E=11 cm or more, thus BA signified that the pebble would fit into a rectangle of 5 x 3 cm) and then given numbers in a sequence within the size grading. The only exception was the AA size group where just the numbers were placed on these small pebbles. For six stations the larger stones were used at the corners and centre of the squares to make the sense of the square easier to see on return visits. One pebble in the EE series formed the centre of each square, with two from the CC series placed at the top left and right corners; one DD or DC stone was placed bottom left, and a CB stone was bottom right. The flake for each square was placed touching the pebble in the top right corner. All the other stones were placed in the centre of the square touching the large EE series pebble. These groupings were chosen so that they were not obvious to a passer-by. For stations 6 and 7 the pebbles would not stay in their positions on the slope even while being laid out, so the pebbles were placed on the top horizontal line of the square with as much care as possible though some slid downhill immediately. The squares were laid out late in the spring season of 1989 with the intention of recording them annually for a few years. They were recorded in 1990, 1992, and lastly in 1999. The Gulf War caused the abandonment of the planned fieldwork in 1991, and the 1999 visit unfortunately was for only one afternoon with armed bodyguards because of security concerns. The detailed results are too lengthy to reproduce here, but summaries of the results at the three year interval 1989-1992 and the ten year data collected in 1999 are given below to support the overall conclusions about the kinds of slopes and rates of artefact movement on these different zones.

Figure 8.30 shows that stations 6, 7 and 8 were all on steep slopes but very different in character. Six was bare clay and caliche and seven was clay with sandstone blocks falling and breaking apart, cascading sand down the slope. Both these stations showed that the assumed rate of stone movement as 'high' was correct. At station 6 the three year data showed that the smaller stones had tended to be lost or move furthest whilst the larger stones had moved but remained visible on what was still a bare >45° slope. The weight of the larger stones and friction seems to keep them on the slope while even heavy rainfall might cause the smaller stones to move. The 10-year results confirmed this overall pattern but the remaining larger stones were now resting by the roots of a thorn tree which could be expected to restrict further movement. In addition the stones had now moved off the steepest bare clay and caliche slope and were now on a medium gradient with odd clumps of grass more akin to station 4. It was also thought that the role of animals might now change since the steep and completely bare slope was not attractive to them whereas now it might become more significant. Station 7 showed that even after one year some of the stones had moved several metres down slope (these tended to be the larger ones) while a cluster were within a metre of their starting positions including one of the smallest stones. Some were now caught behind much larger sandstone blocks and others were beginning to be buried by disintegrating sandstone blocks. In 1992 only one stone was visible, presumably because they were being covered by sandstone blocks and the sands from these. This pattern was confirmed by the cursory search in 1999 when no stones were visible. It is assumed that animals are contributing to this erosion pattern but the erosion seems to be cyclical with periods when the stones are exposed leading to rapid movement until they come to rest against larger blocks which accumulate slope wash behind them until the blocks are dislodged by animals (personal observation) or eventually disintegrate. The 1987 photographs of the slope showed large blocks that were missing the following year or were visible as broken up and disintegrating. In contrast to these two stations, number 8, though on a steep slope, was assessed as having a 'low' rate of stone movement because the animal tracks and sandstone blocks had created a series of miniature terraces. The sandstone blocks were however gradually disintegrating and it was evident that on the slope as a whole some blocks had slipped from their positions presumably due to gradual weathering down of the blocks and the sediments underlying them, aided by animals occasionally dislodging them. The three-year study showed that there was some settling down in the first year but thereafter smaller movements depending upon whether the stones were now wedged between sandstone blocks. The 10-year study confirmed the overall pattern and whilst the cluster was still recognisable, some sandstone blocks had shifted. The largest quartzite stone was still wedged near a sandstone block but the middle range of stones that had been very close to it were now further down the slope with the flake further still. Whilst a 10-year study shows perceptible movement the rate is 'low' and artefacts on such slopes may take longest to move down the profile. Some stones have moved 1 metre in 10 years, some have moved less. So a slope of this kind might contain artefacts for decades, hundreds or perhaps even a few thousands of years, but not tens of thousands of years.

Stations 3 and 4 were assumed to have 'medium' rates of stone movement. Station 3 was mostly bare clay and caliche with occasional sandstone blocks and sparse patches of grass with roots exposed and criss-crossing animal paths, whereas Station 4 was in an area with brown/orange silts and sands with caliche but generally flatter and more open with odd remnants of sandstone blocks. The three-year study at station 3 showed that most of the stones had moved, including one that had moved over a metre. Animals could have dislodged this one causing it to roll although other stones seemed to be moving at a steadier rate. The 10-year study confirmed the overall pattern with all observed stones having moved but the larger ones appear to have moved furthest. Station 4 showed movement of the larger stones, presumably by animals, after only one year. At the three-year point there were more small movements but the sense of the square was still plainly visible. Although an ants nest was building up deposits in the centre these were loose, whereas both nails were now standing above the surface (15 and 22mm) indicating that overall the station was loosing deposits. The 10-year study confirmed the general trend but also showed more movement of the smaller stones since fewer of these were noted than the larger ones, although the flake was still visible. Both stations were confirmed as 'medium' for stone tool movement but affected different sizes of pebbles differently with larger stones travelling further on station 3 whereas small stones were not remaining at station 4. Both effects could be explained by the combination of animals and slope. Larger stones are more likely to be kicked and if on a slope would tend to travel further than if the surface were flat.

Stations 1, 2 and 5 were all thought to have 'low' stone movement. Station 2 was on brown fine sands and a fairly flat surface with very sparse grass tufts with roots showing. Some of the larger stones had moved 10 cm or more after one year, presumably by animals, and after three years all the stones were out of position and even the flake and smaller stones had moved 10 cm or more. One of the nails was 2 mm. above the surface but the other was missing. The low rate of movement was confirmed by the three-year study. However, the ten-year study showed that this flat area was becoming much more broken up. A set of stones large and small were still visible and in close proximity but others were missing, perhaps lost into the dissecting lines which acted as funnels for water runoff and animal movement, both of which would tend to move stones more quickly. Although the assessment of 'low' held, it was a very uneven label and the impression was that the flat 1989 surface had changed in character by 1999, and was entering a more active erosion phase.

Station 1 was chosen to represent a stable flat area, and was assumed to have a 'low' rate of stone movement overall and the lowest rate of all the stations set up. The square was virtually flat and lay just beneath a sandy near vertical small cliff of ca. 1.5 m. The brown sandy sediment was almost covered with green grass with no roots showing. The three year study confirmed the assessment in that after one year the flake was still visible but the whole square was being gradually covered over the by the sands accumulating from the small cliff and after three years only the tops of the larger stones were visible - a clear case of an accumulating site. However, the three-year study conclusion was overset by the 10 year results. The grass cover was now more patchy with roots showing, while the larger stones are being re-exposed. In the space of 10 years the stones have been buried but are starting to be re-exposed. The small cliff line has eroded slightly and whilst the proximity to this feature initially led to accumulating deposits below, the frontier has shifted just enough for the erosion to be active again. This was a complete surprise as were the results from station 5 also on the basal sandy deposits but this time in a more open context.

Station 5 was on fine yellow/orange silty sands with sparse grass with roots exposed. After one year some of the larger stones had moved from the corners slightly but by the end of three years the station had completely changed. The largest stone had moved to one side and other stones from the square had moved metres. Some had been transported into a little outwash where they were now being slightly covered on the upslope side as more deposits washed against them. One nail was 28mm above the surface and the other was 8mm above it. It seems that the flatter area was gradually being lost and that parts of the original square were now being lost as part of this overall erosion. The other factor that could have caused such a rapid change is if a horse chose to roll in this sandy area. The three-year study showed that an area that had been thought to have low rates of stone movement could suddenly change but the cause was not certain. The 10-year study showed that although a set of stones cluster together (zone A in Figure 8. 32), the small depression contains a stream of stones (at B,C and D) that were originally on the flat area near the thorn tree. While smaller stones were still observed at zone A, the outwash area contained the larger stones. It seems likely that it is hillside run off that is effecting the movement of stones down this depression and that smaller stones once coming into the area move down rapidly, which is borne out by finding one reasonably sized stone (BB13) from station 4 (i.e., further uphill than station

5) almost at the kas edge. Furthermore the old nail from station 5 was found completely out of sediment and lying in the direction of slope 1 metre further back from zone A, and this despite the fact that all nails were hammered back flat with the surface in 1992. Again, an area of flat sands assessed as 'low' in stone movement has shown a more rapid and localised pattern.

In general the 10-year study has recovered fewer of the smallest stones. All of the observations suggest that these might be real losses rather than such pebbles being more easily concealed by accumulating sediments. Flakes may however behave differently to these rounded small shapes as a better proportion of the flakes were still visible 10 years later. The sediment loss rates documented by the nails and the small plinths of sediment on which the larger stones were sometimes observed all support the generally high erosion rates.

Overall the squares show that even the flatter areas can erode quickly if the sediment is soft sand. The most stable areas may be the strewn block stepped slopes but these are still eroding. There are cycles of stability and erosion that act over quite short time periods. Square one immediately beside a small sand bluff was gradually covered, but is now being eroded again, the stones initially moved quickly in square 7 as they came to rest on the steep slope, but once in amongst bigger sandstone blocks they were gradually covered as these disintegrated and rolled further down slope. Squares 2, 4 and 5 were comparable at first but once the erosion processes started to be more active the situation could rapidly change, as shown by square 5. Many of the squares showed the movement of larger stones in the first year and it seems likely this is because even in flatter areas they could be kicked by an animal and set moving. This would not happen so easily for smaller flatter pieces such as flakes. In other squares on the flatter areas the flow patterns created in the sands clearly showed water was an agent in moving the sediments on which the stones rested. Some stones were on small plinths of sediment. The soft sands certainly seem to have lost 5 cm or perhaps much more in 10 years. Animal kicks, gradients and whether the surface is a smooth slant or series of steps, the weight and shape of stones, sediment type, rainfall and water flow down slopes will all affect the rates of stone movement as well as very local factors such as the protection offered by a large sandstone block, or thorn trees and their roots. However, the experiments have offered insights into these processes and provided some approximate orders of scale. Not one of the squares remained untouched by the processes of erosion and whilst 10 years is a long time for a research experiment it is a very short time in the scale of years the project was investigating. It may be true that the rates of erosion are partly due to the overgrazing which may be relatively recent but the current rates seem to suggest that in many of the areas in which we found stone tools either the artefacts have come from deposits accumulated on the tops of major sandstones which are now eroding down, and perhaps remaining longest on the block-strewn stepped hillside, whereas the bare earth sands and clays should have virtually no tools since they are where erosion rates are fastest. Even in areas that initially had relatively good grass cover and accumulating deposits, the 10 years of observation showed that erosion was active once more.

In view of the experiments it might be assumed that stones on the basal flatter areas may be there for perhaps tens but not hundreds of years; on the slightly steeper slopes stones might be there for a few to many tens of years; on the steeper slopes stones may move quickly once exposed but become reburied, or become part of a steep but stepped slope which acts as a series of small terraces. This is where the stone tools seem most likely to linger longest. It might take a stone a few hundred, a thousand or even a few thousand years to work its way down a 20 metre slope of this kind into a gully. However, it does not seem likely that even in these stable areas the erosion processes would take many thousands of years. Once stone has been transported by slope erosion in a gully, our observations imply that it is rapidly flushed out of the Pabbi Hills. Our assessment is that most of the stone artefacts on slope surfaces in the Pabbi Hills are highly unlikely to be derived from residues of post-Siwalik material from ca. 400 kya onwards.

SECTION D: Interpreting patterns within the assemblage; discussion and conclusions

In our investigations, we have tried to consider a wide range of explanations of the lithic material that we found. The philosophy underlying the recording of where material was found, and its lithic attributes has been to be as explicit as possible, and to avoid prejudging the nature and significance of what we found. Similarly, we have tried to publish sufficient information for any future team to see where and what we found, and why we assessed it the way we have. Our conclusions are as follows:

By field survey and excavation:

- Although there are no natural lithic outcrops in the Pabbi Hills area, and although stone suitable for tool-making would have been scarce, the Siwalik sequence contains suitable pebbles, deposited as stringers along some stream channels, which would have been available for tool making at the time the Pabbi Hills sequence was laid down. These pebble deposits within the Siwalik sequence could also contain artefacts derived from upstream and washed into the river that laid down the Siwalik sequence of the Pabbi Hills. The dates of stone tools and/or early hominins from 'Ubeidiya, Israel (1.4 Mya; stone tools), Dmanisi, Georgia (1.7 Ma; hominin and tools) Riwat (>1.9Mya; stone tools), Java (1.8 Mya; hominins) and North China (1.35 Mya, stone tools) indicate that hominins were making stone tools in Asia at the time when most of the deposits that we surveyed (i.e. 0.9 – 2.2 Mya) were being laid down

- Although the later Siwalik sequence of the Pabbi Hills has been described as conglomeritic, the evidence does not support this. The only area with a conglomerate was the Sandstone 14 ridge between Kanara and Bawali kas. This area was complex, with at least one later channel within the capping sandstone and possibly some later conglomerate material, but with some other pebbles integral to the Siwalik sequence. Major lithic sites are absent from the main survey areas though smaller archaeological sites with pottery exist on the fringes of the anticline.

More detailed investigations of specific areas within the Pabbi Hills where we had found stone tools on the surface show that:

- 387 and "Pelvis Valley", Kotha kas, showed that the outwash of a steep but stable slope could contain artefacts from a wide range of different ages
- 722, Baroth area, was the largest concentration we found of lithic material. It also contained some pottery, although this could be very recent. Though the stone artefacts were *bona fide* some were relatively sophisticated and could be seen as recent. The excavated artefact found in the sands could have been trampled in because these were relatively soft. A geological fault may also exist nearby. Evidence from this area remains inconclusive, and further investigations are required.
- In the Baroth area, there was also an extensive deposit of sands with pebbles just below Sandstone 12 in the Siwalik sequence. The area in general showed differences in the distribution of natural spreads of stone and concentrations of artefacts
- 218, Choawala kas, showed no obvious lithic source other than the nearby geological strata (1.7 – 1.8 Mya) and these seemed integral to the Siwalik sequence.
- 269, Sohawa kas, contained no obvious lithic source other than the 1.9 – 2.2 Mya-old geological strata which were integral to the Siwalik sequence, and these included quartzites, fossil fragments and mudball clasts of a size commensurate with artefacts. Furthermore there seem to be quartzite cobbles and artefacts eroding from these deposits. Although our excavations failed to find any artefacts in situ, we remain convinced that those found had recently eroded from nearby Siwalik deposits.

In summary, stones are naturally present in the fluvial sediments. Some areas have a mixture of possible sources of the artefacts recovered, but there are places where a detailed study has not shown a source other than the geological strata. The low density of lithic material is also consistent with what would be expected of an area poor in lithic resources, and is similar to that obtained (and widely accepted) from late Pliocene deposits in the Bouri region of Ethiopia.

Our investigations also included investigations of modern processes of stone transport within the survey area, and the type of flaking produced by natural processes. The following conclusions are from these experiments and field observations:

- Fluvial movement in the past may have caused some flaking but this is rarely confusable with intentional fractures and might account for no more than a fraction of 1% of the assemblage.
- Fractures caused by natural falls down vertical slopes onto other hard stones could cause some fractures which might be confused with intentional flaking, but in the survey area the falls are more likely to have been onto softer sediments down less than vertical slopes, both of which reduce the chances of fractures. Shepherds might occasionally throw stones (but only small ones) as part of herding but again these are likely to land on soft sloping sediments, rather than onto hard stone. Falling fractures are estimated to be less than 1% of the assemblage; i.e. negligible to the overall pattern.
- Animals may kick larger stones in motion and larger or more rounded pieces may move further once in motion on slopes than lighter or flatter ones. Conversely, larger stones may shield the deposits beneath from rainfall and the hillside run-off, whereas the smaller stones may move more easily with the soft sediments in such conditions. Though the two may cancel each other out, some winnowing and size sorting might occur from these factors.
- Artefacts in even minor streams will be rapidly moved downstream and will be quickly lost in the main gully systems.
- Different areas and types of slopes will have different rates of erosion, but the stepped boulder strewn slopes are the most stable. Artefacts on such slopes may take longest to move down the profile and out of the Pabbi Hills drainage system. Conversely bare soft sands and silts forming even flat or moderate slopes will have more rapid erosion. Present day erosion rates suggest stones may take only a few hundred years or less to move down very active erosional surfaces, though artefacts on the stable stepped slopes might be more realistically estimated at a thousand or few thousand years.
- Present day erosion rates may be more extreme than those of the past due to overgrazing affecting the vegetation, though the uplift of the anticline, the softness of the deposits, and the monsoon will always have led to erosion.

Given the above it could be expected firstly, that if there are prehistoric artefacts laid down within the Siwalik deposits, these will be eroded out onto modern slopes, and eventually (i.e. within a few hundred or a few thousands of years) end up in gullies. Once there, they would be rapidly transported out of the area. Secondly, the post-Siwalik prehistoric periods will in general be poorly represented in the artefact record, with the earliest post-Siwalik phases being the most rare or non-existent because they would have been washed out of the area. (Interestingly, the fossil assemblages are entirely consistent with an Early Pleistocene or Late Pliocene date, and there is no evidence that any fossil material is derived from the Middle Pleistocene; see chapter 10). The assemblages that should predominate in this residual material scenario are those of the latest prehistoric stone-using periods, i.e. the Neolithic and Chalcolithic, together with any Siwalik age artefacts that have recently been eroded, though the latter may not be very numerous. The character of both these elements can be surmised.

In the period 2.5-0.9 Mya, although a Siwalik river was responsible for the fluvially derived sequence now exposed in the Pabbi Hills, the strata do not represent an area where hominins and animals could only have drowned, but one in which they could have lived. The sequence contains palaeosols and land surfaces, seasonally exposed surfaces, abandoned channels, vast areas of rich fertile overbanks and, of course, active channels so that water would be locally available. This environment is best described from the fossil remains as an open grassland with some woodland and scrub along the stream margins, and many of the fossil sites are terrestrial accumulations of land-based animal remains (e.g. sites 73, 362 and 642; see next chapter and chapter 11). The richest fossil bearing area is that termed 'Sandstone 12', but in fact it is the strata just beneath this sandstone which are rich in fossils; the robust and laterally extensive sandstone which formed *above* these land surfaces has in happy coincidence provided a distinctive geological marker with good scarp exposures of the strata it caps.

The lithic assemblage could in part be the remnants of a hominin toolmaker's use of the landscape. Hominins in this region might be keeping to a Mode 1 (i.e. simple flake and core) assemblage for much of the time span represented, though handaxes are known in Israel (2000 miles to the west) after 1.4 Mya, and nearby at Dina and Jalalpur after 700 kya. Their usage of stone tool might have been to carry artefacts with them, to cache them in the landscape, or to make some or all of them on an ad hoc basis. Given the nature of the stone available it might not be easy to predict the *precise* location of stone, although general areas where it might occur would be obvious. Since the river would sometimes change course or flood, caching might not be a good longer- term strategy but might have been useful on a seasonal basis. The shifting nature of the river would not allow palimpsest sites to accumulate larger numbers of artefacts, and the likely numbers of hominins and artefacts per hominin would always keep the number of artefacts dropped on a land surface low. Rivers in spate might accumulate artefacts washed into the river along with other stones, so it could be expected that relative concentrations of artefacts might be contained within stony riverine deposits, but the proportion of artefacts to natural stones would always be low. The number of toolmakers, the number of tools made and then deposited would thus be epitomised by low numbers of artefacts, widely dispersed.

There are many Mode 1 style assemblage components consistent with an Early Palaeolithic assemblage, but we are aware that these might also have been used in later periods. There are no clear Acheulean-type handaxes or cleavers, unlike areas where Middle Pleistocene deposits are exposed. There are fewer than 10 pieces that could be regarded as definitely Middle Palaeolithic, and not many more from the Mesolithic and Neolithic. If there is a strong Neolithic or Chalcolithic component (e.g. nomad settlements in the hills), it is curious that there are so few retouched objects, e.g. scrapers and blades. The Neolithic finds are consistent with pastoral short-term use of the Pabbi Hills, with the major settlements on the fertile plain. Much of the assemblage remains undateable to a specific period but compatible with a very early, Lower Pleistocene date.

Conclusions

At the start of this chapter, four explanations were offered, singly or in combination, for the lithic material found in the Pabbi Hills each of these can now be taken in turn:

1. *Natural agency.* Less than 1 % of the material is likely to have been caused in this way and any that has will neither confuse the overall patterning of the remaining 99%, nor be mistaken for period-specific types.

2. *Deliberately struck but very recent in origin.* There is no evidence to suggest that much, if any, of the lithic material is from the very recent past except for the chance use of material to hand such as the use of a pebble to surround a hearth, weight the side of a tent down, or hammer in a nail or stake. There is no reason to suggest that any form of knapping strategies have been used in the recent past.

3: *Deliberately struck but derived from reworked residues of post-Siwalik material from ca. 400 kya onwards, i.e. after the anticline was formed and thus Mid-Pleistocene in age onwards.* There seems to be no or very little lithic material from the period immediately after the anticline began to be formed, and relatively little from the most recent stone age periods. Put more precisely, there is no evidence of the Acheulean, and very little that is obviously Middle or Upper Palaeolithic, or even Neolithic. Known later types account for only a few pieces. In particular, the retouched items well-known from the Middle Palaeolithic onwards form only 2.8 % of the assemblage.

4: Deliberately struck and originating from the underlying Upper Siwalik strata dating from about 0.9 to over 2 Mya. Logically, there should be some toolmakers in the area in the time range of the sediment and there should be some material from this source. There are potentially early types present and these form the most numerous part of the dispersed assemblage. There are no obvious accumulations as some archaeologists might expect. But as argued in the discussion, there are many reasons to suspect that material in the braided channel and floodplains of the Upper Siwaliks would result in dispersed and low levels of material. The low numbers overall, dispersed assemblage, and many of the types could fit in with such a source and could account for perhaps a third to two thirds of the assemblage collected. Overall, we see no reason to explain the bulk of the lithic material from the Pabbi Hills as residual items deflated from a previous conglomerate.

Despite the arguments above, we are all too aware that this is a surface assemblage, and it is impossible to date survey material. However, it is interesting to ask what would an assemblage from the earliest palaeolithic look like in this part of the world and in this kind of environment: most likely, the type of dispersed pebble and flake technologies that we have found here. It is not possible to state what date the Pabbi Hills material must represent, but it is reasonable to say that if there were material of this date in this ancient environment, then the resulting assemblage could look much like the one that is documented here.

Acknowledgements

The lithic chapter has of course benefited from the editorial help and challenging discussion over the years of Robin Dennell. Little did I realise the consequences of his invitation to join the team in 1987 and the enjoyment from both the lithics and the fieldwork. He also helped with all of the experimental work and with great temerity stood at the bottom of the section to retrieve the stone cobbles I hurled down in the first falling experiment. Helen Rendell, the late Asif Jah, and Imran Khan have all helped identify rock types and aided the appreciation of clast behaviour and distribution within the sediments. I would also like to thank Bridget Allchin for her expertise on stone tools from the continent. The stone tools were drawn in the field by Eddie Moth in 1986 and myself thereafter. Although I inked up some of the drawings Mike and Sue Rouillard have inked up the majority. Mike was also responsible for several figures in this chapter. Robin Dennell produced the final versions of all the figures, tables and text. Exeter University has provided much technical support and Mike, Sue and Seàn Goddard have all helped with photographic work. Robin Dennell, Eddie Moth, Andrew Chamberlain and Imran Khan all aided the collection of the erosion squares data and Mike Rouillard drew up the data. Most of the stone tool photos were taken by me but Eddie Moth took some as indicated in the captions and drew some of the sections. Hannah Jones took me back to the key 1986 stone find areas and drew specific sketch maps of the Sandstone 14/16 area which proved invaluable in linking the 1986 data into the system. G.S.P. in Islamabad allowed me to use the roof of their building for the falling fracture experiments.

Table 8.1: Stone artefact measurements and data. Items with an overall credibility of <3 (see text) have been omitted.

Key to headings (see text for further definition):
LO: Collecting zone number
TO: Tool number
TYPE: Code for tool-type (see below)
TECH: Techno-type (see Figures 8.1 and 8. 2)
LE: Length (mm.)
WI: Width (mm.)
BR: Breadth (mm.)
WEIG: Weight (gm.)
RO: Rock type
COLOUR: Self-explanatory
FR: Fracture quality
RO: Edge-rounding
DA: Edge damage
SF: Number of scars with flake features
SW: Number of scars without flake features
NO: Number of directions of flaking
PO: Positive scars as % of surface
NE: Negative scars as % of surface
CO: Cortex as % of surface
SIG: Sign
CR: Overall credibility (for more detailed explanation, see pp. 224-228).

Table 8.1: Explanation of Column 3 (TYPE): Definition and codes of different types of flakes, cores and polished/ground stonework

Flake Types	**Code**	**Defining Characteristics**
Knife	FKN	uni- or bi-facially retouched, long, acute edged flake
Blade	FBB	acute edged, parallel sided flake with one or more parallel dorsal ridges; at least three times as long as broad
Flake blade	FFB	parallel or slightly rounded acute edges; at least twice as long as broad with a good length of cutting edge
Rounded, simple flakes	FRS	rounded or oval in plan view with no or few scars on the dorsal surface
Rounded, complex flakes	FRC	approximately round or oval in plan view with complex scars on the dorsal surface
high action core tool (HACT) flake	FHW	flakes which show wear on the platform and proximal dorsal area originating from the use made of the parent piece, a hemispherical tool (see below)
Flake from hammerstone	FHA	one or more surfaces show pecking and/or damage from the parent piece
Polished stone axe flake	FPO	one or more surfaces show polishing; these may originate from a ringstone, rubbing stone or polished stone axe as indicated by the shape and texture of the polished or ground surface
Straight edged flake	FSE	at least one long acute edge; straight in profile and plan view
Scraper	FSC	retouched to form an obtuse rounded or straight edge
Split flakes	FSP	stress fissures on the ventral and side surfaces have the same point of origin, indicating that these flakes split longitudinally as they were detached from the core
Shatter-split flakes	FWS	the blow has resulted in a series of angular facets as the cobble core has shattered or split irregularly rather than flaked; some resemble a segment of orange
Bipolar flakes	FBI	one main bulb on the ventral surface, but the distal ventral surface shows another bulb in the opposite direction
Double bulb flakes	FDB	double bulb of percussion, as occasionally caused by uneven and large hammerstones resulting in two points of impact from one striking event
Irregular flakes	FIR	irregular shape in plan view
Angular flakes	FAN	angular and irregular shape in plan and side views
Core/Core Tool Types		**Defining characteristics**
Chopper	CCH	flakes removed from opposing faces along the same edge, sometimes creating a sinuous profile (term reserved for pieces which are convincing as functioning chopping tools)
Unifacial, uni-directional core	CUD	pebble core with flakes removed from only one face (could be classed as a unifacial chopper on flaking grounds)
Bi-directional core	CBD	pebble core with flakes removed from two opposing faces (could be classed as a bifacial chopper on flaking grounds)
Straight-edged core tool (SECT)	CCL	one or more robust (c 50-80° edge angle), straight or slightly curving fracture surface edges formed with fracture surfaces on both faces, with retouch/damage on one or more faces
Proto or unifacial SECT	CCP	a robust (c 50-80° edge angle), straight or curving edge formed on one fracture surface with retouch/damage along it
High action core tool (HACT)	CHW	core with flakes removed from the circumference, with wear on specific areas of surface, frequently showing further damage from hitting/scraping motion; some show wear facets and affected surfaces may show cracks and the formation of partial Herztian cones
Squared core	CSQ	split cobble, trimmed to make two or more right angles on the platform
Micro-core	CMI	less than 5cms long; less than 2.5 cms wide
Split core	CSP	cobble split in two
Irregular core	CIR	irregular shape
Polyhedron	CPO	core has been extensively reduced leaving an angular shape
Shattered wedge core remnant	CWS	uneven angular fragment remaining from a pebble shattering rather than flaking; flaws or weak spots in the lithic material may be evident
Blade core	CBB	elongated regular core showing the scars of blade removals
Flake blade core	CFB	slightly elongated core showing previous scars where elongated flakes have been removed; neither as long nor as regular as a true blade core
Core scraper	CSC	abrupt angles in side view and curving edge in plan view created via flake removals (much larger than retouch) around the circumference of a flat fracture surface
Biface/handaxe	CAX	roughly oval shape; bifacially worked; symmetrical
Proto-biface/handaxe	CAP	very roughly shaped by removing a few flakes bifacially; vaguely symmetrical
Pebble hammerstone	CHA	unmodified pebble with wear on specific areas of the surface resulting from blows made by the piece impacting onto something else; affected surface may show cracks and formation of partial Herztian cones
Circumferentially flaked simple core (simple discoid)	CCS	a split cobble with flakes removed around one face, either using the cortex as a platform or the split fracture surface (could also be called a simple discoid)
Centripetal complex core, (discoid)	CCC	based on a split cobble; flakes removed around both faces using both the cortex and the fracture face as a platform (could also be called a discoid)
Bipolar core	CBI	convex surface and evidence of reflected force removing flakes as well as the flake scars with origins on the striking platform
Polished/Ground Stone Tool Types		**Defining Characteristics**
Ringstone	CRS	flaked to roughly a disc shape (though usually found broken in half), one or both flatter surfaces may show smoothing with a central depression formed on both faces, resulting in an hourglass profile and occasionally complete perforation. The flaked outer edges appear to be either deliberately lightly smoothed or are worn.
Rubbing stone	CRU	stones with one or more facets of smoothed surfaces
Polished stone axe	PAX	slightly rounded working edge polished to shape; body may be unmodified pebble or smoothed or pecked, elongated shape sometimes tapering at the opposite end to the working edge

Table 8.1: Stone artefact measurements and data. Items with an overall credibility of <3 (see text) have been omitted.

LO	TO	TYPE	TECH	LE	WI	BR	WEIG	RO	COLOUR	FR	RO	DA	SF	SW	NO	PO	NE	CO	SIG	CR
10	1	FFB	F8	70	45	10		QT	GREY	R	SR	?	3	5	3	45	55	0	+	5
10	2	FSE	F6	52	34	14		QT	BROWN	R	F	Y	2	4	4	45	25	30	+	4
10	3	FIR	F2	58	53	22		QT	DGREY	RP	SR	Y	3	0	2	45	10	45	+	3
10	5	CCC	C1B	78	53	23		QT	BROWN	R	SR	0	3	9	4	0	60	40	-	4
10	6	FSE	F8	62	52	14		QT	BROWN	R	F	0	5	6	5	45	55	0	+	5
11	3	CCS	C2A	70	51	37		QT	WHITE	R	SR	0	3	2	2	0	70	30	-	5
11	4	FSE	F7	72	53	20		QT	PINK	RG	VR	0	2	3	3	50	50	0	+	4
18	1	CBD	B3	125	74	62		QT	GREEN	R	F	0	2	1	2	0	25	75	-	3
24	1	FFB	F8	50	28	11		QT	GREEN	R	F	0	3	1	3	50	50	0	+	5
26	1	FIR	F7	34	26	13		QT	BUFF	R	SR	0	1	1	2	45	35	20	+	3
27	1	FSE	F3	52	32	10		QT	BUFF	R	SR	?	6	3	4	30	70	0	+	4
29	1	FSP	F7	37	27	15		QT	GREY	R	SR	?	3	1	3	45	50	5	+	3
30	1	CWS	A1	90	63	42		QT	CREAM	RP	SR	0	1	3	3	0	60	40	-	4
30	2	FIR	F3	58	42	15		QT	BROWN	R	F	0	2	3	3	40	40	20	+	4
36	1	FIR	F2	78	41	39		QT	BROWN	RP	SR	0	2	2	3	35	25	40	+	3
38	1	FAN	F7	80	72	35		QT	PURPLE	VR	SR	0	3	2	5	40	40	20	+	3
39	1	CBD	B4	91	82	63		QT	BUFF	VR	SR	0	1	3	4	0	40	60	-	3
40	1	FFB	F8	76	37	20		QT	BUFF	R	VR	0	1	4	3	45	50	5	+	4
48	1	FIR	F7	45	33	11		QT	GREY	R	SR	0	1	4	3	50	40	10	+	4
48	2	FAN	F3	68	66	25		QT	BUFF	R	R	?	0	5	3	50	20	30	+	3
48	3	FAN	F7	58	55	26		QT	BUFF	R	SR	0	1	1	2	50	10	40	+	4
49	1	FRC	F8	45	30	8		QT	CREAM	R	R	0	3	2	3	50	50	0	+	5
49	2	FSP	F1	50	45	16		QT	BROWN	R	SR	0	1	3	3	10	45	15	+	4
49	3	FAN	F2	82	54	38		QT	BUFF	R	SR	0	2	1	3	50	10	40	+	3
52	1	CCL	B4	90	82	42		QT	MAROON	R	F	0	6	2	5	0	70	30	-	5
52	2	CIR	A1	126	57	50		QT	PINK	R	F	?	0	5	2	0	65	35	-	3
57	1	CIR	A1	100	85	42		QT	PINK	VR	SR	0	0	5	3	0	55	45	-	3
57	2	CIR	A1	82	42	39		QT	PINK	R	F	0	3	10	6	0	100	0	-	3
58	1	FIR	F3	33	31	11		QT	DGREEN	R	SR	Y	5	0	3	40	50	10	+	5
59	1	FIR	F3	60	47	18		QT	GREY	VR	SR	0	1	3	3	40	20	40	+	4
65	1	CSQ	D3	67	53	52		QT	GREEN	VR	SR	?	1	8	5	0	75	25	-	4
65	2	CIR	C2A	85	84	52		QT	BUFF	R	SR	0	0	10	5	0	75	25	-	3
65	3	CIR	A1	79	75	74		QT	BUFF	R	F	0	0	8	5	0	75	25	-	4
65	5	CBD	B3	88	84	58		QT	PGREEN	R	SR	0	2	2	2	45	55	0	-	4
66	1	CFB	E3	67	63	30		QT	PINK	R	SR	0	3	2	3	0	55	45	-	5
66	2	CCS	C2A	63	59	24		QT	PURPLE	R	SR	0	7	0	3	50	10	40	+	5
66	3	FRS	F2	60	60	20		QT	WHITE	R	SR	?	3	1	3	50	25	25	+	5
66	4	CSQ	D2	70	57	45		QT	BROWN	RP	SR	0	3	3	3	0	65	35	-	5
66	5	CWS	A1	72	63	43		QT	GREEN	R	F	0	2	1	2	0	45	55	-	3
66	6	FSP	F2	50	58	19		QT	BROWN	R	F	?	1	2	2	40	10	50	+	3
66	7	FSP	F2	62	42	23		QT	GREY	VR	SR	0	1	2	2	40	20	40	+	3
66	8	CWS	A1	44	41	21		QT	REDGREY	R	F	0	2	3	3	0	100	0	-	3
66	9	CUD	B1	62	48	39		QT	FAWN	RG	F	0	1	0	1	0	30	70	-	3
66	10	CWS	A1	62	29	21		QT	BROWN	R	SR	0	1	5	4	0	80	20	-	3
66	11	CWS	A1	38	31	17		QT	GREY	R	F	0	2	2	2	0	100	0	-	3
66	12	CSP	A3	91	61	33		QT	HONEY	R	F	0	2	2	2	50	10	40	+	3
66	13	FRS	F5	41	47	19		QT	GREY	VR	F	?	2	1	3	45	5	50	+	3
66	14	FSP	F6	75	32	24		QT	BROWN	VR	SR	?	1	3	2	45	30	25	+	3
68	1	CWS	A1	67	52	33		QT	DGREY	R	F	0	2	1	3	0	60	40	-	3
83	1	FRS	F3	86	86	13		QT	GREEN	R	R	?	2	1	3	50	30	20	+	5
91	1	FIR	F8	59	37	14		QT	PPINK	R	F	Y	2	4	4	50	50	0	+	4
91	2	CCC	C4A	68	63	40		QT	BUFF	RP	SR	0	4	5	5	0	75	25	-	5
91	3	CPO	CB4	85	80	70		QT	GREY	R	R	0	3	9	4	0	85	15	-	4
91	5	FDB	F3	62	33	15		QT	DGREY	R	F	0	1	4	4	45	35	20	+	5
91	6	CCC	C2B	90	60	45		QT	GREYBRO	RP	F	0	4	3	4	0	70	30	-	5
91	7	FSE	F7	78	48	13		QT	BROWN	R	SR	Y	8	3	3	50	45	5	+	5
91	8	FFB	F7	68	51	15		QT	GREY	VR	R	Y	2	2	2	50	45	5	+	4
92	1	FAN	F7	76	56	36		QT	MAROON	R	SR	0	3	3	3	50	25	25	+	5
92	4	CFB	E3	71	44	15		QT	BROWN	R	F	0	3	5	3	0	70	30	-	5
92	3	FFB	F7	86	65	18		QT	GREYBUFF	RP	F	0	2	2	2	50	45	5	+	5
93	1	CCC	C3B	100	90	33		QT	GREY	R	SR	?	6	9	7	0	95	5	-	5
93	2	CWS	A1	53	53	20		QT	GREY	R	F	Y	3	2	4	0	80	20	-	3
93	3	FSE	F6	43	34	14		QT	FAWN	R	F	?	2	2	3	45	30	25	+	3

94	1	FIR	F6	78	48	18		QT	PGREY	R	F	0	2	5	4	0	0	5	+ 3
94	2	FIR	F5	68	48	16		QT	BROWN	R	SR	?	1	1	2	45	5	50	+ 5
94	3	CSQ	D2	68	58	54		QT	FAWN	R	SR	?	2	5	3	0	65	35	- 3
94	4	FRS	F1	76	64	16		QT	BROWN	R	SR	?	1	1	2	50	0	50	+ 5
94	5	FSP	F8	45	33	12		QT	GREY	R	SR	?	3	2	3	45	55	0	+ 4
96	1	CFB	E4	103	68	56		QT	BROWN	R	SR	0	3	3	2	0	60	40	- 5
97	1	FRS	F7	48	48	12		QT	DPURPLE	R	R	?	5	1	2	50	45	5	+ 5
97	3	CCC	C3B	94	65	39		QT	GREY	R	SR	0	7	8	7	0	90	10	- 5
97	4	FRS	F7	54	54	16		QT	BUFF	R	SR	0	2	3	3	50	35	15	+ 5
97	2	CHW	C3A	62	55	29		QT	DGREY	RP	SR	?	5	5	5	0	100	0	- 4
104	1	FIR	F3	64	41	20		QT	CREAM	R	R	?	7	2	5	50	35	15	+ 5
104	2	FSE	F2	72	51	19		QT	PINKGREY	R	SR	0	2	2	2	50	25	30	+ 5
104	3	FIR	F3	73	51	18		QT	GREY	RP	SR	?	1	3	3	50	30	20	+ 5
105	1	CPO	C4A	75	60	46		QT	GREY	R	F	0	2	5	4	0	60	40	- 4
105	2	FFB	F7	71	52	14		QT	GREY	R	F	?	8	2	4	45	45	10	+ 5
105	3	FIR	F5	49	41	13		QT	BROWN	R	R	0	6	5	4	50	30	20	+ 4
105	4	FFB	F7	58	32	11		QT	DPURPLE	R	SR	?	6	1	4	50	30	20	+ 5
108	1	CSP	A2	84	65	42		QT	GREY	RP	F	0	0	8	4	0	50	50	- 4
108	2	FFB	F8	50	30	10		QT	FAWN	R	R	?	1	5	2	50	50	0	+ 3
108	3	FIR	F5	59	41	18		QT	DGREY	R	F	0	3	4	4	50	45	5	+ 5
148	1	CBD	B2	96	86	60		QT	BUFF	R	SR	0	4	2	4	0	45	55	- 5
189	1	FIR	F3	66	48	22		QT	GREY	RP	F	R	5	6	3	50	50	0	+ 4
189	2	CWS	A1	65	44	28		QT	PGREY	R	SR	0	2	5	3	45	45	10	- 4
189	3	FSE	F7	60	50	18		QT	PGREY	R	F	0	5	2	3	45	45	10	+ 5
190	1	CIR	B4	99	106	52		QT	CREAM	R	SR	?	5	3	3	0	70	30	- 4
190	2	FRS	F5	72	66	26		QT	HONEY	R	SR	0	3	3	3	40	10	50	+ 3
191	1	FSP	F2	77	36	18		QT	GREY	RP	SR	Y	4	1	1	50	25	25	+ 3
191	2	FSE	F6	78	53	19		QT	MAROON	R	SR	?	5	4	4	45	20	35	+ 4
191	3	CIR	A1	80	88	61		QT	GREY	RP	F	0	2	10	5	0	75	25	- 3
191	4	FSE	F6	86	69	24		QT	CREAM	VR	R	?	1	2	2	45	30	25	+ 4
191	5	FRS	F6	48	48	17		QT	DGREY	R	F	?	10	0	4	40	35	25	+ 4
191	6	FIR	F3	64	76	23		QT	GREY	R	F	?	4	2	5	30	50	20	+ 4
191	7	FIR	F3	74	72	29		QT	GREY	R	F	0	4	4	5	45	45	10	+ 4
191	8	CBD	B3	100	77	46		QT	HONEY	R	F	0	3	1	2	0	30	70	- 3
191	9	CFB	E3	62	108	40		QT	GREY	R	R	0	8	4	4	0	85	15	- 5
192	1	FSE	F6	45	29	14		QT	CREAM	R	F	0	1	2	2	45	30	25	+ 4
192	2	FIR	F6	74	62	31		QT	GREY	R	SR	Y	2	5	6	40	35	25	+ 3
192	3	FIR	F8	26	34	12		QT	CREAM	R	F	0	1	4	3	40	60	0	+ 4
192	4	CWS	A2	88	47	38		QT	GREY	RP	F	Y	2	1	2	0	40	60	- 3
193	1	CCS	C1A	58	52	28		QT	GREY	RP	F	Y	5	3	5	0	55	45	- 5
193	2	CFB	E2	113	87	55		QT	GREY	R	SR	Y	6	1	3	0	45	55	- 5
196	1	CBD	B3	150	75	43		QT	PURPLE	R	SR	?	3	9	5	0	75	25	- 3
196	2	CIR	A1	83	64	43		QT	CREAM	RP	SR	0	1	5	4	0	60	40	- 3
196	3	CWS	A1	120	81	55		QT	GREY	RP	F	0	2	4	5	0	70	30	- 3
196	4	FIR	F3	45	50	13		QT	DGREY	R	F	Y	2	5	5	50	40	10	+ 4
196	5	FSE	F3	52	52	14		QT	GREY	R	F	Y	2	2	3	45	40	15	+ 4
196	6	FFB	F7	89	42	20		QT	GREY	R	F	Y	2	3	3	45	50	5	+ 4
197	1	FSE	F7	55	40	13		QT	HONEY	R	F	?	2	5	2	45	35	20	+ 4
198	1	CIR	A1	177	120	85		QT	CREAM	VR	SR	0	0	8	4	0	70	30	- 3
198	2	FSP	F6	120	65	39		QT	GREY	RP	SR	U	2	2	3	40	35	25	+ 3
198	3	FIR	F6	110	91	45		QT	GREY	VR	SR	U	1	1	2	30	30	40	+ 4
198	4	CIR	C2A	59	58	30		QT	PINKGREY	R	SR	0	2	5	4	0	65	35	- 4
198	5	FRC	F7	87	66	18		QT	BROWN	R	F	?	1	4	4	45	50	5	+ 3
216	1	CUD	B1	100	85	59	521	QT	BUFF	R	SR	0	3	2	3	0	50	50	- 5
216	2	CMI	B4	28	19	15	4	FL	GREY	G	F	?	7	5	5	0	60	40	- 5
218	1	FRC	F3	73	79	26	154	IG	GREEN	R	R	0	3	8	3	40	40	20	+ 4
218	4	FSP	F2	94	36	25	48	QT	GREYBRO	R	F	S	1	2	2	40	20	45	+ 3
218	5	FSP	F3	39	25	12	13	QT	MAROON	R	F	S	1	5	3	40	40	20	+ 3
224	1	FPO		43	32	8	0	IG	GREY	R			0	0	0	0	0	0	5
224	2	FRS	F6	63	70	23	113	QT	BUFF	VR	R	0	1	3	3	50	20	30	+ 4
226	1	CWS	A1	80	58	41	161	QT	GREY	R	SR	S	2	1	3	0	65	35	- 3
248	1	FSP	F2	61	90	28	171	QT	GREYPURP	R	SR	0	1	1	1	40	10	50	+ 4
261	2	CBD	B2	84	75	60	348	QT	BROWN	R	R	0	3	5	4	0	60	40	- 3
269	1	CHW	C3A	94	81	54	354	QT	BROWN	R	SR	0	7	2	5	0	70	30	- 5
276	1	FSP	F2	67	43	25	75	QT	GREY	R	SR	0	0	3	3	40	30	30	- 3
278	1	CCS	C1B	74	70	35	187	QT	GREEN	RP	SR	0	7	0	4	0	50	50	- 5

278	3	FAN	F8	53	41	16	31	QT	GREEN	R	F	0	2	1	3	50	30	20	+ 4
278	6	FRC	F7	52	33	18	14	QT	FAWN	R	SR	?	1	3	2	50	20	30	+ 5
278	7	CCS	C3A	67	65	30	120	QT	GREY	RP	SR	0	2	2	2	0	75	25	- 3
282	1	FRS	F5	73	53	18	85	QT	GREY	R	F	E	1	1	1	50	5	45	+ 3
284	2	CCS	C3A	90	79	47	410	QT	GREY	RP	SR	0	5	2	3	0	50	50	- 4
284	3	FRS	F6	23	20	6	4	FL	PGREY	F	F	Y	2	0	1	50	5	45	+ 5
284	4	FSE	F3	60	51	26	88	QT	DGREY	VR	R	?	2	1	2	45	35	20	+ 4
284	5	CBD	B2	59	48	41	156	QT	GREY	R	SR	0	1	1	2	0	35	65	- 3
284	6	CUD	B1	97	62	53	356	QT	GREYHON	R	SR	0	1	0	1	0	25	75	- 3
305	1	CCS	C1A	63	61	29	124	QT	GREEN	R	SR	?	6	1	3	0	60	40	- 5
307	1	FIR	F2	111	89	61	485	QT	GREEN	R	SR	S	1	1	2	30	40	30	+ 3
319	1	FSE	F3	65	46	21	66	QT	GREENGRE	RP	SR	R	7	0	3	30	50	20	+ 5
321	1	FIR	F2	58	44	19	56	QT	BROWN	R	SR	?	3	1	4	40	30	30	+ 5
322	1	CUD	B1	80	54	41	208	QT	BUFF	R	SR	0	2	0	1	0	30	70	- 3
324	1	FRS	F5	58	43	14	50	QT	CREAM	R	F	0	3	0	3	50	35	15	+ 4
327	1	FIR	F2	69	44	25	68	SL	DGREEN	P	SR	0	0	3	3			20	+ 3
327	2	FIR	F2	75	63	27	46	SL	DGREEN	P	SR	0	0	2	2			40	+ 3
327	3	FIR	F2	53	42	9	29	SL	DGREEN	P	SR	0	0	2	2			20	+ 3
328	1	CCS	C2A	84	74	40	272	QT	PINKBROW	RP	SR	R	7	0	4	35	30	35	- 5
339	4	CMI	B4	34	29	13	14	FL		G	F	E	5	0	4	0	90	10	- 3
340	1	CWS	A1	88	44	43	331	QT	PURPLE	R	R	?	0	4	2	0	50	50	- 3
345	2	FIR	F3	41	50	13	26	QT	GREEN	R	SR	U	1	1	2	50	30	20	+ 3
345	3	FSP	F3	43	26	11	17	QT	MAROON	RP	F	0	1	2	2	0	95	5	+ 3
346	4	FPO		47	46	25	50	IG	BROWN	O									5
346	5	FRS	F6	37	34	10	13	QT	DGREY	R	SR	?	1	2	2	50	20	30	+ 4
346	6	CIR	A1	47	32	20	0	QT	DGREY	R	F	0	0	3	2	0	30	70	- 3
346	8	FIR	F3	50	25	13	20	QT	DBROWN	R	F	0	1	3	2	40	30	30	+ 3
346	15	FIR	F2	70	52	25	99	QT	BROWN	R	F	0	1	1	2	50	25	25	+ 4
346	21	FRS	F7	67	60	23	78	QT	BROWN	R	SR	0	7	1	5	45	35	20	+ 5
347	4	FRS	F2	51	59	27	50	QT	BROWN	R	R	0	1	1	2	45	10	45	+ 3
348	1	CBD	B4	120	72	56	604	QT	FAWN	R	SR	0	3	1	3	0	50	50	- 4
349	1	FRC	F7	64	50	23	120	QT	BROWN	R	R	0	3	1	2	50	35	15	+ 3
349	3	FRS	F7	42	40	12	20	QT	FAWN	R	F	?	3	2	3	50	30	20	+ 5
349	4	FIR	F7	45	38	15	18	QT	HONEY	R	F	?	3	2	3	40	35	15	+ 5
349	5	FSC	F7	23	17	9	4	FL	FAWN	G	F	0	10	2	5	30	55	15	+ 5
349	6	FIR	F5	37	35	11	16	QT	HONEY	R	SR	Y	1	1	2	44	5	50	+ 4
349	7	FSP	F3	29	24	11	6	QT	HONEY	R	SR	Y	2	2	3	15	60	15	+ 4
351	2	FIR	F3	40	34	12	19	QT	GREEN	R	F	0	2	2	2	30	35	15	+ 4
351	1	CMI	B4	30	14	11	6	FL	GREY	G	F	?	0	4	3	0	95	5	- 3
353	1	FSE	F7	66	55	22	93	QT	BUFFGREE	R	SR	R	4	0	2	45	55	5	+ 5
353	4	FRS	F2	48	35	10	20	QT	BROWN	R	SR	?	1	2	3	45	5	50	+ 3
353	5	FIR	F7	65	60	19	73	QT	GREEN	R	SR	E	1	4	3	50	35	15	+ 3
353	6	CBI	B4	66	71	66	378	QT	MAROON	R	F	0	6	0	4	20	20	60	- 5
363	1	CBD	A2	117	70	55	469	QT	BUFF	RP	SR	?	1	2	3	0	45	55	- 3
367	1	CHA		156	129	59	1815	QT	PINK	R			0	0	0	0	0	0	4
374	1	CWS	A1	58	44	19	43	QT	PGREY	RP	SR	0	0	5	3	50	45	5	- 3
376	1	FRS	F3	50	49	14	36	QT	CREAMBU	R	SR	R	2	0	2	50	45	5	+ 5
376	3	FDB	F6	43	39	19	20	QT	BUFF	R	SR	0	2	2	2	40	15	45	+ 3
376	4	FRC	F7	36	28	10	14	QT	PINK	VR	R	0	3	3	3	45	45	10	+ 3
376	5	CBD	A1	46	36	20	2	QT	GREEN	SR	SR	0	1	3	3	0	50	50	- 3
376	8	FRC	F3	43	34	13	19	QT	GREEN	RP	SR	Y	5	2	5	50	50	0	+ 4
376	11	FRS	F2	35	33	19	19	QT	GREYGREE	RP	F	?	1	1	2	45	5	50	+ 3
377	2	FSC	F8	38	31	20	23	QT	CREAM	VR	SR	?	0	5	4	45	55	0	+ 3
377	3	FIR	F2	37	37	19	20	QT	MAROON	R	F	0	1	1	2	50	25	25	+ 3
382	1	FIR	F1	49	55	13	40	QT	GREY	VR	SR	?	0	2	2	50	5	45	+ 3
383	2	FRS	F1	59	71	22	109	QT	BUFF	VR	SR	0	1	1	1	50	0	50	+ 3
382	4	FRS	F1	34	36	10	8	QT	WHITE	R	SR	0	0	1	1	0	0	50	+ 3
383	9	FKN	F3	91	55	32	146	QT	PINKGREY	R	SR	R	8	0	5	40	30	30	+ 5
383	10	CMI	B4	20	21	11	6	FL	FAWN	G	F	0	3	4	3	0	70	30	- 5
384	1	FRC	F6	53	46	12	30	QT	DGREY	R	SR	?	0	4	3	50	25	25	+ 3
384	2	CWS	A1	37	22	10	7	QT	CREAM	R	SR	E	0	4	3	40	60	0	- 3
385	8	CHW	C3A	64	51	48	160	QT	PGREY	R	R	?	4	2	4	0	45	55	- 5
385	9	CSP	A2	103	77	49	236	QT	HONEY	R	R	0	1	2	2	0	50	50	- 3
385	10	FRS	F1	47	42	19	36	QT	MAROON	R	R	?	1	0	1	50	0	50	+ 3
387	1	FIR	F7	38	44	14	17	QT	BROWN	R	F	0	4	3	4	45	50	5	+ 5
387	2	CIR	C2A	61	50	39	135	QT	FAWN	R	F	0	3	4	3	0	60	40	- 3

387	4	FSP	F2	56	47	18	0	QT	BUFF	R	F	0	1	1	2	50	10	40	+ 4
387	8	FIR	F2	37	44	14	20	QT	MAROON	R	SR	Y	5	1	3	50	20	30	+ 5
387	9	FDB	F8	47	41	13	25	QT	MAROON	R	SR	Y	2	3	3	50	45	5	+ 5
387	10	FIR	F2	60	43	22	46	QT	HONEY	RP	SR	Y	0	3	3	40	25	35	+ 3
387	11	FSP	F5	19	41	23	13	QT	GREY	R	SR	Y	1	1	1	0	0	40	+ 3
387	25	FPO		37	43	19	16	QT	DGREY	O									5
389	2	CSQ	D3	58	48	44	157	QT	GREENBUF	R	F	0	1	2	2	0	50	50	- 3
389	7	CWS	A1	54	47	22	51	QT	WHITE	R	F	H	0	3	2			30	- 3
389	8	FRS	F1	55	41	27	77	QT	BUFF	R	F	0	1	2	2	40	10	50	+ 3
389	10	FSP	F3	36	26	11	10	QT	GREY	R	SR	E	0	4	3	0	0	0	+ 3
389	13	FRS	F3	38	56	16	41	QT	BUFF	R	F	0	1	1	2	50	40	10	+ 3
389	14	FRU		30	26	13	18	IG	GREEN				0	0	0	0	0	0	5
389	17	FRS	F7	40	41	11	23	QT	CREAM	VR	SR	?	1	3	4	40	55	5	+ 4
389	18	FIR	F1	47	58	22	62	QT	CREAM	VR	SR	?	1	2	2	40	5	55	+ 3
389	19	CBD	B3	110	76	53	364	QT	CREAM	RP	SR	Y	0	5	4	0	70	30	- 3
393	1	CIR	B4	65	65	69	434	QT	BUFF	VR	F	?	5	5	5	0	65	35	- 4
393	2	FIR	F6	59	56	18	63	QT	PINK	R	SR	?	5	2	4	45	20	35	+ 5
393	3	CBD	B3	63	47	32	115	QT	DGREY	R	SR	?	4	6	5	0	60	40	- 4
393	4	CSP	A2	88	79	39	321	QT	BROWN	R	SR	0	2	1	3	0	45	55	- 3
393	5	FSP	F3	41	40	15	28	QT	GREYPURP	R	SR	?	3	2	2	50	40	10	+ 3
393	8	FSP	F2	65	51	25	107	QT	GREYGREE	R	SR	0	4	0	4	20	35	45	+ 3
393	13	FSC	F3	31	36	13	16	QT	WHITE	R	F	0	6	0	3	50	45	5	+ 5
393	14	CWS	A1	50	26	23	37	QT	GREY	R	F	0	0	5	3	0	0	45	- 3
395	1	FHA	F1	59	46	20	73	QT	WHITE	R	F	?	0	1	1	50	0	50	+ 4
395	4	CBD	B3	97	102	80	650	QT	WHITE	VR	SR	E	0	3	2	0	45	55	- 3
395	5	FRS	F6	46	33	19	30	QT	GREY	R	SR	Y	2	1	3	45	10	45	+ 3
395	6	FRS	F5	41	38	13	26	QT	BUFF	R	F	Y	1	1	2	50	5	45	+ 3
395	11	CHW	B3	75	56	48	255	QT	BUFF	VR	R	0	3	3	3	0	65	35	- 3
397	1	FRS	F6	49	58	19	34	QT	GREY	R	SR	?	2	0	2	0	60	40	+ 3
397	3	FSE	F7	54	41	21	32	QT	PURPLE	VR	SR	0	3	0	2	50	30	20	+ 4
397	5	CIR	B4	92	41	47	1143	QT	PINK	R	SR	0	3	3	4	0	60	40	- 5
397	7	FRS	F1	34	34	11	0	QT	BUFF	R	F	0	1	2	3	45	5	50	+ 3
399	1	FRS	F7	46	36	12	22	QT	CREAM	R	SR	E	0	3	3	50	35	15	+ 3
399	2	FSP	F7	51	47	18	19	QT	CREAM	R	F	E	0	4	2	50	50	0	+ 3
399	4	CUD	B1	71	62	34	178	QT	BROWN	R	F	Y	4	2	4	0	55	45	- 3
399	5	CSP	A2	91	104	61	515	QT	CREAM	R	SR	?	1	2	2	0	50	50	- 3
401	1	FRS	F2	67	70	34	136	QT	CREAM	R	SR	0	1	2	2	50	20	30	+ 3
403	1	FRC	F2	43	75	23	36	QT	PINK	R	F	0	6	0	4	50	25	25	+ 5
403	2	CCS	C2A	84	80	55	302	QT	GREY	RP	SR	0	1	2	3	0	65	35	- 3
403	3	CHW	C3A	65	61	41	180	QT	GREYGREE	R	SR	0	0	5	4	0	70	30	- 4
403	5	FRS	F6	63	55	23	82	QT	WHITE	R	F	0	2	0	2	50	15	35	+ 5
405	1	FIR	F3	65	57	22	99	QT	BUFF	R	F	0	2	3	4	40	40	20	+ 5
405	3	CCS	C2A	71	55	28	119	QT	PURPLE	R	SR	?	7	2	4	0	80	20	- 3
405	4	CCS	C1A	58	52	34	104	BA		R	F	0	0	5	4	0	55	45	- 3
405	5	FSP	F2	50	44	20	54	QT	BUFF	R	F	0	2	1	2	40	20	40	+ 3
405	8	FSE	F3	80	64	31	125	QT	BUFF	RP	F	0	0	4	2	0	0	15	+ 4
405	9	FIR	F3	36	42	20	19	QT	MAROON	R	F	Y	2	1	2	40	50	10	+ 5
409	1	CHA		109	60	55	438	QT	BROWN	R	0	0	0	0	0	0	0	0	0 3
411	2	FSP	F1	21	19	8	4	FL		G	F	E	2	3	2	40	20	40	+ 3
411	3	CUD	B1	95	57	58	64	QT	CREAM	R	F	0	0	2	2	0	45	55	- 3
419	1	FRC	F8	68	48	21	59	QT	GREEN	RP	SR	?	1	6	4	40	60	0	+ 3
426	3	CHW	C1A	70	83	56	377	QT	WHITE	R	R	0	3	2	3	0	60	40	- 4
426	4	CRU		118	78	57	70	QT	PINK	R	SR	H	0	5	5	0	0	45	- 4
426	5	CWS	A1	71	49	30	106	QT	BROWN	R	F	0	1	1	2	0	55	45	- 3
435	1	CCS	C2A	70	59	36	164	QT	GREEN	R	SR	0	4	3	4	0	75	25	- 5
437	2	FIR	F8	53	50	15	34	QT	PINKFAWN	R	F	0	1	9	6	45	55	0	+ 3
438	1	FIR	F7	36	32	9	19	QT	WHITE	R	SR	E	0	4	2	50	50	0	+ 3
440	1	CWS	A1	65	41	42	68	QT	WHITE	R	F	?	2	4	2	0	100	0	- 3
440	2	CSQ	D2	55	71	75	469	QT	BROWN	P	SR	?	5	3	4	0	60	40	- 5
442	2	CCS	C3A	105	102	72	820	QT	WHITE	VR	R	0	0	9	4	0	50	50	- 3
444	1	FRS	F1	55	59	22	91	QT	PINK	R	SR	H	1	1	1	45	5	50	+ 3
445	1	FSP	F2	58	29	11	20	QT	BUFFGREE	R	SR	?	3	1	2	45	50	5	+ 3
446	1	FIR	F6	70	50	18	61	QT	PINK	RP	SR	0	1	3	2	0	70	30	+ 3
446	4	FRS	F2	68	56	27	101	QT	BUFF	R	SR	?	0	4	3	45	25	30	+ 3
446	7	FSE	F2	73	36	15	45	QT	GREY	R	SR	?	1	3	3	35	35	30	+ 3
448	1	CIR	A2	120	82	66	765	QT	BROWN	VR	SR	0	0	5	3	0	75	25	- 3

450	1	FIR	F3	71	59	19	60	QT	ROSE	VR	R	?	1	2	2	50	30	25	+	3
457	1	CBD	B2	65	43	45	126	QT	FAWN	R	F	0	4	0	2	0	60	40	-	4
460	1	CSQ	D3	57	58	53	206	QT	WHITE	R	SR	0	4	3	4	70	30	5	-	5
467	1	CBD	B3	103	83	53	0	QT	BUFF	R	F	0	1	2	3	0	40	60	-	3
473	1	CUD	B1	118	72	64	707	QT	GREY	R	SR	0	1	0	1	0	15	85	-	3
473	2	CBD	B3	60	73	64	383	QT	WHITE	R	SR	?	2	1	1	0	30	70	-	4
473	3	CSP	A1	81	69	43	254	QT	GREY	RP	SR	E	1	4	4	0	60	40	-	3
475	1	CHA	A1	59	44	36	0	QT	WHITE	R	F	H	0	7	4	0	60	40	-	3
475	8	CBD	B2	54	46	36	96	QT	GREYBUFF	R	F	0	1	2	3	0	50	50	-	3
475	9	CWS	A1	60	45	26	96	QT	BUFF	RP	F	0	2	2	3	45	10	45	+	3
481	1	CIR	B4	63	40	41	126	QT	WHITE	RP	F	0	2	3	3	0	80	20	-	4
481	2	FIR	F3	47	36	25	33	QT	PURPLE	R	SR	?	3	3	3	40	45	15	+	4
481	5	CWS	A1	43	85	24	95	QT	BROWN	RP	F	0	1	0	1	0	45	55	-	3
483	1	CBD	B2	149	100	84	1110	QT	GREY	R	SR	0	1	1	2	0	50	50	-	3
487	1	FIR	F2	68	57	21	99	QT	PINKGREY	R	F	0	2	1	3	40	30	30	+	5
489	1	FRS	F2	56	49	10	20	QT	GREYHON	R	SR	?	2	0	2	50	15	50	+	3
489	2	FAN	F7	85	65	40	172	QT	DMAROON	R	R	0	3	2	4	30	50	20	+	4
489	3	CSP	B1	75	56	37	194	QT	ROSE	R	SR	0	2	0	1	0	40	60	-	3
491	2	CUD	B1	94	59	66	270	QT	FAWN	R	SR	0	4	2	4	0	50	50	-	5
491	3	CCS	C1A	74	91	72	497	QT	BROWN	RP	SR	0	3	4	2	0	45	55	-	5
494	1	FPO		65	50	29	102	GR	GREEN				0	0	0	0	0	0		5
494	2	FSP	F3	83	66	24	183	QT	GREY	R	SR	0	2	1	3	40	40	20	+	3
496	1	FIR	F2	51	73	26	100	QT	CREAM	R	SR	0	1	2	3	45	30	25	+	4
497	1	FRS	F1	48	52	13	43	QT	WHITE	R	SR	0	1	0	1	0	50	50	+	3
497	2	FRS	F3	49	42	13	37	QT	WHITE	R	SR	H	0	4	3	45	45	10	+	4
500	6	FIR	F7	64	67	23	97	QT	PINK	VR	SR	?	1	8	4	50	45	5	+	4
500	7	CIR	A2	104	60	37	240	QT	GREY	VR	SR	0	0	4	2	0	55	45	-	3
500	8	CBD	B3	48	52	29	80	QT	GREY	R	F	0	1	1	2	0	30	70	-	3
500	10	FRS	F1	60	64	21	105	QT	GREY	R	F	?	1	0	1	50	0	50	+	4
502	1	CBD	B2	63	69	36	138	QT	BROWN	R	SR	?	2	1	3	0	40	60	-	4
502	2	FRS	F5	47	58	14	43	QT	BUFF	VR	SR	?	1	1	2	10	40	50	+	3
502	3	FIR	F7	48	38	19	30	QT	GREEN	VR	SR	?	0	4	3	40	50	10	+	3
504	1	CSQ	D2	65	62	47	239	QT	BROWN	VR	SR	0	1	2	3	0	60	40	-	3
506	1	FSP	F3	56	36	20	46	QT	BUFF	VR	SR	0	0	5	4	0	0	5	+	3
507	1	CCC	C2B	128	59	70	596	QT	WHITE	R	SR	0	7	5	4	30	65	5	-	5
507	2	CBD	C2A	109	92	47	591	QT	BROWN	R	SR	0	1	2	3	0	55	45	-	3
510	1	FSP	F3	31	34	19	26	QT	PPURPLE	R	F	?	0	4	4	0	0	10	+	3
511	1	CBD	C2A	114	110	39	479	QT	BUFF	RP	SR	?	3	5	6	0	60	40	-	4
511	2	FSC	F7	56	45	34	80	QT	WHITE	R	SR	?	5	3	5	30	45	25	+	4
513	1	CCC	C3B	144	132	78	2001	QT	BUFF	R	SR	0	4	2	4	0	40	60	-	5
515	1	CSQ	D4	65	76	63	359	QT	GREY	R	SR	0	6	3	4	0	50	50	-	5
515	2	FIR	F6	41	41	18	18	QT	GREY	R	SR	0	1	2	2	40	35	25	+	4
515	3	FRC	F3	38	39	23	18	QT	MAROON	R	SR	?	1	6	5	45	45	10	+	4
516	1	CBD	B3	104	61	46	392	QT	BUFF	R	F	0	2	1	3	0	30	70	-	5
516	2	CBD	B3	76	72	53	307	QT	FAWN	R	SR	0	2	2	2	0	40	60	-	5
516	4	FSP	F5	63	55	20	78	QT	WHITE	VR	SR	H	0	4	2	20	70	10	+	3
517	1	FIR	F6	81	75	33	148	IG	GREEN	R	SR	?	3	5	5	50	35	15	+	5
520	2	CWS	A1	85	40	31	111	QT	GREY	R	F	0	2	1	1	0	55	45	-	3
523	1	FSP	F2	47	56	24	88	QT	GREY	R	F	0	1	1	2	30	10	60	+	4
523	2	CCS	C1A	73	71	51	281	QT	BROWN	VR	SR	0	3	3	4	0	45	55	-	3
523	3	CWS	A1	69	53	21	83	QT	GREY	RP	SR	0	0	3	2	0	50	50	-	3
523	5	FIR	F3	51	43	16	41	QT	GREY	RP	F	?	2	3	3	50	30	20	+	3
524	1	CWS	A2	80	41	41	140	QT	GREY	R	SR	0	2	2	2	40	20	40	-	3
524	3	FIR	F1	50	35	18	40	QT	DPINK	R	F	?	1	0	1	45	0	55	+	4
526	1	CIR	C2A	90	72	31	192	QT	BROWN	RP	F	0	5	1	4	40	45	15	-	5
528	2	CIR	C2A	118	86	46	644	QT	WHITE	RP	SR	0	2	2	4	0	60	40	-	3
531	1	CBD	B3	0	0	0	0	QT	FAWN	VR	F	0	1	0	1	0	0	60	-	3
533	1	FSP	F2	60	86	44	272	QT	GREYCRE	R	SR	Y	1	0	1	0	40	60	+	3
533	4	FRS	F1	54	56	18	75	QT	BROWN	VR	SR	H	1	0	1	50	0	50	+	3
533	6	FRS	F1	38	46	19	32	QT	GREY	R	F	0	1	0	1	50	0	50	+	3
600	1	FSE	F7	97	54	20	99	QT	GREY	R	F	?	2	3	4	45	50	5	+	4
605	1	FFB	F7	72	40	18	59	QT	PBROWN	R	SR	?	1	4	2	45	50	5	+	4
605	2	CUD	B1	72	55	48	254	QT	BROWN	R	F	0	2	0	1	0	30	70	-	3
605	4	FFB	F8	59	30	15	22	QT	BROWN	R	F	?	3	5	3	50	40	10	+	4
608	1	CUD	B1	74	67	43	215	QT	PGREY	R	F	0	3	1	2	0	30	70	-	3
611	1	FRS	F2	91	68	40	265	QT	PGREY	R	F	0	2	2	3	40	20	40	+	4

612	1	FIR	F7	66	66	26	86	QT	PGREY	R	SR	?	1	3	2	45	45	10	+ 3
612	2	FIR	F7	36	31	11	12	QT	PINK	R	SR	?	2	3	3	40	55	5	+ 4
614	1	CSP	A2	80	61	37	181	QT	PBROWN	RP	SR	0	0	4	4	0	75	25	- 3
619	1	FFB	F3	59	35	11	25	QT	WHITE	R	R	?	0	4	2	40	50	10	+ 4
625	1	FSE	F7	55	28	14	25	QT	FAWN	R	R	Y	2	2	4	45	50	5	+ 4
625	2	FSP	F2	78	43	37	117	QT	BROWN	RP	F	0	2	0	1	45	0	55	+ 3
625	3	FIR	F3	93	42	27	139	QT	FAWN	R	SR	0	1	3	3	45	35	20	+ 3
625	4	FIR	F6	85	115	33	221	QT	PGREY	R	SR	R	1	2	3	45	45	10	+ 3
625	5	CHW	C3A	48	78	46	212	QT	BROWN	R	SR	Y	2	5	4	0	65	35	- 4
625	6	FIR	F1	53	50	18	40	QT	BROWN	R	SR	Y	0	2	2	45	5	50	+ 3
627	1	FIR	F1	51	59	11	39	QT	GREY	R	F	?	1	1	2	45	5	50	+ 3
627	2	FSE	F7	53	29	20	30	QT	PGREY	R	R	0	2	3	2	45	50	5	+ 4
627	3	CIR	B2	42	74	61	199	QT	PINK	R	R	?	1	5	6	0	85	15	- 3
628	1	CBD	C2A	104	81	46	400	QT	BROWN	R	SR	Y	2	1	2	50	10	40	- 4
629	1	CWS	A2	131	126	79	1000	QT	MAROON	R	C	0	3	6	5	0	80	20	- 3
631	1	CSP	A2	70	75	39	278	QT	GREEN	RP	R	0	1	1	2	10	40	50	- 3
631	2	FSP	F2	44	16	38	34	QT	BROWN	R	F	0	6	3	4	30	35	35	+ 4
631	3	FSP	F2	46	52	18	43	QT	FAWN	R	F	0	2	2	2	30	40	30	+ 4
631	4	FSP	F4	51	53	13	22	QT	FAWN	R	F	0	2	1	2	50	30	20	+ 4
631	5	FSP	F3	36	40	14	11	SS	BROWN	VR	SR	0	0	4	0	50	35	15	+ 3
631	6	FRS	F5	58	46	18	56	QT	FAWN	VR	R	?	5	5	3	45	5	50	+ 3
632	1	CSQ	D4	62	64	58	297	QT	FAWN	R	SR	?	2	5	4	0	80	20	- 4
639	1	CBD	B4	53	41	33	71	QT	GREY	F	F	0	2	1	3	40	25	35	+ 4
639	2	CBI	B2	60	55	28	92	QT	FAWN	RP	F	0	2	1	2	0	30	70	- 4
639	3	FIR	F7	51	69	27	72	QT	GREEN	R	SR	0	3	2	3	45	40	15	+ 4
640	1	CCS	C2A	89	78	37	235	QT	BROWN	R	SR	?	5	1	4	35	30	35	- 5
641	1	FSE	F7	47	28	14	18	QT	PGREY	F	SR	?	3	1	3	45	45	10	+ 5
642	1	CIR	B4	80	77	70	207	QT	GREY	VR	SR	?	2	3	4	0	70	30	- 3
645	1	CCC	C1B	70	72	36	178	QT	PGREY	FP	SR	0	6	7	4	0	50	50	- 5
647	1	FRS	F5	59	50	14	41	QT	PGREY	R	SR	?	2	0	1	45	5	50	+ 4
649	1	CIR	C2A	73	77	39	222	QT	HONEY	RP	SR	0	3	2	4	30	35	35	- 4
649	2	CUD	B1	71	56	35	187	QT	PGREY	RP	R	0	2	1	2	0	30	70	- 4
650	1	FIR	F3	43	48	13	20	QT	BROWN	R	F	?	1	3	3	45	40	15	+ 4
651	1	FIR	F3	37	47	9	21	QT	BROWN	R	F	?	3	1	3	45	55	5	+ 5
651	2	CBI	B2	78	50	34	135	QT	PGREY	RP	F	?	2	1	2	0	40	60	- 4
652	1	FRC	F7	37	29	9	14	QT	GREY	F	SR	?	3	1	2	50	45	5	+ 4
652	2	FWS	F3	96	76	43	215	QT	GREY	RP	SR	0	1	5	3	50	30	20	+ 3
652	3	CWS	A1	58	55	29	79	QT	GREY	RP	SR	0	0	6	3	0	100	0	- 3
652	4	FSE	F3	70	61	20	108	QT	ROSE	R	SR	?	1	5	3	45	20	35	+ 3
652	5	FIR	F6	61	48	14	50	QT	GREY	R	SR	?	1	1	2	50	5	45	+ 3
652	6	FIR	F7	51	43	16	32	QT	BROWN	R	SR	?	0	5	4	35	50	15	+ 3
652	7	FRS	F6	56	59	22	79	QT	GREY	VR	R	0	2	0	2	50	20	30	+ 4
652	9	CWS	A1	85	64	51	239	QT	FAWN	RP	SR	0	2	3	4	0	65	35	- 3
660	1	FIR	F7	57	56	24	92	QT	GREY	VR	SR	?	2	1	2	50	30	20	+ 4
663	1	FSP	F2	83	64	35	185	QT	CREAM	R	SR	0	1	1	2	45	10	45	+ 3
666	1	FSE	F2	99	65	37	232	QT	BROWN	R	SR	?	2	0	2	50	15	35	+ 5
666	2	FSP	F5	103	59	31	167	QT	BROWN	VR	R	0	2	2	2	35	30	35	+ 3
666	3	FSP	F2	48	27	20	31	QT	BROWN	R	SR	?	3	3	3	30	55	15	+ 3
670	1	FSE	F7	42	31	11	16	QT	HONEY	R	SR	0	1	4	3	45	55	0	+ 4
670	2	CWS	A1	53	46	22	60	QT	GREY	R	SR	?	2	3	4	20	70	10	- 3
670	3	FIR	F2	65	49	27	68	QT	GREY	R	SR	?	1	2	3	30	35	35	+ 4
672	1	FSP	F2	56	58	31	68	QT	GREY	F	SR	?	1	1	2	40	10	50	+ 3
672	2	FSP	F7	58	30	14	22	QT	FAWN	R	F	0	1	1	1	50	45	5	+ 3
672	3	FIR	F6	62	41	23	71	QT	WHITE	R	R	0	2	2	2	50	30	20	+ 3
672	4	FIR	F6	107	74	43	227	QT	HONEY	VR	R	?	1	3	3	50	35	15	+ 3
673	1	FIR	F1	51	79	21	96	QT	GREY	VR	SR	0	1	1	2	40	40	20	+ 3
673	2	FIR	F5	96	65	27	179	QT	CREAM	RP	SR	?	1	1	2	50	15	35	+ 3
678	1	FIR	F8	51	52	21	45	QT	PGREY	RP	F	?	1	2	3	50	50	0	+ 4
701	1	CIR	B2	71	55	37	148	QT	GREY	R	SR	0	1	4	3	0	60	40	- 3
702	1	CBI	A1	71	71	42	280	QT	BROWN	RP	SR	0	3	0	2	0	25	75	- 4
703	1	CCS	C2A	66	42	28	100	QT	NR	R	R	?	4	1	2	0	70	30	- 4
710	1	CFB	E3	106	75	61	366	QT	GREY	R	R	0	6	4	4	0	80	20	- 4
710	2	FRS	F1	52	71	26	93	QT	FAWN	R	SR	0	1	0	1	40	0	60	+ 3
710	3	CIR	A1	101	68	38	273	QT	FAWN	RP	SR	?	1	2	2	0	55	45	- 3
710	4	CWS	A2	116	74	58	514	QT	ROSE	VR	R	0	0	5	4	0	60	40	- 3
711	1	FRS	F2	58	60	22	97	QT	BROWN	VR	R	0	1	2	3	45	5	50	+ 4

712	1	CCL	B4	74	63	51	274	QT	ROSE	R	SR	?	2	3	3	0	80	20	- 4
712	3	FAN	F2	75	54	35	125	QT	ROSE	RP	SR	0	0	3	2	45	15	40	+ 3
713	1	FIR	F1	82	70	21	143	QT	WHITE	RP	SR	0	1	1	1	45	10	45	+ 3
713	2	CPO	B4	113	102	88	1014	QT	BROWN	RP	SR	0	2	10	6	0	45	55	- 3
716	1	CIR	A1	52	74	50	214	QT	CREAM	R	SR	0	2	0	2	40	15	45	- 3
716	2	FWS	F5	50	52	31	79	QT	CREAM	RP	R	?	2	2	3	35	15	50	+ 3
716	3	CIR	A1	76	60	48	160	QT	CREAM	RP	0	?	0	3	2	0	50	50	- 3
716	4	FRS	F5	26	39	9	16	QT	CREAM	R	R	0	1	1	2	45	5	50	+ 3
716	6	FIR	F7	36	44	13	21	QT	GREY	R	SR	?	0	7	5	40	55	5	+ 4
716	7	CIR	C2A	32	74	39	241	QT	MAROON	R	R	0	3	5	5	0	60	40	- 4
717	2	FIR	F5	46	28	17	18	QT	BROWN	VR	SR	0	0	4	3	0	0	15	+ 3
718	1	FSP	F6	65	22	10	13	QT	PPINK	R	R	?	1	3	3	0	0	40	+ 3
718	2	FSP	F7	45	28	19	16	QT	PGREY	R	R	0	1	5	4	0	0	5	- 3
719	1	CBD	D2	62	62	51	221	QT	RED	R	SR	0	1	3	3	0	55	45	- 3
722	1	FRS	F7	53	49	19	26	QT	GREYSTRI	R	SR	Y	1	2	3	45	45	15	+ 5
722	2	CCC	C2B	74	55	27	109	QT	BROWN	R	SR	0	7	10	5	0	95	5	- 5
722	3	FSP	F8	50	37	10	12	QT	DGREY	VR	SR	?	1	2	1	50	50	0	+ 3
722	4	FRS	F7	43	38	10	19	QT	GREY	RG	R	Y	1	3	2	45	45	10	+ 3
722	5	FIR	F2	47	32	16	16	QT	MAROON	R	SR	Y	0	4	4	50	40	10	+ 3
722	9	FHW	F6	57	53	29	64	QT	FAWN	R	SR	Y	1	2	3	45	35	20	+ 3
722	43	FIR	F7	0	0	16	24	QT	GREY	RG	R	0	1	2	3	45	40	15	+ 3
722	60	FSP	F8	10	10	2	1	FL	WHITE	G	F	?	1	3	3	45	55	0	+ 4
722	63	CMI	B2	27	18	14	4	FL	FAWN	G	F	0	4	2	3	0	55	45	- 5
722	69	FHW	F8	74	0	24	57	QT	HONEY	VR	SR	?	1	3	4	20	45	35	+ 4
722	70	FHW	F8	54	32	18	26	QT	HONEY	VR	SR	?	1	3	3	50	25	25	+ 3
722	72	FSP	F2	17	9	4	1	FL	FAWN	G	F	?	0	5	3	50	50	0	+ 3
722	73	FRS	F6	0	0	24	43	QT	PINK	RP	R	?	1	2	3	35	30	35	+ 3
722	77	FIR	F3	0	0	10	4	QT	HONEY	R	SR	Y	1	3	3	45	45	10	+ 4
722	79	FIR	F2	0	0	6	1	QT	FAWN	R	SR	0	0	4	4	45	55	0	+ 3
722	86	FRS	F1	23	20	8	1	QT	HONEY	R	SR	0	1	1	2	45	5	50	+ 3
722	91	FRS	F1	32	47	11	12	QT	GREEN	VR	R	Y	1	1	2	50	15	35	+ 3
722	97	FIR	F7	14	21	5	1	FL	FAWN	G	F	?	3	1	3	50	45	5	+ 5
722	100	CCC	C2A	69	52	35	120	QT	PGREY	R	R	0	2	4	4	0	95	5	- 3
722	110	FRS	F2	37	31	9	7	QT	GREEN	R	R	Y	1	1	1	45	35	25	+ 3
722	114	FIR	F2	49	30	13	7	QT	CREAM	R	SR	0	0	4	3	50	50	0	+ 3
722	116	FIR	F2	45	40	13	21	QT	DGREY	VR	F	Y	1	2	3	50	20	30	+ 4
722	125	FHW	F7	52	36	18	26	QT	BROWN	R	R	Y	3	4	4	40	45	15	+ 4
722	128	FIR	F2	35	42	16	14	QT	CREAM	R	R	0	1	2	3	45	35	20	+ 3
722	129	FIR	F6	53	52	21	60	QT	GREEN	VR	SR	Y	1	3	3	45	15	40	+ 3
722	130	FSP	F8	34	23	10	5	QT	MAROON	R	F	0	1	4	3	50	50	0	+ 3
722	135	FSP	F8	22	26	10	2	QT	FAWN	R	SR	0	2	1	3	45	55	0	+ 4
722	136	CWS	A1	45	37	29	61	QT	MAROON	VR	SR	0	3	2	2	0	0	10	- 4
722	139	FIR	F8	43	33	10	10	QT	DGREY	VR	SR	0	0	4	3	50	50	0	+ 3
722	140	FSP	F7	52	38	16	34	QT	GREEN	R	SR	0	3	2	2	45	30	25	+ 3
722	142	FRS	F2	43	34	9	13	QT	BROWN	R	SR	Y	1	0	1	50	0	50	+ 3
722	146	CHW	C4A	87	79	60	444	QT	MAROON	RP	R	0	2	4	5	0	80	20	- 4
722	148	CBD	B2	66	43	39	107	QT	PINK	RP	R	?	1	1	2	0	30	70	- 3
722	154	CBD	B3	97	87	50	583	QT	CREAM	RP	R	0	2	2	3	0	45	65	- 4
722	163	CCC	C4A	101	91	68	568	QT	FAWN	VR	R	0	2	7	7	0	90	10	- 3
722	164	FRS	F7	64	56	15	55	QT	FAWN	R	SR	0	2	5	4	45	35	20	+ 5
722	165	FSC	F2	51	52	20	59	QT	FAWN	R	SR	0	4	6	6	45	25	30	+ 5
722	166	CWS	A1	66	47	32	74	QT	GREY	R	R	0	0	4	3	0	80	20	- 3
722	167	CBD	B3	103	86	49	407	QT	BROWN	R	R	0	1	1	1	0	20	80	- 3
722	168	FSP	F3	63	30	19	19	QT	GREY	R	SR	?	0	4	2	0	0	5	+ 3
722	169	CWS	A1	60	42	35	72	QT	PGREY	RP	R	?	3	3	4	0	70	30	- 4
722	170	FIR	F7	49	62	11	36	QT	GREYSTRI	R	F	Y	3	2	3	45	55	0	+ 5
722	171	CCC	C1B	64	51	28	75	QT	DGREEN	R	SR	?	4	2	6	0	60	40	- 4
722	172	FRS	F1	46	41	15	32	QT	MAROON	R	R	Y	1	0	1	50	0	50	+ 3
722	173	FSE	F8	58	37	15	25	QT	PINK	R	R	0	1	3	4	45	55	0	+ 4
722	174	CBD	B4	54	53	34	92	QT	CREAM	RP	SR	0	0	5	4	0	85	15	- 3
722	176	CWS	A1	46	31	23	29	QT	FAWN	R	SR	0	1	5	5	0	90	10	- 3
722	177	FSP	F1	41	54	17	43	QT	HONEY	R	SR	0	1	2	2	45	5	50	+ 3
722	178	FSE	F2	44	32	12	19	QT	HONEY	R	SR	Y	1	1	1	50	30	20	+ 4
722	179	FIR	F7	36	31	18	8	QT	CREAM	RP	SR	0	0	2	1	50	40	10	+ 3
722	180	FSE	F3	69	41	19	42	QT	BROWN	R	R	0	1	4	2	50	35	15	+ 3
723	1	FRS	F9	62	55	18	60	QT	HONEY	R	R	R	2	0	1	45	45	10	+ 5

724	1	FIR	F1	34	22	6	4	QT	GREY	R	R	Y	2	0	2	50	5	45	+ 3
724	2	FSP	F2	68	60	17	93	QT	CREAM	R	SR	0	2	0	1	50	0	50	+ 3
724	3	FIR	F1	48	47	15	30	QT	CREAM	R	SR	Y	1	0	1	40	0	60	+ 3
724	5	CWS	A1	62	29	29	46	QT	BROWN	RP	R	0	0	3	2	0	60	40	- 3
724	6	FSP	F2	60	47	23	48	QT	GREEN	R	R	?	1	4	2	30	45	25	+ 3
724	7	FIR	F1	54	35	12	22	QT	CREAM	R	R	?	0	1	1	50	0	50	+ 3
724	8	CWS	A1	83	59	32	182	QT	DPURPLE	R	R	0	2	0	1	35	0	65	+ 4
725	1	CUD	B1	84	59	48	317	QT	BROWN	R	SR	0	2	0	1	0	25	75	- 3
725	2	CCS	C3A	67	56	27	96	QT	DGREY	R	SR	0	3	4	5	0	70	30	- 3
725	3	CWS	A1	38	31	14	13	QT	PGREY	R	SR	0	1	2	2	0	0	40	- 3
725	4	CWS	A1	90	64	35	110	QT	MAROON	R	SR	Y	1	6	5	0	70	30	- 4
725	6	FPO		32	33	0	6	QT	DKGREY				0	0	0	0	0	0	5
725	7	CBD	B2	96	78	39	328	QT	FAWN	VR	R	0	4	0	3	0	30	70	- 4
725	8	FRS	F5	34	28	9	12	QT	FAWN	R	SR	?	1	1	2	50	10	40	+ 3
725	9	FIR	F1	30	32	21	25	QT	CREAM	R	SR	?	0	4	3	55	0	45	+ 3
727	1	CSP	A2	81	72	35	210	QT	PURPLE	VR	SR	0	1	3	3	0	60	40	- 3
727	2	CIR	C2A	70	57	29	119	QT	CREAM	R	SR	?	4	3	4	0	60	40	- 4
727	3	FSP	F1	48	39	20	45	QT	BROWN	R	SR	0	2	0	1	45	50	55	+ 3
727	4	FSP	F3	51	29	17	29	QT	CREAM	R	R	0	0	5	3	50	45	5	+ 3
727	5	FSP	F3	52	41	18	42	QT	PGREY	R	R	Y	1	3	3	40	25	35	+ 3
728	1	CCS	C2A	43	65	56	198	QT	FAWN	R	SR	0	4	1	4	0	65	35	- 4
728	2	FRS	F5	62	60	21	94	QT	GREY	VR	R	Y	2	3	2	45	10	45	+ 3
728	3	CHW	C3A	90	80	47	391	QT	BROWN	R	R	0	5	4	6	0	90	10	- 5
728	4	CBD	B4	81	65	49	332	QT	PGREY	VR	R	0	2	5	4	0	60	40	- 3
728	5	CSP	A3	99	78	86	1018	QT	FAWN	RP	R	0	1	5	4	0	60	40	- 3
728	6	CUD	B1	106	84	32	234	QT	GREY	VR	VR	?	1	10	8	0	95	5	- 3
728	7	CBD	B3	95	58	55	295	QT	FAWN	RP	SR	Y	1	5	3	0	30	70	- 3
728	8	FIR	F7	54	44	16	39	QT	HONEY	R	F	Y	2	3	2	45	45	10	+ 4
728	9	FIR	F6	56	41	14	30	QT	PINK	R	SR	Y	1	2	2	45	10	45	+ 4
728	10	CSP	A2	80	64	30	176	QT	BROWN	VR	R	?	0	4	3	0	0	25	- 3
728	11	FSE	F7	64	52	28	87	QT	DMAROON	VR	R	?	0	6	4	0	0	15	+ 3
728	12	CSP	A2	121	77	60	520	QT	CREAM	RP	SR	?	1	2	3	50	20	30	+ 4
728	13	FSE	F3	87	42	17	66	QT	PGREY	R	R	?	0	2	2	0	0	10	+ 3
728	14	FSP	F7	43	28	14	25	QT	CREAM	R	R	0	0	4	3	40	45	15	+ 3
728	15	FDB	F7	47	34	20	35	QT	GREY	R	SR	0	3	1	3	45	50	5	+ 4
728	16	CHW	C3A	72	48	29	91	QT	BROWN	RP	R	?	0	6	4	0	75	25	- 3
728	18	FIR	F5	60	54	33	64	QT	PGREY	R	SR	?	0	2	2	40	5	55	+ 3
730	2	FRS	F3	48	32	11	16	QT	DBROWN	R	SR	Y	2	0	2	50	30	20	+ 4
730	5	FFB	F8	55	32	12	20	QT	PGREY	R	R	Y	1	8	5	45	55	0	+ 5
730	6	FIR	F1	32	50	12	18	QT	GREEN	R	SR	?	1	1	2	45	5	50	+ 3
730	8	FIR	F5	40	56	13	28	QT	PGREY	VR	R	?	1	2	2	50	10	40	+ 3
730	9	CBD	B3	65	58	40	117	QT	PINK	VR	R	?	1	2	3	0	30	70	- 4
730	11	CWS	A1	94	40	22	66	QT	FAWN	R	F	0	1	3	3	50	25	25	- 3
730	12	CUD	B1	52	90	56	52	QT	GREEN	VR	R	?	1	0	1	0	25	75	- 3
731	4	FSP	F3	31	42	10	6	QT	CREAM	VR	R	?	0	2	2	45	5	45	+ 3
731	5	CCC	C3B	63	61	28	102	QT	MAROON	R	SR	?	6	8	7	0	75	25	- 5
731	6	CSQ	D3	42	61	52	193	QT	WHITE	RP	SR	0	2	3	3	0	60	40	- 4
731	7	FSC	F1	63	63	25	70	QT	MAROON	G	SR	?	5	0	2	45	10	45	+ 5
731	8	CHW	C3A	101	67	58	477	QT	HONEY	RP	SR	0	0	5	4	0	75	25	- 3
731	9	FSP	F6	48	33	12	8	QT	PGREY	VR	SR	0	1	0	1	50	5	45	+ 3
731	10	FSP	F6	73	75	37	142	QT	FAWN	VR	SR	0	0	3	2	40	15	45	+ 3
731	16	CCS	C2B	75	52	30	130	QT	PINK	RP	F	?	6	4	6	0	85	15	- 4
731	20	FIR	F1	41	40	15	26	QT	BROWN	VR	SR	0	1	0	1	50	0	50	+ 3
731	24	FSE	F7	45	38	14	25	QT	PINK	R	SR	?	1	4	3	45	45	10	+ 4
731	26	FRS	F1	52	41	13	31	QT	CREAM	VR	SR	Y	1	1	2	45	5	50	+ 3
731	27	FIR	F8	17	22	15	4	QT	PINK	R	SR	Y	1	2	3	40	60	0	+ 3
731	29	FHW	F8	73	48	18	65	QT	FAWN	VR	R	0	0	5	5	45	55	0	+ 3
732	2	FIR	F8	34	23	15	19	QT	FAWN	R	SR	?	1	3	3	45	55	0	+ 4
732	3	FIR	F2	72	44	21	69	QT	CREAM	R	R	Y	2	0	2	50	20	30	+ 5
734	1	FSP	F2	41	20	14	19	QT	GREY	RP	SR	?	1	1	1	45	5	50	+ 3
734	2	FSP	F7	31	29	17	18	QT	HONEY	R	SR	?	2	3	2	50	40	10	+ 3
734	4	CUD	B1	60	48	42	134	QT	FAWN	R	SR	0	1	0	1	0	10	90	- 3
736	1	FRS	F1	55	63	15	54	QT	BROWN	R	SR	Y	2	0	1	50	5	45	+ 3
736	2	FIR	F1	44	32	11	26	QT	PGREY	R	SR	Y	1	0	1	50	0	50	+ 3
736	3	FRS	F1	41	47	14	29	QT	PGREY	R	SR	Y	1	0	1	50	0	50	+ 3
738	1	FIR	F3	75	38	20	69	QT	PGREY	F	SR	0	2	2	3	20	60	20	+ 4

738	2	CSP	A3	52	47	42	147	QT	GREY	R	R	0	1	0	1	0	30	70	- 3
738	3	FRC	F8	32	37	10	18	QT	PGREY	R	R	Y	2	2	3	50	45	5	+ 5
738	4	FIR	F1	55	51	23	67	QT	BROWN	R	R	Y	1	2	3	40	20	40	+ 3
738	6	CWS	A1	66	43	42	82	QT	CREAM	R	SR	0	2	3	2	30	40	30	- 3
738	7	CSP	A1	81	67	57	509	QT	CREAM	RP	SR	0	1	2	2	0	35	65	- 3
739	1	FRS	F1	49	41	15	42	QT	MAROON	R	R	0	1	0	1	50	0	50	+ 3
739	2	CWS	A1	56	31	20	39	QT	FAWN	R	R	0	0	3	2	45	25	30	- 3
739	3	CWS	A1	66	42	31	120	QT	FAWN	R	SR	0	3	1	3	35	35	30	- 3
739	4	CWS	A1	61	43	20	65	QT	FAWN	R	SR	0	2	3	4	0	85	15	- 4
739	6	FRS	F1	67	42	17	46	QT	GREY	VR	SR	?	0	2	2	45	5	50	+ 3
739	8	FSE	F3	55	37	15	35	QT	MAROON	R	SR	?	2	1	2	50	30	20	+ 4
739	9	FIR	F6	30	23	9	2	QT	HONEY	R	SR	?	1	2	2	45	35	20	+ 4
740	2	FSE	F7	58	31	13	19	QT	GREY	R	R	?	0	4	3	50	50	0	+ 3
741	1	CBD	B3	82	67	58	472	QT	BROWN	R	R	Y	2	1	2	0	30	70	- 3
741	2	CSP	A2	108	71	58	516	QT	HONEY	R	R	0	2	0	2	0	25	75	- 3
741	3	CUD	B1	89	95	50	698	QT	GREY	R	R	0	1	0	1	0	10	90	- 3
741	4	CBD	B4	80	55	46	188	QT	BROWN	R	R	0	0	3	3	0	50	50	- 3
741	5	CRS		0	0	0	256	QT	GREEN	O	R	?	0	0	0	0	0	0	5
741	6	FRS	F1	62	34	11	37	QT	GREY	R	SR	?	0	1	1	50	5	45	+ 4
741	7	CWS	A1	58	66	53	146	QT	DGREY	R	SR	0	4	3	4	0	60	40	- 4
741	8	CBD	B2	77	56	20	117	QT	FAWN	R	SR	0	3	0	2	0	70	30	- 4
741	10	FSP	F2	42	30	24	35	QT	FAWN	R	R	0	2	1	3	30	30	40	+ 3
741	11	FIR	F6	33	32	15	23	QT	FAWN	R	R	0	2	1	3	40	20	40	+ 5
741	12	CBD	B4	65	61	50	234	QT	BROWN	RP	SR	0	1	1	2	0	60	40	- 3
741	13	FRS	F2	44	38	19	32	QT	MAROON	R	SR	0	2	2	3	40	35	25	+ 4
743	1	CBD	B2	73	45	41	180	QT	MAROON	R	SR	0	2	1	3	0	40	60	- 3
744	1	CBD	B4	121	89	62	786	QT	HONEY	RP	SR	Y	2	1	2	0	40	60	- 3
744	3	CWS	A2	79	50	32	123	QT	HONEY	R	R	?	2	2	3	0	80	20	- 4
744	4	CCC	C1B	68	53	30	128	QT	GREEN	RP	SR	0	3	3	3	0	60	40	- 4
744	6	CWS	A1	27	41	18	27	QT	GREEN	VR	R	0	1	1	2	0	60	40	- 3
744	7	CWS	A1	65	53	41	116	QT	FAWN	R	SR	Y	2	0	2	0	35	45	- 3
745	1	FRS	F1	63	45	15	34	QT	HONEY	R	SR	Y	1	0	1	50	0	50	+ 3
745	2	CBD	B4	80	51	36	156	QT	BROWN	R	CR	0	2	3	3	0	70	30	- 4
745	3	FIR	F2	68	35	20	52	QT	FAWN	R	R	0	3	0	1	45	10	45	+ 4
746	1	CBD	B4	181	91	80	1000	QT	FAWN	RP	R	Y	2	3	3	0	60	40	- 3
746	2	FIR	F2	79	62	35	156	QT	BROWN	RP	R	Y	1	1	1	45	20	35	+ 4
746	3	FRS	F1	52	39	18	47	QT	FAWN	RP	R	Y	1	0	1	45	0	55	+ 3
746	4	FSP	F3	40	33	21	16	QT	CREAM	VR	R	?	1	3	3	40	50	10	+ 3
747	1	FRS	F1	68	76	28	135	QT	HONEY	R	SR	0	1	0	1	40	0	60	+ 3
747	3	CCL	B3	110	86	73	812	QT	BROWN	R	R	?	2	4	2	0	15	85	- 4
749	1	FRS	F9	66	62	20	81	QT	BLACK	VR	SR	Y	2	1	2	50	30	20	+ 4
749	2	FRS	F9	63	55	23	66	QT	BLACK	VR	SR	Y	2	0	2	45	30	25	+ 4
750	1	FIR	F8	30	42	13	20	QT	FAWN	R	SR	?	3	6	4	45	55	0	+ 4
751	1	FRS	F5	28	31	8	7	QT	RED	R	SR	Y	1	1	2	45	5	50	+ 4
751	2	FHW	F7	47	33	10	14	QT	BROWN	R	SR	Y	0	3	2	45	10	45	+ 3
799	2	FPO					30	ME	DGREEN										5
799	1	CRS					202	IG	PGREEN										5
804	1	CBD	B3	171	81	74	1185	QT	FAWN	R	SR	?	3	2	2		20	80	- 3
808	1	CHW	B3	90	59	58	452	QT	BROWN	VR	R	0	2	2	2		40	60	- 3
812	1	CCL	B1	109	74	81	870	QT	BROWN	R	SR	0	4	2	2		85	15	- 3
816	1	CCL	C2A	83	55	35	166	QT	BROWN	VR	SR	0	3	0	2		55	45	- 3
816	2	FDB	F2	84	52	48	204	QT	FAWN	VR	SR	0	3	2	2	45	25	30	+ 3
817	1	CCS	C2A	98	76	62	356	QT	CREAM	RP	SR	0	2	4	4	65	0	35	- 3
817	2	CCS	C3A	74	72	45	228	QT	PBROWN	RP	R	0	2	2	2	0	55	45	- 3
818	2	FRS	F2	42	36	16	22	QT	BROWN	R	SR	?	2	1	1	35	50	15	+ 4
818	3	FFB	F7	62	47	32	50	QT	PGREY	RP	R	?	7	5	5	45	50	5	+ 4
818	1	FSP	F6	75	49	18	63	QT	FAWN	RP	SR	0	4	0	1	45	15	40	+ 3
818	4	FAN	F2	158	103	51	705	QT	BROWN	VR	R	Y	2	0	1	45	10	45	+ 3
818	5	FRS	F2	77	72	33	220	QT	GREYBRO	R	R	?	3	1	2	45	10	45	+ 3
824	1	CIR	A1	62	49	40	160	QT	GREY	R	R	?	2	3	5	0	60	40	- 3
828	1	CHW	C3B	63	57	45	116	QT	GREYBRO	RP	R	0	6	3	5	0	65	35	- 4
828	2	FRU	F1	56	50	21	90	BA	DGREY	VR	SR	H	0	8					+ 5
830	1	CHW	C2B	81	69	47	264	QT	PGREY	VR	SR	0	4	5	7	0	80	20	- 4
830	2	CBD	B3	81	51	30	100	QT	HONEY	RP	SR	0	4	6	4	0	55	45	- 4
830	3	CAP	B4	109	75	39	304	QT	FAWN	RP	R	?	2	6	3	0	60	40	- 3
831	1	CWS	A1	80	60	35	124	QT	GREY	RP	SR	?	3	0	3	0	55	45	- 4

831	2	CIR	A1	74	71	60	278	QT	PINKFAWN	RP	R	?	2	0	2	0	65	35	- 3
832	1	CWS	A2	54	45	26	60	QT	GREY	RP	SR	0	2	2	2	0	55	45	- 3
834	1	FSE	F4	48	42	22	22	QT	DGREY	R	SR	Y	1	4	2	45	50	5	+ 3
834	2	CWS	A1	113	55	50	366	QT	REDBROW	R	R	Y	4	1	3	0	40	60	- 3
835	1	FSE	F7	69	54	21	106	QT	FAWN	R	R	?	2	2	3	45	45	10	+ 5
835	2	FIR	F7	46	34	27	22	QT	MAROON	VR	SR	0	4	0	3	45	50	5	+ 4
835	3	CCS	C2A	86	74	46	254	QT	PGREY	RP	R	0	4	2	4	0	55	45	- 4
835	4	FAN	F2	95	54	32	170	QT	GREEN	VR	R	?	5	3	3	55	25	20	+ 4
835	5	FAN	F3	77	62	37	116	QT	GREY	VR	R	?	2	4	2	45	45	10	+ 3
835	6	FSP	F2	56	54	15	48	QT	GREEN	RP	R	Y	1	1	1	45	10	45	+ 3
835	7	CWS	A1	76	68	39	218	BA	DGREY	RV	SR	0	4	4	4	0	55	45	- 3
835	8	CWS	A1	118	93	80	830	QT	GREY	R	R	0	1	2	2	0	50	50	- 3
840	1	CMI	B4	22	26	11	4	FL	GREY	F	F	Y	8	3	5	0	100	0	+ 5
841	1	CBD	B3	69	57	41	174	QT	FAWN	R	SR	0	4	0	2	0	40	60	- 4
841	2	FSE	F3	64	39	19	34	QT	GREY	VR	SR	?	1	1	1	50	20	30	+ 3
841	3	CWS	A1	65	45	43	112	QT	DGREY	VR	R	0	1	0	1	0	40	60	- 3

Table 8. 2: The number of stone artefacts found grouped by age of underlying deposit showing the low density overall.

	Number of stone artefacts per collecting zone														
Age band groups:	**1**	**2**	**3**	**4**	**5**	**6**	**7**	**8**	**9**	**12**	**13**	**14**	**17**	**51**	**TOTAL**
Younger than SST 12		5	9	4	3	1	1		1						36
SST12 and just older	41	15	10	4	3	5	3	2		1	1	1	1	1	88
Older than SST 12	43	18	11	6	2	3		3	1						87
Total	96	38	30	14	8	9	4	5	2	1	1	1	1	1	211
Percentage	45.5	18.0	14.2	6.6	3.8	4.3	1.9	2.4	0.9	0.5	0.5	0.5	0.5	0.5	

Table 8. 3: Summary of the lithic assemblage from the Pabbi Hills by type, retouch, and lithology.

Codes	**Artefact Types**	**Total No.**	**%**	**Retouched Pieces**	**Quartzite**	**Flint**	**Other Rock**
CAP	proto-biface	1	0.1		1		
CCL	straight edged core tool (SECT)	5	0.8	3	5		
CFB	flake blade core	6	1.0		6		
CHW	high action core tool (HACT)	14	2.3		14		
CMI	microcore	6	1.0			6	
CSQ	squared core	10	1.6		10		
CBI	bipolar core	4	0.7		4		
CWS	shattered wedge core fragment	51	8.4		50		1
CIR	irregular core	30	4.9		30		
CSP	split cobble	16	2.6		16		
CUD	unifacial core	16	2.6		16		
CBD	bidirectional/bifacial core	48	7.9		48		
CCS	simple disc core (discoid)	23	3.8		22		1
CCC	complex disc core (discoid)	14	2.3		14		
CPO	polyhedron	3	0.5		3		
	Cores/core tools total	**247**	**40.7**	**3**	**239**	**6**	**2**
FWS	shatter-split flake	2	0.3		2		
FAN	angular flake	11	1.8		11		
FIR	irregular flake (decortication flakes)	110 (21)	18.1	1	105	1	4
FSP	split flake (decortication flakes)	64 (4)	10.5		60	3	1
FDB	double bulb flake	5	0.8		5		
FRS	rounded simple flake (decortication flakes)	76 (39)	12.5	1	75	1	
FRC	rounded complex flake	13	2.1		12		1

Table 8. 3: Summary of the lithic assemblage from the Pabbi Hills by type, retouch, and lithology.

Codes	Artefact Types	Total No.	%	Retouched Pieces	Quartzite	Flint	Other Rock
FSE	straight-edged flake	37	6.1	3	37		
FFB	flake blade	14	2.3	2	14		
FSC	scraper	6	1.0	6	5	1	
FHW	HACT flake	6	1.0		6		
FKN	knife	1	0.2	1	1		
	flakes/flake tools total	345	56.8	14	333	6	6
	chipped stone total	592	97.5	17	572	12	8
CRS	ring stone	2	0.3		1		1
CRU	rubbing stone	1	0.2		1		
FRU	rubbing stone flake	2	0.3				2
FPO	polished stone fragment	6	1.0		2		4
	ground/polished stone total	11	1.8	0	4	0	7
FHA	hammerstone flake	1	0.2		1		
CHA	hammerstone	3	0.5		3		
	hammerstones total	4	0.7	0	4	0	0
	TOTAL	**607**		**17**	**580**	**12**	**15**
	% in assemblage			2.8	95.6	2.0	2.5

Table 8. 4: Technological types for all quartzite flake artefacts

	Techno-type	No. with cortex platform surface	%	Techno-type	No. with fracture platform surface	%	Totals	%
decortication primary flakes	F1	43	24.7	F5	23	14.5	66	19.8
secondary flakes >25% cortex	F2	67	38.5	F6	36	22.6	103	30.9
secondary flakes < 25% cortex	F3	59	33.9	F7	71	44.7	130	39.0
tertiary flakes no cortex	F4	2	1.1	F8	29	18.2	31	9.3
salami slices	F9	3	1.7	-	-	-	3	0.9
Totals		**174**	**52.3**		**159**	**47.7**	**333**	**99.9**

Table 8.5: The technotypes for all cores and core tools (Types A-C)

Cores and core tools	Row total	A1	A2	A3	B1	B2	B3	B4	C1A	C1B	C2A	C2B	C3A	C3B	C4A	C4B
Fragments																
CSP	16	2	10	3	1											
CWS	51	45	6													
CIR	30	13	2			2		5			8					
Pebble cores																
CUD	16				16											
CBD	48	1	1			11	21	10			3					
CAP	1							1								
CCL	5				1		1	2			1					
CBI	4	1				2		1								

Table 8.5: The technotypes for all cores and core tools (Types A-C) continued:

Cores and core tools	Row total	A1	A2	A3	B1	B2	B3	B4	C1A	C1B	C2A	C2B	C3A	C3B	C4A	C4B
Disc cores																
CCS	22								5		11	1	5			
CCC	14									4	1	3		4	2	
CPO	3							1							1	1
CHW	14						2		1			1	8	1	1	
Long-flake cores																
CSQ	10															
CMI	6					1		5								
CFB	6															
TOTAL		**62**	**19**	**3**	**18**	**16**	**24**	**25**	**6**	**4**	**24**	**5**	**13**	**5**	**4**	**1**
Strategy totals	**246**	**84 As**			**83 Bs**				**62 Cs**							

Table 8.5 continued: The technotypes for all cores and core tools (Types D and E)

Cores and core tools	D2	D3	D4	E2	E3	E4
Fragments						
CSP						
CWS						
CIR						
Pebble cores						
CUD						
CBD	1					
CAP						
CCL						
CBI						
Disc cores						
CCS						
CCC						
CPO						
CHW						
Long-flake cores						
CSQ	4	4	2			
CMI						
CFB				1	4	1
TOTAL	**5**	**4**	**2**	**1**	**4**	**1**
Strategy totals	**11 Ds**			**6 Es**		

Table 8. 6: The types of artefacts grouped by the age of the geological strata on which they were found.

Artefact type code	Age band 1 (0.9 – 1.2 Mya)	%	Age band 2 (1.2 – 1.4 Mya)	%	Age band 3 (1.7 – 2.2 Mya)	%	Total
CAP	1	1.0					1
CCL			5	1.6			5
CHW	2	2.0	6	2.0	6	3.0	14
FHW			6	2.0			6
CUD			9	2.9	7	3.5	16
CBD	8	7.8	24	7.8	16	8.1	48
FRS	9	8.8	38	12.4	29	14.6	76
CCS	2	2.0	8	2.6	13	6.6	23

Table 8. 6 continued: The types of artefacts grouped by the age of the geological strata on which they were found.

CCC	5	4.9	7	2.3	2	1.0	14
FRC	1	1.0	3	1.0	9	4.5	13
CPO	2	2.0	1	0.3			3
CBI			3	1.0	1	0.5	4
CFB	4	3.9	2	0.7			6
FFB	6	5.9	8	2.6			14
CMI	1	1.0	1	0.3	4	2.0	6
CSQ	2	2.0	3	1.0	5	2.5	10
FKN					1	0.5	1
FSC			2	0.7	4	2.0	6
Irregular pieces							
CWS	7	6.9	30	9.8	14	7.1	51
FWS			2	0.7			2
CIR	8	7.8	16	5.2	6	3.0	30
FIR	19	18.6	59	19.2	32	16.2	110
CSP	1	1.0	10	3.3	5	2.5	16
FSP	7	6.9	37	12.1	20	10.1	64
FDB	1	1.0	2	0.7	2	1.0	5
FAN	1	1.0	6	2.0	4	2.0	11
FSE	12	11.8	17	5.5	8	4.0	37
polished/ground pieces							
CRS	1	1.0	1	0.3			2
CRU					1	0.5	1
FRU					2	1.0	2
FPO	2	2.0	1	0.3	3	1.5	6
Hammerstones							
CHA					3	1.5	3
FHA					1	0.5	1
Total	**102**		**307**		**198**		**607**

Table 8.7: The techno-types of flakes and cores grouped by the age of the geological strata on which they were found.

TECHNO-TYPE	Age band 1 No.	%	Age band 2 No.	%	Age band 3 No.	%	Totals
n.d. ignored			2		4		6
Cores							
A1	9	21.4	36	29.0	17	23.0	62
A2	4	9.5	10	8.1	5	6.8	19
A3			3	2.4			3
B1			10	8.1	8	10.8	18
B2	1	2.4	9	7.3	5	6.8	15
B3	7	16.7	10	8.1	7	9.5	24
B4	1	2.4	14	11.3	5	6.8	20
C1A	1	2.4			4	5.4	5
C1B	1	2.4	3	2.4	1	1.4	5
C2A	4	9.5	12	9.7	8	10.8	24
C2B	2	4.8	2	1.6	1	1.4	5
C3A	1	2.4	6	4.8	6	8.1	13
C3B	2	4.8	1	0.8	2	2.7	5
C4A	2	4.8	2	1.6			4
CB4	1	2.4					1
D2	2	4.8	2	1.6	1	1.4	5
D3			2	1.6	2	2.7	4
D4					2	2.7	2
E2	1	2.4					1
E3	2	4.8	2	1.6			4
E4	1	2.4					1

Table 8.7 continued: The techno-types of flakes and cores grouped by the age of the geological strata on which they were found.

Core totals	**42**		**124**		**74**		**240**
Flakes							
F1	3	5.4	28	15.9	13	12.7	44
F2	5	8.9	35	19.9	27	26.5	67
F3	10	17.9	25	14.2	24	23.5	59
F4			1	0.6	1	1.0	2
F5	5	8.9	13	7.4	6	5.9	24
F6	11	19.6	16	9.1	9	8.8	36
F7	16	28.6	37	21.0	17	16.7	70
F8	6	10.7	18	10.2	5	4.9	29
F9			3	1.7			3
Flake totals	**56**		**176**		**102**		**334**

Table 8.8: The percentage of cortex remaining for flakes grouped by the age of the geological strata on which they were found.

Flake CORTEX %	**Age band 1 No.**	**%**	**Age band 2 No.**	**%**	**Age band 3 No.**	**%**	**Totals**
0	4	13.3	9	8.2	6	7.5	19
5	6	20.0	10	9.1	7	8.8	23
10	4	13.3	13	11.8	7	8.8	24
15	2	6.7	8	7.3	6	7.5	16
20	1	3.3	11	10.0	13	16.3	25
25	4	13.3	2	1.8	7	8.8	13
30	2	6.7	5	4.5	4	5.0	11
35			7	6.3	3	3.8	10
40	2	6.7	7	6.3	4	5.0	13
45			11	10	12	15.0	23
50	4	13.3	19	17.3	10	12.5	33
55	1	3.3	4	3.6			5
60			3	2.7	1	1.3	4
65			1	0.9			1
Total flakes	**30**		**110**		**80**		**220**

Table 8. 9: The percentage of cortex remaining for cores grouped by the age of the geological strata on which they were found.

Core CORTEX %	Age band 1 (0.9 – 1.2 Ma)	%	Age band 2 (1.2 – 1.4 Ma)	%	Age band 3 (1.7 – 2.2 Ma)	%	**Row total**
0/5	2	7.1	4	5.0	5	7.2	11
10/15	3	10.7	6	7.5			9
20/25	2	7.1	12	15.0	5	7.2	19
30/35	5	17.9	14	17.5	8	11.6	27
40/45	8	28.6	23	28.8	19	27.5	50
50/55			5	6.3	21	30.4	26
60/65	5	17.9	6	7.5	7	10.1	18
70/75	2	7.1	8	10.0	3	4.3	13
80/85/90	1	3.6	2	2.5	1	1.4	4
Total cores	**28**		**80**		**69**		**177**

Table 8. 10: The occurrence of stone in the 269 area as indicated on Figure 8.24

No.	Length	Breadth	Thickness	Type	Rating	Details
269.1	94	81	54 mm	quartzite, brown	-5	collected in '87
261.1	190	112	95 mm	quartzite, grey	-2	collected in '87
261.2	84	75	60 mm	quartzite, brown	-4	collected '99, marked on Figure 8.24
273.1	40	36	15 mm	quartzite, fawn	-2	collected in '87
273.2	33	28	11 mm	quartzite, pink	+2	collected in '87
273.3	34	31	16 mm	quartzite, brown	-2	collected in '87
A	43	29	22 mm	quartzite, cream	0	found '89, marked on Figure 8.24
B	37	29	20 mm	quartzite, cream	0	found '89, marked on Figure 8.24
C	100	48	37 mm	quartzite, brown	0	found '89, marked on Figure 8.24
D	7	4	2 cm	igneous, grey	0	found '89, marked on Figure 8.24
E	5	4	3 cm	quartzite, pale grey	0	found '89, marked on Figure 8.24
F	6	6	3 cm	quartzite, honey	0	found '89, marked on Figure 8.24
G	7	4	3 cm	quartzite, cream	0	found '89, marked on Figure 8.24
H	3	2	1 cm	quartzite, n.d.	0	found '89, marked on Figure 8.24
I	6	3.5	2 cm	quartzite, fawn		found '89, marked on Figure 8.24
J	7.5	5	4 cm	quartzite, fawn	-1	found '89, marked on Figure 8.24
K	7	5	4 cm	quartzite, brown	0	found '89, marked on Figure 8.24
L	196	73	46 mm	quartzite, cream	0	found '99, marked on Figure 8.24 near 261.2
M	41	32	25 mm	quartzite, cream	0	found '99, marked on Figure 8.24 near 261.2
N	88	74	57 mm	quartzite, white rose	1	found '99, marked on Figure 8.24 c. 10 m from 261.2
O	110	84	65 mm	quartzite, brown	0	found '99, marked on Fig. 8.27, on spur south of 261.2
P	51	29	21 mm	quartzite, white grey	0	found '99, marked on Fig. 8.27, on pink clay between 269 and spur
Q	45	29	22 mm	quartzite, white grey	0	found '99, marked on Fig. 8.27, on pink clay between 269 and spur
R	96	56	65 mm	quartzite, brown	0	found '99, marked on Figure 8.24, down 269 gully

Table 8. 11: Summary of the pieces created when 18 nodules fractured, showing the numbers that were potentially confusable with intentional flaking (credibility rating 1-3, probably geological; 4-5 probably/certainly intentional).

Credibility	Cores	Flakes	Total
1	2	45	47
2	7	19	26
3	14	8	22
4	0	0	0
5	0	0	0

Table 8. 12: Details of erosion experiments of stone along gully bases

Size	Stream gully laid out in 1989	Recovered in 1990	Main gully laid out in 1989	Recovered in 1990
AA < 3 x 3 cm	29	1	31	0
BA < 5 x 3 cm	10	2	10	0
BB < 5 x 5 cm	12	0	16	0
CB < 7 x 5 cm	14	0	13	0
CC <7 x7 cm	1	0	2	0
DD <9 x 9 cm	0	0	1	0
EE > 11x >9 cm	4	0	5	1
Total	**70**	**3**	**78**	**1**

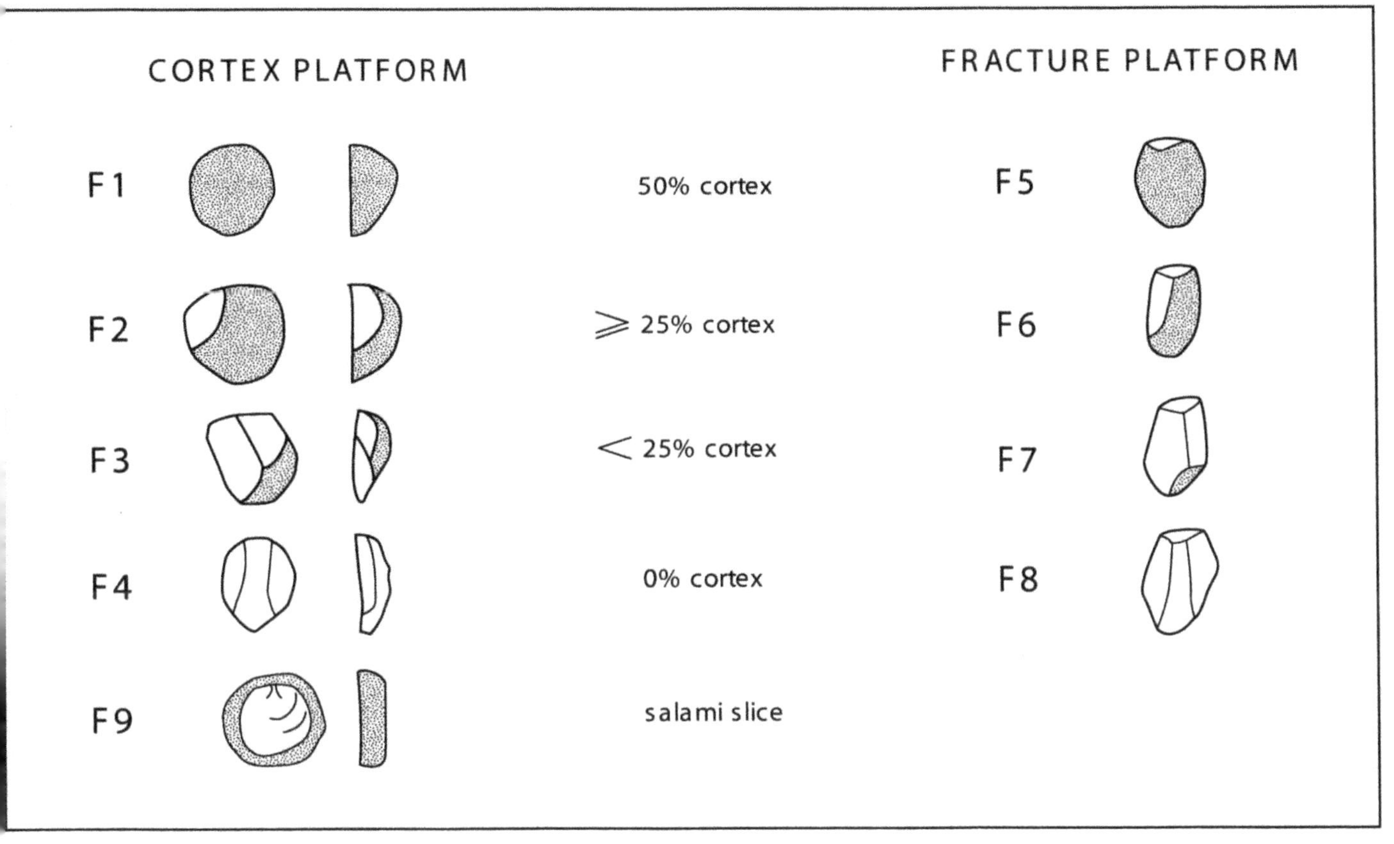

Figure 8. 1: Flake techno-types defined by the amount of cortex as a % of total surface area and the presence or absence of cortex on the striking platform.

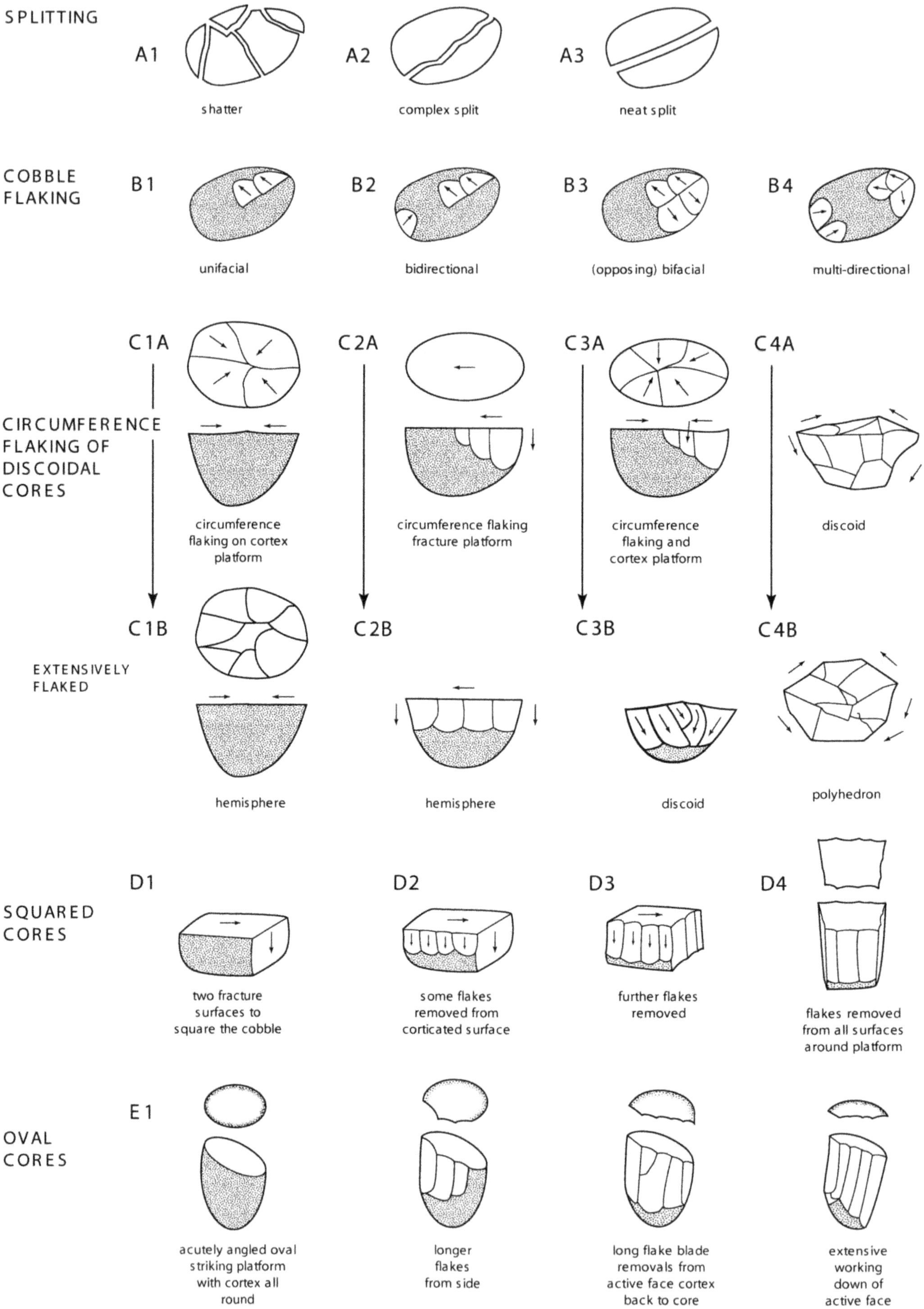

Figure 8.2: Core technotypes defined by flaking strategy and sequence

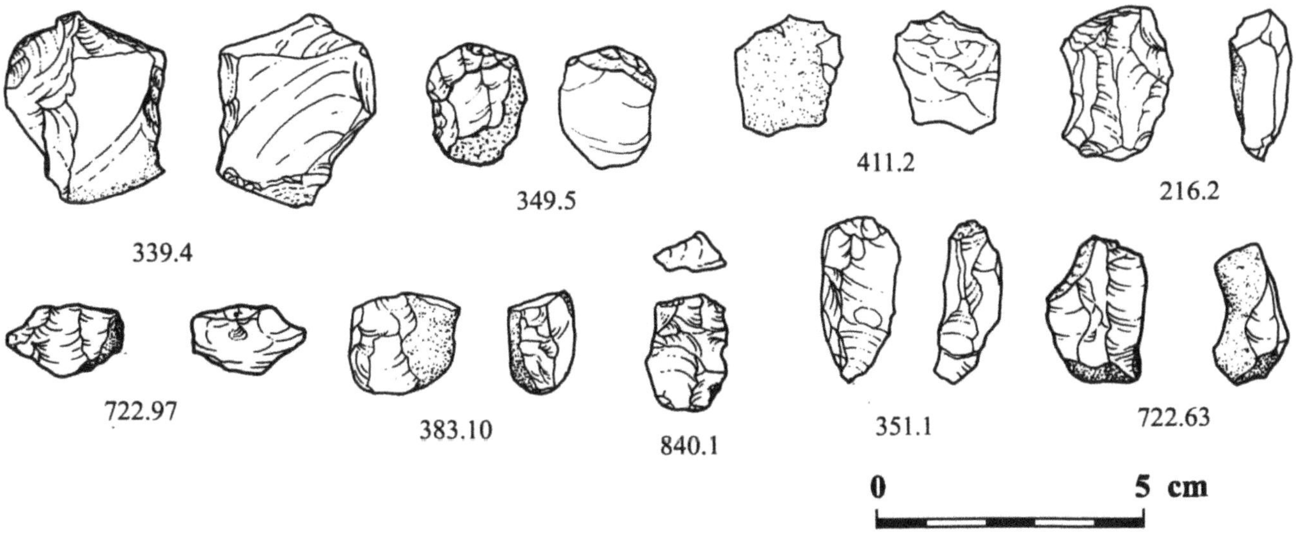

Figure 8.3: Flint artefacts, including micro-cores, flakes and scraper.

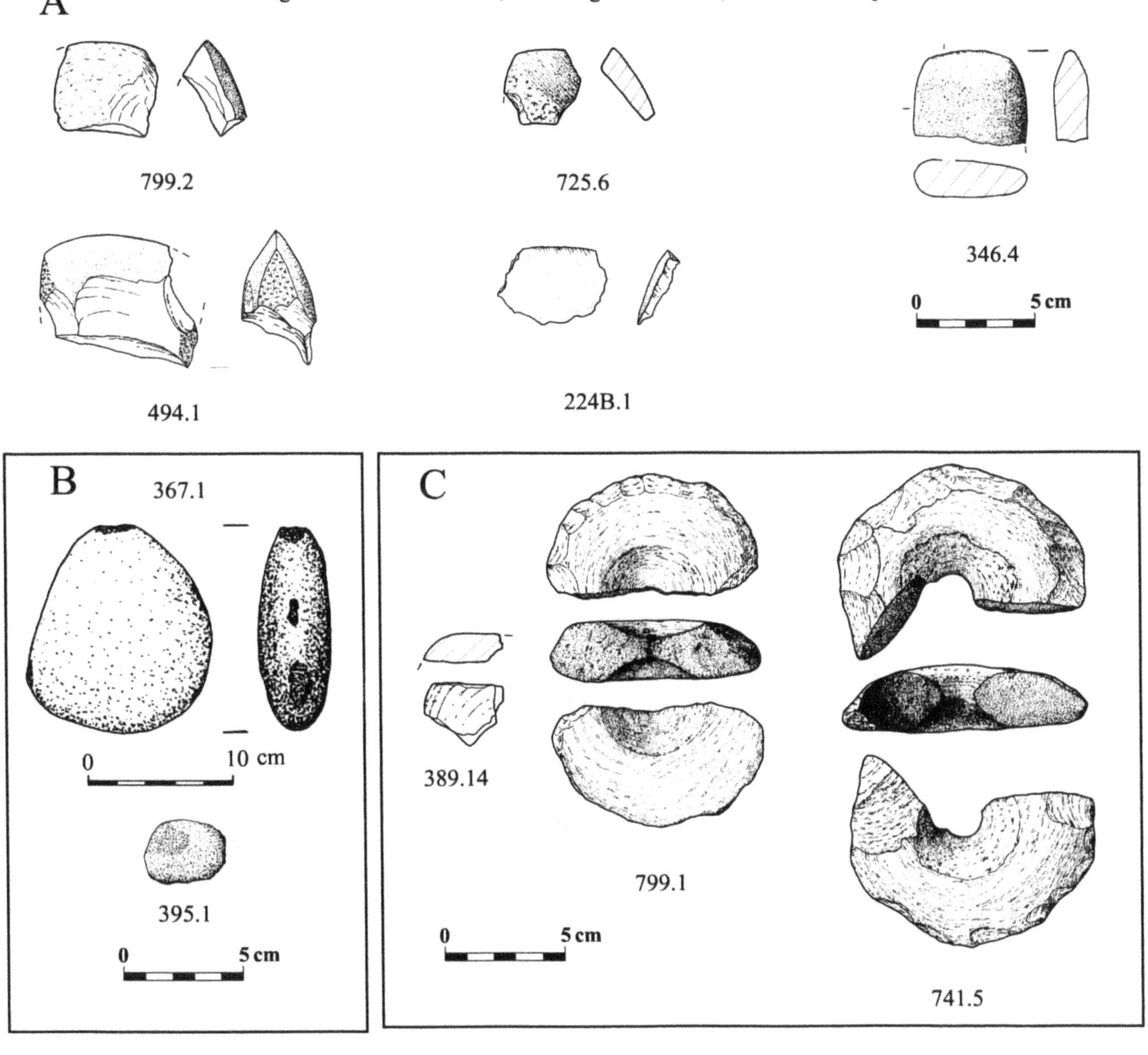

Figure 8.4: Hammerstones, ground and polished stone objects including axes (A), hammerstone (B) and ringstones (C).

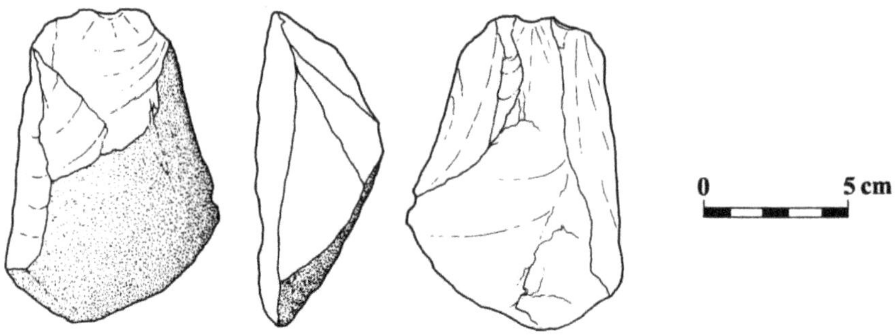

830.1

Figure 8.5: Proto-biface

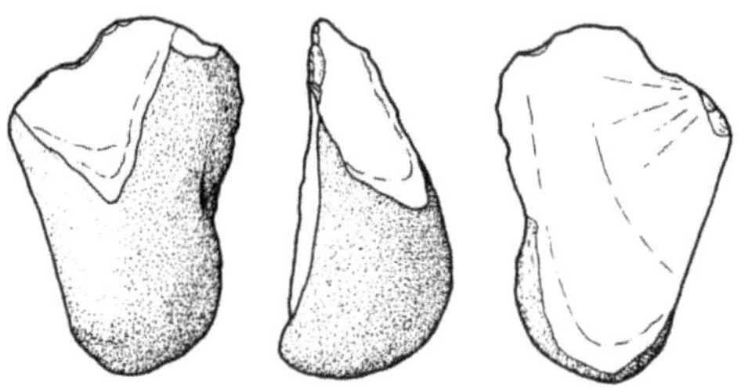

816.1

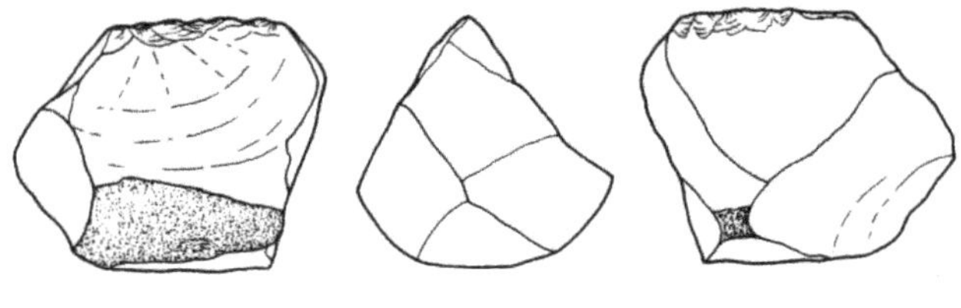

712.1

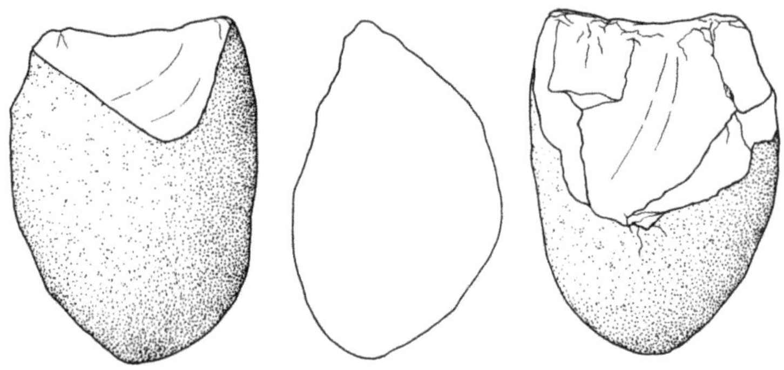

747.3

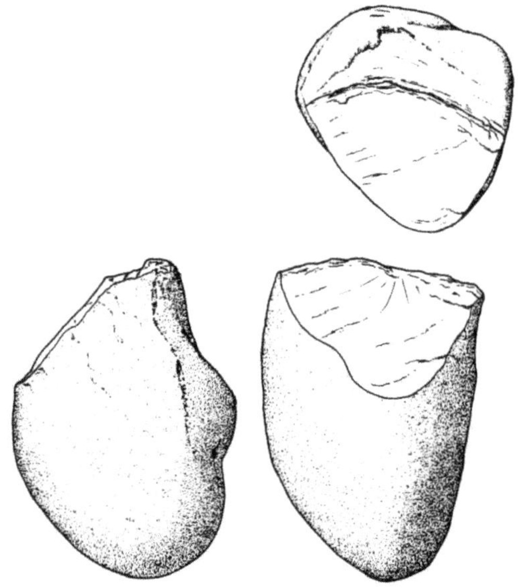

812.1

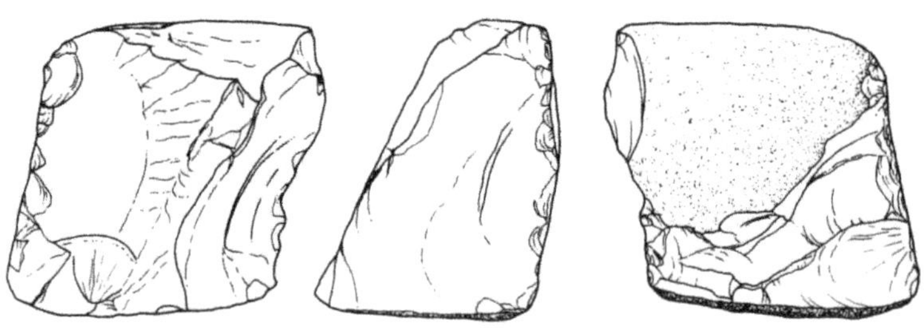

51.1

Figure 8.6: Straight-edged tools (SECT's) , most on cores with damage along the straight edge suggesting a cleaver-like function.

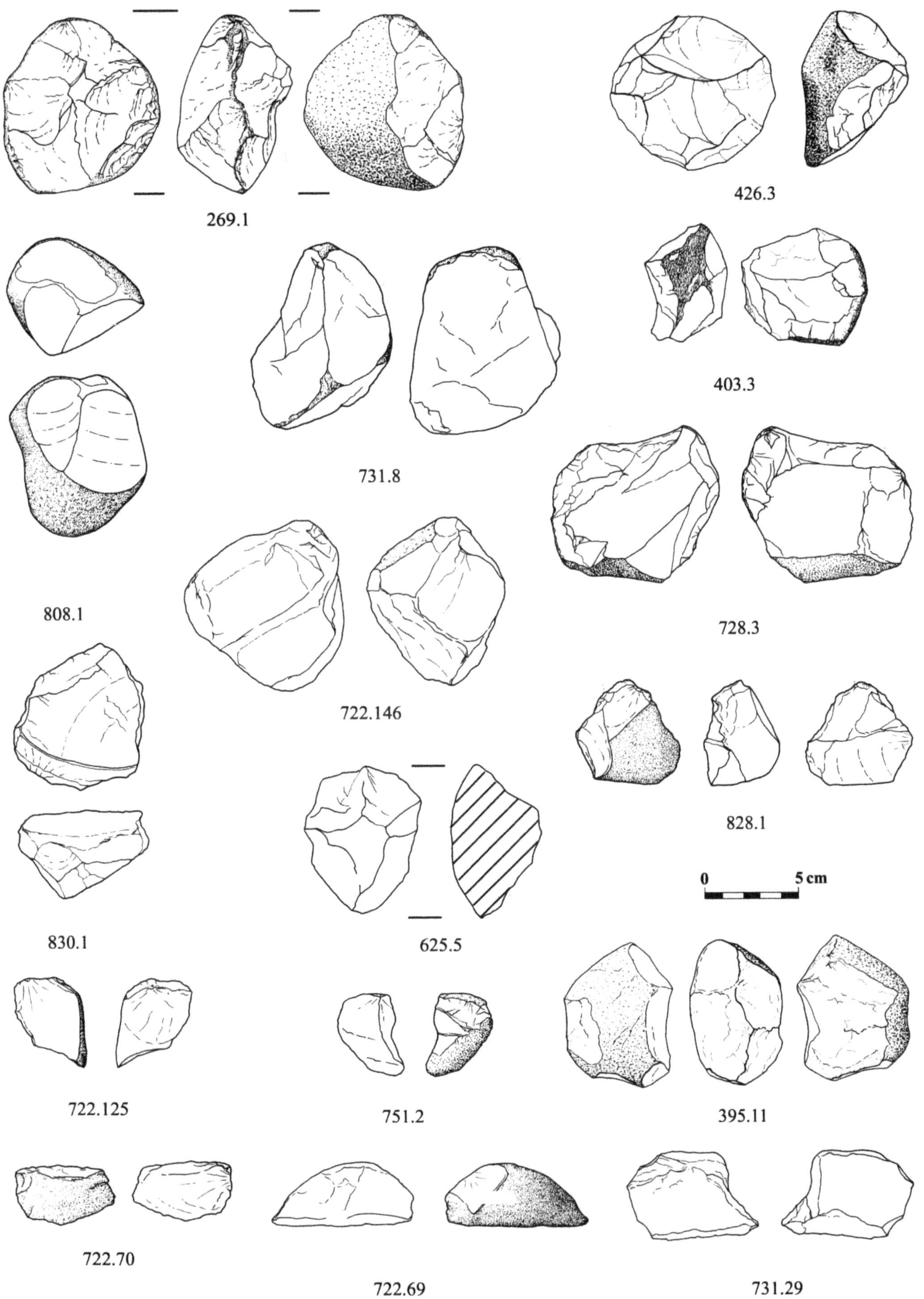

Figure 8.7: High action core tools (HACTs) and HACT flakes made on a variety of sinuous or straight edges.

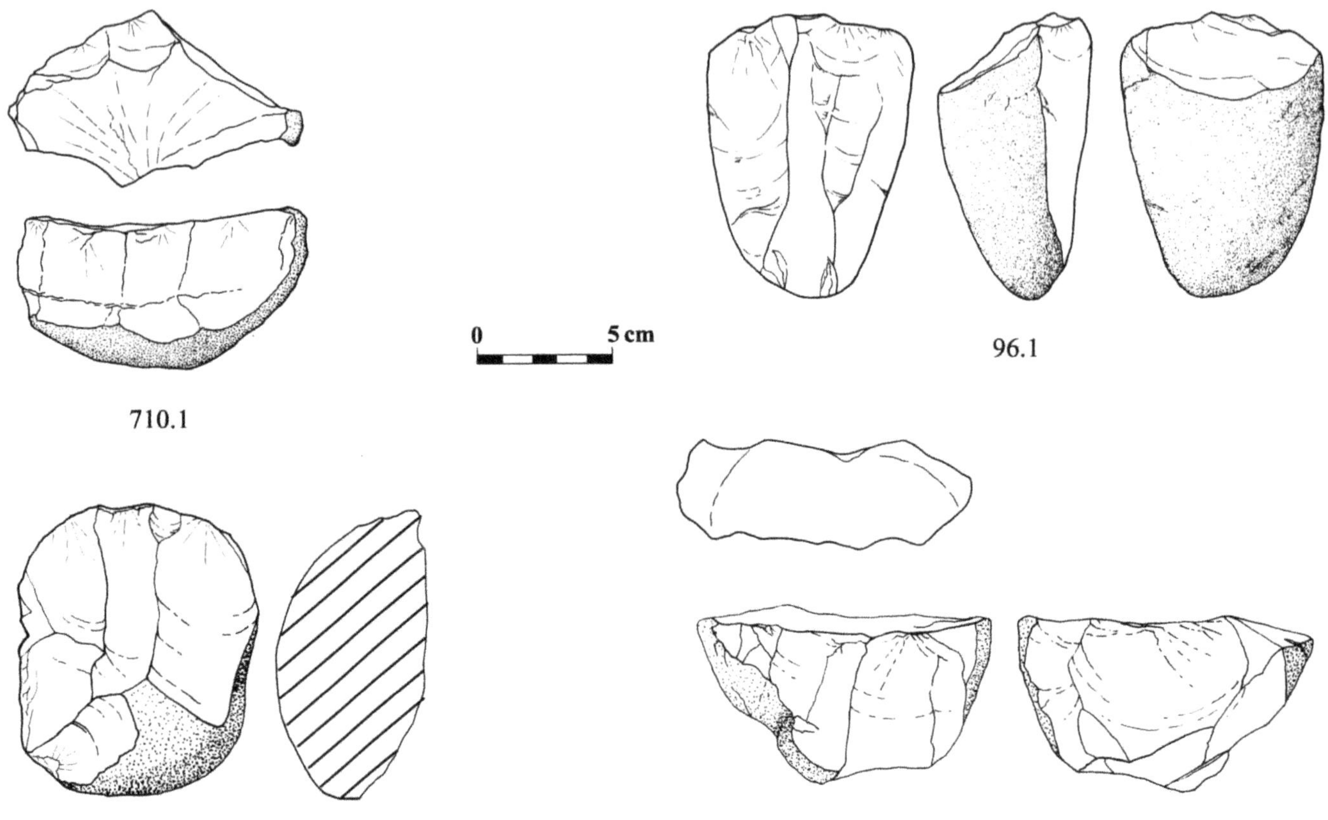

Figure 8.8: Flake-blade cores

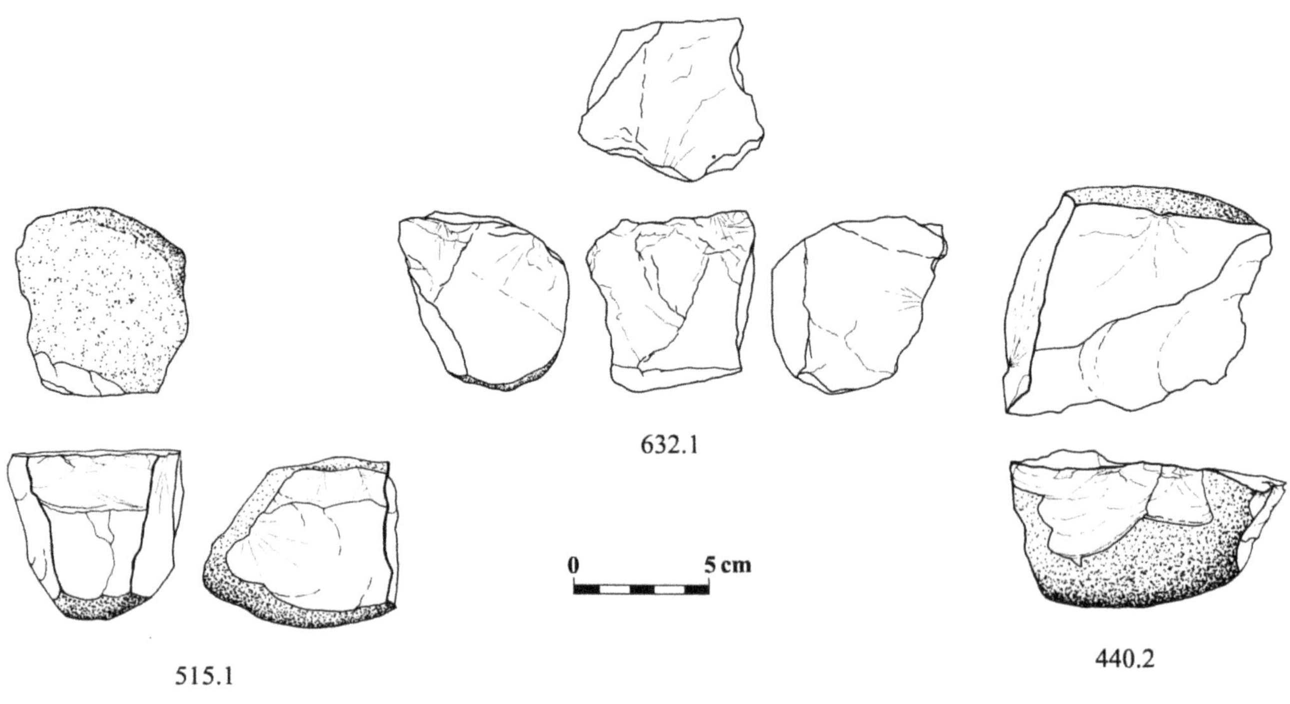

Figure 8.9: Squared cores

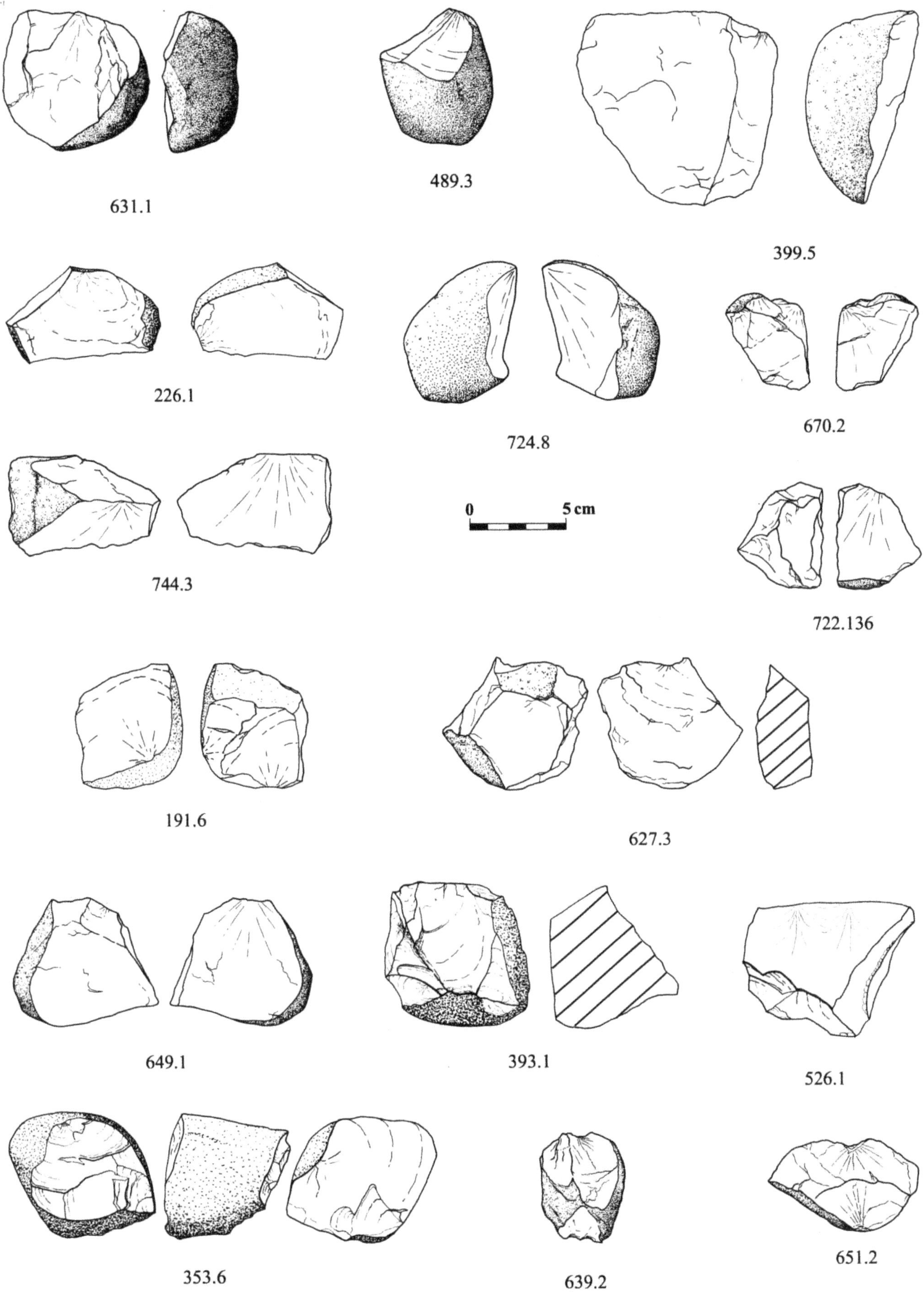

Figure 8.10: Split cobbles, shattered wedge fragments, irregular and bipolar cores as evidence of core reduction problems.

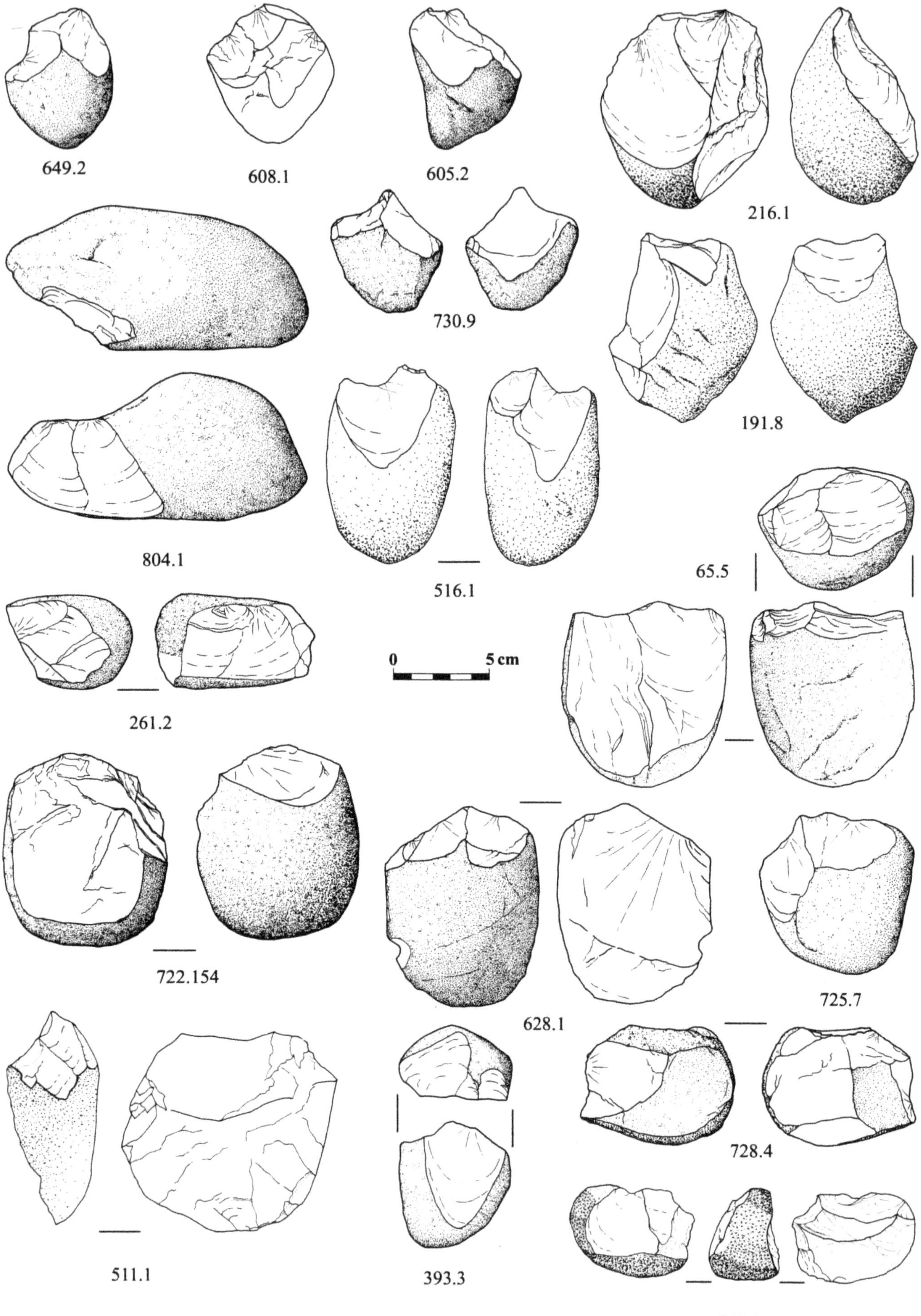

Figure 8.11: Pebbles cores showing uni- and bi-directional flaking

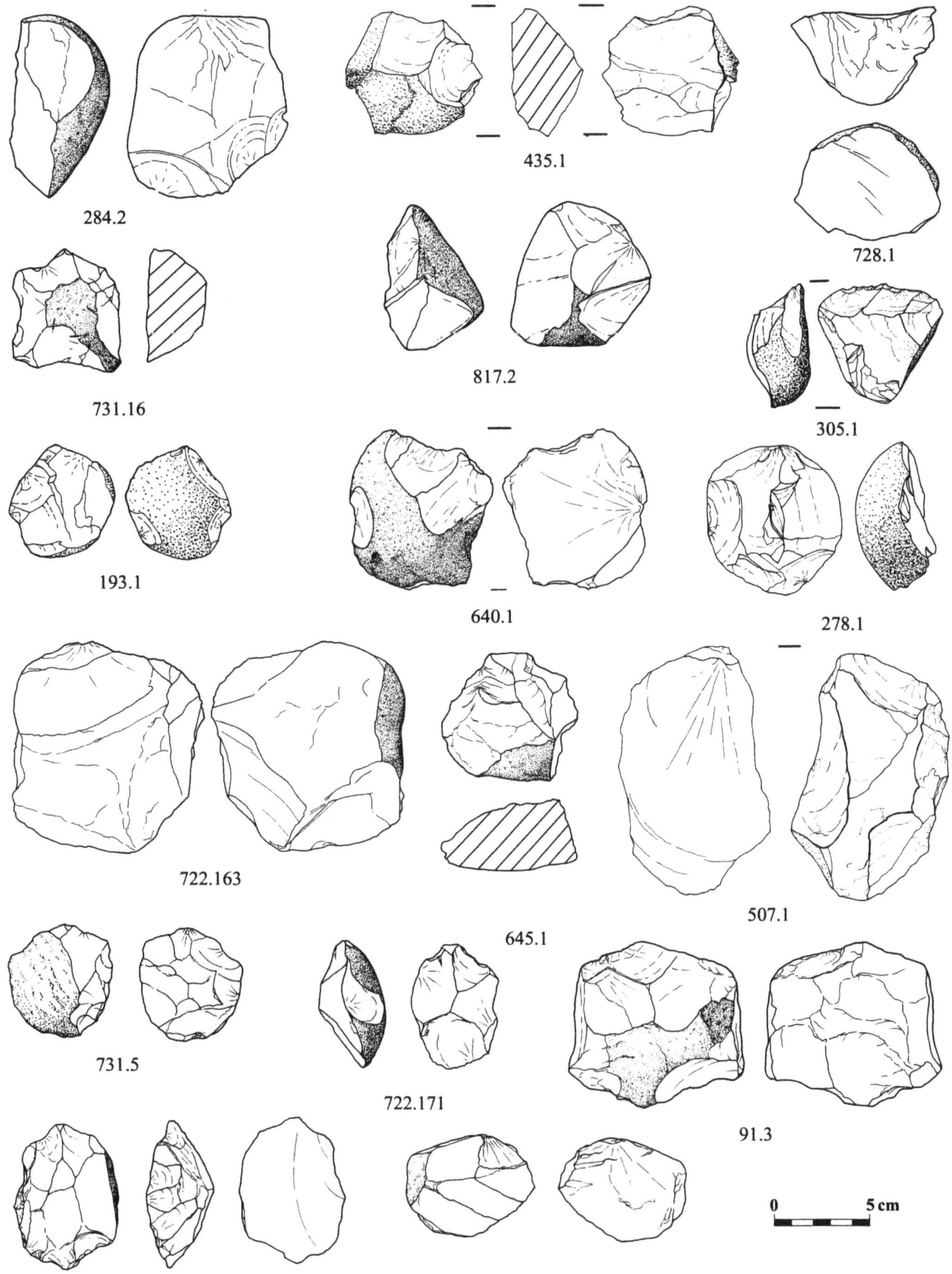

Figure 8.12: Simple and complex disc cores, and polyhedral cores.

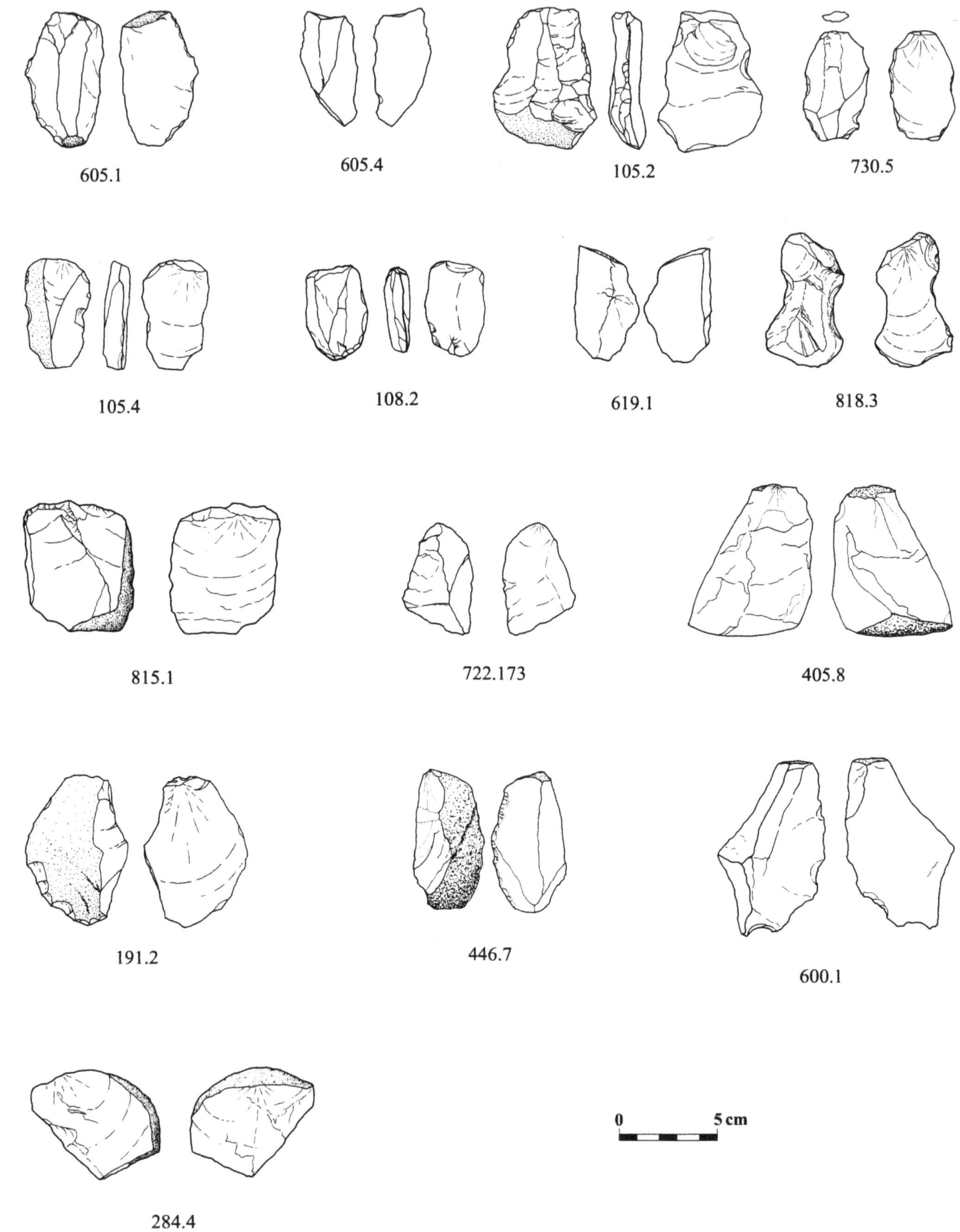

Figure 8.13: Flake blades and straight-edged flakes

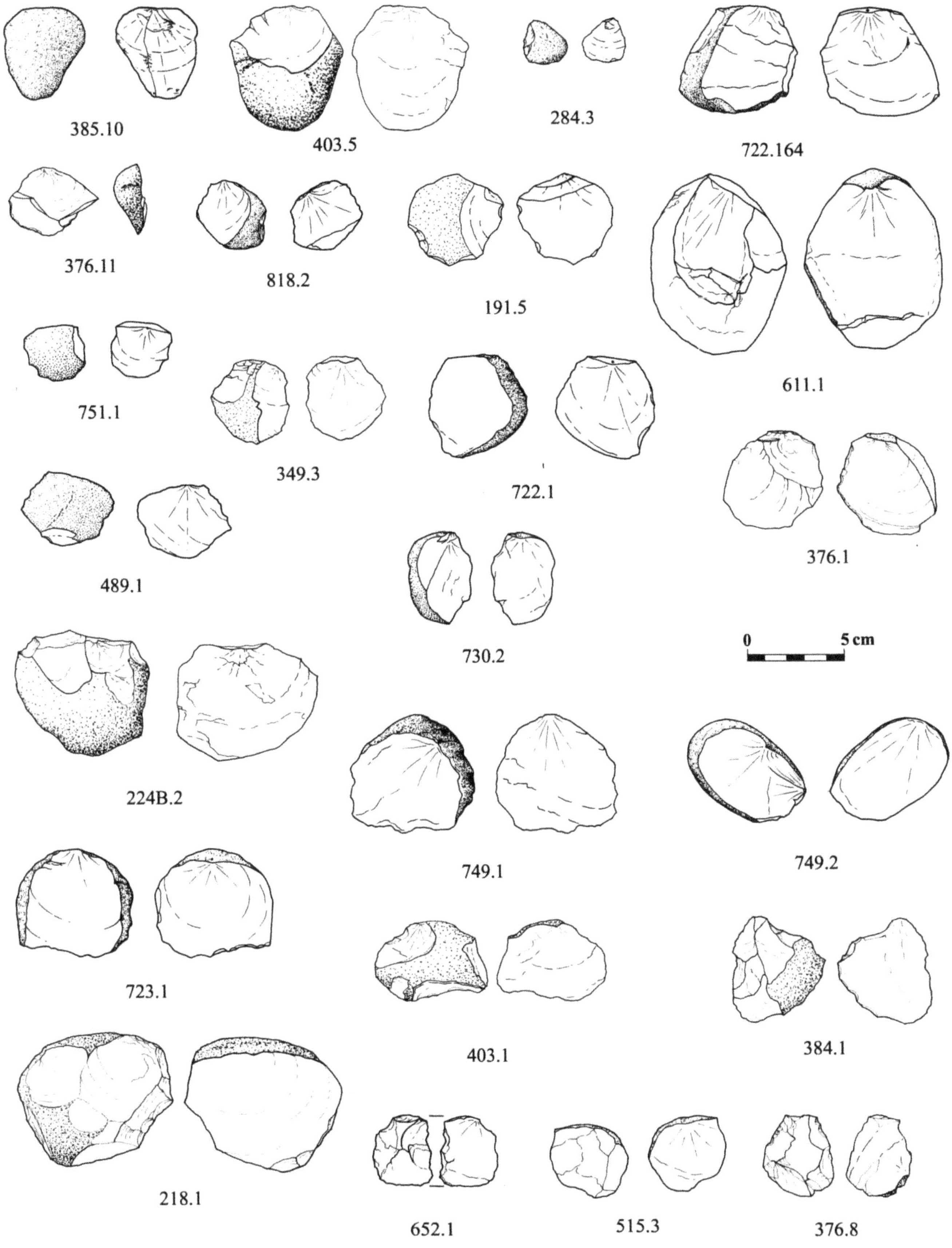

Figure 8.14: Simple and complex round flakes, and salami slices.

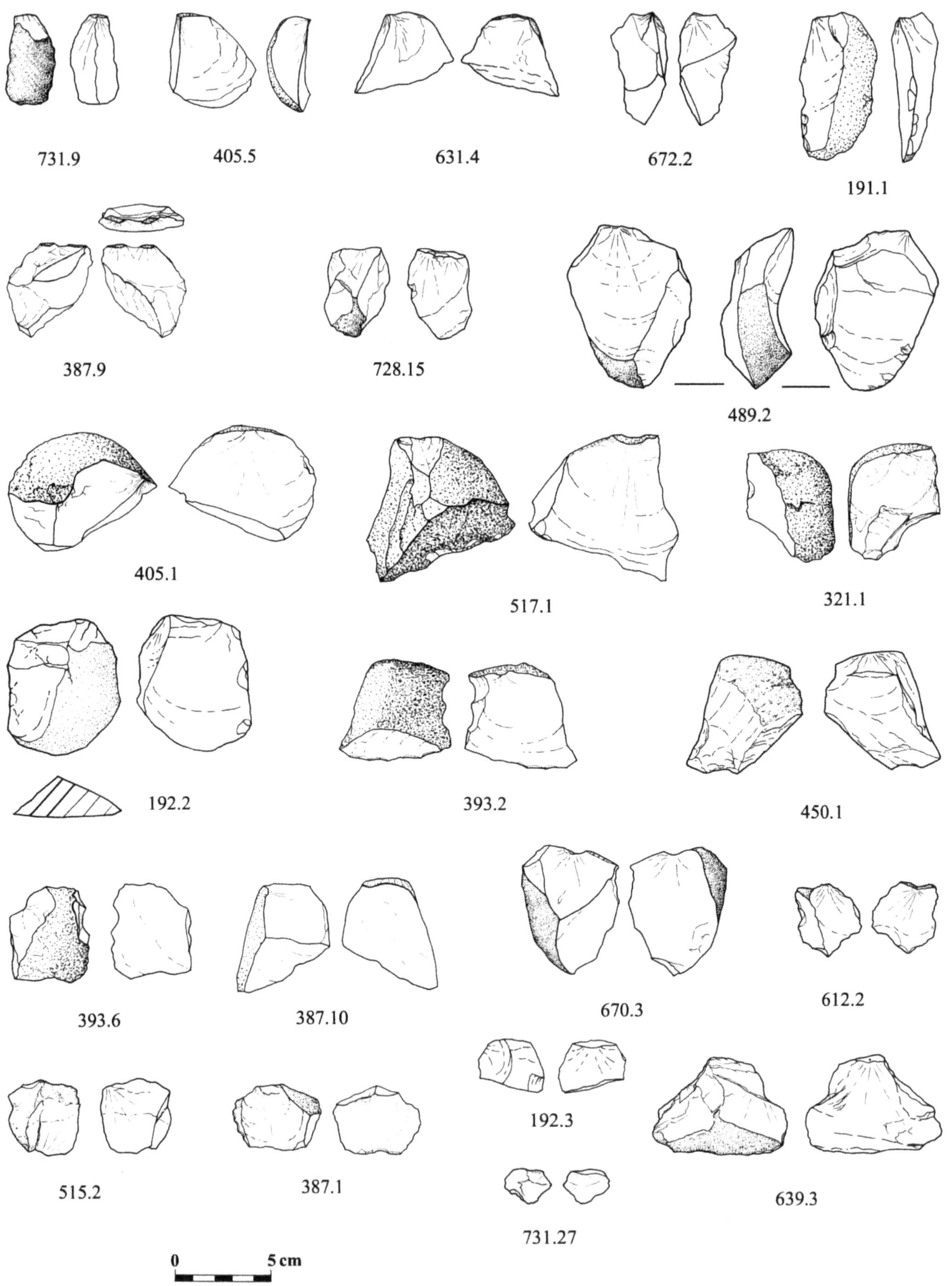

Figure 8.15: Split and double-bulb flakes as evidence of knapping problems, and angular flakes.

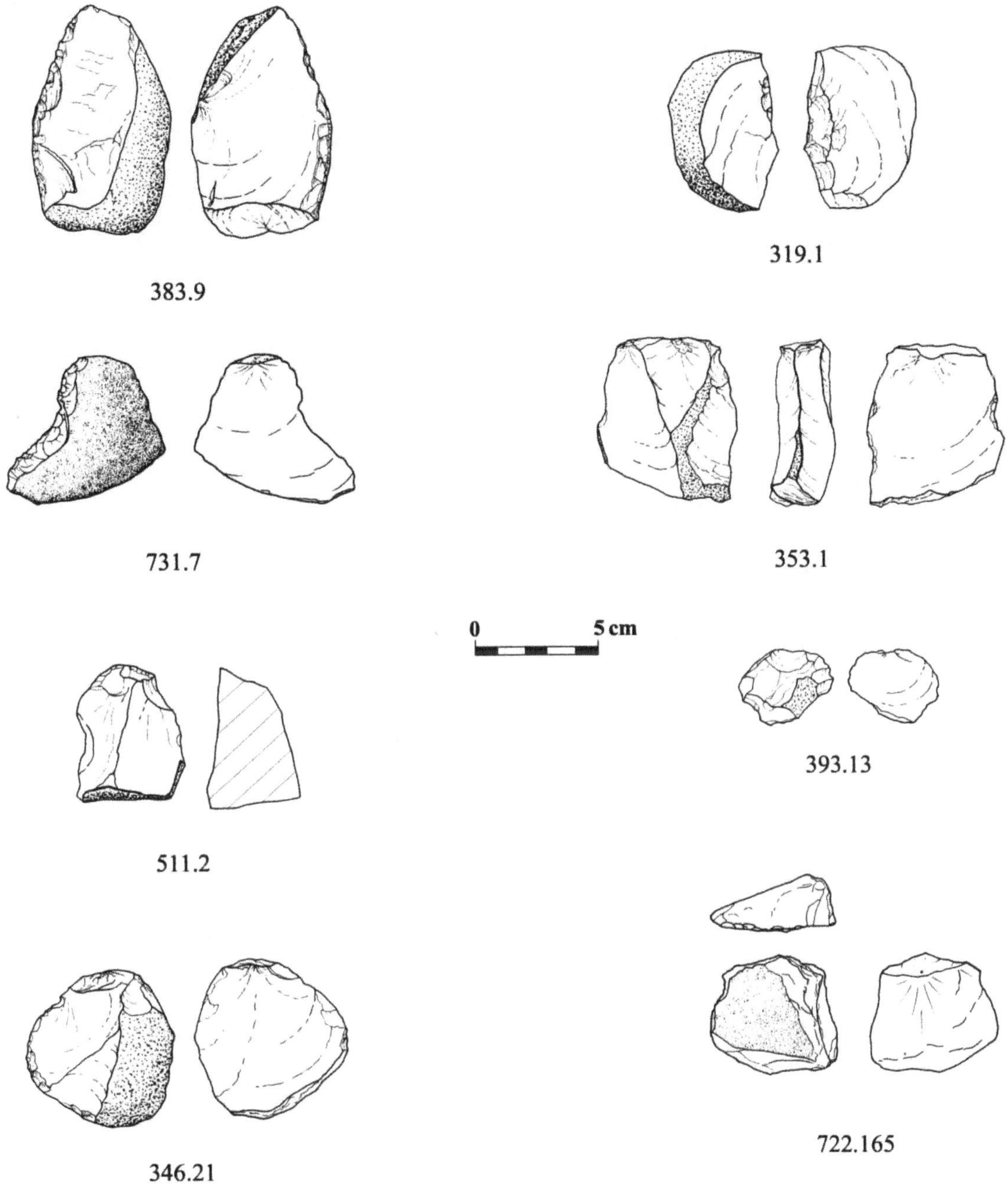

Figure 8.16: Retouched flakes, including knife and scrapers

Figure 8. 17: The conglomerate capping Sandstone 14, here exposed as a consolidated slab and showing the size of the pebbles.

Figure 8. 18: The sands with pebble stringers running along the top of the Sandstone 14 scarp slope.

Figure 8. 19: Laterally-restricted conglomerate layer at the base of sands of Sandstone 14, showing the 50 cm section cut back into the hillside (scale = 20 cm).

Figure 8. 20: The small spur of locality 218 being searched for stones, and the arrow marking where an artefact was recovered in 1987.

Figure 8. 21: The conglomerate and pebble exposures in the Baroth area mapped against artefacts from the collecting zones.

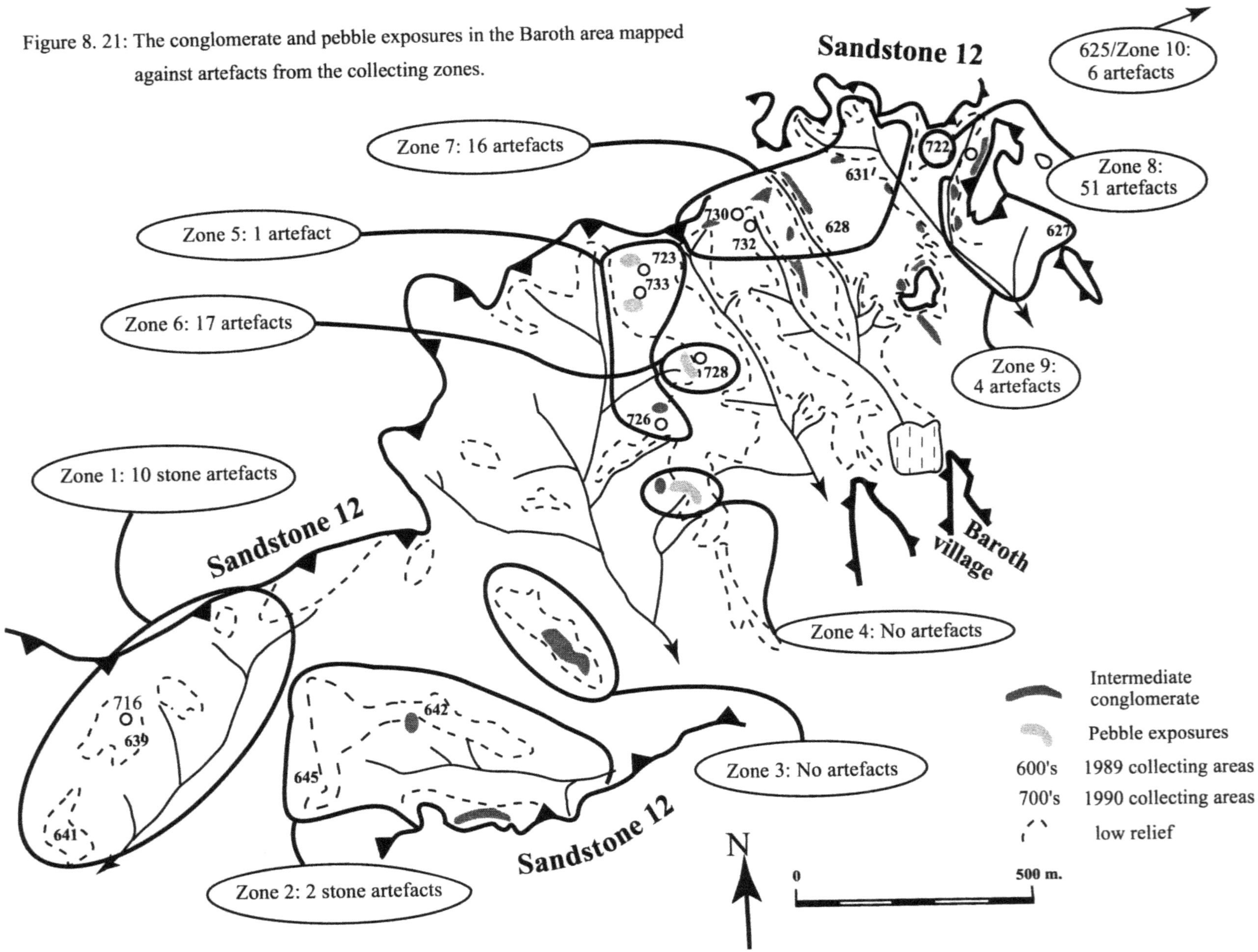

Figure 8. 22: The 269 stone and fossil bone in close proximity as found in 1987. The fossil refitted with other fragments further upslope.

Figure 8. 23: The general area of 269 looking North, showing the main scarp in the background, and the 'horseshoe' containing the stone artefact, with person standing at the level of the path in the area of excavations.

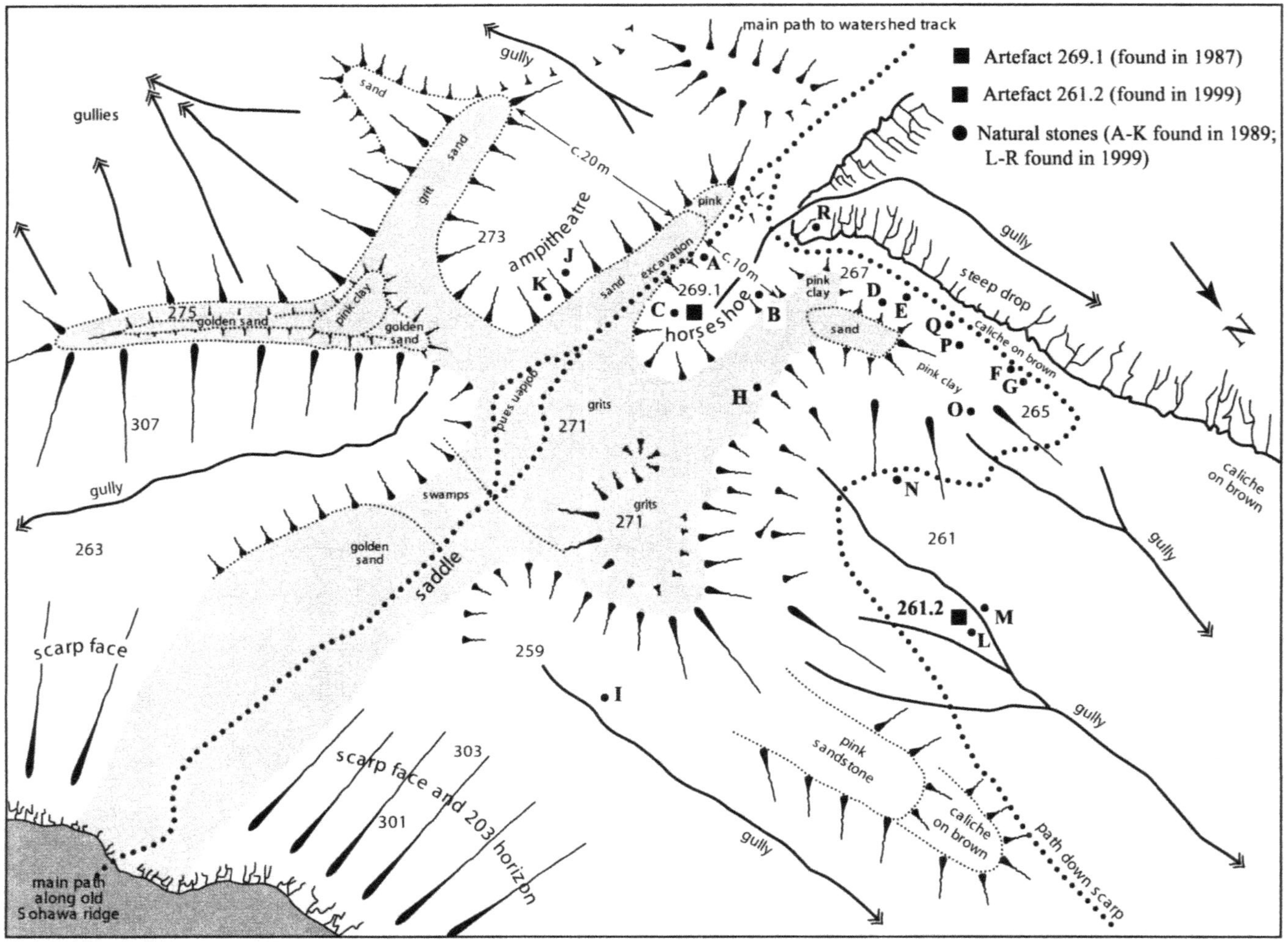

Figure 8.24: Map of the 269 area, showing also collecting areas, and natural and flaked stone

Figure 8. 25: The 269 horseshoe area, showing the level of the path, with soft sands above and steep slope below

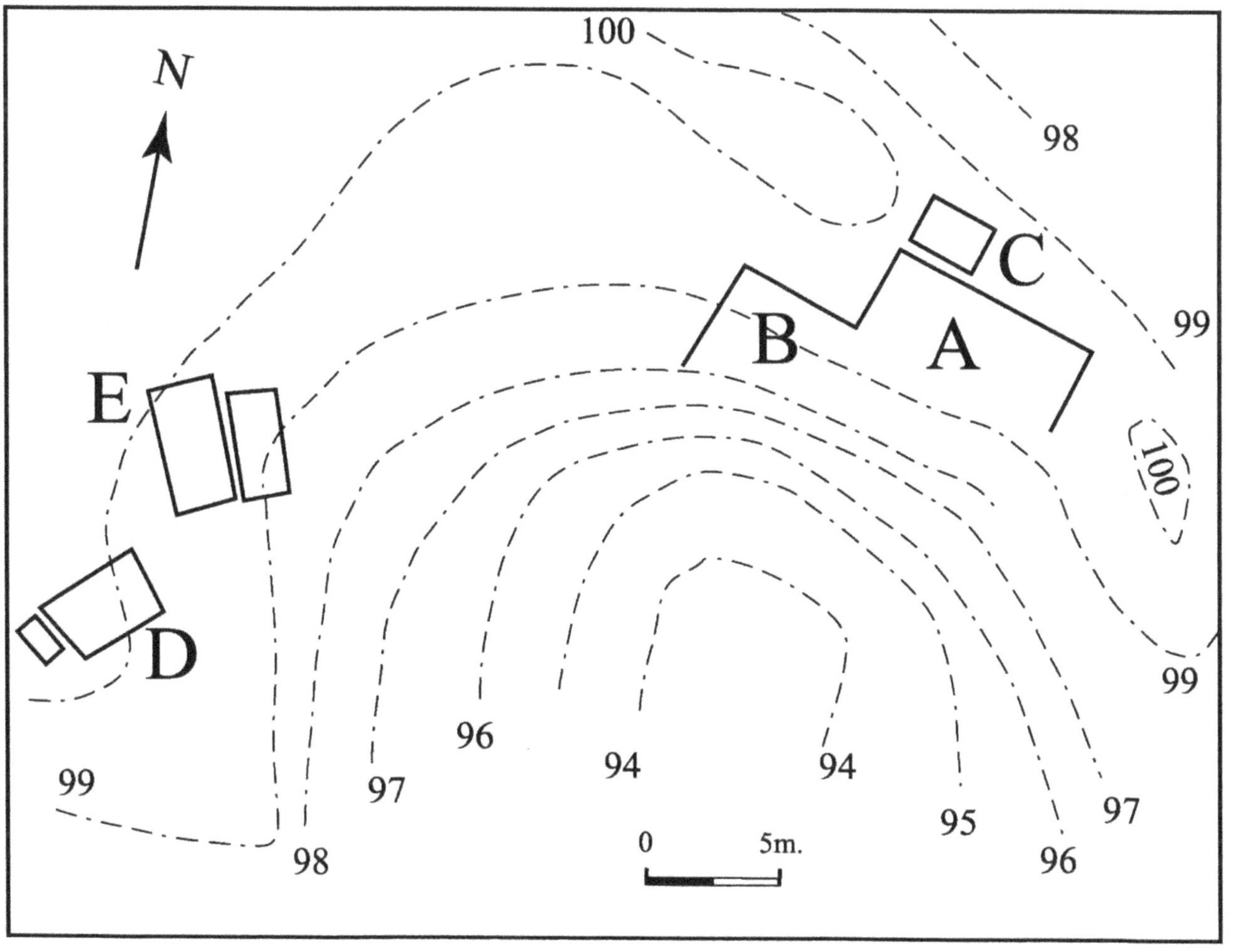

Figure 8. 26: Contour map of the 269 horseshoe area and excavation trenches.

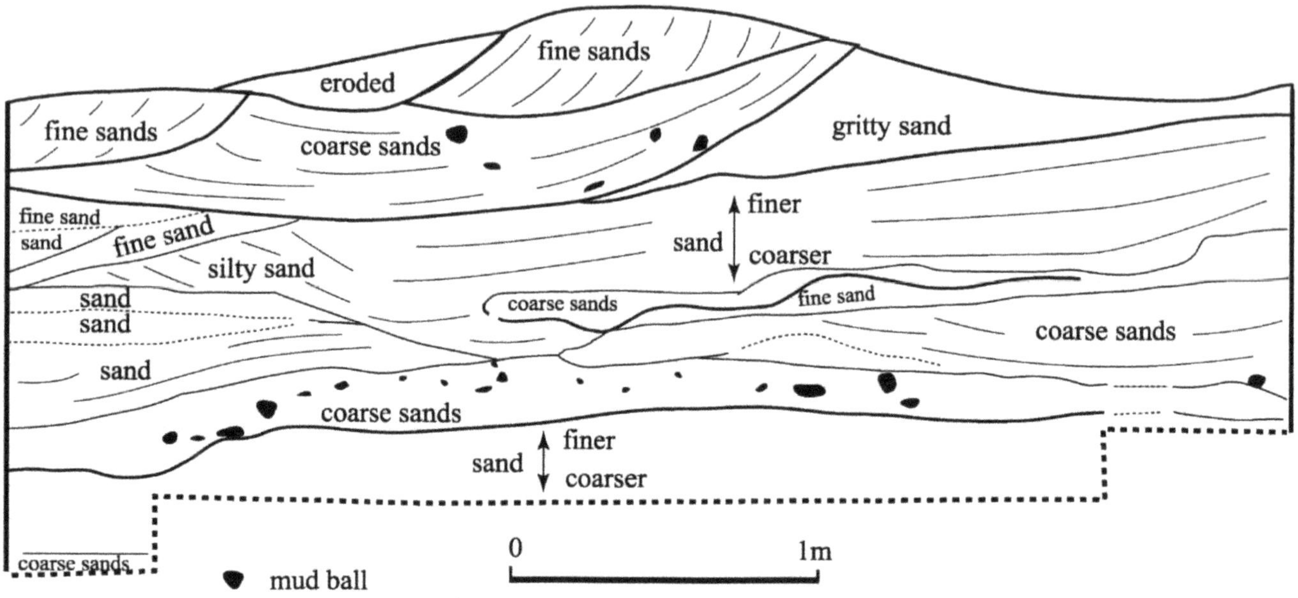

Figure 8.27: The north section of Trench A, site 269

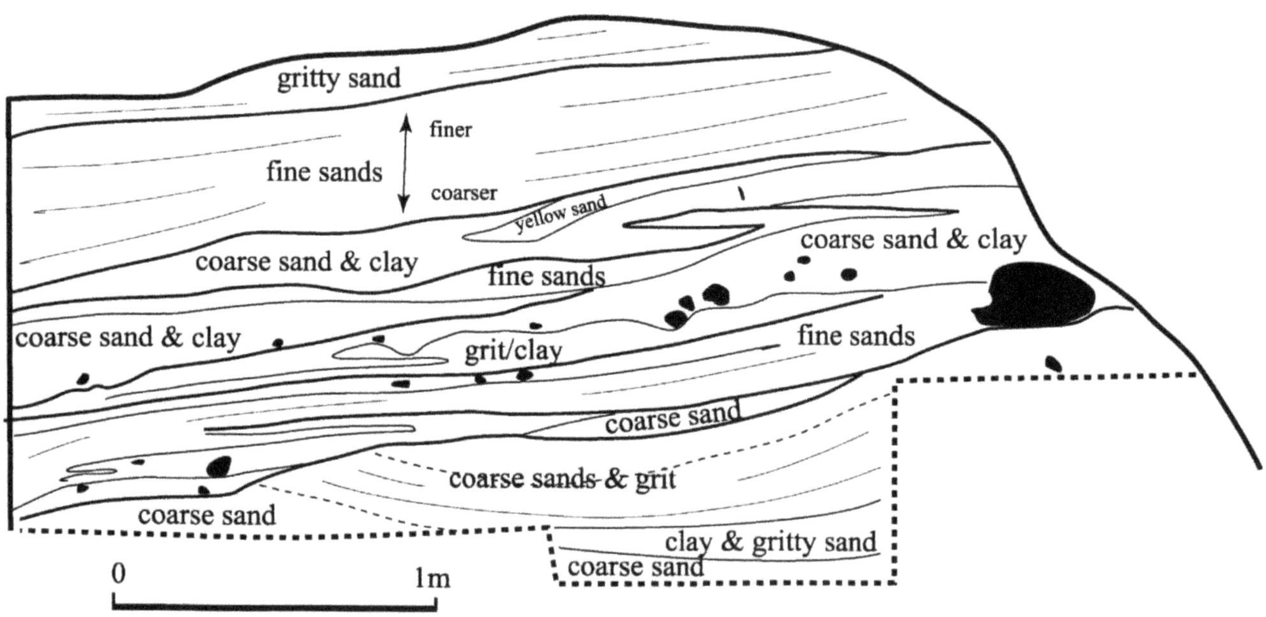

Figure 8. 28: The east section of Trench A, site 269

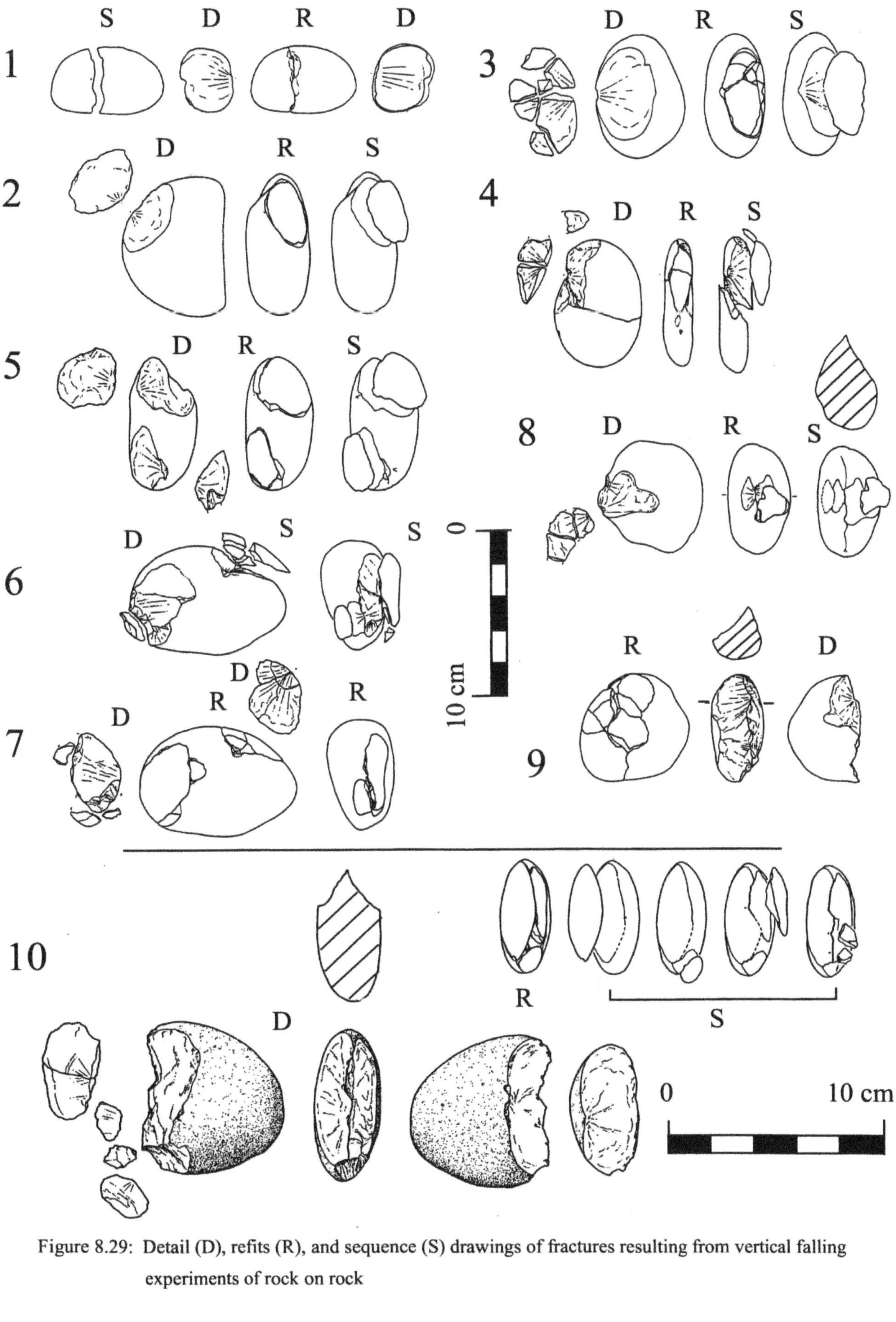

Figure 8.29: Detail (D), refits (R), and sequence (S) drawings of fractures resulting from vertical falling experiments of rock on rock

Figure 8. 30: Hillside on the west bank of the west fork of Kotha Kas (1999) indicating the layout of the erosion experiments. Flag Hill can be seen in the background.

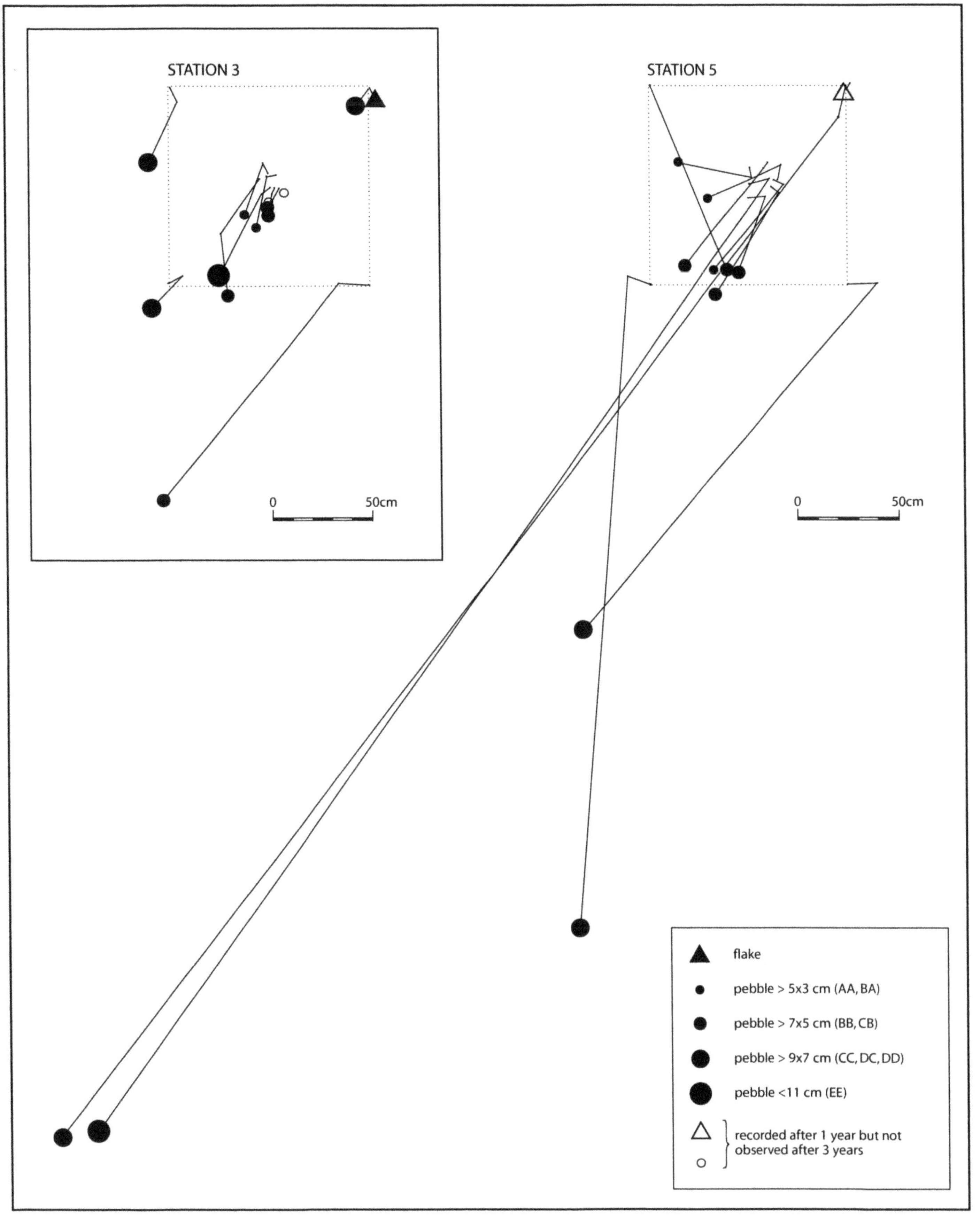

Figure 8.31: The movement of stones from 1989 to 1990 and 1992 in two of the erosion experiment stations.

Figure 8. 32: Erosion and movement of stones at experiment station 5 in 1999 after 10 years, showing the spread of material away from the original cluster.

Figure 8. 33: Erosion and movement of stones at experiment station 4 in 1999 after 10 years, showing the thorn bush roots resisting erosion to form a cone above the present land surface.

Chapter 9: The excavations of fossil localities 73, 362 and 642

Three particularly rich fossil localities were excavated: 73, found in 1986; 362, found in 1987; and 642, found in 1989 (see Figure 9. 1) Localities 73 and 642 were in the Sandstone 12 complex, and thus between 1.2 – 1.4 Mya-old; locality 362 was older, and in deposits that dated to the Olduvai Event, ca. 1.7 – 1.9 Mya-ago. There were several reasons for excavating these sites: they were eroding rapidly, and in the case of localities 73 and 362, were unlikely to survive for much longer; they offered excellent prospects of obtaining complete, and even articulated, specimens for the Geological Survey of Pakistan, in contrast to the fragmented and isolated specimens found in surface surveys; they offered opportunities of investigating the processes by which dense fossil accumulations were formed; and they offered an outside chance that they might include the remains of a hominid. Although hominid remains were not found, the excavations were worthwhile in producing excellent reference material, and much information on site formation processes, particularly by carnivores.

The excavation of locality 73 in 1986 showed the problems encountered in excavating this type of fossil accumulation. First, the sand matrix in which the fossils were embedded was very hard to excavate. Sometimes it was possible to soften it with a small amount of water, but it was usually necessary to use hammers and cold chisels to remove most of the matrix, followed by dental picks and fine brushes to expose the surface of the fossil. Many of the fragile bones were heavily cracked due to bioturbation and compaction, and fragile ones were given a coating of 70% PVA (Polyvinyl acetate adhesive) solution in the field. Some specimens required further treatment with PVA once back in the field base. Secondly, some specimens were very fragile, often needed consolidating before removal, and had to be carefully exposed on a block of matrix before being lifted as part of a plaster block, but without endangering other specimens around and underneath them. Thirdly, many fossils were in contact with their neighbours, and this often made it very difficult to isolate each specimen prior to removal. For these reasons, it was often necessary to isolate blocks of fossils that could then be removed together in a plaster jacket. Plastering was done by covering the surface of the block with tissue paper, and then strips of Gypsona, a locally-available plaster-of-Paris 2"-or 4"-wide bandage which set quickly once wetted. (Its main use is for making plaster casts for broken limbs). As each block had to be carried back to the vehicle (or, in the case of locality 73, to the field base at Banni Bangla over an hour's walk away over rough terrain), it was necessary to ensure that each block was sufficiently robust to withstand the inevitable vibration, yet light enough to be carried reasonably easily by one person.

The recording of material during and after the excavation was problematic. Sometimes, only small areas of specimens were exposed prior to removal, especially if they were being removed as part of a block. In some cases, it later transpired that pieces that were recorded in the field as separate specimens were actually part of the same bone. (Examples are specimens 30 and 31 from locality 73). Often, it was not possible to give accurate spatial co-ordinates for specimens, or draw their entire shape on plans because they dipped under other specimens. Additionally, a bone might be numbered during the excavation, lifted as a block, but later renumbered if the original number had faded or become illegible; great care had to be taken to ensure that specimens were not recorded twice. Most blocks also contained other specimens that were not visible during the excavation; these could be allocated new numbers, but their spatial co-ordinates remain imprecise, except by reference to the original location of the main specimen that formed the main component of a block.

In retrospect, locality 73 was the most difficult of the three fossil excavations. It was the most difficult to reach, being totally inaccessible by vehicle, and accessible only by an hour's walk through difficult terrain. In hot weather, sufficient water for the day had to be carried as well as food and equipment at the start of the day. Many days ended with a heavy load of plastered packages of fossils to carry back. Fossil specimens were more densely packed, and far more numerous, than at locality 362, even though the two localities were roughly the same size. Although site 642 was a much larger fossil accumulation, fossils were also more dispersed, and it was usually possible to expose and remove specimens individually.

Each site report follows the same format. The core of the report comprises the plans of each "level" (i.e. the surface of the fossil accumulation as the excavation unfolded), and the inventory of finds. The text of each report summarises the results, and the excavators' assessment of the processes by which the fossils accumulated. Those primarily responsible for each excavation (myself for site 73, and Beech and Anwar for localities 362 and 642) then summarise the results, and present their assessment of the main factors that resulted in the accumulation of the fossils. Ros Coard's own assessment of each site, undertaken as part of her Ph.D., (Coard, 1993) is presented as a separate contribution at the end of the chapter.

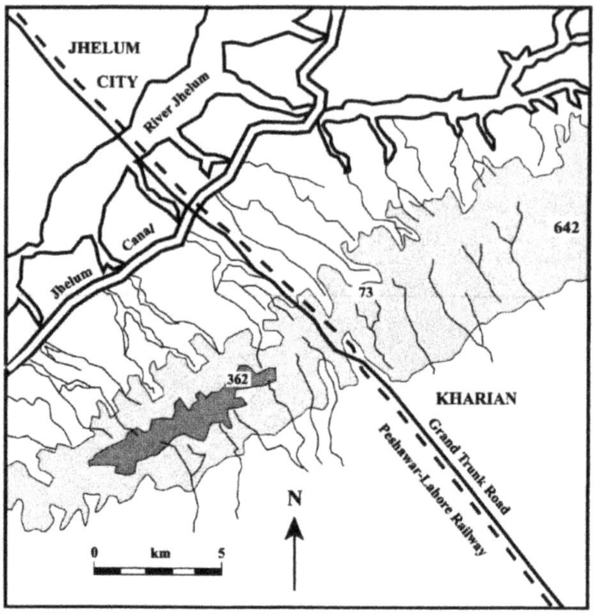

Figure 9.1: Location of localities 73, 362 and 642. Localities 73 and 642 are between 1.2 and 1.4 Mya-old; locality 362 is ca. 1.7 – 1.9 Mya-old.

Locality 73

Site 73 was the first of the three fossil localities that were excavated in the Pabbi Hills. As described in chapter 5, it was discovered early in the first field season of 1986, when over 600 specimens were collected from the top and sides of the small spur from which fossils were eroding. These included numerous limb and dental specimens of bovids, cervids, and equids, as well as a couple of pig teeth. Carnivore remains were common, and were mainly of hyaena, with bear, a canid, a large cat and a herpestid also represented. (In terms of carnivore remains, this locality contained more than any other found in the Pabbi Hills). As narrated in chapter 5, there was also a carnivore cranium and complete rhinoceros scapula exposed on the surface of the main fossil deposit when the locality was first discovered.

There were several reasons why it was decided to excavate it: preservation was excellent; a wide range of taxa were present; it was small and easily defined as an excavation; and its rate of erosion was such that it would clearly not survive much longer. A small excavation was thus conducted in 1986 (Feb. 17 - March 6th), and a longer, larger-scale operation mounted in 1987 (March 9 - April 6th). Figures 9.3–9.5 show the distribution of material at the beginning, middle and near the end of the 1987 excavation. Tables 9.4 and 9.5 give details of the material excavated in 1986 and 1987 respectively.

Methodology

The 1986 excavation developed out of the surface collecting. The initial collecting areas were defined as 73A (north flank), 73B (south flank) and 73C (top of spur) (see chapter 5, Figure 5.7 and Tables 5. 10 – 15). As erosion was recent and on-going, the sub-surfaces of each collecting zone were also sieved. Thus the sub-surface of 73A was 73X; 73B sub-surface was 73Y; and the sub-surface of 73C was 73Z. The top of the spur was then excavated, and numbered 73EX1 onwards. In 1986, the numbers used to record specimens that were drawn and/or lifted ran from 1- 166; in 1987, the excavated specimens were numbered from 167 to 386. Eight sets of specimens were lifted as plaster blocks in 1986, and 27 in 1987. These contained numerous specimens that were not visible, or were only partly exposed, during the excavation. These were later given numbers from 400-848, and their original approximate location can be determined by the position of the main specimen(s) that formed the nucleus of a given block.

Sedimentary context

A detailed section of the entire sequence of deposits at locality 73 has already been presented as part of Rendell's sedimentological investigations of Sandstone 12 (see Figure 5. 2). Figure 9. 2 shows a schematic section of the deposits at site 73. These are:

Unit A (117cm.): a fine grained soft, homogeneous yellowish-grey (5Y7/2) sandstone, with a "salt and pepper" appearance due to black (probably of iron or manganese) clasts with white translucent, slightly rounded SiO_2 clasts.

Unit B (44 cm.) This is unbedded, and differentially eroded from both A above and C below. It is a very fine-grained sand that is almost a siltstone. Its overall colour was grayish-orange (10YR7/4) with "veins" of moderate brown (5YR 4/4) iron colouring that may indicate bioturbations and burrowing by insects. Jay Quade (pers. com.) suggested that this may be a weak soil horizon.

Unit C (93 cm.): this had the same texture as Unit A, but lacked the "salt and pepper" appearance. Its overall colour was a light brown, 5YR 5/6. This unit contained the fossil accumulation. The bone distribution was predominantly north to south, dipping down slightly towards the north, with a very clear eastern limit and an almost continuous parallel western limit. Fossil was found in the top 20-30 cm. of the excavated area of this unit. Weak, dipping bedding planes could be seen in this unit, and the fossils appeared to lie on top of one of them

Unit D (23 cm.). This was a homogeneous, unbedded, slightly clayey, grayish orange (10YR 7/4) sand, and may denote a weak soil unit (Quade, pers. com.).

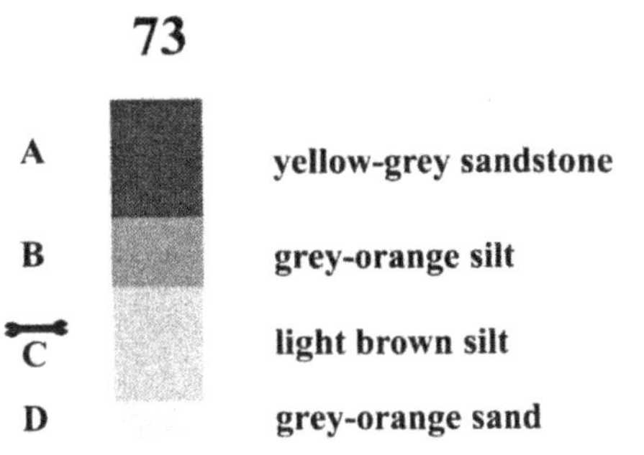

Figure 9. 2: Schematic section of the local profile at locality 73.

Results

Overall, some 350 specimens were excavated, and a further 500 pieces were found in the plaster blocks. The abundance of each taxon recognised from the excavation is shown in Table 9.1, along with the probable number of individuals present. This information needs also to be considered in conjunction with the surface and sub-surface collections from 73A-C, and 73X-Z. In general terms, the excavated part of the locality contained the remains of one rhino, pig and horse, perhaps four bovids and three cervids; adult and young hyaena, a felid, a small canid and possibly a bird and rodent. Additionally, the surface collections contained specimens of a bear and an herpestid.

Table 9. 1: Principal taxa represented in the excavated assemblage from locality 73.

Taxon	No. of specimens	No. of individuals
Rhinoceros sivalensis	5	1
Equus sivalensis	53	3
Bovid	31	3
Bovid (large)	6	1
Cervid	43	3
Suid	20	1
Hyaenid	3	1
Canid	2	1
Felid	10	1
Carnivora indet.	13	?

The following carnivore remains are also represented (those in bold are from the excavated sample), and are discussed by Turner later in this volume:

Crocuta crocuta: 73X GB 56, 73X GB 57, **73EX410**
Pachycrocuta: 73C GB 133
Hyaenidae indet: 73A GB 202, 73B GB 184, 73B GB520, 73A GB 328, 73BB GB 22
Panthera uncia: **73EX238**; **73EX GB 58, 73EX 568**, 73X GB51, 73C 203, 73C 420, 73C 233, 73C 382
Felidae indet: 73B GB 407/181; 73B GB, 6 fragments
Canis 73X 36, 59, 20; 73C GB 63
Herpestidae: 73A 52
Ursidae: 73B GB 515

Skeletal part representation

Many specimens were articulated, notably the hind legs of a pig and horse (see contents of blocks 9 and 10), horse foot (block 19) and part of a vertebral column (block 25). As the contents of other blocks indicate, however, pieces of several animals were often mixed up together (see e.g. blocks 5 and 10). Limb-bones were also much commoner than cranial pieces: unlike localities 362 and 642, there were no well-preserved cranial specimens from site 73, although there were several good dental and jaw specimens. The lack o cranial material, particularly of bovids, impede identification of much of this material.

The herbivores are represented overwhelmingly by lim segments, whereas the carnivores are primarily identifie from dental and foot specimens. Details are as follows:

Rhinoceros is represented by a partial forelimb (specimen 5, scapula; 403 humerus) and part of a foot (52 cuneiform; 10 astragalus; 177 metapodial) from a minimum of one individual.

Equus is overwhelmingly represented by limb fragments from a minimum of two individuals. The limb fragments are mainl from the lower limb; there are very few trunk and head elements, as shown below:

upper fore-limb: specimen 270 scapula; 57 proximal humerus; 357 complete humerus + proximal radius; 240 left radius + ulna; 379, 427 radius fragments.

lower fore-limb: specimen 78, 198, 366 lateral metapodial; 131 metapodial; 53, 138, 230, 360 proximal metacarpal; 35 metacarpal; 356 complete metacarpal; 504 metacarpal fragment

feet: specimens 25, 139, 204, 250, 581, 606 1st phalange; 44 proximal phalange; 173 distal 1st phalange; 285, 369, 849 2n phalange, complete; 273 proximal 3rd phalange; 539, 555 3rd phalange fragment; 618 3rd phalange, complete

lower hind limb: specimens 263, 850 distal tibia; 331 tibia shaft; 189 metatarsal; 220 complete metatarsal; 569 right metatarsa + tarsals; 610 metatarsal shaft + lateral metapodial; 247, 269 astragalus fragments; 321, 341, 845 astragalus complete; 84 calcaneum complete; 847 cuboid complete; 857 carpal
trunk: specimen 156 axis, 186 atlas
head: specimen 239 left mandible; 605 mandible fragment; 280 left maxilla; 363 maxilla fragment

One medium-sized bovid is represented by a variety of limb segments:

upper fore-limb: specimen 57 distal humerus; 30/31, 222 proximal radius; 534 radius
lower fore-limb: specimens 7, 20, 579 distal metacarpal; 239/310 metacarpal;
upper hind-limb: specimens 226, 268 distal tibia; 287 tibia shaft
lower hind-limb: specimen 188 metatarsal + navicular + magnum; 332 navicular-cuboid; 15 astragalus; 212, 351 complet astragalus; 272 distal calcaneum; 86 pisiform
feet: specimen 82, 227 complete 1st phalanges; 848 proximal 1st phalange
trunk: specimen 190 atlas

A large bovid is also represented by part of a fore- and hind-limb:

122 left complete radius + part of ulna (117); 763 proximal ulna; 286 proximal and distal metatarsal; 191 complete 1st phalange; 295 complete 2nd phalange

Cervids are also largely represented by head and limb segments:

Upper fore-limb: 79, 206 complete scapula; 158 scapula fragment; 290 complete radius; 299 distal radius
lower fore-limb: 55 right metacarpal, complete; no number and 608, proximal metacarpal; 426, 607 distal metacarpal
Upper hind-limb: 242 complete left femur; 27, 330/811 complete tibia; 34 proximal tibia; 128 distal tibia; 262 tibia shaft
lower hind-limb: 21, 333, 613 proximal metatarsal; 218 complete astragalus
feet: 51, 83 cuneiform; 160 sesamoid; 43 complete 1st phalange; 275 complete 2nd phalange; 159 3rd phalange; 545 phalange fragment
head: 40 PM_3; 61, 236, 803 PM; 311 M3; 94 upper incisors; 96 mandible with dm1-3, M1-2, M3 erupting; 296 left mandible with dm1-3, M1 erupting; 219/223 right mandible; 369 left mandible with dm2-3; 313 mandible with P2-4, M1-3; 182 left maxilla M1-3, P4 erupting; 235 maxilla M1-3; 570 maxilla with M1-2, M3 erupting; 867 mandible fragment

(A number of limb-bone specimens were identified as bovid or cervid:
245 distal scapula; 210 proximal radius; 4 distal femur; 224, 234 distal femur; 604 distal metacarpal; 308 astragalus).

The **suid** remains comprise one fore-limb + neck, and one hind-limb from one individual:
314 mandible ; 33/449 scapula; 118 humerus; 428, 611 humerus shaft; 609 ulna
110, 737; femur; 323 distal tibia; 355, 854 lateral metapodial; 615 proximal lateral metapodial
196 calcaneum; 853 astragalus; 326, 855, 856 tarsals; 243, 543 lateral 3rd phalanges

Ribs and vertebrae were recorded as indeterminates. Both were rare when considered against the range of taxa present. Only 47 rib fragments were recovered from the excavation, and 24 vertebra fragments, including one find of a sacrum and three articulated vertebrae (281). Only two indeterminate tooth fragments were recorded from the excavation.

Age structure
This is discussed by Coard later in this chapter

Bone weathering
Most bones in both the surface and excavated assemblages had very little surface modification through weathering, and this implies that most were buried shortly after death. Details are shown in Table 9. 2; comparative data from other localities are presented in chapter 10.

Table 9. 2: Bone surface weathering at locality 73.

Weathering stage	1	2	2a	2b	3	4	5	Total
Locality 73 surface	27		**46**	29	19	6	4	131
Locality 73, excavation	12		**140**	59	43	8	2	264

Taphonomy

As research has repeatedly shown over the last 20 years, it is rarely possible to single out one set of processes that unambiguously led to the formation of a fossil deposit. Locality 73 was no exception, and the issues involved are best illustrated by a summary of a workshop held at the excavation by team members on 23/03/87:

"Much discussion concerned fluvial versus carnivore activity. The mélange of sediments (including grey silt-sand and red clay, and a clay coated ball) amongst bones suggest to one participant that sediments had been redeposited/reworked fluvially (they might, for example, have been hill wash). The fresh nature of bones suggested rapid covering by sediments with no sub-aerial erosion. There was also very little evidence of any gnawing by carnivores, and most bones were not broken. It was pointed out that carnivores such as hyaenas might not have damaged the bones if meat was plentiful. The presence of coprolites (observed in 1986) also suggested that material had not been moved significantly. This point was supported by the fresh, unrolled state of bones. Two participants observed that the bedding crack against which the bones lie along the eastern margin of the concentration became roughly horizontal to the west, and may mark the base of the fossil horizon. A possible scenario might thus have been a depression or gully that was utilised by hyaenas, and later infilled by hillwash (causing some orientation of bones). Alternatively, it might have been a burrow (with bones lined against edge), followed by collapse of the roof, leading to a mélange of deposits. Another possibility is that some carcasses/partial carcasses were deposited fluvially and then visited by carnivores - i.e. it appears to be a mixture of carnivores and fluvial activity"

To summarise, the following points suggested that carnivores were the main accumulating agency:

1) diversity of species present, particularly of carnivores
2) number of individuals present
3) presence of gnaw marks (but very few seen in 1987)
4) presence of carnivores
5) presence of young hyaena
6) presence of hyaena coprolites (but not in 1987)
7) the freshness of the bones

The following points suggested that fluvial transport was primarily responsible:

1) mixed sand, silt and clay and nodules among bones
2) the alignment of some bones, although this could be due to carnivores or the configuration of a burrow
3) the presence of a mud-ball (specimen 260, see Figure 9. 4

The following points argued against fluvial transport as primary agency:

1) bones were fresh and unrolled
2) a mixture of hydrodynamic groups of elements w present (vide Voorhies, Behrensmeyer etc.) but this mig have been complicated by the presence of bloated carcass rather than defleshed individual bones
3) the presence of articulated units of bones

As can be seen, the team members expressed a variety viewpoints. My own assessment is that carnivores we primarily responsible for the fossil accumulation, but th there had also been some fluvial activity at the site, hence t presence of a mud-ball, and the mélange of sedimen amongst the bones. Coard presents her assessment, a comes to broadly the same conclusion, with perhaps a great emphasis on carnivore activity. As hyaenids are the bo accumulators par excellence amongst carnivores, it probable that they accumulated most of the prey remain and then faced competition from other scavengers, notab the felid, canid and ursid Their remains would suggest th the competition for feeding sometimes had leth consequences.

Table 9. 3 lists the size of complete skeletal elements fro the excavation. Further comparative data are presented chapter 11.

Table 9. 3: Measurements of limb-bones from locality 73

Specimen	Taxon	Element	Length (mm)	A-P (mm)	Width (mm)
7	Bovid	Metacarpal (distal)		34.0	48.0
15	Bovid	Astragalus	89.0	59.0	47.0
212	Bovid	Astragalus	93.0	60.5	51.0
351	Bovid	Astragalus	58.0	39.0	30.0
30	Bovid	Radius (proximal)		34.0	103.0
82	Bovid	1st phalange	151	32.0	24.5
286	Bovid (large)	Metatarsal	250	56.0 (prox)	75.0 (prox)
188	Bovid	Metatarsal	280	48.0 (prox)	48.0 (prox)
188	Bovid	Metataral		32.6 (prox)	49.0 (distal)
226	Bovid	Distal tibia		64.0	88.0
122	Bovid (large)	Radius	354.0	51.0 (prox)	91.0 (prox)
122	Bovid (large)	Radius			93.0 (distal)
191	Bovid (large)	1st phalange	77.0	40.0 (prox)	37.0 (prox)
191					37.0 (distal)
295	Bovid (large)	2nd phalange	49	36.0 (prox)	
43	Cervid	1st phalange	67.5	29.0 (prox)	23.0 (prox)
275	Cervid	2nd phalange	38.5	25.0 (prox)	19.0 (prox)
275	Cervid	2nd phalange		25.0 (distal)	15.0 (distal)
55	Cervid	Metacarpal (proximal)	248.0	29.0 (prox)	43.8 (prox)

Table 9. 3 continued: Measurements of limb-bones from locality 73

608	Cervid	Metacarpal (proximal)		31.0 (prox)	42.0 (prox)
55	Cervid	Metacarpal (distal)		29.0 (distal)	42.3 (distal)
426	Cervid	Metacarpal (distal)		26.5 (distal)	41.5 (distal)
607	Cervid	Metacarpal (distal)		28.0 (distal)	44.0 (distal)
21	Cervid	Metatarsal (proximal)		39.0 (prox)	41.0 (prox)
613	Cervid	Metatarsal (proximal)		35.0 (prox)	
330	Cervid	Tibia	270.0		
128	Cervid	Tibia (distal)		41.0	54.0
218	Cervid	Astragalus	61.3	40.9	34.6
290	Cervid	Radius	294.0	33.0 (prox)	63.0 (prox)
299	Cervid	Radius (distal)			57.0 (distal)
210	Cervid/bovid	Radius + ulna	300	36.0 (prox)	79.0 (prox)
210	Cervid/bovid	Radius + ulna			68.0 (distal
53	Equid	Metacarpal (proximal)		31.0 (prox)	52.0 (prox)
138	Equid	Metacarpal (proximal)			49.5 (prox)
356	Equid	Metacarpal	232	29.0 (prox)	49.0 (prox)
356				33.0 (distal)	43.0 (distal)
360	Equid	Metacarpal	220		51.0 (prox.W)
139	Equid	1st phalange	75.0	29.0 (prox)	
139				22.0 (distal)	40.0 (distal
204	Equid	1st phalange	82.0	34.0 (prox)	45.0 (prox)
204					38.0 (distal)
250	Equid	1st phalange	83.0	34.0 (prox)	47.0 (prox)
250					38.0
606	Equid	1st phalange (split)	82.0		
285	Equid	2nd phalange	43.7	29.3 (prox)	44.7 (prox)
369	Equid	2nd phalange	31.0	28.0 (prox)	45.0 (prox)
849	Equid	2nd phalange	39.0	32.0 (prox)	48.0 (prox)
189	Equid	Metacarpal	270	48.0 (prox)	42.5 (prox)
189				33.0 (distal)	43.6 (distal)
220	Equid	Metatarsal		35.5 (prox)	40.0 (prox)
220				30.0 (distal)	36.0 (distal)
569	Equid	Metatarsal	252.0		33.0 (prox)
				26.0 (distal)	36.0 (distal)
240	Equid	Radius + ulna	308	40.0 (prox)	66.0 (distal)
263	Equid	Tibia (distal)		41.0 (distal)	59.0 (distal)
196	Suid	Calcaneum	92.0		35.5

The following figures show the distribution of fossil specimens at various stages of the excavation. The highly localised distribution of material is particularly evident from the middle and lowest part of the excavation, with no fossil at all on either side of a very tight clustering of specimens.

Figure 9. 3: Plan of excavation of upper part of locality 73 (based on plans by Eddie Moth and others)

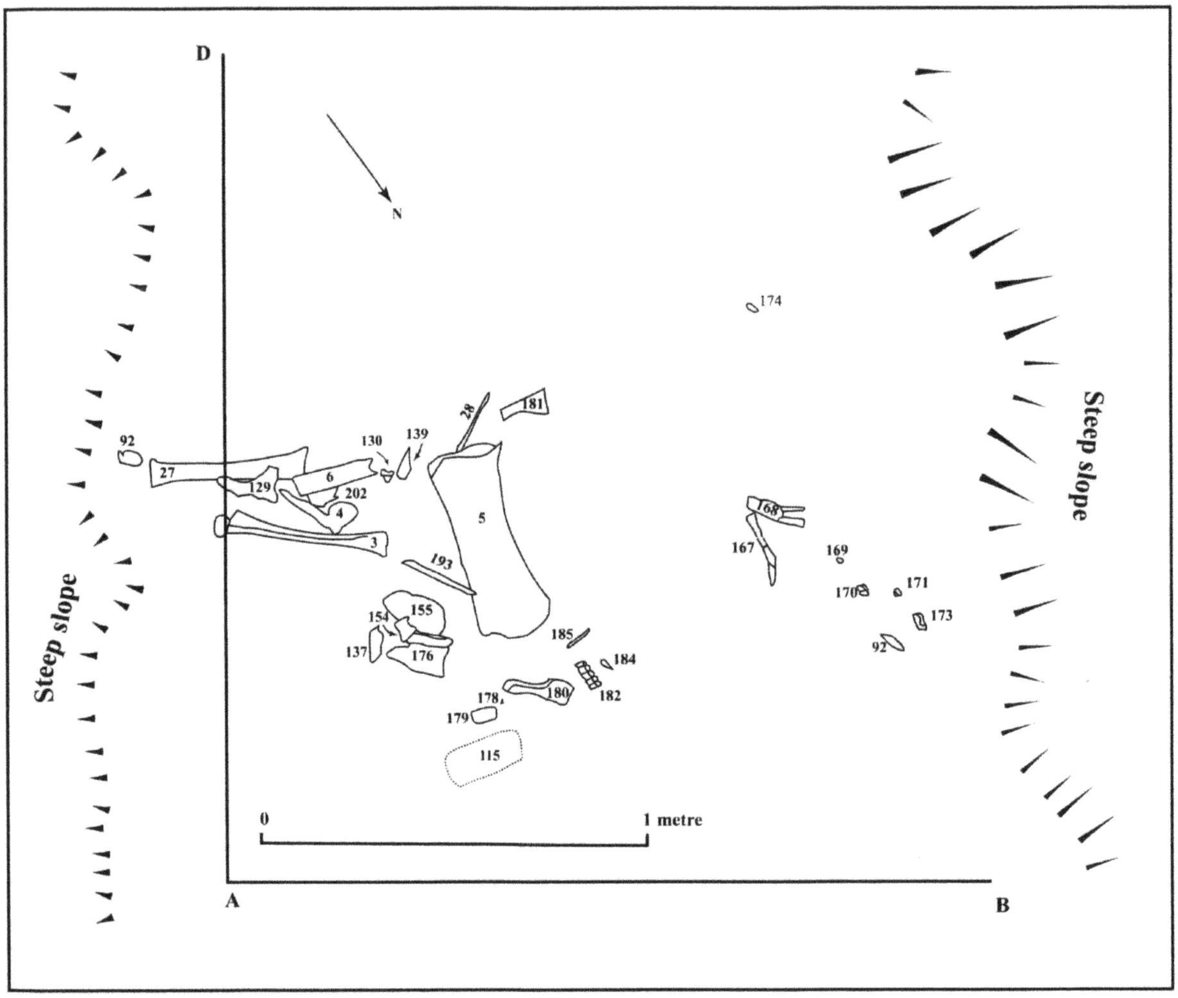

Figure 9. 4: Plan of excavation of middle part of locality 73 (based on plans by Eddie Moth and others)

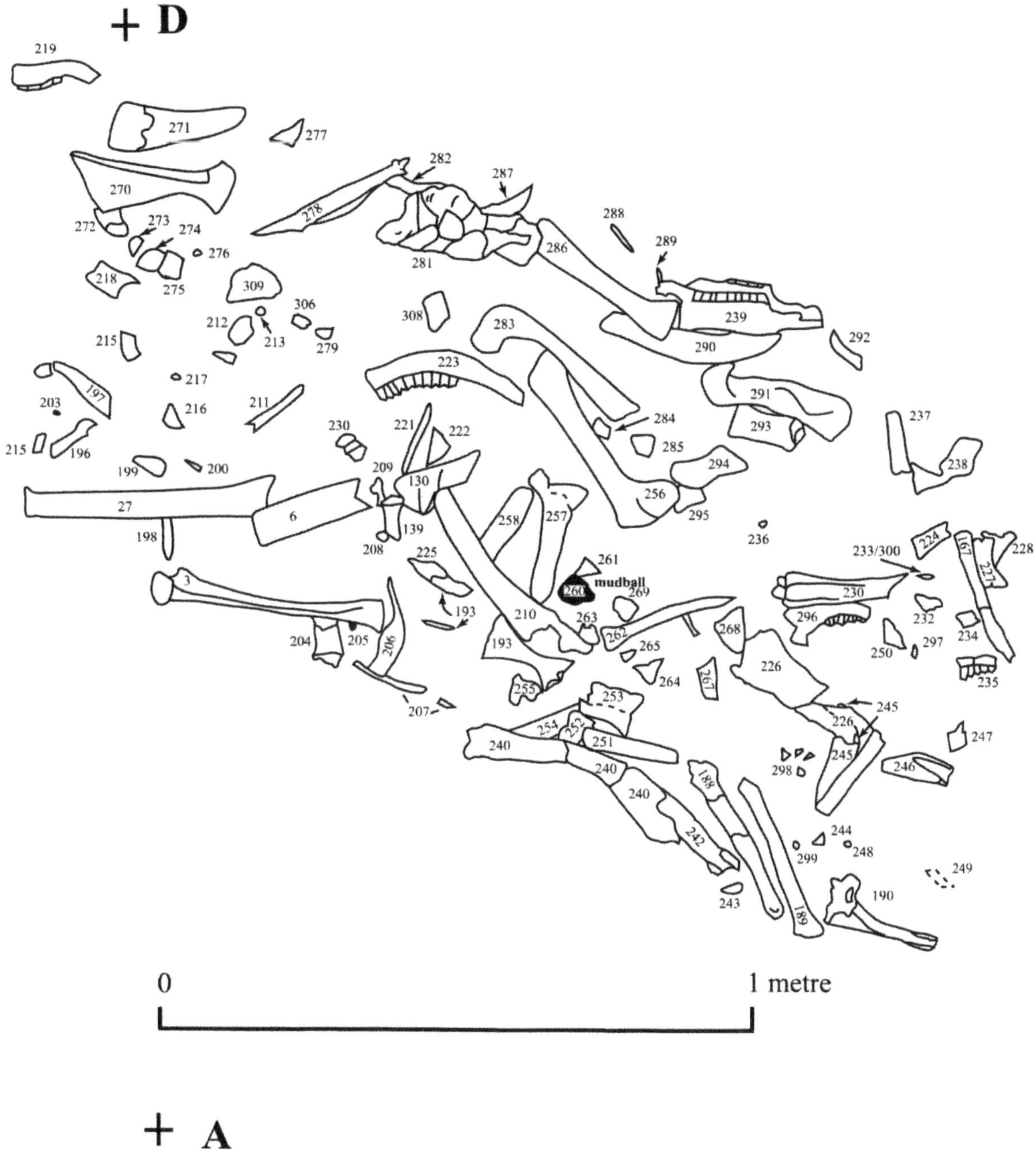

Figure 9. 5: Plan of excavation of bottom part of locality 73 (based on plans by Eddie Moth and others)

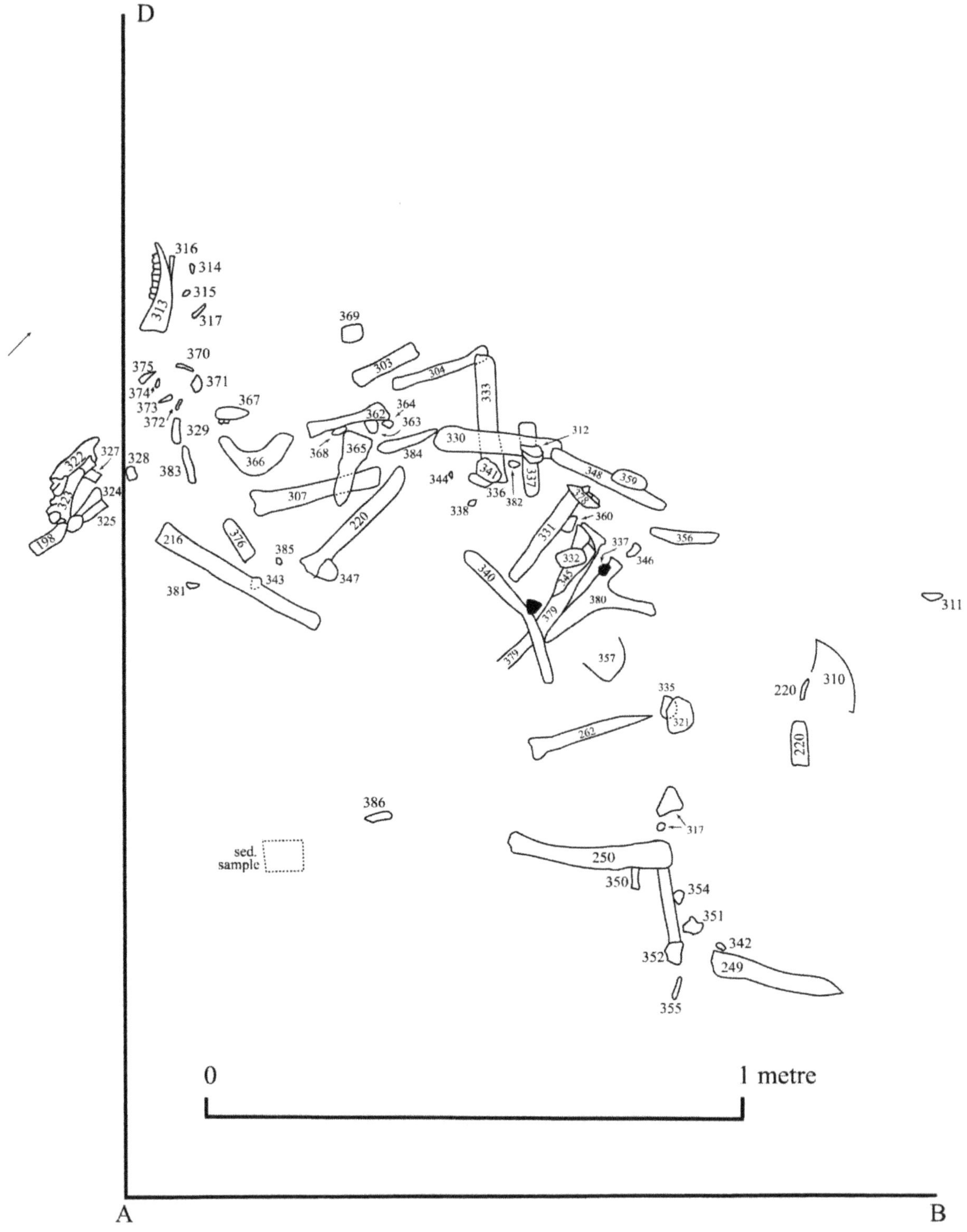

Table 9. 4: Fossil specimens from the excavation of locality 73, 1986

Specimen	New No.	Block	W	Description	N
0				Indeterminate fragments	138
0				Indeterminate shaft fragments	207
0				Canid 2nd phalange, complete	1
0				Carnivora 1st phalange, complete	1
2				Cervid m1/2 upper	1
3	503	20	2a	Cervid proximal radius + ulna, complete	0
4	507	20	2a	Bovid/Cervid distal femur	0
5				*Rhinoceros* scapula, complete	1
7			2a	Bovid distal metacarpal	1
8		ac	2a	Cervid ulna, complete	1
9		ac	2a	Indeterminate vertebra spine	1
10, 12		ac	2a	Indeterminate (small) rib	2
11				Indeterminate epiphysis	1
12				Carnivora carnassial tooth fragment	1
14		z	3	Indeterminate shaft (removed with 34)	1
15			1	Bovid astragalus, complete	1
18			1	Carnivora 2nd phalange, complete	1
19			2a	Indeterminate rib (removed with 41)	1
20				Bovid distal metacarpal	1
21			2b	Cervid proximal metatarsal	1
22	738	24A		Indeterminate condyle	0
24			2a	Indeterminate rib, 19 x 8 mm.	1
24			2a	Suid lateral 3rd phalange	1
25	606	24B		Equid 1st phalange	0
26				Indeterminate fragments	2
27		13	2a	Cervid tibia, complete; lifted 1987 with 29 and 198	1
28				Indeterminate rib 10 cm. long; lifted 1987	1
29		13	2a	Indeterminate pelvis, lifted 1987	1
30,31		b	2b	Bovid proximal radius + ulna; lifted with 466	2
32			2b	*Rhinoceros* cuneiform	1
33				Suid scapula, left; no glenoid	1
34		z	3	Cervid proximal tibia, left; removed with 14	1
36				Indeterminate rib fragments	2
36			2b	Artiodactyla (medium) proximal radius	1
38				Indeterminate (large) fibula, complete	1
40			1	Cervid tooth, PM_3; no wear	1
41			2a	Ungulata pelvis (acetabulum+ ischium); removed with 19; crushed	1
43		a	4	Cervid 1st phalange, complete	1
43		a	2a	Indeterminate rib	1
44		a		Equid? proximal phalange	1
45		a		Indeterminate rib	1
47			2a	Indeterminate distal scapula (no glenoid); see also 446-448	1
48				Indeterminate fragment (removed with 12)	1
49				Indeterminate flakes, discarded	2
51			1	Cervid cuneiform; complete	1
52				Indeterminate/Carnivore rib	1
53		a	2b	*Equus* proximal metacarpal	1
55				Cervid metacarpal, right, complete	1
56			2a	*Equus* humerus shaft	1
57		x	2b	Bovid distal humerus, split	1
58				Felidae (*Panthera uncia*) mandibular ramus; P4 root, M1 broken	1
58				Indeterminate rib 31 x 20 mm. (part of set 33,449)	1
59		a		Indeterminate rib; see also 43,45,53,67,121	1

60		a		Indeterminate rib (part of 58,59)	1
61			4	Cervid mandible premolar	1
62				Indeterminate shaft	1
65				Indeterminate phalange	1
67		a		Indeterminate fragment	1
70		a,a		Indeterminate fragment; same bone as 90	1
71				Indeterminate fragment	1
73				Indeterminate fragment; goes with 107,108	1
75				Hyaenid metatarsal, complete	1
76, 92, 106				Indeterminate rib	3
78			2a	Equid lateral metacarpal	1
79			2a	Cervid scapula, complete	1
82			2	Bovid 1st phalange, complete	1
83			1	Cervid cuneiform, complete	1
85				Ungulata cuboid, complete; removed with 61	1
86				Bovid pisiform, complete	1
89				Carnivore (small) ribs	3
90		a,a	2a	Artiodactyla radius/ulna shaft; see also 70, 436-445	1
93				Indeterminate fragment	1
94				Cervid maxillary incisors 1-2	1
95		x	2a	Indeterminate cranium; with 57, 97	1
96			2a	Cervid mandible (DM1-3,M1,M2 errupting)	1
97		x	2b	Indeterminate (large) rib (13 x 14 mm.); with 57 and 95	1
100				Carnivore (small) 3rd phalange; ? small canid	2
101				Indeterminate shaft	1
104		a,d		Indeterminate cervical vertebra	1
107			2b	*Rhinoceros* astragalus, complete	1
108				Indeterminate fragment; goes with 73,107	1
110	428			Suid femur; goes with 120	0
111				Artydactyla atlas	1
112, 116				Indeterminate fragments	2
117				Bovid (large) left ulna; goes with 122	1
118				Indeterminate fragment in silt block	1
118				Indeterminate vertebra	1
118				Suid humerus	1
120				Indeterminate shaft; goes with 110	4
121		a	2a	Indeterminate vertebra centrum	1
122			2a	Bovid (large) radius, left and complete + part of ulna; goes with 117	1
123				Indeterminate fragments	2
124			3	Bovid proximal radius or metacarpal; carnivore damage	1
125				Indeterminate fragment; part of 96	1
125				Carnivora small canine fragments	2
126, 137				Indeterminate rib	2
127		a,b		Indeterminate sacrum; goes with 131,455-460	1
128			2a	Cervid distal tibia; almost complete	1
130			2a	Artiodactyla (medium) distal humerus; spiral fracture; gnawed	1
131		a,b		Equid metapodial, complete; see also 127, 455-60	1
134				Indeterminate tooth spall	1
135			4	Indeterminate fragment	1
138			2a	Bovid proximal metacarpal (split) + carpal	1
139			2b	Equid 1st phalange, complete	1
143				Indeterminate flake	1
148		26		Indeterminate fragment	1
149-153, 157				Numbered but not exposed; not removed in 1986	1
154		26		Indeterminate vertebra centrum; not removed in 1986;	1
155		26		Indeterminate vertebra fragment; not removed in 1986	1
156		25	2b	Equid axis, complete; not removed in 1986	1

157				Indeterminate axis from under scapula 5	1
158				Cervid carpal; goes with 159,160	1
159				Cervid 3rd phalange; goes with 158,160	1
160				Cervid sesamoid; goes with 158,159	1
163				Indeterminate cuboid	1
164			2b	Carnivora incisor, worn; in matrix of 13	1
165				Indeterminate splinter; in matrix of 13	1
166				Indeterminate splinter; in matrix of 44	1
Total					**472**

Table 9. 5: Fossil specimens from the excavation of locality 73, 1987

Specimen	New No	Block	W	Description	N.
		11		Indeterminate fragments	3
		20		Equid carpals	3
		12		Indeterminate fragments	4
		3		Indeterminate rib	1
				Cervid proximal metacarpal	1
		12		Indeterminate fragments	7
				Indeterminate fragment	1
		10		Indeterminate fragments	14
		20		Indeterminate shaft	1
		20		Indeterminate rib	1
167		6		Artiodactyla distal metapodial	1
168				Artiodactyla (small) pisiform	1
169				Indeterminate fragments	3
170				Indeterminate rib	1
170				Artiodactyla carpal	1
171			2a	Indeterminate epiphysis fragment	1
172			2b	Indeterminate shaft	1
173			2a	*Equus* distal 1st phalage	1
174				Indeterminate tooth fragment	1
175				This number not used	0
176		26	2b	Bovid/Equid distal humerus	1
177		26	3	*Rhinoceros* metapodial	1
178			1	Carnivora 2nd phalange,	1
178				Carnivora 3rd phalange	1
179				Artiodactyla 1st phalange	1
180				Bovid/Equid distal humerus, complete; spiral fracture	1
181		24B	2a	Artiodactyla (medium) scapula glenoid	1
182			2b	Cervid left maxilla, M1-M3,P4 errupting	1
183			2b	Indeterminate splinter	1
184				Indeterminate shaft	1
185				Indeterminate rib fragment	1
186		25		Equid atlas, complete	1
187				Indeterminate long bone fragments near scapula (5)	9
188				Bovid metatarsal, navicular and magnum	1
189				*Equus* metatarsal, complete	1
190		11	2a	Bovid atlas	1
191		25	2b	Bovid (large) 1st phalange, complete	1
192			2b	Carnivore calcaneum	1
193		19		Indeterminate scapula fragments	5
194		3	2a	Artiodactyla femur; carnivore damage	1
195			3	Artiodactyla/Cervid scapula fragments	3
196		9		Suid calcaneum, complete	1
197			2a	Artiodactyla (medium) distal humerus	1
198		13	2a	Equid lateral metapodial, complete	1

199				Indeterminate sesamoid	1
200				Canid canine	1
201			4	Indeterminate shaft	1
202		20		Artiodactyla metacarpal	1
203		20		*Panthera uncia* left lower DM4 1	1
204		20	2a	*Equus* 1st phalange, complete; carnivore damage	1
205				No data in daybook; not used?	0
206		19	2b	Cervid scapula, complete; carnivore damage	1
207		19		Indeterminate rib	1
208			3	Indeterminate fragment	1
209		23		Indeterminate calcaneum	1
210		16	2b	Cervid/Bovid proximal radius and ulna, complete	1
211			2a	Indeterminate/Carnivore tibia shaft	1
212			1	Bovid astragalus, complete	1
213			2a	Canid canine	1
214			2b	Indeterminate patella	1
215				Indeterminate carpal/tarsal	1
216		17	2b	Indeterminate shaft	1
217			3	Indeterminate fragments	3
218			1	Cervid astragalus, complete	1
219			2a	Cervid right mandible with P4,(P2,3 broken)	1
220			2a	*Equus* metatarsal, complete	1
221		23		Long bone fragment	1
222		23		Bovid proximal radius	1
223		2	3	Cervid right mandible, complete with P2-M3	1
224			2a	Bovid/Cervid distal fibula	1
225		19	2b	Indeterminate vertebra	1
226		1	3	Bovid distal tibia	1
227		6		Bovid 1st phalange, complete	1
228		6		Artiodactyla pelvis (acetabulum+ ischium)	1
229		6		Indeterminate shaft	1
230		7		Equid proximal metacarpal + 190mm. of shaft	1
231			5	Indeterminate fragment	1
232		5		Artiodactyla (medium) navicular-cuboid	1
233			2a	Felid right lower incisor 3	1
234			2b	Bovid/Cervid distal fibula	1
235		6		Cervid maxilla, with M1-3	1
236				Cervid premolar	1
237		5		Indeterminate shaft	1
238				*Panthera uncia* mandibular ramus with DM3 root	1
238		6		Bovid metacarpal; part of 310	1
239	567	10	2a	*Equus* left mandible, complete with P2-4,M1,M2 errupting	1
240		14	2a	Equid left radius + ulna, complete	1
241				Indeterminate carpal/tarsal	1
242		24B	2b	Cervid left femur, complete	1
243			2b	Indeterminate shaft; bite mark	1
244		12		Indeterminate shaft + 832-843	1
245		12	2a	Artiodactyla/Cervid distal scapula	1
246		12	2b	Indeterminate shaft	1
247		12	2a	Equid astragalus	1
248		11	2a	Indeterminate shaft, squashed	1
249	400	11	3	Indeterminate shaft	1
250		7	2a	Equid 1st phalange, complete	1
251		24B	3	Indeterminate/Suid tibia shaft	1
252		18		Indeterminate condyle	1
253		18	2a	Indeterminate atlas	1
254		24B	2a	Indeterminate/Artiodactyla radius, split	1

255	405		2a	Indeterminate carpal	1
256	425	8	3	Bovid/*Equus* humerus, complete	1
257		23		Indeterminate humerus	1
258		23		Long bone fragment (fits ulna 763)	1
259		23	2a	Artiodactyla thoracic vertebra	1
260				Clay ball 70 x 58 x 25 mm.	1
261		24A		Indeterminate cranial fragment	1
262		24A		Cervid tibia shaft	1
263		24A	2a	Equid distal tibia, complete	1
264				Indeterminate fragment	1
265			2a	Indeterminate fragment	1
266			3	Indeterminate ribs	7
267		24A		Long bone fragment	1
268		24A		Bovid distal tibia	1
269			1	Equid astragalus fragment, right	1
270		5	2a	Equid scapula glenoid	1
271		5	4	Indeterminate shaft	1
272	541	5	2b	Bovid distal calcaneum	1
273		5	3	*Equus* proximal 3rd phalange	1
274			2a	Indeterminate carpal/tarsal	1
275			2a	Cervid 2nd phalange, complete	1
276			2b	Indeterminate carpal	1
277			2a	Artiodactyla vertebra centrum	1
278		4		Indeterminate (large) rib; carnivore damage	1
279		5	2a	Indeterminate patella	1
280			2a	*Equus* left maxilla	1
281		3	2b	Artiodactyla sacrum + 3 lumbar vertebrae	1
282		3	2b	Indeterminate shaft; carnivore damage	1
283	401	8		Bovid/Cervid distal femur + shaft (300mm) ; carnivore damage	1
284		8		Equid/Bovid distal phalange	1
285			2a	*Equus* 2nd phalange, complete	1
286		10	2b	Bovid (large) proximal metatarsal, complete	1
286		3	2a	Bovid distal metatarsal, complete; carnivore damage	1
287			2b	Bovid tibia shaft	1
288		10	2a	Indeterminate rib	1
289		10		Felid lower canine	1
290		10	2b	Cervid radius, complete	1
291		10		Artiodactyla/Suid distal humerus	1
292				Long bone fragment	1
293		10		Artiodactyla calcaneum	3
294			2b	Indeterminate vertebra fragment	1
295				Bovid (large) 2nd phalange, complete	1
296		7	2a	Cervid left mandible with dm1-3, M1 errupting	1
297				Indeterminate carpal	1
298		10		Indeterminate fragments	3
299		11	2a	Cervid distal radius	1
300				Felid (*Panthera*) 1st incisor	1
301				Indeterminate patella	1
302				Indeterminate incisor	1
303		3	2a	Artiodactyla metatarsal shaft	1
304				daybook says bovid metatarsal	1
305		12		Indeterminate ~~shaft~~ rib	1
306				Indeterminate femur shaft	1
307			2b	Artiodactyla proximal metatarsal	4
308			2a	Bovid/Cervid left astragalus	1
309				Indeterminate pelvis fragments	5
310		6	3	Ungulata scapula fragment	1

311			2a	Cervid 3rd upper molar, fesh	1
312				Indeterminate proximal ulna	1
313		5	2a	Cervid left mandible, complete with P2-4,M1-3	1
314		5		Suid L. and R. mandible, complete with I1-M2,M3 errupting	2
315		5		Indeterminate fragment	1
316		5		Long bone fragment	1
317		5		Long bone fragment	1
318				Long bone fragment	1
319		24A		Indeterminate cranial fragment	1
320			2a	Indeterminate fragment; carnivore damage	4
321			2a	*Equus* astragalus, complete	1
322		9		Indeterminate shaft	1
323		9		Suid distal tibia, complete + shaft	1
324		9		Metatarsal fragment	1
325			2a	Indeterminate proximal rib fragments	3
325		9		Suid lateral metapodial	1
326		9		Suid tarsal	1
327		9		Indeterminate fragment	4
328			2a	Indeterminate carpal	1
329		21	3	Indeterminate fragment	1
330		15	2b	Cervid tibia, complete + piece 811; carnivore damage	1
331		20		Equid tibia	1
332			2a	Bovid navicular-cuboid	1
333		15	2b	Cervid proximal metatarsal	1
334				Not fossil	0
335				Indeterminate rib	1
336				Indeterminate fragment	1
337			2a	Artiodactyla cuneiform	1
338				Indeterminate fragment	1
339			2b	Indeterminate fragment	1
340				Long bone fragment	1
341			2a	*Equus* astragalus, complete	1
342			2a	Indeterminate condyle	1
343				Artiodactyla distal phalange	1
344				Felid left incisor 3	1
345				Long bone fragment	1
346				Indeterminate fragment	1
347			1	Indeterminate carpal, complete	1
348, 349		10		Indeterminate radius-ulna	1
350		24A		*Equus* metacarpal	1
351			1	Bovid astragalus, complete	1
352			2a	Artiodactyla carpal/tarsal, complete	1
353, 354				Indeterminate metapodial fragments	2
355		11		Indeterminate fragment	1
356		24A	2a	Equid metacarpal, complete; carnivore damage	1
357		22	4	Equid humerus, complete + proximal radius: previously 261	1
358		10	2a	Indeterminate rib	2
359				Indeterminate fragment	1
360		20		Equid proximal m/carpal + lateral m/podials; carnivore damage	1
361		21		Indeterminate fragment	1
362		21	2a	Artiodactyla femur shaft	1
363		21	2b	Equid navicular	1
364				Indeterminate carpal	1
365		21	3	Indeterminate shaft	1
366		21		Equid lateral metapodial	1
367		21		Cervid left mandible with DM2,3	1
368		21	3	Indeterminate fragment	1

369			2a	*Equus* 2nd phalange, complete	1
370				Indeterminate fragments	3
371				Indeterminate fragments	2
372				Indeterminate fragments	6
373				Indeterminate fragments	3
375			2a	Indeterminate shaft	1
376			2a	Indeterminate shaft	1
377-378		15		Long bone fragment	1
379		20		Equid radius with 260mm. of shaft	1
380		20		Indeterminate shaft 130 mm.	1
381			1	Indeterminate rib 13x5 mm.	1
381				Artiodactyla distal fibula, complete	1
382				Felid left maxilla with I3 + I2	1
383			2a	Indeterminate shaft; carnivore damage	1
384				Indeterminate fragments	13
385			2a	Artiodactyla premolar	1
386				Indeterminate rib	1
Total					**323**
				Post-excavation numbers:	
400	249	11	3	Indeterminate shaft	1
401	Ex 283			Indeterminate vertebrae	1
403			3	*Rhinoceros* humerus	1
404			4	Indeterminate shaft	1
405	Ex 255		2a	Indeterminate carpal/tarsal	1
406			2a	Indeterminate carpal	1
407				Indeterminate fragment	1
409			3	Hyaenid cranium	1
410				Hyaenid (*Crocuta crocuta*) right maxilla with C,P2-4	1
411				Indeterminate fragments	8
419			2a	Indeterminate scapula fragment	1
420				Felid lower incisor 1/2	1
421				Indeterminate fragment	1
425	256	8	3	Bovid/*Equus* humerus, complete	1
426				Cervid distal metacarpal	1
427				*Equus* radius	1
428	Ex 110		2a	Suid humerus shaft, tooth marks	1
429				Indeterminate shaft fragments	22
430-433		'x'		Small fragments from 1986 block 'x' with 57,95,97	4
435		12		Carnivora 1st phalange, complete	1
436			2a	Artiodacyla scapula fragment	1
437-445	Ex 1986	'a,a'		Indeterminate fragments; block 'a,a' with 73C90ex	9
446-448	Ex 1986			Indeterminate fragments assoc. with 47	3
449				Suid scapula (part of 33)	1
455-460	Ex 1986		3	Indeterminate flakes (part of 127)	6
461, 462	Ex 1986	a,b		Indeterminate fragments	2
463	Ex 1986	a,b	2b	Indeterminate/carnivore calcaneum shaft with 127, 131.	1
464		a,b	2a	Indeterminate fragment with 127, 131	1
465		'a,d'		Indeterminate scapula; goes with 104 -ADD	1
466		b		Indeterminate fragment (goes with 31)	1
467-499		a,d		Indeterminate fragments (part of 104)	33
470				Indeterminate vertebra fragment	1
500		20		Indeterminate cuboid	1
501		20		Artiodactyla carpal	1
502		20		Indeterminate cuboid	1
503		20	2a	Cervid radius and proximal ulna; carnivore damage	1
504		20		*Equus* metacarpal + carpal	1
505		20		Indeterminate vertebra	1

506		20		Indeterminate vertebra	1
507		20	2a	Bovid/Cervid distal femur, carnivore damage, spiral fracture	1
508		20		Carnivora metapodial	1
509-512		20		Indeterminate shaft	4
513		20		Indeterminate fragment	1
514		20		Artiodactyla metapodial	1
515		20		Indeterminate fragment	1
531		20	2b	Indeterminate rib	1
532		20		Indeterminate rib	1
533		20		Indeterminate fragment	1
534		22		Bovid radius	1
535		22		Indeterminate fragment	1
536		5	2a	Artiodactyla navicular, complete	1
537		5	2a	Indeterminate cranial fragment	1
538		5	2a	Artiodactyla vertebra	1
539		5	2a	Equid 3rd phalange fragment	1
540		5	2a	Indeterminate shaft	1
541	272	5	2b	Bovid distal calcaneum	1
542		5	2a	Indeterminate shaft	1
543		5	2a	Suid lateral 3rd phalange	1
545		5		Cervid phalange	1
546		5		Indeterminate fragment	1
547		21		Indeterminate metapodial	1
549		21		Indeterminate fragment	5
552		21		Indeterminate metapodial	1
554		21		Artiodactyla vertebrae	1
555		21	2b	Equid 3rd phalange, split	1
556-566		21		Indeterminate fragments.	11
567	239	10	2a	*Equus* left mandible, complete with P2-4,M1,M2 errupting	1
568		10		*Panthera uncia* mandible and carnassial	1
569		10	2a	*Equus* right metatarsal, complete + tarsals etc	1
569		10		*Equus* magnum; part of metataral set	1
569		10		*Equus* external cuneiform; part of metataral set	1
569		10		*Equus* pisiform; part of metataral set	1
570		10	2a	Cervid maxilla, with M1-2, M3 errupting	1
571		10	2a	Indeterminate rib 17 x 7.5 mm.	1
572		10	2a	Indeterminate rib, split	2
573		10	2a	Indeterminate vertebra fragment	1
574		10		Indeterminate vertebral flange	1
575		10	2b	Indeterminate shaft	1
576		10		Indeterminate fragment	1
578		10	2b	Artiodactyla/Suid proximal humerus	1
579		10	2a	Bovid distal metatarsal, split; carnivore damage	1
580		10		Artiodactyla proximal ulna; carnivore damage	1
581		10	2a	*Equus* 1st phalange. complete	1
582		10	2b	Artiodactyla carpal	1
583		10	2b	Artiodactyla carpal	1
584		10		Indeterminate fragment	1
585		10	2b	Artiodactyla mandible fragment	1
586		10		Indeterminate shaft; carnivore damage	1
587		10	2a	Indeterminate shaft fragment	1
601		10		Indeterminate rib	1
602		10		Indeterminate fragments, very small	1
603		10		Artiodactyla phalange	1
604		10	2a	Bovid/Cervid distal metapodial condyle	1
605		10	2a	Equid mandible with I1-3	1
606	Ex 25	24B	2a	*Equus* 1st phalange, split	1

607		24B	2a	Cervid distal metacarpal, complete	1
608		24B	2a	Cervid proximal metacarpal, complete	1
609		24B		Suid ulna	1
610		24B	3	Equid metatarsal shaft +1 lateral metapodial	1
611		24B	3	Suid humerus shaft	1
612		24B	2b	Indeterminate shaft	1
613		24B	2a	Cervid proximal metatarsal, split	1
614		24B		Artiodactyla/Suid vertebra centrum	1
615		24B	2b	Suid proximal lateral metapodial	1
616		24B	2a	Artiodactyla/Suid calcaneum fragment	1
617		24B		Indeterminate tibia	1
618		24B	2b	*Equus* 3rd phalange, complete	1
619		24B	2a	Rodent (large) incisor	1
620		24B	2a	Indeterminate carpal, complete	1
621		24B	2b	Indeterminate carpal	1
622		24B	2b	Artiodactyla distal fibula	1
623		24B		Indeterminate carpal	1
624		24B		Indeterminate fragment	1
625-736		24B		Indeterminate scraps	111
737		24B		Suid femur fragment	1
738	Ex 22	24A	2a	Indeterminate condyle, split	1
739		24A		Indeterminate fragments	21
760		11		Indeterminate fragment	1
761		24A	2a	Artiodactyla metatarsal shaft	1
762		24A		Indeterminate condyle	1
763		23	2a	Bovid (large) proximal ulna (fits 258)	1
764		11		Indeterminate calcaneum	1
765		23		Artiodactyla cuboid	1
791		11	3	Indeterminate shaft	1
792		17		Indeterminate sesamoid	1
793, 794		17		Indeterminate fragments fitting 216	1
795-801		14		Indeterminate fragments (goes with 240)	7
802		23		Indeterminate shaft	1
803		23		Cervid upper premolar, little wear	1
804		23		Artiodactyla molar fragment	1
805		23		Indeterminate carpal	1
806		23		Indeterminate fragment	1
811		15		Fragment, fits 330.	1
812		15		Indeterminate fragment	1
813		15		Indeterminate fragment	1
814		15	3	Indeterminate rib 25x13 mm.	1
817-831		15		Indeterminate fragments	15
832-843		12		Fragments fitting 244	1
845		19	2a	Equid astraglus, complete	1
846		19	2a	Equid calcaneum, complete	1
847		19	2a	Equid cuboid, complete	1
848		19	2a	Bovid proximal 1st phalange	1
849		19	2a	Equid 2nd Phalange, complete	1
850		19	2a	Equid distal tibia, complete	1
851		3		Indeterminate shaft	1
852		9		Indeterminate metatarsal fragment	1
853		9		Suid astragalus, complete	1
854		9		Suid lateral metapodial	1
855		9		Suid tarsal	1
856		9		Suid tarsal	1
857		7		Equid carpal	1
858		7		Indeterminate/Equid carpal	1

859		26	2a	Indeterminate vertebra centrum epiphysis	1
860		26	2a	Indeterminate shaft (?bird bone)	1
862		25		Indeterminate (Large) Shaft	1
863		25	2a	Aves? long bone shaft	1
864		2		Indeterminate cuboid	1
865, 866		23		Indeterminate fragments (v. small)	2
867		7		Cervid mandible fragment	1
868				Indeterminate fragments	3
872				Indeterminate fragments beneath scapula 5, 1986	13
885				Indeterminate fragment	1
886				Indeterminate fragment below 148	1
887				Indeterminate fragments around 157	12
899		11	2a	Artiodactyla distal humerus	1
Total					**756**

Notes:
1986: In order to avoid double-counting, bones recorded in 1986 but renumbered when lifted in 1987 are returned as zero in Table 9.4 unde their original number, and counted against their new, 1987 or post-excavation number in Table 9.5. An example is specimen 3, which became no. 503 in 1987. Some specimens have the same number (e.g. 118), usually because they were removed as one item and then found to be composite.

3 and 4 were partly exposed in 1986, but lifted in 1987 as part of block 20, and renumbered 503 radius/ulna and 507 femur.
27 and 29 were lifted in 1987 as part of block 13; 27 and 29 were lifted in 1987 as part of block 13, with bone 198.
Bone number 30 was deleted when it turned out to be part of no. 31. Cervid specimen no. 43 is gnawed.
1986 blocks:
Block 'a': 43 cervid 1st phalange, 44 equid 1st phalange, 45, 53 equid proximal metacarpal, 59,60 (part), 67, 121
Block 'a,a'', 1986: 70, 90 artiodactyl radius-ulna, 437-445: 70 and 90 were initially numbered as 2 bones, then only 90 was used when it wa evident that they were both bits of the same bone. Scraps 437-445 were added when the block was opened in 1987
Block 'a,b': 127,131, equid metapodial, 461,462,463,464
Block 'a,c': 8 cervid ulna, 9, 10
Block 'a,d': 104, 465, 467-499.
Block 'b': 30 bovid radius-ulna, 31; 31 is now replaced by 30, as they were both from the same bone. 466 was added in 1988.
Block 'x': 57 bovid distal humerus, 95 and 97 were lifted as a plastered block. Nos. 430-433 were added in 1987
Block 'z': 14, 34 cervid proximal tibia

1987:
Carnivore damaged: 204, 206, 243, 244, 250, 251, 278, 282, 283, 291, 327, 330, 376,377, 380, 381, 383, 428, 503, 507, 509, 510, 511, 512 540, 569, 579, 580, 586, 817
290, 333, squashed; 306: small tooth marks. No. 130 Equid metapodial - gnawed, with opposing puncture marks.

286 was lifted as part of blocks 3 and 10

Principal contents of 1987 blocks

Block 1 comprises 226 bovid distal tibia, 369 and scraps, 370-390.
Block 2 consists of cervid mandible 223 plus cuboid 864.
Block 3 comprises sacrum 281, long bone fragment 282, half of bovid metatarsal 286 (the other half is in block 10), 303, bovid metatarsal and 851.
Block 4 contains a large indeterminate rib 278
Block 5 is very complex: 232, 270 equid scapula, 271, 272 bovid calcaneum, 273 and 539 equid 3rd phalanges, 279, 313 cervid mandible, 314 suid mandible, 315-317, 536-546. Bovid skull fragment 537 was partly destroyed to salvage pig mandible and scapula.
Block 6 comprises: 167 artiodactyl metapodial, 227 bovid 1st phalange, 228 artiodactyl pelvis, 229, 235 cervid maxilla, 238 felid ramus, an 310 ungulate scapula.
Block 7 comprises: 230 equid metacarpal, 250 equid 1st phalange and 296 cervid mandible, plus carpals 857 and 898.
Block 8 comprises: 256 bovid/equid femur, 283 bovid/cervid femur, 284 bovid/equid distal phalange and 336.
Block 9 comprises a suid lower hind-limb: 196 suid calcaneum, 322, 323 suid distal tibia, 324 metatarsal, 325 and 854 suid lateral metapodials, 326 suid tarsal, 327 fragment plus 852 metatarsal, 853 suid astragalus, 855, 856 suid tarsals.
Block 10: 239/567 equid mandible, 286 bovid metatarsal, 288, 289 felid canine, 290 cervid radius, 291 artiodactyl/suid humerus, 293 artiodactyl calcaneum, 298, 312, 348-349 radius-ulna, 358, 568A felid mandible, 569 equid metatarsal plus tarsals, 570 cervid maxilla, 579

bovid metatarsal, 581 equid 1st phalange, and 605 equid mandible. There were also numerous minor elements. Note also that half of 286 is in block 3, and the equid mandible 239/567 were in separate pieces.

Block 11 comprises: 190 bovid atlas, 248, 249, 299 cervid distal radius, 355, and new numbers 760, 764, 791 and 899.

Block 12 comprises: 244, 245 artiodactyl scapula, 246, 247, 305 rib plus fitting fragments 832-843, 435 carnivore phalange, and 844.

Block 13 comprises: 27 cervid tibia, 29 pelvis, and 198 equid lateral metapodial. Nos. 27 and 29 are from 1986.

Block 14 comprises 240 radius-ulna and fragments 795-801.

Block 15: 330 cervid tibia, 333 cervid proximal metatarsal, 377, 378 and fragments 811-814, 817-831

Block 16 comprises only 210, a cervid/bovid proximal radius.

Block 17 comprises 216 plus fitting fragments 793, 794; plus sesamoid 792.

Block 18: 252 and 253 atlas.

Block 19: 193 scapula fragments, 206 cervid scapula, 207 rib, 225 vertebra; plus 845 equid astragalus, 846 equid calcaneum, 847 equid cuboid, 848 bovid proximal phalange, 849 equid 2nd phalange, and 850 equid distal tibia. Equid specimens 845,846, 847 and 850 are articulated.

Block 20: This comprises two groups: 3,4 (radius-ulna from 1986 and renumbered 503 and 507), 201, 202, 203 felid tooth, 204 equid 1st phalange; and also 331 equid tibia, 360 equid proximal metacarpal, 379 equid radius, 380, 503 cervid proximal radius, 504 equid metacarpal, 508 carnivore metapodial, 531-535.

Block 21 comprises: 329, 361, 362 artiodactyl femur, 363 equid navicular, 365, 366 equid lateral metapodial, 367 cervid mandible, 368, 371 (renumbered as original number had rubbed off), 555 equid 3rd phalange plus numerous lesser pieces.

Block 22 comprises 357 equid humerus plus proximal radius (previously 261), 534 bovid radius and 535.

Block 23 comprises 209, 221, 222 bovid proximal radius, 257, 258, 259 artiodactyl vertebra, 763 (fits 258 radius/ulna), 802-807 (left in block); 763, 765; 865,866 (small indeterminate fragments).

Block 24A comprises 22 (renumbered 738), 260, 262, 264, 267, 268 bovid distal tibia, 319, 350 equid metacarpal, 353, 356 equid metacarpal, 761 (originally 313) plus scraps 739-760 and 762.

Block 24B comprises 181 artiodactyl scapula, 242 cervid femur, 251 artiodactyl tibia, 254 artiodactyl radius, 25 (from 1986; now 606) equid 1st phalange, 607-8 cervid metacarpal, 609 suid ulna, 610 equid metatarsal, 611 suid humerus, 613 cervid proximal metatarsal, 615 suid lateral metapodial, 616 artiodactyl/suid calcaneum, 618 equid 3rd phalange and various lesser elements. Numbers 606-624 were for individual bones; 625-737 were just scraps, and so not used on individual fragments

Block 25: 156 (from 1986, equid axis), articulates with 186 atlas; there is also 191 bovid 1st phalange, 862 and ?bird shaft 863.

Block 26: 148, 154, 155, 176 bovid/equid distal humerus, 177 *Rhinoceros* metapodial; 859, and 860, a possible bird bone.

The excavation of site 362

Mohammad Anwar (Assistant Director, Geological Survey of Pakistan) and Mark Beech (Senior Resident Archaeologist, Abu Dhabi Islands Archaeological Survey, PO Box 45553, Abu Dhabi, United Arab Emirates; formerly Department of Archaeology and Prehistory, University of Sheffield, U.K.)

Introduction

Locality 362 was found during the 1987 field season by Helen Wilson, a member of the British Archaeological Mission team (see chapter 6, Figure 6. 7, this volume). The fauna collected from the surface of the locality in 1987 included Bovid, Tragulid, Cervid, Rhinoceratid, Giraffid and a Proboscidean (see Table 6.6). The bones appeared to come from an area of about one square metre lying within orange-brown sandy silt with calcium carbonate concretions. This sandy silt graded downward into medium grained sandstone and had a sharp upper contact with a purple-pink mudstone. The mudstone had sharp contact with the overlying medium grained sandstone. According to palaeomagnetic studies by Dr. Helen Rendell, the orange-brown sandy silts from this locality date fall within the Olduvai event (ca. 1.67 - 1.87 Ma) (see this volume, Figure 6.3).

In 1987, all bone from this locality and immediately down slope was collected, and the site was protected with cloth and a layer of earth to retard further erosion. In 1989, it was apparent from the amount of material that had eroded in only two years that the site would not survive much longer. It was thus decided to excavate it on account of its diverse fauna, good fossil preservation and dense concentration of material within a defined area. The main objectives of the excavation were to recover *in situ* fossil material, and to record it systematically in order to aid understanding of how it accumulated, and to obtain good fossil specimens for the Geological Survey of Pakistan that could serve as research and reference material for future workers.

Methodology

Before beginning the excavation, the fossil material that had eroded from the surface since the site was covered in 1987 was collected. This material was confined to the gully running northwards, down slope from the site, and the distance of the main pieces from the site was recorded to provide an indication of the rate at which the site was being eroded. All the specimens collected were cleaned and marked as described below, and are catalogued in Table 9. 6 (end of report). This material complements that found in the 1987 survey, and summarised in Table 6. 6.

The excavation of the site by the two authors began on 11th March, 1989 and lasted three weeks. A trench 2.6 m. long and 1.46 m. wide was cut into the slope below the sandstone ridge to provide a working platform either side of the bone distribution visible in the slope surface. A permanent site datum was established by fixing a red-painted 6" nail into the surface of the sandstone ridge, c. 1 metre west of the trench, and a datum level below it was established with the aid of a theodolite. This datum level was given an arbitrary height of 100 metres, and the permanent site datum is therefore located at 102.14 metres.

Deposits were removed in horizontal spits down to the level of the bone concentration. The western side of the trench did not contain any bones, and was excavated to c. 40 cm below the surface of the bone distribution to provide greater accessibility and a work space for the excavation of the bones. Each sedimentary unit was recorded, and the north, east and west facing sections (see Figure 9. 6) were drawn at a scale of 1:20.

As most bones were concentrated within a small area, their position was recorded by establishing a one metre square grid. All co-ordinates taken on the bones, i.e. northings and eastings, were measured in relation to this grid square, the depth being measured down from the 100 metre temporary site datum. All co-ordinates were measured to the central axis point of each particular bone. For larger bone fragments, a note was made of the orientation and the angle of dip of the bones. A series of scale 1:10 plans were made of the bone distribution at each appropriate level (see Figures 9. 7 - 10). Bones were carefully excavated with dental picks and brushes. As the sediments were very hard, it was sometimes necessary to soften them with a small amount of water. Many of the fragile bones were heavily cracked due to bioturbation, and fragile ones were given a coating of 70% PVA (Polyvinyl acetate adhesive) solution in the field. Some specimens required further treatment with PVA once back in the field base.

Each specimen was cleaned manually in the field and at the field base with dental picks and toothbrushes, and marked with the site number, followed by GB (indicating G for Geological Survey of Pakistan (GSP) and B for British Archaeological Mission to Pakistan), and the specimen catalogue number. Table 9. 7 (end of report) lists each specimen by number, provisional taxon, anatomical element, co-ordinates, orientation, angle of dip, along with other relevant information. Once cleaned, marked, and recorded, all bone material was stored at GSP, Islamabad.

Results: Surface Collection of the Locality and Gully

Table 9. 6 (end of report) lists the fossil collected in 1989 from the surface of the locality and the gully running downslope from the site. Over 75 specimens were collected, of which more than two thirds were diagnostic; i.e. identifiable to at least anatomical element and/or genus. Provisional identification of this material indicates the presence of Bovid, Proboscidean, *Sivatherium* and *Rhinoceros*.

Table 9. 8: Comparison of the surface and excavated fossil assemblages from site 362.

Taxon	% of species in surface collections	% of species in excavation
Bovid	25	15
Carnivore	--	11
Sivatherium	1	2
Rhinoceratid	5	1
Proboscidean	1	<1
Suid	--	<1
Cervid	1	--
Tragulid	<1	--
Bird	--	1

It is interesting to compare the overall composition of the species represented at site 362 from its surface collection, and from the excavation. The general features are shown in Table 9. 8. As indicated, the surface collection provided a reasonably accurate indication of the importance of bovids, giraffids, Proboscidean and rhino. However, it failed to indicate the presence of carnivore (11% of the excavated total), pig and bird. Conversely, there was no trace of cervids in the excavated material, although these were present in small amounts in the surface collection. However, it should be noted that the identifications of the Bovidae in the excavated assemblage are very provisional, and more detailed research may show that cervids are present in small amounts. In conclusion, it is suggested that surface assemblages give a reasonable indication of the taxa in a fossil accumulation, but are unlikely to show the full range of types, or the rarer ones.

Erosion rates

It was obvious that locality 362 was eroding very quickly. Seventy five specimens had been eroded from their original context and washed down slope in just two years between 1987 and 1989. This suggests that the 217 specimens found in 1987 might have resulted from ca. 6 years' erosion. At this rate of erosion, it is unlikely that the 175 bones remaining in the deposit would have remained in situ for much more than 10 years.

The Excavation: Stratigraphic sequence

The sedimentary sequence of the site is shown in Figure 9. 6, and consisted of (from top to bottom): loose sandstone (context 1), mudstone (context 2), sandy silt (context 3), and sandy silt with abundant fossils and increased mineralised concretions (context 4). Underlying deposits were not investigated.

Context one is a grey coloured, medium grained, c. 10 cm thick, bioturbated loose sandstone that has a well defined contact with the underlying mudstone. Context two is purple-pink, c. 40 cm thick, bioturbated, moderately compacted mudstone that has a sharp, uneven contact with the lower sandy silt unit. Context three is an orange-brown, c. 50 cm thick, bioturbated, sandy silt with occasional concretions. The unit is well compacted and has very occasional round-oval mudstone inclusions (c. 5 cm in diameter).

Context four contains the fossil accumulation, and is an orange-brown sandy silt which grades downwards into a medium-grained sandstone. It has abundant concretions which increase in size with depth from c. 5 mm - 2 cm in diameter. Mudstone inclusions are also present. Roots in this unit are marked by mudstone and sandstone infilling. Fresh roots could be seen protruding out of the surface fractures of many of the bones. The bone bed extended in a roughly rectangular strip from the centre of the northern part of the trench into its south-east corner. It was approximately 1.20 m long, c. 75 cm wide and had a uniform thickness of c. 30 cm.

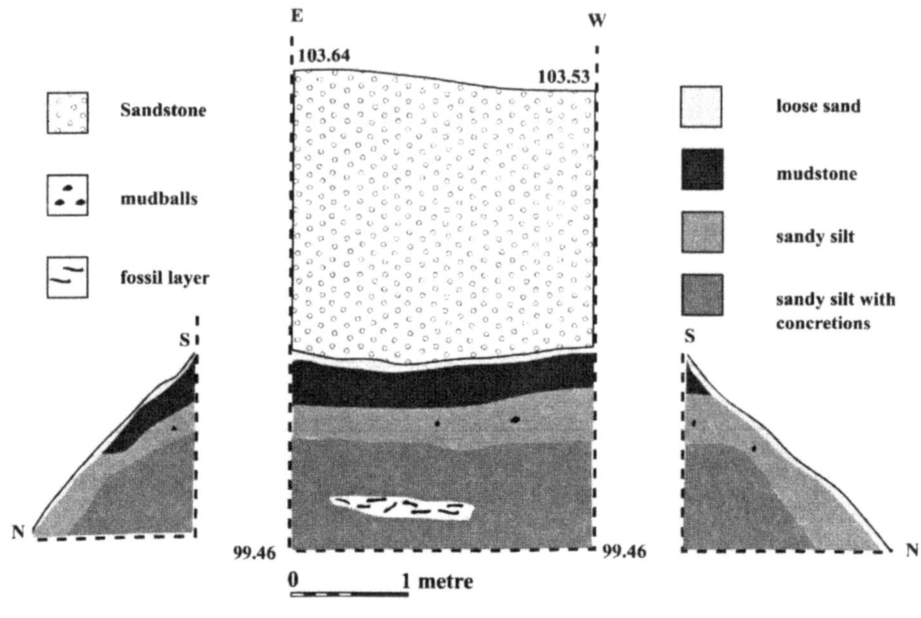

Figure 9. 6: Section of locality 362. (Based on field records by Mark Beech and Eddie Moth).

The Fauna

A total of 179 specimens were excavated, of which almost half were diagnostic i.e. identifiable to genus or family. As Table 1 shows, the provisionally identified fauna consists (in decreasing numerical importance) primarily of Bovid, followed by carnivore (*Canis*), *Sivatherium*, Gazelle, *Rhino*, Proboscidean, a large bird (?eagle size/type) and ?Suid.

Nature of the Fossil Accumulation

Three particular groups of bones were articulated: a Bovid humerus-radius-ulna (specimen numbers X21, X28 and X29), a *Sivatherium* humerus-radius-ulna (specimen numbers X19, X25, X27 and X136) (see Figure 9. 7), and a canid partial skeleton (see Figure 9. 10). Many of the other bones appeared to come from only a few individuals.

The major part of the skeleton of an adult male carnivore (*Canis cautleyi*; see Turner, this volume) was represented within the deposits (jaw (X119/X124), canines (X88/X104/X126), atlas (X92), humerus (X38), radius (X55), ulna (X41), os penis (X75), a pair of tibias (X125 (left) and X56 (right)), calcaneum (X105), astragalus (X100), metapodials (X84 / X93 / X154/ X158), carpals - tarsals (X103 / X105 / X112 / X160) and phalanges (X83/X175/X178). In addition many of the smaller ribs may have belonged to this particular animal, as no other species of small body size have yet been identified within the assemblage.

A pair of horns attached to a fragment of the crania (X12) and two maxillaries (X46-left maxilla, X51-right maxilla) all appear to belong to the same individual. The slender horns are set quite close together and project backwards from the skull at about 45 degrees, the tips of the horns curling upwards. The size and form of the horns suggest that the specimen is some sort of gazelle-type creature within the family of Bovidae.

A complete bird ulna (X15/X16) and almost complete carpometacarpus (X66) appear to belong to the same bird (? of eagle size/type).

Taphonomic factors

Most of the long bones were transversely fractured. Longitudinal and diagonal fractures are also present. Features such as step-fracturing, polishing, and sun-cracking were not seen on bone surfaces. A few bones exhibit traces of carnivore activity. Two specimens came from the gully collect: a pelvis shaft fragment (specimen G11/1) had several puncture marks and striations, and a Bovid distal humerus (specimen G16) had gnawing marks on its surface.

The best evidence of carnivore activity came from one of the excavated specimens: a Proboscidean calcaneum (specimen number X62), which had one very deep puncture mark and at least three other shallower marks. It is worth noting that the deep puncture mark is of a size comparable with the carnivore canines recovered from the site (specimens X88 and X104).

Discussion: 1) Sediments

The sediments at site 362 are fine-grained, poorly sorted, and do not contain silt layers. Sedimentary structures are not preserved, perhaps due to the effects of bioturbation. The sandy silt grades downward into medium grained sandstone. In fluvial rocks, fine sediments generally represent lake, overbank or abandoned channel deposits. There are various ways of explaining the accumulation of the sediments at site 362:

1. Lake or depression: Deposition in lakes/depressions is usually marked by oscillatory ripple marks, and the presence of freshwater invertebrate fossils (bivalves and gastropods). In the case of site 362, no such ripple marks were present, and no freshwater invertebrate fossils were recovered during the excavation. It is worth also noting that no crocodile or turtle fossils are present in the recorded fauna.

2. Overbank deposit: Overbank deposits are generally recognised by the presence of silt layers, ripple marks and well sorted sediments. As mentioned above, the sediments at site 362 are poorly sorted and do not have any ripple marks or contain any silt layers.

3. Abandoned channel: Abandoned channel deposits are generally characterized by an upward decrease in particle size. This feature is present in the sediments at site 362, the sediments becoming increasingly finer upwards through the profile.

In the light of the above observations with regard to the sedimentary evidence, it would seem that the site may have existed within an abandoned channel system.

2) Bones

The absence of sun-cracking on bone surfaces would appear to indicate that the bones were not exposed to the sun for any great length of time. The material would therefore appear to have been either rapidly buried, or transported to where the sun could not affect them.

The excavation of articulated and semi-articulated limbs, as well as several complete fragile specimens e.g. the bird bones, the carnivore penis bone, and many ribs, suggests a negligible amount of transportation. The lack of roundness or step-fracturing to any of the specimens would also appear to confirm this hypothesis.

The presence of a large, adult, male carnivore (*Canis cautleyi*) within the assemblage may explain the accumulation of the bone material. Its skeleton was, however, found at the base of the fossil deposits, so it clearly cannot have been the primary collector. As noted already, several specimens exhibit traces of carnivore puncture marks and striations. The distribution of the bones in an elongated spread, with a north-west/south-east orientation may possibly represent the remnants of bones deposited in a carnivore passage-way or burrow.

In summary, the sedimentary evidence suggests that sediments accumulated within an abandoned channel system. The fossil evidence points to some sort of carnivore accumulation. A model may therefore be proposed as follows:

1. An abandoned channel slowly accumulates increasingly finer sediments, and eventually dries out.

2. A carnivore makes its den/cache of bone within this abandoned channel, either by burrowing, or by using a narrow gully near its bank.

3. The den/cache subsequently goes out of use, perhaps with the roof/passage-way collapsing, leading to the rapid burial of the bones.

Conclusions

With reference to our initial objectives, set out in the introduction to this report:

1. *In-situ* fossil bone material was recovered from site 362 and recorded in detail. The bones appear to represent a carnivore accumulation, which resulted in the preservation of some excellent specimens of Bovid, a carnivore (*Canis cautleyi*), *Sivatherium*, Gazelle, Rhino, some sort of bird of ?eagle size/type and a Suid.

2. Sedimentological evidence suggests that this site was formed on the edge of an abandoned river channel. Further study of the sedimentary history of the deposits associated with this site is recommended to gain a fuller understanding of the environmental setting of this site.

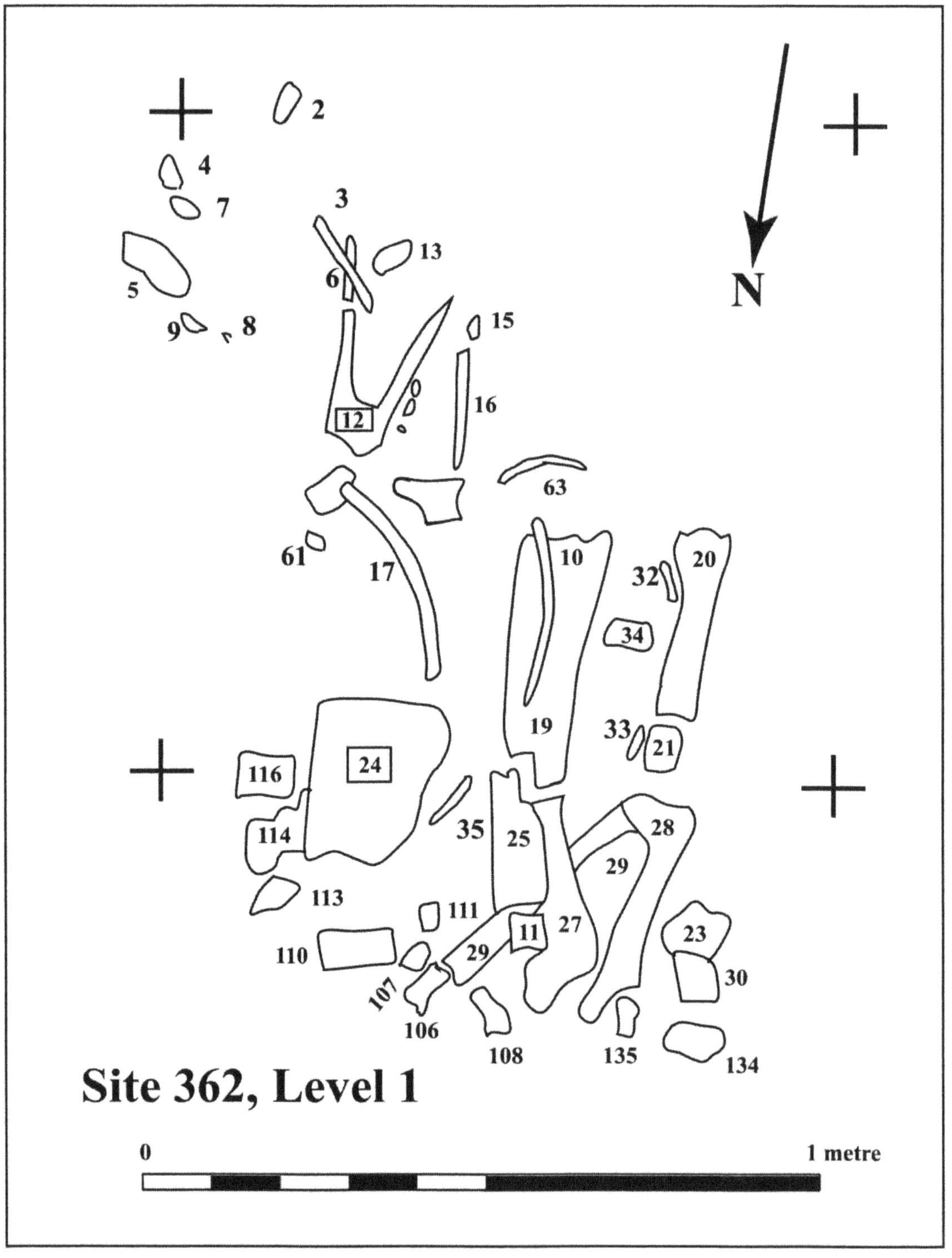

Figure 9. 7: Plan of level 1, locality 362 (based on field records by the authors and Eddie Moth).

The most important finds shown here are the bovid skull/horns (12), a radius-ulna of a bird (15,16), an articulated radius-ulna of *Sivatherium* (19, 25, 27), and a bovid humerus and radius-ulna (28, 29).

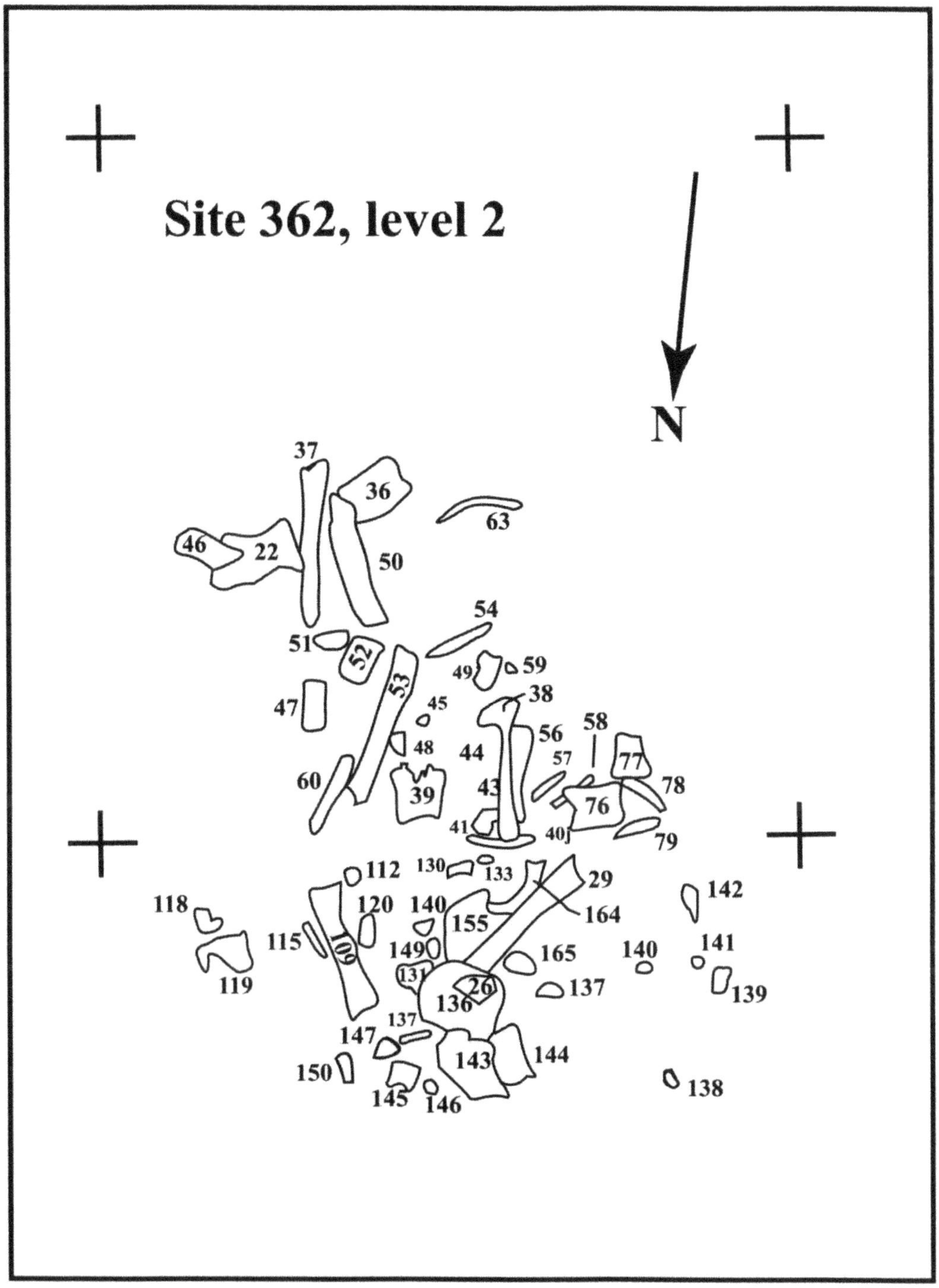

Figure 9. 8: Plan of level 2, locality 362 (based on field records by the authors and Eddie Moth).

The *Sivatherium* proximal humerus (136) in this level articulates with the radius-ulna in level 1 (19, 25 and 27).

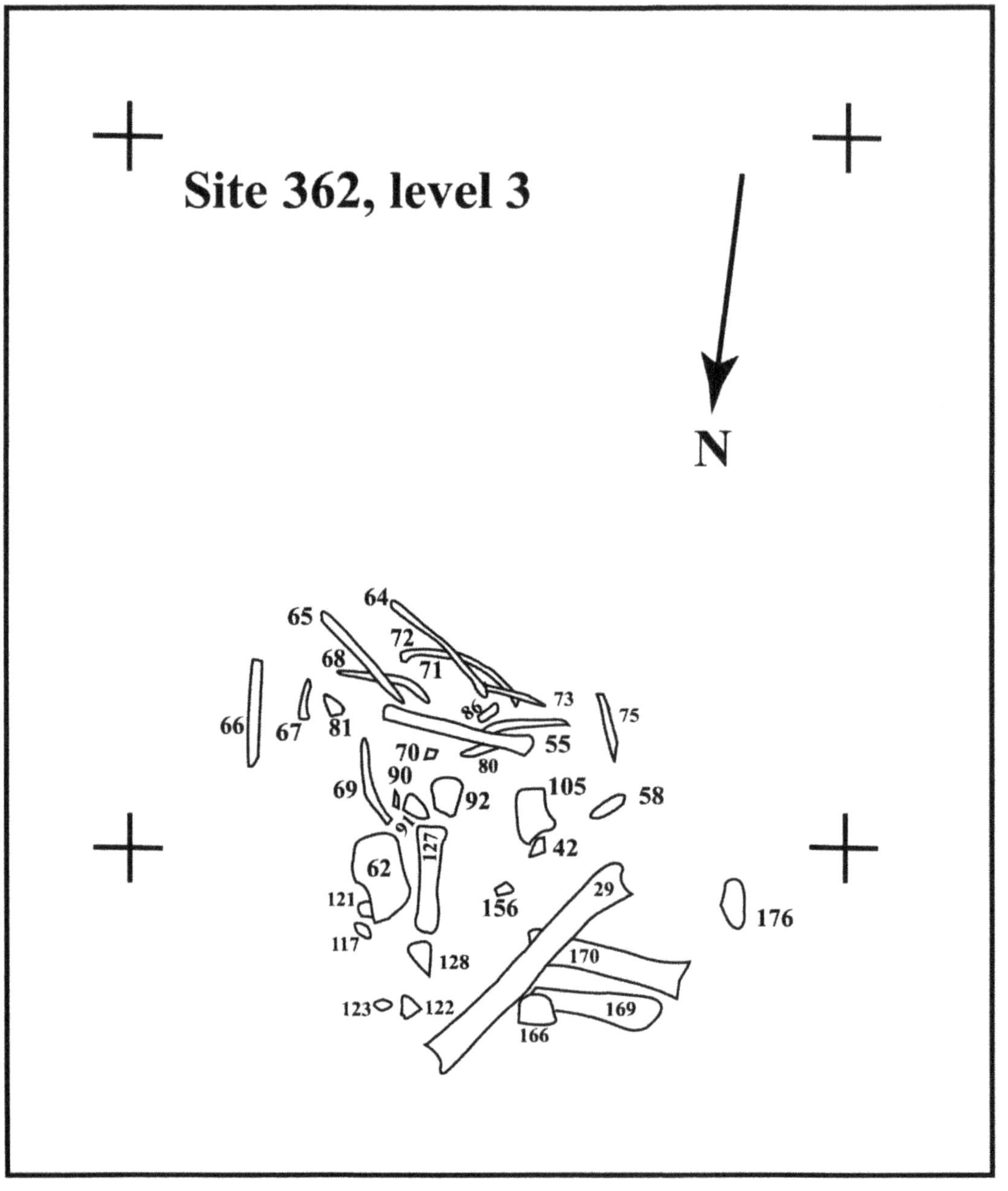

Figure 9. 9: Plan of level 3, locality 362 (based on field records by the authors and Eddie Moth).

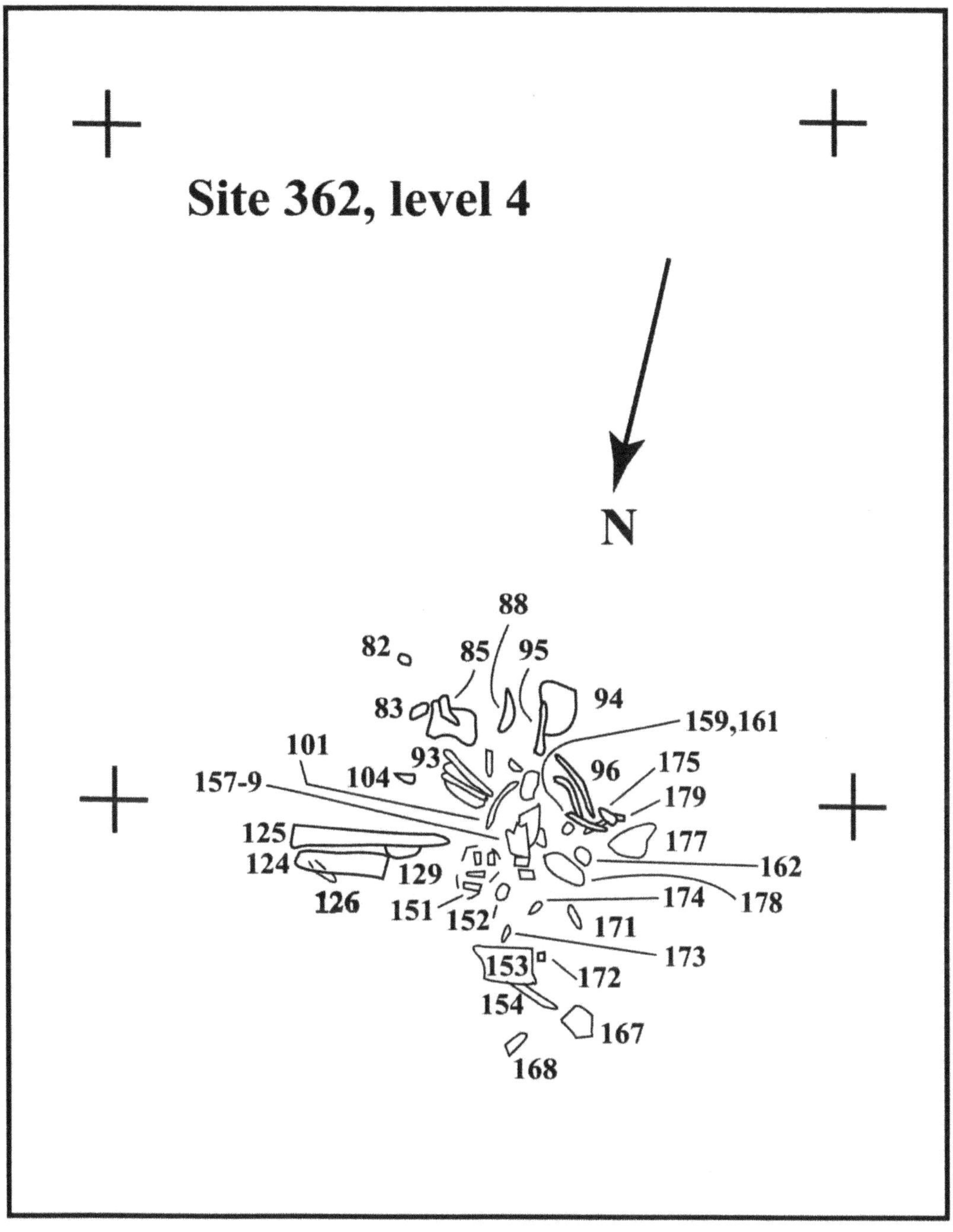

Figure 9. 10: Plan of level 4, locality 362 (based on field records by the authors and Eddie Moth). The canid partial skeleton was the main find in this part of the excavation.

Table 9. 6: Fossil material collected in 1989 from the gully down slope of locality 362

Key: lbf = long-bone fragment

Spec. No.	Distance down gully	Taxon	Element	Notes
GB 1	49 m	Large mammal	Indeterminate	
GB 2/1	148 m	Medium mammal	Indeterminate	
GB 2/2	248 m	Medium mammal	Indeterminate	
GB 3	47.5 m	Medium mammal	Indeterminate	
GB 4	46.6 m	Medium mammal	skull fragment	
GB 5	44.4 m	Bovid	Atlas	
GB 6	44.3 m	Small mammal	Indeterminate	
GB 7	43.3 m	Bovid	Jaw fragment with broken teeth	
GB 8	41.7 m	Bovid	Right distal metacarpal fragment	
GB 9	41.1m	Large mammal	Indeterminate	
GB 10	40.0 m	Medium mammal	Vertebra fragment	
GB 11/1	38.4 m	Medium mammal	Pelvis shaft	carnivore puncture marks and striations
GB 11/2	38.4 m	Medium mammal	Indeterminate	
GB 12	36.3 m	Bovid	Right distal tibia	
GB 13	36.0 m	Bovid	Right distal humerus	
GB 14	35.9 m	Bovid	Right distal tibia	
GB 15	33.4 m	Medium mammal	Indeterminate	
GB 16	33.5 m	Bovid	Left distal humerus	? chewing marks
GB 17	32.8 m	Medium mammal	Indeterminate	
GB 18	31.3 m	Medium mammal	Skull fragment	
GB 19	31.0 m	Bovid	Left proximal radius	
GB 20	30.5	Medium mammal	Indeterminate	
GB 21	29.7 m	Large mammal	Indeterminate	
GB 22	28.2 m	Medium mammal	Indeterminate	
GB 23	27.9 m	Medium mammal	Indeterminate (+ 4 other small ones)	
GB 24	25.3 m	Medium mammal	Indeterminate	
GB 25/1	23.3-23.6m	?Bovid	Pelvis (ilium)	
GB 25/2	23.3-23.6m	Medium mammal	Pelvis fragment	
GB 25/3	23.3-23.6m	Medium mammal	Indeterminate lbf	
GB 25/4	23.3-23.6m	Medium mammal	Vertebra fragment	
GB 25/5	23.3-23.6m	Large mammal	Proximal tibia + 1 indeterminate	
GB 26	23.1 m	Bovid	Left proximal radius	
GB 27	22.1 m	Bovid	Right distal humerus	
GB 28	20.7 m	Large mammal	Indeterminate	
GB 29/1	20.5 m	?Bovid	Distal radius	
GB 29/2	20.5 m	Large mammal	?vertebra	
GB 30/1	20.0 m	?*Sivatherium*	Left metatarsal shaft	
GB 30/2	20.0 m	Medium mammal	Pelvis fragment	
GB 31/1	19.6 m	?Bovid	Right radius-ulna fragment	
GB 31/2	19.6 m	Large mammal	Indeterminate lbf	
GB 32	18.95 m	Bovid	Right maxilla with tooth fragment	
GB 33	18.6 m	Bovid	Right proximal radius-ulna	
GB 34/1	14.6 m	?Proboscidean	astragalus	
GB 34/2	14.6 m	Medium mammal	Indeterminate	
GB 34/3	14.6 m	*Rhinoceros*	Right distal tibia	
GB 35/1	14.2 m	Medium mammal	Indeterminate	
GB 35/2	14.2 m	Medium mammal	Indeterminate	
GB 36/1	14.1 m	Bovid	Astragalus	
GB 36/2	14.1 m	Medium mammal	Indeterminate	
GB 37/1	14.2 m	Medium mammal	Indeterminate	
GB 37/2	14.2 m	?Bovid	Radius midshaft	
GB 37/3	14.2 m	Medium mammal	Indeterminate	

GB 38	13.2 m	Medium mammal	Indeterminate	
GB 39/1	12.1 m	Large mammal	Vertebra fragment	
GB 39/2	12.1 m	Medium mammal	Indeterminate	
GB 39/3	12.1 m	Medium mammal	Indeterminate	
GB 39/4	12.1 m	Medium mammal	Indeterminate	
GB 40	12.8 m	Large mammal	Scaphoid	
GB 41	13.5 m	Medium mammal	Indeterminate	
GB 42	14.0 m	Bovid	Humerus midshaft	
GB 43/1	12.9 m	Medium mammal	Indeterminate	
GB 43/2	12.9 m	Medium mammal	Indeterminate	
GB 44	12.2 m	Bovid	Right distal tibia	
GB 45	12.0 m	Bovid	Right distal scapula	
GB 46	10.8 m	Bovid	Right distal tibia	
GB 47/1	11.2 m	?Bovid	Carpal	
GB 47/2	11.2 m	?Bovid	Radius midshaft	
GB 48	10.9 m	Medium mammal	?vertebra	
GB 49	11.6 m	?Bovid	Proximal metatarsal	
GB 50/1	11.3 m	Medium mammal	Indeterminate	
GB 50/2	11.3 m	Medium mammal	Indeterminate	
GB 50/3	11.3 m	Medium mammal	Indeterminate	
GB 51	9.6 m	Small mammal	Indeterminate	
GB 52	9.6 m	?Rhino	Carpal	
GB 53/1	8.5 m	?Bovid	Metatarsal	
GB 53/2	8.5 m	Medium mammal	Indeterminate	
GB 54	8.6 m	?Bovid	? phalange+ 2 Indeterminates	
GB 55	7.5 m	Large mammal	Indeterminate	
GB 56/1	7.0 m	Medium mammal	Rib fragment	
GB 56/2	7.0 m	Medium mammal	Indeterminate	
GB 56/3	7.0 m	Medium mammal	Indeterminate	& 7 other indeterminates
GB 57	6.1 m	Bovid	1st phalange, left	
GB 58/1	6.0 m	Large mammal	Indeterminate	
GB 58/2	6.0 m	Medium mammal	Indeterminate	& 1 other indeterminate
GB 59	6.2 m	Bovid (Medium mammal)	Jaw fragment	& 1 other indeterminate
GB 60/1	7.2 m	?Proboscidean	?metapodial	
GB 60/2	7.2 m	Large mammal	Indeterminate lbf	
GB 60/3	7.2 m	Medium mammal	Indeterminate lbf	
GB 60/4	7.2 m	Bovid	left distal unfused metatarsal	
GB 60/5	7.2 m	Large mammal	Indeterminate lbf	
GB 61	2.5 m	Bovid (Medium mammal)	Right jaw fragment + PM fragment	& 1 other indeterminate
GB 62/1	1.0 m	Bovid	Right proximal metatarsal	
GB 62/2	1.0 m	?Bovid &	Metatarsal	probably part of 62/1
		(Medium mammal)	3 Indeterminate	
GB 63	0.3 m	?	Indeterminate	
GB 64/1	0 m	Medium mammal	?vertebra fragment	
GB 64/2	0 m	Medium mammal	Indeterminate	& 3 other indeterminates
GB 65	50 m	Bovid	Right jaw fragment with M3	
GB 66	0.25 m	Medium mammal	Indeterminate lbf	
GB 67	0.58 m	Medium mammal	Indeterminate lbf	
GB 68	0.10 m	?Bovid	Distal metacarpal	
GB 69	0.09 m	?	Indeterminate	
GB 70	0.30 m	Medium mammal	Indeterminate lbf	
GB 71	0.28 m	Medium mammal	Indeterminate lbf	
GB 72	0.15 m	?	Indeterminate	
GB 73	0.20 m	Large mammal	2 Indeterminate lbf	
GB 74	0.25 m	Small mammal	1 rib	
			1 Indeterminate	
GB 75	0.30 m	Medium mammal	2 ribs	
		?	1 Indeterminate	

Table 9. 7: Catalogue of site 362 fossil specimens, excavated 1989

No.	Taxon	Element	East	North	Depth	Orientation/Notes
X1	Indeterminate	Fragment	0.85	0.05	100.42	
X2	Indeterminate	Fragment	0.82	0.02	99.9	N/S
X3	Indeterminate	Rib	0.75	0.24	99.85	N 60° W; horizontal
X4	Bovid	2nd phalange	1.0	0.1	99.83	NW/SE, horizontal
X5	Indeterminate/Suid	Pelvis	0.97	0.23	99.91(t)	Acetabulum W. = 75mm
X6	Bovid	Bovid horn; joins 12	0.75	0.18	99.87(t)	
					99.84(m)	
X7	Indeterminate	Vertebra	0.98	0.15	99.86	NW/SE; horizontal
X8	Indeterminate	Fragment	0.92	0.34	99.85	
X9	Indeterminate	Fragment	0.99	0.31	99.86	
X10	Indeterminate	Rib	0.44	0.6	99.88	N/S; horizontal
X11	*Sivatherium* (part of 19)	Indeterminate	0.44	1.1	99.92	NW/SE; horizontal
X12	Bovid/Tragulid	Skull/horns; joins 6	0.60(t)	0.27(t)	99.84	NE/SW, horns horizontal
			0.77(m)	0.46(m)		
X13	Indeterminate	Carpal	0.70	0.28	99.86	NE/SW; horizontal.
X14	Indeterminate	Vertebra	0.65	0.63	99.82	N/S, dipping north.
X15	Bird (large)	Distal ulna; part of 16	0.54	0.39	99.85	N/S; almost horizontal
X16	Bird (large)	Ulna; part of 15	0.58	0.52	99.82	N/S, dipping slightly north
X17	Indeterminate	Rib	0.69	0.64	99.86	NW/SE; dipping west
X18	Indeterminate	Carpal?	0.83	0.55	99.81	NE/SW; horizontal
X19	*Sivatherium*	Radius; joins 25 and 27	0.45(N)	1.01(N)	99.86	N/S; horizontal
			0.40(S)	0.62(S)		
X20	Bovid	Right tibia	0.26(N)	0.89(N)	99.84	N/S; horizontal
			0.23(S) 0.61(S)			
X21	Bovid	Ulna; goes with 28, 29	0.27	0.96	99.81	N/S; horizontal
X22	Indeterminate	Vertebra	0.73	0.57	99.83	NE/SW; horizontal
X23	Indeterminate	Carpal? Tarsal?	0.20	1.22	99.85	
X24	Indeterminate	Scapula?	0.75	1.0	99.85	NE/SW; almost horizontal
X25	*Sivatherium*	Proximal radius	0.50	1.10	99.87	NE/SW; horizontal
X26	Indeterminate	Epiphysis	0.48	1.20	99.83	Horizontal
X27	*Sivatherium*	Ulna; belongs with 19,25	0.40	1.18	99.85	NE/SW; horizontal
X28	Bovid	Right humerus; goes with 29	0.27	1.17	99.83	N/S; proximal end dips north
X29	Bovid	radius-ulna; goes with 28	0.23	1.09	99.86	Distal end dips steeply to NE
X30	Bovid	Left humerus; goes with 28	0.20	1.23	99.81	NW/SE: dips slightly towards NW
X31	Indeterminate	Rib fragment	0.66	0.43	99.83	
X32	Indeterminate	Rib fragment	0.28	0.66	99.82	N/S; horizontal.
X33	Indeterminate	Rib fragment	0.35	0.93	99.80	N/S; horizontal
X34	Bovid	2nd phalange, complete	0.34	0.77	99.81	E/W; horizontal
X35	Indeterminate	Rib; ?part of 3	0.59	1.03	99.83	NE/SW; horizontal
X36	Bovid	Astragalus, left, complete	0.62	0.48	99.85	NE/SW; horizontal
X37	Indeterminate	Tibia, left	0.69	0.54	99.83	N/S; horizontal
X38	*Canis cautleyi*	Humerus, left	0.42	0.90	99.80	NW/SE; horizontal,
X39	Bovid	Metacarpal	0.54	0.92	99.81	N/S; horizontal
X40	Indeterminate	Rib	0.44	1.0	99.79	E/W; horizontal
X41	*Canis cautleyi*	Ulna	0.45	0.97	99.78	E/W; horizontal
X42	Indeterminate	Fragment	0.42	0.99	99.77	N/S; horizontal
X43	*Rhinoceros*	Carpal	0.47	0.93	99.82	Horizontal
X44	Indeterminate	Sesamoid; goes with 43	0.49	0.90	99.81	
X45	Indeterminate	Vertebra	0.53	0.84	99.78	
X46	Bovid/Tragulid	Maxilla; goes with 12	0.79	0.64	99.85	NW/SE, slight dip to NW
X47	Bovid	1st phalange, complete	0.69	0.81	99.80	N/S; horizontal
X48	Indeterminate	carpal	0.56	0.86	99.79	N/S, horizontal.
X49	Bovid	2nd phalange, complete	0.44	0.78	99.8	NE/SW, horizontal
X50	Indeterminate	rib	0.63	0.62	99.81	NW/SE, horizontal

X51	Indeterminate	Vertebra	0.64	0.71	99.82	E/W, dipping 30° east
X52	Bovid	1st phalange, complete	0.61	0.73	99.83	NE/SW; horizontal
X53	Bovid	Metatarsal, right	0.60	0.82	99.81	NE/SW: horizontal
X54	Indeterminate	Rib	0.47	0.72	99.77	NE/SW; horizontal
X55	*Canis cautleyi*	Radius, left, complete	0.49	0.85	99.76	NW/SE; horizontal
X56	*Canis cautleyi*	Tibia, right proximal	0.41	0.90	99.78	N/S; horizontal
X57	Indeterminate	Rib	0.38	0.94	99.80	
X58	Indeterminate	Rib	0.37	0.95	99.80	E.W; horizontal.
X59	Indeterminate	Fragment	0.41	0.76	99.80	E/W; horizontal
X60	Indeterminate	Rib	0.72	0.91	99.82	NE/SW; horizontal.
X61	Indeterminate	Rib	0.79	0.65	99.83	NW/SE; horizontal
X62	Proboscidean	Calcaneum	0.66	1.0	99.84	NW/SE; horizontal
X63	Indeterminate	Rib	0.45	0.51	99.79	E/W; horizontal.
X64	Indeterminate	Rib	0.57	0.71	99.79	NW/SE; horizontal
X65	Indeterminate	Rib	0.67	0.73	99.80	NW/SE, horizontal
X66	Bird (Large)	Carpometacarpus; goes with 16	0.82	0.82	99.78	N/S; horizontal
X67	Indeterminate	Rib	0.75	0.80	99.77	NE/SW, horizontal
X68	Indeterminate	Rib	0.62	0.77	99.76	E/W; horizontal
X69	Indeterminate	Rib	0.66	0.91	99.77	N/S; slight dip to north
X70	Indeterminate	Rib	0.57	0.87	99.78	E/W; horizontal
X71	Indeterminate	Tooth fragment	0.54	0.80	99.75	horizontal
X72	Indeterminate	Rib	0.51	0.75	99.79	NW/SE; slight dip to NW
X73	Indeterminate	Rib	0.48	0.79	99.80	NW/SE; slight dip to NW
X74	Indeterminate	Rib	0.43	0.83	99.78	E/W; horizontal
X75	*Canis cautleyi*	Os penis; L = 9 cm	0.33	0.84	99.80	NW/SE; horizontal
X76	Indeterminate	Vertebra	0.29	0.95	99.84	Horizontal
X77	Indeterminate	?pelvis	0.26	0.91	99.84	Horizontal
X78	Indeterminate	Rib	0.23	0.97	99.83	Horizontal
X79	Indeterminate	Vertebra	0.25	1.0	99.81	Horizontal
X80	Indeterminate	Fragment	0.50	0.85	99.75	Horizontal
X81	Indeterminate	Skull fragment	0.71	0.80	99.77	Horizontal
X82	Indeterminate	Fragment	0.60	0.79	99.74	Horizontal
X83	Carnivore	Phalange	0.57	0.86	99.72	NE/SW; horizontal
X84	Indeterminate	?vertebra	0.55	0.86	99.77	Horizontal
X85	Indeterminate	Rib	0.54	0.86	99.80	Horizontal
X86	Indeterminate	Rib	0.50	0.84	99.81	Horizontal
X87	Indeterminate	Fragment	0.49	0.88	99.80	
X88	Carnivore	Canine	0.45	0.89	99.81	NE/SW; horizontal
X89	Indeterminate	Rib	0.46	0.94	99.82	Horizontal
X90	Indeterminate	Rib	0.61	0.93	99.80	Horizontal
X91	Indeterminate	Fragment	0.59	0.96	99.80	Horizontal
X92	Carnivore	?atlas	0.54	0.94	99.80	NE/SW; slight dip to SW
X93	*Canis cautleyi*	Metatarsals I – IV; opp. foot to 158	0.50	0.99	99.81	NW/SE; horizontal
X94	*Rhinoceros*	Tooth fragment	0.37	0.88	99.81	
X95	Indeterminate	Rib	0.40	0.90	99.81	N/S; horizontal
X96	Indeterminate	Rib	0.35	0.95	99.81	NW/SE, horizontal.
X97	Indeterminate	Rib	0.35	1.0	99.81	NW/SE, horizontal
X98	Indeterminate	Fragment	0.35	1.01	99.82	Horizontal
X99	Indeterminate	Fragments	0.33	1.0	99.82	Horizontal
X100	*Canis cautley*	Astragalus, right	0.42	0.98	99.80	N/S; horizontal
X101	Indeterminate	Rib	0.45	0.97	99.80	NE/SW; horizontal
X102	Indeterminate	Rib	0.46	0.90	99.80	NW/SE; horizontal
X103	Indeterminate	Fragment	0.45	0.94	99.80	Horizontal
X104	*Canis cautleyi*	Canine	0.59	0.96	99.80	Horizontal
X105	*Canis cautleyi*	Calcaneum + 3 carpals /tarsals	0.41	0.92	99.80	Horizontal
X106	Bovid	1st phalange, complete	0.50	1.29	99.75	NE/SW, slight dip to NE
X107	Bovid	1st phalange, proximal fragment	0.60	1.23	99.75	Horizontal
X108	Bovid	2nd phalange, complete	0.50	1.34	99.74	NW/SE, slight dip to NW
X109	Bovid	Proximal radius-ulna	0.68	1.10	99.75	NW/SE; horizontal
X110	*Sivatherium*	1st phalange	0.70	1.20	99.77	E/W; horizontal
X111	Indeterminate	Carpal/tarsal	0.60	1.19	99.75	Horizontal

X112	Indeterminate	Sesamoid	0.68	1.02	99.73	Horizontal
X113	Indeterminate	Fragment	0.80	1.17	99.76	Horizontal
X114	Indeterminate	Lumbar vertebra	0.81	1.10	99.77	Dipping slightly to east
X115	Indeterminate	Rib	0.66	1.16	99.71	NW/SE; horizontal
X116	Bovid	Navicular cuboid	0.80	1.02	99.75	Horizontal
X117	Indeterminate	Carpal/tarsal	0.66	1.10	99.74	Horizontal
X118	Indeterminate	Fragment	0.86	1.12	99.75	Horizontal
X119	*Canis cautleyi*	Mandible ramus, right; goes with 124	0.82	1.16	99.74	NW/SE; horizontal
X120	Indeterminate	Fragment	0.62	1.13	99.76	
X121	Indeterminate	Fragment	0.66	1.09	99.76	
X122	Indeterminate	?carpal/tarsal	0.60	1.23	99.75	Horizontal
X123	Indeterminate	Rib	0.63	1.22	99.75	
X124	*Canis cautleyi*	Mandible, right	0.68	1.07	99.76	E/W; horizontal
X125	*Canis cautleyi*	Tibia,left; opp. leg to 56	0.67	1.05	99.76	E/W; horizontal
X126	*Canis cautleyi*	?incisor, upper	0.71	1.09	99.75	NE/SW; horizontal
X127	Bovid	Metacarpal, right prox.	0.59	1.05	99.79	N/S; horizontal
X128	Indeterminate	Fragment	0.58	1.15	99.79	
X129	*Canis cautleyi*	Jaw fragment plus ribs	0.60	1.05	99.74	
X130	Indeterminate	Phalange	0.48	1.03	99.72	E/W; horizontal
X131	Indeterminate	?carpal/tarsal	0.55	1.19	99.75	Horizontal
X132	Indeterminate	Rib	0.55	1.27	99.76	Horizontal
X133	Indeterminate	Rib	0.46	1.03	99.73	Horizontal
X134	Bovid	Distal femur, unfused epiphysis	0.20	1.37	99.78	Horizontal
X135	Indeterminate	Fragment	0.30	1.20	99.80	slight dip to north
X136	*Sivatherium*	Humerus, right; goes with 19,25,27	0.49	1.22	99.85	Horizontal
X137	Indeterminate	Fragments	0.36	1.22	99.83	Horizontal
X138	Indeterminate	Fragment	0.20	1.39	99.72	Slopewash
X139	Indeterminate	Fragment	0.12	1.20	99.76	horizontal, ? slopewash
X140	Indeterminate	Fragment	0.23	1.20	99.83	Horizontal
X141	Indeterminate	Fragment	0.16	1.19	99.77	Horizontal
X142	Indeterminate	Fragment	0.15	1.09	99.78	
X143	Bovid	Tibia, right	0.47	1.33	99.77	NW/SE; horizontal
X144	Indeterminate	calcaneum, proximal fragment	0.42	1.30	99.75	NE/SW, slight dip to NE
X145	Indeterminate	?carpal/tarsal	0.58	1.33	99.72	Horizontal
X146	Indeterminate	Fragments	0.53	1.35	99.73	Horizontal
X147	Indeterminate	?carpal/tarsal	0.58	1.30	99.80	Horizontal
X148	Indeterminate	?sesamoid	0.53	1.12	99.82	Horizontal
X149	Indeterminate	2 ?sesamoids	0.52	1.15	99.81	Horizontal
X150	Indeterminate	?carpal/tarsal	0.64	1.32	99.79	Horizontal
X151	Indeterminate	ibs	0.49	1.08	99.80	Horizontal
X152	Indeterminate	?carpal/tarsal	0.45	1.12	99.80	
X153	Bovid	Astragalus, right side	0.44	1.23	99.78	E/W; horizontal
X154	Indeterminate	Rib	0.43	1.25	99.76	NW/SE; horizontal.
X155	Indeterminate	Fragment	0.48	1.12	99.80	NE/SW; horizontal
X156	Indeterminate	?sesamoid	0.48	1.06	99.80	
X157	Indeterminate	Fragment	0.42	1.04	99.81	N/S; horizontal
X158	*Canis cautleyi*	Foot bone group	0.42	1.04	99.81	
X159	Indeterminate	?rib	0.40	1.03	99.72	Vertical
X160	Indeterminate	?carpal/tarsal	0.42	1.09	99.72	Horizontal
X161	Indeterminate	Ribs	0.38	1.00	99.74	Horizontal
X162	Indeterminate	?rib	0.32	1.04	99.75	E/W; horizontal
X163	Indeterminate	Ribs	0.38	1.08	99.75	
X164	Indeterminate	Pelvis	0.38	1.07	99.81	N/S; horizontal
X165	Bovid	3rd phalange	0.40	1.18	99.80	NE/SW
X166	Bovid	Calcaneum; right, fragment	0.43	1.23	99.80	Horizontal
X167	Indeterminate	Fragments	0.33	1.30	99.82	E/W; horizontal
X168	Indeterminate	Rib	0.42	1.33	99.77	NE/SW; slight dip to NE
X169	Indeterminate	Fragment	0.35	1.23	99.84	
X170	Bovid	Metatarsal, right, prox.	0.32	1.17	99.84	E/W; horizontal
X171	Indeterminate	Rib	0.32	1.14	99.82	NW/SE; horizontal
X172	Indeterminate	Rib	0.39	1.20	99.81	N/S, horizontal

X173	Indeterminate	Rib	0.44	1.18	99.80	NE/SW, horizontal
X174	Indeterminate	Rib	0.40	1.18	99.82	NE/SW; horizontal
X175	Carnivore	Phalange fragment	0.32	1.0	99.78	
X176	Indeterminate	Fragment	0.16	1.08	99.86	Horizontal
X177	Indeterminate	?vertebra	0.27	1.04	99.88	
X178	Indeterminate	Fragment	0.33	1.07	99.85	
X179	Indeterminate	Rib	0.31	1.01	99.86	E/W; horizontal

Notes:

Column 6 (Depth): t = top, m = midpoint
X5: t=top,b=bottom, NW/SE. Suid left acetabulum/ischium/ilium fragment
X16 : Length = c. 26 cm.
X19 : *Sivatherium*.radius; goes with (X25) and ulna (X27). Length = 59 cm; Dist.radius breadth = 12.5 cm
X20 : Bovid, Right, almost complete. Distal breadth = 28 mm
X21 :Bovid ulna, lying with humerus, Joins with X28 and X29.
X28 :Bovid, right side, articulates with X29; distal breadth = 95 mm
X29 :Bovid, almost complete, articulates with X28. Length radius = 33 cm, Dist.radius breadth = 88 mm.
X30 :Bovid, Left side shaft fragment ?opposite leg to X28.
X38 :Canis, Left side, almost complete, Distal breadth = 43 mm
X39 : Bovid, metacarpal; Dist. breadth = 76 mm
X41 :*Canis*, Left Prox. fragment, articulates with X55?.
X46 :Belongs with X12, ? gazelle, left side, permanent dentition, worn.
X53 :Bovid, right side, Distal breadth = 68 mm.
X55 :*Canis*, left side, Complete, Length = 21.8 cm, Prox. breadth = 24 mm, Dist. breadth = 31 mm
X56 : *Canis*, Right proximal. Opposite leg to X125.
X62 : Proboscidean, one deep carnivore puncture mark and at least three shallower ones. Length = 17.5 cm.
X66 : Bird bone, may belong to same bird as spec.no.X16?, eagle size/type??
X83 : Carnivore phalanx, Length = 19 mm.
X100 : *Canis*, right side, Length = 40 mm.
X105 : Length calcaneum = 63 mm
X110 : *Sivatherium*; Length = 12.5 cm, Prox. breadth = 56 mm.; Distal breadth = 54 mm
X124 : *Canis*: X119 probably belongs with this; right side, adult dentition: Incisors 1-3, canine and premolars 1-3, all = worn.
X125 : Canis, left side, complete, opposite leg to X56, Length = 23.5 cm.
X127 : Bovid, right side, Prox. breadth = 67 mm
X129 : Group of bones: cluster of ?ribs plus *Canis* jaw fragment (joins with X124), and one carpal/tarsal.
X143 : Bovid, right side, Distal breadth = 70 mm
X158 : Group of bones taken out as one, ?*Canis* foot, i.e. metapodials, carpals/tarsals etc, opposite foot is ? X93
X165 : Articular surface dipping to SW,. ? Bovid, almost complete.
X170 : Bovid, right side, Prox. breadth = 56 mm.

The excavation of fossil locality 642

Mark Beech (Senior Resident Archaeologist, Abu Dhabi Islands Archaeological Survey, PO Box 45553, Abu Dhabi, United Arab Emirates) and Mohammad Anwar (Assistant Director, Geological Survey of Pakistan)

Introduction

As narrated in chapter 7, this volume, locality 642 was discovered in the 1989 field survey, and was the largest fossil locality found in the Pabbi Hills. As already shown in Tables 7. 5- 10, over 4,200 specimens were collected here in 1989. It was decided to excavate this site in 1990 in order to retrieve well-preserved *in-situ* remains of taxonomically important animals, to investigate the taphonomic history of this locality, and to see if hominids had been involved in its formation.

Methods

The 1990 spring field season lasted for seven weeks, work taking place on the site from the 30th January until the 24th March. The first two days were spent carrying out a systematic surface collection of the area surrounding the locality. The results are listed in Tables 9. 9 – 11. The surface collection used the same collect zones as in 1989; i.e. as zones 1A, 1B, 1C, 2A, 2B, 3A and 3B. In addition, *in-situ* bone material, or clusters of surface fragments within these zones were given find spot numbers (Table 9. 12), the location of which was recorded (see Figure 9. 11). Measurements of specimens are given in Table 9. 13. As shown, over 1000 new specimens were found that had eroded to the surface during the year following the 1989 collection. This would imply a very fast rate of erosion for this locality, as the 4,200 specimens found in 1989 might have resulted from no more than 5 years' erosion.

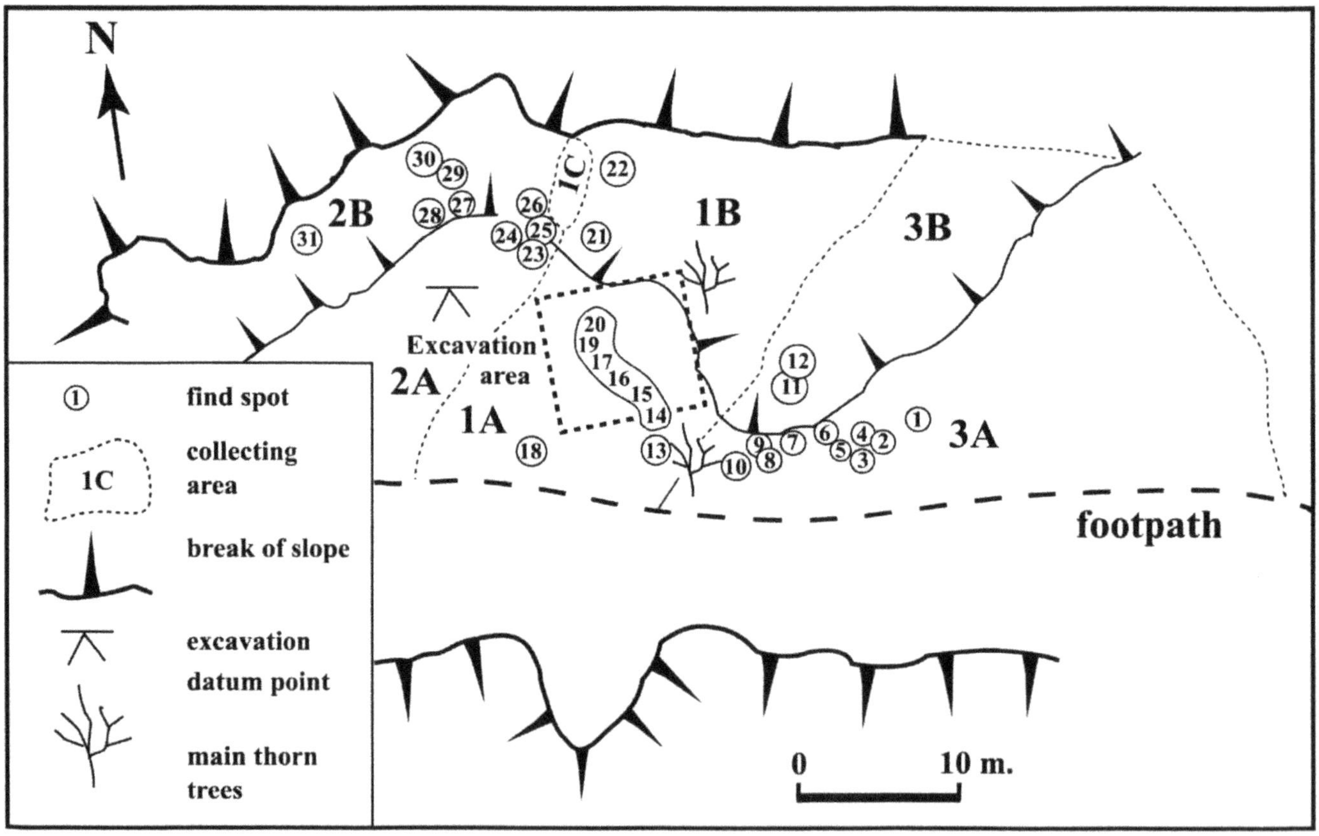

Figure 9. 11: Collecting zones, datum point, and find-spots at locality 642, 1990. The datum point here is the same as for the excavation grid, shown in Figure 9. 12. Based on field records by Mark Beech and Malik Sikander.

Figure 9. 12 shows the 8 metre square grid that was laid out on the site. The location of this gird was determined by the following factors:
(1) there appeared to be a clear line of *in-situ* bone material eroding out of the slope.
(2) the siting of the trench was immediately above the area where an almost complete *Equus sivalensis* skull and many c.f. *Proamphibos lachrymans* and c.f. *Hemibos triqueticornis* horncores had been found in 1989.
(3) it appeared that a greater depth of sediment was preserved in that particular area of the site, and the adjacent areas of the site appeared more eroded, partially due to the presence of the nearby footpath. The area selected therefore was potentially less disturbed and contained a thicker fossiliferous horizon.

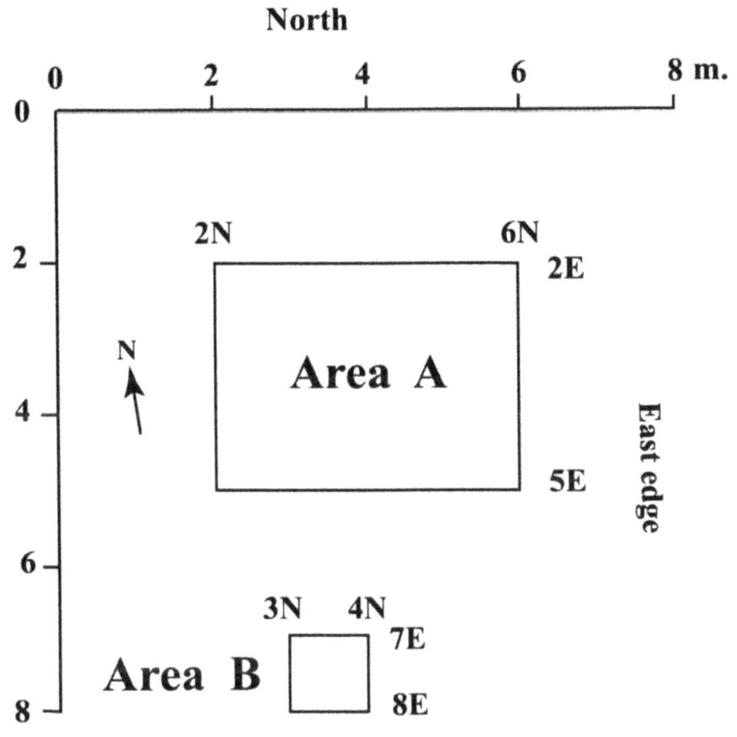

Figure 9. 12: Excavation grid and areas excavated at locality 642. The zero points are at the datum point shown in Figure 9. 11.

The area within the 8 metre square grid was firstly cleared of all loose surface material (extracting all thorn bushes, grass material and loose 1-3 cm diameter quartzite pebbles which were scattered over the general area). A 2cm spit of sediment was then removed from the entire area. It soon became apparent that due to the hardness of the sediment at the locality, it would not be possible to tackle the entire area in the time available. It was therefore decided to continue by excavating a strip one metre wide across the site (2-3 metres east/0-8 metres north), just above the line of visible *in-situ* material (running north-south) eroding out of the slope surface.

Excavation soon concentrated upon two principle areas (A and B) (see Figure 9. 12). Area A became the main area of excavation, and area B a subsidiary one. It seemed worthwhile to excavate area B because of the abundance of bones that were eroding out of this part of the site. However, the greater effort soon became concentrated on area A when it became clear that the quality and general preservation of specimens found there were better than in area B.

All fossil specimens were recorded three dimensionally within the superimposed site grid. All grid references taken were measured from the central axis point of each bone. In addition, the orientation was noted for all bones with a long axis, as well as the angle of dip. A full catalogue of these finds is contained in Table 9. 14. Relevant measurements are shown in Table 9. 15. All specimens were planned at a scale of 1:10 (cm) at each appropriate level (if it was not possible to fit a bone onto the existing plan then a new plan was commenced). These are shown in Figures 9. 13 – 15 for area A, and Figures 9. 16 – 19 for Area B. Sections were drawn of each profile, but were generally uninformative beyond showing that fossil material was contained within brown silty-sands. The most informative section is shown in Figure 9. 20.

Results

The following were the main species were identified at site 642 in 1990:

c.f. *Hemibos triquetricornis* (Falconer, 1868).
The identification of this species is problematic. It has pear shaped to circular cross section in its horncores, and its horns diverge at up to 90 degrees. Further studies need to be carried out on this particular material.

c.f. *Damalops palaeindicus*: This species usually has horncores which are sub-triangular in cross section and which diverge at up to 45 degrees. Again further studies need to be carried out on this particular material. (This was initially identified as *Proamphibos lachrymans* (Pilgrim, 1939), but later revised in the light of Coard's study of the bovid material.

Pachycrocuta sivalensis (Matthew, 1929). It appears to be this species from the relative size of the teeth.

Equus sivalensis (Falconer and Cautley, 1868).
The diagnostic features of this species are as for *Equus* but it has a deep jaw (as contrasted with *Hipparion*). In addition, the protocone of its premolars are very large in its 2nd molar. Generally speaking all the ostoemetric data appears to show a homogenous group of horses, all of which can be classified as *Equus sivalensis*.

Rhinoceros sivalensis (Falconer and Cautley, 1868).

The complete skull excavated from site 642 (specimen 1602) had the following diagnostic features characteristic of *Rhinoceros sivalensis*: a saddle shaped skull; a single horn; molars with parastyle, crochet absent, and concave ectolophs.

Further comparative data are presented in chapter 20, and Coard's assessment of the taphonomic history of 642 is presented after this report.

Table 9. 9: Fossil specimens from the surface of locality 642, zone 1, 1990

Zone	Specimen	W	Taxon	Element	Part	Other	N
1	88	4	Carnivora (Small)	Maxilla+Teeth	C,P2,P4 sockets		1
1A	8,17	5	Artiodactyl	Tooth	Split		2
1A	19	2b	Artio (Medium)	Humerus	Distal	Shaft	1
1A	12-16	3	Bovid	Horncore	Base		5
1A	17,18	3	Bovid	Horncore	Tine		2
1A	1	4	Bovid (Medium)	Mandible	PM3,PM4,M1 split	Worn, left	1
1A	2	3	Bovid (Medium)	Tooth, M1/2	Mod. wear	Complete, upper	1
1A	4	4	Bovid (Medium)	Tooth, Molar	Mod. wear	Split, lower	1
1A	5	5	Bovid (Medium)	M1/2	Complete	Worn, lower	1
1A	10	3	Bovid (Medium)	Mandible+ Teeth	Fresh, Right	P2 fresh,P3 broken	1
1A	20	3	Bovid (Medium)	Radius	Proximal	Complete	1
1A	24,25	3	Bovid (Medium)	Metapodial	Distal barrel		2
1A	3	4	Equid	Tooth	Complete	Mod. wear; PM2 or M3	1
1A			Indeterminate	Fragments			21
1B	0		Artiodactyl	Tooth, PM	Complete	Worn	1
1B	0	5	Artiodactyl	Tooth	Split	Mod. wear	1
1B	0	5	Artiodactyl	Tooth, PM	No crowns		1
1B	0	4	Artiodactyl	Tooth	Split	Mod. wear	1
1B	3	4	Artiodactyl	Tooth, PM	Fragment	Lower	1
1B	7	5	Artiodactyl	Mandible	+ Teeth	1 broken, 1 root	1
1B	12,19,24, 28	4	Artiodactyl	Tooth	Split	No crowns	4
1B	20	5	Artiodactyl	Mandible	+ Tooth	Root	1
1B	83	4	Artiodactyl	Metatarsal	Shaft		1
1B	90	4	Artiodactyl	Tibia	Distal	Complete	1
1B	114	5	Artiodactyl	Metapodial	Distal barrel		1
1B	5	5	Artio (Large)	Mandible	+ Teeth	2 broken, 1 root	1
1B	0	4	Artio (Medium)	Tooth	Mod. wear	Crowns broken	1
1B	95,98	5	Artio (Medium)	Metapodial	Distal barrel		2
1B	94	4	Artio (Medium)	Humerus	Distal	Complete	1
1B	1	5	Bovid	Mandible	M1,M2	Worn	1
1B	4	4	Bovid	Tooth	PM Complete	Worn, Lower	1
1B	17	4	Bovid	Tooth	Complete, upper	Mod. wear, PM	1
1B	56	4	Bovid	Horncore			22
1B	77,78,84	4	Bovid	Horncore			3
1B	109	3	Bovid	Fibula	Distal		1
1B	0	4	Bovid (Large)	Tooth, PM	Complete	Mod. wear	1
1B	2	4	Bovid (Large)	Molar	Split	Worn	1

Table 9. 9 continued: Fossil specimens from the surface of locality 642, zone 1, 1990

1B	14	4	Bovid (Large)	Tooth, PM	Complete	Worn, Upper	1
1B	21,23	3	Bovid (Large)	Tooth	Split	Worn	2
1B	26	3	Bovid (Large)	Tooth	Complete	PM, lower	1
1B	89	5	Bovid (Medium)	Metacarpal	Distal		1
1B	78	3	Bovid (Medium)	Metatarsal	Proximal	Complete	1
1B	0	4	Bovid (Medium)	Molar	Mod. wear	Split	1
1B	0	4	Bovid/Cervid	Maxilla	+ Teeth	M1-3, mod. wear	1
1B	108	4	Carnivora	Radius	Proximal	Split	1
1B	82	3	Equid	Metacarpal	Distal	Complete	1
1B	101	5	Rhino	Tibia+Astrag	Distal		1
1B			Indeterminate	Fragments			192
1C	1	4	Artiodactyl	Mandible	+ Tooth	Split	1
1C	8,9,20, 17,44		Artiodactyl	Tooth	Split	No crowns	5
1C	13		Artiodactyl	Tooth, PM	Split, lower	Mod. wear	1
1C	0,18,32		Artiodactyl	Tooth	Split		3
1C	23,24,26,47		Artiodactyl	Tooth	Split	Worn	4
1C	71	4	Artiodactyl	Pelvis	Ichium		1
1C	73	4	Artiodactyl	Pelvis	Acetabulum		1
1C	89	5	Artio (Medium)	Metapodial	Distal barrel		1
1C	5	4	Bovid	Mandible	+ Teeth	PM split, M broken	1
1C	6	3	Bovid	Mandible	+ Tooth	Mod. wearM1/2	1
1C	22		Bovid	Tooth	Split	Mod. wear	1
1C	35	4	Bovid	Horncore			26
1C	10	4	Bovid (Large)	Tooth	Erupting	PM, Upper	1
1C	12	4	Bovid (Large)	Tooth PM	Complete	Worn, Upper	1
1C	31	4	Bovid (Large)	Tooth	Mod. wear	PM	1
1C	75	4	Bovid (Large)	Metacarpal	Proximal	Complete	1
1C	0	5	Bovid (Medium)	Radius	Proximal	Complete	1
1C	4	3	Bovid (Medium)	PM	Worn	Split, Upper	1
1C	65, 68,70,72	3	Bovid (Medium)	Humerus	Distal	Complete	4
1C	69	4	Bovid (Medium)	Radius	Proximal	Complete	1
1C	74	4	Bovid (Medium)	Metacarpal	Distal	Complete	1
1C	77	4	Bovid (Medium)	Metacarpal	Proximal	Complete	1
1C	82	4	Bovid (Medium)	Metatarsal	Shaft		1
1C	84	4	Bovid (Medium)	Metacarpal	Shaft		1
1C	96,97	4	Bovid (Medium)	1st Phalange	Proximal	Complete	2
1C	16, 27, 30	3	Bovid (Medium)	PM	Complete	Worn	3
1C	67,94	4	Carnivora	Shaft			2
1C	66	4	Cervid	Antler	Pedicel		1
1C	34	4	Hyaenid	Tooth			1
1C	100	4	*Rhinoceros*	Carpal	Complete		1
1C			Indeterminate	Fragments			113
Total							**487**

Notes: 1C 16 is upper, 1C 30, 65, 70 and 72 are left; 1C68 is right.

Table 9. 10: Fossil specimens from the surface of locality 642, zone 2, 1990

Zone	Specimen	W	Taxon	Element	Part	Other	N
2B	15	3	Artiodactyl	Ulna	Proximal		1
2B	1	3	Artio (Medium)	Tooth	Complete		1
2B	2	4	Artio (Medium)	Tooth	Split		1
2B	16	3	Artio (Medium)	3rd Phalange	Proximal		1
2B	17	3	Artio (Medium)	Femur	Distal	Split	1
2B	14	4	Artio(Medium)	Radius	Proximal	Split	1
2B	12	4	Bovid	Horncore			1
2B	3	2b	Bovid (Large)	Tooth	Split	Fresh	1
2B	11	4	Bovid (Large)	Horncore	Tine		1
2B	4	4	Bovid (Medium)	PM Lower	Complete	Mod. wear	1
2B	5	4	Bovid (Medium)	PM Upper	Split	Mod. wear	1
2B	20	5	Bovid (Medium)	Metacarpal	Proximal	Complete	1
2B	22	4	Cervid	1st Phalange	Complete		1
2B	0,6,25	5	Indeterminate	Fragments			84
Total							**97**

Table 9. 11: Fossil specimens from the surface of locality 642, zone 3, 1990

Zone	Specimen		Taxon	Element	Part	Other	N
3B	20	4	Artio (Large)	Tooth	Split		1
3B	8, 10, 13	5	Artio (Medium)	Tooth	Split		3
3B	18	4	Artio (Medium)	Mandible	+ Tooth Split		1
3B	33	3	Artio (Medium)	Metatarsal	Shaft		1
3B	35	3	Artio (Medium)	Metatarsal	Proximal	Split	1
3B	46	4	Artio (Medium)	Humerus	Distal	Split	1
3B	91	3	Artio (Medium)	Humerus	Distal	Complete	1
3B	92	4	Artio (Medium)	3rd Phalange	Proximal		1
3B	93	4	Artio (Medium)	Radius	Proximal	Complete	1
3B	97	5	Artio (Medium)	Metapodial	Distal barrel		1
3B	23,24,30-32,33,34	5	Bovid	Horncore			7
3B	25, 26	4	Bovid	Horncore	Tine		2
3B	28	4	Bovid	Horncore			1
3B	67-84*	3	Bovid	Horncore	Curved		18
3B	16	4	Bovid (Large)	Tooth, lower	Complete	Worn	1
3B	46	4	Bovid (Large)	Calcaneum	Shaft		1
3B	52	4	Bovid (Large)	Tooth	Molar, Upper	Complete	1
3B	56	4	Bovid (Large)	Split	Tooth		1
3B	61	4	Bovid (Large)	Tooth	Split		1
3B	96	3	Bovid (Large)	Calcaneum	Distal	Shaft	1
3B	44	3	Bovid (Medium)	Astragalus	Complete		1
3B	39	3	Bovid (Medium)	Humerus	Distal	Complete	1
3B	3	4	Bovid (Medium)	Mandible	+ Tooth	M1/2	1
3B	51	4	Bovid (Medium)	Mandible	+ M2, M3	Right	1
3B	53	5	Bovid (Medium)	Mandible	+ Tooth socket		1
3B	54	4	Bovid (Medium)	Mandible	+ M3	Right	1
3B	57	5	Bovid (Medium)	Mandible	+ Molar	Split	1
3B	1	5	Bovid (Medium)	Mandible + M3	Split		1
3B	36,37	3	Bovid (Medium)	Metatarsal	Proximal	Complete	2
3B	38	5	Bovid (Medium)	Molar Split	Worn		1
3B	62	5	Bovid (Medium)	PM, Upper	Complete	Worn	1
3B	47	4	Bovid (Medium)	Radius	Proximal	Complete	1
3B	21	4	Bovid (Medium)	Split	Worn	Molar	1
3B	64	5	Bovid (Medium)	Split M3	Right, upper	Tooth	1
3B	46	44	Bovid (Medium)	Tibia	Distal	Complete	1
3B	0, 15	3	Bovid (Medium)	Tooth	PM, Upper	Complete, worn	2
3B	11,17,59,66	5	Bovid (Medium)	Tooth	Split	Worn	4
3B	39	4	Bovid (Medium)	Tooth	Split	Mod. wear	1
3B	102	5	Bovid (Small)	2nd Phalange	Complete	? *Gazella*	1
3B	95	4	Bovid/Cervid	Metatarsal	Proximal	Complete	1
3B	2	5	Equid	Mandible	+ Tooth Split		1
3B	22	4	Equid	Tooth, lower	Complete, worn	Molar	1
3B	30,94	4	Equid	Metacarpal	Distal	Complete	2
3B	48	3	Rhino	Carpal	Complete		1
3B	4,60,63,65	5	Ungulata	Tooth	Split		4
3B	5	4	Ungulata/Rhino?	Tooth	Split		1
3A			Indeterminate	Fragments			133
Total							**214**

Note: * includes 3 tines and 2 curved pieces

Table 9. 12: Fossil specimens from find spots on the surface of locality 642, 1990

Zone	Find Spot	Specimen	W	Taxa	Element	Part	Other	N
3B	FS11	0	5	Artiodactyl	Tooth	Fragment		1
3B	FS11	5	3	Bovid (Medium)	Mandible, right	+ M1/2, erupting	Complete	1
3B	FS11	6	3	Bovid (Medium)	Mandible	+ Tooth	Split	1
3B	FS11	0		Indeterminate	Fragments			7
3B	FS12	6	5	Artio (Medium)	Maxilla	+Tooth	Worn, split	1
3B	FS12	various		Bovid	Horncore			8
3B	FS12	5	5	Bovid (Medium)	Mandible	M1/2	Worn	1
3B	FS12	4	5	Bovid (Medium)	M3, lower	Complete	Worn	1
3B	FS12	12	4	Bovid (Medium)	Molar, upper	Complete	Mod. wear	1
1A	FS13	0	5	Bovid	Horncore			4
1A	FS13	0	4	Indeterminate	Tooth	Fragments		50
1A	FS14	0	4	Bovid	Horncore			1
1A	FS14	2	4	Bovid (Medium)	Tooth	Split		1
1A	FS14	3	3	Bovid (Medium)	Maxilla	M3, mod. wear	Split	1
1A	FS14	0		Indeterminate	Fragments			8
1A	FS15	1	3	Artiodactyl	PM, lower	Complete	Worn	1
1A	FS15	2	3	Artiodactyl	Tooth			1
1A	FS15	3	4	Bovid (Small)	Maxilla	PM		1
1A	FS15	0	5	Indeterminate	Fragments			23
2A	FS17	2	4	Artio (Medium)	Tooth	PM	Fragment	1
2A	FS17	1	5	Equid	Tooth	Split	Worn	1
2A	FS17	0	4	Indeterminate	Tooth	Fragments		4
2A	FS18	1,2	3	Artiodactyl	Tooth	Fragment		2
2A	FS18	0	5	Indeterminate	Fragments			4
1A	FS19	3	4	Bovid	Horncore	Fragment		1
1A	FS19	1	4	Bovid (Medium)	Mandible	+ Teeth		1
1A	FS19	2	3	Bovid (Medium)	PM	Upper	Split	1
1A	FS19	14	5	Indeterminate	Fragments			12
1A	FS20	4	3	Artio (Medium)	Mandible	2 premolars	Erupting	1
1A	FS20	5	5	Bovid	Mandible	Tooth,split	Worn	1
1A	FS20	3	4	Indet/Carnivore	Radius	Distal		1
1A	FS20	1,2	5	Indeterminate	Fragments			2
1B	FS21	1	5	Artiodactyl	Tooth	Spall		1
1B	FS21	2	4	Bovid (Medium)	Molar, lower	Worn	Split	1
1B	FS21	3	3	Bovid	Horncore			1
1B	FS21	4,5	4	Indeterminate	Fragments			2
1B	FS22	3	3	Bovid	Horncore	Tine		1
1B	FS22	0	4	Indeterminate	Fragments			5
2B	FS23	1	5	Artio/Bovid	Mandible	+ Tooth		1
2B	FS23	3	4	Bovid (Medium)	Mandible	1 split	1 broken	1
2B	FS23	2	3	Bovid (Small)	Mandible, left	Tooth	1 split, broken	1
2B	FS23	19	4	Bovid (Small)	Metacarpal	Proximal	Complete	1
2B	FS23	0	5	Indeterminate	Fragments			140
1B	FS25	7	3	Bovid (Small)	Humerus	Distal	Complete	1
1B	FS25			Indeterminate	Fragments			6

Table 9. 12 continued: Specimens from find spots on the surface of locality 642, 1990

1B	FS26	1	4	Artio (Medium)	Tooth	Split	No crowns	1
1B	FS26	2	4	Artio (Medium)	PM	Complete	Worn	1
1B	FS26	3	5	Bovid	Horncore			1
1B	FS26	4	3	Bovid (Medium)	1st Phalange	Proximal		1
1B	FS26	5	3	Bovid (Medium)	2nd Phalange	Proximal		1
1B	FS26			Indeterminate	Fragments			5
2B	FS28	16	4	Artio (Small)	Calcaneum	Distal		1
2B	FS28	10	5	Bovid	Horncore			1
2B	FS28	1	4	Bovid (Medium)	Mandible	+ Tooth	Worn, split	1
2B	FS28	4	4	Bovid (Medium)	Tooth	Split	Errupting	1
2B	FS28	5	5	Bovid (Medium)	PM	Complete	Worn	1
2B	FS28	7,8	4	Bovid (Medium)	PM, lower	Complete	Worn	2
2B	FS28	14	4	Bovid (Medium)	2nd Phalange	Complete		1
2B	FS28	18	4	Bovid (Medium)	Metatarsal	Proximal	Complete	1
2B	FS28	18		Carnivore	Foot*			1
2B	FS28	0		Indeterminate	Fragments			38
2B	FS29	4	4	Artiodactyl	Tooth	Root		1
2B	FS29	5	4	Artio (Medium)	PM, lower	Complete	Worn	1
2B	FS29	6-9	5	Bovid	Horncore	Fragment		4
2B	FS29	2	4	Bovid (Medium)	PM, lower	Complete	Worn	1
2B	FS29	3	4	Bovid (Medium)	Molar, lower	No crowns	Split	1
2B	FS29	11	3	Bovid (Medium)	Metatarsal	Proximal	Complete	1
2B	FS29	0	5	Indeterminate	Fragments			41
2B	FS30	1,3	4	Artiodactyl	Tooth	Spall		2
2B	FS30	2	5	Artio (Medium)	PM, lower	Complete	Worn	1
2B	FS30	7	3	Artio (Medium)	Calcaneum	Distal	Shaft	1
2B	FS30	4	5	Bovid (Medium)	PM2, lower	Complete	Worn	1
2B	FS30	0	5	Indeterminate	Fragments			25
3A	FS4	1,2	5	Artiodactyl	Tooth	Spall		2
3A	FS4	0	3	Bovid	Horncore			1
3A	FS4	0	5	Indeterminate	Fragments			13
3A	FS9	1	3	Artiodactyl	PM	Complete	Lower	1
3A	FS9	8	4	Bovid (Medium)	Metatarsal	Shaft		1
3A	FS9	9	4	Bovid (Medium)	Metapodial	Distal barrel		1
3A	FS9	0		Indeterminate	Fragments			26
2A	XX	3	3	Bovid	Horncore			1
2A	XX	1	4	Ungulata	Mandible	+ Tooth	No crowns	1
2A	XX	2	5	Ungulata	Tooth	Split	No crowns	1
2A	XX	0	5	Indeterminate	Fragments			8
Total								**570**

Notes: The following find spots had only indeterminate fragments: Find spot 6, (zone 3A), 5 fragments; find spot 12 (zone 3B), 16 fragments; find spot 16 (zone 2A), 8 fragments; find spot 24 (zone 1B, 22 fragments; find spot 31 (zone 2B), 16 fragments.

FS28/18: carnivore foot - proximal metacarpals 1-5 fused together as a block

Notes on Table 9.13 (next page): Prox = Proximal; Dist = Distal; AP = anterior-posterior, W = Width

3B/51 Bovid Medium: M3, length = 30mm; M2 = 21.0mm.; FS25 1B/7:Bovid Small: cf 642/1/275; FS26 1B/4 Bovid: cf. 26/5; FS28 2B/14: cf. 30GB16; FS28 2B/18: squashed laterally; 1C/6: cf 3B/54; IC/65 and 1C/68; these may pair 1C70 and 1C/72; 3B/44 Astragalus: L= 45.0mm.; W = 30.0 mm.; Th. = 25.0mm.

Table 9. 13: Measurements (in mm.) of specimens from the surface collection of locality 642, 1990

Zone	Specimen	Taxon	Element	Length	Prox AP	Prox W	Dist AP	Dist W
3B	91	Artio (Medium)	Humerus, distal					42.0
1B	94	Artio (Medium)	Humerus, distal					51.0
1C	89	Artio (Medium)	Metapodial distal barrel				22.0	20.0
1B	95	Artio (Medium)	Metapodial, distal barrel, unfused				21.0	13.5
1B	98	Artio (Medium)	Metapodial, distal barrel				24.0	16.0
3B	97	Artio (Medium)	Metapodial, distal barrel				22.5	17.0
3B	35	Artio (Medium)	Metatarsal, proximal, split		29.0			
3B	93	Artio (Medium)	Radius, proximal		24.0	39.0		
1B	90	Artio (Medium)	Tibia, distal				31.0	39.0
1C	75	Bovid (Large)	Metacarpal, proximal		43.0	69.0		
3B	52	Bovid (Large)	Molar, upper	26.0				
705	1	Bovid (Large)	Radius, proximal	50.0	102.0			
3B	44	Bovid (Medium)	Astragalus, complete	45.0				
1C	6	Bovid (Medium)	Mandible + M1/2, mod.wear; right	22.0				
FS11	5	Bovid (Medium)	Mandible + M1/2 erupting	25.0				
3B	1	Bovid (Medium)	Mandible + M3	32.0				
3B	3	Bovid (Medium)	Mandible + M1/2, little wear	22.0				
3B	51	Bovid (Medium)	Mandible + M2,M3 *	30.0				
3B	54	Bovid (Medium)	Mandible + M3, little wear	28.0				
1A	2	Bovid (Medium)	M1/2, upper, mod. wear	22.0				
1A	5	Bovid (Medium)	M1/2, lower, worn	21.0				
FS12	4	Bovid (Medium)	M3, complete, worn	28.0				
FS12	12	Bovid (Medium)	Molar, complete, mod.wear	22.0				
FS26	4	Bovid (Medium)	1st Phalange, proximal		26.0	18.0		
1C	96	Bovid (Medium)	1st Phalange, proximal		25.0	19.0		
1C	97	Bovid (Medium)	1st Phalange, proximal		25.0	19.0		
FS28	14	Bovid (Medium)	2nd Phalange		21.0	16.5		
1C	70	Bovid (Medium)	Humerus, distal, left					47.0
1C	72	Bovid (Medium)	Humerus, distal, left					47.0
1C	65	Bovid (Medium)	Humerus, distal, right					50.0
1C	68	Bovid (Medium)	Humerus, distal, right					48.0
3B	39	Bovid (Medium)	Humerus, distal					48.0
1C	74	Bovid (Medium)	Metacarpal, distal				24.0	36.0
705	3	Bovid (Medium)	Metacarpal, distal				26.0	38.0
1C	77	Bovid (Medium)	Metacarpal, proximal		21.0	36.0		
2B	20	Bovid (Medium)	Metacarpal, proximal		20.0	32.5		
FS23	19	Bovid (Small)	Metacarpal, proximal		14.0	21.0		
FS9	9	Bovid (Medium)	Metapodial, distal barrel				25.0	20.0
1A	24	Bovid (Medium)	Metapodial, distal barrel				26.5	21.0
FS28	18	Bovid (Medium)	Metatarsal, proximal		41.0	39.0		
FS29	11	Bovid (Medium)	Metatarsal, proximal		30.0	29.0		
1B	78	Bovid (Medium)	Metatarsal, proximal		32.0	31.0		
3B	36	Bovid (Medium)	Metatarsal, proximal		32.0	36.0		
1B	89	Bovid (Medium)	Metacarpal, distal				23.0	34.0
1C	0	Bovid (Medium)	Radius, proximal		26.0	51.0		
1C	69	Bovid (Medium)	Radius, proximal		26.0	52.0		
3B	47	Bovid (Medium)	Radius, proximal		27.0	38.5		
3B	46	Bovid (Medium)	Tibia, distal				26.5	38.0
FS25	7	Bovid (Small)	Humerus, distal				22.5	26.0
3B	102	Bovid (Small)	2nd Phalange, complete	25.0	16.5	12.0	14.0	10.0
3B	95	Bovid/Cervid	Metatarsal, proximal		28.0	28.0		
2B	22	Cervid	1st Phalange, complete	62.0	24.0	19.0		
1B	82	Equid	Metacarpal, distal				33.0	45.0
3B	30	Equid	Metacarpal, distal				28.0	43.0
3B	94	Equid	Metacarpal, distal				34.0	46.0

Table 9. 14: Catalogue of locality 642 fossil specimens, excavated in 1990

Spec No.	Description	East	North	Depth	Orientation
1001-1010*	Bovid mandible and maxilla, + various long bone fragments	458	268	99.36	
1011	c.f. *Damalops* distal metatarsal	425	220	99.32	NW/SE, horizontal
1012-1017*	vertebra centrum (MM) + fragments	448	275	99.34	N/S
1018-1028*	*Damalops* maxilla, vertebra + fragments	446	275	99.44	NW/SE
1029-1034*	Vertebra centrum (MM)+ fragments	440	279	99.44	NE/SW
1035-1040*	*Damalops* mandible+ 2 cervical vertebra fragments (MM) + lbf	425	298	99.24	E/W
1041	*Damalops* horncore	400	298	99.24	E/W; horizontal
1042	Long bone fragment (LM)	442	700	99.63	NE/SW; horizontal
1043	Rib fragment (LM)	443	685	99.61	NE/SW
1044	*Damalops* horncore	460	630	99.66	
1045	Rib fragment (MM)	461	605	99.67	N/S; horizontal
1046	c.f. *Damalops* 1st phalange + smaller one, proximal unfused	505	656	99.36	E/W; horizontal
1047	Long bone fragment (MM)	512	685	99.34	N/S; horizontal
1048	Long bone fragment (MM)	518	684	99.34	E/W
1049	*Damalops* upper molar	518	712	99.26	E/W; horizontal
1050	c.f. *Damalops*, 2nd phalange	591	733	98.76	E/W; horizontal
1051	Long bone fragment (MM)	692	730	98.36	N/S
1052	*Damalops* mandible; 3 vertebra frags. (MM) + 2 long bone fragments	702	682	98.41	
1053	Thoracic vertbra (MM)	665	650	98.46	
1054	c.f. *Damalops* distal scapula	740	618	98.51	NW/SE; horizontal
1055	*Damalops* 2 upper molars + pelvis fragment	761	615	98.48	
1056	c.f. *Hemibos* distal metacarpal	768	535	98.41	N/S; horizontal
1057	cf. *Damalops*, ?tibia shaft	750	500	98.46	E/W
1058	*Damalops* horncore	675	540	98.47	N/S; horizontal
1059	*Damalops* horncore	616	285	99.21	
1060	*Damalops* horncore	660	142	98.76	
1061	*Damalops* horncore	722	221	98.56	
1062	*Rhinoceros* mandible, juvenile	245	335	99.47	
1063	Not used				
1064	*Damalops* PM + fragments	770	300	98.88	
1065	*Pachycrocuta* PM upper + Bovid (*Hemibos* and *Damalops)* fragments	425	220	99.41	
1066	*Damalops* mandible and prox. tibia, & other fragments (MM)	410	313	99.48	
1067	Long bone fragment (MM)	410	313	99.47	N/S; horizontal
1068-1075	*Damalops* maxilla and distal radius, + long bone fragments (MM)	425	220	99.38	N/S; horizontal
1076	c.f. *Damalops* distal metatarsal	787	302	98.88	NW/SE
1077	*Damalops* mandible	785	297	98.88	N/S
1078	*Damalops* horncore and tibia, and rib (MM)	780	306	98.87	NE/SW
1079	*Damalops* horncore and tibia shaft, c.f. *Damalops*	750	313	98.91	NW/SE
1080	c.f. *Damalops* humerus shaft	755	307	98.90	NW/SE
1081	c.f. *Damalops* humerus, proximal radius-ulna, + distal tibia, atlas, axis + 2 ribs	800	301	98.84	NW/SE
1082	c.f. *Damalops* proximal ulna	783	317	98.87	N/S
1083	c.f. *Damalops* mandible	780	312	98.86	NE/SW
1084	Bovid (c.f. *Damalops*) axis	761	316	98.92	
1085	Vertebra fragment	800	307	98.86	
1086	c.f. *Damalops* mandible fragment	795	310	98.86	
1087	c.f. *Damalops* mandible fragment	799	319	98.86	
1088	c.f. *Damalops* proximal radius	793	317	98.85	
1089	Rib fragment (MM)	789	290	98.83	N/S
1090	Vertebra centrum (MM)	795	303	98.83	
1091	Bovid (c.f. *Damalop*) atlas fragment	797	315	98.36	
1092	Fragment	777	304	98.87	
1093	Vertebra fragment	791	305	98.84	E/W
1094	Vertebra centrum (MM)	788	306	98.84	E/W
1095	Vertebra centrum (MM)	785	307	98.84	E/W
1096	*Rhinoceros* mandible fragment	215	325	99.31	
1097	c.f. *Damalops* mandible fragment	796	305	98.84	N/S
1098	Vertebra centrum (MM)	743	305	98.94	
1099	Vertebra centrum (MM)	737	312	98.94	
1100	*Damalops* mandible	396	316	99.54	

1101	Vertebra fragment	245	564	99.71	
1102	Bovid horncore (large)	255	562	99.81	N/S; horizontal
1103	Bovid horncore fragment	254	557	99.84	
1104	Fragment	249	550	99.84	N/S; horizontal
1105	c.f. *Damalops* proximal radius	285	553	99.82	N/S; horizontal
1106	*Damalops* horncore	295	556	99.82	N/S; horizontal
1107	Long bone fragment)	287	548	99.82	
1108	Long bone fragment (LM	290	547	99.82	
1109	*Damalops* mandible	305	545	99.79	N/S; horizontal
1110	*Equus* maxilla	295	512	99.76	N/S; horizontal
1111	*Damalops* horncore	305	530	99.78	E/W; horizontal
1112	Fragment	236	487	99.90	
1113	*Rhinoceros* tibia shaft	815	322	98.80	NW/SE
1114	Rib fragment (SM)	818	320	98.79	NW/SE
1115	c.f. *Damalops* radius-ulna fragment	824	317	98.82	NW/SE
1116	*Damalops* skull & horncore	775	285	98.86	N/S
1117	*Damalops* maxilla	773	302	98.86	N/S
1118	Proximal rib fragment (MM)	762	297	98.24	NE/SW
1119	*Damalops* mandible	295	525	99.78	NW/SE; horizontal
1120	Fragment	304	532	99.79	
1121	*Damalops* horncore	295	530	99.81	E/W; horizontal
1122	Fragment	755	325	98.87	NW/SE
1123	*Damalops* horncore	307	536	99.82	E/W; horizontal
1124	*Damalops* horncore	314	540	99.82	E/W; horizontal
1125	*Damalops* horncore	320	530	99.80	N/S; horizontal
1126	Rib fragment (MM)	346	521	99.73	NW/SE; horizontal
1127	Bovid tooth fragment	353	518	99.72	
1128	Rib fragment (MM)	355	507	99.71	N/S; horizontal
1129	Fragment	352	502	99.70	
1130	*Pachycrocuta* 1st phalange	356	498	99.70	E/W; horizontal
1131	*Damalops* horncore	355	595	99.74	E/W; horizontal
1132	Bovid atlas, c.f. *Damalops*	822	320	98.80	
1133	Rib fragment (LM)	818	315	98.76	NW/SE
1134	c.f. *Damalops* distal metapodial	800	316	98.81	E/W
1135	c.f. *Damalops* metatarsal	827	317	98.82	NNE/SSW
1136	*Damalops* mandible	884	320	98.83	E/W
1137	Fragment	230	490	99.61	
1138	c.f. *Damalops* distal metapodial	270	437	99.67	
1139	Fragment	273	435	99.66	
1140	Long bone fragment (LM)	256	553	99.84	
1141	Bovid mandible + molar fragment	289	436	99.62	
1142	Tooth fragment	335	365	99.51	
1143	c.f. *Damalops* 2nd phalange	786	312	98.84	
1144	Fragments	774	315	98.85	
1145	c.f. *Damalops* horncore	271	256	99.54	NW/SE
1146	c.f. *Damalops* horncores (pair)	345	365	99.53	E/W
1147	Bovid horncore	445	294	99.40	E/W; horizontal
1148	Long bone fragments (MM)	345	368	99.51	
1149	Rib and long bone	400	288	99.50	E/W; horizontal
1150	Fragments	412	295	99.50	
1151	Bovid horncore tip	287	272	99.48	N/S; horizontal
1152-1161*		831	312	98.72	
1152-1154	3 articulated cervical vertebrae (MM)				
1155	*Damalops* mandible				
1156	c.f. *Damalops* metapodial				
1157	c.f. *Damalops* humerus				
1158-1161	Fragments				
1162	Bovid (large) carpal	430	295	99.36	
1163	c.f. *Hemibos* humerus	410	375	99.57	E/W; horizontal
1164	c.f. *Hemibos* horncore	246	560	99.87	E/W; horizontal
1165	c.f. *Hemibos* distal humerus	287	338	99.85	NW/SE; dip 30°
1166	Fragment	301	503	99.78	
1167	Long bone fragment (MM)	305	607	99.78	E/W
1168	c.f. *Damalops* metapodial fragment	307	607	99.78	E/W
1169	Bovid (medium) horncore	262	538	99.80	

1170	Bovid humerus & femur shafts	976	303	98.84	
1171	Bovid atlas	763	312	98.88	NW/SE
1172	Bovid 1st cervical vertebra	762	317	98.88	NW/SE
1173	Bovid 2nd cervical vertebra	758	323	98.89	NW/SE
1174	Bovid 3rd cervical vertebra	754	328	98.90	NW/SE
1175	Bovid 4th cervical vertebra	753	335	98.91	NW/SE
1176	Bovid 5th cervical vertebra	752	340	98.91	NW/SE
1177	Rib fragment (MM)	748	317	98.91	NW/SE
1178	Rib fragment (MM)	747	322	98.93	NW/SE
1179	Rib fragment (MM)	745	325	98.91	NW/SE
1180	Rib fragment (MM)	745	330	98.89	NW/SE
1181	*Hemibos* horncore & skull	772	310	98.86	NE/SW
1182	Rib fragments (SM)	425	298	99.28	
1183	Rib fragments (SM)	440	299	99.28	N/S
1184	c.f. *Damalops* humerus shaft	764	327	98.91	N/S
1185	c.f. *Damalops* distal radius	773	320	98.95	E/W
1186	Rib frament (MM)	764	333	98.91	NE/SW
1187	c.f. *Damalops* carpal	783	313	98.90	
1188	Fragments	232	488	99.69	
1189	c.f. *Hemibos* astragalus	227	429	99.64	NW/SE; horizontal
1190	Fragment	250	464	99.60	
1191	Bovid tooth fragment	263	472	99.60	
1192	Fragment	265	453	99.56	
1193	Fragment	276	472	99.57	
1194	Fragment	277	482	99.57	
1195	Fragment	289	446	99.58	
1196	*Rhinoceros* tooth fragments	375	240	99.38	
1197	Long bone fragment (SM)	757	312	98.81	
1198	c.f. *Damalops* lower incisor	749	312	98.82	
1199	Vertebra fragment (MM)	790	295	98.81	NW/SE
1200	Rib fragment (MM)	772	293	98.83	N/S
1201	Fragment	770	332	98.82	
1202	Fragment	771	333	98.82	
1203	Rib fragment (SM)	737	325	99.91	
1204	c.f. *Damalops* tooth fragment	737	327	99.91	
1205	c.f. *Damalops* metacarpal	770	335	98.84	NE/SW
1206	*Damalops* mandible	380	298	99.43	NE/SW; horizontal
1207	c.f. *Hemibos* scapula	372	375	99.46	NE/SW; horizontal
1208	Vertebra fragment	385	389	99.46	
1209	Mandible	372	375	99.46	
1210	Scapula fragment (part of 1207)	372	375	99.46	
1211	Mandible fragment (part of 1209)	372	377	99.47	
1212	*Damalops* mandible	380	375	99.39	
1213	Rib fragment	774	327	98.88	
1214	c.f. *Damalops* carpal/tarsal	788	312	98.82	NW/SE
1215	*Equus* mandible	367	307	99.44	NW/SE; horizontal
1216	c.f. *Damalops* mandible	340	390	99.53	
1217	*Damalops* mandible	385	387	99.48	N/S; horizontal
1218	*Damalops* mandible	387	380	99.48	N/S; horizontal
1219	Long bone fragment (MM)	362	398	99.51	
1220	Bovid horncore (tip)	302	388	99.48	
1221	Fragments	422	299	99.30	
1222	Fragments	231	413	99.50	
1223	c.f. *Damalops* distal metapodial	265	420	99.55	E/W; horizontal
1224	Number cancelled				
1225	Proximal rib (MM)	782	321	98.86	E/W
1226	Bovid (med.) sesamoid	782	341	98.93	
1227	Fragment	782	342	98.83	
1228	Rib fragment (SM)	379	416	99.81	N/S
1229	Long bone fragment (MM)	435	388	99.59	
1230	c.f. *Damalops* distal femur	500	492	99.58	N/S; horizontal
1231	*Damalops* mandible	286	238	99.56	E/W; horizontal
1232-1237*	c.f. *Damalops* mandible	282	238	99.51	E/W; horizontal
1238	c.f. *Damalops* prox. radio-ulna	800	333	98.80	E/W
1239	c.f. *Damalops* distal humerus	800	340	98.80	E/W

1240	?Carnivore (small) femur shaft	790	325	98.81	N/S
1241	c.f. *Damalops* lower incisor	793	329	98.80	
1242-1246*	c.f. *Damalops* tibia + fragments	265	229	99.57	NE/SW; horizontal
1247	Metapodial fragment	270	241	99.47	
1248	c.f. *Damalops* 3rd phalange	272	239	99.42	
1249	Cervical vertebra (MM)	276	255	99.50	
1250	c.f. *Damalops* lower premolar	370	248	99.14	
1251	*Damalops* mandible	274	234	99.37	
1252	Proximal rib (MM)	350	380	99.47	
1253	c.f. *Damalops* distal metapodial	305	380	99.48	
1254	c.f. *Hemibos* navicular-cuboid	258	238	99.44	Horizontal
1255	c.f. *Damalops* metatarsal	270	242	99.47	NE/SW; horizontal
1256	Rib fragment (SM)	330	278	99.47	
1257	c.f. *Damalops* distal scapula	345	246	99.44	
1258	c.f. *Damalops* upper molar	345	240	99.43	
1259	Fragment	350	240	99.65	
1260	Mandible fragment	396	468	99.65	NE/SW
1261	Scapula fragment (MM)	396	479	99.69	E/W
1262	Long bone fragment (SM)	382	474	99.66	E/W
1263	Fragment (assoc. with 1255)	270	242	99.47	
1264	*Pachycrocuta* canine	442	388	99.56	
1265	c.f. *Damalops* upper premolar	447	379	99.53	
1266	*Rhinoceros* tooth fragment	355	265	99.39	
1267	Fragment	337	272	99.39	
1268-1270	Rib fragments (MM)	270	238	99.37	Horizontal
1271	c.f. *Damalops* metacarpal	772	322	98.81	NE/SW
1272	Long bone fragment	781	323	98.82	N/S
1273	Cervical vertebra	786	322	98.78	NNE/SSW
1274	Cervical vertebra	787	327	98.79	NNE/SSW
1275	Cervical vertebra	788	333	98.80	NNE/SSW
1276	Cervical vertebra	791	339	98.81	NNE/SSW
1277	Vertebra	800	327	98.84	
1278	Vertebra	798	337	98.82	
1279	Metapodial	800	320	98.78	E/W
1280	Thoracic vertebra	792	316	98.79	
1281	Fragment	797	319	98.79	
1282	Vertebra	794	306	98.79	
1283	Fragment	790	314	98.80	
1284	*Equus* distal scapula	350	390	99.49	
1285	Rib	397	345	99.43	N/S; horizontal
1286	Tooth	387	345	99.42	
1287**	*Pachycrocuta* mandible	390	305	99.46	N/S; horizontal
1288	c.f. *Damalops* mandible fragment (ramus)	380	315	99.42	
1289	Teeth	355	260	99.45	
1290	Tooth	368	280	99.49	
1291	Vertebra	795	331	98.80	
1292	Vertebra	797	326	98.82	N/S
1293	Vertebra	798	330	98.78	N/S
1294	Long bone fragment	800	336	98.79	NW/SE
1295	c.f. *Damalops* mandible fragment				
1296	Rib	385	287	99.74	
1297	Horncore	432	349	99.53	
1298	3rd molar	452	356	99.50	
1299	Tooth fragment	453	356	99.50	
1300	Metapodial	448	374	99.52	NE/SW; horizontal
1301	Tooth fragment	345	285	99.49	
1302	Vertebra	796	335	98.82	
1303	Rib	279	241	99.43	
1304	Fragment	375	323	99.45	
1305	Fragment	460	360	99.50	Horizontal
1306	Tooth	290	260	99.48	
1307	Vertebra	800	313	98.81	
1308-1310*	2 vertebrae + ribs	800	306	98.83	N/S
1311	Vertebra	800	300	98.80	
1312	Rib	800	295	98.80	NW/SE

1313	Vertebra	812	307	98.88	
1314	*c.f. Damalops* mandible	803	323	98.78	
1315	Fragment	789	306	98.81	
1316	Fragment	793	350	98.84	
1317	Rib	387	476	99.64	
1318-1323*	Vertebrae	810	305	98.76	
1324	*Equus* upper molar	281	295	99.46	
1325	Fragments	340	270	99.44	
1326	Rib	365	280	99.40	
1327	Fragment	366	280	99.57	
1328	Sesamoid	795	331	98.80	
1329	Bovid horncore	439	385	99.55	N/S; horizontal
1330	Long bone fragment	279	302	99.43	N/S; horizontal
1331	*Hemibos* humerus (attached to 1472, 1473)	275	555	99.89	E/W; horizontal
1332-1335*	Teeth	255	306	99.43	N/S
1336	Rib	226	212	99.41	
1337-1347*	*Damalops* mandible; contents unclear	235	225	99.49	NE/SW
1348	Long bone fragments	326	371	99.39	
1349	Fragments	325	345	99.38	
1350	Rib	369	403	99.55	NW/SE
1351	*Damalops* mandible	355	435	99.56	E/W; horizontal
1352	Rib	394	447	99.57	E/W
1353	Long bone fragment	391	444	99.55	E/W
1354	Atlas	409	384	99.50	
1355	Tooth fragment	411	386	99.51	
1356	Long bone fragment	258	557	99.84	
1357	Long bone fragment	252	557	99.84	
1358	Long bone fragment	259	553	99.84	
1359	Long bone fragment	258	548	99.83	
1360	Scapula fragment	257	543	99.80	N/S; horizontal
1361	Fragment	265	532	99.81	
1362	Bovid horncore	266	526	99.93	N/S; horizontal
1363	Bovid horncore	268	524	99.84	NE/SW; horizontal
1364	Fragment	270	517	99.81	
1365	Long bone fragment	274	519	99.79	N/S
1366	Vertebra	274	525	99.82	
1367	Fragment	235	564	99.82	
1368	Fragment	235	558	99.80	
1369	Pelvis fragment	236	552	99.78	N/S; horizontal
1370	Fragment	233	550	99.79	
1371	Long bone fragment	227	552	99.80	E/W; horizontal
1372	Fragment	289	510	99.78	
1373	Long bone fragment	322	383	99.44	E/W; horizontal
1374	Rib	340	385	99.45	
1375	Rib	377	375	99.33	
1376	*Rhinoceros* tooth	249	284	99.43	
1377	Bovid horncore	253	276	99.42	NE/SW; horizontal
1378	Long bone fragment	375	382	99.32	E/W
1379	Long bone fragment	244	556	99.77	N/S; horizontal
1380	Vertebra fragments	258	547	99.78	Horizontal
1381	Fragments	268	546	99.78	
1382	Fragment	296	542	99.78	
1383	Fragments	296	505	99.78	
1384	Fragment	274	509	99.74	
1385	Fragment	276	510	99.74	
1386	Fragment	280	512	99.73	
1387	Fragment	286	509	99.72	
1388	Fragment	277	520	99.77	
1389	*Damalops* horncore	265	287	99.45	NW/SE; horizontal
1390**	*Pachycrocuta* maxilla	272	285	99.42	NW/SE; horizontal
1391-1393*	Part of horncore	265	287	99.45	
1394	Long bone fragment	280	280	99.47	E/W; horizontal
1395	Teeth	287	292	99.43	
1396	Tibia	262	260	99.45	NW/SW; horizontal
1397-1400	Maxilla and teeth	267	275	99.53	E/W; horizontal

1401	Fragment	310	406	99.49	
1402	*Damalops* mandible	350	443	99.55	NE/SW; horizontal
1403	Rib	370	470	99.52	E/W
1404	*Damalops* mandible	385	479	99.48	NE/SW; horizontal
1405	Rib	350	450	99.50	N/S
1406	Rib	351	448	99.51	NE/SW
1407	Radius fragment	205	579	99.91	NE/SW; horizontal
1408	Long bone fragment	257	581	99.90	NE/SW
1409	Fragment	263	562	99.79	NE/SW
1410	*Equus* scapula fragment	254	535	99.72	NW/SE; horizontal
1411	Fragment	225	565	99.73	N/S
1412	Long bone fragment	355	345	99.68	E/W
1413	c.f. *Hemibos* distal humerus	242	543	99.63	
1414	Fragment	238	545	99.61	
1415	*Damalops* mandinble fragment	346	378	99.33	N/S; horizontal
1416	*Damalops* mandible fragment	280	270	99.44	NNW/SSE; horizontal
1417	*Damalops palaeindicus* complete skull				
1418	*Damalops* mandible	444	360	99.65	N/S; horizontal
1419	Scapula fragment	335	370	99.34	E/W; horizontal
1420	Rib	315	368	99.33	E/W; horizontal
1421	Rib	322	380	99.31	NE/SW; horizontal
1422	Rib	332	380	99.32	NW/SE; horizontal
1423	Rib	331	379	99.35	N/S; horizontal
1424	Long bone fragment	315	373	99.33	E/W; horizontal
1425	Rib	345	385	99.31	E/W; horizontal
1426	Rib	345	382	99.32	NE/SW; horizontal
1427	Rib	340	363	99.32	E/W; horizontal
1428	Rib	334	360	99.29	N/S; horizontal
1429	Scapula fragment	355	353	99.32	E/W; horizontal
1430	Maxilla	338	348	99.34	N/S; horizontal
1431	c.f. *Damalops* mandible	300	525	99.84	NE/SW; horizontal
1432	Fragment	205	561	99.68	E/W
1433	Fragment	209	562	99.67	
1434	Fragment	340	370	99.32	E/W; horizontal
1435	Rib	353	540	99.74	
1436	Bovid horncore	344	588	99.76	
1437	Rib	292	502	99.75	E/W; horizontal
1438	Fragment	285	516	99.70	
1439-1441	Fragments	235	523	99.60	
1442	Tibia	282	253	99.48	NW/SE; horizontal
1443	Bovid horncore	292	282	99.39	N/S; horizontal
1444	Bovid, pair of horns	285	283	99.42	N/S; horizontal
1445	Fragment	285	553	99.85	E/W
1446	Bovid horncore	275	536	99.76	NW/SE; horizontal
1447	Vertebra	313	380	99.25	N/S
1448	Bovid horncore	266	555	99.72	N/S
1449	Fragment	271	541	99.70	N/S; horizontal
1450	c.f. *Hemibos* proximal metatarsal	267	533	99.68	E/W; dips west 30°
1451	Fragment	276	288	99.43	E/W; horizontal
1452	Fragment	308	589	99.75	
1453	Rib	310	572	99.72	NE/SW
1454	Skull fragment	308	576	99.73	
1455	Long bone fragment	315	550	99.74	E/W
1456	Canine fragment	316.	571	99.72	E/W; horizontal
1457	Scapula fragment	322	571	99.72	E/W; horizontal
1458	Vertebra	320	559	99.75	
1459	Metatarsal	328	570	99.72	NE/SW; horizontal
1460	Pelvis fragment	327	580	99.73	N/S; horizontal
1461	Humerus fragment	332	570	99.75	NE/SW; horizontal
1462	Rib	336	589	99.74	
1463	Rib	345	577	99.73	NW/SE; horizontal
1464	Rib	347	584	99.71	N/S; horizontal
1465	Rib	348	582	99.70	N/S; horizontal
1467-1471*	Block	280	275	99.47	
1472	Rib (attached to 1331)	292	554	99.87	E/W; horizontal

1473	Bovid horncore (attached to 1331)	292	556	99.87	E/W; horizontal
1474	*Damalops* mandible	360	285	99.20	E/W; horizontal
1475	Vertebra	347	278	99.10	
1476	Bovid teeth	357	283	99.30	
1477	Rib	348	283	99.05	E/W; horizontal
1478	Tibia	355	286	99.05	E/W; horizontal
1479	Humerus	357	284	99.05	E/W; horizontal
1480	Rib	367	282	99.05	
1481	Maxilla	365	292	99.40	E/W; horizontal
1482	Fragment	360	287	99.05	
1483	Tooth fragment	353	292	99.20	
1484	Long bone fragment	233	237	99.37	N/S
1485	Rib	270	250	99.33	E/W; horizontal
1486	*Equus* distal metapodial	417	330	99.52	E/W; horizontal
1487	*Damalops* mandible	436	336	99.51	
1488	*Damalops* mandible	435	337	99.48	NW/SE; horizontal
1489	Rib	437	327	99.50	E/W
1490	Rib	327	592	99.76	NE/SW, dips 15°
1491	*Damalops* mandible	322	576	99.73	NE/SW
1492	Vertebra	323	588	99.75	NE/SW; horizontal
1493	Vertebra	324	589	99.73	Horizontal
1494	Distal femur	333	567	99.77	NW/SE; dips 10°
1495	Distal femur	334	562	99.74	NW/SE; horizontal
1496	Rib	333	569	99.77	NW/SE; dips NW 10°
1497	Metatarsal	336	575	99.74	NW/SE; horizontal
1498	Rib	335	556	99.76	N/S; horizontal
1499	Scapula	327	511	99.72	E/W; horizontal
1500	Long bone fragment	370	540	99.75	NE/SW; horizontal
1501	Rib	324	589	99.75	NE/SW; horizontal
1501	Rib	324	589	99.76	NW/SE; dips 20°
1502	Metapodial fragment	440	344	99.66	NE/SW
1503	Rib	330	365	99.28	NE/SW; horizontal
1504	Rib	335	345	99.26	E/W; horizontal
1505	Rib	343	588	99.74	NW/SE; horizontal
1506	Fragment	339	578	99.72	NW/SE; horizontal
1507	*Damalops* mandible	275	260	99.46	NW/SE; horizontal
1508	Metapodial fragment	292	255	99.53	E/W; horizontal
1509	Rib	290	252	99.51	N/S; horizontal
1510	Bovid horncore	274	543	99.83	NW/SE; horizontal
1511-1515*	?Vertebra	335	383	99.25	
1516	Rib	324	592	99.73	NW/SE; horizontal
1517	Rib	340	355	99.28	N/S; horizontal
1518	Rib	338	355	99.25	NE/SW; dips E. 10°
1519	Rib	317	357	99.22	E/W; dips E. 10°
1520	Bovid tooth	337	567	99.74	NE/SW; horizontal
1521	Rib	315	365	99.25	E/W; horizontal
1522	*Damalops* mandible	292	260	99.41	E/W, dips W. 18°
1523	Long bone fragment	285	267	99.41	NW/SE; dips NW 30°
1524	Metapodial fragment	295	263	99.40	E/W; dips W. 45°
1525	*Damalops* mandible	280	270	99.42	NNW/SSE; horizontal
1526	Fragment	290	554	99.83	
1527	Bovid horncore	285	549	99.85	E/W; horizontal
1528	*Rhinoceros* metapodial	288	540	99.87	NNW/SSE; dips 30°
1529	Bovid metatarsal	280	540	99.84	E/W; horizontal
1530	*Rhinoceros* ulna	295	535	99.85	N/S; horizontal
1531	Metapodial	298	542	99.84	NNW/SSE; horizontal
1532	*Equus* mandible	285	537	99.82	N/S; horizontal
1533	Rib	363	362	99.23	E/W; horizontal
1534	c.f. *Damalops* mandible	368	340	99.24	N/S; horizontal
1535	Scapula fragment	375	355	99.28	E/W; horizontal
1536	Scapula fragment	360	375	99.25	N/S; horizontal
1537	*Equus* tooth	323	355	99.19	N/S; horizontal
1538	Long bone fragment	328	355	99.20	NW/SE; horizontal
1539	Rib	325	590	99.73	E/W; horizontal
1540	Long bone + rib fragments	330	340	99.26	

1541	Rib fragments	362	348	99.28	NE/SW; dips SW 20°
1542	Metatarsal block	290	273	99.42	NE/SW; dips 45°
1543-1548*	No data				
1542	*Rhinoceros* tooth fragment				
1549	Fragment	333	372	99.24	
1550	Long bone fragment	345	364	99.21	E/W; horizontal
1551	Long bone fragment	365	385	99.22	E/W; horizontal
1552	Rib	378	377	99.27	NE/SW; dips SW 10°
1553-1557*	Block				
1554	*Equus* distal metapodial				
1558 + 1559-1561*	1558 = c.f. *Damalops* mandible 1559-61 = Metapodial, teeth etc.	263	265	99.39	
1562	Bovid horncore	285	528	99.81	E/W; horizontal
1563	Bovid horncore	278	542	99.82	N/S; horizontal
1564	*Equus* incisor	284	543	99.80	N/S; horizontal
1565	Fragment	263	255	99.40	Horizontal
1566	Fragment	263	255	99.40	
1567	c.f. *Hemibos* prox. metacarpal	407	310	99.54	E/W; dips E. 10°
1568	Bovid metatarsal	395	305	99.53	NE/SW; horizontal
1569	Long bone fragment	415	310	99.52	E/W; dips E. 10°
1570	?Metatarsal	415	323	99.54	E/W; horizontal
1571	Cervical vertebra	430	329	99.54	
1572	Bovid molar	402	302	99.50	NE/SW; horizontal
1573	Tooth fragment	410	308	99.51	
1574	Bovid horncore	326	547	99.75	NE/SW; horizontal
1575	Bovid tooth	308	504	99.56	E/W; horizontal
1576	*Damalops* mandible	275	535	99.77	E/W; dips 30°
1577	Long bone fragment	405	323	99.58	
1578	Fragment	260	218	99.45	N/S; horizontal
1579	Atlas	402	291	99.51	Bovid
1580	Fragment	283	535	99.72	Horizontal
1581	Long bone fragment	261	536	99.72	E/W; horizontal
1582	*Equus* incisor	280	540	99.71	E/W; horizontal
1583	Long bone fragment	275	534	99.72	E/W; horizontal
1584	Long bone fragment	273	537	99.72	E/W; horizontal
1585	Rib	406	289	99.50	
1586	Long bone fragment	282	539	99.71	E/W; horizontal
1587	Bovid humerus	299	558	99.80	N/S; horizontal
1588	Metapodial	301	564	99.77	N/S; horizontal
1589	*Damalops* mandible	406	312	99.49	
1590	Bovid atlas	270	542	99.68	NW/SE; horizontal
1591	Fragment	295	550	99.66	Horizontal
1592*	c.f. *Hemibos* radius				
1593	Fragment	298	261	99.40	NE/SW; horizontal
1594	*Equus* incisor	305	255	99.34	N/S; horizontal
1595-1596*	No details				
1598	Rib	392	310	99.44	NE/SW; dips SW 15°
1599	Rib	392	310	99.44	NE/SW; dips SW 15°
1600	Rib	392	310	99.44	NE/SW; dips SW 15°
1601	Bovid metatarsal				
1602	*Rhinoceros*, complete skull	330	525	99.62	NW/SE, slight dip
1603	Rib	392	310	99.44	NE/SW; dips SW 15°
1604	Fragment	330	236	99.43	E/W; horizontal
1605	Fragment	343	547	99.57	Horizontal
1606	Vertebra fragment	328	566	99.64	N/S; horizontal
1607	Bovid teeth	304	528	99.50	N/S; horizontal
1608	Rib	385	298	99.44	E/W; horizontal
1609	*Equus*, complete skull	400	300	99.42	Dips SW 10°
1610	Fragment	299	247	99.34	Horizontal
1611	Scapula fragment	342	568	99.59	E/W; horizontal
1612	Long bone fragment	353	577	99.66	NE/SW; dips SW 15°
1613	*Rhinoceros*, prox. humerus unfused epiphysis	361	558	99.71	Horizontal
1614	Distal humerus	284	270	99.39	ENE/WSW; horizontal
1615	Incisor	292	270	99.40	

1616	Fragments	202	564	99.69	
1617	Rib	290	263	99.40	ENE/WSW; horiz.
1618	Rib	275	272	99.45	ENE/WSW; horiz.
1619	Not identified	435	380	99.64	NNW/SSE; dips SSE 10°
1620	Metapodial	297	265	99.38	E/W; horizontal
1621	Rib	297	265	99.38	E/W; horizontal
1622	Rib	297	265	99.38	E/W; horizontal
1623-1626*	Contents unclear	254	255	99.39	Horizontal
1627	Bovid pelvis	382	581	99.70	NE/SW; dips SW 10°
1628	*Rhinoceros* radius	406	565	99.65	NW/SE; dips 10°
1629	*Rhinoceros* metapodial	385	550	99.69	NW/SE; horizontal
1630	Bovid horncore	392	567	99.67	E/W; horizontal
1631	Bovid tooth	351	589	99.67	E/W; horizontal
1632-1636*	*Rhinoceros* mandible (juvenile)	430	410	99.65	NE/SW; horizontal
1637	Part of 1632				NW/SE; dips NW 10°
1638	Rib	426	380	99.63	E/W; dips W. 10°
1639-1644*	Scapula	255	255	99.36	NW/SE; dips 30°
1645-1649*	Long bone fragment	450	345	99.51	NE/SW; horizontal
1650-1652*	Scapula block	250	248	99.39	NW/SE; dips 20°
1653	Fragments	440	330	99.52	
1654	Fragment	249	244	99.38	NW/SE; horizontal
1655	Metatarsal	440	350	99.58	NE/SW;dips SW 5°
1656	? Metapodial	445	350	99.58	Dips SW 5°
1657	Vertebral column	390	600	99.68	NE/SW; horizontal
1658	?Vertebra	405	467	99.27	E/W; horizontal
1659-1664*	*Damalops* mandible	450	405	99.63	NE/SW; horizontal
1665	Tooth	416	457	99.76	N/S; horizontal
1666	Fragment	413	457	99.76	N/S; horizontal
1667	Bovid maxilla	226	220	99.41	NW/SE; dips SE
1668	*Pachycrocuta* canine	440	381	99.64	
1669	Rib	405	458	99.77	E/W; horizontal
1670	Rib	404	455	99.74	E/W; horizontal
1671	Rib	406	464	99.73	E/W; horizontal
1672	Bovid incisor	408	464	99.73	E/W; horizontal
1673	Rib	405	475	99.73	NE/SW; horizontal
1674	Rib	410	473	99.73	N/S; horizontal
1675	Rib	411	474	99.73	N/S; horizontal
1676	? Skull fragment	413	459	99.74	Horizontal
1677	Rib	420	465	99.72	E/W; horizontal
1678	Rib	421	460	99.72	E/W; horizontal
1679	Rib	407	468	99.72	NW/SE; horizontal
1680	Bovid radius	375	513	99.61	NE/SW; dips 15°
1681	Rib	393	559	99.62	NW/SE; horizontal
1682	Rib	390	586	99.57	E/W; dips E. 10°
1683	Scapula fragment	387	556	99.64	N/S; horizontal
1684	*Damalops* mandible	388	530	99.59	NE/SW; dips 15°
1685	Long bone fragment	392	524	99.51	NE/SW; dips NE 15°
1686	Bovid tooth	379	527	99.61	NE/SW; horizontal
1687	Bovid horncore	379	527	99.64	NE/SW; dips NE 10°
1688	Bovid horncore	383	518	99.69	E/W; dips W. 15°
1689	Bovid horncore	333	512	99.64	N/S; horizontal
1690	Bovid horncore	366	516	99.62	NE/SW; horizontal
1691	Bovid (small) maxilla	313	539	99.47	N/S; horizontal
1692-1696*	Metapodial	300	524	99.43	NW/SE; horizontal
1697	Bovid horncore (medium twist)	382	510	99.60	NE/SW horizontal
1698	Rib (SM)	424	460	99.69	NW/SE horizontal
1699	Bovid tooth	434	340	99.56	NW/SE horizontal
1700	Bovid (small) mandible	433	725	99.83	E-W horizontal
1701	Bovid atlas	405	395	99.64	
1702	Bovid (small) upper molar	415	398	99.62	N/S horizontal

1703	Rib (MM)	425	325	99.56	NE/SW horizontal
1704	Bone fragments concreted to 1703	420	325	99.56	
1705	Indeterminate fragment	403	330	99.57	
1706	Rib (MM)	400	442	99.62	E/W horizontal
1707	Bovid horncore	570	540	99.77	E/W, dips 5°
1708	Bovid horncore (medium twist)	589	538	99.72	E/W horizontal
1709	Bovid, pair of horncores + skull fragment, medium twist	563	525	99.75	N/S horizontal
1710	Rib (MM)	405	345	99.56	
1711	Rib (MM)	407	345	99.57	NW/SE horizontal
1712	Rib (MM)	400	346	99.56	N/S horizontal
1713	Rib (MM)	420	343	99.55	E/W horizontal
1714	Indeterminate mandible	423	355	99.62	N/S horizontal
1715	Bovid (small) proximal metapodial	415	435	99.57	NE/SW horizontal
1716	Bovid (small) proximal metatarsal	430	360	99.61	NNW/SSE, horizontal
1717	Bovid (small) proximal metapodial	420	360	99.61	E/W, horizontal
1718	Bovid (small) metapodial	428	356	99.60	NNW/SSE, horizontal
1719	Rib (MM)	422	346	99.61	N/S horizontal
1720	Rib (MM)	419	345	99.61	N/S horizontal
1721	Bovid (small) scapula	415	372	99.63	E/W horizontal
1722	Bovid (small) scapula	418	355	99.62	N/S horizontal
1723	Rib	408	367	99.58	E/W horizontal
1724	Bovid metacarpal, complete	312	524	99.49	NW/SE horizontal
1725	Long bone fragment	325	538	99.54	NE/SW, dips 10°
1726	Rib	326	535	99.56	NE/SW horizontal
1727	Rib	332	570	99.61	NE/SW horizontal
1728	Bovid (small) mandible (possible carnivore damage)	345	525	99.57	NW/SE, dips 5°
1729	Ulna fragment, LM	350	527	99.58	E/W horizontal
1730	Small carnivore (? Hyaenid) femur fragment	365	542	99.69	NE/SW horizontal
1731	Long bone fragment	367	524	99.68	N/S, dips 5°
1732	Bovid horncore, medium twist	367	533	99.65	NW/SE horizontal
1733	*Rhinoceros* tooth fragment	366	530	99.61	NW/SE horizontal
1734-1737*	Block; bovid horncore, ribs etc.	387	520	99.65	Horizontal
1738	Indeterminate fragment	327	530	99.48	NW/SE horizontal
1739	*Bovid (small) mandible*	415	450	99.75	E/W horizontal
1740	Metatarsal	410	377	99.63	WNW/ESE horizontal
1741	Bovid distal metatarsal; section cleaning in Area B	845	345	99.88	
1742	Indeterminate vertebra	845	342	98.86	
1743	Indeterminate atlas	845	338	98.84	
1744-1747	Fragments with 1741-1743				
1748	*Rhinoceros* humerus	378	522	99.49	NE/SW; horizontal
1749	Bovid horncore	556	542	99.77	E/W; horizontal
1750	Bovid horncore	570	542	99.76	E/W; horizontal
1751	Indeterminate fragment	573	542	99.75	Horizontal
1752	Bovid (small) atlas fragment	584	542	99.70	Horizontal
1753	Bovid (small) tibia fragment	580	534	99.73	E/W, dips 15°
1754	Rib	575	575	99.72	E/W; horizontal
1755	Bovid (small) mandible	583	537	99.70	NE/SW; horizontal
1756	*Rhinoceros* tooth fragment	388	530	99.51	NW/SE; horizontal
1757	*Equus* metapodial, complete	580	534	99.71	E/W; horizontal
1758	Indeterminate tibia fragment; part of block 1207	386	385	99.46	E/W; horizontal
1759	Indeterminate rib; attached to 1209; part of block 1207	372	375	99.44	

Notes: Specimen numbers marked by * were removed as plastered blocks. MM = medium mammal; LM = large mammal; SM = small mammal. Lbf = long bone fragment. Most indeterminate fragments are probably bovid, given that this is the commonest taxon in the assemblage. *Pachycrocuta*** = discussed in Alan Turner's report (this volume). The other *Pachycrocuta* specimens are extrapolated from these. Identifications of *Damalops* and *Hemibos* are based primarily on cranial, dental and horn-core specimens. Post-cranial identifications of each are primarily on the basis of size, and indicated as c.f. *Damalops* or *Hemibos*. Specimen 1700 was found eroding to the surface outside the main area of excavation.

Table 9. 15: Measurements (in mm.) of specimens from the excavation of locality 642, 1990

N.	Taxa	Element	Details	Length	ProxAP	ProxW	DistAP	DW
1011	Bovid	Metatarsal	Distal					39.6
1042	c.f. Damalops	1st phalange		59.2		20.0		
1050	Bovid	2nd phalange		33.0				
1065	Bovid	Tibia	Distal					38.3
1065	Bovid	Astragalus		46.0		29.1		
1076	Bovid	Metacarpal	Distal					36.6
1078	Bovid	Tibia		304				37.9
1081	c.f. Damalops	Radius	Proximal			52.3		
1081	cf. Damalops	Tibia	Distal					39.0
1088	Bovid	Radius	Proximal			50.8		
1135	Bovid	Metatarsal		254				
1205	Bovid	Metacarpal	BT 39.5?	261		39.7		
1223	Bovid	Distal	Metapodial					34.5
1253	Bovid	Metapodial	Distal					33.2
1255	Bovid	Metatarsal		258				
1724	Bovid	Metacarpal	Complete	265		40.0		39.8

Notes: ProxAP = proximal anterior-posterior; ProxW = proximal width; DistAP = distal anterior-posterior; DW = distal width. MM = medium-sized mammal; LM = large mammal; SM = small mammal.

1046: Two 1st phalanges of c.f. *Damalops*; larger one – Length = 59.2mm.; proximal breadth = 20.0; smaller one, unfused, proximal
1054: Bovid distal scapula: glenoid length = 52.3; possible carnivore damage.
1056: possible carnivore puncture mark.
1065: group of bones including maxillary premolar (*Pachycrocuta*); two horncores, mandible, four tooth fragment;s:,distal tibia; astragalus (lateral length = 46.0, medial length = 44.7, distal breadth = 29.1) c.f. *Damalops*; calcaneum c.f. *Hemibos triquetricornis*; and four vertebra fragments, long bone fragment (large mammal) and fragments.
1066: group of bones incl. mandible and proximal tibia c.f. *Damalops*; vertebra centrum fragment (MM), two rib fragments (MM) and fragments.
1079: possible carnivore damage to horncore
1081: Articulated distal humerus, proximal radio-ulna, c.f. *Damalops* (prox. radius) and, distal tibia, c.f. *Damalops*; and, atlas, axis and two ribs (MM)
1096: mandible fragment (belongs with 1062), *Rhinoceros sivalensis*
1134: Bovid distal metapodial, possible carnivore damage
1135: Bovid metatarsal; possible carnivore damage; GL = 254.0
1155: Bovid mandible; possible carnivore puncture mark
1171: atlas, articulated with vertebrae 1172-1176*
1205: Bovid metacarpal GL = 261.0, Bp = 39.7, BT = 39.5mm.
1217: Bovid mandible. Tooth row GL = 98.1mm.
1223: Bovid distal metapodial; BT (est) = 34.5mm.
1238: Bovid proximal radio-ulna; possible carnivore damage; articulated with 1239*
1239: Bovid distal humerus; articulated with 1238*
1271: Bovid metacarpal; BT = 36.9, BT = 37.6mm.
1273: cervical vertebrae articulated with 1274-1276
1472: rib (attached to proximal humerus no.1331)
1473: horncore (attached to proximal humerus no.1331)
1525: mandible under 1416
1540: long bone and rib fragments; disturbed by goat.
1616: fragments from section cleaning

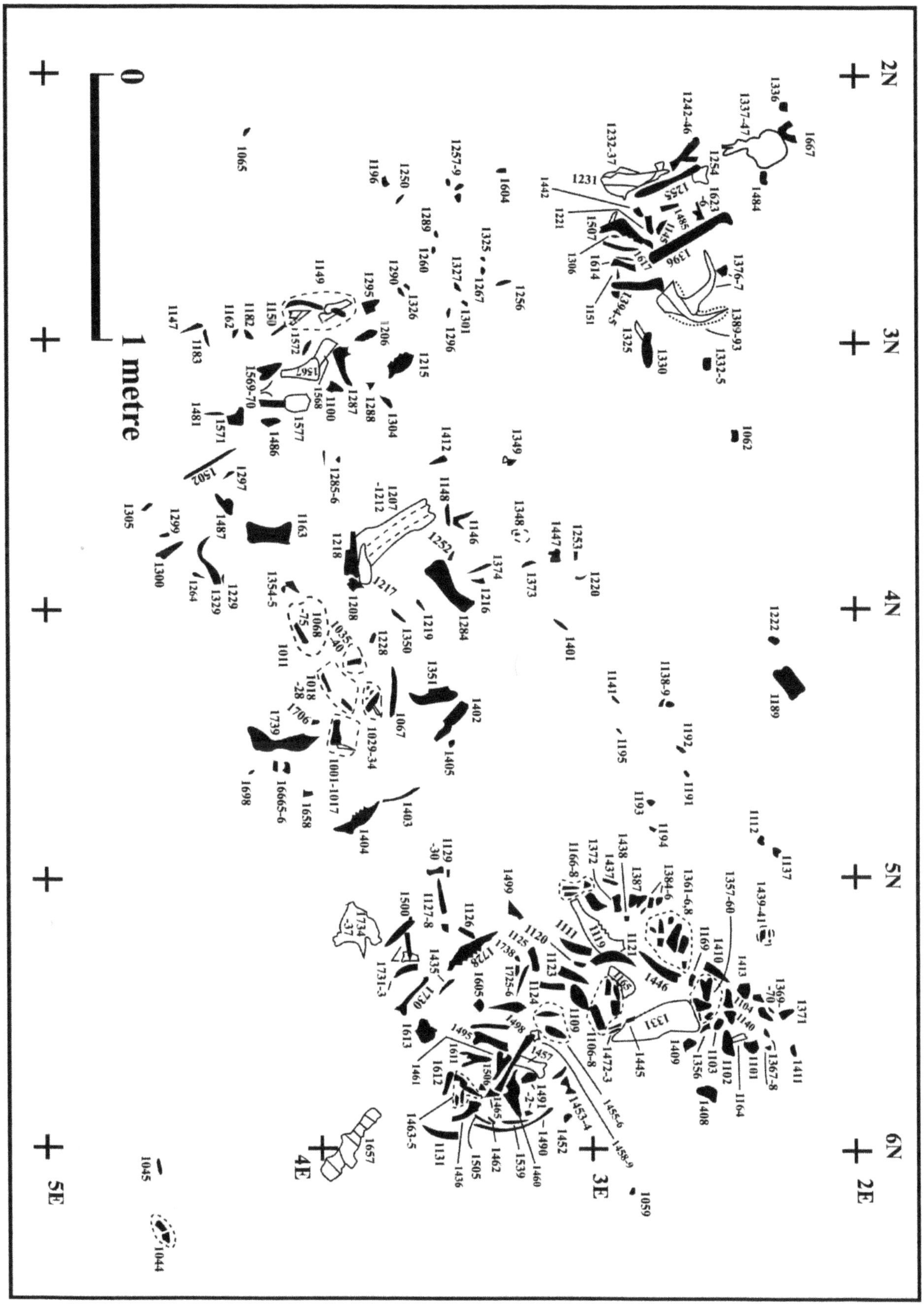

Figure 9. 13: Plan of excavation of locality 642, area A, level 1 (redrawn from original by Mark Beech and team).

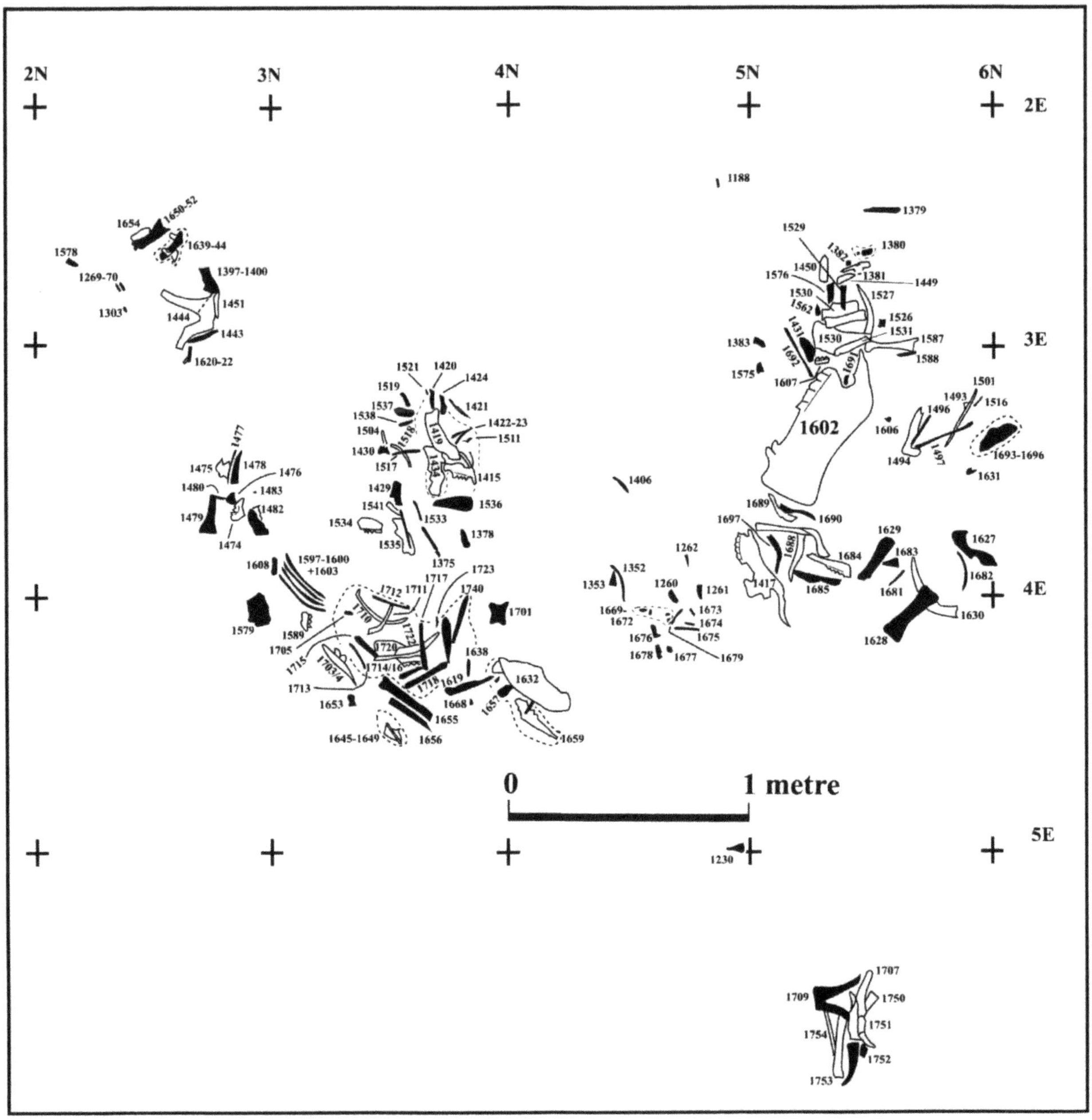

Figure 9. 14: Plan of excavation of locality 642, area A, level 2 (redrawn from original by Mark Beech and team).

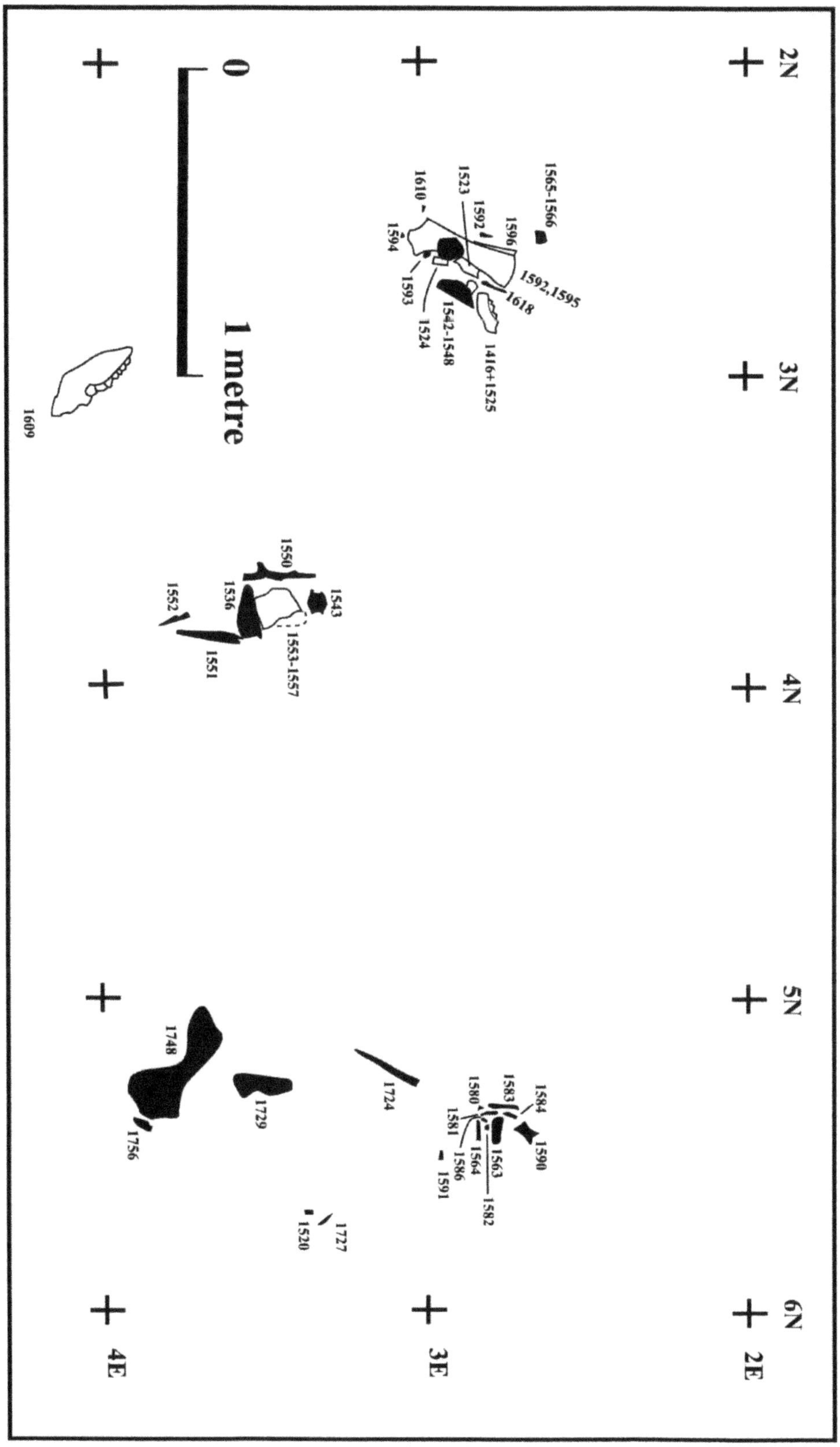

Figure 9. 15: Plan of excavation of locality 642, area A, level 3 (redrawn from original by Mark Beech and team).

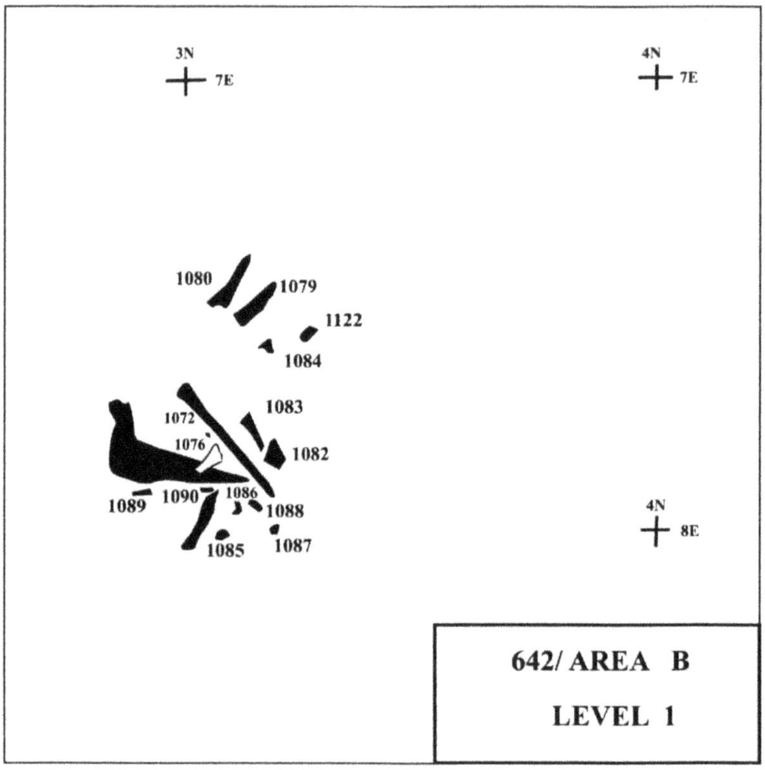

Figure 9. 16: Plan of excavation of locality 642, area B, level 1 (redrawn from original by Mark Beech and team).

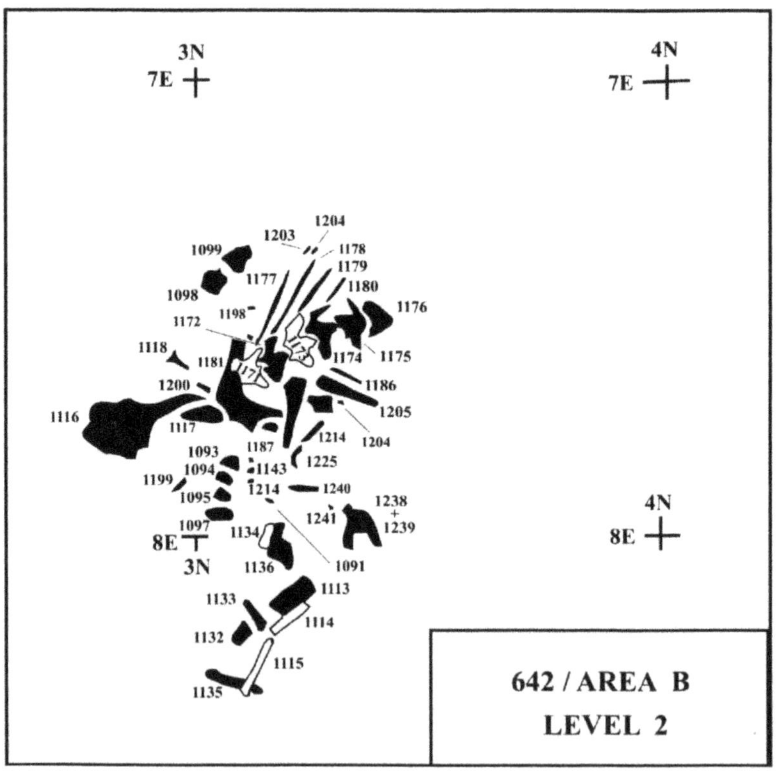

Figure 9. 17: Plan of excavation of locality 642, area B, level 2 (redrawn from original by Mark Beech and team).

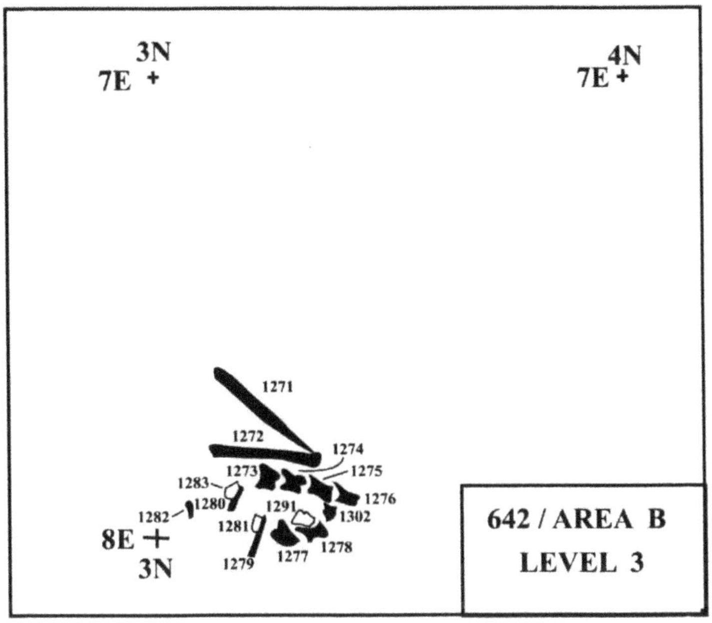

Figure 9. 18: Plan of excavation of locality 642, area B, level 3 (redrawn from original by Mark Beech and team).

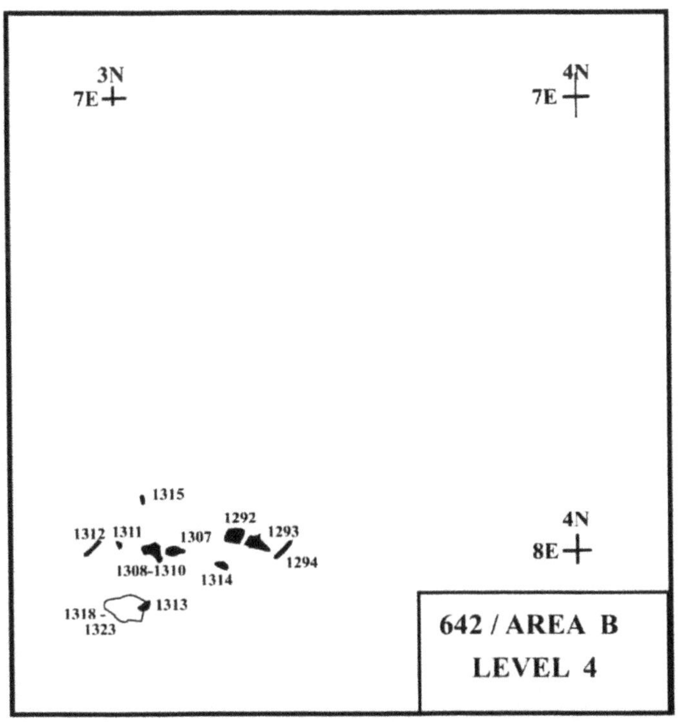

Figure 9. 19: Plan of excavation of locality 642, area B, level 4 (redrawn from original by Mark Beech and team).

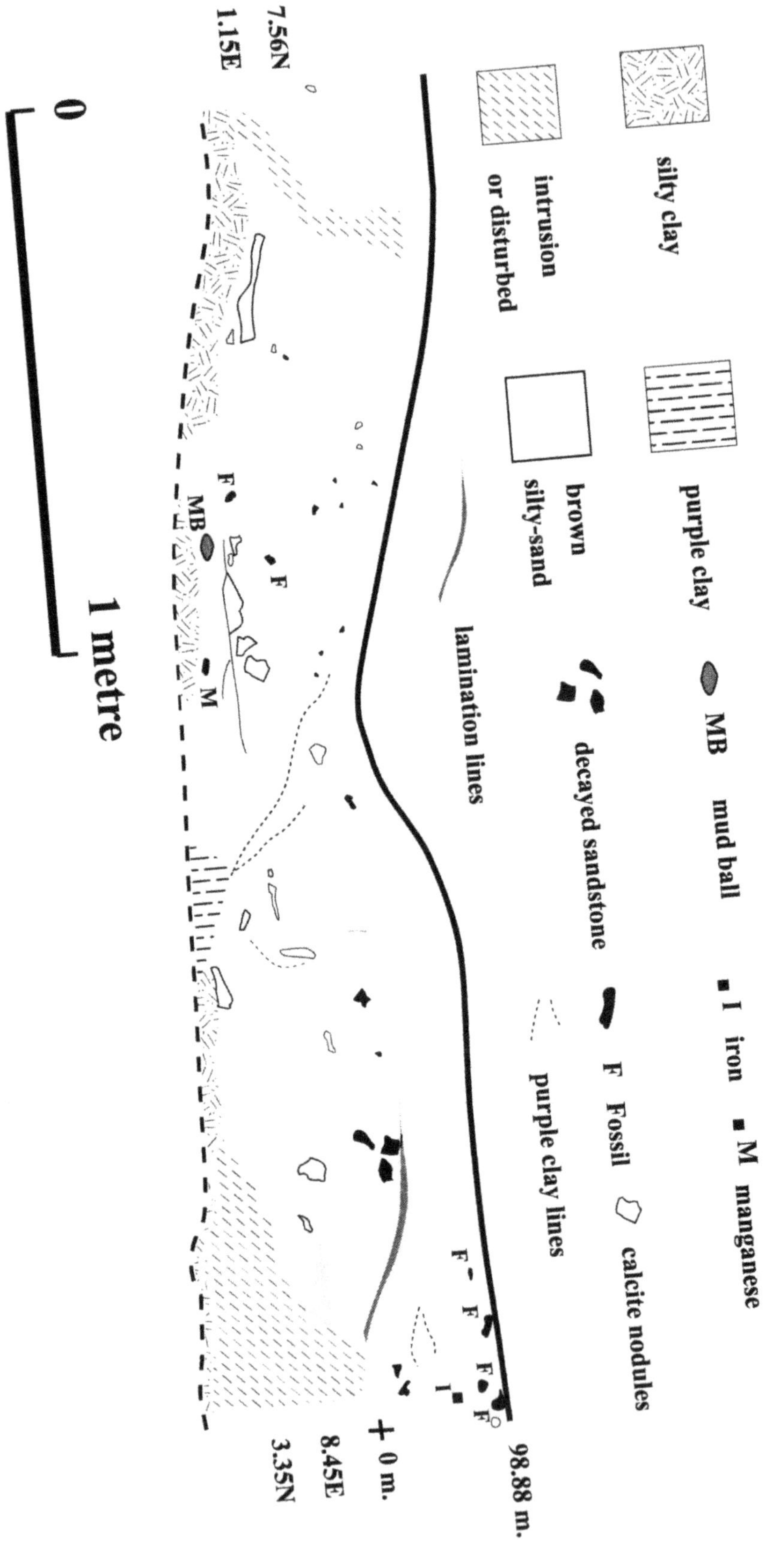

Figure 9. 20: East section of locality 642, Area B (redrawn from an original by Ros Coard)

Taphonomic observations of localities 73, 362 and 642, Pabbi Hills, Pakistan

Ros Coard, Dept. of Archaeology, University of Wales, Lampeter, Ceredigion, Wales, SA48 7ED.

Observations at a distance: an interpretation of locality 73

Locality 73 comprised of an extraordinarily dense concentration of fossil remains, many of which were complete and well preserved. The presence of gnawing marks on the bone, and the large numbers of carnivores present, suggested that it may have been accumulated by carnivores, or at least modified by them. The herbivorous taxa present at locality 73 comprise Bovid, Cervid, Equid, Suid and Rhinocerid: the carnivores present are Hyaenid, Canid, Felid, Ursid, and a Herpestid.

Population profiles

Although not all material has been closely examined to determine age profiles for all taxa, those that have been determined include the cervids, equids, and suid.

There are at least nine cervids represented, and they indicate a wide age range from juveniles through to senile adults. The known ageing systems of extant cervids are not always easy to apply to extinct forms. Age estimates are based on eruption and wear rates and are largely adapted from Spinage (1973). The material includes four juveniles still with deciduous dentition, four adults with permanent dentition (M_3 erupted) and a single senile adult with extremely worn adult dentition.

At least three equids are present. Based on the fusion of the surviving epiphyses they are all at least 2-2.5 years of age and one 3.5 years of age (1969). The dentition of two of the mandibles shows the M_2 erupting and as they are from a left and a right mandible, they are probably from the same individual. This would be consistent with the estimated age range from the bone fusion of around 2-2.5 years of age. Two of the equids are taller and more slender than the third, which is shorter and more robust in the measurements of the long bones, podials and metapodials. All measurements were compared to Walker (1985) and indicate that these equids are larger than the extant zebra of Africa.

Two very different sizes of Bovid are represented by some cranial and post-cranial material, mainly limb bones. One of the bovids is very large with all measurable indices in excess of the largest extant African bovids of buffalo and giant eland (Walker 1985). The second form is smaller and more slender, with few measurable indices but falling within a medium size range.

A single suid is present and is well represented by dentition and post-cranial bones. There are two virtually complete mandibles, a left and a right, both with the M_3 erupting. The eruption rate would indicate an individual around 2 years of age (quoted in Hillson 1986). A virtually complete leg, from the distal tibia to the tarsals, is present, as are isolated bones from both front limbs. All of the bones showing fusion would support an age estimate of around 2 years (Silver 1969).

The rhinoceros is represented by post-cranial material only. It includes a minimum of two limbs, at least one front and one rear limb, both from the left side of the individual, and possibly from the same individual. In appearance it shows stronger affinities to the present day black rather than the white rhino of Africa.

A single hyaenid is represented by cranial remains and dentition. A smaller carnivore, a canid, is represented by dentition and post-cranial remains, including limb bones and metatarsals, tarsals (from a left and right rear limb) and phalanges. The permanent dentition in extant canids emerges over a short period of time and usually within the first year (see Hillson 1986). As the 3rd molar is recorded as emerging this would indicate an individual who is not yet fully adult and in their first year. There is no duplication of any bone and as the right and left tarsals are consistent in size it would suggest that a single individual is present.

Skeletal part representation

There is a high degree of articulation present with at least nine groups of bone either articulated or anatomically adjacent to each other. Of the nine groups only one consists of vertebrae (three lumbar vertebrae and sacrum), and all of the rest are lower limb bone combinations. These vary from the presence of the distal tibia or radius/ulna down to the podials or metapodials, or just combinations of podials with metapodials. This pattern of predominantly lower limb articulation is more suggestive of carnivore predation and is consistent with reported and observed carnivore denning activity (Haynes 1988, Hill 1980a). The articulations are associated predominantly with the equids, with 5 of the 9 articulated groups coming from this taxon. The remaining groups are associated with either the suid or bovid, or are indeterminate, such as the articulated vertebrae.

Weathering stages

Even though many of the bones were either in sediment blocks or encrusted with sediment, it was still possible to assign weathering stages to some of them. It had been noted in the field that the bones were in a state of good preservation and initially the bone surface texture certainly supported these observations. Behrensmeyer (1978) related the degree of deterioration of the bone surface texture to the length of time a bone had been exposed and assigned stages to periods of exposure. Of the bones that could be assigned a weathering stage under Behrensmeyer's scheme, 23 recorded a stage 1, the majority (N- 208) recorded a stage 2. These stages would indicate a reasonably rapid burial. Bones indicating longer periods of exposure recorded stage 3 (N=39) and only 8 bones recorded the higher stages of 4 or 5. (This assessment is in keeping with Dennell's Table 9.2, which was compiled independently). These stages would indicate a much longer period of exposure, possibly tens of years. The only pattern that could be observed across all the bones is that most of the stage 1 bones tend to be the small blocky bones, such as podials or phalanges. It is known that different skeletal elements do weather at different rates (Behrensmeyer 1978, Lyman and Fox 1989) and this may be what is observed with the small robust bones. The weathering stages did not appear to be more consistent with any individual taxa.

Bone breakage and damage

Carnivore damage is very evident, with some typical damage being gnawed epiphyses of long bones and the gnawing often extends into the shaft of the bone. This type of damage is present to a large extent and is very typical of carnivore behaviour (Cruz-Uribe 1991, Haynes 1988, Hill 1980b). Spiral fractures also accompany many of these carnivore damaged bones, particularly on the long bones and flake fragments. Equally evident is the damage left on the bone surface caused by carnivore teeth ,which includes evidence of chewing, pitting, striations, puncture marks and even whole tooth row outlines. This type of damage is typically on mandibles but can also be seen on long bones and long bone fragments. The damage is not restricted to any particular taxa but is observed particularly on the suid and cervid remains.

From the carnivore damage present it is evident that at least two sizes of carnivore were active at the site. The first includes a large bone-crunching individual which had the ability to inflict a considerable amount of damage to the bones of different sized animals and different skeletal parts. This damage could be attributed to the Hyaenid who is present at the site. The second is a much smaller carnivore inflicting the less destructive damage in the form of striations and small pitting marks. The bones of both sized carnivores are recorded at the site but it may also possible that some of the observed damage is caused by juveniles of the larger bone crunching form.

Mode of accumulation

The skeletal part representation can be used as an indicator of the mode of accumulation with the suggestion that carnivores tend to remove selected bones from a large carcass, most often limb bones or scavenged cranial bones (Cruz-Uribe 1991). There is some evidence that for the larger animals post -cranial bones do predominate. This is supported by the remains of the rhinoceros where only the bones from the appendicular skeleton have survived and include a scapula, a fragment of humerus, distal femur, an astragalus and two phalanges, for example. Most of the bones are limb bones, and particularly so for the large bovid. The cervids, who are slightly smaller in size range, are better represented by mandibular and cranial remains. The single suid is also represented by dentition as well as post cranial limb bones. Certainly the predominance of articulated limb bones would support the suggestion that elements were selectively removed from carcasses and brought to the site.

The differences in the survivorship of certain anatomical elements can also used as an indicator of the mode of accumulation, and can be seen in a cursory analysis of the metapodials of the typical prey species. Forty seven metapodia of prey species were recorded from the 1986 excavations. Of these 15 entries were not specific about which part had survived. Of the 32 metapodia that were listed as either having the distal, proximal or shaft surviving, the shafts were better represented (N=15) than the proximal ends (N=9) or the distal ends (N=8). This is consistent with the bone destruction pattern and skeletal part representation observed by Marean and Spencer (1991). They observed that the distal metapodia were the least likely to survive and shaft pieces the most likely, after intensive feeding by hyaenas.

There are many features of both the skeletal remains and their context that would suggest that locality 73 is primarily a product of carnivore activity. The degree of articulation and pattern of the articulated elements, the bone breakage and bone damage patterns along with the skeletal part representation, are all consistent with the fossil assemblage being accumulated by large bone crunching carnivores. The site lists do not always give sufficient detail on the survivorship of certain anatomical parts, whether they were bone shafts or ends for example to undertake a higher degree of analysis. Certainly this level of analysis would need to be undertaken before it would be possible to say that the assemblage was a product of scavenging or hunting behaviour. It is entirely possible that both forms of behaviour were occurring. The rhinoceros remains may be a product of scavenging, hence the fewer skeletal elements from the limbs bones only, compared to the suid, which is much better represented with a range of bones from across the skeleton and some still in articulation. If the suid was scavenged, the carnivore was certainly early in the access sequence to the carcass as the pattern would be similar had the animal been hunted. The compact nature of the deposit, the dense concentration of bones and their spatial patterning would also support the initial suggestion that Locality 73 does represent a carnivore accumulation and most probably denning activity.

A taphonomic analysis of fossil locality 362

Locality 362 was the second of three fossil localities excavated by the British Archaeological Mission to Pakistan and the Geological survey of Pakistan, and has its own distinct character. It is the oldest of the three, with an age of ca. 1.7 – 1.9 Mya, and was similar to locality 73 in terms of its size and compactness. As seen below, it probably shared a similar taphonomic history in that it was primarily a carnivore accumulation.

Sedimentary context

The fossils at locality 362 are located within a discontinuous horizon of orange-brown sandy silts containing calcium carbonate concretions. In general, few depositional structures are preserved in the sedimentary sequence throughout the Pabbi Hills, and this is certainly the case for Locality 362. The original sediment has clearly been subject to extensive post-depositional modification. This is indicated by the absence of any clearly distinguishable internal sedimentary structures, the development of calcium carbonate nodules, and evidence of bioturbation by plants. The sedimentary sequence is characterised by sharp and well-defined bed contacts with overlying and underlying deposits. Although few internal sedimentary structures are evident, there is a fining upward of the sediment within the orange-brown sandy silts, and a more gradational contact with a similar sandy silt deposit. This overlying sandy deposit has markedly fewer calcium carbonate nodules. The Locality 362 sediments are generally hard and well cemented. All of the Locality 362 sedimentary sequence is heavily bioturbated with both modern roots and palaeoroot casts being visible. Bioturbation may be one reason why sedimentary structures do not preserve well in the Pabbi depositional sequence.

The sandy units that make up part of the sedimentary sequence at Locality 362 are evident across a large section of the Pabbi Hills (Jenkinson et al. 1989). The sand units, evident at Locality 362, are spatially variable and vary in thickness from 2-8 m. They are fluvial in origin and are generally characterised by a fining upwards in their deposition. The discontinuity and spatial variability are indicative of the lateral migration of the depositional environment. They most probably represent abandoned channels or tributaries, or minor channels to the main palaeo-river system. Jenkinson et al. (1989) suggests that such deposits could be interpreted as crevasse-splays/levee deposits. It is primarily within these sand units that most of the fossil localities occur in the Pabbi Hills.

Description of locality 362

The excavated area of Locality 362 was small, compact and spatially discrete. The bones were concentrated in a small area, little larger than 1m^2. The fossil deposit was uniformly 30 cm deep. The concentration of the bones appeared to be even more marked as the surrounding sediments were virtually devoid of any fossil material. This made the assigning of the surface collect material that much easier, as there was simply nowhere else the eroded fossils could have come from. For the purposes of this study, 132 bones or bone fragments that had eroded down slope from the site, and were part of the sample collected between 1987 and 1989 were analysed. The majority of these bones, around 91 or 69%, were indeterminate bone fragments or could only be assigned to a class size of mammal. Only 59 (45%), of the bones from the surface collection were identifiable to anatomical part and/or species. Although the period of erosion for these bones was only one or two seasons at most, many of the bones were unidentifiable. Once exposed to the destructive forces of the weather and climate, fossil deteriorates very rapidly in the Pabbi Hills. Only 39 (31%) of the bones collected from the surface could be reasonably identified to genus or species level. This is in marked contrast to the quality and hence the number of identifiable bones recovered from the excavation.

In total 223 complete and fragmentary bones were recovered from the excavated fossil locality. (The site catalogue (Table 9.7) records 179 entries; however, some of these included several specimens, such as X31 (rib fragments), X93 and X158 (canid feet)). Of these 121, or 54%, were identifiable to taxonomic level. A further 21 bones (9%) were identifiable to anatomical part, mostly rib and vertebrae fragments. 102 bones, around 46%, were indeterminate or unidentifiable small bone fragments and, at best, could only be assigned to a class size of mammal. Identifications, unless otherwise stated, are provisional as much of the material has yet to be compared to known taxonomic types. Although this report considers a large proportion of the bones found at the site, much of the analysis concentrates on the fossil specimens that were identifiable to a taxonomic level.

The range and type of species

Around six taxonomic groups have been positively identified, Canidae, Bovidae, a Proboscidean, Rhinoceratidae, Giraffidae and Aves, although within the Bovidae, there are possibly up to three different forms. Given that the area from which the bones were derived was only ca. $1m^2$, this appears to be a much more diverse range than perhaps one would initially expect. The animals represented at locality 362 are predominantly large forms and such a range of large forms is intriguing. This may simply be explained by preferential survival of large robust forms, or, it may be indicative of other factors.
The species present at Locality 362 consist predominantly of large terrestrial mammals. The very largest of the animals with an estimated body weight of over 820 kg include the *Sivatherium*, the rhinoceros and a Proboscidean. Other animals include, a large bovid (provisionally identified as *Hemibos triqueticornis*). The smaller mammals include a medium sized bovid and another type represented by a skull fragment with both horns largely intact. The remaining mammals consist of a single carnivore, a canid (identified as *Canis cautleyi*, see Turner, this volume) and a bird that is estimated to be about the equivalent size of a large bird of prey, possibly eagle-sized. As a generality, a single, individual animal represents most of the taxa. The fauna could be summarised as consisting of predominantly large mammals and, with the exception of the canid, a restricted range of anatomical parts. The assemblage mainly consists of post-cranial remains and specifically limbs bones.

The fauna

Table 9. 18 lists the number of all of the diagnostic fossils that were identifiable to either genus or species level and include those that were collected from the surface or recovered during the excavation. Table 9. 16 shows the total number of taxonomically identified fossils from Locality 362 and their relative abundance as a percentage, for both the excavated and surface collect material. The carnivore remains with 67 bones or 55% of the fauna, dominate the excavated assemblage. The next most numerically important taxon from the excavation is the large bovid with 34 bones, 28% of the fauna The remaining species are represented by between one and eight bones, and constitute less than 10% of the excavated fauna each.

The surface collection is dominated by bovids. Collectively the bovid remains constitute 82% of the fauna. The large bovid is the most abundant species here, with 20 bones or 51% of the fauna. The remaining identified material consists of between one and two bones (3-5% of the fauna) for the *Sivatherium*, rhinoceros and a Proboscidean. Taxonomically the canid and the *Sivatherium* are the most interesting fossils at the site.

The canid material has been extensively examined and clearly represents a single adult male. This individual is represented by mandibular and vertebral material, a virtually complete front-limb, two lower rear-limbs and an os penis bone. The excavation report initially identified this material as belonging to *Crocuta sivalensis*, Mathew, 1929, with, perhaps, *Crocuta sivalensis*, Falconer, 1887 in mind. On re-examination, this material was identified as Canidae and not Hyaenidae. The dental formula and dental morphology shows clear affinities to canids and not to the hyaenids. The post-cranial material is also far too gracile to be classed as Hyaenidae. The mandible has been further examined and is described elsewhere (Turner, this volume) as belonging to a large form of *Canis*.

The Locality 362 canid is of interest because it is unusually large for an Early Pleistocene date (Turner, personal communication). Turner (chapter 10, this volume) suggests that there is some affinity between the Locality 362 canid and *Canis lupus* Linné, the extant true wolf. Although large for this species, Turner also suggests affinities with an extinct canid, *Canis cautleyi* (Bose, 1880). Bose (1880) reported on two *C. cautleyi* specimens and noted their likeness to the living Indian wolf. The stratigraphic locations of these two specimens were not clearly indicated by Bose, other than that they were found in Siwalik deposits, with a possible mid-Pliocene date, although Bose did acknowledge the possibility of them being more recent.

There is no duplication of any anatomical element and the size and fusion of the canid bones suggests that it is consistent with all of the bones coming from the same individual. The preservation of this fossil material is excellent. Much of the material consists of either anatomically adjacent bones or fully articulated remains. The articulation is predominantly in the lower limbs, with the feet varying in degrees of articulation from a whole foot (tarsals, metatarsals and first, second and even third phalanges) to two carpal bones (scaphoid and lunar). Other material is anatomically adjacent, such as the humerus, ulna and radius.

The *Sivatherium* is of taxonomic interest as it is rare type of giant Pleistocene giraffe. The Sivatheriinae, a sub-family of the Giraffidae, contains a variety of fossil forms, all of which are now extinct. The Sivatheriinae are widely, although not abundantly, represented in Pleistocene deposits across Asia and present at some South African cave sites of a similar date (Brain, 1981). All the material at Locality 362 is post-cranial; examples of cranial material would be extremely useful in determining which particular form is present in the Locality 362 deposits. The Sivatheriinae in the Pabbi deposits are most probably *Sivatherium giganteum*, Falconer and Courtley, 1835. Although the cranial material for this species is more extensively described (Lydekker, 1878, Arif and de Vos, 1989), the post-cranial specimens at Locality 362 fit the general description of a large, heavy *Sivatherium* (Colbert, 1935), not elongated in the limb.

Skeletal part representation.

Table 9. 18 shows the number of each skeletal elements recovered and identified for all the fossil material at Locality 362. The skeletal part representation of the animals present at Locality 362 is interesting. With the exception of the canid, the fauna could be categorised as comprising a few individuals, each represented by a restricted range of anatomical parts. There is also a difference in skeletal part representation of the large versus the smaller mammals. Most of the large animals consist of limb bones and in the case of the *Sivatherium* just a single limb plus two metatarsal shafts. The Proboscidean consists largely of bones attributable to a single foot, including a metapodial, an astragalus and a calcaneum, plus a distal humerus fragment. The large bovid is dominated by limb bones with only three vertebrae representing the axial skeleton. The rhinoceros bones are dissimilar in pattern in that they are isolated elements from disparate parts of the skeleton, consisting of fragmentary teeth, a fragment of tibia and two carpal bones and fragments of vertebra, radius, and phalange. The only cranial material for the larger animals are the isolated teeth from the large bovid, the rhinoceros, and a piece of proboscidean tusk and molar.

In contrast, the smaller animals show a different pattern. Although equally represented by a restricted range of elements, the gazelle-type or antilope consists of cranial material only, with no post-cranial bones evident at all. A wider range of anatomical parts, including some cranial and post-cranial bones represents the canid. The bird bones are attributable to a single wing, consisting of the anatomically adjacent ulna and carpus-metacarpus. Although the skeletal part representation differs there is little to suggest that more than one individual for most of the species represented at Locality 362.

Number of individuals present

The majority of the material at Locality 362 is attributable to a single individual or part thereof. This conclusion is supported by very limited duplication of bones and the occurrence of both articulated bones and anatomically adjacent bones. The NISP (number of individual specimens present) and MNI (minimum number of individuals) are presented in Table 9. 17. The canid is numerically dominant but all its bones can be attributed to a single individual.

Not only is a single individual represented but also most of skeletal elements come from a restricted part of the skeleton. The *Sivatherium* is represented by a single front-limb and is attributable to a single individual. The Proboscidean is possibly a single individual, represented by the bones of a single, rear foot. The small gazelle type bovid is represented by the skull and horns but is again only attributable to one individual. Although represented by isolated cranial and post-cranial bones, the rhinoceros could still be attributed to a single individual. The wing bones of a single wing indicate the presence of only one bird.

The exception to this pattern is the large bovid, which is represented by limb bones only. The duplication of the bones and different fusion rates suggests that they are attributable to at least three individuals.

Age structure

The age structure of the Locality 362 population indicates that most of the animals were fully adult when they died. The only juvenile is that of a large bovid, indicated by an unfused femur and metatarsal bones. The epiphysial fusion of the long bones and/or the presence of permanent dentition indicate that, the other animals were fully adult at the time of their inclusion into the fossil assemblage. As so little dentition is present at the site, it is difficult to be precise about the age at death of the individuals. The dentition of the gazelle is permanent and worn, which would suggest a mature individual (Spinage, 1973). The remaining teeth are mainly isolated or fragmented and, as such, their usefulness as an age determinant is limited.

Articulation

The presence of articulation and anatomically adjacent bones is a dominant feature of several of the animals at Locality 362. The canid remains show the greatest degree of articulation with many of the portions of the lower limbs, and particularly the feet, being fully articulated. The remaining canid bones are disarticulated but are clearly associated, with the majority being anatomically adjacent bones. Several other animals also have evidence of articulation. Articulated limb bones, particularly the radius and ulna, are present for the large bovid. Also present are the anatomically adjacent bones of the humerus, which articulate with the with the radio-ulna. Articulated bones and anatomically adjacent bones are present for the smaller bovids also. The *Sivatherium* is represented almost entirely by a single front-limb with the anatomically adjacent bones of the humerus down to carpals and phalanges being present (plus the two metatarsal fragments). The anatomically adjacent bones of a single wing represent the bird.

Carnivore damage

Several of the bones from both the excavation and the surface collection show damage that is typical of a large carnivore (Binford, 1981, Haynes, 1980, 1983, Hill, 1980a). The damage ranges from a single puncture mark to several puncture marks, striations, gnaw marks, spiral fractures, depressed fractures and evidence of gnawed ends to the long bones. The number of bones affected by carnivore activity is not great, around 10% for the whole assemblage. The excavators supposed that the canid present at the site was the main agent of carnivore damage. It is possible that a large canid could inflict much of the damage observed at the site. The creature would have to be of a reasonable size, as bones as large and dense as a proboscidean calcaneum has at least four large and deep puncture marks in it. One of the puncture marked bones has the same dimensions as one of the carnivore teeth, demonstrated by the fitting of an excavated canid tooth into a puncture marked bone. Whilst not direct evidence that a carnivore accumulated these bones, it is evidence that they were modified by a carnivore of considerable size.

What is clear from a subsequent examination of the material is that even the carnivore at the site is heavily carnivore-damaged. The ends of some of the limb bones are missing as a result of chewing and there are puncture and gnaw marks on the bones. It is most likely that a large carnivore did produce this damage. There is evidence, in the form of finer and smaller striations, of a second and much smaller modifier of bones. The canid mandible has several groups of fine striations on it. The striations are concentrated below the lower canine and range from 4 mm-20 mm long. The irregular striations have more similarity to a small carnivore gnawing the bones than the more uniform and parallel striations of a rodent (Brain, 1981). A smaller carnivore, either a juvenile of the canid species already at the site or a different species not preserved at the site, is more likely to have produced these marks.

It is probable that carnivores accumulated the assemblage and, possibly, carnivores similar to that observed at the site. This is by no means a certainty, as other large predators are known of in the area and in similarly aged deposits. These other large predators, such as *Pachycrocuta brevirostris*, are also known as accumulators and crushers of bone and they are present at other fossil localities in the Pabbi Hills. Hyaenas are also known to be cannibalistic (Balestra 1962, Kruuk 1972).

There is good evidence that carnivores were the main agents of accumulation of the assemblage. A carnivore of the size of that present at the site would be capable of accumulating the bones of large animals. The skeletal part representation is also consistent with a carnivore-accumulated deposit. Limb bones only or cranial bones represent most of the animals at Locality 362, and this pattern is in keeping with that of large carnivores and possibly scavengers. The schlepp effect or preferentially bringing back limb bones or a head, would explain this patterning (Perkins and Daly, 1968, Marean et al., 1992). There is little evidence that the locality was a den. Firstly, there is no carnivore juvenile material present. The only possibly evidence of juveniles is the fine striations made on the large canid mandible, but this could equally be attributed to other agents, such as other small carnivores. Secondly, coprolites, which could also suggest denning activity, are not recorded at the site. Thirdly, there is no evidence of a den as such. No den-like structures are evident in the sediments. This is not so surprising as the Pabbi Hills sediments rarely contain any evidence of any form of structures, and even bedding structures are not always obvious. A den made in a depression on the ground could be a possibility, and certainly the bones are spatially confined, but if this was the case, then no evidence of such a depression remains.
Certainly the high representation of limbs bones and the spatial confinement of the bones would suggest a location where the bones were brought from elsewhere.

The presence of the bird wing is unusual and could represent an unfortunate individual who got in the way during a frenzied feeding session. Intense competition over food resources is certainly well documented in modern environments and is evident in other palaeo-faunal assemblages (Brain, 1981, Kruuk, 1972, Potts, 1988).

Fluvial modification

Although clearly deposited in a fluvial environment, it is not thought that fluvial processes have played a major role in the accumulation of this assemblage. Factors that suggest the assemblage was <u>not</u> accumulated by water include the body size of the animals, the skeletal part representation, the presence of articulated bones, and the compact spatial distribution of the bones (Badgley and Behrensmeyer, 1980, Behrensmeyer, 1982, 1991), and the complete absence of turtle, crocodilian remains; and the lack of orientation on the bones. Most of the bones are limb bones of some of the largest land animals. Many of these limb bones are articulated or anatomically adjacent suggesting they have come from the same limb, and such bones are not traditionally considered as having a high hydraulic potential (Badgley and Behrensmeyer, 1980, Behrensmeyer, 1982, 1991). Although some articulated bones have shown experimentally to have a much greater hydraulic potential than previously thought (Coard and Dennell, 1995, Coard, 1999), little is known about how such large bones behave in fluvial environments and in particular, real fluvial environments. Generally, it is only the larger species and the more robust bones from the smaller animals that have survived at the site. Given the presence of mainly the larger bones of some of the larger land mammals, then this pattern might suggest that this assemblage represents a lag deposit.

It could be argued that the lack of small bones or even small species could suggest that hydraulic forces winnowed out the smaller, lighter elements. Equally, it could be that these small bones are simply too fragile to have survived the deposition and fossilisation processes. The survival of the very fragile bird wing may argue against a winnowing or survivorship hypothesis, as may the preservation of the smaller canid bones. There is, however, other evidence that could explain the survival of so many of the canid bones.

The distribution of the canid skeletal material is interesting and relevant to a winnowing hypothesis. Within the excavated area, the canid skeleton is situated in the lower levels of the fossil deposit, with no evidence of its presence in the upper levels. It may well be that the canid, being lower down in the sedimentary sequence, was protected from most of the effects of water activity by the volume of bones deposited on top of it or by the sediment. Certainly, the canid skeleton was sufficiently protected to enable some of the smallest bones to survive. It is thus possible that the bones in the upper and less protected levels were subject to fluvial modification in the form of winnowing.

Had all the smaller, lighter bones of all the animals present been winnowed out then one might have expected a more dispersed distribution with a series of bones trailing downstream. This is simply not the case. As has been noted, the surrounding sediments were devoid of any fossil material and the fossil deposit is compact and spatially discrete. Post-depositional forces have not acted to disperse these bones more widely.

If fluvial forces have played a part in the accumulation of these bones, then it is suggested that the majority of the bones were deposited when they were in an articulated or semi-articulated state. Most of the bones could have come from the same animal and the same limb of that animal. Many of the bones are either articulated, anatomically adjacent or from the same side. The major exception is the rhinoceros, where bones from different parts of the skeleton are present, e.g. two teeth, one distal tibia and one carpal. It is thus possible that animal skeletons in various states of articulation could enter a fluvial environment and their hydraulic potential may not necessarily have been impeded (Coard, 1999).

Weathering

All of the fossils from the excavation were examined in an attempt to assess the period of exposure prior to burial and the period over which the assemblage had accumulated. Weathering stages were assigned to 181 of the excavated bones, the vast majority of which (N=160) fell in the 0 to 1 stages of weathering (Behrensmeyer, 1978), indicating a relatively short exposure period and one not above a few seasons. However, it is likely that the more weathered bone may give a more realistic assessment of exposure (Lyman and Fox, 1989). 21 bones were recorded in the higher categories of weathering stages evident at the site, that is fossils in Behrensmeyer's stage 2 or 3 (Behrensmeyer, 1978). This may indicate an exposure of nearer three years or so. It is difficult to assess if the bones have been exposed for a longer period as it may just be that one or two of the bones were exposed for longer or were exposed for the same period but were subjected to more extreme weathering agents. As the overwhelming majority of the bones indicate a much shorter exposure period then it is possible that the bulk of the bones were relatively protected in a short period of time. It is difficult to draw many conclusions from this data but it may be more a more accurate reflection of events if differential weathering is envisaged of an assemblage that was protected within a season or two at most.

Palaeoecological indicators

Most of the species consisted of mainly large to very large animals, or class size D to E as determined by Potts (1988), and included *Sivatherium*, rhinoceros and a proboscidean. At Locality 362, these large forms constituted around 10% of the taxonomically identified fauna and significantly more if the large bovid is included. With the inclusion of the large bovid, the large forms represent over a third (38%) of the fauna. Compared to other Pleistocene sites, one may have expected the ratio of bovids to other large forms (rhinoceros, giraffe, elephant etc.) to be significantly higher (Brain, 1981, Potts, 1988). Potts (1988:185) states that, in Bed I assemblages at Olduvai, such large animals tend to be around 1-3% of the number of specimens present at the sites. This, he suggests, reflects the low frequencies with which they are present on modern savannahs. As the frequency with which they occur is greater than this at Locality 362, it may suggest one of three scenarios. The first is that some agency was preferentially collecting large forms at the site, the second, that the ecological conditions were such that large forms were more abundant and represented a greater than normal proportion of the palaeocommunity; or thirdly, that destructive forces preferentially acted upon the smaller and more fragile bones. It is not felt that this is likely as the very small and fragile bones of the canid are evident at the site.

Western (1980) suggests that there is a close relationship between the number of animals with a large body size and greater biological productivity of an environment, with a positive correlation between increased rainfall and numbers of large bodied animals. This line of reasoning applied to the Locality 362 would indicate that the environment was stable, with higher precipitation levels and a high plant biomass. At Olduvai Gorge, however, the contrary seems to have occurred. There, these large, rarer forms were usually present in higher frequencies when moisture levels were low (Potts 1988). Unfortunately, there is little independent environmental evidence for most of the Pabbi Hills sequence with which to make statements that are more definitive. Pollen has not been successfully preserved and/or extracted. Equally, there are few independent indicators for the immediate area, such as beetle or micro-faunal evidence. However, Quade et al. (1993) showed that the flood-plain environments of the Pabbi Hills had a C4 vegetation, implying a predominantly grassland vegetation.

The Giraffidae, proboscideans and rhinocerii, tend to be regarded as browsers and would indicate the local presence of trees and shrub or bush lands within a savannah environment. The less elongate nature of the *Sivatherium* would lend itself more to a browsing and more closed habitat than that of the extant giraffes. Churcher (1978) suggests a grazing or low browsing habit of foraging with an adaptation to open woodland for at least some types of *Sivatherium*. There is a complete absence of any aquatic taxa (notably turtle and crocodile that are very common at other fluvial localities in the Pabbi Hills), and equally, the diagnostic types of an open savannah, such as the equids, are also absent.

Conclusion

The taphonomic indicators gained from the fossils at Locality 362 suggest that the sedimentary depositional environment was not active at the time of the deposition of the fossils. The presence of articulated bones, the presence of anatomically associated bones, the compact and clustered nature of the assemblage and the evidence of carnivore activity, all suggest a stable depositional environment. Equally, the sediments indicate a period of some stability also. The presence of calcium carbonate nodules and palaeo-root casts possibly indicates a weakly developed land surface. The depositional environment is more typical of a channel margin or floodplain area or an abandoned channel area within a larger fluvial environment. The taphonomic history of the assemblage suggests that the fauna is autochthonous to the local channel environment. Although there may be some evidence for fluvial modification to the assemblage, the assemblage shows clear evidence of modification, and probable accumulation by carnivores.

Table 9.16: Number and % of Individual Specimens (NISP) from the excavation and surface collections of locality 362

	Excavation		Surface collection	
Genus/Species	**NISP**	**%**	**NISP**	**%**
Canis	67	55	0	0
Large bovid	34	28	20	33
Sivatherium	6	7	10	16
Gazelle-type	4	2	0	
Aves	3	2	0	
Medium Bovid	2	2	12	20
Rhinoceros	2	2	12	20
Proboscidean	0	0	6	10
Total	121	100	60	100

Table 9.17: Number of Individual Specimens and Minimum Number of Individuals from locality 362

Genus/Species	NISP	MNI
Canis	67	1
Large bovid	54	3
Sivatherium	16	1
Medium Bovid	14	1
Rhinoceros	14	1
Proboscidean	7	1
Aves	3	1

Table 9.18 Number of individual specimens present for all mammal taxa at Site 362, N= 352.

Skeletal Part	Canis	Large Bovid	*Sivatherium*	Medium Bovid	Rhinoceros	Proboscidean	*Gazella*	S.M.	M.M.	L.M.	Indeterminate fragments
Skull	-	-	-	-	-	-	1	-	2	-	1
Horncore	-	-	-	-	-	-	1	-	-	-	-
Mandible	3	-	-	-	-	-	-	-	-	1	-
Maxilla	-	-	-	-	-	-	2	-	-	-	-
Tooth	3	1	-	4	2	-	-	-	-	-	-
Vertebra	-	3	-	-	-	-	-	1	4	2	3
Atlas	1	1	-	-	-	-	-	-	-	-	-
Rib	-	-	-	-	-	-	-	3	2	-	47
Scapula	-	1	-	-	-	-	-	-	-	1	-
Pelvis	-	1	-	1	-	-	-	-	3	-	2
Humerus: Proximal	1	-	1	-	-	-	-	-	-	-	-
Shaft	-	3	-	-	-	-	-	-	-	-	-
Distal	-	4	-	-	-	-	-	-	-	-	-
Radio-ulna: Complete	1	-	-	-	-	-	-	-	-	-	-
Proximal	-	4	1	-	-	-	-	-	-	-	1
Shaft	-	3	1	2	-	-	-	-	-	-	-
Distal shaft	-	1	1	-	-	-	-	-	-	-	-
Distal	-	1	-	-	-	-	-	-	-	-	-
Ulna	1	-	-	-	-	-	-	-	-	-	-
Femur: Distal	-	1	-	-	-	-	-	-	-	-	-
Tibia: Complete	1	1	-	-	-	-	-	-	-	-	-
Proximal	1	-	-	-	-	-	-	-	-	-	-
Distal	-	5	-	1	1	-	-	-	-	-	-
Metacarpal: Complete	2	-	-	-	-	-	-	-	-	-	-
Proximal	-	1	-	-	-	-	-	-	-	-	-
Distal	4	2	-	1	-	-	-	-	-	-	-
Metatarsal: Complete	8	1	-	-	-	-	-	-	-	-	-
Proximal	-	1	-	1	-	-	-	-	1	-	-
Shaft	-	-	1	1	-	-	-	-	1	-	-
Distal shaft	-	1	-	-	-	-	-	-	-	-	-
Distal	-	-	-	1	-	-	-	-	-	-	-
Metapodia: Complete	-	-	-	-	-	1	-	-	-	-	-
Shaft	-	-	-	-	-	-	-	-	1	-	-
Carpal	2	-	2	-	1	-	-	-	-	1	-
Tarsal	9	2	-	-	-	-	-	-	-	1	-
Calcaneum	-	2	-	-	-	1	-	-	-	-	-
Astragalus	1	2	-	1	-	1	-	-	-	-	-
Podials	2	-	-	-	-	-	-	-	-	-	8
Sesamoid	-	-	-	-	-	-	-	-	-	1	6
Phalanges 1	9	5	1	1	-	-	-	-	-	-	1
2	9	4	-	-	-	-	-	-	-	-	-
3	8	3	-	-	-	-	-	-	-	-	-
Os penis	1	-	-	-	-	-	-	-	-	-	-
Long bone fragments	-	-	-	-	-	-	-	-	-	-	-
Indeterminate	-	-	-	-	-	-	-	2	29	4	52
TOTAL	**67**	**54**	**9**	**14**	**4**	**3**	**4**	**6**	**50**	**17**	**124**

SM = small mammal; MM = medium-sized mammal; LM = large mammal

A Taphonomic Study of Site 642: A Plio-Pleistocene fossil assemblage

Fossil locality 642 was originally located on a small stream channel that was peripheral to any main or permanent river course. Figure 9.20 shows a representative section through the channel fill. There is evidence of smaller channels, indicated by mud balls and calcium carbonate concretions, which may suggest a relatively rapid accumulation of the sands, 57% by weight of which falls into a fine sand category. The fossils accumulated on a land surface characterised by flash-flood conditions where pluses of sand were deposited, possibly as a suspended load, and subsequently became dry and desiccated. The brown silty-sand overlies silty clays and fine purple clays which, although rather homogeneous in appearance, contain lamination lines (or a weakly developed cross-bedding), suggesting some alteration to the stream flow. Dessication cracks are evident by their subsequent infilling of finer purple clay and along with decayed sandstone nodules, suggest periods of drying out (see Figure 9. 20). The immediate environment of the site was probably a loop in a braided channel system where the banks of the stream were breached and the sands deposited as a suspended load type deposit at the bend of the loop.

The fossil assemblage consists of eroded fossils collected from the surface and well preserved fossil remains recovered by excavation. Although some of the fossils from the surface collections are identifiable to taxon, most are categorised by size as their poor preservation enables little else. The material recovered from the excavation shows a good level of preservation and enables a higher degree of analysis. The total number of fossils from Site 642 that were examined in this study was 5739. Just over 1444 bones, or 25.2 % of the assemblage, are identifiable to taxon. The remaining 4295 bones (74.8 %) are taxonomically indeterminate or can only be classed as a large, medium or small sized mammal. Although the percentage of indeterminate specimens is high, it is not uncommon for sites of a similar age (Brain 1981, Potts 1988). The list of species present at Site 642 and the minimum number of individual specimens present (NISP) for each taxon is presented in Table 9.19.

Table 9.20 shows the skeletal part breakdown and the frequency for each skeletal element for the fossils classified to a taxonomic level, and Table 9.21 shows similar information for the size classed and the indeterminate material. Table 9.22 shows the NISP and estimate of the minimum number of individuals (MNI) for some of the taxa at Site 642.

Of the fossils that are assigned to a species, the bovids dominate, comprising 90.4% of the total. *Damalops palaeindicus* is the most abundant single species in terms of the NISP (N = 465) and also in terms of the MNI, with twenty individuals. *Equus sivalensis* is the next most abundant species in terms of NISP (N = 63), but only two individuals can be recognised from this material. A large bovid, *Hemibos triqueticornis* is only represented by 20 bones but at least three individuals can be recognised. A very small bovid that is probably a gazelle is represented by some foot bones and a few dental pieces. All other taxa are represented by few bones, from which only one or two individuals can be recognised.

Re-assignation of two of the species should be noted, that of *Pachycrocuta brevirostis* (Aymard, 1846) that was listed in the initial site report as *Crocuta sivalensis* but revised on morphological as well as metrical grounds (see Turner, this volume). The material listed as *Proamphibos lachrymans* (Pilgrim 1939) in the initial site report was also wrongly attributed, and has been re-identified as *Damalops palaeindicus* Falconer (Pilgrim 1939), following my own investigations. (These changes have been incorporated into Table 9. 14, the catalogue of finds from the site).

A note on the identification of *Damalops palaeindicus* (Falconer).

From a preliminary study of the remains listed as *Proamphibos lachrymans*, including mandibles and a near-complete cranium (specimen GB 642 1417), evidence arose casting doubt on the identification. Major differences in the horncores, skull and dentition are evident: for example, the skull is not as elongate but is more slender in the Site 642 specimen than that of Pilgrim's (1939) description of *Proamphibos lachrymans*. Equally, the cranial vault is narrow and higher. The horncores, although not complete on the 642 specimen, show that that they leave the skull at a less acute angle than those described by Pilgrim for *Proamphibos lachrymans* (see Figure 9. 21). With regards to the dentition, the occlusal surfaces show stronger affinities with the tribe Alcelaphini rather than the Bovini, the tribe to which *Proamphibos lachrymans* is attributed. The teeth are strongly hypsodont with a relatively short premolar to molar row and the absence of basal pillars and goat folds on the lower molars would suggest affinities with the Alcelaphini (Gentry 1978). Alcelaphini are a medium to large bovid, now largely confined to Africa, but the Pinjor Formation of the Siwaliks does contain related species (Gentry 1978). The Site 642 remains show a strong affinity to *Damalops palaeindicus* (Falconer), a Pleistocene bovid that is well documented in the Pinjor Formation and in Tadjikistan, and with close relatives at Hadar (East Africa) (Gentry 1981) and Langebaanweg (South Africa) (Gentry 1980). The Site 642 specimens are also similar to the *Damalops palaeindicus* specimens described and illustrated by Dmitrieva (1977: 99) from Tadjikistan.

Population profile of ***Damalops palaeindicus*** **and other taxa**

A population profile based on dentition was determined for *Damalops palaeindicus*, the most extensive and complete species represented. Age profiles for other taxa were difficult to determine as good dentition did not survive. The estimates of age are based on published data of eruption rates and wear stages, namely those of Spinage (1967), Atwell (1980) and Atwell and Jeffery (1987). Although these ageing methods are not ideal for determining the age at death of an individual, in that they cannot provide an absolute age, the eruption rates, at least, are fairly standardised for the bovine family. The mandibles were grouped into three broad categories based on the presence of the deciduous molars (juvenile), the eruption of the M3 (prime adult), heavily worn cusps and exposed dentine (senile adult). The range for *Damalops palaeindicus* is wide, with juveniles to very old individuals being represented, but also remarkably even with six juveniles, seven prime adults and seven senile adults.

Of the other taxa at Site 642, *Rhinoceros sivalensis* is a juvenile. Due to wear on the dentition, the *Pachycrocuta brevirostis* individual is thought to represent a more mature adult (A. Turner, pers. comm.). Of the equids, one at least, is thought to be young, possibly a sub-adult with the permanent dentition erupting. At least one juvenile is represented in the *Hemibos triqueticornis* material, based on the fusion of long bones.

Determination of the size of the Bovidae remains

Measurements were taken of the adult bovine material and compared to those published by Walker (1985) and then assigned to a weight class (Vrba 1980). The smallest type of herbivore is the gazelle, whose remains are distinct in that they are very small, relatively complete and even articulated. Measurements suggest a Body Weight Class a (up to 26 kg). Bovids in the Weight Class b (27 – 124 kg), a small to medium bovid are also represented. The *Damalops palaeindicus* remains fall in the Weight Class c (125 – 342 kg), consisting of medium to larger bovids with measurements within the modern hartebeest to wildebeest range. The largest bovine present, *Hemibos triqueticornis*, is in the Weight Class d (>343 kg), with measurements either within, or exceeding those of modern African buffalo (i.e. >500 kg).

The modification processes identified at Site 642.

The modification processes at Site 642 include fluvial activity, weathering, carnivore activity, trampling and root damage, bone deformation and damage caused by the drying out and desiccation of the site.

Fluvial modification: Bone orientation and clustering

From a cursory glance at the site, bones lay in clusters and pronounced orientations were not readily apparent. However, flume experiments demonstrate that the orientation and clustering of bone is severely affected when bones are transported as anatomical units (Coard 1999, Coard and Dennell 1995). The orientation for the majority of the bones had been recorded during the excavation, but the clustering of bones masked any preferred orientation. However, for the isolated bones where their orientation had been recorded (N=97), a preferred orientation was more evident. Of 97 isolated bones, 54% lay in a North-South alignment. These bones are predominantly from the uppermost levels, suggesting that they had been affected by a current. This preferred orientation decreased in the lower levels where an East-West orientation became more pronounced. Some of the observed clustering of bone can also be attributed to fluvial action. Clustering of bone is suggested as a feature of hydraulic transport and sorting, especially if bones are articulated or of a similar hydraulic potential (Coard 1999). Fluvial action may have moved bones with similar hydraulic potential to slack water areas or more protected reaches. This may represent a similar pattern to that observed by Haynes (1980b), where bone piling occurred when bones were deposited in slack water areas. Clustering of similar anatomical parts, as can be seen with horn cores at the site would not be unexpected in a fluvial accumulation, as similar bones have the same hydraulic potential and would behave in a similar way.

Winnowing

Behrensmeyer (1975) suggested a "teeth to vertebrae ratio" as an indicator of winnowing on the basis of teeth being the densest element and vertebrae one of the least dense, and, therefore, most likely to be transported out. An assemblage dominated by a high ratio of teeth to vertebrae is more suggestive of a lag deposit. In mammals, the ratio of teeth to vertebrae is roughly equal. For *Damalops palaeindicus* there are 7.4 teeth for every vertebra (based on loose teeth only). For the medium sized mammals the ratio is still high, with 5.4 teeth to every vertebra. For the largest mammals, 32 teeth are recorded but no vertebrae are present. Such a high ratio of teeth surviving over several class sizes of animal does suggest a high degree of winnowing. To a lesser extent, a very small number of bones showed rounding and smoothing, which is indicative of water modification but not necessarily of water transport.

Carnivore damage

This form of damage is distinctive and well documented (Binford 1981, Brain 1981, Haynes 1980a, 1980b, 1983, 1985). Recorded damage at site 642 includes puncture marks, bite marks leaving the tooth row outline, gnawing damage and breakage patterns. The mandibles often have the ascending and horizontal ramus removed exposing the tooth roots. Metapodia suffered extensive gnawing to the distal ends and puncture marks to the shaft or proximal ends. Horn cores suffered distinctive gnawing damage, often covering the length of the horn core. Spiral fractures are evident on long bones, as are other forms of fracture, but not all of these are unequivocally due to carnivores. These types of damage are not observed on all of the mandibles, horncores, metapodia etc., many remain complete and unaffected.

In all, less than 10% of the fauna is classed as unequivocally carnivore damaged. This proportion of damaged bones is perhaps not significant, as the extent of the damage inflicted by carnivores is dependent on such factors as prey vulnerability, size of feeding groups, predator prey ratios etc. (Haynes 1980a, 1980b, 1983). The carnivore damaged bones range across most sizes of animals, including the Rhinoceros and *Pachycrocuta brevirostis* itself.

The most likely carnivore to produce the damage would be a hyaenid. The puncture marks are large and deep, and are present on the largest animals, including *Pachycrocuta brevirostis*, which does suggest a powerful bone crunching animal. Such cannibalistic behaviour is not uncommon for hyaenas (Balestra 1962, Kruuk 1972) even though Kurten (1968: 66) regards these gigantic hyaenas to be less of a carrion feeder and more of an active hunter than other hyaenid species.

(The gazelle remains are unusual in terms of the way they have been damaged. The post-cranial remains (see Table 10. 15) are derived from one animal, and include two distal humeri, one distal radius, one distal and one proximal tibia, four proximal and distal metapodia, three 1st and two 2nd phalanges. None of these bones shows obvious carnivore damage such as puncture marks, tooth scour or pit marks. Although spiral fractures are present (N=3), other fracture types not so indicative of carnivore predation are also present. Although not obviously displaying clear or diagnostic signs of carnivore damage, one interpretation is that these remains represent the remains of a single consumption event, whereby the main body of the carcass was consumed, leaving these fragments of bone relatively untouched. The spatially-restricted distribution of the remains (see chapter 7) may also support a single consumption event, with little post-depositional disturbance).

A small carnivore with sharp teeth is also evident from the damage inflicted to some of the bones. The striations are fine and non-uniform in direction which is suggestive of small carnivore chewing (P. Halstead, pers.comm.), and not the highly characteristic parallel broad groves produced by rodent gnawing (Brain 1981).

From this evidence at least two sizes of carnivores are present at the site, a large powerful bone-crunching animal and a much smaller carnivore inflicting finer-striation damage. Both these animals are evidenced by their skeletal remains. Two coprolites have been recovered and may well be the result of the larger carnivore.

Bone splinters that have been unequivocally produced as a product of bone crunching or as those digested by carnivores have not been recovered. Bones produced as a product of feeding are described by Potts (1988:100-101) and are quite distinctive, triangular flakes of bone, displaying both the cortical bone and the bone cavity. The vast majority of the Site 642 bone fragments are sections of bone ranging from a quarter of the bones circumference to a full shaft piece and do not all show distinctive carnivore damage. They generally lack the serrated jaw outline of the ends of the bone splinters (Potts 1988: 100, Binford 1981: 173), do not show the bulb of force (Potts 1988: 100), or chipped bone edges caused by teeth (Haynes 1980b: 80).

It is possible that such diagnostic splinters were present, but were winnowed out by subsequent fluvial action. Such action occurred, but equally other small fragments have remained. It is possible that splinters were totally consumed by carnivores and removed them from the site by regurgitation and/or defecation, but as coprolites have been recovered this clearly was not always the case at Site 642.

Weathering

Once exposed to the air the surface of a bone progressively undergoes changes in its texture. These changes, due to weathering, have been related to the time a bone has been exposed, and such, can give an estimate of the time that lapsed between death and deposition (Behrensmeyer 1978, see also Lyman and Fox 1989 for a cautionary note). Although many of the bones are encrusted with calcite weathering stages were assigned to as many of them as possible. Just over 89% of the bones in the excavated assemblage fall in the Behrensmeyer stage 2 or below category indicating a rapid, or at most three years, exposure before deposition. Around 12% of the bones are classed in the more advanced weathering stages of 3 or 4, and this could represent up to 15 years exposure.

No pattern relating to either anatomical element or position on the site is evident from the weathering patterns but it does appear to be related to particular genera. Most of the bones assigned to Stage 3 or 4 come from the largest genera, namely the Rhinoceros, *Hemibos*/large bovid, *Equus*, and a few in the medium mammal size range. This indicates something "taphonomically" distinct about the larger and more robust genera. Evidence does suggest that the larger the animal the more resistant it is to weathering (Behrensmeyer 1978, Gifford 1981, Lyman and Fox 1989) and that the more heavily constructed the bone is, the more weather resistant it is (Gifford, 1981). It may be that larger species were present as a lag-type deposit and had lain on the surface for several years. Deposition of the smaller animals and the majority of the *Damalops palaeindicus* occurred at a later stage and their deposition was fairly rapid. It is possible the bones of the more robust animals acted as a trap for the later inclusions.

It may be that all of the animals died in a single catastrophe, were all buried together but the larger bones stuck out of the sediment and remained exposed longer. This would be plausible except that small bones as well as large bones show advanced stages of weathering. Equally, large bones showing differential weathering stages across the bone, as may have been expected had part of it been exposed, are very rare but a single tibia shaft of *Rhinoceros sivalensis* does show such a pattern. The possibility of the bones being reworked, and hence showing mixed weathering stages, is excluded because in such reworked assemblages articulation, especially to the degree it is present at Site 642, is unlikely. Some bones do show a differential weathering pattern on the anterior to posterior surface, indicating that one surface had been less protected. A rapid deposition and limited reworking of the sediment at least is supported to some extent by the presence of mud balls and the poorly developed palaeosols at the site.

The anatomical part representation of the largest animals is reminiscent of a lag deposit, particularly if they had been transported as dry and articulated carcasses (Coard 1999, Coard and Dennell 1995). Skeletal parts such parts as crania, mandibles, metapodia and scapula are present whereas the more transportable vertebrae, phalanges etc. are not. The distribution of the *Rhinoceros* bones suggests that the skeleton disarticulated *in situ*. The humerus, ulna, radius, metapodia, skull, and teeth fragments are all in close proximity as a major cluster of bones and trail in a NE direction, roughly in line with the palaeostream.

It is possible that the animals were deposited on the landscape, subject to disarticulation and weathering, and during varying degrees of disarticulation, the more transportable bones were winnowed out. This fluvial action may have caused subsequent depositional events with some of the material being transported in and then trapped, which may explain why the small and highly transportable tragulid bones and more the transportable elements of other species are also present.

Root damage, bone deformation and desiccation damage.

Damage caused by palaeoroots is that of dendritic pattern of lines and an etched bone-surface appearance. Also root casts are evident, where the root has most probably decayed and the root cavity has calcified leaving a cast. The other main form of damage is that of roots splitting the bone. This can be seen in long bones, where roots have grown into the bone cavity and then split the bone, and in the mandibles, where the horizontal ramus has split away from the main body of the mandible. This process is depicted in Behrensmeyer (1978: Fig. 5). The resulting damage is a fracture that is difficult to distinguish from carnivore or other types of damage.

Bone deformation is generally termed plastic deformation (Haynes 1980b, Shipman 1981), and the compression of larger bones and the near-complete *Damalops palaeindicus* skull may be due to sediment compression, as one whole side of the crania has been pushed forwards. The near complete *Equus* skull (GB 642 1609) shows similar compression damage. Metapodia are also affected, particularly on articular ends where the bone appear to have been softened and then compressed. Such compression is usually localised. Although there are many forms of deformation (Shipman 1981) and as many reasons for deformation, such localised damage may be due to trampling (Haynes 1980b). Deformation due to sediment filling cracks in the bone and then pushing the bone apart, is particularly evident. The distortion due to deformation is not great enough to obscure the recognition of the bones, but it does render what would have been complete bones useless for measurements.

The desiccation cracks consist of lines about 1 cm wide, infilled with fine purple pink clay, and run for up to three metres across the site in a NW/SE direction. Other smaller lines are present, some almost at right angles to the main lines. The cracks appear to be unjoined, random and straight with regular crack edges (Allen 1987). Coloured and textured deposits of a similar nature are present on some of the bones, and the site note book records mud balls of fine clay up to "fist size" being present. The presence of the desiccation cracks suggests episodes of drying out with subsequent deposition of much finer material and may represent "flash type" depositions (J. Hutt, pers. comm.). Here, pulses of sand are deposited by flash flooding, leaving small ponded areas to become dried and cracked. The desiccation cracks are known to be younger than the deposition of the bone because some bones are bisected by the cracks and are displaced by them.

Trampling

Some *Damalops palaeindicus* metapodia show differential weathering at the proximal as opposed to the distal end, suggesting that the less weathered end was pushed down into the protective sediment. This change in declination is suggestive of displacement due to trampling, as has been observed experimentally (Gifford Gonzalez et al. 1985), and in other fossil localities (Hill and Walker 1972). Displacement of conjoining bones is a feature of trampling (Olsen and Shipman 1988), and is evident here. Conjoining bones were sometimes displaced up to 25 cm apart with there being no obvious signs of carnivore damage to account for the displacement.

Conclusions

The intrinsic morphology of bone dictates the nature of a break as much as the force acting upon it. It is for this reason that it is so difficult to identify the perpetrator of the damage, and "..it is more difficult to assign an unambiguous meaning to aggregate patterns of traces and remains in assemblages especially when evidence exists for the action of two or more agents" (Gifford-Gonzalez, 1991:226). When such taphonomic overprinting occurs, it is often only possible to say what factors can be identified as contributory. Taphonomic overprinting is evident on a grand scale for some of the bones at Site 642, whereas other bones remain unscathed.

The two potential accumulators at Site 642 are fluvial processes and carnivore behaviour, by either hunting or scavenging. Fluvial processes are known to have modified the site but may not have been responsible for the transportation of all the bones. The stream competence, if the bed load remained stable, was not enough to transport the remains of the larger species but it could have been sufficient to transport the more highly transportable elements or species. The stream competence was enough to winnow out the more transportable bones of many of the species represented. Fluvial processes are responsible for the orientation and the clustering of bones. It is entirely possible that larger anatomical elements and, indeed, whole animals could have been transported as part of the suspended bed load.

Carnivores, however, are responsible for considerable damage to some of the bone. This damage affects most taxa, including the carnivores themselves. To what extent carnivores accumulated the assemblage, as opposed to scavenged from it, is more difficult to determine. Haynes (1988) states that mass-death accumulations and serial-predation sites often share many important characteristics, such as abundant bone deposits, presence of many taxa/ individuals, degree of gnawing, the presence of different weathering stages etc. All of these features are present at Site 642 to a greater or lesser extent.

There is some evidence to suggest that this site does not fall readily into a carnivore accumulation category, and this can be seen in the survivorship of some of the articular ends of long bones. Marean et al. (1992), noted the frequencies of articular ends to shafts of the femur, tibia and metatarsals after they had been ravaged by intensive feeding by hyaenas. They noted for the femur the greatest change in abundance was to the distal end. This contrasts with the Site 642 where the distal end, N= 9 is better represented than the proximal end, N=2, or shafts, N=5. This is equally the case for the tibia, where the articular ends changed the most in the Marean et al. (1992) study ,and the shaft the least after carnivore ravaging. However, at Site 642 there is a good survival of tibia distal ends, N= 24, the second highest of the long bones, but fewer shafts, N=10. The metatarsals data did concur with Marean et al's observations in that the distal ends did not survive well in either study, but there was little difference between the survival pattern of the shafts and proximal ends, with 24 proximal ends and 23 shafts, which was not a pattern observed by Marean et al. These figures are based on all species but the pattern is evident within a single species also. Neither the relative abundance of shafts nor the survivorship of certain articular ends when compared with shafts fitted as well as Marean et al. (1992) suggest they should had they been simply ravaged by carnivores.

The Site 642 data are more comparable with mass-deaths or scavenged sites rather than predated or denning sites. The bones are not widely scattered as in the case of predation nor does the distribution fit that of denning where bones decrease in size away from the den entrance (Haynes 1988, Hill 1978, 1980a). Although it is possible that subsequent fluvial modification would distort any previous patterns of activity. Equally, the common articulated elements at Site 642 are vertebrae and ribs, and distal humeri and proximal radio-ulna rather than the skulls or lower leg segments (Haynes 1988, Hill 1980a). The information suggests that Site 642 is not just a carnivore accumulated or ravaged site. More than one factor was responsible for the observed patterning, which involves a combination of fluvial and carnivore modification processes. Although evidence of carnivore modification is evident on some bones, many remained unaffected. This, along with the observed pattering of the fossils, the skeletal part representation, and known 'signatures" of carnivore accumulations do not support carnivores as the only modification agents, nor does it support a denning hypothesis.

Fig.9. 21: Comparison of *Damalops palaeindicus* (Falconer) GB 642 1717, left, and *Proamphibos lachrymans* (reproduced from Pilgrim 1939). Scale 1/8 actual size.

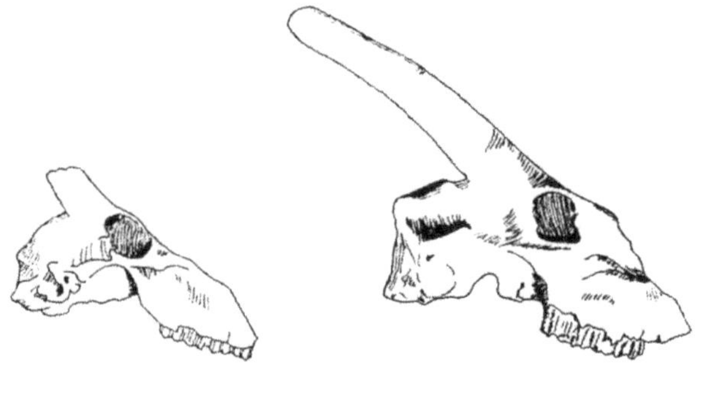

References:

Allen. J. R. L. (1987) Desiccation of mud in the temperate intertidal zone: Studies from the Severn Estuary and eastern England. *Philosophical Transactions of the Royal Society London*. B315:12-156.

Atwell, C. A. M. (1980) Age determination of the blue wildebeest, *C. taurinus* in Zululand. *South African Journal of Zoology* 15: 121-130.

Atwell, C. A. M. and Jeffery, R. C. V. (1987) Aspects of molariform tooth attrition in eland and wildebeest. *South African Journal of Wildlife Research* 11: 31-34.

Badgley, C. and Behrensmeyer, A. K. (1980) Paleoecology of Middle Siwalik sediments and faunas, northern Pakistan. *Paleogeography, Paleoclimatology, Paleoecology*. 30: 133-155.

Balestra, F. A. (1962) The man-eating hyaenas of Mlanje. *African Wildlife Journal*. 16: 25-27.

Behrensmeyer, A. K. (1975) The taphonomy and paleoecology of Plio-Pleistocene vertebrate assemblages east of Lake Rudolf, Kenya. *Bulletin of Comparative Zoology* 146 (10): 473-578.

Behrensmeyer, A. K. (1978) Taphonomic and ecologic information from weathering. *Paleobiology* 4 (2): 150-162.

Behrensmeyer, A. K. (1982) Time resolution in fluvial vertebrate assemblages. *Paleobiology* 8 (3): 211-227.

Behrensmeyer, A. K. (1991) Terrestrial vertebrate accumulations. In (Eds.) P. A. Allison and D. E. G. Briggs *Taphonomy: Realising the Data Locked in the Fossil Record, 291-335*. New York: Plenum Press.

Binford, L. R. (1981) *Bones: Ancient Men And Modern Myths*. New York: Academic Press.

Bose, P. N. (1880) Undescribed fossil Carnivores from the Sivâlik Hills in the collection of the British Museum. *Quarterly Journal of the Geological Society of London* 16: 119-136.

Brain, C. K. (1981) *The Hunters or the Hunted*? Chicago: University of Chicago Press.

Churcher, C. S. (1978) Giraffidae. In eds. V. J. Maglio and H. B. S. Cooke. *Evolution of African Mammals*, 509-535 Cambridge: Harvard University Press.

Coard, R. (1999) One bone, two bones, dry bones, wet bones. *Journal of Archaeological Science* 26 (11): 1369-1375.

Coard, R. and Dennell, R. (1995) A taphonomic study of some articulated remains from an artificial fluvial environment: An experimental study. *Journal of Archaeological Science* 22: 441-448.

Colbert, E. H. (1935) Siwalik mammals in the American Museum of Natural History. *Transactions of the American Philosophical Society* 16:339-348.

Cruz-Uribe, K. 1991 Distinguishing hyena from hominid bone accumulations. *Journal of Field Archaeology*. 8: 467-486

Dmitrieva, E. L. 1977 Tajikstan's and India's fossil Aleclaphinae. *Journal of the Palaeontological Society of India* 20: 97-101.

Gentry, A. W. (1978) Bovidae. In (Eds.) V. J. Maglio and H. B. S. Cooke. *Evolution of African Mammals*. Cambridge, Mass.: Harvard University Press.

Gentry, A.W. (1980) Fossil Bovidae (Mammalia) from Langebaanweg, South Africa. *Annals of the South African Museum* 79 (8): 213-337.

Gentry, A.W. (1981) Notes on Bovidae (Mammalia) from the Hadar Formation, and from Amado and Geraru, Ethiopia. *Kirtlandia* 33: 3-30.

Gifford, D. (1981) Taphonomy and paleoecology: a critical review of archaeology's sister disciplines. In (Ed.) M. B. Schiffer *Advances in Archaeological Method and Theory* Volume 4. New York: Academic Press, pp. 365-438.

Gifford-Gonzalez, D. (1991) Bones are not enough: Analogues, knowledge and interpretative strategies in zooarchaeology. *Journal of Anthropological Archaeology* 10: 215-254.

Gifford-Gonzalez, D., Damrosh, P. R., Pryor, J. and Thunen, R. L. (1985) The third dimension in site structure: an experiment in trampling and vertical displacement. *American Antiquity* 50 (4): 803-818.

Haynes, G. (1980a) Evidence of carnivore gnawing on Pleistocene and recent mammalian bones. *Paleobiology* 6 (3): 341-351.

Haynes, G. (1980b) Prey bones and predators: Potential ecologic information from analysis of bone sites. *Ossa* 7: 75-97.

Haynes, G. (1983) A guide for differentiating mammalian carnivore taxa responsible for gnaw damage to herbivore limb bones. *Paleobiology* 9 (2): 173-182.

Haynes, G. (1985) On watering holes, mineral licks, death and predation. In (Eds) J. I. Mead and D. J. Meltzer *Environments and extinctions: Man in Late Glacial North America*. University of Maine, Orano, pp. 53-71.

Haynes, G. (1988) Mass deaths and serial predation: Comparative taphonomic studies of modern large death sites. *Journal of Archaeological Science* 15: 219-235.

Hill, A. (1978) Hyaenas, bones and fossil man. *Kenya, Past and Present* 9: 1-3.

Hill, A. (1980a) A modern hyaena den in Amboseli National Park. In *Proceedings of the 8th Pan-African Congress of Prehistory and Quaternary Studies*, ed. R.E. Leakey and B.A. Ogot, 137-138.

Hill, A. (1980b) Early postmortem damage to the remains of some contemporary East African mammals. In (Eds.) A. K. Behrensmeyer and A. Hill *Fossils in the Making,* 133-152. Chicago: University of Chicago Press.

Hill, A. and Walker, A. (1972) Procedures in vertebrate taphonomy; notes on a Ugandan Miocene fossil locality. *Journal of the Geological Society of London* 128: 399-406.

Hillson, S. (1986) *Teeth*. Cambridge: Cambridge University Press

Jenkinson, R. D. S., Rendell, H. M., Dennell, R. W., Jan, M. A. and Sutherland, S. A. (1989) Upper Siwalik palaeoenvironments and palaeoecology in the Pabbi Hills, northern Pakistan. *Zeitschrift fur Geomophologie.* 33(4): 417-428.

Kruuk, H. (1972) *The Spotted Hyena*. Chicago: University of Chicago Press.

Kurten, B. (1968) *Pleistocene Mammals Of Europe*. London: Weidenfeld & Nicholson.

Lydekker, R. (1878) Crania of ruminants. *Palaeontologica Indica. Memoirs of the Geological Survey of India* 10 (3): 88-171.

Lyman, R. L. and Fox, G. L. (1989) A critical evaluation of bone weathering as an indication of bone assemblage formation. *Journal of Archaeological Science* 16: 293-317.

Marean, C. W. and L. M. Spencer (1991) Impact of carnivore ravaging on zooarchaeological measures of element abundance. *American Antiquity* 56 (4): 645-658.

Marean, C. W., Spencer, L. M. Blumenschine, R. J. and Capaldo, S. D. (1992) Captive hyaena bone choice and destruction, the Schlepp effect and Olduvai archeofaunas. *Journal of Archaeological Science* 19 (1): 101-121.

Olsen, S. L. and Shipman, P. (1988) Surface modification on bone: Trampling versus butchery. *Journal of Archaeological Science* 15: 535-553.

Perkins, D. and Daly, P (1968) A hunters village in Neolithic Turkey. *Scientific American* 219 (5): 96-106.

Pilgrim, G. E. (1939) The fossil Bovidae of India. *Palaeontologia Indica* 26 (1): 1-356.

Potts, R. (1988) *Early Hominid Activity at Olduvai*. New York: Aldine de Gruyter.

Shipman, P. (1981) *Life History of a Fossil*. Cambridge, Mass.: Harvard University Press.

Silver, I. A. (1969) The ageing of domestic animals. In (eds) D. Brothwell and E. Higgs, *Science in Archaeology*. pp 283-302. London: Thames and Hudson.

Spinage, C. A. (1967) Ageing the Uganda Defassa waterbuck. *East African Wildlife Journal* 5: 1-17.

Spinage, C. A. (1973) A review of the age determinations of mammals by means of teeth, with special reference to Africa. *East African Wildlife Journal* 11:165-187.

Vrba, E. (1980) The significance of bovid remains as indicators of environment and predation patterns. In (Eds.) A. K. Behrensmeyer and A. Hill *Fossils in the Making*. Chicago: University of Chicago Press, pp. 247-271.

Walker, R. (1985) *A Guide to Post-Cranial Bones of East African Animals*. Norwich: Hylochoerus Press.

Western, D. (1980) Linking the ecology of past and present communities. In (Eds.) A. K. Behrensmeyer and A. Hill *Fossils in the Making*, 41-54. Chicago: University of Chicago Press.

Table 9. 19: Number of Individual Specimens from the excavation and surface collections of locality 642.

Taxon	**642 excavation**	**1990 surface collection**	**1989 surface collection**
Damalops palaeindicus	245	220	-
Hemibos triqueticornis	19	1	-
Equus sivalensis	14	11	38
Rhinoceros sivalensis	18	1	15
Cervus punjabicus	1	1	1
Gazella	1	-	22
Pachycrocuta brevirostris	5	1	1
Carnivore	3	-	1
Bovid (general)	41	149	251
Bovid (small)	-	8	-
Bovid (medium)	-	-	316
Bovid (large)	-	2	49
Suid	-	-	5
Large mammal	18	5	18
Medium mammal	109	34	112
Small mammal	59	3	17
Indeterminate	443	1214	2263
Total	**980**	**1650**	**3109**

Table 9. 20: Skeletal part break down for all taxa at Site 642

Skeletal Part	*Damalops paleindicus*	*Hemibos triquetricornis*	Tragulid	Bovid Medium	Bovid Large	*Equus sivalensis*	*Rhinoceros sivalensis*	*Pachycrocuta brevirostris*	Carnivore	Suid	*Cervus punjabicus*
Cranial	4	-	-	2	-	1	1	-	-	-	-
Horncore	139	3	-	247	6	-	-	-	-	-	1
Mandible	75	-	1	92	7	12	3	2	-	3	-
Maxilla	14	-	1	4	2	3	-	1	1	-	1
Tooth	74	1	3	282	13	24	7	3	-	1	-
Atlas	5	-	-	7	-	-	-	-	-	-	-
Axis	2	-	-	6	-	-	-	-	-	-	-
Vertebrae	7	-	-	11	-	-	-	-	-	-	-
Ribs	-	-	-	-	-	-	1	-	-	-	-
Scapula	15	3	-	1	-	2	-	-	-	-	-
Pelvis	4	-	-	-	1	-	-	-	1	-	-
Humerus											
Proximal	1	-	-	-	-	1	1	-	-	-	-
Shaft	4	1	-	3	-	-	-	-	-	-	-
Distal shaft	2	2	-	-	-	-	-	-	-	-	-
Distal	10	1	1	7	4	-	-	-	-	-	-
Complete	0	-	-	-	-	-	1	-	-	-	-
Radio-ulna											
Proximal	7	1	-	5	-	-	1	-	-	-	-
Shaft	6	-	-	4	-	-	-	-	-	-	-
Distal shaft	2	-	-	-	-	-	-	-	-	-	-
Distal	3	-	-	5	-	-	-	-	-	-	-
Complete	1	-	-	-	-	-	1	-	-	-	-
Ulna Proximal	2	1	-	-	1	-	1	-	-	-	-
Femur											
Proximal	1	-	-	-	-	-	1	-	-	-	-
Shaft	-	-	-	-	-	-	-	-	2	-	-
Distal	3	-	-	1	1	-	1	-	-	-	-
Tibia											
Proximal	-	-	-	1	1	-	-	-	-	-	-
Proximal shaft	2	-	-	-	-	-	-	-	-	-	-
Shaft	5	-	-	-	-	-	1	-	-	-	-
Distal shaft	2	-	-	-	-	-	-	-	-	-	-
Distal	1	-	1	11	2	4	-	-	-	-	-
Complete	2	-	-	-	-	-	-	-	-	-	-
Metacarpals											
Proximal	5	-	2	10	-	-	-	-	-	-	-
Proximal shaft	1	1	-	-	-	-	-	-	-	-	-
Shaft	4	-	-	2	-	-	-	-	-	-	-
Distal shaft	1	-	-	-	-	-	-	-	-	-	-
Distal	3	-	1	18	1	4	-	-	-	-	-
Complete	5	-	-	-	-	-	-	-	-	-	-

Metatarsal											
Proximal	6	1	2	12	-	-	1	-	-	-	-
Shaft	6	-	-	7	1	-	-	-	-	-	-
Distal shaft	2	-	-	-	-	-	-	-	-	-	-
Distal	3+1	-	-	2	-	1	-	-	-	-	-
Complete	6	-	-	-	-	-	-	-	-	-	-
Metapodial											
Proximal	4	1	-	6	-	-	-	-	-	-	-
Shaft	7	-	-	-	-	-	-	-	-	-	-
Distal	16	1	6	4	3	5	1	-	-	-	-
Complete	-	-	-	-	-	-	2	-	-	-	-
Carpals	2	1	2	-	-	-	4	-	-	-	-
Calcaneus	-	1	-	-	4	1	-	-	-	-	-
Astragalus	2	1	-	1	2	2	3	-	-	-	-
Tarsals	1	1	-	-	-	-	2	-	-	-	-
Sesamoid	2	-	-	-	-	-	-	-	-	-	-
Phalanges											
1st	6	-	3	2	1	1	-	1	-	-	1
2nd	3	-	-	2	1	-	-	-	-	1	-
3rd	1	-	-	-	-	-	-	-	-	-	-
Fragments											
Shaft	-	-	-	-	-	-	1	-	2	-	-
Articular End	-	-	-	2	-	-	-	-	-	-	-
Totals	**477**	**20**	**23**	**757**	**51**	**63**	**34**	**7**	**6**	**5**	**3**

Table 9. 21: Skeletal part breakdown for the size classed and indeterminate material at Site 642

Skeletal Part	**Small Mammal**	**Medium Mammal**	**Large Mammal**	**Indeterminate**	**Small Indeterminate**	**Medium Indeterminate**	**Large Indeterminate**
Cranial	-	2	-	14	-	1	-
Mandible	6	30	3	43	4	7	3
Maxilla	4	3	2	6	1	2	-
Tooth	3	108	3	55	3	26	-
Atlas	-	3	-	1	-	-	-
Axis	-	3	-	1	-	-	-
Vertebrae	4	45	3	127	-	32	1
Sacrum	-	-	-	1	-	-	-
Ribs	50	51	1	158	-	7	-
Scapula	1	2	1	6	1	2	-
Pelvis	-	3	-	20	1	2	4
Humerus							
Proximal	-	-	-	-	-	1	1
Shaft	-	-	2	-	-	1	-
Distal	-	6	1	3	-	-	-

Radius							
Proximal	-	3	1	2	-	-	2
Distal	2	3	1	2	-	-	1
Ulna Proximal	-	-	1	1	-	-	-
Femur							
Shaft	-	-	-	3	-	-	-
Distal	-	3	-	-	-	-	-
Tibia							
Proximal	-	-	1	1	-	1	-
Shaft	-	-	-	-	-	-	-
Distal	-	5	-	-	-	-	-
Metacarpal							
Proximal	-	-	-	1	-	-	-
Shaft	-	3	-	-	-	-	-
Metatarsal							
Proximal	1	-	-	-	-	-	-
Shaft	2	5	-	1	-	-	-
Metapodial							
Proximal	-	9	-	-	-	2	-
Shaft	-	5	-	2	-	-	-
Distal	-	4	-	2	-	-	2
Carpals	-	1	-	-	2	3	1
Calcaneus	1	5	-	1	-	-	-
Astragalus	-	2	1	-	-	-	-
Tarsals	-	-	1	12	-	1	-
Phalanges							
1st	2	1	1	-	-	-	-
2nd	1	1	-	-	-	-	-
3rd	-	1	-	-	-	-	-
Fragments							
Articular End	-	1	2	20	-	8	-
Shaft	4	24	16	601	-	-	4
Indeterminate	-	1	-	2612	-	-	-
Totals	**81**	**331**	**41**	**3705**	**12**	**96**	**20**

Table 9. 22: Number of Individual Specimens (NISP) and Minimum Number of Individuals (MNI) for some taxa at locality 642.

Taxon	NISP	MNI
Damalops palaeindicus	465	20
Hemibos triqueticornis	20	3
Equus sivalensis	63	2
Rhinoceros sivalensis	34	2
Cervus punjabicus	3	1
Pachycrocuta brevirostris	7	1
Carnivore	4	2
Gazelle	23	2
Suid	5	1

Chapter 10: The fossil vertebrate record of the Pabbi Hills: palaeontology and biostratigraphy

Introduction

Early in this study (chapter 2), it was pointed out that there is little precision concerning the provenance of Upper Siwalik fossil specimens. Important finds have often been described with little detail of where they found, or their likely position within the Pinjor or Tatrot Formations. Additionally, the Pinjor Stage of the Upper Siwaliks lasted as long as the Villafranchian of Europe, or the entire Olduvai sequence in East Africa, yet is still treated as a largely undifferentiated unit, lasting from ca. 2.5 to 0.7 Mya. Consequently, there is little clear evidence of the chronological span of many taxa, or of any extinctions or immigrations that might have occurred during the Early Pleistocene. A second limitation is the lack of information on variation in the size of populations of each taxon, as the emphasis has been on individual specimens, rather than groups of them. Thirdly, most information so far published on Upper Siwalik vertebrates has been on their cranial and dental features, with less attention paid to post-cranial evidence.

In chapter 3, the methodology underlying the main surveys of 1986-89 was outlined. This was strongly influenced by the work of Barry et al. (1980) at Chinji, who had tried to grade the quality of information regarding the provenance of key specimens. Ideally, they pointed out, their spatial and stratigraphic provenance within a geological formation should be known to within 10m horizontally and 3m vertically, and also linked to a specific set of deposits. Their own work at Chinji showed the benefits of linking specimens with precise spatial and stratigraphic co-ordinates to palaeomagnetic sections. The potential of this type of investigation was also shown by our predecessors' research in the Pabbi Hills (Keller et al. 1977, Opdyke et al. 1979), which succeeded in tying several taxa to a palaeomagnetic time-frame, as was also done by the Dutch-Pakistani team at Mangla-Samwal (Hussain et al. 1992).

The field surveys of 1986, 1987 and 1989, and the excavations of the three fossil localities that were described in previous chapters produced over 40,000 specimens, including non-diagnostics. All material from those fossil occurrences classed as localities or find-spots can be located to within 10m horizontally and 3 m vertically, and dated palaeomagnetically to within 200-300 kya. Some scatters (perhaps 30%) are also of this quality. Most of the remaining material (most scatters and the general collections) can be located to within 100-200 m spatially and 15 m vertically, comparable to Barry et al's Grades 2-4. Much of this could probably be upgraded with subsequent visits and GPS readings. Table 10.1 shows the results of the fieldwork in terms of quality of information about provenance and dating, and grading in terms of the scheme suggested by Barry et al. (1980). This information should therefore have high biostratigraphic value.

Additionally, the fossil material from the Pabbi Hills is largely derived from two time-bands: half from Sandstone 12-times (1.2 - 1.4 Mya), and half from deposits between 1.7 and 2.2 Mya-old. Each is roughly the same size (ca. 20,000 specimens), and samples the same range of depositional environments: channel bars, abandoned channels, and overbanks. Consequently, the absence of a taxon from either time band is likely to be genuine unless it was extremely rare, and is unlikely to be an artefact of inadequate sampling or small sample size. The Pabbi Hills collections should thus be useful in documenting any extinctions or immigrations of large mammals during the Late Pliocene and Early Pleistocene. Further work is needed to provide comparable information for the periods 0.9-1.2 and 1.4–1.7 Mya.

Table 10.1: The Pabbi Hills fossil material in terms of provenance, dating and grading

Barry et al. (1980) criteria		**Location known to within 10m spatially and 3m stratigraphically**	**Known to within <100-200m spatially and 3-15m stratigraphically**			**Very limited stratigraphic value**
Barry et al. (1980) grading		**Grade 1**	**Grades 2 – 4**			**Grades 5-6**
Age (Mya)	**Time-span (kya)**	**Localities**	**General collections and scatters**		**Total**	**Gullies**
0.9 – 1.2	300	-	255 + 882	1137	1137	
1.2 – 1.4	200	13521	2160+ 7002	9162	22683	
1.4 – 1.7	300	-	672 + 384	1056	1056	270
1.7 – 1.9	200	4330	2677 + 2356	5033	9363	
1.9 – 2.2	300	2423	3121 + 4016	7137	9560	
Total		**20274**	**8885 + 14640**	**23525**	**43799**	

(Figures complied from Tables 5. 2, 6. 1 and 7. 1. Excavated data and a few find-spots are excluded).

This chapter will describe the most important specimens from each major group of mammals, and attempt to relate them to the genera and species already identified in the Upper Siwaliks. (The carnivores are treated separately in an appendix at the end of this chapter by Alan Turner). The chapter will then consider biostratigraphic issues: what evidence is there for local extinctions and immigrations in the million-year-record of the Pabbi Hills?

Part 1: Fossil remains from the Pabbi Hills

i) Artiodactyla

The most diagnostic parts of the artiodactyl skeleton for taxonomic purposes are the teeth, and for bovids, antelopes and cervids, the horn-cores or antler. Many Siwalik artiodactyls (as well as other types of mammals) are defined by their skulls and maxillary dentition. Unfortunately, complete crania are rarely found, so it is not easy to link different types of teeth with different types of horns or antlers, especially if several types are present. However, some of the material from the excavated fossil localities does permit some reasonably confident linkage between bovid dentitions and horn-cores. Different types of Artiodactyls are difficult to recognise from post-cranial elements (particularly if they are fragmented, as is the norm in surface collections), but these are nevertheless useful in showing the overall size range and numbers of sub-groups within this large and diverse grouping. In some cases (notably locality 642), post-cranial remains can be linked to specific types of teeth and horn-core

Bovidae: Teeth

The main pre-war studies of Siwalik bovids were by Pilgrim (1937, 1939) and Colbert (1935). Gentry (1978) provides an essential overview of the evolution of African bovids , with some reference to Siwalik material. Various papers by Nanda (e.g. Nanda 1979), and Gaur (1982) on some of the Indian Upper Siwalik bovids are also essential reading. It is, however, both difficult and unwise to identify fragmented fossil material on the basis of second-hand information, and thus identifications have been cautious. As the bovids are primarily of African origin, the Siwalik bovids require re-assessment in the light of African material, particularly as they appear to have more resemblances to African than other Eurasian forms (see Gentry 1978: 565). With these remarks in mid, we can consider the main types of bovids evidenced in the Pabbi Hills.

Type 1: *Damalops palaeindicus* Falconer.

In a previous chapter, Coard showed that the most of the bovid material from locality 642 was derived from the medium-sized *Damalops palaeindicus*. This type was initially identified as *Proamphibos lachrymans*, which was also the main type of bovid recognised in Opdyke et al.'s (1979) analysis of their material from the Pabbi Hills. In the opinion of Gentry (1978: 557), *Damalops palaeindicus* is most likely related to the *Alcelaphus – Rabaticeras – Damaliscus* group of alcelaphines, with hartebeest as a modern representative. It is medium-sized in comparison to other bovids in the Pabbi Hills collections. The M_3 is usually ca. 25 - 30mm. long. There is no cingulum; the buccal side is slightly concave, and the lingual side slightly convex. The infundibula are shallow, the central cavities are complicated, and the cusps are strongly hypsodont. In occlusal view the molars are rather selenodont. The crown enamel is smooth and thin (ca. 0.6 - 0.8mm thick on the lingual side). The mesial end on the lingual side of the smaller, and usually younger specimens, has a moderate pillar, as does the fold between the mesial and medial cusp. Goat folds are absent in the lower molars, and the M_3 is widest mesially. The premolar rows are short.

Good examples are a near-complete cranium (specimen GB 642EX1417); 642EX 1351, a left mandible with the P2-M3, all moderately worn, 642EX1404, another left mandible, and a right mandible from the bovid skeleton at 636. Mandibles are marked by deep horizontal rami. Teeth size are given below:

Table 10. 2: Size of Type 1 bovid teeth (*Damalops palaeindicus*)

Specimen	P2	P3	P4	M1	M2	M3
642XGB1351	6.5 x 4.2	10.0 x 6.3	12.5 x 8.0	14.0 x 10.0	20.9 x 11.9	29.7 x 11.5
642XGB1404	5.5 x 5.0	7.5 x 6.0	11.9 x 8.0	12.3 x 7.7 estimated	20.0 x 10.6	28.7 x 10.4

(The M1 of 642XGB1404 has broken crowns, hence the estimated width).

Third molar measurements of bovids in this group are listed below:

Table 10. 3: Size of Type 1 third molars (*Damalops palaeindicus*)

Specimen	Age (Mya)	Element	Crown length	Max.width
14AGB2	1.2 - 1.4	M_3, right, very worn	27.0	12.0
14AGB387	1.2 - 1.4	Left mandible + M_3, unworn	25.8	10.8
20GB4	1.2 - 1.4	Right M_3, unworn	25.8	10.3

Table 10. 3 continued: Size of Type 1 third molars (*Damalops palaeindicus*)

23GB38	1.2 - 1.4	Left mandible + M_3, mod. wear	25.0	10.0
44GB6	1.2 - 1.4	Right M_3, little wear	24.3	7.6
57GB5	1.2 - 1.4	Right M_3, mod. wear	23.9	8.9
68B GB1587.1	1.2 - 1.4	Left mandible + M_3, unworn	27.8	12.2
128BGB137	1.2 - 1.4	Left mandible + M_2, M_3, mod.wear	28.0	10.0
134GB7	1.2 – 1.4	M_3, heavy wear	25.0	9.0
267GB5	1.9 - 2.2	Left M_3, mod/heavy wear	27.0	10.0
273GB1005	1.9 - 2.2	Right M_3, mod. wear	22.0	8.0
420GB10	>1.7	Left M_3 left, mod. wear	25.0	13.5
440GB2	?1.7 - 1.9*	Left mandible + M_3, worn	24.0	7.8
452GB13	?1.7 - 1.9*	Right mandible + M_3M3, little wear	31.0	12.3
*503GB8	1.7 - 1.9	Right M_3, moderate wear	28.0	11.0
524GB1	0.9 - 1.2	Right mandible + M_3, worn	27.2	10.2
529GB21/1	1.9 - 2.2	Left mandible +M_3, unworn	25.2	8.0
815GB4	1.2 - 1.4	M_3Lower M3, left, moderate. wear	26.0	9.1
815GB5	1.2 - 1.4	M_3Lower M3, right, moderate. wear	22.0	6.8
634GB4	1.2 - 1.4	M_3 left, moderate. wear	28.4	11.0
		Average	**25.9**	**9.9**
Other teeth:				
14AGB193	1.2 - 1.4	Left mandible +dm_3	17.0	8.0
642XGB1418	1.2- 1.4	Right mandible + dm_3	22.0	8.5
648GB6	1.2 - 1.4	Left mandible + dm_3, M_1 erupting	25.4 (dm3)	7.1 (dm3)
648GB6	1.2 - 1.4	Left mandible+ dm_3	25.1	7.8
14AGB190	1.2 - 1.4	Right maxilla, M^3	26.5	11.2
14GB303	1.2 - 1.4	M^3, right, worn	16.0	12.0
35GB24	1.2 - 1.4	Left M^3, unworn	19.5	12.2
68B GB1587.2	1.2 - 1.4	Right maxilla + M^3, mod. wear	16.0	12.0
642, recollect 1999	1.2 - 1.4	Complete cranium (!) + Maxilla: M^3	22.0	12.8
437C GB47	1.9 - 2.2	Right M^3, worn	19.0	8.8

Comments: This group may contain two types of bovid, as some molars have a smoother enamel, with no cuspids between the 1st and 2nd cusp. This group is called Type 1B; Locality 610 had the main examples.

Type 1B: this is similar to the other Type 1 bovids, but the molars are rounder on both the buccal and lingual edges, less bunched-up distally to mesially, with no cuspid between the 1st and 2nd cusps. They are also a bit flatter, and less wide buccal-lingually. This type is tentatively identified as an alcelaphine. The crown length and width are similar to those of *Damalops palaeindicus*.

Table 10. 4: Third molar measurements of Type 1B alcelaphine-type teeth

Specimen	**Age**	**Element**	**Crown length**	**Crown width**
610AGB10	1.2 - 1.4	Lower M3, right, mod. wear	28.0	--
610CGB1	1.2 - 1.4	Lower M3, right, mod. wear	29.0	10.0
610CGB3	1.2 - 1.4	Lower M3, right, mod. wear	22.0	10.0
610CGB5	1.2 - 1.4	Lower M3, left, mod. wear	32.0	11.0
610CGB9	1.2 - 1.4	Lower M3, right, mod. wear	38.0	11.0
610CGB51	1.2 - 1.4	Lower M3, left, mod. wear	32.0	--

Note: 133GB409 M1/2 lower, 23 x 11mm may also be of this type.

Type 2: *Hemibos triqueticornis*. This type is a large bovid, similar in size to a modern eland or African buffalo, with teeth that are strongly hypsodont and selenodont. The enamel is slightly rugose and usually ca 0.8 - 1.0mm. thick but up to 1.5mm thick (as with, for example 630GB33); there is no cingulum. The cusps are strongly folded buccally. On the buccal side, there is a basal pillar between each cusp. The specific identification of *Hemibos* as *H. triqueticornis* is on the basis of its probably horn morphology, which is described below (see Type B horns).

Table 10. 5: Size of Type 2 bovid (*Hemibos triqueticornis*) third molars

Specimen	Age	Element	Max. length	Max. width
68 recollection	1.2 - 1.4	Left lower M_3M3, worn	40.2	14.0
107GB4	0.9 - 1.2	Right M_3, unworn	49.0	14.5
203GB6	1.9 - 2.2	Right M_3, worn	39.0	14.5
203GB39	1.9 - 2.2	Right M_3 lower, heavy wear	45.0	17.0
204GB8	1.7 - 1.9	Right mandible +P4-M3, unworn: **M_3**	44.0	14.8
261GB2001	1.9 - 2.2	Right lower M_3, mod. wear	42.0	18.0
361GB15	1.9 - 2.2	Lower M_3, right, worn	40.0	15.5
362G GB65	1.7 - 1.9	Left mandible + M_3, mod. wear	38.0	12.0
379GB1	1.9 - 2.2	Right mandible + M_3, unworn	40.8	15.0
391GB	1.9 - 2.2	Right mandible + M_3, worn	40.0	15.0
393GB3	1.9 - 2.2	M_3, worn	44.0	16.0
398GB1	1.9 - 2.2	Right mandible, M1-3: M_3, worn	36.0	15.5
427A	1.9 - 2.2	Left mandible + M_3	39.0	16.0
435GB6	1.9 - 2.2	Right M_3, heavy wear	40.5	15.0
451AGB15	1.9 - 2.2	Left mandible + M3, worn	38.2	12.6
495GB8	1.7 - 1.9	Right mandible + M_3, mod. wear	38.0	13.0
531GB26	1.9 - 2.2	Left mandible + M_3 , little wear	40.5	16.5
630GB33	1.2 - 1.4	Left mandible+ M_3	42.0	17.3
747GB1	1.2 - 1.4??	Left mandible + M_3, mod. wear	45.0	14.8
		Average	**41.1**	**15.1**
362FGB51	1.7 - 1.9	Left mandible + dm_3	33.0	7.5

Upper teeth:

Specimen	Age	Tooth	Length	Width
271GB13 or 15	1.9 - 2.2	M^3, no wear	31.0	18.0
341GB18 or 28	1.9 - 2.2	M^3, worn	32.0	24.0
361GB12	1.9 - 2.2	M^3	34.0	22.6
373GB41	1.9 - 2.2	M^3, very worn	30.0	16.0
460GB1	1.7 - 1.9	M^3, left, moderate wear	31.8	21.5
704GB2	1.2 - 1.4	M^3, moderate wear	28.0	18.0
628GB1	1.2 - 1.4	Maxilla P^2- M^3, moderate wear	34.0	25.0
		Average	**31.5**	**20.7**
Mirpur 16589		Maxilla + M^3	28.7	18.0

The best example of the upper dentition of this type of bovid is 628GB1, a left maxilla with the P2, P3, M1-M3, all moderately worn. The P^2 is 18.0 x 19.0mm; the P^3, 18.0 x 19.0; the P^4, missing; the M^1 27.0 x 27.0; and the M^2, 30 x 27mm. The best example of the upper dentition is probably the excellent cranium 16589 from Mangla/Samwal, Mirpur (Hussain et al. 1992); the size of the M_3 is given above. (The size of the other teeth in the right maxilla are as follows: P2, 21.0 x 13.8mm; P3, 17.8 x 15.0mm; P4, 17.2 x 16.8mm; M1, 17.2 x 16.8mm.; M2, 16.2 x 20.0mm.)

Some specimens (e.g. 373GB41) had smoother and slightly thicker enamel on the buccal side, and less or no fluting on the buccal side. However, there were not enough specimens to suggest that this group should be split.

Type 3

This is the smallest bovid-type, perhaps comparable in size to a modern dik-dik or small gazelle. It is cautiously identified as one of the Antilopini, or "small- to medium-sized antelopes that generally live in arid areas" (Gentry 1978: 559). The best examples are of a right mandible from the surface of locality 642 and a lower M3 from 610. As Table 10. 6 shows, a few other dental specimens of a very small artiodactyl were also found during the field surveys. Post-cranial material attributed to whatever animal had Type 3 teeth is presented below (see Tables 10.12, 10.13, 10. 15, Figure 10. 2). As Table 10. 6 shows, the length of the M_2 and M_3 from 610 and 642 are identical to those illustrated by Pilgrim (1937: 807) for *Gazella lydekkeri* from the Dhok Pathan Formation, and considerably smaller than that of the Pinjor Stage *Antilope subtort*a (ibid, 815). As shown, the M_2 is also identical to that of a modern gazelle.

Table 10. 6: Size of Type 3 (Gazelle-like) teeth. Measurements (*) of *Gazella* and *Antilope* are from Pilgrim (1937: 807, 815), and (**) from Gaur (1987: Table 3.27).

Specimen	Age	Element	Crown length	Max. width
113	1.2 - 1.4	Upper PM	6.5	7.0
285GB39	1.9 - 2.2	$M_{1/2}$, worn	10.6	8.0
452GB29	1.7 - 1.9	$P^{3/4}$, worn	10.2	7.0
489Brecollect	1.7 - 1.9	Mandible fragment + $M_{1/2}$	13.0	
489J	1.7 - 1.9	Right mandible +M_2	11.8	6.8
610GB41	1.2 - 1.4	Right M_3, no wear	17.5	--
642GB2112	1.2 - 1.4	Right mandible + M_2	14.0	
642GB2112	1.2 - 1.4	Right mandible + M_3	17.5	
*Gazella**	Dhok Pathan	M_3	17.5	7.0
*Antilope**	Pinjor	M_3	24.5	11.5
*Gazella***	Pinjor/PUA 84/11	M_2	12.5	6.8
*Gazella***	modern	M_3	11.8	6.8
*G. lydekkeri***	Dhok Pathan	M_3	11.5	7.0

Type 4: This type is a very large bovid indeed. Only three examples were found. Specimen 481E GB39 (1.7 - 1.9 Mya) comprised a lower M3, broken across the mesial cusp, and with the crowns largely encrusted in yellow silt. Its length was 53mm., with an estimated length of 65 mm. The crown width was 23 mm., and the enamel was 2-3 mm. thick. The second specimen is 125GB2 (1.2 - 1.4 Mya): a right mandible with the M3; the enamel is smooth; the cusps are selenodont, with the buccal cusps slightly lower than the lingual ones. 125GB3 is the conjoining ramus, with a sandstone matrix. The M3 is 59mm. long, and 23mm. wide. The third specimen was 668GB6 + 668GB1 (1.2 - 1.4 Mya): 2 joining pieces of a right mandible, with a broken M3. The M3 is broken at the mesial, forward end. The tooth is moderately worn. The mandible is very robust, and 36mm wide at the exposed root of the M3. The M3 is 19mm wide at the base.

(This does not seem to be either *Sivatherium* (cf the Dutch specimen 15913 maxilla, in which the enamel is very rugose, and the buccal cusps very low. *Sivatherium* teeth also have very deep fossae between the lingual and buccal cusps. According to the criteria set out by Nanda (1978a), this specimen does not appear to be *Camelus* either).

Bovid horns

The main evidence on the shape and size of bovid horn-cores are those excavated at localities 362 and 642, supplemented by some material found on the surface during the other surveys. The following types can be recognised.

Type A: *Damalops palaeindicus*

The horns in this group have a circular or slightly oval cross-section, and are twisted slightly outwards at the tip. The type specimen is 642X 1146 (see Plate X). The basal diameter is 36mm A-P (anterior-posterior), and 32mm transversely. Other specimens are 642EX1116, 642EX 1377, 642EX1389 and 642EX1417. Table 10.6 lists the size of specimen 642EX1116. A simple indicator of the degree of twisting in this specimen can be obtained by the ratio of total length along the horn to the direct length from tip to base. In this case, it is 1.22.

Table 10. 7: Size of Type A horns (*Damalops palaeindicus*)

Measurement (mm)	642XGB116
Direct length along horn, tip to base	180
Tape length along horn	220
Basal diameter, anterior - posterior	32
Basal diameter, transversely	35

Type B: *Hemibos triqueticornis* Rütimeyer

Pilgrim (1939: 259-269) recognised three types of *Hemibos* – *H. triqueticornis, H. acuticornis* and *H. antelopinus* – and provided detailed descriptions. Each was distinguished primarily on the basis of the angle at which the horns diverge outwards and backwards. In *H. triqueticornis* – the type identified here - the horns diverge at >90°, and are tilted back considerably above the plane of the face, following criteria set out in Nanda (1979). The best examples from the Pabbi Hills are 642EX1164 and 642EX1181. There is also the superlative cranium 16859, found by the Dutch-GSP group at Mangla-Samwal, Mirpur, and identified as *Hemibos* (de Vos, pers. com.). This specimen is especially important it comprises the complete cranium, including the entire maxillary dentition as well as both horns. These are straight; triangular in cross-section, with the keel at the back. Table 10. 7 lists specimens from the Pabbi Hills that are probably of the same type; note the absence of examples from the 1986 assemblage.

Table 10. 8: Type B horn-cores (*Hemibos triqueticornis*):

Specimen	Age	Element
203GB5	1.9 - 2.2	Horn-core
220GB1	1.7 - 1.9	Horn-core + part of cranium
243GB5	1.4 - 1.7	Cranium + horns
312GB6	1.9 - 2.2	Horn-core tine
355GB3	1.9 - 2.2	Horn-core
445BGB23	1.9 - 2.2	Horn core tine
451CGB1	1.9 - 2.2	Horn-core, complete; 42mm A-P, 30mm transverse
487GB48	1.7 - 1.9	Horn-core
620GB1	1.2 - 1.4	Cranium + horns

Type C

This is larger and less twisted than Type A. Its affiliation is unknown, but superficially resembles one of the Tragelaphines, or bush-buck/kudu tribe. The type example is 362EX12 (see Plate XX). The horns are more or less circular in cross-section, and there is no keel. The horns are only slightly twisted. In this specimen, the tape-length from tine to skull is 25cm., but 24cm if measured directly. The horns are 4cm apart at the base. The twisting ratio is 1.04.

362X GB12 can probably be associated with two mandibles. One is 362G GB65 (see Table 8.6), a right mandible with a moderately worn M3, 38mm long and 12mm wide. The other is 362E GB2, a right mandible, which has the dm3 and M1, erupting. The dm3 is 34.8mm long and 10mm wide.

Other specimens: 529GB9 (1.9 - 2.2 Mya) A right, complete horn, with part of the cranium. The basal diameter is 47mm A-P and 41mm transversely. The tape length from tip to skull is 30 cm, and the direct length is 29cm. Twisting ratio = 1.03.

368GB4. (1.7 - 1.9 Mya) A horn with part of the cranium preserved. The horn is slightly flattened. The diameter is 47 mm A-P., and 36.5mm transversely.

252AGB6 and 252AGB8. (1.7 - 1.9 Mya) A pair of horns , broken, plus part of the frontal. Basal horn diameter is 43 by 43mm. The gap between the horns at the base is 3.5cm.

14AGB257. (1.2 - 1.4 Mya) A right horn with part of the skull. Max. Basal diameter 41 by 41mm.

522BGB11. (0.9 - 1.2 Mya) A horn plus pedicel; 35 mm AP and 32mm transverse.

Linking the bovid teeth and horns

Which bovid teeth are associated with these types of horn-cores? Given that the vast majority of horn-core fragments found in surface collections are too fragmentary to indicate the original overall shape and size, the small number of excavated and reasonably complete specimens from localities 362 and 642 are especially useful, as these can sometimes be linked to the teeth. Table 10. 9 (end of chapter) summarises the most obvious and parsimonious way of associating horn-core to dentitions.

As is evident, there is room for considerable improvement. Unfortunately, progress depends upon the excavation of well-preserved bovid horn-cores that can be confidently linked to dentitions. As indicated above, we have yet to discover the type of horn-core(s) linked to teeth of Type 1B (alcelaphine) teeth, the very small Type 3, or the very large Type 4 bovids. It is also apparent that although teeth are more commonly preserved than horn-cores, they under-represent the full range of bovid taxa.

ii) Cervidae

A large amount of dental, antler and post-cranial material was found that could be classed as cervid. Almost all was found in surface collections, and it was not possible to link different types of antler to either post-cranial or dental material. Some post-cranial material attributed to cervids may have belonged to a gracile type of bovid. This is particularly likely for locality 14, found right at the start of the survey. It was only after the fieldwork had been completed and locality 642 had been analysed that many of the dentitions identified as cervid in 1986 were re-identified as a gracile bovid. As the taxonomy of Siwalik cervids is both complex and probably in need of revision (Opdyke et al. 1979), it is not proposed to attempt identifications of the Pabbi Hills cervid material beyond the family level.

The principal cervid dental data are listed below. The teeth have a pronounced cingulum, and the enamel is slightly rugose. The crowns are hypsodont. The enamel is ca. 1.5 - 2.0 mm. thick. On the basis of size, these teeth can be interpreted as showing one type of cervid, albeit of fairly varying size. As seen shortly, however, there are two types of antler. Note that the specimens from locality 73 are largely immature or sub-adult, and thus smaller than the fully adult individuals; unfortunately, no antler was preserved here.

Table 10. 10: Size of cervid teeth.

a) third molars

Specimen	Age	Element	Crown length	Max. width
73AGB100	1.2 - 1.4	Left mandible + M2,M3, fresh	29.0 (M3)	10.8 (M3)
73AGB289	1.2 - 1.4	Left mandible +M3, unworn	45.0	18.5
485	1.7 - 1.9	Left mandible +P3-M3, worn	36.5 (M3)	18.9
68BGB327	1.2 - 1.4	Left mandible+ dm3	30.0	12.0
GB73C219	1.2 - 1.4	Right mandible +P4	28.5 (dm3)	9 (dm3)
73XGB9	1.2 - 1.4	dm3	38	8.5
491D	1.7 - 1.9	Left mandible +dm3	31.0	14.0
73AGB147	1.2 - 1.4	Right maxilla + M3, worn	22.0	17.0

b) other teeth

Specimen	Age	Element	Crown length	Max. width
491D	1.7 - 1.9	Left mandible +M1	30.6	8.5
68BGB23	1.2 - 1.4	Left mandible +M1, unworn	23.6	
68BGB23	1.2 - 1.4	Left mandible +M2, unworn	25.5	
73AGB100	1.2 - 1.4	Left mandible + M2,M3, fresh	25 (M2)	13 (M2)
73EX182	1.2 - 1.4	Left maxilla+P2,P3,M1,M2, unworn	25 (M2)	12 (M2)
73AGB147	1.2 - 1.4	Right maxilla + M2, worn	19.0	15.0
73EXGB2	1.2 - 1.4	Upper M1/2	24	10
73BGB9	1.2 - 1.4	Upper M1/2, fresh	28	13
676GB2		Right maxilla + M1	23.0	25.0
676GB2		Right maxilla + M2	26.0	25.0

Cervid antler

It is assumed that antler fragments are derived from cervids, although a small amount may have belonged to *Sivatherium*, particularly in the earlier horizons. There appear to be two types. One is branched at the pedicel, with at least one smaller branch near the top of the main fork; the antler itself has a roughly circular cross-section. The best example is specimen 695GB1, found by a shepherd in 1989 between locality 269 and the watershed track, and likely to be pre-Olduvai in date. The other is more palmate, with a more oval cross-section. Details of both types of antler are listed in Table 10. 11 at the end of this chapter.

In modern terms, the former is similar to *Cervus*, and the other to *Dama*. The example from 695 is broadly similar to one of *Cervus simplicidens* shown by Sahni and Kahn (1988: Figure 168) from the Tatrot-Pinjor zones (see Figure 10: 1). *C. punjabiensis* (see Colbert 1935: Figure 149) has a similar but more elaborate type of antler; this example came from the Upper Siwaliks immediately below the Boulder Conglomerate, and is thus probably from the Pinjor Formation.

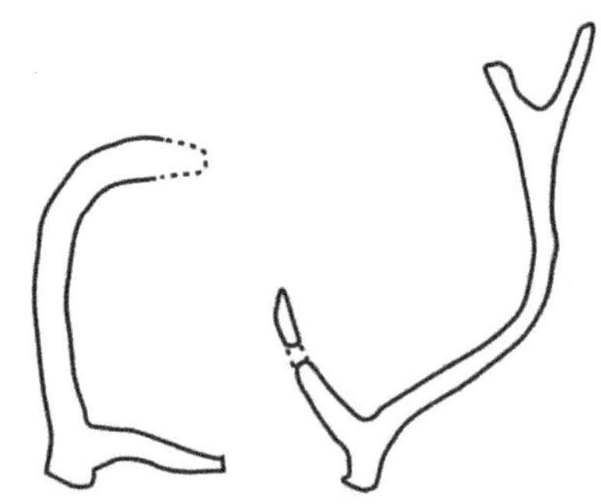

Figure 10. 1: Cervid antler. The Tatrot-Pinjor example is from Sahni and Kahn (1988, Figure 168)

The palmate type is harder to identify. It is necessary to rely only on specimens that show the pedicel, as the upper branches of *C. simplicidens* and *C. punjabiensis* flatten out and look palmate. However, Table 10.11 lists several palmate examples with the pedicel, and it is thus likely that a second type of cervid is present.

Artiodactyla (Bovid/Cervid) post-cranial remains

A large amount of post-cranial material was found that could be classed as bovid or cervid, and most of the remains identified as Artiodactyl were probably bovid or cervid. Measurements of distal elements are listed in Table 10. 12, and proximal ones in Table 10.13. Table 10. 14 provides measurements of astraguli.

Although few of the post-cranial bovid and cervid remains can be linked to specific types of dentitions, horns and antler, they are nevertheless a useful source of data. First, they give some indication of the likely body size and range. Secondly, they can also provide some information on the likely community structure. As example, the size range of elements such as the distal metapodia and astragali can provide some indication of the probable ratio of large- to medium- to small-bovids and cervids (see Figures 10. 2 and 3). Although these ratios are subject to a variety of distorting factors - such as the predatory preferences of *Pachycrocuta*, or differential taphonomic histories - these should be ironed out in samples as large and diverse as those from the Pabbi Hills. (This line of enquiry will be explored in the next chapter).

Finally, the post-cranial data supplements dental evidence for different types of bovid, particularly when the dental evidence is sparse. A good example is the evidence for the smallest bovid recognised from dentitions and which may have been a small antelope. As indicated above, the dental evidence for this taxon in the Pabbi Hills collections is very slight. However, the post-cranial evidence for this animal is much greater (see Table 10. 15), particularly at locality 642, where several remains of the fore- and hind-limbs of one individual were found. As Figure 10.3 shows, its distal metacarpals form a small but distinct group from those identified as *Damalops*, and thus these are unlikely to represent a juvenile form of the latter. Relevant details are given in Tables 10. 12 – 14.

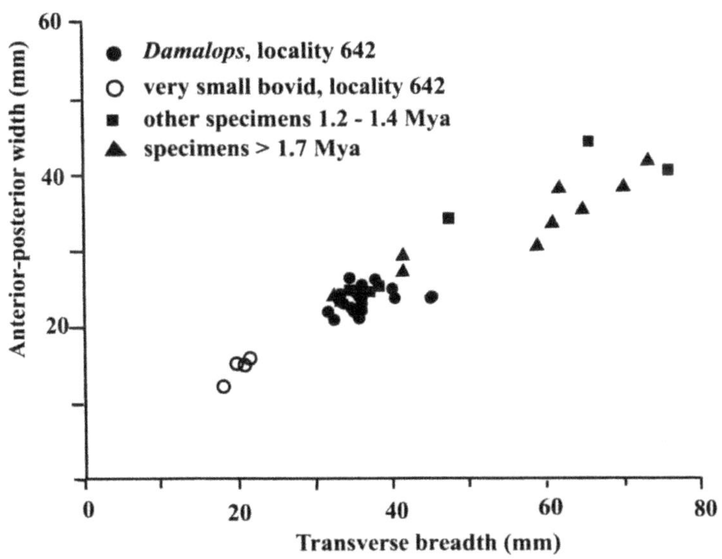

Figure 10. 2: Scattergram of bovid distal metacarpals

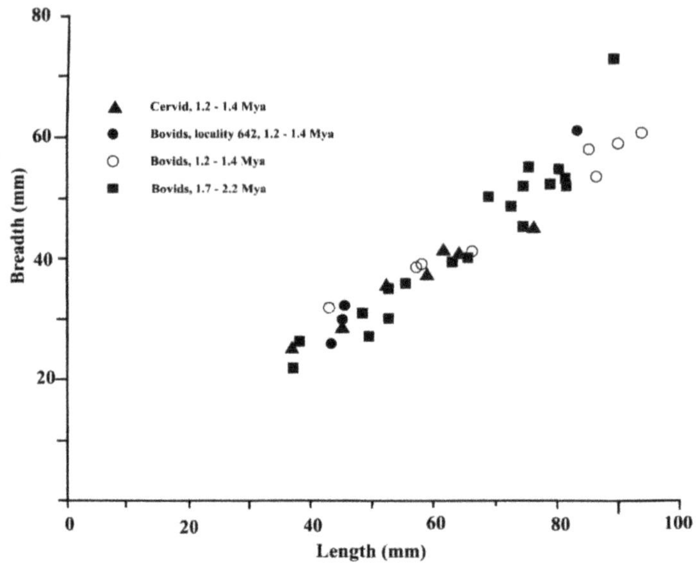

Figure 10. 3: Scattergram of bovid and cervid astraguli

iii) Giraffidae

The basic papers on fossil giraffids are by Pilgrim (1911), Churcher (1978), and Arif and de Vos (1989). Colbert (1935: 339-345) also provides detailed descriptions, and recent examples of *Sivatherium giganteum* from Mangla-Samwal are described by Arif and de Vos (1989). Siwalik giraffids are probably ripe for a detailed taxonomic review. Nevertheless, a few dental specimens were found that can be referred to *Sivatherium giganteum,* a large, 4-horned and probably short-necked giraffe. According to Churcher (1978: 530), Sivatheres were browsers that probably lived in woodland sufficiently open for them to move without obstruction because of their size. Because of their short neck, they probably fed on low woody plants and ground-level vegetation. The cheek teeth are easily recognised by their size; by very rugose enamel; and by very deep infundibula between the lingual and buccal cusps. The best example was specimen 378GB1 (1.7 – 1.9 Mya), an excellent upper M1 or M2, 51mm long, by 49mm wide. This compares with 50 x 50mm for an M^1 and 55 x 52 for a M^2, listed in Colbert (1935: 345). A good comparable example is shown in Plate 36, Figure 2, of Badam (1979). Table 10. 16 lists the specimens attributed to *Sivatherium giganteum;* further measurements can be found at the end of Table 8.7. The length of 1st phalange 362X110 was 125.0mm., comparable to that of a modern giraffe, length ca. 132.0mm.

Table 10. 16: *Sivatherium giganteum* tooth and post-cranial remains
a) teeth

Fossil occurrence	Age	Element
264GB1	1.7 - 1.9	Upper premolar
265GB61	1.9 - 2.2	Molar fragment
355GB15	1.9 - 2.2	Upper molar fragment
378GB1	1.7 – 1.9	M1/2 upper, 51 x 49 mm.
412GB2	1.7 - 1.9	Spall
440GB1	1.7 - 1.9	Molar crown fragment; deep fosse, rugose
459GB8	1.9 - 2.2	Rugose fragment, thick enamel
475GB19	1.9 - 2.2	Rugose fragment, thick enamel
477GB19	1.9 - 2.2	Spall
481B49	1.7 - 1.9	Mandible, P2-4, M1 roots

b) Giraffid post-cranial material.

Specimen	Age	Element	Anterior-Posterior	Width
293GB2001	1.9 - 2.2	Proximal radius	84.0	165.0
299GB1	1.9 - 2.2	Unciform		
352GB2	1.7 - 1.9	Proximal 1st phalange	59.0	59.0
353GB2	1.9 – 2.2	Proximal 1st phalange	57.0	61.0
362GB100	1.7 - 1.9	Metacarpal		
362B13	1.7 - 1.9	Unciform		
362XGB19, 25, 27	1.7 - 1.9	Radius and ulna, complete		125.0 (distal)

Table 10. 16 continued: *Sivatherium giganteum* tooth and post-cranial remains

362B1	1.7 - 1.9	Distal metacarpal	61.0	112.0
362C/0	1.7 - 1.9	Distal radius		110.0
362GB30/1	1.7 - 1.9	Metatarsal shaft		
362EX110	1.7 - 1.9	Proximal 1st phalange	56.0	55.0
481EGB2	1.7 - 1.9	Distal humerus	88.0	115.0
535GB2001	1.9 - 2.2	Navicular-cuboid		
535GB2	1.9 - 2.2	Proximal 1st phalange	57.0	61.0

iv) Suidae

Suid remains are not plentiful in Upper Siwalik deposits, but some useful additional new material has been obtained which considerably amplifies the scant record obtained to date. These include two superb crania found eroding from the surface (fossil occurrence 525, found in 1987, and a specimen found at fossil occurrence 343 in 1999), a pair of mandibles and an articulated rear leg excavated at locality 73, and several dentitions. There appear to be two types, one larger (e.g. 666), and the others (e.g. 525) smaller. However, there are major size differences in suids between males and females, young and old, and so the observed differences may be due to age and sex. This material awaits detailed study by a suid specialist. If there are indeed two types, previous studies suggest that they should be attributable to *Sus falconeri* (related to the modern wart-hog), and *Potamochoerus palaeindicus* (Colbert 1935: 256-266).

The principal dental evidence is summarised below. (Note: maximum lengths of isolated 3rd molars teeth are not taken along the crown, as the posterior end can be difficult to define, but along where the tooth emerged from the mandible. Otherwise, these individuals may appear to derive from much larger animals than those whose teeth were still retained within the mandible or maxilla.).

a) mandibles
73EX314 (1.2 – 1.4 Mya) A pair of mandibles of a young suid. Both mandibles have the I1-2, C, dm1, dm2, dm3, M2 and M3. The dm2 of the left mandible has snapped off, showing the permanent tooth erupting underneath. The M2 and M3 are unworn. The left M2 has lost its buccal anterior cusp. The right mandible has the dm1; and mandible has broken along the dm3, which thus looks like two separate teeth.

Left dm3 = 20.0 x 10.0 mm.; Right dm3 = 21.0 x 11.0 mm.;
Left M2 = 22.0 x 14.0 mm.; Right M2 = 22.0 x 14.5 mm.;
Left M3 = 24.0 x 16.8 mm.; Right M3 = 28.0 x 17.0

495GB7 (1.7 – 1.9 Mya). A left lower M3, with part of the mandible. It is broken anteriorly. The enamel is slightly rugose. This is moderately worn.

610/1/GB1 (1.2 – 1.4 Mya) A right mandible fragment with the M3, unworn. Length in mandible = 48.0 mm.; Total length = 60.0 mm.; width = 20.0 mm.; Length = >51.0 mm.; Width = 31.0mm.

b) maxillae: The main specimens are as follows:

343GB2001 (1.9 – 2.2 Mya). A superb cranium, similar to 525GB1 (see below); found in 1999

525GB1 (1.9 – 2.2 Mya). A magnificent specimen of a complete cranium, slightly distorted through compression. It has a clay matrix and requires further conservation. This is an old individual, as the teeth are very worn. The tooth measurements are as follows: M1 = 27.0 x 23.0 mm.; M2 = 28.0 x 28.0 mm.; M3 = 27.8 x 23.0 mm.; M1-M3 = 81.5 mm.; Length of M1+M2+M3 = 82.8 mm.; Overcrowding index = 98.4%

610/4/1/GB1 (1.2 – 1.4 Mya) A right maxilla with the P2, P3, P4 and M1. The P2-4 have lost their crown enamel, and have very flat occlusal surfaces. The M1 has good enamel, and is little worn. Subsequent to discovery, the specimen broke, revealing the erupting M2 inside the maxilla. Length (M1) = 22.0 mm.; width (M1) = 15.0 mm.;

642GB3/258 (1.2 – 1.4 Mya). A left maxilla with the M2 and M3. The M2 is worn, and the M3 is unworn. M2 length = 25.6 mm.; M2 width = 17.8 mm.; M3 length = 43.0 mm.; M3 width = 18.0mm.

660GB10 (1.2 – 1.4 Mya). A maxillary palate with the M2, in poor condition. The tooth is moderately worn, and the occlusal surface is crinkly. M2 length = 35.5mm.; width = 24.0mm.

666GB1 (1.2 – 1.4 Mya). A left upper M3, with part of the upper M2 and part of the maxillary palate. The teeth are very worn; the occlusal side is slightly concave, and the enamel is rugose. M3 length = 53.0mm.; width = 28.0mm.

666GB104. (1.2 – 1.4 Mya) A maxilla with the M3. The tooth is very worn, and the occlusal surface is slightly concave. M3 length = 52.0mm.; width = 29.0mm.

Note that the M3 of this specimen is much smaller than 68BGB280, 387GB17 and 666GB1, but is similar in size to 68BGB1003, 68BGB2006, 208GB2001 and 383GB2001.

c) isolated teeth. These are tabulated below.

Table 10: 17. Isolated suid tooth measurements

Specimen	Age	Details	Length	Width
16GB8	1.2 - 1.4	A lower M3 fragment, very worn.	-	-
38GB15.	1.2 - 1.4	A lower molar, possibly a dm2.	13.2	8.0
68BGB280	1.2 - 1.4	A left, very worn, lower M3.	62.5	33.0
68GB999.	1.2 - 1.4	A right lower M3, in poor condition	49.0	20.0
68BGB324	1.2 - 1.4	Ca. half an upper molar fragment, worn	40 (est)	17.0
68BGB1003	1.2 - 1.4	A right upper M3, with little wear	54.0	29.0
68GB2006	1.2 - 1.4	An upper M3, very worn	>49.0	27.0
68CGB24.	1.2 - 1.4	A right lower M3, broken anteriorily, and moderately worn.	>54.0	22.0
379GB10		the posterior part of a lower M3		>22.0
383GB2001		upper molar, (?M3), broken anteriorly. Cf 68GB2006	>35.0	22.5
387GB17		upper M3, very worn	57.0	27.0
435GB8		lower dm3, little wear, and part of mandible	23.8	9.0
463GB1		broken molar fragment, very worn		25.0
610/3/GB1	1.2 - 1.4	lower PM	14.0	7.5
208GB2001		right upper M3, little wear	34.0	20.0

Notes: **68BGB1003**: Crown length = 30.0mm.; crown width = 22.1mm. The anterior end is broken. The occlusal surface is broken, and it is lower on the buccal side. The enamel is rugose, and the sides are crinkled. This specimen is very like 387GB17.
68GB2006: crown length = 32.5mm.; crown width = 21.0mm. The enamel is rugose, with a wrinkly occlusal surface. There are two distinct pillars for the cusps.
387GB17. The anterior end is broken. The occlusal surface is broken, and it is lower on the buccal side. The enamel is rugose, and the sides are crinkled.

v) Perissodactyla: Equidae

A substantial amount of equid remains were found during the surveys, and useful specimens were also recovered from the excavations of localities 73 and 642. These are especially useful as each contained the remains of several horses, and provide information on the size of fore- and hind-limbs as well as dentitions. Another useful group of finds came from Sandstone 12 (fossil occurrences 30 and 34, 16 and 17, and 109). Find-spot 689 (hairpin area near Baroth) also produced a good pair of equid mandibles, and an excellent pair was also excavated in 1989 at find-spot 693 in the 203 horizon (near and slightly above 269). Another superb pair, with all teeth present from the incisors to rear molars was excavated from locality 109 (Sandstone 12) in 1999: these three finds await conservation, as does cranium 642EX1609 from locality 642. Principal dental material is listed below. Later in this chapter, we shall examine i) whether there is any evidence for *Hipparion* (the 3-toed equid); ii) whether the equids are homogeneous and iii) their relationship to the equids found in earlier Upper Siwaliks, notably at Mangla-Samwal.

Details of the main equid mandibles are summarised in Table 10. 18, and isolated teeth in Tables 10. 19 and 10. 20 at the end of the chapter. Table 10. 21 lists some comparative measurements, and show that the Pabbi Hills equids are comparable in size to those found elsewhere in S.W. and S. Asia.

Table 10. 18: Details of equid mandibles

Specimen	P3	P4	M1	M2	M3
26GB1-3, left	26.0 x 15.0	26.0 x 15.0	- x 15.0	23.0 x 15.0	29.0 x 13.0
73GB5, left					31.4
693GB1, left		25.0 x 16.0	28.0 x 16.0	28.0 x 16.0	32.0 x 15.0
689GB2, right		28.0 x 15.0	27.5 x 16.0	27.0 x 16.0	33.0 x 16.0
642/2A/GB64, right					34.0
683GB1-6	Left mandible +M2, M3. Worn, good condition. Shallow linguaflexid, shallow U-shaped ectoflexid M3; part of right mandible				28.0 x 13.5
693GB1, 693GB2	Left and right P4, M1-3. Very worn, good condition. All teeth hollow except for main cingula.				

Notes: There is also another complete cranium (642EX1609) that needs cleaning. The teeth (P3-M2 on both sides) are too encrusted to draw or measure accurately, but the protocone is joined to the rest of the cusps. The excavations of locality 642 also produced an equid maxilla (642EX1110), and two mandibles (642EX1215 and 1532) that also require cleaning.

Table 10. 21: Comparative specimens of equid teeth. Sources: M17030 and 17040 from Mirpur are from Hussain et al. (1992: 73). The Dmanisi measurements are from Vekua 1995, Tables 23 and 24. The CASG and WIF/A measurements are from Nanda, 1978b, Tables 2 and 3, and the last three are from Badam 1979, Table 11.

Specimen	Tooth	Length	Width
M17030	Lower M3	32.7	13.7
M17040	Lower M3	28.9	13.8
CASG342	*E. sivalensis*, upper M3	26.0	27.0
CASG344	*E. sivalensis*, upper M3	28.0	27.0
CASG346	*E. sivalensis*, upper M3	31.0	31.0
CASG347	*E. sivalensis*, upper M3	31.0	30.0
WIF/A117	*E. sivalensis*, upper M3	28.0	31.0
WIF/A119	*E. sivalensis*, lower M3	35.0	15.0
Dmanisi	*E. cf. stenonis*, upper M3	33.4	28.5
Dmanisi	*E. cf. stenonis*, lower M3	32.0	15.0
Dmanisi	*E. cf. stenonis*, lower M3	33.0	13.7
Dmanisi	*E. cf. stenonis*, lower M3	33.0	13.7
Dmanisi	*E. aff. altidens*, upper M3	29 - 35	13 – 15.5
Dmanisi	*E. aff. altidens*, upper M3	23 - 26	23 – 24.5
A/601	*E. sivalensis*, lower M3	28	8
A/602	*E. sivalensis*, lower M3	28	14
B/40	*E. sivalensis*, lower M3	35	16

Hipparion or *Equus?*

One of the key issues in studies of Upper Siwalik biostratigraphy concerns the timing of the first appearance of *Equus*, and the last appearance of *Hipparion*, the tridactyl (3-toed) horse. Lindsay et al. (1980) showed that *Equus* first appeared in South Asia ca. 2.5 Mya, most probably from North America. According to Barry et al. (1982), the earliest specimen of *Equus* in the Upper Siwaliks is from locality D68 at Hasnot, just below the Gauss-Matuyama palaeomagnetic boundary (2.48 Mya or 2.7 Mya on the revised Kent time-scale). At Mangla-Samwal, Mirpur, *Equus* also appears at around this time (Hussain et al. 1992). Its first appearance in Europe and East Africa was placed somewhat later, at ca. 1.8 Mya. The appearance of *Equus* is thought to coincide with the immigration of new taxa, notably *Elephas* and *Leptobos* (Barry's *Elephas-Leptobos-Equus* horizon).

The date at which *Hipparion* becomes extinct remains problematic. According to Barry et al. (1982) its last appearance was during the *Elephas planifrons* Interval-Zone (3.3 - 2.7 Mya); in other words, it was probably replaced rapidly by the immigrant *Equus*. Opdyke et al. (1979) however claim that the last appearance of *Hipparion* is just above the Olduvai Event, ca. 1.5 Mya, on the basis of material from the Pabbi Hills. This find is very anomalous, as *Hipparion* is not evidenced at Mangla-Samwal, Jalalapur, Rhotas, or any other Upper Swialik section that covers the period between 1.5 and 2.5 Mya old. In India, *Hipparion* is not thought to extend into the Pinjor Zone, i.e. beyond 2.5 Mya (see Nanda 1997: 182-3). One possibility is that the Pabbi Hills specimen was reworked; or (unlikely) misidentified. Given the large number of equid remains from the Pabbi Hills, the absence of *Hipparion* in earlier sections is unlikely to be an artefact of inadequate sampling. (Sadly, the specimen cannot now be traced).

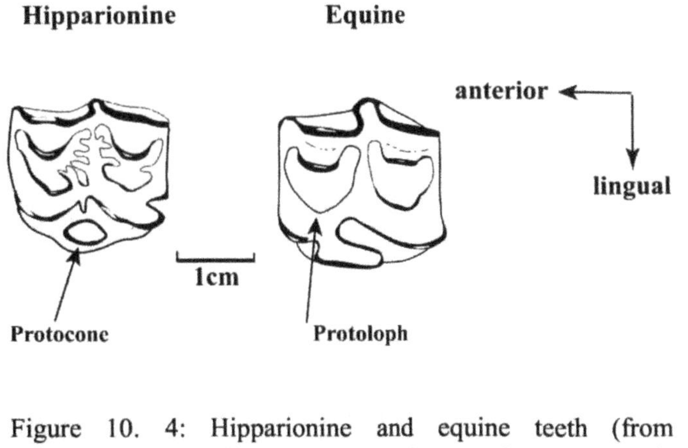

Figure 10. 4: Hipparionine and equine teeth (from MacFadden 1992, Fig. 5.18). Note the way the protocone is distinct in *Hipparion* but not in *Equus*.

Somewhat perversely, the main criterion for distinguishing *Hipparion* from *Equus* is dental rather than pedal. In *Hipparion*, the protocone of the molars and premolars is separate from the main cusp, whereas it is joined in *Equus* (see Figure 10. 4). MacFadden (1992: 109) notes that in old specimens of *Equus*, the protocone may appear to be distinct as a result of wear. This may explain the anomalously late sighting of *Hipparion* by Opdyke et al. (1979). Figure 10. 5 (end of chapter) shows the cusps of those equid teeth from the Pabbi Hills that were sufficiently complete and clear to be drawn. As can be seen, none shows a distinct protocone, and all show the cusp morphology of *Equus*. This evidence is interpreted as indicating that *Hipparion* was either extinct or extremely rare after 2.5 Mya.

Equid post-cranial specimens

Very little information is available on the post-cranial skeleton of early Pleistocene equids, particularly from Asia. Measurements of complete specimens are listed at the end of the chapter in Table 10. 22, with metapodial measurements from Mirpur for comparison. Table 10. 23 shows other post-cranial measurements, and Figure 10. 6 illustrates the size of distal metapodia from the Pabbi Hills and Mirpur.

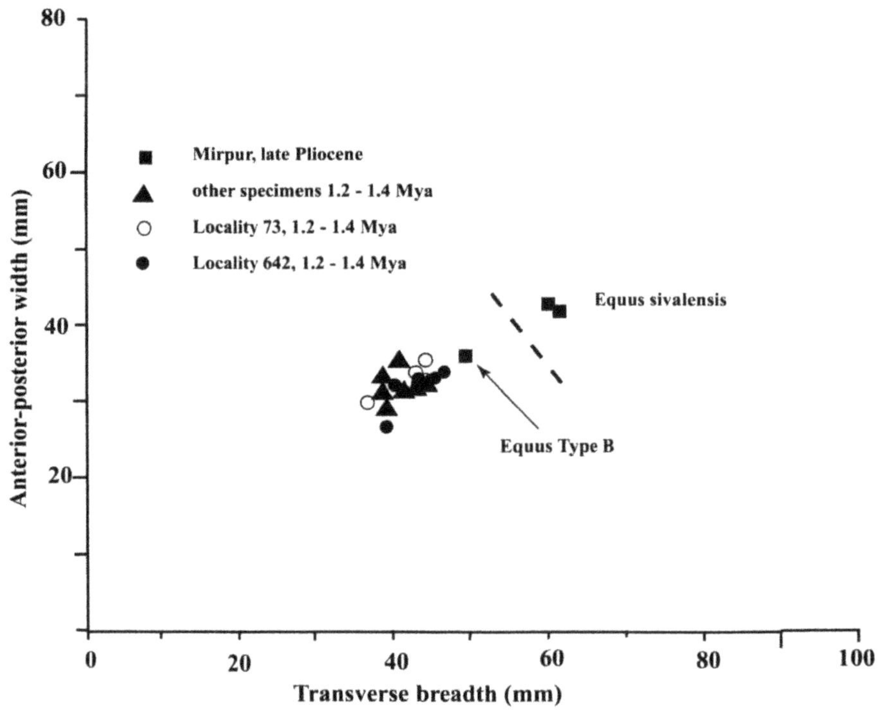

Figure 10. 6: Scattergram of equid distal metapodia from the Pabbi Hills and Mirpur

Discussion

Equus sivalensis is the main type of equid recognised in the Upper Siwaliks, particularly in India, and thus it seems reasonable to attribute the Pabbi Hills equid specimens to this taxon. However, Hussain et al. (1992) recognised two types of *Equus* in their early Upper Siwalik section at Mangla-Samwal. Most of these specimens were identified as *Equus sivalensis* on the basis of size and morphology. A smaller type was also recognised at locality 8460, and provisionally called *Equus sp. A*. Interestingly, the equid post-cranial material from the Pabbi Hills is distinctly smaller than those remains identified at Mirpur as *E. sivalensis*, but also slightly smaller and more gracile than that called *Equus sp. A*. by Hussain et al. (1992) (See Table 10.22 and Figure 10. 5).

This issue cannot be resolved with the evidence available, but various explanations can be offered. One is that *E. sivalensis sensu strictu* has a short time-range and was later replaced by an equid which was smaller post-cranially. Alternatively, *Equus sivalensis* may have become smaller during the Late Pliocene and Early Pleistocene. Another possibility is that other Upper Siwalik equid material that is normally classified on dental grounds as *E. sivalensis* may belong to a smaller species that has yet to be formally designated. Finally, the larger type of equid at Mirpur might derive from *E. cautleyi*, which Sahni and Kahn (1988: 51) state is larger than *E. sivalensis*. Given these uncertainties, it seems prudent to refer to the Pabbi Hills equids as *Equus cf. sivalensis*.

i) Rhinoceritidae

Although these are rare, some useful material was obtained that probably relates to *Rhinoceros sivalensis*. These include a complete cranium (642EX1602), two mandibles (341GB33 and 642EX1062), a scapula (73EX5) and numerous limb and teeth specimens. These are listed in Table 10.24 at the end of the chapter.

ii) Proboscidea

Useful papers on Siwalik Proboscidea are by Nanda (1976, 1988), and Nanda and Corvinus (2000). Sahni and Kahn (1988) show some useful photographs and descriptions. Two types of Proboscidean are represented in the Pabbi Hills. The first, and commoner, is *Elephas hysudricus,* and the other is a species of *Stegadon*. There is no evidence of *E. planifrons*, which conforms earlier suggestions by Hussain et al. (1992) that it may have become locally extinct by 2.7 Mya.

According to Nanda (1976: 16), *Elephas hysudricus* is distinguished from *E. planifrons* by:

i) presence of medially expanded plates in *E. planifrons*
ii) greater thickness of enamel (3.5 - 4.0mm) in *E. planifrons*, compared with only 2.0 - 3.0mm in *E. hysudricus.*
iii) almost complete absence of plications in the enamel of *E. planifrons*, whereas the enamel is highly plicated and also smooth in *E. hysudricus*
iv) lower lamellar frequency (ca. 4 - 4.5 in *E. planifrons*, and 5.5 - 6.5 in *E. hysudricus*)

The following Pabbi Hills specimens (see Figure 10. 7, Table 10. 25) concur with these criteria.

Table 10. 25: Principal *Elephas hysudricus* specimens:

Specimen	Age (Mya)	Details
12GB1*	1.2 - 1.4	Right mandible + molar, 5 plates
32GB1*	1.2 - 1.4	Rear part of molar, 3 rows of conelets; cf. 265GB2001
57GB72**	1.2 - 1.4	Plate fragment
83GB1	1.2 - 1.4	Juvenile molar, 4 plates, 64 x 31mm
238GB4	1.7 - 1.9	Molar fragment
265GB2001	1.9 - 2.2	Molar fragments, 4 plates
265GB7	1.9 - 2.2	Molar fragment, 3 plates; enamel thickness 2.0 - 2.3 mm
293GB2	1.9 - 2.2	Premolar fragment, 3 plates
302GB6	1.9 - 2.2	Molar fragment
375AGB3	1.9 - 2.2	Molar fragment
386GB10	?1.7 - 1.9	Molar fragment
448GB1	?1.7 - 1.9	Molar fragment, 5 plates, worn. Valleys filled with cement.
535	1.9 - 2.2	Complete maxilla

Notes: * = gully collections early in the 1986 season. 12GB1 was in the vicinity of Sandstone 12. ** a large sandstone block with Proboscidean skull fragments, left in situ.

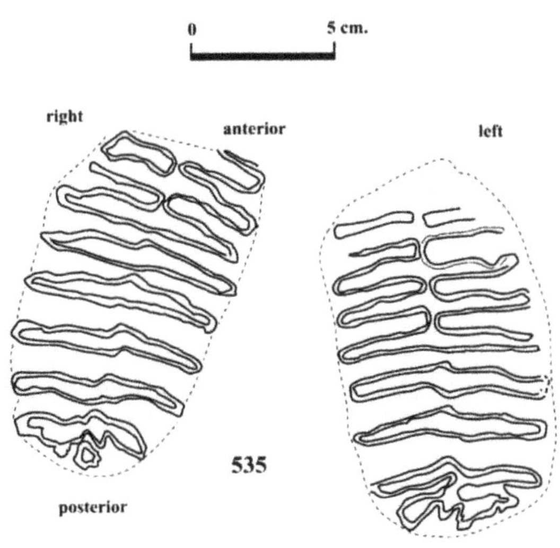

Figure 10. 7: *Elephas* molar plate details of 535GB1.

Stegodon:

Nanda (1988) lists the following criteria for the plates on a tooth of the Stegodont, *S. kathiensis*:

"**Plate 1**: On the posterior-most plate, there is a lingual cone to expose the enamel. Buccal to this cone, the plate forms a figure, and dentine is exposed. The enamel is folded (?). The anterior enamel slopes gently anteriorily, unlike the posterior enamel, which is steep.
Plate 2: Much more worn and forms a full figure. The plate is convex anteriorily.
Plate 3: Much more worn. In the medial part, the anterior and posterior enamel of the plate are widely separated, giving it a greater anterior-posterior width than in other species. The valleys between plates 2 - 3, and 3 - 4, are blocked medially as enamel touches.
Plates 4,5: Much more worn. Valley between 4 and 5 is almost blocked. The A-P width of plate 5 is 33mm."
The best specimen of *Stegodon* was X507, a complete pair of mandibles ca. 1.9 - 2.2 Mya-old. On the right mandible, the tooth is 174mm long and 83mm wide. There are 7 ridges. The lamellar frequency is estimated at 4.5. The enamel thickness is 5.5 - 9.0mm, and the crown height above the mandible is 47mm. (see Figure 10. 4). A second specimen (also kindly identified by A.C. Nanda) was a molar fragment 139GB56 (ca. 1.4 - 1.7 Mya-old). The length is 180mm., and the lammelar frequency is estimated at 4.0. It can also be compared to specimen 17264 from Mirpur. Comparable examples of *Stegodon* teeth are shown in Sahni and Kahn (1993, Figures 1, 8, 19, 22 and 23).

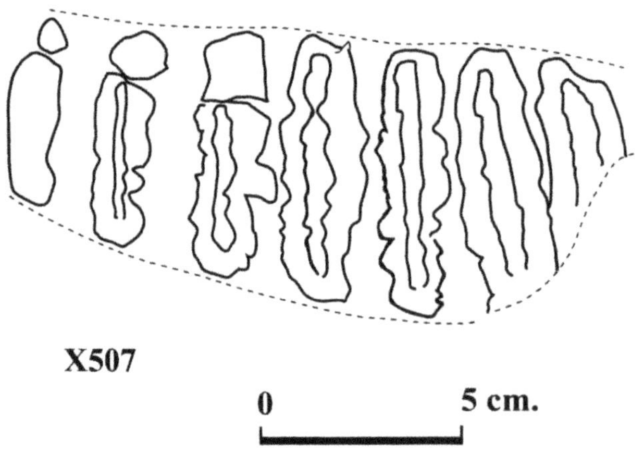

Figure 10.8: Detail of *Stegodon* tooth X507

viii) other taxa

Anthracoceridae: These are very rare in the Upper Siwaliks. Recently, Steensma and Hussain (1992) described a maxillary fragment of *Merycopotamus dissimilis* from Mangla-Samwal, dated at 2.40 ± 0.20 and 2.65 ± 0. 21 Mya b.p. This is the first well-dated specimen of this type, which is known from Pakistan, India, Burma and Java. Although no dental specimens of anthracotheres were found in the Pabbi Hills, two metacarpal fragments and a distal ulna from occurrence 529 (1.9 - 2.2 Mya; Table 6. 26) were identified by A. Currant (Natural History Museum, London) as anthracocere.

Hippopotamidae: 17 specimens were referred to this taxon (see Table 10. 26), and in two cases, the identifications were confirmed in the UK against comparative material. All came from the early part of the sequence, and apart from two tusk fragments from scatter 241 (1.4 – 1.7 Mya) and locality 68 (1.2 – 1.4 Mya) that were initially listed as Hippotamid but are more likely to be suid, none was found in deposits younger than 1.7 Mya-old. It is likely that these belong to *Hexaprotodon sivalensis*, a type known elsewhere from the Upper Siwaliks (see e.g. Nanda 1978c). This genus is identified chiefly from its six incisors, although Matthew (1929: 555) stated "I cannot see any very strong reason for according full generic status to *Hexaprotodon*. Except for the incisors, it is in no way different from the modern species" of *Hippotamus*". The genus has persisted as independent of *Hippopotamus*, however. *Hexaprotodon* is somewhat smaller, and probably related to the pygmy hippo, now found only in West Africa (Jablonski 2004). According to Jablonski (2004: 121), the last definite appearance of *Hexaprotodon sivalensis* in the Pakistani Siwaliks is 3.5 Mya, although it may have persisted until 2.2 Mya (Barry et al. 2002). The above data indicate that it persisted longer, up to, but not beyond the Olduvai Event.

Table 10. 26. Hippopotamid specimens from the Pabbi Hills

Occurrence	Specimen	Age (Mya)	Element	Other
244	1, 3, 9	1.7 - 1.9	Mandible	
244	7	1.7 - 1.9	Tooth	
248*	8	1.7 - 1.9	Tooth	Tusk
347*	15	1.9 - 2.2	Mandible	Premolar
385C	3	1.9 - 2.2	Radius	
393	5	1.9 - 2.2	Mandible	
413A	1	1.7 - 1.9	Tooth	Tusk
437B	7	1.9 - 2.2	2nd Phalange	Complete, mud skin matrix
437D	11	1.9 - 2.2	Mandible	No teeth
437D	23	1.9 - 2.2	Ramus	
437D	23	1.9 - 2.2	Unciform	
443	1	1.9 - 2.2	Mandible	Fragment
443	4	1.9 - 2.2	Mandible	Fragment
443	2	1.9 - 2.2	Mandible	Tooth root+1 broken tooth
443	5	1.9 - 2.2	Phalange 1/2	Fragment
443	10	1.9 - 2.2	Tooth	Fragment
489B	46	1.7 - 1.9	Femur	Distal

Notes: * = identification confirmed in the Natural History Museum, London

Chelonia: Almost all the evidence for turtles consisted of scute fragments. Two types were recognised: a smooth, and a dimpled type that can be sub-divided into a fine and coarse type. There was also a remarkably well-preserved cranium from scatter 492 (see Table 6. 27). It was not possible to identify this material to generic level without further study by a specialist. According to J. Head (formerly, Southern Methodist University, USA), the dimpled type was a Caroletocheyid, and the other, a Trionychid.

.

Crocodile: This taxon was represented by numerous teeth, carapace fragments, and some post-cranial material, all of which awaits a specialist's attention. The carapace fragments can be recognised by their deep netted surface. (Some of the egg-shell fragments attributed to *Struthio* may have been crocodilian). The standard paper on Siwalik crocodiles is by Patnaik and Schleich (1993), who examined material from two Upper Siwalik localities in Himachal Pradesh, India. They found evidence of both *Crocodylus cf. palustris* and *Gavialis cf. gangeticus*. Their choice of habitat is broadly similar: "Tropical woodlands with wide river systems, deep permanent water channels and associated riverine shore areas ranging from rocks, sand banks, grassbanks are to expect for *Gavialis*, while *Crocodylus palus*tris seems to be more restricted to hot swamp areas ranging from small riverine forests with side channels, streams to ponds, lakes and slower river parts of similar tropical climates" (Patnaik and Schleich 1993: 93). It is likely that both crocodile and gavials would have been present in the Pabbi Hills region in the Late Pliocene and Early Pleistocene.

Molluscs: Samples of fossil molluscs were collected from several fossil occurrences, particularly in Kotha Kas in 1987. I am indebted to Dr. Katsumi Takayasu (ReCCLE, Shimane University, Nishikawatsu-cho, Matsue, Japan) and Damayanti Gurung from Nepal. In their opinion, they are all typical of a large, slow-moving river with clean, shallow water, typically <5 m. deep, and with abundant aquatic vegetation. They suggested that non-polluted sections of the Ganges would be a suitable analogue. It is unsurprising that turtle and crocodile remains were common at most of these occurrences. The following table lists their identifications:

Bellamya sp.: fossil occurrence 500, 664
Indonaia sp.: fossil occurrence 347
Lamellidens sp.: fossil occurrences 347, 371, 389, 489B, 489C, 489E, 506, 669
Parreysia sp: fossil occurrences 297, 389, 427, 445, 450, 451, 459, 467, 481, 485, 489B, 489E, 491D, 507
Viviparidea: fossil occurrence 500

Primates: In 1999, a small mandibular fragment was found during a re-visit to locality 491 that has been provisionally identified as a macaque. If it is, it was the only Upper Siwalik primate discovered in the 20th century. Four other specimens were reported in the 19th century (Barry 1987) and were probably Cercopithecidae.

Other finds: coprolites: These were found at several fossil occurrences: 34, 79, 113, 124, 437, 452, 463, and 654 (1 each); 83 (2 specimens); 14 (3 examples); and 15 at 666. Some were also found at locality 73 during the excavation. These were typically spherical, and not unlike those illustrated in Larkin et al. (2000, Figure 6a and 6b).

Interestingly, these are not the elongated, segmented type shown by Brain (1981: Figure 58). Of those that were measured, 113GB3 was 37 x 25mm, and 654GB5 was 35 x 33mm.: both are within the size ranges of recent and fossil hyaena coprolites listed in Larkin et al. (2000). Further analysis of these specimens would be desirable.

Part 2: an overview of the faunal remains from the Pabbi Hills

In discussing the fossil mammalian material collected from the Pabbi Hills, we need to consider two themes that were raised in the introductory chapters of this study. The first was that of faunal change within the Upper Siwaliks. It was noted in chapter 2 that Hussain et al. (1992) have identified an *E. hysudricus* faunal zone that begins at Mangla-Samwal ca. 2.7 Mya, but which is currently left open-ended, and without further subdivision. Although the Pabbi Hills cannot illuminate any trends between 2.7 and 2.2 Mya, it can indicate if any changes occurred between 2.2 and 0.9 Mya ago. Additionally, the biostratigraphic scheme proposed by Opdyke et al. (1979) needs re-assessment in the light of the new data presented here.

The second issue is whether the Pabbi Hills data are consistent with those from East Africa in showing a general trend towards more open conditions between 1.8 and 1.6 Mya ago. Each can be taken in turn.

1) Continuity and extinction in the Pabbi Hills sequence

As stated at the beginning of this chapter, approximately half the fossil material from the Pabbi Hills is derived from Sandstone 12-times (1.2 - 1.4 Mya), and the other half from deposits between 1.7 and 2.2 Mya-old. Each is roughly the same size (ca. 20,000 specimens), and sampled the same range of depositional environments: channel bars, abandoned channels, and overbanks. Consequently, the absence of a taxon from one time band is likely to be genuine unless it was extremely rare, and unlikely to be an artefact of inadequate sampling or small sample size. In contrast, the very limited amount of material from deposits between 1.4 and 1.7 Mya and 0.9 - 1.2 Mya means that the absence of a taxon from these units probably reflects the small amount of material relating to those time-bands. Further data collection could reasonably be expected to increase the range of taxa in these time-bands.

The main trends are summarised in Figure 10. 9. Overall, there appears to be a considerable amount of continuity in the range of large mammals in the Pabbi Hills between 2.2 and 1.2 Mya. The medium-sized bovid *Damalops palaeindicus*, and equids were the commonest ungulates. *Rhinoceros sivalensis, Equus cf. sivalensis, Elephas hysudricus* and *Stegodon* the only members of their genus, and there were probably two types of pig. The range of bovids has probably been under-estimated because of the limited number of well-preserved horn cores. In addition to *Damalops*, there were probably two others of the same size: an alcelaphine (seen at 610) and the bovid represented at locality 362. Additionally, there was a larger bovid, *Hemibos triqueticornis* and the even larger and much rarer one with Type 4 teeth. The very small bovid that is tentatively identified as an small antelope/gazelle is probably under-represented, but the numerous post-cranial remains at locality 642 confirms its existence in the Upper Siwaliks as late as 1.2 – 1.4 Mya.

The main changes appear to be ones of extinction after 1.7 Mya. Although rare, *Sivatherium giganteum* is evidenced at several fossil occurrences in the Pabbi Hills, but all are between 1.7 and 2.2 Mya-old. Its remains are robust, and easily seen by virtue of their large size. Its absence from the whole of the Sandstone 12 assemblage, both from 1986 and 1989, is thus probably genuine, and it may reasonably be expected to have become locally extinct by 1.4 Mya, and perhaps shortly after 1.7 Mya. However, Ranga Rao et al. (1995) report *Sivatherium giganteum* from the base of the Brunhes (0.63-0.73 Mya) at Patiali Rao, Pinjor. This find appears very anomalous in the light of the data presented here, and further work is needed to resolve this discrepancy.

Other taxa are also evidenced before, but not after, 1.7 Mya ago, most notably a small primate, *Canis cautleyi* (found in the excavated deposits of locality 362, see chapter 8 and Turner's report, this volume) and a very large bovid. All appear to have been extremely rare. The small primate is represented by only one specimen, found at 481 in 1999, by which time over 40,000 specimens had been collected in the Pabbi Hills since 1986. Locality 481 itself was first discovered in 1987, and revisited on several occasions thereafter. *Canis cautleyi* is represented only at locality 362, at which several parts of one skeleton are preserved. Although carnivore remains are rare in the Pabbi Hills (only 120?) in total, over half were found at locality 73, and the absence of this canid from this assemblage is probably genuine. Other taxa known only in the early part of the Pabbi Hills sequence (i.e. 1.7 - 2.2 Mya) include anthracoceres, hippopotamids, and possibly the medium-sized bovid with twisted horns, known only from locality 362. *Stegodon* may also have been extinct before Sandstone 12 times: its remains were very rare, and the youngest specimen (139GB56) is older than 1.4 Mya. According to Opdyke et al. (1979), however, at least one specimen of *Stegodon* is dated to the early Brunhes, ca. 0.7 mya-ago, although no details are given about provenance or identification. This find is additionally anomalous because it is the only taxon that they report from this period.

A few taxa are presented in assemblages between 1.2 and 1.4 Mya-old, but absent from older ones. These are *Ursus* (one vertebra, and one tooth), and snow leopard (one tooth); (see Turner's report, end of this chapter). As the teeth of these animals were found at locality 73, we can assume that they were autochthonous to the area, and not the result of downstream transport from elsewhere. The significance of their presence in the Pabbi Hills area by 1.2 - 1.4 Mya is hard to assess. Both are usually associated with upland/montane regions. Their presence may indicate cooler conditions, or that the uplift and forefront advance of the Karakourum had brought the Pabbi Hills within the annual migratory range of these animals. The fossil record of these animals is very sparse, however.

By the time of Sandstone 12, it is assumed that *Sivatherium giganteum, Canis cautleyi*, a small primate, anthracoceres and hippopotamids, and possibly *Stegodon*, had become locally extinct. This may indicate the expansion of grassland at the expense of woodland. This assumes that *Sivatherium* and the very large bovid were browsers, and that the small primate was arboreal. Future research may test this hypothesis by, for example, examining the C3/4 ratios in the dental enamel of herbivores from the Pabbi Hills. Another possibility worth exploring would be to see if gavials replaced crocodiles if the hot, swamp habitats of the latter became rarer during the Early Pleistocene.

2) The *Elephas hysudricus* faunal zone

In chapter 2, we noted that Hussain et al. (1992) had identified an *E. hysudricus* faunal zone, beginning at 2.7 Mya, but left open-ended as to duration or subsequent change. We can now suggest that this can be subdivided into an *E. hysudrics-Sivatherium* part up to 1.7 Mya, which is defined by the local extinction of *S. giganteum*, hippopotamids, anthracoceres, and also small primates, and a large canid. The period between 1.7 and 0.9 can be defined by *E. hysudricus-Ursus-Panthera*, although the FAD of both *Ursus* and *Panthera* may be extended back in time with further exploration. The LAD (last appearance date) of *E. hysudricus* is currently unclear. In Middle Pleistocene times, it is replaced by the modern *E. namadicus*, the appearance of which may be part of the Galerian faunal turnover that affected Europe and southern Asia. According to Nanda (1997), the end-point of *E. hysudricus* may be as late as 0.6 Mya.

3) Opdyke et al's (1979) Upper Siwalik biostratigraphy reassessed

Opdyke et al. (1979) made a major contribution in attempting to establish a reliable biostratigraphy for the Upper Siwaliks that was based on specimens that could be dated by reference to a palaeomagnetic time-scale. Their work considered material from several sections from both the Tatrot and Pinjor stages. In considering the Pinjor Stage, their main attention was on the transition from the Tatrot, and the first appearance of key taxa such as *Equus* and *Elephas*. Since their investigations, important new material has been published from Mangla-Samwal, Mirpur, and our research in the Pabbi Hills has of course produced a great deal of new material. In the light of these new data, the following revisions and additions are proposed to the scheme they proposed in 1979. Figures 10. 9 and 10. 10 provide schematic overviews of each scheme.

Bovids: we propose that *Damalops palaeindicus* is substituted for the material they identified as *Proamphibos*, and *Hemibos* for *Bubalus*. This takes in account material published by Dimitrieva (1977) that was not mentioned by Opdyke et al. (ibid.), and the discovery of the *Hemibos* specimen 16589 at Mirpur. We would also extend the range of *Damalops* (alias *Proamphibos*) from mid- to late-Matuyama times. The existence and time-range of *Gazella* is confirmed. Additionally, at least three other types of bovid are recognised: a medium-sized alcelaphine (locality 610, 1.2 – 1.4 Mya); a possible tragelaphine (locality 362, 1.7 – 1.9 Mya), and a very large bovid (1.9 – 1.2 Mya).

Cervids: two types of recognised on the basis of their antler, and their time range is extended slightly into the late Matuyama.

Giraffids: their presence in only the early part of the Pabbi Hills sequence is confirmed, with *Sivatherium giganteum* as probably the only member of this group.

Suidae: new material from the Pabbi Hills supplements the very sparse record for Upper Siwalik suids presented by Opdyke et al. (ibid.). There are probably two types that may be *Sus* and *Potamochoerus*, but these need confirmation.

Equids: the chronological range of *Equus* (probably *sivalensis*) is extended from mid- to late Matuyama times. More importantly, no evidence of *Hipparion* was found, and we consider it probable that it became extinct or was extremely rare by 2.5 Mya.

Rhinoceros: its time-range is extended into late Matuyama times

Elephas. The presence of one type of *Elephas* (most likely *E. hysudriscus*) is confirmed, as is its time-range throughout the Matuyama.
Stegodon: our data indicate that it was present in the early, but not the later, part of the Pabbi Hills sequence. The persistence of *Stegodon* into the Middle Pleistocene, as suggested by Opdyke et al. (1979), needs confirmation.

Hippopotamids/*Hexaprotodon*: in Opdyke et al.'s (ibid) scheme, its presence in the Matuyama chron was very tentative. We confirm its presence in the early Pleistocene prior to 1.7 Mya, but it seems unlikely that it persisted long thereafter.

Carnivores: these were very sparsely represented in Opdyke et al.'s (ibid) scheme. The new data indicate two types of hyaena (*Crocuta* and *Pachycrocuta*), not one, and possibly also *Hyaenictis* or *Leecyaena*. We suggest that *Panthera* should be substituted for *Felis*, and *Megantereon* for *Machairodus*. A third type (perhaps *Dinofelis*) may also be present. Two types of canid are recognised rather than one. Two types of canid are recognised rather than one. The larger one (*Canis cautleyi*) may have become extinct after 1.7 Mya. An ursid was also present in late Matuyama times.

Anthracoceres: these are very rare in the Upper Siwaliks, and were not reported by Opdyke et al. (ibid.) . New data indicate that these persisted into the early Matuyama, but were probably extinct by 1.9 Mya.

4) Faunal turnover in South Asia compared with that in East Africa

In the first part of this study, we noted that the East African faunal record (currently the best available for the period) indicates a series of complex, small and cumulative changes between 3.0 and 1.0 Mya ago. DeMenocal (1995) suggests there were three shifts towards more arid conditions, ca. 1.8, 1.7 and 1.0 Mya-ago. In general, conditions were more wooded before 1.6 Mya, and grasslands were commoner thereafter. Contra Vrba (1996), the radiation of hominids after 2.5 Mya ago cannot be linked to a marked increase in more open environments, although such a proposition appears more reasonable in discussions of *Homo ergaster* after 1.6 Mya ago.

The Pabbi Hills data generally support these conclusions. Overall faunal change between 2.2 and 1.2 Mya ago appears to have been slight, but extinctions noted above may be attributed to a reduction of tree-cover. This is consistent with analyses of soil carbonates that shows an overwhelmingly grassland environment in the Pabbi Hills after 1.7 Mya (Quade et al. 1993). What is perhaps surprising is the stability of the faunal communities and grassland-dominated landscapes in this region: despite at least a dozen glacial-interglacial cycles every 40 kya during the Lower Pleistocene, the overall characteristics of the landscape appear to have been very stable apart from those associated with tree-cover.

Table 10. 9: Suggested linkage of bovid teeth and horn-cores

Horn-type / **Teeth**	**Type A (small, twisted)**	**Type B (straight, large)**	**Type C (large, less twisted)**	**Type D?**	**Type E?**
Type 1 teeth (small) 14, 610 **642X1351**, 642X**1404** 362X46 (smaller)	*Damalops* 642X1377 642X1166 642X1181				
Type 1B teeth (small) 610AGB10, 610C1,3,5,9,51					
Type 2 teeth (large) 628GB1, Mirpur 16589		*Hemibos* 642EX1181, Mirpur 16589			
Type 2 teeth (large) 362G GB65			362EXGB12		
Type 3 (very small) 642				*Gazella*	
Type 4 (very large) 481EGB39, 125GB2, 668GB1, 6					Unknown

Figure 10. 5: Equine molar occlusal morphology from the Pabbi Hills

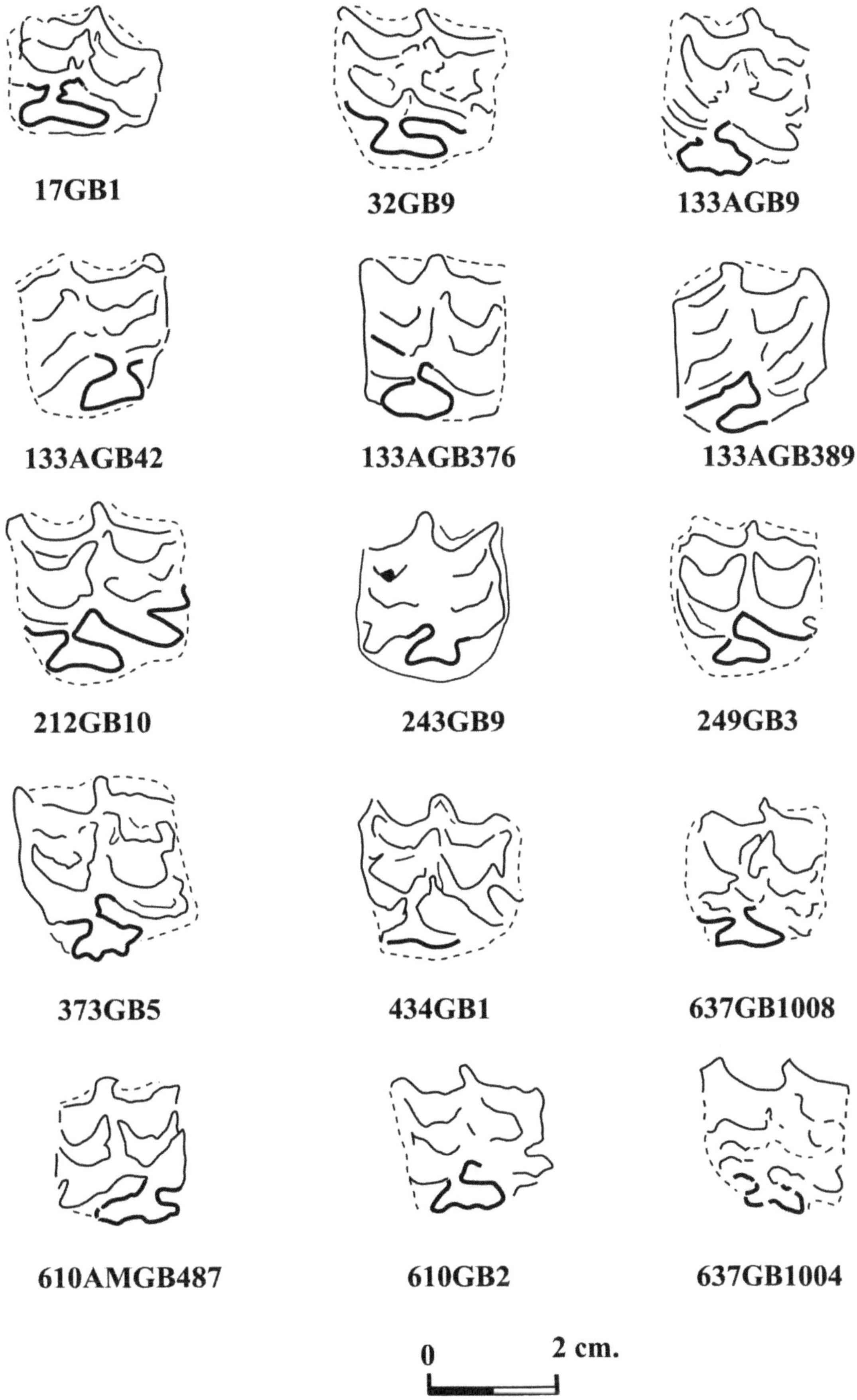

Figure 10. 9: Overview of the main biostratigraphic features of the Pabbi Hills. Note that the absence of many taxa from the periods 1.4 – 1.7 and 0.9 – 1.2 Mya is almost certainly due to the small number of fossil specimens from those periods.

Age range	**2.2 – 1.9**	**1.9 – 1.7**	**1.7 – 1.4**	**1.4 – 1.2**	**1.2 – 0.9**
Number of fossil specimens	**9560**	**9363**	**1056**	**22683**	**1137**
Taxon					
Hemibos triqueticonis	x	x	x	**X (642)**	x
Damalops palaeindicus	x	x		**X (642)**	x
Medium-sized alcelaphine				**X (610)**	
Medium-sized bovid at locality 362		**X (362)**	? ?	?	
Very large bovid		x		x	
Very small bovid	x	x		**X**	
cf. *Cervus*	x	x		X	x
cf. *Dama*	x	x		X	x
Sivatherium giganteum	X	**X (362)**	Probably extinct		
Suid, large	**X (343, 525)**	x		**X (73)**	
Suid, small	x	x		X (666)	
Equus cf. sivalensis	X	X	X	**X (73, 642)**	X
Rhinoceros sivalensis	**X (341)**	X		**X (642)**	
Elephas hysudriscus	**X (535)**	x		x	
Stegodon	**X (507)**		X		
Anthracoceres	X (529)	Probably extinct			
Hippotamids	x	x	Probably extinct		
Crocodylus/Gavialis	X	X	X	X	X
Chelonids	X	X	X	X	X
Cf. Cercopithecidae		**X (491)**	Probably extinct		
Crocuta crocuta				**X (73)**	
Pachycrocuta brevirostris				**X (642)**	
Hyaenidae	X	X		X	
Large *Felis*, not leopard	X			X	
Cf. *Panthera uncia*				**X (73)**	
Megantereon cultridens				**X (674)**	
Cf. *Canis cautleyi*		**X (362)**	Probably extinct		
Canis, small	x	x		x	
Herpestid				**X (73)**	
Ursid				**X (73)**	

Notes: X = common evidence; **X** = excellent examples, often in an excavated context; x = other evidence. The numbers after X denotes the number of the main fossil locality where it was found, or where the main specimen was found.

Figure 10. 10: Overview of the main biostratigraphic features of the Matuyama Chron (2.5 – 0.7 Mya) as suggested by Opdyke et al. (1979, Figure 21). This scheme relied heavily upon material from the Pabbi Hills.

Age range	**2.5 – 2.2**	**2.2 – 1.9**	**1.9 – 1.7**	**1.7 – 1.4**	**1.4 – 1.2**	**1.2 – 0.9**	**Early Brunhes**
Taxon							
Proamphibos	x	x		x			
Bubalus		x	x	x			x
Gazella	x	x			x		
cf. *Cervus*	x	x	x	x	x		
Giraffidae	? x	x					
Sivatherium giganteum			x				
Suidae	x		x				
Sus	?			x			
Equus	x	x	x	x			
Hipparion	x			x			
Rhinoceros			x				
Elephas sp.	x	x	x	x	x		
Stegodon	x	x	x			x	x
Hippotamids/*Hexaprotodon*	x				?		
Hyaenidae			x				
Machairodus			x				
Felis			x				
Canis			x				
Viverrid			x				

Notes: According to Keller et al. (1979: 196), the oldest occurrence of *Equus* is ca. 1.8 Mya, rather later than shown in Opdyke et al. (1979). Isolated records of shrews, rabbits, rats and weasel, all ca. 1.5 Mya, have been omitted as these were not found in our investigations.

Table 10. 11: Types and provenance of cervid antler: a) branched fragments:

Number	Age	Details
209GB6	1.9 - 2.2	Fork at base, pedicel; dia. 63 x 62mm. Main branch dia. 41 x 48mm. Shed
242GB3	1.7 - 1.9	Pedicel, shed, sandstone matrix. Dia. 65 x 59mm
261GB7	1.9 - 2.2	Fork at base, pedicel, dia. 65 x 56mm.
281GB4	1.9 - 2.2	Fork at base, pedicel, dia. 30 x 25mm. Diameter base of branch, 23 x 21mm
333GB1	?1.7 - 1.9	Small tine at base, pedicel, dia. 38 x 33mm.Main branch 25mm.
355GB5	1.9 - 2.2	Fork at base, pedicel, dia. 61 x 48mm. Diameter base of branch, 61 x 48mm
405GB2	?1.7 - 1.9	Fork at base, pedicel, shed; dia. 59 x 49mm.
419GB2	1.7 - 1.9	Fork at base, pedicel, dia. 48 x >42mm. Diameter of branch, 40 x 35mm
429GB1	1.9 - 2.2	Fork at base, pedicel, broken.
437FGB5	1.9 - 2.2	Fork at base, pedicel, dia. 32 x 32mm.
489CGB13	1.7 - 1.9	Fork at base, pedicel; dia. 46 x 39mm. Main branch dia. 27 x 27mm.
624GB30	1.2 - 1.4	Fork at base, pedicel, dia. 77 x 67mm.
628GB1	1.2 - 1.4	Fork at base, pedicel, dia. 68 x 58mm. Diameter base of branch, 43 x 35mm; shed
695GB1	1.9 - 2.2	Pedicel dia.: 54 x 37mm; main branch dia. 30 x 28. Smooth surface. 310mm long
734GB1	1.2 - 1.4	Part pedicel and side branch (642 area; found in 1990)

b) palmate fragments:

Number	Age	Details
68BGB339	1.2 - 1.4	Large fragment
83CGB6	1.2 - 1.4	Fork at base, pedicel in part, strong ribbing
203GB17	1.9 - 2.2	Fragment
218GB2	1.7 - 1.9	Fragment
240GB4+7	1.7 - 1.9	Fork at base, pedicel; dia. 60 x 49mm
258GB1	1.7 - 1.9	Fragment
280GB13	1.7 - 1.9	Fork at base, pedicel; dia. 60 x 49mm
280GB9	1.7 - 1.9	Fragment
288GB1	1.9 - 2.2	Fragment
345GB4	1.9 - 2.2	Fragment
359GB2	1.9 - 2.2	Has pedicel, dia. 60 x 55mm
359GB3	1.9 - 2.2	Fork at base, pedicel; dia. 58 x 48mm
368GB1	?1.7 - 1.9	Fragment
371GB7	1.9 - 2.2	Fragment
387GB12	1.9 - 2.2	Fragment
445GB26	1.9 - 2.2	Fragment
485GB2001	1.7 - 1.9	Fork at base, pedicel; dia. 45 x 38; re-collection
489AGB1	1.7 - 1.9	Pedicel and side branch; pedicel dia. 68 x 58mm. Not shed
716GB1	1.2 - 1.4	Fragment
716GB3	1.2 - 1.4	Pedicel, 61 x 58mm.; 642 area; found in 1990

Table 10.12: Distal measurements for bovid and cervid post-cranial bones.

Occurrence	No.	Age	Taxon	Element	A-P	Width
128C	18	1.2-1.4	Bovid	1st Phalange	0.0	12.0
437E	9	2.2-1.9	Bovid	1st Phalange	0.0	26.0
489E	2	1.7-1.9	Bovid	1st Phalange	13.0	14.5
642/IS1	1	1.2-1.4	Bovid	1st Phalange	0.0	16.0
489E	1	1.7-1.9	Bovid	2nd Phalange	13.5	16.0
815	33	1.2–1.4	Bovid	2nd Phalange	37.0	35.0
156A	5	>1.4	Bovid	Humerus	0.0	70.0
298	3	2.2-1.9	Bovid	Humerus	0.0	46.0
362D	14	1.7-1.9	Bovid	Humerus	0.0	75.0
368	37	?1.7-1.9	Bovid	Humerus	0.0	86.0
485	17	1.7-1.9	Bovid	Humerus	0.0	84.0
642/3	222	1.2-1.4	Bovid	Humerus	0.0	48.5
73EX	7	1.2-1.4	Bovid	Metacarpal	34.0	48.0
73A	12	1.2-1.4	Bovid	Metacarpal	0.0	36.0
73B	12	1.2-1.4	Bovid	Metacarpal	0.0	38.5
301*	2	2.2-1.9	Bovid	Metacarpal	41.5	74.0
343	12	2.2-1.9	Bovid	Metacarpal	38.0	62.0

386	2	?1.7-1.9	Bovid	Metacarpal	24.0	33.0
388	11	1.7-1.9	Bovid	Metacarpal	33.5	61.0
406	7	?1.7-1.9	Bovid	Metacarpal	38.0	70.0
437C	10	2.2-1.9	Bovid	Metacarpal	27.0	42.0
489B	34	1.7-1.9	Bovid	Metacarpal	0.0	71.0
497	9	1.7-1.9	Bovid	Metacarpal	29.0	42.0
642/2	189	1.2-1.4	Bovid	Metacarpal	24.0	36.5
642/1B (1990)	89	1.2-1.4	Bovid	Metacarpal	23.0	34.0
642/1C (1990)	74	1.2-1.4	Bovid	Metacarpal	24.0	36.0
642/SE1	8	1.2-1.4	Bovid	Metacarpal	22.5	35.0
642 recollect	3	1.2-1.4	Bovid	Metacarpal	26.0	38.0
642/16		1.2-1.4	Bovid	Metacarpal	23.5	45.5
674	1	1.2-1.4	Bovid	Metacarpal	40.5	76.5
491D	1	1.7-1.9	Bovid	Metacarpal, unfused	34.0	47.0
73EX	286	1.2–1.4	Bovid	Metatarsal	42.0	78.0
73A	105	1.2-1.4	Bovid	Radius	0.0	67.0
200	1	1.7-1.9	Bovid	Radius	0.0	80.0
245	5	1.4-?1.7	Bovid	Radius	31.0	40.5
250	3	1.7-1.9	Bovid	Radius	0.0	79.0
362C	8	1.7-1.9	Bovid	Radius	0.0	83.0
408	4	?1.7-1.9	Bovid	Radius	0.0	46.0
445F	6	2.2-1.9	Bovid	Radius	0.0	48.0
481G	12	1.7-1.9	Bovid	Radius	46.0	57.0
489C	12	1.7-1.9	Bovid	Radius	46.0	57.0
507	31	2.2-1.9	Bovid	Radius	34.0	60.0
481E	18	1.7-1.9	Bovid	Radius/Ulna	63.0	86.0
279	1	2.2-1.9	Bovid	Tibia	53.0	71.0
397	13	2.2-1.9	Bovid	Tibia	30.0	40.0
445B	46	2.2-1.9	Bovid	Tibia	70.0	86.0
481B	6	1.7-1.9	Bovid	Tibia	32.0	38.0
481B	2	1.7-1.9	Bovid	Tibia	42.0	59.0
489B	6	1.7-1.9	Bovid	Tibia	32.0	38.0
642/1	33	1.2-1.4	Bovid (Large)	1st Phalange	0.0	39.0
642/1	67	1.2-1.4	Bovid (Large)	Femur	0.0	52.0
642	2	1.2-1.4	Bovid (Large)	Humerus	0.0	98.0
614	8	1.2-1.4	Bovid (Large)	Metacarpal	44.0	66.0
362G	9	1.7-1.9	Bovid (Large)	Metapodial	35.0	65.0
73EX	286	1.2-1.4	Bovid (Large)	Metatarsal	42.0	78.0
362D	2	1.7-1.9	Bovid (Large)	Radius	0.0	75.0
362F	10	1.7-1.9	Bovid (Large)	Radius	0.0	85.0
642SE1	1	1.2-1.4	Bovid (Large)	Radius	0.0	91.0
642/3	862	1.2-1.4	Bovid (Large)	Tibia	0.0	83.0
636	130	1.2-1.4	Bovid (Medium)	2nd Phalange	17.0	11.0
642/3**	600	1.2-1.4	Bovid (Medium)	Tibia	32.0	42.0
636	106	1.2-1.4	Bovid (Medium)	Humerus	25.0	52.0
642/3	114	1.2-1.4	Bovid (Medium)	Humerus	0.0	47.0
642/3**	190	1.2-1.4	Bovid (Medium)	Humerus	0.0	52.0
642/3	216	1.2-1.4	Bovid (Medium)	Humerus	0.0	49.0
666	89	1.2-1.4	Bovid (Medium)	Humerus	0.0	62.0
642X	25	1.2-1.4	Bovid (Medium)	Metacarpal	22.0	32.0
636	114	1.2-1.4	Bovid (Medium)	Metacarpal	25.0	37.0
636	104	1.2-1.4	Bovid (Medium)	Metacarpal	26.0	38.0
642	8	1.2-1.4	Bovid (Medium)	Metacarpal	22.5	35.0
642/3	9	1.2-1.4	Bovid (Medium)	Metacarpal	25.0	40.0
642/3	16	1.2-1.4	Bovid (Medium)	Metacarpal	23.5	45.5
642/3	53	1.2-1.4	Bovid (Medium)	Metacarpal	22.0	36.5
642/3	102	1.2-1.4	Bovid (Medium)	Metacarpal	23.0	33.5
642/1	181	1.2-1.4	Bovid (Medium)	Metacarpal	24.0	36.0
642/3	223	1.2-1.4	Bovid (Medium)	Metacarpal	24.0	33.5
642/3	247	1.2-1.4	Bovid (Medium)	Metacarpal	25.0	36.5
642/1	260	1.2-1.4	Bovid (Medium)	Metacarpal	23.0	36.0
642/1 Cluster 2	399	1.2-1.4	Bovid (Medium)	Metacarpal	24.0	36.0
642/1 Cluster 2	430	1.2-1.4	Bovid (Medium)	Metacarpal	23.5	40.0
642/1 Cluster 2	431	1.2-1.4	Bovid (Medium)	Metacarpal	24.0	36.0

642/1	598	1.2-1.4	Bovid (Medium)	Metacarpal	25.0	36.0
642/3	699	1.2-1.4	Bovid (Medium)	Metacarpal	26.0	35.0
642/3	855	1.2-1.4	Bovid (Medium)	Metacarpal	36.0	21.0
642/3	868	1.2-1.4	Bovid (Medium)	Metacarpal	24.0	36.0
642/1C (1990)	74	1.2-1.4	Bovid (Medium)	Metacarpal	24.0	36.0
642/1B (1990)	89	1.2-1.4	Bovid (Medium)	Metacarpal	23.0	34.0
642 (1990)	3	1.2-1.4	Bovid (Medium)	Metacarpal	26.0	38.0
642/3	16	1.2-1.4	Bovid (Medium)	Metacarpal	23.5	45.5
655	3	1.2-1.4	Bovid (Medium)	Metacarpal	24.0	34.0
675	3	1.2-1.4	Bovid (Medium)	Metacarpal	25.0	35.0
642/3	20	1.2-1.4	Bovid (Medium)	Metacarpal, unfused	25.0	39.0
642/3	10	1.2-1.4	Bovid (Medium)	Radius	31.0	41.0
642/3	168	1.2-1.4	Bovid (Medium)	Radius	32.0	42.5
642/3	192	1.2-1.4	Bovid (Medium)	Radius	29.0	45.0
642/3	768	1.2-1.4	Bovid (Medium)	Radius	29.0	45.0
636	44	1.2-1.4	Bovid (Medium)	Tibia	30.0	42.0
642/3/Pt22	13	1.2-1.4	Bovid (Medium)	Tibia	29.5	38.0
642/IS3	38	1.2-1.4	Bovid (Medium)	Tibia	33.0	45.0
642/1	70	1.2-1.4	Bovid (Medium)	Tibia	30.0	41.0
642/3	157	1.2-1.4	Bovid (Medium)	Tibia	29.5	39.0
642/1	208	1.2-1.4	Bovid (Medium)	Tibia	33.0	38.5
642/3	246	1.2-1.4	Bovid (Medium)	Tibia	32.0	41.0
642/3	346	1.2-1.4	Bovid (Medium)	Tibia	30.0	41.0
642/3	623	1.2-1.4	Bovid (Medium)	Tibia	38.0	48.0
642/3	721	1.2-1.4	Bovid (Medium)	Tibia	24.0	40.0
642/1	275	1.2-1.4	Bovid (Small)	Humerus	0.0	22.5
642/1	199	1.2-1.4	Bovid (Small)	Metacarpal	16.0	22.0
642/1	276	1.2-1.4	Bovid (Small)	Metacarpal	16.0	21.0
642/2	135	1.2-1.4	Bovid (Small)	Metapodial	12.0	18.5
642/2	166	1.2-1.4	Bovid (Small)	Metapodial	15.0	20.0
642/2	212	1.2-1.4	Bovid (Small)	Metapodial	15.0	21.0
642/3/Pt.22	9	1.2-1.4	Bovid (Small)	Metapodial barrel	15.5	9.0
642/2	187	1.2-1.4	Bovid (Small)	Tibia	18.0	22.5
664	23	1.2-1.4	Bovid/Cervid	2nd Phalange, unf.	20.0	16.0
68C	1	1.2-1.4	Bovid/Cervid	Metacarpal	25.0	33.8
664	32	1.2-1.4	Bovid/Cervid	Tibia	29.0	39.0
74	74	1.2-1.4	Cervid	2nd Phalange	0.0	20.0
57	68	1.2-1.4	Cervid	Humerus	0.0	44.0
128B	62	1.2-1.4	Cervid	Humerus	0.0	43.4
18	2	1.2-1.4	Cervid	Metacarpal	21.0	29.5
68A	258	1.2-1.4	Cervid	Metacarpal	24.0	29.0
73	8	1.2-1.4	Cervid	Metacarpal	29.0	44.0
481B	5	1.7-1.9	Cervid	Metacarpal	34.0	50.0
247	16	1.4-?1.7	Cervid	Metatarsal	28.0	42.0
14A	25	1.2-1.4	Cervid	Radius	45.0	78.0
91	64	0.9	Cervid	Tibia	41.0	52.0
133B	33	1.2-1.4	Cervid	Tibia	32.0	39.0
343	12	2.2-1.9	Cervid/Bovid	Metacarpal	27.0	41.0

Notes: A-P = anterior-posterior. * Complete, both proximal and distal ends preserved. 301GB1/2, length 257 mm.; 343/11, length = 262; 343/12, length 240 mm. **: 642\3\190 distal humerus is also attached to the proximal radius and ulna, and 642/3/600 tibia is attached to the astragalus. 815GB33 was found in 1999 on the scarp of Sandstone 12 near locality 14.The length was 54 mm.

Table 10. 13: Bovid and cervid proximal post-cranial measurements

Place	Zone	No.	Taxon	Element	A-P	Width
14	A	26	Bovid	1st Phalange	24.5	19.5
124		116	Bovid (Large)	1st Phalange	36.0	40.0
437	F	2	Bovid	1st Phalange	38.0	37.5
481	B	33	Bovid	1st Phalange	28.0	20.0
489	G	26	Bovid	1st Phalange	35.5	33.0
73	A	176	Bovid	1st Phalange	41.0	35.0
489	E	10	Bovid	2nd Phalange	28.0	21.0
362	A	15	Bovid	Humerus	55.0	60.0

73	A	20	Bovid	Metacarpal	50.0	80.0
214	R	2	Bovid	Metacarpal	44.0	69.5
301		1	Bovid	Metacarpal	42.0	70.0
362	A	8	Bovid	Metacarpal	44.0	69.0
467		1	Bovid	Metacarpal	44.5	72.0
515		1	Bovid	Metacarpal	37.0	58.0
17		2	Bovid	Metatarsal	51.0	50.0
73	Y	1	Bovid	Metatarsal	43.0	40.0
73	EX	286	Bovid	Metatarsal	56.0	75.0
247		15	Bovid	Metatarsal	40.0	48.0
337		3	Bovid	Metatarsal	35.0	39.0
481		8	Bovid	Metatarsal	44.0	44.5
491	C	11	Bovid	Metatarsal	42.0	41.0
642	3	15	Bovid	Metatarsal	33.0	32.0
642	X	6	Bovid	Metatarsal	35.0	35.0
24		1	Bovid	Radius	33.5	67.2
223		1	Bovid	Radius	49.0	86.0
259		2	Bovid	Radius	35.0	60.0
288		2	Bovid	Radius	52.0	91.0
382		1	Bovid	Radius	50.0	86.0
445	E	13	Bovid	Radius	22.5	43.0
481	B	1	Bovid	Radius	25.0	48.0
491	C	5	Bovid	Radius	44.0	88.0
73	EX	30	Bovid	Radius + ulna	34.0	103.0
200		2	Bovid	Tibia	0.0	76.0
642	3	64	Bovid (Large)	1st Phalange	43.0	40.0
815		33	Bovid (Large)	2nd Phalange	47.0	41.0
128	C	16	Bovid (Large)	Metacarpal	0.0	73.0
254		3	Bovid (Large)	Metacarpal	42.0	71.0
630		37	Bovid (Large)	Metacarpal	50.0	76.5
73	EX	286	Bovid (Large)	Metatarsal	56.0	75.0
602		1	Bovid (Large)	Metatarsal	57.0	57.0
630		38	Bovid (Large)	Radius	48.0	74.0
630		29	Bovid (Large)	Radius, split	0.0	106.0
128	C	30	Bovid (Medium)	1st Phalange	21.0	16.0
636		26	Bovid (Medium)	Metacarpal	25.0	38.0
642		0	Bovid (Medium)	Metacarpal	25.0	38.0
642		17	Bovid (Medium)	Metacarpal	24.0	35.0
642	1,Clust.2	439	Bovid (Medium)	Metacarpal	24.5	36.0
642	1	124	Bovid (Medium)	Metacarpal	27.0	40.0
642	3	611	Bovid (Medium)	Metacarpal	24.0	40.0
642	3	260	Bovid (Medium)	Metacarpal	24.0	35.0
642	3	745	Bovid (Medium)	Metacarpal	21.0	38.0
642	X	30	Bovid (Medium)	Metatarsal	33.0	33.0
636		118	Bovid (Medium)	Metatarsal	33.0	32.5j
642	1	277	Bovid (Medium)	Metatarsal	34.0	31.0
642	1,Clust.2	400	Bovid (Medium)	Metatarsal	37.0	42.0
642	1	251	Bovid (Medium)	Metatarsal	31.0	28.0
642	1	248	Bovid (Medium)	Metatarsal	33.0	33.0
642	1	203	Bovid (Medium)	Metatarsal	33.0	33.0
642	3	870	Bovid (Medium)	Metatarsal	32.0	29.0
642	3	622	Bovid (Medium)	Metatarsal	35.0	29.0
642	3	758	Bovid (Medium)	Metatarsal	33.0	32.0
642	3	836	Bovid (Medium)	Metatarsal	32.5	29.0
642	3	103	Bovid (Medium)	Metatarsal	33.0	0.0
642	3	210	Bovid (Medium)	Metatarsal	40.0	41.0
656		24	Bovid (Medium)	Metatarsal	33.0	31.5
636		36	Bovid (Medium)	Radius	27.0	52.0
642	1	266	Bovid (Medium)	Radius	24.5	50.0
642	1	185	Bovid (Medium)	Radius	24.0	45.0
642	3, Pt.22	12	Bovid (Medium)	Radius	26.0	51.0
642	3	251	Bovid (Medium)	Radius	27.0	46.0
642	3	132	Bovid (Medium)	Radius	27.5	54.0
642	3	126	Bovid (Medium)	Radius	26.0	52.5

642	3	167	Bovid (Medium)	Radius	27.0	50.0
655		31	Bovid (Medium)	Radius	20.5	43.0
666		91	Bovid (Medium)	Radius, split	26.0	0.0
636		84	Bovid (Medium)	Tibia	64.0	0.0
642	2	135	Bovid (Small)	1st Phalange	15.0	0.0
642	2	291	Bovid (Small)	1st Phalange	14.0	9.5
642	2	197	Bovid (Small)	Metacarpal	15.0	20.5
642	1	364	Bovid (Small)	Metatarsal	21.0	19.0
642	3	690	Bovid (Small)	Metatarsal	20.0	19.0
642	3	810	Bovid (Small)	Tibia	27.0	33.0
642	Pt.20	9	Bovid/Cervid	1st Phalange	23.0	17.0
642	3	92	Bovid/Cervid	Metatarsal	34.0	30.0
73	C	162	Cervid	1st Phalange, epiph.	19.0	14.0
73	C	93	Cervid	2nd Phalange	0.0	16.5
14	A	75	Cervid	Metacarpal	23.0	34.5
68	B	309	Cervid	Metacarpal	33.0	48.0
73	A	106	Cervid	Metacarpal	32.0	51.0
73	EX1987	0	Cervid	Metacarpal	31.0	42.0
73	EX	608	Cervid	Metacarpal	31.0	42.0
83	E	4	Cervid	Metacarpal	23.0	35.0
119		2	Cervid	Metacarpal	39.0	0.0
57		6	Cervid	Metatarsal	32.0	32.0
73	A	107	Cervid	Metatarsal	38.0	38.0
73	EX	613	Cervid	Metatarsal, split	35.0	0.0
116	C	18	Cervid	Metatarsal	29.0	25.0
122		1	Cervid	Metatarsal	35.0	35.0
31		1	Cervid	Radius	46.6	77.6
44		4	Cervid	Radius	21.3	37.5
217		5	Cervid	Radius	18.0	0.0
128	B	44	Cervid	Tibia, unf. epipysis	48.0	73.2
73	EX	210	Cervid/Bovid	Radius + ulna	36.0	79.0

Note: 642GB2/197 is attached to the carpals. 815GB33 was found in 1999 on the scarp of Sandstone 12 near locality 14.The length was 54 mm.

Table 10. 14: Astraguli measurements from the Pabbi Hills

Place	No.	Taxon	Completeness	Length	Width	Thick ness	Age
14B	29	Bovid	Complete	85.0	58.0	0.0	1.2-1.4
47	8	Bovid	Complete	41.0	25.0	0.0	1.2-1.4
73EX	15	Bovid	Complete	89.0	59.0	47.0	1.2-1.4
73EX	212	Bovid	Complete	93.0	60.5	0.0	1.2-1.4
73EX	351	Bovid	Complete	58.0	39.0	30.0	1.2-1.4
73A	7	Bovid	Complete	66.0	41.0	0.0	1.2-1.4
83A	2	Bovid	Complete	86.0	53.0	0.0	1.2-1.4
250	8	Bovid	Complete	81.0	53.0	43.0	1.7-1.9
311	1	Bovid	Complete	63.0	39.0	32.0	?1.7-1.9
326	1	Bovid	Complete	39.0	24.0	21.5	>0.5
345	1	Bovid	Complete	75.0	55.0	42.0	2.2-1.9
362D	12	Bovid	Complete	65.0	40.0	36.0	1.7-1.9
362D	9	Bovid	Complete	74.0	45.0	40.0	1.7-1.9
362G	2	Bovid	Complete	89.0	73.0	63.0	1.7-1.9
362A	4	Bovid	Split	72.0	0.0	42.0	1.7-1.9
393	4	Bovid	Complete	44.0	27.0	24.5	2.2-1.9
399	11	Bovid	Complete	37.0	22.0	22.0	2.2-1.9
403	1	Bovid	Complete	47.0	30.0	26.0	?1.7-1.9
404	4	Bovid	Split	78.0	0.0	40.0	?1.7-1.9
412	5	Bovid	Split	62.0	0.0	40.0	?1.7-1.9
437C	29	Bovid	Split	40.0	0.0	0.0	2.2-1.9
485	8	Bovid	Complete	55.0	36.0	29.0	1.7-1.9
485	18	Bovid	Complete	68.0	50.0	37.0	1.7-1.9
489C	26	Bovid	Complete	81.0	52.0	45.0	1.7-1.9
489D	1	Bovid	Complete	80.0	53.0	46.0	1.7-1.9

489B	13	Bovid	Complete	38.0	26.0	22.5	1.7-1.9
489B	9	Bovid	Complete	52.0	35.0	31.0	1.7-1.9
491A	0	Bovid	Complete	72.0	49.0	39.0	1.7-1.9
491C	7	Bovid	Complete	74.0	52.0	44.0	1.7-1.9
529	12	Bovid	Complete	48.0	31.0	30.0	2.2-1.9
481E	41	Bovid (Large)	Complete	80.0	55.0	39.0	1.7-1.9
642/1	64	Bovid (Large)	Complete	83.0	61.0	51.0	1.2-1.4
610A	9	Bovid (Medium)	Split	0.0	29.0	26.0	1.2-1.4
636	0	Bovid (Medium)	Complete	43.0	31.5	25.0	1.2-1.4
642	44	Bovid (Medium)	Complete	45.0	30.0	25.0	1.2-1.4
642/1	48	Bovid (Medium)	Complete	43.0	26.0	24.0	1.2-1.4
642	0	Bovid (Medium)	Complete	45.0	32.0	26.5	1.2-1.4
73EX	308	Bovid/Cervid	Complete	57.0	39.0	31.0	1.2-1.4
14A	1	Cervid	Complete	36.0	25.0	0.0	1.2-1.4
14A	87	Cervid	Complete	45.0	29.0	0.0	1.2-1.4
36	1	Cervid	Split	76.0	45.0	0.0	1.2-1.4
45	3	Cervid	Complete	52.0	35.0	0.0	1.2-1.4
73EX	218	Cervid	Complete	61.3	40.9	34.6	1.2-1.4
73BB	1	Cervid	Complete	58.5	37.0	0.0	1.2-1.4
83A	3	Cervid	Complete	64.0	41.0	0.0	1.2-1.4
235	13	Cervid	Split	39.0	26.0	0.0	1.4-1.7
237	1	Proboscidean	Split	120.0	0.0	80.0	?1.7-1.9
156A	4	*Rhinoceros*	Split	88.0	69.0	0.0	>1.4

Table 10. 15: Specimens from locality 642 attributed to *Gazella*.

Year	No.	Zone	Specimen	Sub-area	Element	Part	Other
1989	642	2	135		1st Phalange	Proximal+dist. metapodial	
1989	642	2	291		1st Phalange	Proximal	Complete
1989	642	2	255		1st phalange	Distal	
1989	642	2	273		2nd phalange	Proximal	
1990	642	3	102		2nd Phalange	Complete	
1990	642	2	16	FS28	Calcaneum	Distal	
1989	642	1	275		Humerus	Distal	Complete
1990	642	1	7	FS25	Humerus	Distal	Complete
1989	642	2	224		Mandible	+ 1 tooth	
1990	642	2	2	FS23	Mandible	+ Tooth	1 split, 1 broken
1990	642	EX	1700,1702		Mandible	+ Molar fragment	
1989	642	1	372		Maxilla	+ Teeth	Incisor 1,2
1990	642	1	3	FS15	Maxilla	+ Tooth	PM
1989	642	1	199, 276		Metacarpal	Distal	Complete
1989	642	2	197		Metacarpal	Proximal + carpals	Complete
1990	642	2	19	FS23	Metacarpal	Proximal	Complete
1989	642	3	9	Pt22	Metapodial	Distal barrel	
1989	642	2	135		Metapodial	Distal +2 1st phalanges + sesamoid	Fused
1989	642	2	166, 212		Metapodial	Distal	Complete
1989	642	3	297		Metapodial	Shaft	
1989	642	1	364		Metatarsal	Proximal	Complete
1989	642	3	690		Metatarsal	Proximal	Complete
1989	642	1	311		Metatarsal	Shaft	
1989	642	3	34	Pt22	Metatarsal	Shaft	
1989	642	1	169		Molar	Split	Worn
1989	642	2	19		Navicular-cuboid		
1989	642	1	262, 303		Radius	Distal	Complete
1989	642	2	187		Tibia	Distal	Complete
1989	642	3	810		Tibia	Proximal	Complete
1989	642	3	6, 7	Pt22	Tooth	Complete	PM
1987	254		7		Vertebra	Centrum	
Other specimens attributed to *Gazella*							
Year	**No.**	**Zone**	**Specimen**	**Age (Mya)**	**Element**	**Part**	**Other**
1987	284	A	3, 4	1.7 – 1.9	Tooth	Fragment	
1987	481	B	24	1.7 – 1.9	Vertebra	Lumbar	Complete

1987	481	B	12	1.7 – 1.9	Metatarsal	Shaft	
1987	261		6	1.9 – 2.2	Horncore		
1987	285		39	1.9 – 2.2	M1/2, lower	Fragment	Worn
1987	327	A	11	1.7 – 1.9	Molar	Split	Worn
1987	414		2	1.7 – 1.9	PM2/3	Upper	Complete
1987	427	D	45	1.9 – 2.2	M1/2, lower	Lit. wear	Complete
1987	433		9	1.9 – 2.2	Molar	Split	Mod.wear
1987	435		112	1.9 – 2.2	PM	Complete	Worn
1987	437		4	1.9 – 2.2	M1/2, upper	Complete	Little wear
1987	437	C	47	1.9 – 2.2	M1/2, upper	Complete	Little wear
1987	522	A	2	0.9 – 1.2	M1/2	Complete	
1987	489	D	70	1.7 – 1.9	M1/2, lower	Split	Mod.wear
1987	489	J	35	1.7 – 1.9	Mandible	Right	M1/2,broken
1987	489	E	187	1.7 – 1.9	Calcaneum	Fragment	
1999	815		39	1.2 – 1.4	Calcaneum	Shaft	
1999	815		41	1.2 – 1.4	Astragalus	Complete	
1999	815		27	1.2 – 1.4	Scapula	Glenoid	
1999	815		31	1.2 – 1.4	Metatarsal	Proximal	
1999	815		32	1.2 – 1.4	1st Phalange	Proximal	
1999	815		15	1.2 – 1.4	Tibia	Distal	
1999	815		17	1.2 – 1.4	Pelvis		Acetabulum
1999	815		26	1.2 – 1.4	Metacarpal	Proximal	
1999	837		3	1.2 – 1.4	Metapodial	Distal	Barrel

Table 10. 19: Equid lower molars/premolars from the Pabbi Hills

Specimen	Tooth	Length	Width
69GB4	Lower M3	31.0	
341GB0	Right lower M3, worn	30.0	
632GB5	Lower M3, no wear	28.0	
353GB8	Right lower M3, worn	36.0	
353GB8	Lower M3, clear cusps, good condition, mod. wear	34.0	14.0
610AGB5	Lower M3; moderate wear	31.0	13.0
677GB11	Lower M3, worn	25.0	11.0
	Other teeth		
107GB3	M2 right, moderate wear, fair condition; sand in crowns	27.0	12.0
327CGB7	Lower P3/4 or M1/2. Worn, moderate condition	26.0	14.5
368GB12	Molar, clear cusps, good condition. V-shaped linguaflexid, ectoflexid deep but partly obscured.	25.0	15.0
407GB1	Lower M3, slight wear, good condition	25.0	14.0
500GB3	Molar, slight wear, good condition	29.5	16.0
610AGB4	Molar, worn, poor condition, crowns partly obscured	25.5	17.0
610GB8	Lower P2, mod. wear	30.5	10.5
610/9GB2	Lower M3 or P3. Worn, poor condition	25.0	17.0
642/1GB 136	P3 right, worn, good condition	34.0	15.0
665GB2	Moderate wear and condition. Fairly deep ectoflexid but clear isthsmus	27.0	17.0
683GB2	M2	23.5	15.5
683GB3	Molar, slight wear, good condition	27.0	18.0
721GB2	Very poor condition, very worn	26.0	15.5
721GB3	Ditto	24.0	14.5

Table 10. 20: Equid upper molars/premolars from the Pabbi Hills

Specimen	Tooth	Length	Width
133AGB42	P3 right; mod. wear, poor surface condition	26.8	26.7
133GB376	Molar, worn, poor surface condition	24.0	29.0
133GB389	Molar, worn, poor surface condition	26.0	28.0
17GB1	Molar, worn, good condition	26.0	22.0
212GB10	Molar, slightly worn, good condition	30.0	30.0
243GB9	Molar, mod.wear, sand matrix on crowns	25.9	26.0
249GB3	Molar, slight wear, good surface condition	25.6	23.2
32GB9	Molar, slight wear, good surface condition. Protocone and isthsmus clear	28.5	27.0
371GB18	Molar, slight wear, crowns partly obscured by grit	20.5	23.3
373GB5	Upper M1/2. Very clear surface, mod. wear	28.0	32.0

434GB1	Molar, slightly worn, good condition, sand matrix	27.0	26.0
610AGB7	PM/Molar, worn, fair condition	21.0	25.5
610GB2	PM/Molar, worn, fair condition	25.0	25.0
637GB1004	Poor condition but no distinct protocone	26.0	26.0
637GB1008	Poor condition but no distinct protocone	24.5	25.0
83GB511	Upper M3	26.0	20.0
133GB14	Upper M3	26.0	
133BGB72	Upper M3	33.0	
243GB12	Upper M3, worn	29.0	23.5

Table 10.22: Equid complete post-cranial specimens

Specimen	Element	Length	Prox. AP	Prox.W.	Dist. AP	Dist.W.
73EX139	1st Phalange	75.0	29.0	not taken	22.0	40.0
73EX204	1st Phalange	82.0	34.0	45.0	not taken	38.0
73EX581	1st Phalange	68.0	30.0	41.0		
73EX250	1st Phalange	83.0	34.0	47.0		38.0
660GB70	1st Phalange	71.0	31.0	38.0	not taken	33.0
73BB2	2nd Phalange	37.0	23.0	45.0	22.0	not taken
73EX285	2nd Phalange	43.7	29.3	44.7		
73EX240	Radius + ulna	308.0	40.0			66.0
73EX189	Metatarsal	273.0	42.0	42.0	33.0	43.0
73EX220	Metatarsal	not taken	35.5	40.0	30.0	36.0
73EX356	Metacarpal	232.0	30.0	44.0	33.0	43.0
73EX569	Metatarsal	252.0	not taken	33.0	26.0	36.0
642EX1757	Metapodial	267.0				
665GB	Metacarpal	260.0	36.0	41.6	29.5	39.0
Mirpur, no num.	Metacarpal	268.0	44.0	69.0*	43.0	60.0
Mirpur 16195	Metacarpal	262.0	41.0	70.0*	42.0	61.0
Mirpur 14042	Metatarsal	289.0	44.0	50.0	35.6	49.5

Notes: * Measurement includes the lateral metapodials. 73EX569 is right, and has the tarsal bones attached; and 73EX240 is left. The proximal and distal ends of 642EX1757 were not accessible to measurement.

Table 10. 23: Additional equid post-cranial measurements

Specimen	Age	Element	AP	Width
30GB9	1.2 - 1.4	Distal metacarpal	31.0	40.5
30GB65	1.2 - 1.4	Distal metacarpal	35.0	41.0
68GB23	1.2 - 1.4	Distal metacarpal	33.0	38.5
73GB104	1.2 - 1.4	Distal metacarpal	33.0	44.0
346GB5	?1.7 – 1.9	Distal metacarpal	33.3	40.0
642/1B GB82	1.2 - 1.4	Distal metacarpal	33.0	45.0
642/3B GB30	1.2 - 1.4	Distal metacarpal	28.0	43.0
642/3B GB94	1.2 - 1.4	Distal metacarpal	34.0	46.0
642/3GB222	1.2 - 1.4	Distal metacarpal	32.0	42.0
642GB129	1.2 - 1.4	Distal metacarpal	27.0	39.0
642GB3/IS6	1.2 - 1.4	Distal metacarpal	33.0	43.0
677GB8		Distal metacarpal	38.0	44.0
Average			**32.5**	**42.2**
68CGB23	1.2 - 1.4	Distal metapodial	32.0	43.5
73AGB109	1.2 - 1.4	Distal metapodial	35.0	44.0
392GB2	not listed	Distal metapodial	29.0	39.7
660GB4	1.2 - 1.4	Distal metapodial	31.5	39.0
660GB6	1.2 - 1.4	Distal metapodial	32.0	42.0
30GB64	1.2 - 1.4	Proximal 1st. phalange	33.2	47.7
276GB14	1.7 - 1.9	Proximal 1st. phalange	28.0	43.5
29GB1	1.2 - 1.4	Proximal 2nd phalange	31.5	

488GB1	1.4 - 1.7	Proximal 2nd phalange	23.0	39.0
642/1GB196	1.2 - 1.4	Proximal 2nd phalange	29.0	43.0
73BB2	1.2 - 1.4	Proximal 2nd phalange	23.0	45.0
73EX369	1.2 - 1.4	Proximal 2nd phalange	28.0	45.0
73EX849	1.2 - 1.4	Proximal 2nd phalange	32.0	48.0
73EX53	1.2 - 1.4	Proximal metacarpal	31.0	52.0
73EX360	1.2 - 1.4	Proximal metacarpal		51.0
133BGB31	1.2 - 1.4	Proximal metacarpal	25.6	38.7
642/2GB127	1.2 - 1.4	Proximal metacarpal	41.0	47.0
660GB3	1.2 - 1.4	Proximal metacarpal	42.0	46.5
660GB15	1.2 - 1.4	Proximal metacarpal	30.0	46.5
642/3GB27	1.2 - 1.4	Proximal metacarpal	41.0	45.0
642/1GB400	1.2 - 1.4	Proximal metacarpal	37.0	43.0
642/1GB183	1.2 - 1.4	Proximal metatarsal	37.0	40.0
610/BGB29	1.2 - 1.4	Proximal metatarsal	33.0	34.0
34GB1	1.2 - 1.4	Proximal radius, unfused	40.0	60.0
247GB5	1.4 - 1.7	Distal radius	35.5	72.5
16GB11	1.2 - 1.4	Distal tibia	50.0	76.0
17GB3	1.2 - 1.4	Distal tibia	47.5	66.0
73EX263	1.2 - 1.4	Distal tibia	41.0	59.0
225GB7	1.4 - 1.7	Distal tibia, sand matrix	48.0	71.0
642/1GB55	1.2 - 1.4	Distal tibia	48.0	62.0
642/3GB80	1.2 - 1.4	Distal tibia	45.0	63.0
642/3GB864	1.2 - 1.4	Distal tibia	43.0	67.0
642/3GB290	1.2 - 1.4	Distal tibia	45.0	60.0

Notes: Distal width of 642GB196 = 36.5mm. Occurrence 346 (near locality 362) has pottery but the fossils are Upper Siwalik. 642/2/129 retains the lateral metapodials.

Table 10. 24: *Rhinoceros sivalensis* specimens from the Pabbi Hills

Occurrence	**Specimen**	**Element**	**Type**	**Age (Mya)**
19	5	Molar	Split	1.2 - 1.4
49	17	Tooth	Molar	1.2 - 1.4
57	53-55	Molar	Fragments	1.2 - 1.4
74	35	Metacarpal	Complete	1.2 - 1.4
127	1	Patella		1.2 - 1.4
156A	4	Astragulus	Split	?1.4 - 1.7
156A	1	Vertebra	Centrum	?1.4 - 1.7
156A	3	Vertebra	Atlas	?1.4 - 1.7
156B	188	Metacarpal		?1.4 - 1.7
156B	104, 194	Molar	Fragments	?1.4 - 1.7
73C	9	Fibula		1.2 - 1.4
73C	713	Metatarsal		1.2 - 1.4
73C	38	Radius		1.2 - 1.4
73C	74	2nd Metatarsal		1.2 - 1.4
73E	7	2nd Phalange	Complete	1.2 - 1.4
73X	5	Calcaneum	Shaft	1.2 - 1.4
Excavation, locality 73				
73EX, 1986	107	Astragalus	Complete	1.2 - 1.4
73EX, 1987	177	Metapodial		1.2 - 1.4
73EX, 1986	32	Cuneiform		1.2 - 1.4
73EX, 1986	5	Scapula	Complete	1.2 - 1.4
73EX, 1986	403	Humerus		1.2 - 1.4
1987 survey				
234	1	Metapodial	Distal	>0.5
273	7	Metapodial	Distal	1.9 - 2.2
339	20	Metatarsal		1.9 - 2.2
339	14	Molar	Worn	1.9 - 2.2
341	33	Mandible	Complete	1.9 - 2.2

343	8	Tibia	Distal	1.9 - 2.2
348	8	Tibia	Distal	1.7 - 1.9
362A	6	Astragulus	Complete	1.7 - 1.9
362A	1	Unciform		1.7 - 1.9
362F	31,56,70	Tooth	Fragment	1.7 - 1.9
362F	2	Astragulus		1.7 - 1.9
362F	1, 8	Radius	Distal	1.7 - 1.9
362F	11	Vertebra		1.7 - 1.9
362G	3	Astragulus	Complete	1.7 - 1.9
362G	0	Phalange		1.7 - 1.9
362G	0	Tibia		1.7 - 1.9
362, 1989	GB 52, X43	Carpal		1.7 - 1.9
362, 1989	GB 34/3	Tibia ,	Distal, right	1.7 - 1.9
368	14, 16	Molars		?1.7 - 1.9
372	3	Ulna	Distal	?1.7 - 1.9
375A	2	Scapula	Glenoid	1.9 - 2.2
375A	8	Premolar	Upper	1.9 - 2.2
413B	26	Humerus		1.7 - 1.9
413B	0,5,6,9-11,20-23	Molar	Fragments	1.7 - 1.9
437D	45	Navicular-cuboid	Complete	1.9 - 2.2
457	1	Molar	Fragment	1.9 - 2.2
477	20	Molar	Fragment	1.9 - 2.2
489B	12	Metacarpal	Distal	1.7 - 1.9
491C	59	Molar, lower	Fragment	1.7 - 1.9
507	1	Metapodial		1.9 - 2.2
507	23	Scapula		1.9 - 2.2
507	0	Molar	Fragment	1.9 - 2.2
517	1	Molar	Split,worn	1.9 - 2.2
531	2,8,20	Vertebra	Cervical	1.9 - 2.2
531	16	Molar	Fragment	1.9 - 2.2
642, surface				
642/1	30	Astragalus	Complete, left	1.2 - 1.4
642/3	50, 69	Astragalus	Complete	1.2 - 1.4
642/3	16	Carpal	Complete	1.2 - 1.4
642/1	337 Clust. 1	Carpal	Complete	1.2 - 1.4
642/2	17	Carpal	Complete	1.2 - 1.4
642/1	329	Femur	Proximal	1.2 - 1.4
642/2	1	Femur	Distal	1.2 - 1.4
642/2	117	Metapodial	Proximal	1.2 - 1.4
642/3	32	Metapodial	Proximal	1.2 - 1.4
642/3	70	Metapodial	Distal	1.2 - 1.4
642/3	113	Navicular-cuboid		1.2 - 1.4
642/3	863	Radius	Distal	1.2 - 1.4
642/1	15	Tooth	Fragment, upper	1.2 - 1.4
1989				
640	3, 8	Tooth	Fragments	1.2 - 1.4
650	25	Tooth	Fragment	1.2 - 1.4
642 excavation				
642	1062	Mandible,	Juvenile	1.2 - 1.4
642	1096	Mandible	Fragment	1.2 - 1.4
642	1113	Tibia shaft		1.2 - 1.4
642	1196, 1266, 1542,1733, 1756	Tooth	Fragments	1.2 - 1.4
642	1376	Tooth		1.2 - 1.4
642	1528	Metapodial		1.2 - 1.4
642	1530	Ulna		1.2 - 1.4
642	1602	Cranium	Complete	1.2 - 1.4
642	1613	Prox. humerus,	Unfused epiphysis	1.2 - 1.4
642	1628	Radius		1.2 - 1.4
642	1629	Metapodial		1.2 - 1.4
642	1632-1636	Mandible	Juvenile	1.2 - 1.4
642	1748	Humerus		1.2 - 1.4

Notes: 57GB54 molar is lower. 413B - 11 fragments of one tooth.

Appendix 1: Vertebrate Fossils: Carnivora

Alan Turner
School of Biological and Earth Sciences, Liverpool John Moores University, Liverpool L3 3AF, UK

Introduction

This paper describes and discusses the new specimens of Carnivora recovered from Plio-Pleistocene deposits of the Upper Siwalik Formation of northern Pakistan by the British Archaeological Mission to Pakistan during the past decade (Dennell 1993, 1999).

Identifications have been undertaken in comparison with other material from the Siwaliks and elsewhere held in the collections of the Natural History Museum, London. Unfortunately, much of the new material is fragmentary and most of the older specimens from the region are in need of considerable taxonomic revision. Identification of the newer fossils beyond the level of family is therefore often difficult. This report should therefore be seen as a preliminary assessment.

Measurements of some of the specimens, quoted in millimetres, are given in the various tables (10. 27 – 10. 31) together with a selection of measurements of other material used in comparisons.

Dating

Specimens are referred to by their unique catalogue number, prefixed by a number that gives the find locality. The find locality numbers 73 and 642 are excavated assemblages from deposits dated to 1.2 - 1.4 Ma. Locality number 362 is an excavated assemblage from deposits dated to 1. 7 - 1. 9 Ma. The remaining specimens come from surface finds. The likely age of other fossil occurrences that contained carnivore remains is given elsewhere in this volume.

Material from the region in existing collections is usually at best no more than vaguely provenanced, as a glance at the discussion provided by Pilgrim (1932) will show. Although the sample dealt with here is small, it therefore provides one very important bonus in establishing the presence of a number of species in the region within two very specific time bands.

Systematic descriptions

Family **HYAENIDAE** : Genus ***CROCUTA*** Kaup, 1828: ***Crocuta crocuta*** (Erxleben, 1777)

MATERIAL

73X GB 56, rear portion of a right horizontal mandibular ramus with P/4 and M/1 in place; 73X GB 57, a left anterior mandible fragment with P/2 in place and a portion of the P/3; 68B GB 1034, an isolated right P/4, moderately worn and with a broken anterior edge; 73 EX 410, a right maxilla of an aged individual with the canine to P4/ in place, broken P1/, no evidence concerning the presence or absence of M1/

REMARKS

The first specimen is from an aged individual. Although the teeth are very worn, the shape and proportions of the carnassial are typical of *Crocuta:* a long and relatively narrow tooth with a short talonid. The second specimen is not necessarily from the same individual, but the stage of wear is similar and the morphology of the tooth and the general appearance suggest allocation to the same taxon. The enamel of the isolated P/4 is fairly smooth and the tooth appears to have been reasonably hypsodont. On the basis of size and morphology allocation of this tooth to *Crocuta* would be reasonable.

The right maxilla is from an old animal, and the wear on the teeth hampers identification. However, the metastyle of the P4/ is relatively long, and allocation to this species therefore seems secure.

Genus *PACHYCROCUTA* Kretzoi, 1938: ***Pachycrocuta brevirostris*** (Aymard, 1846)

MATERIAL

642 1287, posterior portion of a left horizontal mandibular ramus with P/4 and M/1 in place; 642 1390, a left maxilla with the root of P1/, complete P2/, P3/, P4/ and broken M1/ in place; 73C GB ?12, posterior fragment of a left P4/ with parastyle and protocone missing; 291 GB 9, a distal metapodial fragment; 214B GB 19, a broken left P3/; 68B G8 314, a posterior fragment of the horizontal ramus of a mandible of a larger hyaenid, preserving the anterior part of the masseteric fossa; 642 GB I, a right horizontal mandibular ramus containing P/3 and P/4 and with a separated M/1 that clearly belongs to the specimen

REMARKS

Pachycrocuta brevirostris, commonly referred to as a gigantic hyaena, was perhaps the largest member of the family with a body size approaching that of a lion (Turner and Antón, 1996; Turner, 2001). The distal metapodial fragment is referred to the species on the basis of size, but all the other dental specimens show clear morphological as well as metrical characters that put identification beyond doubt. The upper carnassials in particular are high crowned and generally massive in appearance, with distinctively rugose enamel, a high and somewhat backwardly-raked paracone, large protocone and short metastyle.

Although no measurable landmarks are preserved, the posterior mandible fragment is clearly too flat bucco-lingually and too deep dorso-ventrally to be of a large cat, and on size and overall morphology it appears to belong to this species rather than to *Crocuta crocuta*.

Pachycrocuta* cf *P. brevirostris
73C GB 133. The worn root of a large canine. This tooth lacks the root bulge more typical of a felid canine. Referral to this species is based on size rather than on any particular features.

HYAENIDAE genus and species indet.
Several specimens fall into this category. They are either completely unidentifiable beyond the family level or unidentifiable with the comparative material to hand, although in some cases certain identifications can be ruled out. I shall deal with these items individually.

73A GB 202. A left horizontal mandibular ramus fragment with a broken P/2, worn and broken P/3 and a very worn and broken P/4. The tooth enamel is black, smooth and almost polished in appearance. The P/3 and P/4 are, respectively, about 20 and 22 mm in total length. The specimen is too small to belong to *P. brevirostris*, and the size and general smoothness of the enamel would be more appropriate to *Crocuta*.

315 GB 1 and 2 (joined specimens). This was found in a gully cutting through deposits that probably span the Olduvai Event, ca. 1.7 - 1.9 Mya. It comprises a right horizontal ramus, with broken off crowns of the canine, P/2, P/3 and P/4 and a trace of the anterior surface of the M/1 alveolus. The measurements given are close approximations to the actual tooth dimensions, although necessarily slightly smaller. It is evident that all teeth were in wear, and the small size of the dentition and mandible therefore rules out allocation to either *Crocuta* or *Pachycrocuta*. Although shallow, the mandible is quite robust.

437D GB 26. This was found in a rich locality ca 1. 9 - 2.2 Ma old. It comprises a left anterior horizontal mandibular ramus with the rear of the canine alveolus and the roots and broken crowns of P/2, P/3 and the anterior half of P/4. The specimen is of similar size and morphology to 315 GB 1/2.

214A GB 1. This was found in a small but rich locality ca 1.7 - 1.9 Ma old. It consists of a right P2/ in a fragment of maxilla with the anterior portion of the P3/ alveolus behind. There is a separate fragment of right maxilla and the anterior portion of the jugal bone with a broken and worn P4/ lacking the parastyle and paracone and with the roots of the M1/. Both teeth are very worn, and could easily come from the same individual.

Inter-dental wear on the rear border of the P2/ makes measurement difficult, but the tooth is evidently short and square. The carnassial is of modest length, with a well-developed protocone set at almost right angles to the axis of the tooth and a moderately elongated metastyle in line with the axis. The angle of the protocone, and the absence of buccal flaring in the metastyle, rule out allocation to *Pachycrocuta*, while the roots of what seems to a have been a fairly large M1/ rule out allocation to *Crocuta*.

608 GB 29. An isolated right P/2, slightly worn. This a fairly large tooth, from an animal the size of *Crocuta* or *Pachycrocuta*.

388 B 9. This was found on the surface of deposits ca 1.7 - 1.9 Ma old. It consists of the distal portion of a right radius of a hyaena. This specimen is of moderate, *Crocuta*-like size, but taxonomically uninformative.

445B GB 46 F51. This was found in a rich locality ca 1.9 - 2.2 Ma old, and is the distal fragment a first phalanx of a moderately large hyaenid.

73B GB 184. A right anterior mandibular ramus fragment of a juvenile hyaenid with the canine half erupted. Although the canine is partly obscured by the mandible, it is high crowned and somewhat slender, and therefore unlike a typical *Crocuta*.

73B GB 520. The tip of a left upper canine.

343 GB 10. This was found in a medium-sized fossil occurrence ca. 1.9 - 2.2 Ma old, and is the broken and somewhat eroded distal end of a right humerus, consisting of the articulation and a small portion of the epicondylar region. The specimen is of a moderate-sized hyaena, with a maximum width across the epicondylar region of ca. 47 mm and an articulation width of ca. 36 mm, but is taxonomically undiagnostic.

214 GB 1. Seven fragments. Three may be identified as respectively the left zygomatic process of a temporal with fused jugal, a fragment of right zygomatic arch at the junction of the temporal and jugal and a portion of a left mandibular condyle with a fragment of the ascending ramus, all of moderately large animal.

73A GB 328. A left calcaneus. The specimen is similar, but not identical, to the calcaneus in *Crocuta*, with a total length of 60.8 mm.

73BB GB 22. The proximal half of a first phalanx.

666 GB 26. A right ulna, missing the distal third of its total length. It is of similar size to the ulna of *Crocuta crocuta*. The likely age is 1.2 - 1.4 ma old.

214 GB 2001-5, 2007-8. Various pieces, seemingly from one individual. Only two items are noteworthy, 2005, a worn left M/1 with a broken talonid, and 2002, a left maxillary fragment with a very worn P2/. The M/1 does not appear to be of *Crocuta*, and is more probably *Pachycrocuta*.

610 GB 27. Fossil locality 610 was one of the richest found in the Pabbi Hills, and is ca. 1.2 - 1.4 ma old. This find was a right upper canine.

610B GB 24. A broken and worn anterior fragment of a right M/1.

Family **FELIDAE:** Genus ***PANTHERA*** Oken, 1816: ***Panthera*** cf ***P. uncia*** (Schreber, 1775)

MATERIAL

73 EX 238. A right horizontal mandibular ramus of a medium-sized juvenile felid with the roots of DM/3, a moderately worn DM/4 and M/ 1 erupting and in initial stages of wear on both the protoconid and the paraconid; 73 EX GB 58, a fragment of right horizontal mandibular ramus with the anterior alveolus and rear root of the P/4 and a heavily damaged M/1; 73 EX 568, a left horizontal mandibular ramus of a medium-sized juvenile felid with an almost erupted M1 and an unworn canine in place and what appear to be broken roots of DM/3 and DM/4; 73 X GB 51, a fragment of the crown of a left upper canine with the base and tip missing; 73B GB ?505, an isolated, unworn right P3/, broken, eroded and partially covered in matrix; 73C 203, an isolated left DM4, broken and moderately worn; 642 IC 34; an isolated left P/4; 73C 300, an isolated I/1; 73C420, an isolated lower I/2; 73C344, an isolated lower left I/3; 73C 233, an isolated lower right I/3; 73C 382, joined left I2/ and I3/.

REMARKS

The more complete mandibular specimens, 238 and 568, are clearly not of the machairodont genus *Megantereon*. In the case of specimen 238 the M/1 does not have a shortened paraconid, the angle between paraconid and protoconid of the tooth is not acute, and there is no sign of any flange developing at the antero-ventral border of the mandible. Specimen 568, with its canine in place, is clearly of a pantherine cat. The morphology of the DM/4s suggests allocation to *Panthera* rather than to *Acinonyx*, the cheetah, since the teeth lack the posterior fourth cusp seen in deciduous molars of the latter species. In so far as comparisons are possible, the heavily damaged mandibular specimen, 58, is very similar to the two more complete mandibles.

The upper third molar, although too broken to permit measurements, is clearly not a specimen of *Acinonyx* in view of the presence of an antero-internal ridge on the crown absent in the latter taxon.

Specific allocation is more difficult. All of the specimens are from cats the size of a leopard, *Panthera pardus* L, and one possibility to consider is the snow leopard, *Panthera uncia*. Among the collection of fossil felid remains from the Siwaliks held in the Natural History Museum in London is at least one mandible, 16537a of similar size to the specimen and with the shortened diastema characteristic of the snow leopard among cats of the genus *Panthera* (unpublished data of the author: see Table 10. 29).

Genus ***MEGANTEREON*** Croizet and Jobert, 1828: ***Megantereon cultridens*** (Cuvier, 1824, partim)

MATERIAL

674 GB 21. A broken right horizontal mandibular ramus containing worn and somewhat broken P/3, P/4 and M/1.

REMARKS

This specimen is entirely typical of the genus *Megantereon*, often referred to as the dirk-toothed cat, an animal variously distributed in Africa, Eurasia and North America from earlier Pliocene to Middle Pleistocene times (Turner, 1987). The tooth row is curved outwards (lingually) in occlusal view. The third premolar is little more than a peg, while the P/4 exhibits the characteristic narrowness, hypsodonty, well-developed anterior and posterior cusps and backwardly raked appearance. The carnassial has a short paraconid and a marked talonid cusp behind the protoconid. The preserved portion of the mandible behind the carnassial shows that the ascending ramus was low.

The new specimen is similar in size to others from the Siwaliks in the Natural History Museum collections (Table 10. 29), and larger than most specimens from many other localities (Turner, 1987: Table 1). The size differences have been the basis for specific differentiation of the Siwalik representatives (Petter and Howell, 1982), but there is no overwhelming argument for the existence of more than one widely distributed species (Turner, 1987

FELIDAE genus and species indet.

608 GB 129. A left second phalanx, broken proximally. The offset distal articulation identifies this specimen as felid, but it is rather larger and longer than comparable items in leopard-sized animals.

291 GB 37. A left pisiform of a larger felid. Based on comparisons with extant taxa this specimen is not from an animal of the genus *Panthera*. Neither does it compare at

all with the specimen of *Homotherium* from the French locality of Senèze illustrated by Ballesio (1963: fig. 30).

445 GB 2. A left astragalus. This bone is from a larger felid, the size of a smaller lion, *Panthera leo L.*, but it does not match the astragalus of either *Panthera* or *Homotherium*. This specimen is probably ca. 1.9 - 2.2 My old.

291 GB 18/6/1. A left innominate, missing the illiac crest and much of the ischium and pubis. The specimen is from a larger, lion-sized animal.

291 GB 16. A left ectocuneiform of a medium sized felid. The morphology of the specimen exhibits no clear agreement with the corresponding element in *Panthera*, *Acinonyx* or *Homotherium*.

73B GB 407/181. A right fourth metacarpal, formerly broken into two. Morphologically, the specimen could be from a species of *Panthera*, although it is large for a leopard.

290 GB 8. A distal lateral metapodial of a felid similar in size to that represented by 73B GB 407/181. The likely age for this specimen is 1.9 - 2.2 Mya.

73B GB. Six miscellaneous fragments. 61, the anterior portion of a felid DM4/; unnumbered, the broken metastyle of a larger cat P4/; 62, an indeterminate lower incisor; 73, 84, and 86 three small phalanx/metapodial fragments.

642 GB 28/18. Broken and squashed proximal halves of four metapodia. They appear to be of a medium-sized cat, but their condition precludes further identification.

Family **CANIDAE:** Genus cf. ***CANIS*** Linnaeus, 1758

MATERIAL

362 124, a semi-complete skeleton of a large canid preserving much of the lower dentition (see Figure 10. 11); 54 GB 1, an edentulous fragment of the right horizontal mandibular ramus of a small canid; 73X 36 59 and 20, a right calcaneus and distal metapodial fragment of a small canid; 508A F48 GB1, a first phalanx of a small canid; 73C GB 63, three associated broken metapodia of a small canid; 203 GB 5, a broken left femur of what appears to be a small canid.

REMARKS

All of the smaller canid material is undiagnostic, and is simply referred here to cf *Canis* for completeness. The larger specimen, which is approximately the size of a modern wolf, *Canis lupus* L, is more remarkable.

The dentition of this larger specimen shows *Canis*-like morphology in the structure of the lower carnassial, with a marked entoconid in addition to the main hypoconid of the talonid. The metaconid is also strongly developed. This morphology is very similar to that of the Upper Pleistocene and extant Holarctic *Canis lupus,* and in marked contrast to that seen in Natural History Museum specimens of the large European hunting dog, *Canis (Xenocyon) lycaonoides* (Kretzoi, 1938), from the British Middle Pleistocene site of Westbury-sub-Mendip (Turner, 1999).

Although it is somewhat larger, the new specimen bears a considerable resemblance to material previously collected from the Siwaliks during the earlier part of last century and now held in the Natural History Museum in London. Two mandibles, referred to *Canis cautleyi* Bose, 1880, are of particular relevance, and their measurements are given for comparison in the tables.

Family **HERPESTIDAE:** Genus and species indet.

MATERIAL

73A 52. A right posterior mandible fragment retaining most of a broken M/1 and the alveolus for M/2. The broken molar exhibits the trenchant cusps typical of a herpestid.

REMARKS

Barry (1983) discusses the presence of the common mongoose genus *Herpestes* (mongooses were then then referred to the Viverridae as a subfamily, Herpestinea) in the Siwaliks but based on material from Miocene-age deposits. This material could perhaps be referred to that genus, but without certainty.

Family **URSIDAE:** Genus and species indet.

73B GB 515. An isolated right M1/ of an ursid, somewhat larger than a small specimen of the extant European brown bear, *Ursus arctos* Linnaeus with a length of 19.0 mm and a breadth of 14.6 mm.

703 GB 2. A left distal calcaneus of a small ursid. This was found in 1990 in the same area as locality 642, and is probably 1.2 - 1.4 My old.

Discussion

Although it is clear that much revision of the existing collections of Carnivora from the Siwaliks is needed, the literature is already extensive and confusing (see, for example, Lydekker, 1884; Matthew, 1929, Pilgrim, 1932 and Gaur, 1987). The hyaenas in particular, as is usual in deposits that include fragmentary Pliocene material, provide something of a taxonomic nightmare, and I shall deal with these first.

The most recent detailed discussion of the Siwalik hyaenas has been undertaken by de Vos *et al.* (1987) in the course of identifying two new finds from deposits to the east of those examined by the British Archaeological Mission, although a brief summary is provided by Gaur (1987). De Vos *et al.* provide an extensive summary of the relevant literature, and conclude (p. 366) with a listing of likely species that includes *Pachycrocuta brevirostris*, *Crocuta sivalensis* and possibly *Crocuta crocuta*. In my view, the presence of *P. brevirostris* is clearly attested in the region - indeed it is virtually impossible to mistake this animal (Turner, 2001, Turner and Antón, 1996), but there is no basis for distinguishing *Crocuta sivalensis* from *Crocuta crocuta* and I would place all the Siwalik specimens of *Crocuta* in the latter species (Turner, 1990). The larger hyaena specimens of the British Archaeological Mission sample can then be placed with confidence in these two taxa.

The indeterminate portion of the hyaenid sample comes mostly from smaller animals. In terms of the listing of taxa provided by de Vos *et al.* (1987: 366), the most likely candidate, based on size and inferred stratigraphic range, is what the latter authors refer to as *Hyaenictis bosei* (following Matthew, 1929), and Werdelin and Solounias (1991) refer to *Leecyaena boisei*. The holotype is the skull M37133 in the Natural History Museum, London, comparative measurements of which are given in Table 10. 28 where it may be seen that the holotype it is somewhat smaller than the new maxillary fragment 214A GB 1.

So far as the Felidae are concerned, it is hard to add to the remarks already made. It is clear from the existing collections at the Natural History Museum (Turner, unpublished data), as well as from wider considerations of other samples (Ficcarelli, 1979; Petter and Howell, 1982; Turner, 1984, 1987) that the genera *Megantereon*, *Homotherium*, *Acinonyx* and *Panthera* were certainly present in later Pliocene and earlier Pleistocene deposits in the Siwaliks. Gaur (1987) also suggests the presence of the genus *Machairodus*, but this is unlikely in Upper Siwalik deposits and the material to which he refers is more probably of the genus *Homotherium*. Only *Panthera* cf *P. uncia* and *Megantereon cultridens* have been identified in the new material discussed here, but the unidentified portion of the sample clearly contains at least one other larger taxon although, as discussed, this does not appear to be *Homotherium*. One further intriguing possibility for allocation of these specimens is raised by the presence in the Natural History Museum collection of several fragmentary specimens, plus a complete felid skull, M32148 collected near Pinjor (but otherwise unprovenanced) and presented to the museum in 1977. With its flattened canines, shortened and wide muzzle, domed skull and short, wide zygomata, this specimen is clearly a species of *Dinofelis*, and Werdelin and Lewis (2001) have recently referred it to the species *D. cristata*.

The relationship of *Canis cautleyi* to other canids is currently unclear. European canids referred to the genus *Canis* are generally much smaller until the Upper Pleistocene (Turner, 1999), and the presence of such a large species in the Siwaliks at this time is clearly a matter for further investigation. So far as the smaller *Canis* is concerned, Gaur (1987) refers to two small canid species, the already known *Sivacyon curviplatus* Falconer, 1868 and a newly identified species, *Canis pinjorensis* Gaur, 1987. In the absence of specimens for comparison, the fragmentary material available here cannot be referred with any confidence to either. For the ursid specimens, I can find no later compilation than that of Pilgrim (1932). To attempt the identification of such isolated material on the basis of so dated a treatment would be pointless.

Acknowledgements

I am grateful to Robin Dennell for his invitation to work on the material discussed here, and to Jerry Hooker and Andrew Currant for access to collections at the Natural History Museum. This research was undertaken as part of a larger project supported by a grant from The Leverhulm Trust.

References

Barry, J.C. 1983. *Herpestes* (Viverridae, Carnivora) from the Miocene of Pakistan. *Journal of Paleontology* **57**: 150-156.

Ballesio, R. 1963. Monogaphie d'un machairodus du gisement Villafranchien de Senèze: *Homotherium crenatidens* Fabrini. *Travaux du Laboratoire de Géologie de la Facultée des Sciences de l'Université de Lyon,* **9**: 129 pp.

Dennell, R.W., Hurcombe, L., Jenkinson, R. and Rendell, H. (1990). Preliminary results of the Palaeolithic Programme of the British Archaeological Mission to Pakistan, 1983 - 1987. In: *South Asian Archaeology 1987*, ed. Taddei, M., pp. 17-30.

de Vos, J., Leinders, J.J.M. and Hussain, S.T. (1987). A historical review of the Siwalik Hyaenidae (Mammalia, Carnivora) and description of two new finds from the Upper Siwalik of Pakistan. *Proceedings of the Koninklijke Nederlandse Akademie van Wetenschappen* **B 90**: 333-369.

Ficcarelli, G. (1979). The Villafranchian machairodonts of Tuscany. *Paleontographica Italica* **71**: 17-26.

Gaur, R. 1987. *Environment and Ecology of Early Man in Northwest India*. Delhi: B.R. Publishing Co.

Lydekker, R. (1884). Siwalik and Narbada Carnivora. *Memoirs of the Geological Survey of India Series X*, Vol **II** (6): 178 pp. plus plates.

Matthew, W.D. (1929). Critical observations upon Siwalik Mammals. *Bulletin of the American Museum of Natural History* **56**: 437-560.

Petter, G. and Howell, F.C. (1982). Un felide machairodonte des formations plio-pleistocenes des Siwaliks: *Megantereon falconeri* Pomel/= *M. sivalensis* (F. et. C.). *Academie des Sciences, Comptes Rendus Hebdomadaires des Seances*, Serie III, **295**: 53-56.

Pilgrim, G.E. (1932). The fossil Carnivora of India. *Palaeontologia Indica* **18**: 1-232.

Turner, A. (1984). *Panthera crassidens*, Broom, 1948. The cat that never was? *South African Journal of Science* **80**: 227-233.

Turner, A. (1987). *Megantereon cultridens* (Cuvier) (Mammalia, Felidae, Machairodontinae) from Plio-Pleistocene deposits in Africa and Eurasia, with comments on dispersal and the possibility of New World Origin. *Journal of Paleontology* **61**: 1256-1268.

Turner, A. (1990). The evolution of the guild of larger terrestrial carnivores in the Plio-Pleistocene of Africa. *Geobios* **23**: 349-368.

Turner, A. (1999). A. Larger carnivores (Mammalia, Carnivora) from Westbury-sub-Mendip, Somerset. In: *Westbury Cave: The Natural History Museum Excavations 1976-1984*, ed. P. Andrews, J. Cook, A. Currant & C. Stringer, 175-193. Bristol: Western Academic & Specialist Press.

Turner, A. (2001). Remains of *Pachycrocuta brevirostris* (Mammalia, Hyaenidae) from the Lower Pleistocene site of Untermassfeld near Meiningen. In: *Das Pleistozän von Untermassfeld in Thüringen*, Vol 2, ed. R.-D. Kahlke, 673-690. Bonn: Rudolf Harbelt Verlag.

Turner, A. and Antón, M. 1996. The giant hyaena, *Pachycrocuta brevirostris* (Mammalian, Carnivora, Hyaenidae). *Geobios* **29**: 455-468.

Werdelin, L. and Lewis, M.E. 2001. A revision of the genus *Dinofelis* (Mammalia, Felidae). *Zoological Journal of the Linnean Society* **132**: 147-258.

Werdelin, L. and Solounias, N. 1991. The Hyaenidae: taxonomy, systematics and evolution. *Fossils and Strata* **30**: 1-104.

Figure 10. 11: Mandible of *Canis cautleyi* from locality 362 (photograph by Ros Coard).

Table 10. 27: Siwalik Hyaenidae: lower dentitions

P/2		P/3		P/4			M/1				DEPTH		B	
L	B	L	B	L	B	Lp	L	B	Lt	M/c	A	P	P/3	
Crocuta crocuta														
British Archaeological Mission specimens														
-	-	-	-	a22	13.4	-	a27	12.2	3.5	-	-	47.8	-	73X GB 56
15.4	10.4	-	-	-	-	-	-	-	-		-	-	-	73X GB 57
-	-	-	-	-	15	13.0	-	-	-		-	-	-	68 GB 1034
15.1	10.8	a20	15.2	-	-	-	-	-	-		-	-	-	M16584
-	-	-	-	22.2	13.1	12.1	29.0	11.6	5.0	-	-	-	-	M1553
Pachycrocuta brevirostris														
British Archaeological Mission specimens														
-	-	-	-	25.6	16.4	13.4	29.5	15.1	5.9	+	-	48.3	22.8	642 1287
-	-	23.5	16.4	26.1	16.8	12.7	29.3	14.3	6.1	-	-	-	23.3	642 GB I
Hyaenidae indet.														
13.7	9.4	18.1	11.8	20.6	-	10.5	-	-	-		37.1		19.1	315 GB 1/2
14	10	17	12	-	-	-	-	-	-		37.0	-	19.2	437D GB 26
18.6	12.4	-	-	-	-	-	-	-	-	-	-	-	-	608 GB 29

L and **B**, length and breadth of tooth; **Lp**, length of paraconid; **Lt**, length of talonid; **Mc**, clear presence (+) or absence (-) of metaconid; **DEPTH A** and **P**, depth of mandible anterior to P/2 and posterior to M/1; **B P/3**, breadth of mandible below P/3

Table 10. 28: Siwalik Hyaenidae: upper dentitions

P1/	P2/		P3/				P4/			
B	L	B	L	B	L	Ba	Bbl	Lp	Lm	
Pachycrocuta brevirostris										
British Archaeological Mission specimens										
root	20.5	14.2	25.1	18.2	40.5	25.0	13.9	14.1	16.0	642 1390
Natural History Museum specimens										
-	-	-	-	-	40.0	23.5	13.2	13.3	16.2	M37139a
Crocuta crocuta										
British Archaeological Mission specimens										
-	17.3	11.5	20.7	15.3	37.5	20.5	11.6	-	15.3	73 EX 410
Natural History Museum specimens										
-	-		21.4	16.6	40.6	20.2	10.9	14.6	18.2	M37139
Hyaenidae indet										
15.3	10.8	-	-	-	-	-	-	-	-	214 GB 2002
-	15+	11.8	-	-	36.8	20.0	11.3	-	15.2	214A GB 1
'Leecyaena bosei'										
Natural History Museum specimen										
-	14.8	9.4	18.0	12.8	a30	18.0	10.8	11.1	-	M37133

L and **B** as Table 10. 27; **Ba**, anterior breadth of P4/ at protocone; **Bbl**, breadth of P4/ blade; **Lp**, length of paracone, **Lm**, length of P4/ metastyle

Table 10. 29: Siwalik and Modern Felidae: lower dentitions

C		DM/4		P/3		P/4		M/1		
L	B	L	B	L	B	L	B	L	B	
Siwalik Fossil specimens										
Cf. *Panthera uncia*										
British Archaeological Mission specimens										
12.8	9.4	-	-	-	-	-	-	18.5	9.8	73 EX 568
-	-	14.8	5.0	-	-	-	-	17.7	9.0	73 EX 238
-	-	-	-	-	-	-	-	18.9	9.1	73 EX GB 58
-	-	14.5	4.8	-	-	-	-	-	-	73 C 203
						19.4	9.5	-	-	642 1C 34
Natural History Museum specimens										
				13.5	6.3	17.0	8.3	18.7	9.1	M16537a
Megantereon megantereon										
British Archaeological Mission specimens										
-	-	-	-	8.1	5.0	21.5	8.8	25.5	12.2	674 GB 21
Natural History Museum specimens										
-	-	-	-	-	-	20.7	9.7	24.5	10.3	16557
-	-	-	-	9.0	5.5	21.0	9.5	22.7	9.8	16554
Natural History Museum modern specimen										
Panthera uncia										
13.5	9.9	-	-	13.4	7.7	18.5	8.1	19.1	8.9	1963.2.25.1

Measurements as Table 10. 27

Table 10. 30: Siwalik Felidae: metapodia

Metacarpal 4					
TL	PML	PAP	SW	DW	
cf *Panthera uncia*					
97.1	15.0	16.6	9.9	15.2	73 GB 407/181

TL, total length; **PML**, **PAP**, medio-lateral and antero-posterior width of proximal end; **SW**, minimum shaft width; **DW**, width of distal epicondyle

Table 10. 31: Siwalik Canidae: lower dentitions

C		P/3		P/4		M/1			M2		P/1-M/3	
L	B	L	B	L	B	L	B	Lt	L	B		
Canis cf *C. cautleyi*												
British Archaeological Mission specimens												
14.0	9.7	15.0	7.3	17.5	9.0	31.5	12.7	8.5	-	-	102	362 124
Canis cautleyi												
Natural History Museum specimens												
-	-	-	-	-	-	a27	10.7	7.2	11.1	7.6	-	M40182
-	-	-	-	-	-	25.2	10.0	6.3	10.6	7.6	-	M40181

Measurement as Table 10. 27, plus **P/1-M/3**, length of tooth row

Chapter 11. The taphonomy of the fossil record of the Pabbi Hills

The surveys of 1986-90 collected over 40,000 vertebrate fossils from both floodplain and channel deposits, varying in age from 2.2 to 0.9 Mya-old. Approximately half were derived from Sandstone 12, and are thus ca. 1.2 - 1.4 Mya-old, and almost as many (found mainly in 1987) were 1.7 - 2.2 Mya in age. The Pabbi Hills collections thus comprise a large set of material from the middle part of the Early Pleistocene, and an equally large set from late Pliocene. As noted, particular attention was paid to recording each occurrence in terms of areal extent, number of specimens, stratigraphic level, topographic location and aspect. This chapter will examine the main features of the fossil record of a million-year-long fluvial system. The main factors to consider are: the distribution of fossils across the landscape; their sedimentary context; the type of fossil occurrences that were found; the type of animals and skeletal elements that are represented in each type of fossil occurrence; the likely degree of destruction of skeletal elements before burial and since re-exposure; and the problems of sampling rare taxa. The chapter ends with an overview of the general characteristics of the vertebrate fossil record from this type of palaeo-landscape, and considers why the evidence for hominids is so slight.

These issues have not so far been explored in studies of the Upper Siwaliks in Pakistan. Some similar work has been done in India, most notably in an excellent study by Gaur (1987), whose analysis was sadly limited by the very small size of his sample, which comprised <500 specimens from both the Tatrot and Pinjor Formations.

1) What was found where: the distribution of fossils across the landscape

Given that over 40,000 specimens were found in over 500 fossil occurrences (with an average of 87 specimens per occurrence), it may seem perverse to observe that fossils are generally very rare in the Pabbi Hills. Yet substantial areas produced hardly any specimens. As examples, Death Valley (occurrences 262, 264 et al.; chap. 6) produced only 65 specimens (including undiagnostics) from several thousands of square metres of exposures; likewise, along Sandstone 12, there were several long stretches of scarp face from which very little was found in 1986 and 1989.

The non-random nature of fossil distribution across the Pabbi Hills is best demonstrated by the simple observation that over one-third of all fossil found from a total of over 500 occurrences was derived from just 10 localities, and half of all fossil found came from 20 occurrences (see Table 11. 1).

Table 11. 1: The 20 most productive fossil occurrences in the Pabbi Hills

Number	Type of occurrence	Age (Mya)	No. of specimens
642	Locality	1.2 - 1.4	4220
610	Locality	1.2 - 1.4	3229
73	Locality	1.2 - 1.4	1673
83	Locality	1.2 - 1.4	1563
437	Locality	1.9 - 2.2	1346
489	Locality	1.7 - 1.9	1330
481	Locality	1.7 - 1.9	1287
133	Locality	1.2 - 1.4	820
435	Locality	1.9 - 2.2	695
14	Locality	1.2 - 1.4	693
68	Locality	1.2 - 1.4	672
491	Locality	1.7 - 1.9	630
664	Scatter	1.2 - 1.4	629
128	Locality	1.2 - 1.4	586
451	Scatter	1.9 - 2.2	466
375	Scatter	1.9 - 2.2	421
427	Locality	1.9 - 2.2	382
487	Locality	1.7 - 1.9	345
523	General collection	1.9 - 2.2	341
445	Scatter/Locality?	1.9 - 2.2	337
Total			**21665**

The following table shows the cumulative totals of the most productive localities as a proportion of the total number of fossils recovered:

Table 11. 2: The number and % of fossils from the most productive localities in the Pabbi Hills

Fossil source	No. of specimens	% of total
The 10 most productive localities	16228	37.1
The 20 most productive localities	21665	49.5
The 30 most productive localities	23986	54.8
The other 472 occurrences	19865	45.4
Total, all occurrences	**43761**	**100**

Table 11.3 shows the number of fossil occurrences found in each major field season in terms of the number of specimens from each. This shows even more clearly the clustered distribution of fossils: over half the fossil occurrences produced less than 25 specimens.

Table 11. 3: Size of fossil occurrences from each field season

N of occurrences	1986	1987	1989	Total	% of total
N of occurrences <25	72	148	42	262	52.2
N of occurrences >25, <50	19	60	13	92	18.3
N of occurrences >50, <100	11	45	7	63	12.5
N of occurrences >100, <250	13	31	8	52	10.4
N of occurrences >250, <500	7	9	3	19	3.8
N of occurrences >500, <1000	5	2	1	8	1.6
N of occurrences >1000	1	3	2	6	1.2
Total no. of occurrences	**128**	**298**	**76**	**502**	**100**
Total no. of fossils	**11710**	**20053**	**11734**	**43497**	
Average no./occurrence	**92.9**	**68.1**	**155.2**	**87.4**	

Notes: Gully collections are omitted from these figures. Note that the 1989 data are skewed by two very large localities, 610 and 642. If these are omitted, the average number per occurrence from that year was 4285/74 = 57.9.

With this brief overview in mind, we can now consider the main types of fossil occurrences.

2) The sedimentary context of fossils

When collecting fossil material from erosional surfaces, it is clearly difficult to identify precisely its original sedimentary context. Indeed, it can be impossible if material has rolled down slope, or been left "winnowed" on a deflated surface. Precise information on the sedimentary context of fossils is available for the excavated localities (73, 362 and 642); for various find-spots; and for specimens that still retain some of their original matrix. Tables 11. 4 and 11. 5 summarise the type of sediment still adhering to fossils collected in 1986, 1987 and 1989.

Table 11. 4. Matrix on fossil specimens by type of fossil occurrence and numbers of specimens (NISP) (in parentheses).

Matrix	Localities	Scatters	General collections	Total
Black clay	0	0	1 (1)	1 (1)
Pink/red clay	0	0	2 (6)	2 (6)
Clay skin	3 (5)	8 (38)	11 (21)	22 (64)
Silt	8 (25)	10 (14)	3 (4)	21 (43)
Grey sand	0	1 (1)	1 (1)	2 (2)
Sand (unspecified)	10 (19)	26 (29)	39 (39)	75 (87)
Sand (coarse)	1 (1)	1 (1)	1 (1)	3 (3)
Sand (fine)	6 (8)	8 (8)	4 (4)	18 (20)
Sand, grey	0	1 (1)	2 (2)	3 (3)
Sand, orange	1 (1)	0 0	2 (2)	3 (3)
Sandstone	0 0	6 (18)	27 (37)	33 (55)
Grit	4 (4)	16 (19)	9 (9)	29 (32)
Grit/pebbles	0	1 (5)	1 (1)	2 (6)

Table 11. 4 continued:

Pebbles	0	1 (1)	0	1 (1)
Total	**65 (127)**	**78 (133)**	**90 (113)**	**215 (326)**

Table 11. 5: Combined totals of matrix adhering to fossils by numbers of occurrences and specimens.

Matrix	No. of occurrences	% of occurrences	No. of specimens	% of specimens
Clay	3	1.4	7	2.1
Clay skin	22	10.2	64	19.6
Silt	21	9.8	43	13.2
Fine sand	18	8.4	20	6.1
Sand	86	40.0	98	30.1
Sandstone	33	15.3	55	16.9
Grit	31	14.4	38	11.7
Pebbles	1	0.5	1	0.3
Total	**215**	**100.0**	**326**	**100.0**

Because some occurrences produced several specimens with the same type of matrix, it is more useful to consider the number of occurrences associated with different types of matrix than the total number of specimens. The matrix on specimens from both localities and scatters is most commonly sand, followed by silt/fine sand, and then grit, indicating that both localities and scatters were primarily associated with the same type of sedimentary context, namely gently flowing water. The general collections are somewhat different, as the main matrix recorded on specimens from them was sand, followed by sandstone and then specimens with clay skins. The last type of matrix is interesting in that it indicates reworking: that is to say, fossil initially deposited in clay were then re-incorporated into sands and silts after the clay crust o the bone had hardened. Interestingly, only 4% of t specimens with matrix from localities had clay skir compared with 19% from the general collections. Th difference adds weight to the view that the amount of r working was least in the localities, and greatest in the gener collections. Overall, general collections were more strong associated with either active channels with greater strea velocity, or included material that had been re-deposited. Th would further suggest that the fossil recorded from gener collections are unlikely to represent material from localiti and scatters that has simply been dispersed over a larger ar since re-exposure.

3) The main types of fossil occurrences

As noted in previous chapters, the largest and most productive concentrations of fossil in the Pabbi Hills were defined as localities; smaller and more dispersed ones were classed as scatters, and areas where both the number and density of fossils were low were categorised as general collections. Although these categories overlap to a small extent (for example, a few of the largest scatters contained more fossils than some of the smaller localities), these categories remained a heuristically useful way of recording fossil occurrences in the field.

Table 11. 6 summarises the main types of fossil occurrences in the Pabbi Hills. In general terms, there are two main types of localities, scatters and general collections: those in which the remains of turtle are absent or very rare, and those in whi turtle is common. The remains of crocodile (usually teeth) a also absent in the former, but sometimes present in the latt In contrast, the remains of mammalian carnivores a primarily (but not exclusively) associated with tho occurrences in which the remains of turtle and crocodile a absent. In the absence of detailed contextual sedimentologic data, these indicators allow a simple distinction between lan based and riverine assemblages. (Some scatters and gener collections contained only a few pieces of turtle scu typically one or two pieces, and are not easily defined as lan based or fluvial; however, their contribution to the over results is slight, and can be omitted).

Table 11. 6: The main types of fossil occurrences in the Pabbi Hills

Type of fossil occurrence	Land-based assemblages: those lacking turtle and crocodile	Riverine assemblages: those in which turtle and crocodile are common
Type 1: Localities	Type 1a: Concentrated (e.g. 68, 73, 214, 362, 610, 642); fossils very numerous, with several taxa and skeletal elements present, and often articulated.	Type 1b: Localised (e.g. 83, 489); fossils are numerous, and several taxa and skeletal elements are present, but rarely articulated
Type 2: Scatters - Type A	Localised; remains of >1 animal, each with a few elements	Localised; remains of >1 animal, each with a few elements
Type 3: Scatters - Type B	Localised: remains of one animal only (e.g. 120, 636)	Localised: outcrops of turtle scute
Type 4: General collections	Dispersed, few taxa or skeletal elements, and sometimes containing scatters or localities	Dispersed, few taxa or skeletal elements, and sometimes containing scatters or localities

Each type can now be more fully described:

Localities. When discovered, the surfaces of these typically contain between 250 - 5000 specimens (including undiagnostics).

a) Type 1a: numerically large, concentrated, lacking turtle and crocodile. The localities in this set are: 68, 73, 133, 610 and 642 from Sandstone 12, and 214 and 362 from older units. Of these, locality 68 suffered most from problems of preservation; 133 and 610 had very high fragmentation rates (see below, Table 11. 22), and had probably been exposed the longest; 73 had the widest range of taxa (especially carnivores), and 642 was by far the largest of the localities. The taxa represented at each are summarised in Table 11. 7 below, and shown in greater detail in Tables 11. 8 and 11. 9 (end of chapter).

These are very well-defined spatially: in the case of locality 73, as small as 2 x 1m, or as bone pockets ca. 1m wide at the top of steep scarp slopes, with the fragments scattered down slope, as at 214 and 362. The largest, 642, covered the top of a low hill, and was perhaps 20 x 20m square. All contain abundant remains of ungulates (usually medium-sized), and sometimes larger ones (e.g. *Sivatherium* at 362, and *Rhinoceros* at 73 and 642). Sub-surface remains at the three localities in this group that were excavated (73, 362 and 642) were often articulated. The commonest prey species are usually medium-sized bovids, cervids and equids (except at 214 and 362, where equids were absent). Although the primary prey is similar, there are distinct differences in their frequencies. The main prey species at 68 were bovids and cervids (roughly equally), followed by suids; cervids, equids and bovids, plus some rhinoceros and suids, were the main types at 73; bovids, equids and cervids were the main types at 133 and 610; and bovids and equids at 642. *Rhinoceros* was represented only at 73 and 642, and suids at all except 133. Locality 214 was the only one to lack evidence of both rhinoceros and suids. Elephant is attested only at 73 and 133, but even there, very weakly indeed. The bovids at these localities are overwhelmingly medium-sized, with a few larger ones, and a very small number of small ones at 642. The most atypical locality was 362, which was the only one to contain the remains of a large canid, *Canis cautleyi,* and a large bird (neither represented in the surface collection), as well as the giraffid *Sivatherium*.

All localities contain evidence of at least one mammalian carnivore, usually a hyaenid such as *Pachycrocuta*, and it is likely that different predators, with different feeding habits and preferences, were involved in the accumulation and modification of these assemblages. (That is not to say they were the sole agency of accumulation; as the report on the material excavated from locality 642 makes clear, carnivore as well as fluvial agencies were involved). Locality 362 was unusual, in that the only carnivore present was *Canis cautleyi*, represented by a partial skeleton. Locality 73 was truly exceptional in that *Pachycrocuta brevirostris*, *Crocuta, Panthera,* and *Ursus* were present, as well as a small canid, an herpestid and another large felid: overall, locality 73 contributed 65 of the 83 carnivore specimens from this type of locality.

Table 11. 7: Taxa represented at large, concentrated (Type 1a) localities:

Locality number	Bovids	Cervid	Suids	Equids	Elephas	Rhino	Hyaena	Other
68	x	x	x	x	-	-	x	Carnivore
73	xx	xx	x	x	-	x	x	Herpestid, felids, ursid, canid
133	xx	x	-	xx	-	-	x	
214	xx	x		-	x		x	Carnivore, and trace of fish and turtle
362	xxx	x	-	-	x	xx		*Canis*, bird, *Sivatherium*
610	xxx	trace	x	x	-	-	x	Canid, *Gazella*
642	xxx	trace	x	x	-	x	x	*Gazella*

Note: xxx = very common; xx = common; x = present.

b) Type 1b: numerically large, dispersed, with abundant remains of turtle. Examples: localities 14, 83, 124, 128, 156, 427 489, 491

The localities in this sub-set include some of the richest in the Pabbi Hills, notably 437, 481 and 489. These are usually much more extensive than the Type 1a locality; and often extend for 25 - 100m along an eroding scarp face. Some, such as 489 and 491, had probably been exposed for several decades; others, such as locality 14 that are on steep, freshly-eroding scarp faces, were being actively eroded at the time of discovery. The sedimentary context of this type is usually a channel bar (e.g. 83, 427, 489, 491) or sandy, shallow water (e.g. 124, 128). Their generalised characteristics are shown in Table 11. 10. Turtle and crocodile remains are common, and those of mammalian carnvores are usually absent. Pigs and rhin tended to be rarer than at the Type 1a localities, but *Elepha* remains were commoner. Overall, these localities contain th largest range of taxa of any in the Pabbi Hills, with th exception of site 73. However, although they contain a divers fauna, the actual number of diagnostic mammalian specimen tends to be small, and usually smaller than that in the Type 1 localities. As examples, locality 437, the richest in this grou of localities, had only 107 diagnostics, compared with 549 a locality 73 (excluding the excavated material).

Table 11. 10: Taxa represented at large, dispersed (Type 1b) localities:

Locality	Bovids	Cervids	Suids	Equids	Rhino	Elephas	Turtle	Crocodile	Other
14	xxx	xx	x	x		x	xx	x	
83	x	xx		x		xx	xx	x	
124	xx	xx					xx	x	Carnivore
128	xx	xx		x			xx	x	Carnivore
156	xx	x		x	x	x	x	x	
427	x						xxx	x	
435	xxx	x	x			x	xxx	x	
437	xxx			x					Carnivores, hippo
481	xxx	x				x			Carnivore, *Sivatherium*
487	xxx	x	x			x	xx		Carnivore
489							x	x	Primate
491	xxx	x				x	x	x	Giraffid, Carnivore
666			x				x	x	

Note: xxx = very common; xx = common; x = present.

Table 11. 11 (end of chapter) shows the taxa represented in these assemblages >1.6 Mya-old. Bovids - predominantly medium-sized - are the commonest type, followed by cervids/tragulids. Suids, giraffids, equids and rhinoceros and elephant are also represented at some, as are crocodile and a few mammalian carnivores. A mandible of what may hav been a small primate was also found at 489 in 1999.

The taxa represented in this type of assemblage fro Sandstone 12 are shown in Table 11. 12 (end of chapter). O the five localities in this set, 14 and 83 were on the scarp slop of Sandstone 12; 124 and 128 on sandy outcrops in the badlands in front of Sandstone 12; and 666, in a gully in front of the Sandstone 12 unit. Overall, cervids are the commonest animal, followed by bovids and equids. Proboscideans are over-represented, in that 800 of the 845 Proboscidean tooth fragments are from locality 83. Carnivores are very rare, with only one specimen identified as a canid.

) **Type 2: small heterogenous scatters**. These were found in areas where collecting produced modest amounts of fossil, often in small local concentrations. Sometimes it was evident that material was eroding locally from the surface. Adjacent scatters of this kind along the same stratigraphic unit were probably sampling the same fossiliferous layer, and possibly the same fossil occurrence. Examples are probably the set of scatters (nos. 30-38) along Sandstone 12, and those along the 203 horizon. In other cases, scatters may lie in areas where material has recently accumulated through erosion, as along the sides and head of some recent gullies, particularly those incising into scarp slopes: examples were probably scatters 415, 419 and 423 in Kotha Kas (see chapter 6).

Most scatters contained a few elements of more than one type of animal, and occasionally a few pieces from one taxon (and possibly the same animal), such as an equid distal metacarpal and equid proximal 1st phalange at scatter no. 30. Although few scatters were particularly spectacular, the scatters contributed a substantial amount of information (see Table 11. 13) on the range of animals present, and their metrical attributes.

As with localities, small dispersed scatters can be broadly divided into those with several (>5) scute remains, and those without. Table 11. 8 includes data on the taxa present in those fossil occurrences > 1.6 Mya-old that lack evidence of turtle or crocodile. These scatters show a similar predominance of bovids, mostly medium-sized, as well as evidence for giraffids, rhinoceros and Proboscideans. The last-mentioned are represented mainly by tooth and tusk fragments, and are probably over-represented. Scatters of this time-range also contain evidence of suids, hippo, and equids, all of which were absent from both 214 and 362. Carnivore remains are absent, however, from these scatters.

Table 11. 11 includes data on the taxa represented in heterogenous scatters >1.6 Mya-old and containing evidence of turtle and/or crocodile. The main difference between these and scatters with no evidence of turtle or crocodile is that suids are absent, but Hippopotamids and carnivores are sometimes present.

The scatters and general collections in turtle-rich areas have the same type of animals as in the localities, apart from suids, which are not presented in the general collections. However, these are rare in the much larger set of scatters and localities, and their absence from the much smaller set of general collections could be fortuitous.

The turtle-free scatters from Sandstone 12 are very impoverished by comparison with the turtle-free localities from this horizon (see Table 11. 9), in that the scatters lacked evidence of suids and rhinoceros. Additionally, the scatters had only one carnivore specimen (38GB22). In contrast, scatters containing turtle remains had slightly more evidence of suids and rhinoceros (see Table 11. 12).

Table 11. 13: Overall composition of scatters

Year	Total	N diags	N teeth	Bovids +Cervids	Equids	Elephas	Rhino	Suid	Carnivore	Crocodile
1986	3969	1693	463	193	33	271	3	5	2	19
1987	7098	1806	555	459	35	296	21	9	12	16
1989	2741	724	151	177	8	0	1	0	1	4
	13808	**4223**	**1169**	**829**	**76**	**567**	**25**	**14**	**15**	**39**

Type 3: small homogenous scatters. Examples are the substantial parts of elephant carcasses (usually skulls, as near 57, 610; the elephant femur from 664 etc.) or a single skeleton, notably of one bovid at 636, or the partial remains of a large felid at 291. Some concentrations of turtle scute are probably other examples of this type of scatter. These probably represent isolated death events in which a carcass was buried by sediment shortly after death and before it was dispersed by scavengers and water transport. The bovid remains at 636 may represent a bloated carcass that became entrained in sediment before it disarticulated. Concentrations of ostrich egg-shell (as at 120) can also be regarded as individual death events.

Type 4: low density, low frequency. This category includes the general collections and also some of the smaller associated scatters. Fossil remains were mainly tooth fragments and undiagnostic pieces, often very fragmented, abraded and weathered, with only the occasional good specimen. General collections show a similar predominance to the scatters of bovids, mostly medium-sized, as well as evidence for equids, rhinoceros and Proboscideans. The last-mentioned are represented mainly by tooth and tusk fragments, and are probably over-represented. Carnivore remains are very rare.

These collections can be regarded as the "background noise and probably comprises 20% of the total fossil found in th surveys. Their general features are shown belov

Table 11. 14: Overall composition of general collections

Year	Total	N diagnostics	N teeth	Bovids +Cervids	Equids	Elephas	Rhino	Suid	Carnivore	Crocodile
1986	1579	569	237	75	20	128	6	0	1	5
1987	6436	1405	640	453	46	320	17	9	7	5
1989	1234	265	172	126	22	0	2	0	0	3
	9542	**2239**	**1049**	**654**	**88**	**448**	**25**	**9**	**8**	**13**

Note: only 82 of the 1049 teeth were complete.

4) Type of animals

This section considers the range of taxa represented in the Pabbi Hills collections, their probable relative abundance, body si and mortality structure.

i) range of taxa and relative abundance: Table 11. 15 (end of chapter) summarises the type of animals represented in localities, scatters and general collections. As is evident, some taxa are commonly found in a variety of contexts, but some are more restricted in terms of where they were found. A complicating factor is that some taxa are also likely to be over-represented by total numbers of fragments: for example, tooth and tusk fragments account for most of the specimens attributed to Proboscideans. Even so, bovids, cervids, equids, *Rhinoceros* and Proboscideans (probably *Elephas*) are commonly found in localities, scatters and general collection Carnivores are much more restricted: most are found localities, and we have already seen that around half we derived from locality 73 alone.

ii) body size: The likely body size of the animals represente in the Pabbi Hills collections is shown below (Table 11. 16) As can be seen, most have a probable body weight of betwe 100 and 500 kg.

Table 11. 16: Probable body size of taxa represented in the Pabbi Hills

Size class	Animal type	Probable adult body weight (kg)
Extremely large	Rhino, Hippo, *Hexaprotodon,* Elephant, *Stegadon*	5,000 - 10,000
Very large	*Sivatherium*, largest bovids	1,000 - 5,000
Large	Bovids	500 - 1000
Large medium	Bovids, equids	250 - 500
Medium	Bovids, pigs, large felids, hyaenids	100-250
Small	Bovids	50 - 100
Very small	Bovids, large canids	25 - 50
Extremely small	*Gazella*, primate, small canid	<25

iii) age structure: Unfortunately, there is virtually no information on the age at death of Upper Siwalik taxa. Table 11. 17 (end of chapter) summarises the evidence for immature specimens in the surface collections from the Pabbi Hills. As can be seen, very few specimens were found in which fusion was not complete. These were mostly bovids or cervids (it can be assumed that most of the immature artiodactyla specimens were also bovid or cervid), and only one immature pig and horse was recorded. Of the 54 records, only 6 were from general collections; 15 were from scatters, and the remaining 33 records were from localities. As 55% of the fossils four came from localities, it is perhaps not surprising that the contained roughly the same proportion of immature specime (61%). What is interesting is that the proportion of immatu specimens in excavated assemblages seems to be higher th in the surface collections. At locality 362, for example, the was an immature large bovid; at locality 73, perhaps 4 cervid 1 equid, 1 suid and a hyaenid were immature, and at locali 642, one-third of the *Damalops*, 1 *Hemibos*, an equid and rhinoceros were also immature. This would imply that

Figure 11. 1: Fossil collecting along steep slopes of Sandstone 12 (people at top as scale). Here, erosion rates are very high, and any fossil material found is likely to have been exposed very recently.

Figure 11. 2: Survey in an area of moderate slopes, high erosion rates, and little ground cover

Figure 11. 3: Sandstone 12, near locality 83. Here, erosion rates are variable, with some slopes actively eroding and others more stable. The scarp face has receded a few metres in places in the last century.

Figure 11. 4: A general view from Kanarawala kas down Sandstone 12, showing variable but high rates of erosion, and considerable incision into the scarp face.

Figure 11. 5: Sohawa kad headwaters (near the 203 horizon). The depth of gullying, steepness of many slopes, and high rates of erosion are very evident. Once exposed, fossil is likely to be rapidly transported down slope and into the main gullies.

Figure 11. 6: Kotha Kas. Here, slope angles are more moderate, with more flat areas and shallower gullies. Erosion is still moderate to high; there is also a recent face collapse in the gully in the foreground.

immature specimens had a lesser chance of survival after they had been eroded to the surface. Given that the Pabbi Hills collections are also weighted towards large animals (see Table 11. 16), it is worth considering what information has been lo before and after fossils were exposed.

5) Information loss: weathering, fragmentation and disarticulation

Fossil material was collected from a wide variety of erosional surfaces. As mentioned several times already, these were mostly ones that were experiencing high rates of erosion. Some examples of the types of surfaces from which fossil was collected are shown in Figures 11. 1 -6. Figure 11. 1(hillllside) shows an example of the steepest slopes (here, along Sandstone 12) that were surveyed; here, erosion is very active, and fossil found at the top of these slopes has probably been exposed very recently. Localities 14, 30, and 109 along Sandstone 12, and 214 and 362 in Choawala Kas, were some of the best discoveries in this type of situation. Figure 11. 2 shows a more common situation – a moderate slope, with very little ground vegetation, and little to prevent erosion and run-off by rain. Two representative examples of Sandstone 12 are shown in Figures 11. 3 and 11. 4. The first, near locality 83, shows that erosive rates are quite variable: some slopes are eroding rapidly, whilst others are more stable, with some grass and sandstone blocks to impede erosion. Nevertheless, the scarp face is receding, as evident by trigometric points that would have been a safe distance from the edge when placed in the early 20th century, but are now very near the edge, or even now on the slopes below. The general view along Sandstone 12 (Figure 11. 4) shows the numerous gullies cutting into the current scarp face, the variety of slope angles along the face, and the variable but generally high rates of erosion along it.

On the south side of the G.T. road, erosion rates were ve high in the headwaters of Choawala and Sohawa kas; Figure 20) shows the dramatic extent of gullying in Choawala Ka and a similar scene is shown for Sohawa Kas (near 269 ar the 203 horizon) in Figure 11. 5. Kotha Kas was rath different (Figure 11. 6); here, the landscape was gentler, with higher proportion of gentler slopes and flat surfaces. Erosic was still high, and it was apparent in 1999 that a considerab amount of fossil had been exposed since last visited in 1989.

As Table 11.18 indicates, fossil vertebrate assemblages fro modern erosional surfaces are the product of numero processes that have occurred since the death of the animals which they were once a part. Establishing how muc weathering, fragmentation and dispersal occurred before ar during burial, and subsequent to re-exposure, is a difficult b essential aspect of taphonomic investigations. In particular, v need to know how long material has been exposed, how mu deterioration it has experienced since re-exposure, and how f it may have been transported from its original location.

As shown below, fossil vertebrate remains that are collect from modern erosional surfaces are the product of numero processes that have occurred since the death of the animals which they were once part:

Table 11. 18: Taphonomic processes and bone fragmentation.

Stage	Process	Consequence
Death	Decomposition	
	Consumption	bone gnawing, fragmentation, dispersal
	Weathering	bone/tooth surface modification, further fragmentation
	Trampling	bone/tooth surface modification, further fragmentation and dispersal
Burial	in situ or transported	destructive if high energy; dispersal, further fragmentation
	diagenetic changes	sometimes adverse on bone preservation
	potential re-deposition through e.g. stream action	further dispersal, fragmentation and time-averaging
Re-exposure	sub-surface damage from roots	further fragmentation
	renewed weathering	further fragmentation
	renewed trampling	further fragmentation
	rain, slope transport	further fragmentation and dispersal; loss of context
	Wind and rain	deflation, loss of context

The principal consequences of these post-mortem processes are weathering, fragmentation and dispersal, and establishing how much occurred before and during burial, as well as subsequent to re-exposure, is a difficult but essential aspect of taphonomic investigations.

i) weathering data: As indicated in Table 11. 19 (end chapter), the degree of surface weathering of a large numb of fossil specimens from both the surveys and excavations w recorded, based on the scheme proposed by Behrensmey (1978) for modern carcass remains. Lyman and Fox (198 have discussed the problems associated with interpreting bo weathering, and argue that "accumulation and exposu histories are not readily inferred from assemblages of weathered bones" (ibid., p. 293). Bone weathering is a multiple signal that reflects sub-aerial exposure after death (as measured by Benhrensmeyer (1978), but also sub-surface deterioration as well as further deterioration after re-exposure. Bone weathering stages are thus best studied in conjunction with other types of data, notably fragmentation rates.

The data from the excavated assemblages provides thc clearest indication of how much weathering occurred before burial. Generally, the state of preservation of all taxa in the excavated assemblages from localities 73, 362 and 642 was excellent (typically stage 2), implying that bone was not exposed to weathering for more than a few months between death and burial. If these assemblages resulted from carnivores removing carcasses or portions of them to dens, they may of course be atypical of bone weathering elsewhere on the palaeolandscape. However, other, partially-exposed, bones were excavated during the survey. Examples were the artiodactyl pelvis from Pelvis Valley (fossil occurrence 387), a complete pair of equid mandibles at occurrence 109 in 1998, another pair near 269, and a rhinoceros forelimb in 1998. Almost all these were as little weathered as the material in the three excavated localities.

The surface condition of specimens from the surface collections provides a compound signal, of weathering before burial, diagenetic changes during burial, and further surface damage after re-exposure from its parent deposit. If the quality of preservation in sub-surface material is also known, the extent of weathering on specimens found on the surface can give some indication of how long it has been exposed. One would expect bones that had been re-exposed to show a high degree of recent weathering due to the length of re-exposure. Once bone is re-exposed, it quickly deteriorates; often the surface detail is poor, cracked, flaked, and the colour changes to dark grey or black. Recently-exposed fossil tends to look much fresher, with better surface detail, and less encrustation.

As examples, the surface condition of specimens found on the surface at locality 73 and 362 was mainly that of Behrensmeyer's stages 1 and 2a; at the time of discovery, these bone pockets were clearly being actively eroded from their soft matrix. In contrast, the surface condition of bone at locality 642 was generally very poor, and was lying on a flat surface that was being eroded much more slowly. Fossils were also heavily encrusted with cemented sand that was difficult to remove and which often obscured all surface detail.

Surface changes to bone after re-exposure are also linked to fragmentation, as generally, the worse the surface condition of fossil bone, the more fragmcnted the material. The clearest examples are from the excavated localities (73, 362 and 642), all of which contained complete and often articulated limb segments as well as complete crania, in contrast to the fragmented material on the surface.

The surface condition of specimens from the scatters and general collections tended to be much the same as from the localities, apart from those along Sandstone 12. Here, however, it is worth noting that this unit is very extensive, and in general, the surface condition of specimens was generally better in areas of rapid erosion, implying that the longer the period of re-exposure, the more fossils deteriorate.

ii) fragmentation: Patterns of bone and tooth fragmentation provide a number of opportunities for ascertaining how much information has been lost before and during burial, as well as since re-exposure. An additional question is whether the degree of fragmentation is related to the type of fossil occurrence.

a) gross fragmentation rates: One simple indicator of fragmentation is the ratio of diagnostic (to either taxa or skeletal part) to indeterminate specimens in individual or sets of assemblages (excluding turtle and crocodilean scute, and egg-shell fragments). Two initial questions are whether the amount of fragmentation varies first between localities, scatters and general collections, and secondly, between fluvial and land-based assemblages, i.e. between those with or without large amounts of scute/crocodilean fragments.

Tables 11. 20 and 11. 21 show the overall level of fragmentation in fluvial and non-fluvial assemblages:

Table 11. 20: Proportions of diagnostic and indeterminate specimens from non-fluvial fossil occurrences, based on all surface data, 1986, 1987 and 1989.

Type of fossil occurrence	Localities	Scatters	General collections
Total number of specimens excluding turtle, crocodile and egg	10317	4277	3605
Total number of diagnostics	3090	936	897
Total number of indeterminates	7227	3341	2708
% diagnostic of total	**30.0%**	**28.0%**	**24.9%**

Table 11. 21: Proportions of diagnostic and indeterminate specimens from fluvial fossil occurrences (i.e. fossil occurrences with >5 specimens of turtle/crocodile), based on all surface data, 1986, 1987 and 1989.

Type of fossil occurrence	Localities	Scatters	General collections
Total number of specimens excluding turtle, crocodile and egg	7953	4156	1869
Total numer of diagnostics	2507	998	406
Total number of indeterminates	5446	3158	1463
% diagnostic of total	**31.5%**	**24.0%**	**21.7%**

In each case, localities have the highest proportion of diagnostic specimens, and general collections, the lowest. Interestingly, the ratio of diagnostics: indeterminates is almost identical in both the land-based and fluvial localities. This might suggest that mammalian carnivores did not fragment bone material more than fluvial agents.

In greater detail, simple indicators of fragmentation can also indicate the length of re-exposure, on the grounds that similar types of fossil occurrences should have the same degree of fragmentation. As a simple example, the following table shows the fragmentation at non-fluvial occurrences >1.6 Mya old.

Table 11. 22: Fragmentation at non-fluvial occurrences >1.6 Mya-old (derived from Table 11: 8):

Type of fossil occurrence	Locality 214	Locality 362	Scatters	General collections
Total	234	218	2673	2132
Diagnostics	26	109	485	393
% of diagnostics	**11.1%**	**50.0%**	**18.0%**	**17.8%**

The scatters and general collections (providing most of the data in this set) indicate that on average, ca. 20% of fossil specimens were diagnostic. Locality 214 had been more or less completely eroded, and only 11% of specimens were diagnostic. In contrast, half the specimens were diagnostic at locality 362, which was discovered at an early stage of erosion.

The same pattern is evident in later, non-fluvial occurrences from Sandstone 12:

Table 11. 23: Fragmentation at non-fluvial fossil occurrences from Sandstone 12 (1.2 - 1.4 Mya-old) (derived from Table 11: 9)

Localities	68	73	133	610	642	Scatters	General collections
Total	672	1673	820	3229	4220	1310	1017
Diagnostics	199	515	217	388	1073	242	279
% diagnostics	**29.6%**	**30.1%**	**26.5%**	**12.0%**	**25.4%**	**18.5%**	**27.4%**

In this set, Locality 73, found when being actively eroded, has the highest proportion of diagnostics (but not as high as at locality 362), and Locality 610 the worst, and longest-exposed, and in a comparable setting to 214, and with the same proportion of diagnostics. (As shown in Table 11. 19, the surface conditions of material from locality 610, and to some extent, 214, was worse than at, for example, localities 14, 73 and 128).

The next two tables show the fragmentation at fluvial fossil occurrences. The Sandstone 12 set (Table 11. 24) consists of localities either on steep scarp slopes (14, 124, 83) and/or on very soft sediments (14, 124, 128) where erosion was rapid. Overall, the proportion of diagnostics was high at locality 14, consistently good (15-20%) in the scatters and most localities. At locality 83, the proportion of diagnostics was very low if Proboscidean tooth fragments are excluded; here, the surface condition of bone was also not as good as at, for example, 14 or 128 (see Table 11. 19).

Table 11. 24: Fragmentation at fluvial fossil occurrences from Sandstone 12 (1.2 - 1.4 Mya-old)

Localities	14	83	124	128	666	Scatters	General collections
Total	693	1563	137	586	279	1498	558
Diagnostics	184	885 (61)*	24	55	46	307	70
% diagnostics	26.6%	60% (4.1%)*	17.5%	15.2%	16.5%	20.5%	12.5%

otes: The figures in parentheses for locality 83 are those excluding Proboscidean tooth plate fragments. Scatters: 17, 19, 23, 30, 33, 34, 35, 6, 38, 42, 47, 48, 114, 117, 122, 125, 127, 140, 628, 650, 655, 656, 664. (N=22). General collections: 49, 51, 74, 119, 126, 132, 144, 672. ‹=8)

ı contrast, the older set (Table 11. 25) was in Kotha Kas, and had probably been exposed for some time. As can be seen, agmentation rates here are much higher.

able 11. 25: Fragmentation at fluvial fossil occurrences >1.6 Mya old

ocality	156	427	435	437	481	487	489	491	Total
)verall total	287	382	695	1346	1287	345	1330	630	6302
ıdeterminates	219	249	514	1019	1124	234	864	439	4662
Ion-indeterminates	68	133	116	327	47	61	278	96	1126
)iagnostics less turtle, roc, egg	61	9	59	106	115	50	182	94	676
6 diagnostics	21.4%	3.5%	10.3%	8.4%	9.3%	17.6%	17.4%	17.6%	12.7%

Iote: By way of comparison, the % of diagnostic specimens from scatters and general collections > 1.6 Mya-old and with >5 pieces of scute vas 18.5% and 17.2% respectively. (Scatters: 244, 248, 254, 260, 265, 271, 275, 284, 285, 413, 415, 433, 445, 446, 451, 466, 452, 454, 485, 15. (N= 19); General collections: 203, 216, 236, 259, 276, 406, 479, 483, 497, 499, 503, 507. (N=12)).

These data indicate the degree of fragmentation was very ariable, and overall, was generally worse in localities than in catters and general collections. With the exception of locality 56, the localities in this set were in Kotha Kas, and preservation was best at numbers 487, 489 and 491, which were also still actively eroding.

) limb-bone fragmentation

Klein and Cruz-Uribe (1984), and especially Marean (1991), ave drawn attention to the problems of post-depositional lamage to bone, and consequent loss of information. Marean ibid.) has developed the most detailed methodology for scertaining post-depositional loss of information. In articular, he stresses the need to identify criteria that are not ffected by either carnivores or hominids; that should be ndependent of the bone transport behaviour of bone collectors; and independent of the calculation procedure for measuring element abundance. He proposes that carpals and tarsals are the most appropriate indicator, and that a "completeness index" should be devised, based on the numbers of complete and fragmented specimens. He was dealing with carefully controlled excavated samples, unlike the surface collections studied here. In the field surveys, carpals/tarsals were either collected intact, or not at all, simply because fragments of pisiforms, carpals etc. would have been almost impossible to spot. However, the principles he suggests are reasonable, and because carpals/tarsals are poor in marrow, they are unlikely to be splintered by carnivores, or split by hominids. Klein and Cruz-Uribe (ibid.) also propose that the ratio of carpals/tarsals (as "small compact bones") to other post-cranial elements can also be a useful indicator of bone survival.

If Marean is correct, land-based assemblages - i.e. ones where carnivores could have had easy access to carcasses prior to burial - should show more evidence of limb-bone fragmentation than fluvial ones. If that is the case, the proportion of identifiable limb fragments (i.e. easily broken by carnivores) relative to carpals/tarsals (i.e. immune to carnivore damage) should be higher at feeding locales than at other types of land-based assemblages (scatters, general collections); and also higher than in fluvial assemblages.

Tables 11. 26 and 11. 27 summarise these measures of limb-bone fragmentation in different types of non-fluvial and fluvial fossil occurrences.

Table 11. 26: Limb-bone fragmentation in non-fluvial fossil occurrences (all surface data, 1986, 1987 and 1989).

Elements	Localities	Scatters	General collections
Carpals/tarsals	48	15	20
Fore-limb (F-L) total	244	90	55
Hind-limb (H-L) total	248	73	64
Fore-limb + hind-limb	492	163	119
Carpals/tarsals: F-L + H-L	**1: 10.3**	**1: 10.9**	**1: 6.0**

Note: Fore-limb = scapula, humerus, radius, ulna and metacarpal; hind-limb = pelvis, femur, tibia, fibula, astragalus, calcaneum and metatarsa
Carpals/tarsals include counts for scaphoid, navicular-cuboid, pisiform, unciform, and magnum.

Table 11. 27: Limb-bone fragmentation in fluvial fossil occurrences (all surface data, 1986, 1987 and 1989).

Elements	**Localities**	**Scatters**	**General collections**
Carpals/tarsals	60	28	14
Fore-limb total	171	75	43
Hind-limb total	191	94	34
Fore-limb + hind-limb	362	169	77
Carpals/tarsals: F-L + H-L	1: 6.0	1: 6.0	1: 5.5

Table 11. 26 indicates that the proportion of carpals/tarsals etc. to limb-bone fragments in non-fluvial assemblages is about the same in localities and scatters (ca. 1:10), and higher in general collections (1:6). The proportions in fluvial assemblages are similar, irrespective of type of occurrence, and higher than in non-fluvial localities and scatters (see Table 11. 27). That is to say, as expected, limb-bones are generally less fragmented in fluvial than in non-fluvial assemblages.

The higher rate of limb-bone attrition in fluvial assemblage and in non-fluvial general collections, can be attributed several factors. An obvious one is the effect of fluvi transport on skeletons, and ensuing fragmentation. Anoth might be the feeding behaviour of fluvial predators, notab crocodile. Regarding the high proportion of compact, durab podials to more fragile limb-bones in all general collections, is likely that prolonged dispersal, exposure, and reworking h favoured preservation of compact podials over more fragi limb-bones

c) tooth and jaw preservation

Additional indicators of information loss might also be the % of complete but isolated teeth to teeth fragments, and the % of complete, isolated teeth to mandible and maxilla fragments. The latter are problematic as some jaw specimens had no teeth, whilst others had several; nevertheless, in a perfectly preserved assemblage, one would expect all permanent teeth to be found in their respective mandibles or maxillae. Thus th ratio of isolated teeth to mandibular/maxillary fragments ma give some indication of post-mortem fragmentation.
We can first consider the overall patterns in tooth/ja preservation in non-fluvial and fluvial occurrences:

Table 11. 28: Jaw and tooth preservation in non-fluvial fossil occurrences (all surface data, 1986, 1987 and 1989).

Elements	**Localities**	**Scatters**	**General collections**
Tooth fragments	1061	380	395
Complete teeth	225	28	48
Mandibles + maxillae	308	31	42
% teeth that are complete	17.5	6.9	10.8
Ratio of complete teeth: jaw fragments	**1: 1.37**	**1: 1.11**	**1: 0.9**

Table 11. 29: Jaw and tooth preservation in fluvial fossil occurrences (all surface data, 1986, 1987 and 1989).

Elements	**Localities**	**Scatters**	**General collections**
Tooth fragments	317	346	125
Complete teeth	58	31	19
Mandibles + maxillae	65	28	6
% of teeth that are complete	15.5	8.2	13.2
Ratio of complete teeth: jaw fragments	**1: 1.1**	**1: 0.9**	**1: 0.3**

In both types of assemblages, localities have the highest proportions of complete teeth preserved relative to teeth fragments, and scatters the lowest. There is no obvio explanation as to why scatters should have the lowest %

complete isolated teeth relative to all teeth fragments, especially when the ratio of isolated teeth: jaw fragments in scatters is intermediate between that of localities and general collections. The overall picture of tooth and jaw fragmentation, however, is similar to that shown in Tables 10: 19 and 10:20 regarding the proportion of diagnostic to indeterminate specimens: the larger and more localised the assemblage, the less the degree of fragmentation.

d) teeth and vertebra preservation

Behrensmeyer (1975a) discussed at length the transport potential of different skeletal elements, and showed how these differences can result in the "winnowing" of an assemblage through hydraulic transport, whether in stream channels or surface run-off. The ratio of teeth to vertebrae in an assemblage can provide an indication of how much "winnowing" took place, as vertebrae would be preferentially removed. The following table shows tooth: vertebrae ratios in different types of assemblages:

Table 11. 30: Tooth and vertebrae preservation in different types of fossil occurrences (all surface data, 1986, 1987 and 1989).

Type of fossil occurrence	Teeth (all)	Teeth (complete)	Vertebrae	Complete teeth: vertebrae
>1.6 Mya-old				
Localities	217	42	210	1: 5.0
Scatters	162	7	31	1: 4.4
General collections	92	8	11*	1: 1.4
Type of fossil occurrence	**Teeth (all)**	**Teeth (complete)**	**Vertebrae**	**Complete teeth: vertebrae**
1.2 - 1.4 Mya-old				
Localities	957**	16	53	1: 3.3
Scatters	152	14	13	1: 0.9
General collections	42	8	1	1: 0.125

Notes: * 3 of these were vertebral flanges. If these are ignored, the total is 8, leaving the ratio of complete teeth to vertebrae at parity.
** distorted by 800 fragments of Proboscidean tooth fragments from locality 83.

The overall pattern is that irrespective of the age of the fossil occurrence, trunk parts are best represented at localities and least well in the general collections. This pattern is consistent with a greater degree of winnowing and attrition of body parts in the general collections than in the localities. Interestingly, the values from scatters are intermediate between those from localities and general collections.

Overview:

We can now consider the questions asked at the start of this section:

i) how long material has been exposed?

It was very apparent during the survey that a large proportion of the overall fossil assemblage had not been exposed for very long; i.e. a few years, or perhaps a couple of decades. This is particularly true of several major localities, notably 14, 68, 73, 124, 128, 214, and 362. In the case of localities 73 and 362, the bone in the surface assemblages was generally as fresh as that in the excavated assemblage. Most of the other major localities (427, 435, 437, 481, 489, 491 and 610) had probably been eroding for longer, and judging by the surface condition of the bone, localities 133, 156 and 666 had been exposed the longest. Many of the scatters were also very recent; examples are nos. 30-34, 109, and those new ones observed in 1999. It would probably be a fair estimate that at least half the total number of specimens found (i.e. those from the localities and many if not most of the scatters) had not been exposed for than a few years/decades.

ii) how much deterioration has occurred since re-exposure? This is related to be above question. In general, the longer fossil is exposed, the more it is likely to fragment and suffer surface deterioration. In this process, may bones will simply become unrecognisable in respect of taxon and skeletal element. In general, as is evident from the weathering and fragmentation data, the survival of material tended to be best at the major localities, and worst in the general collections.

iii) how far may fossil have been transported from its original location?

There are numerous instances where it was evident that fossil had not moved more than a few metres (typically, down slope) since re-exposure. This is true of the localities and probably most of the scatters. In some cases, it was obvious that movement had been very local as well as recent: as example, the four conjoining pieces of fossil from 269 (see chapter 8). The least movement of material was observed at locality 642, where most of the surface collection came from the flat surface forming the top of the present low hill: i.e. it had simply been washed out of its matrix by rain. (This may also explain why the surface condition of the fossil at this locality was the worst of all the localities; once exposed, the entire surface area was vulnerable to erosion and trampling, and fossil has no chance of being re-buried (even temporarily) during down slope transport). The greatest movement was observed at 214 and 362, both of which originated near the top of steep slopes, down which some fossil had travelled several metres. As noted in chapter 6, some "scatters" were local concentrations of material washed down small gullies and deposited at the base of scarp slopes. Transport of fossil found on modern erosional surfaces is likely to be very small and typically down-slope until it enters a major drainage gully, at which point it would be sluiced out of the area.

The overall impression of the Pabbi Hills fossil assemblage is that it is mostly very recent, in the sense that it has only recently been exposed. This is certainly true of the main fossil localities, and probably most of the scatters and even some of the general collections. Given the high rates of erosion, and the softness of most of the sediments, it is difficult to envisage fossil surviving re-exposure for more than a few centuries at most before experiencing either total disintegration, or transport into one of the major gullies. In these respects, the Pabbi Hills is probably similar to other Siwalik exposures.

The above implies that on time-scales of <100 years, there is very little information loss subsequent to re-exposure, in the sense that the fossil found on erosional surfaces will probably be an accurate sample of the original assemblage. (This is why, with careful collecting and recording, Siwalik fossil assemblages tend to be so fine-grained). On longer time-scales, information loss is likely to be substantial, if not total, as very little is little to survive exposure or remain in the area. This point is borne out by the complete absence of any faunal items that could be attributed to the Middle and Upper Pleistocene. In general, the fossil evidence is entirely consistent with the erosional data presented in chapter 8 by Hurcombe, and implies – independently of her evidence – that fossils found on modern erosional surfaces are derived from them, and are not lag elements from younger deposits now eroded away.

6) Type of skeletal elements

Skeletal element totals from localities, scatters and general collections are summarised in Tables 11: 31 - 33, which show totals for land-based and fluvial assemblages > and < 1.6 Mya-old. Given what has already been mentioned about the effects of weathering, fragmentation and length of exposure, it is probably unwise to search for significant differences between each set of data. Nevertheless, it is interesting that the % of diagnostic specimens from scatters is virtually the same as from general collections (21-22%), and much lower than from the well-preserved locality 362 (59%).

Nevertheless, some general trends may be noted if data are aggregated in the matter used by Behrensmeyer (1975b), in her study of carcass disarticulation in the Amboseli Basin East Africa. She found that carcass remains were thinly distributed, and might amount to an average of only 4.4 remains per sq. km. over a 10,000 year period. Her results are summarised in Figure 11. 7 below.

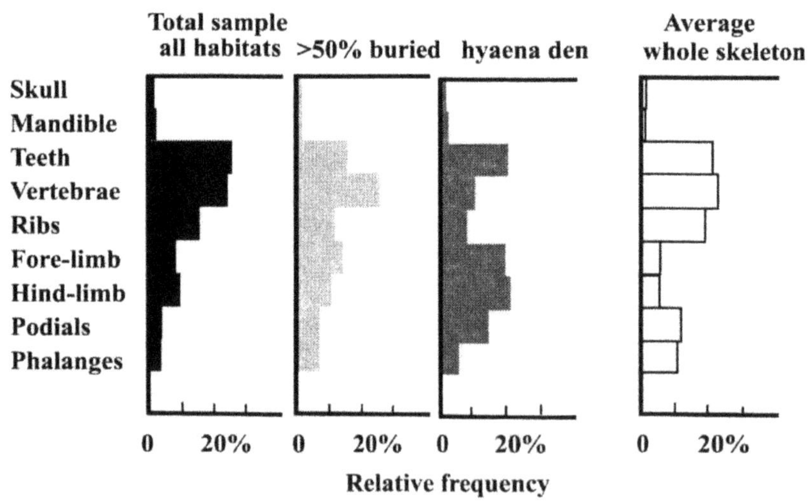

Figure 11. 7: The relative frequencies of skeletal parts in different modern situations in the Amboseli Park, East Africa. After Behrensmeyer 1975b: Figure 5).

Behrensmeyer (1975b: 101) showed that "the combined effects of carnivory, trampling., and weathering have significantly altered the (original) bone assemblage in the loss of ribs, podials, and phalanges". These effects are relatively minor in the sample of bones already half buried. In the hyaena den, however, selective transport of body parts and the effects of gnawing result in increases in the frequencies of fore- and hind-limb parts and podials, and some loss or destruction of ribs and vertebrae. Since her study, a considerable literature has accumulated on body-part representation in different kind of modern and ancient settings (see e.g. Lyman 1994 and references).

Table 11. 31 shows carcass part representation in land-based fossil assemblages (i.e. with <5 pieces of turtle/crocodilean remains). The main differences between localities and general collections is the increasing proportion of cranial fragments and vertebrae, but decreasing proportion of mandibles/maxillae and ribs. The frequency of body-parts is otherwise fairly evenly distributed across all three types.

Table 11. 31: Carcass part representation in land-based fossil occurrences (all surface data from 1986, 1987 and 1989).

Type of fossil occurrence	**Localities**	**Scatters**	**General collections**
Carcass part	**NISP (%)**	**NISP (%)**	**NISP (%)**
Cranium	49 (3.3)	34 (8.1)	24 (6.6)
Mandible + Maxilla (1)	308 (20.4)	31 (7.4)	42 (11.6)
Teeth, complete	225 (14.9)	28 (6.6)	48 (13.3)
Vertebrae	117 (7.7)	57 (13.6)	53 (14.7)
Ribs	202 (13.4)	75 (17.9)	33 (9.1)
Fore-limb	244 (16.1)	90 (21.4)	55 (15.2)
Hind-Limb	248 (16.4)	73 (17.4)	64 (17.7)
Carpals/tarsals	48 (3.2)	15 (3.6)	20 (5.5)
Phalanges	71 (4.7)	17 (4.0)	22 (6.1)
Total	**1512 (100.1)**	**420 (100)**	**361 (99.8)**

The frequency of body parts in fluvial fossil occurrences (i.e. with several pieces of turtle scute) is shown below.

Table 11. 32: Carcass part representation in fluvial fossil occurrences (all surface data from 1986, 1987 and 1989).

Type of fossil occurrence	**Localities**	**Scatters**	**General collections**
Carcass part	**NISP (%)**	**NISP (%)**	**NISP (%)**
Cranium	107 (8.1)	69 (13.7)	46 (22.1)
Mandible + Maxilla (1)	65 (4.9)	28 (5.6)	6 (2.9)
Teeth, complete	58 (4.4)	31 (6.1)	19 (9.1)
Vertebrae	263 (20.0)	49 (9.7)	13 (6.3)
Ribs	326 (24.8)	77 (15.2)	23 (11.1)
Fore-limb	171 (13.0)	75 (14.9)	43 (20.7)
Hind-Limb	191 (14.5)	94 (18.6)	34 (16.3)
Carpals/tarsals	60 (4.6)	28 (5.5)	14 (6.7)
Phalanges	75 (5.7)	54 (10.7)	10 (4.8)
Total	**1316 (100)**	**505 (100)**	**208 (100)**

The main differences between localities, scatters and general collections is the increasing proportion of cranial fragments and forelimbs, and decreasing proportion of vertebrae and ribs.

We may also note the following, broader groups (see Tables 11. 33 and 11. 34, derived from the previous tables):

Table 11. 33: Head and trunk parts in land-based occurrences.

Type of fossil occurrence	**Localities**	**Scatters**	**General collections**
Carcass part	**NISP (%)**	**NISP (%)**	**NISP (%)**
Head (cranium, teeth +mandibles/maxillae)	582 (38.6%)	93 (22.1%)	114 (31.5%
Trunk (vertebrae +ribs)	319 (21.1%)	132 (31.5%)	86 (23.8%)

Table 11. 34: Head and trunk parts in fluvial occurrences.

Type of fossil occurrence	Localities	Scatters	General collections
Carcass part	**NISP (%)**	**NISP (%)**	**NISP (%)**
Head (cranium, teeth +mandibles/maxillae)	230 (17.5)	128 (25.4)	71 (34.1)
Trunk (vertebrae +ribs)	589 (44.8)	126 (24.9)	36 (17.4)

Table 11. 33 suggests that in land-based assemblages, localities have a slightly higher proportion of head parts and slightly fewer trunk elements, although the differences are unlikely to be significant. In fluvial assemblages (Table 11. 34), the differences between localities, scatters and general collections is much clearer: localities have a far smaller proportion of head parts, and a much higher proportion of trunk parts. These contrasts between fluvial and non-fluvi localities may be attributed to the processes by which the assemblages were accumulated. The high proportion of hea parts in the non-fluvial assemblages is consistent wi predators transporting those elements away from kill- feeding-sites, whereas the high proportion of trunk elements fluvial assemblages is consistent with stream transport.

7) Carcass dispersal

It is obvious that substantial parts of individual skeletons are preserved in the major localities, particularly those associated with carnivores. (Indeed, that was one of the prime reasons for excavating three of them). Examples are the suid leg from locality 73, the partial skeleton of a large canid at locality 362, and articulated segments of herbivores from locality 642. The same type of results would probably accrue from excavating the larger, more extensive fluvial localities such as 489. What is difficult to establish is the area over which individu carcasses were distributed prior to burial, particularly alo stream channels. In some instances, individual carcasses we buried with very little dispersal, the best example being t remains of one bovid from occurrence 636. In other cases, considerable amount of disarticulation may have occurre with carcass parts dispersed over several hundreds of metres.

8) The sampling of rare taxa: needles and haystacks

As stated at the beginning of this volume, one of the major objectives of the fieldwork in the Pabbi Hills was to find hominid remains. Although 40,000 fossils were collected, no hominid remains were found, and it is relevant to ask whether their absence is genuine, or the product of inadequate sampling. In order to address this question, it is first necessary to consider it in conjunction with the rarest taxa that are represented in the Pabbi Hills.

Table 11. 35 indicates the numbers of each taxon, their number as a % of the total of fossils found, and the number of fossils that had to be collected before evidence of a particular taxon was likely to be found. As is evident, some groups are extremely rare, whether at the level of orders, families or genera. Birds and primates are extremely rare, for example, as are families such as Anthracoceridae, Ursidae, Viverridae, and genera such as *Megantereon* and *Stegodon*, each being represented by only one or two specimens each, out of a total of >40,000. In contrast, one in very 20 fossils found was identified as Proboscidea or Bovid, and and *Equus*, one of the commonest genera, occurred once in every 135 fossils. (The same relative frequencies occur if indeterminate specimens a excluded).

It is unfortunately impossible to ascertain whether the figur cited above are typical of other comparable palaeolandscape as numbers of fossils found of a particular group are rare given, and total number of fossils found, almost never. The is unfortunately no comparable data from other Upper Siwal data sets, and so the information presented here provid perhaps the first indication of the relative abundance of ea taxon in the Pinjor Formation. In the absence of comparab data, it might also serve as a useful indicator of ta abundance (or rarity) in other Early Pleistocene fluvial palae landscapes.

The reasons why particular types of animal might be rare a complex, but have some relevance as to why hominid remai were not found. Key factors are problems of identificatio fragmentation; body size and skeletal fragility, where anima lived and died, and the amount of time spent in the field. Ea can be taken in turn.

Table 11. 35: The relative abundance of various fossil taxa in the Pabbi Hills

Taxon	Total	Proportion of total number of fossil found	Average number needed to find 1 specimen
Orders			
Chelonia	3125	7.8%	13
Carnivora	65	0.1625%	615
Aves	2	0.005%	20,000
Primates	1	0.025%	>40,000
Families			
Bovidae	2443	6.1075%	16
Elephantidae	1996	4.99%	20
Cervidae	760	1.9%	53
Suidae	71	0.1775%	564
Giraffidae	25	0.0625%	1,333
Hyaenidae*	20	0.05%	2,000
Canidae*	20	0.05%	2,000
Felidae*	12	0.03%	3,333
Ursidae*	3	0.0075%	13,333
Anthracoceridae	3	0.0075%	13,333
Herpestidae*	1	0.0025%	40,000
Genera			
Equus sivalensis	296	0.74%	135
*Rhinoceros sivalensis**	87	0.2175	460
cf *Gazella*	49	0.1225	816
Hippotamus/Hexaprotodon	25	0.0625%	1,600
*Sivatherium**	23	0.0575%	1,739
*Panthera**	12	0.03%	3,333
*Canis**	9	0.0225%	4,444
*Pachycrocuta**	8	0.02%	5,000
*Crocuta**	4		10,000
*Ursus**	3	0.0075%	13,333
Stegodon	2	0.005%	20,000
Megantereon	1	0.0025%	40,000

Notes: * = includes excavated data. *Panthera* was found only at localities 73 and 642; *Megantereon* only at 674; Ursidae at 73, 703 and 714, the latter two in 1999 in the Baroth area). Anthracoceres (cf. *Merycopotamus*) - 2 metacarpal and 1 ulna fragment from 529. *Stegodon*: 2 teeth from 139 and X507.

i) Problems of identification

The most confident identifications are obviously those based on complete skeletons. As these are extremely rare, the next most reliable are those based on complete, well-preserved adult skulls and dentitions, preferably supplemented by associated post-cranial and cranial material. Isolated teeth and individual post-cranial elements, especially if fragmented, are often insufficient for an identification to be made beyond family level. In particular, much of the bovid and cervid post-cranial material is very hard to distinguish from each other, even though each group might be distinguished by individual cheek teeth or type of horn/antler. As seen in the previous chapter, *Damalops palaeindicus* and *Hemibos triqueticornis* were two types of bovid that could be identified to specific level on the basis of dentitions and associated horn-cores. Given that most of the Pabbi Hills material consists of isolated and fragmented material, most identifications of groups such as elephants, bovids, cervids, suids and hippopotamids were confined to the family level. As example, only a few complete and well-preserved dentitions allowed Proboscidean specimens to be identified as *Elephas* or *Stegodon*. In a few cases, identifications could be made to a specific level because only one species is known to have existed within a particular genus and family. The clearest examples were *Sivatherium giganteum, Rhinoceros sivalensis,* and *Equus sivalensis.*

ii) Food chains, family size and fossil preservation

As prey species are far commoner than their predators, it is unsurprising that bovids, equids and cervids are far commoner than the hyaenids, felids and canids that ate them. Regarding the herbivores, bovids are an extremely large and diverse group, especially when compared with equids and cervids, and thus it is unsurprising that their remains are ca. three and eight times commoner than those of the smaller cervid and equid families respectively. (At the generic level, of course, some bovids may have been rarer than cervids or equids). The largest herbivores tend to be rare because of their low birth rates and enormous feeding requirements, and thus the numbers of rhinoceros, hippotamid and elephant remains (excluding fragmented tooth and tusk fragments) are probably indicative of their rarity relative to, for example, equids and cervids, even if each family had an equally small number of genera. Omnivores such as pigs and bear that do not ingest large amounts of grass are usually rare, and this too seems to be reflected in the numbers of their fossils that were found.

Social structure may also have been important. The commonest types of animal tended to be ones that were probably gregarious, such as the larger bovids, the equids, hyaenids and canids. Animals that were solitary or formed smaller social groupings such as rhinoceros, the hippopotamids and the felids tended to be rarer. Body size and skeletal fragility were also important.

iii) Body size and skeletal fragility

In general, animals with large, robust skeletons should have a better chance of entering the fossil record than ones with skeletons that area small and fragile. As example, the very small gazelle-like bovid that probably weighed <25 kg may have been commoner than the small number of its fossil remains might suggest. Given its small size, its bones are much more likely to have been destroyed by a scavenger or predator, or fragmented through trampling or exposure to the elements. (The same point would apply equally a small primate or a bird). As noted already in this chapter, most of the adult mammals represented in the Pabbi Hills were probably >100 kg in body weight, and thus animals within the weight-range of hominids are also under-represented. On the other hand, very large animals may be disadvantaged as candidates for becoming fossils. Unless they die in or near an active stream channel or other large body of water, they are unlikely to be buried rapidly, and because their skeletons are so large they will be exposed to weathering for much longer. Re-exposure onto a modern surface will also take longer, and that too increases the chances of fragmentation, and so lessens the chances of accurate identification. (As example, during the fieldwork, hundreds and probably thousands of large bone fragments were found that were probably once part of an elephant, yet very few complete limb elements were found). In general, fossil preservation away from active channels tends to favour those animals sufficiently robust to withstand prolonged weathering and small enough to be fairly rapidly buried.

iv) Where animals lived and died

It was obvious during the field surveys that fossils of some groups were found at virtually any fossil occurrence (e.g. bovids), and some were most likely to occur in specific types of locales. The most restricted were turtle and crocodile, which unsurprisingly were not found in carnivore localities away from rivers. We have also noted that Proboscidean remains were common almost anywhere except carnivore localities such as 73, 362, 620 and 642, even if these otherwise contained a very wide range of taxa. The carnivores in particular occurred most at localities - especially site 73 - and were very rare outside them. Unsurprisingly, arboreal taxa are almost invisible; in particular, small primates, and birds.

v) Field time - the moral of the primate and 491

It was noted earlier in this chapter than half the fossils found came from just 20 fossil occurrences out of the 500 that were recorded. Those 20 major fossil occurrences also contained evidence of almost every family found in the Pabbi Hills, and every genus except *Megantereon* (fossil occurrence 674), *Stegadon* (X507 and 139) and *Anthracoceres* (fossil occurrence 529). This suggests that once these large and rich localities are found, the chances of finding rare taxa are improved considerably. (This is not surprising: a good comparable example are the key localities of FLK and FKL NN at Olduvai, which have yielded most of the hominid remains, other vertebrate fossils, and stone tools from Bed and lower Bed 2). Thus the best prospects of finding another primate fossil, or even one of a hominid, are likely to be realised at a locality such as 481 or 642.

An unfortunate reality in a country such as Pakistan is that field-time is limited to the length if time that foreign teams can spend searching for fossils: neither the funds nor the trained personnel are locally available for this type of work. In contrast with, for example, Kenya, indigenous counterparts of Kimoya Kimeu and his team do not exist, and thus th coverage of major fossil-bearing localities or strata is much less frequent. Even though each major field season in the Pabbi Hills lasted ca. 10 weeks and involved several field workers, the total time spent was almost certainly inadequate for locating the remains of the rarest taxa, including hominids.

Given that hominid remains tend to be extremely rare, and field seasons relatively short, it might be argued that the search for them should have over-ridden any other considerations. Only once they were found, it might be argued, was there any point to investigating the faunal communities of which they were a part. As should be clear from the rest of this volume, this approach was never followed: the field teams never were "hominid gangs", as on various other projects. The policy of the field work was to collect specimens of all taxa, and to build up as complete a record as possible of the biostratigraphy and spatial patterning of fossils in different types of context. To do otherwise would have been a high-risk strategy that risked not only finding no hominid remains, but also compromising the quality of other information that was obtained. (Additionally, the logistic support from the Geological Survey of Pakistan was given in order to build up a vertebrate fossil reference collection).

We may also note that some of these localities were extremely localised, particularly those that lacked turtle remains. Numbers 68, 73, 214, and 362, for example, were very small bone pockets that could easily have been missed only 50 - 100m away. They also tended to occur in areas that were otherwise very poor in fossil; this is particularly true of the two largest localities, 610 and 642. The turtle-rich localities tended to be more extensive, and often in areas that were generally productive. These observations lead to the pessimistic conclusion that even areas that appear poor in fossils need to be searched thoroughly in case they contain an extremely rich occurrence.

9) The absence of hominins: real or a sampling problem?

Various reasons have been put forward as to why hominin remains are (so far) absent from the Pabbi Hills fauna despite the excellence of fossil preservation. They would have been rare; they weighed less than most of the animals for which there is fossil evidence; their skeletons are fragile; they do not appear to have been eaten by carnivores or crocodiles at the few localities so far discovered; and as non-herbivores, they are also under-represented in a fauna overwhelmingly dominated by herbivores. The two best chances at present of finding a hominin fossil in the Pabbi Hills would appear to be at locality 642, which is not only the largest carnivore site, but also still mostly intact; and locality 481, where one specimen of what may have been a small primate was found in 1999, some 12 years after the locality was first discovered.

Nevertheless, stone tools were found in the course of fossil collecting, even if they were not found in large concentrations or their primary context. As related by Hurcombe (this volume), there are good reasons for arguing that many of these are more likely to have eroded from the fossil-bearing deposits on which they were found than to have been derived from post-Siwalik Middle or Upper Pleistocene contexts. As a hominin is capable of flaking a far greater number of stone tools in its lifetime than the number of bones in its body, the stone tools imply - but do not prove - that hominins were present, even if their remains have not been found.

As indicated in Chapter 1, there are several reasons why we might expect to find hominid remains in the Pabbi Hills or other comparable Upper Siwalik deposits in South Asia. The first is that there is now especially good evidence of hominid remains and associated stone tools from Dmanisi, Georgia, dated to 1.8 Mya and thus comparable in age to the earlier part of the Pabbi Hills sequence. In Asian terms, Dmanisi is only a stone's throw from the Pabbi Hills (or, more accurately, 1200 km), and thus it would be odd if hominids did not inhabit South Asia at that time. Secondly, the new dates for the hominid remains from Sangiran and Mojokerto (Swisher et al. 1994) suggest that *H. erectus* had colonised S.E. Asia by 1.8 Mya ago, even if others (e.g. Hyodo et al. 2002) contest this; see Dennell (2003). A third pertinent factor is Xiaochangliang, North China, where stone tools have been dated at 1.35 Mya (Zhou et al. 2001), which is within the time-range of Sandstone 12. Finally, our earlier research in the Soan Valley indicated that the earliest stone tools at Riwat have a minimum age of 1.9 Mya (Rendell et al. 1989). We might therefore reasonably attribute the absence of hominid remains in the Pabbi Hills to inadequate sampling. If so, more fieldwork should eventually result in the discovery of hominid remains.

Matters might not be so simple, however. Recently, Dennell (2003) has argued that even though *Homo erectus* was distributed across southern Asia in the Early Pleistocene, it does not appear to have been a successful coloniser, judging by the half-million to one-million-year gaps in many regional records. Even though a good case can be made for arguing that hominids had dispersed across southern Asia by 1.8 Mya, and even though fossil preservation in the Pabbi Hills was excellent, there may also be grounds for arguing that they would have been rare, and often absent, from the type of palaeo-landscape represented by the Pabbi Hills. In other words, the absence of large concentrations of stone tools, the absence of hominid remains, and the patchy record of artefacts across the palaeo-landscape could be seen as significant pieces of information about the behaviour of *H. erectus* in large-scale fluvial landscapes.

Homo erectus, rivers and stone

A basic truism in studies of human evolution is that the ability to flake stone was one of the major landmarks in our evolution. Stone tools opened up possibilities for accessing meat and bone-marrow, and thus had far-reaching dietary consequences. They also enabled hominins to make a wide range of implements from wood, and thus probably increased the quantities and types of plant and other types of foods that they could eat. Bipedal hominins that could make stone tools eventually developed larger brains because of their ability to acquire large amounts of animal protein; and larger brains increased the range of opportunities and solutions for early hominins. Their dispersal out of Africa and across southern Asia is thus seen as a consequence of bipedalism, tool-making and larger brains.

On this scenario, tool-making is seen as one of the "Mag Wands" that empowered early hominins. However, the Pab Hills data suggest a counter-argument, that the ability to mak stone tools may have been as much a ball-and-chain as source of empowerment. By 1.8 Mya, African hominins we probably technologically dependent on stone tools. Data fro Olduvai, Koobi and elsewhere indicate that the stone th preferred to use was basalt, lava, and quartzite; ve importantly, we also know that the stone they used wa obtained locally. At Olduvai (Beds I and II), for exampl most artefacts were made from stone obtainable within 2 – km, and occasionally 8 – 10 km (Hay 1976: 183 quoted Isaac 1989: 171). The distances over which stone was obtain appear similar throughout the Early Palaeolithic. How then d hominins cope with environments in which suitable stone w either absent or extremely rare?

Fluvial landscapes: resource-rich, stone-poor

The sedimentary history of the Pabbi Hills is essentially one of sand, silt, and clay. As is evident from the numerous sections presented in earlier chapters, most of the sediments were fine sands and silts that probably represent overbanks and abandoned channels. Large stones and cobbles were absent from every section prior to Sandstone 12. They were not entirely absent from the area surveyed: stringers of pebbles were noted in several places (see Hurcombe chapter), and one was partially excavated, at locality 269. In the Sandstone 12 horizon (1.2 – 1.4 Mya), stones were absent from the areas surveyed in 1986 and 1989, except for the Intermediate Conglomerate in the Baroth area, and this was the only place where stone would have been reasonably abundant, and then for only a relatively short period between 1.2 and 1.4 Mya. As indicated in the chapter by Hurcombe, most of the flaked stone found that year and in 1990 also came from this area.

The main type of stone in pebble stringers and the Intermediate Conglomerate was quartzite, which would have been transported as part of the bed load of the rivers flowing from the Himalayas and Karakorum. The forefront of these mountains would have been further away from the Pabbi Hills than today, given that the 3000m.-high range of the Pir Panjal did not exist when Sandstone 12 was deposited (Burbank and Raynolds 1984:118). There is no indication of any alternative and nearer source of stone suitable for making tools. The flood-plain associated with the river channels that flowed through the Pabbi Hills would have been wide and flat, wi no obvious high ground eastwards towards Lahore southwards towards the modern Indus. West of the mode Jhelum, most of the modern relief, particularly around Dina Jalalpur and Rohtas (see Figure 2. 3) is a product of t Middle Pleistocene, as the final Upper Siwalik fluvial deposi date from the early Brunhes. Consequently, the Pabbi Hi flood plain would have been essentially stone-free apart fro the occasional stringer or larger deposit along the main riv channels.

It was noted in chapter 2 that the monsoon regime of Sou Asia has existed since the late Miocene. Consequently, t rivers that flowed southwards from the Himalayas a Karakorum would have fluctuated considerably in their wat levels throughout the year. They are typically at their lowest winter, and then start to rise in spring when snow starts to m in the mountains. They then rise to their maximum lev during the summer monsoon, between June and Septemb when most of the annual rainfall occurs. At this time of yea large areas of the flood plain adjacent to the active channe would be inundated if it were not for the extensive series drainage canals along the modern river system. The like consequence of this variation in the depth and width of riv channels is that the few sources of workable stone would ha been inaccessible for much of the year. Figure 11. 8 sugges how this might have affected hominids.

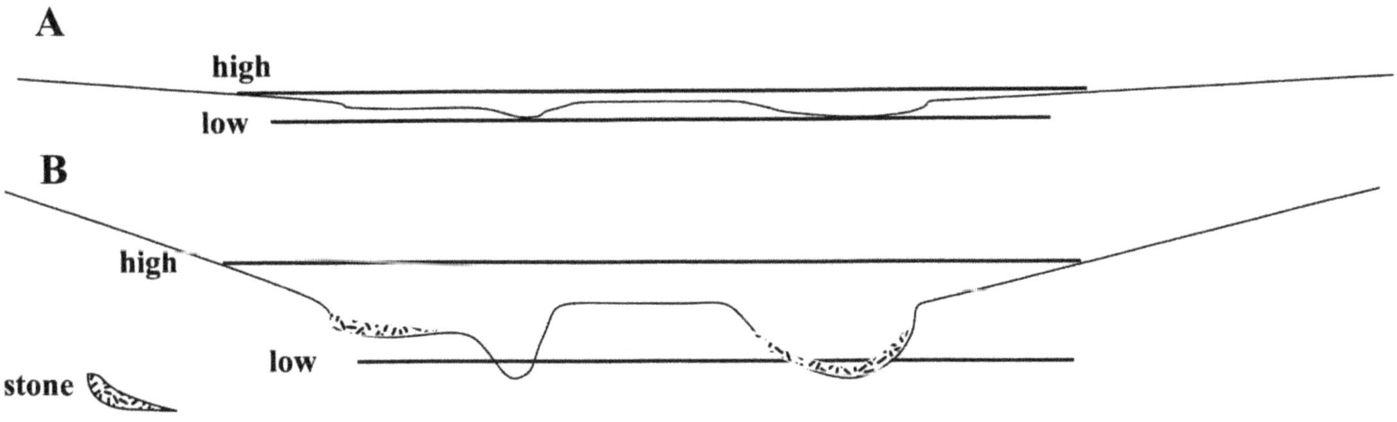

Figure 11. 8: The effect of river-levels upon the availability of stone. In this scenario, the only stone suitable for flaking is in river channels. In B, the vertical scale has been considerably exaggerated in order to show the profile of two river channels. In one, a pebble layer has been exposed through down-cutting, and in the other, stone is part of the current bed-load. Both are accessible only when river levels are low, typically in mid-winter. Following the influx of melt-water in late spring, and the summer monsoon, the sources of stone would be under –water, and thus inaccessible.

This figure shows a schematic cross-section of a flood-plain and two channels. In one channel, an earlier pebble layer has been re-exposed, whereas the other channel is currently carrying stone as part of its bed load. Both would be accessible during the later part of the dry season, between January and March. Between April – when rivers start to rise following the influx of melt-water from the Karakorum mountains – and December – when rivers are still falling after the monsoon – these sources of stone would be inaccessible. Given the absence of other alternative sources of stone within 10 km, hominids would be unable to exploit the animal and plant resources of the flood-plain for a large part of the year. The absence of large concentrations of stone tools, the absence of hominid remains, and the patchy record of artefacts across the palaeo-landscapes of the Pabbi Hills could therefore reflect the consequences of stone being available for only a restricted part of the year.

Overview

Although coarse-grained, the material from the Pabbi Hills provides a great deal of new information on the fossil record of a late Pliocene and early Pleistocene fluvial palaeo-landscape. The total of >40,000 specimens, from >500 collecting points or areas, considerably exceeds the sample of ca. 300 specimens from the KBS horizon collected by Stern (1993), or that of ca. 400 specimens from the Tatrot and Pinjor Formations studied by Gaur (1987).

The Pabbi Hills between one and two million years ago would have comprised a large flood plain, through which an antecedent of the Jhelum flowed. The active channel would probably have been wide, with several channel bars, and subject to major seasonal differences between the winter, dry season, and the summer, monsoonal season. Soil carbonate analyses (Quade 1994) indicate that tree-cover would have been slight, and largely confined to river margins. Faunal data suggest that this tree cover might have been thinner after 1.8 Mya ago. The landscape was probably similar to that already observed for northern India at this time: "a dominant grassland vegetation, with some trees and bushes, and a highly seasonal climate" (Gaur 1987: 198).

The fossil occurrences recorded in the Pabbi Hills represent material accumulated under a wide variety of circumstances, and over a variety of time-spans. Some of these fossil occurrences were probably "moments in time": e.g. the death and burial of a bovid at 636, and the breakage of a few ostrich eggs at 120. The land-based localities (e.g. 68, 73, 362 and 642) were probably accumulated by hyaenids over a few seasons; the smallest such as locality 362 may have been a single-season accumulation, and the largest (642), the outcome of perhaps a decade. The fluvial localities (e.g. 83, 489, 491) probably took longer to form, and perhaps span a hundred to a few thousand years, depending upon the stability of the stream channel. According to Behrensmeyer (1988: 194), channel-lag assemblages might span 10^2 – 10^4 yr., whilst channel-fill ones might result over a shorter period. As the largest 20 localities account for ca. 50% of fossil found in the Pabbi Hills, it is also likely that the same proportion represents a very small fraction of the total time-span of deposits in the one-kilometre thick, million-year sequence of the Pabbi Hills. The same is likely for other Upper Siwalik sequences: Gaur (1987: 31), for example, commented that Pinjor fossils in India "often occur concentrated in pockets", which implies that most are found in localities, as here. Raynolds and Johnson (1985: 310) point out that because Upper Siwalik fluvial sequences are strongly cyclical, with a range in the Jhelum area of 30-50,000 years, "[fossil] collections are likely to sample fossil populations at 10^3 to 10^5 yr increments". The biostratigraphic implications of this point were noted in the previous chapter.

The spatially-extensive but low-density general collections that constitute some 20% of the total fossil from the Pabbi Hills are most likely to be a palimpsest of numerous smaller episodes of animal death, disarticulation and burial. It is probably this fraction that has been the most subject to time-averaging and reworking. General collections from near the base of scarp slopes are also likely to be composite, as they could easily consist of specimens derived from other sedimentary units further upslope. The remaining 30% of fossil are largely from scatters and the larger general collections. Their temporal spans and accumulating agents are harder to identity, but are probably diverse. Some scatters may perhaps an occasional lion, and very rarely a bear and small dog. Along the margins of the river, scatters of egg-shell might show former nesting sites of ostrich; turtles and crocodiles would also have been common.

If any animal had been so interested, it might have observed occasional scraps of carcass and bone as it walked across the floodplain or along the river margin. There might occasionally have been evidence away from the river margins of where carnivores had fed; small to large clusters of partially-articulated and scavenged remains of bovids, equids and various other animals, and sometimes the remains of have been the initial stages of exposure and erosion of a much larger locality. Others that were contiguous and originating from the same stratigraphic level might also have part of a much larger locality. As noted, other contiguous ones might simply represent areas where fossil is accumulated by present-day erosion. Scatters might also represent small, discrete events, or areas of repeated, episodic accumulation.

Taken together, these different types of fossil accumulation provide an indication of a flood-plain in the Late Pliocene and between 1.2 - 1.4 Mya-ago. As noted in chapter 10, different types of animals would soon have been encountered. The commonest were medium-sized bovids. Before 1.8 Mya-ago very large ones might have been glimpsed, as well as a large short-necked giraffe. Occasionally, very small bovids, <0.50 m high, might also have been seen. Herds of small horses would have been common, and different types of tragulid and some deer. Larger animals too would have been spotted *Elephas hysudricus*, more frequently than *Stegadon*, *Rhinoceros* and Hippopotamids by the river edges, and both large and small pigs, although rarely. Carnivores would also have been seen: a very large hyaena, the size of a small pony, carnivores that had died there. A walk during the dry season, when the river was low, to a channel bar might also have revealed pieces of animal carcass scattered along the surface, perhaps over several hundred metres, and sometimes mixed up with the remains of turtle, ostrich shell, showing that other types of creature also frequented these areas. Just very occasionally, pieces of flaked stone might have been encountered, made from quartzite pebbles stone obtained from occasional stringers in the channel bed, and visible only at this time of year. As for the hominids themselves, they might have been occasional residents, depending upon where and when workable stone was available.

Table 11. 8: Taxa represented at land based fossil occurrences >1.6 Mya old and turtle-free

Taxa	214	362	214 and 362	Scatters	General collections
Artiodactyla (general)	6	32	38	102	769
Bovids	9	46	54	129	108
Cervid/Tragulids	2	3	6	53	56
Giraffids	0	8	8	3	6
Suids	0	0		6	4
Hippo	0	0	0	1	2
Equids	0	0	0	22	12
Rhinoceros	0	13	13	6	8
Proboscidean	2	5	7	155 (83)	108 (70)
Carnivora (general)	1	1	2	0	1
Hyaenideae	3	0	3	0	0
Felidae	0	0	0	0	2
Canidae	0	0	0	0	1
Ursidae		0	0	0	0
(All carnivores	**4**	**1**	**5**	**0**	**4**
Crocodylus	0	0	0	0	0
Turtle	2	0	2	0	0
Eggshell	0	0	0	4	14
Indeterminates	208	109	317	2192	1753
Diagnostics (non-indet)	26	109	134	485	393
less turtle, eggshell **	**26**	**109**	**132**	**481**	**379**
Average	n/a	n/a	225.5	46.2	41.3
Total	**234**	**218**	**451**	**2677**	**2146**

Scatters: nos: 210, 212, 215, 218, 223, 242, 252, 269, 278, 282, 298, 289, 301, 304, 306, 314, 321, 329, 330, 337, 340, 343, 344, 345, 354, 356, 363, 365, 367, 373, 375, 376, 377, 379,385, 387, 389, 391, 402, 409, 416, 420, 423, 441, 546, 457, 458, 460, 471, 472, 474, 482, 484, 477, 493, 505, 509, 527, 529. (N = 58).

General collections: nos. 208, 220, 246, 258, 264, 313, 323, 339, 341, 347, 351, 353, 355, 357, 359, 371, 4381, 383, 393, 397, 399, 408, 410, 411, 414, 417, 418, 421, 428, 430, 432, 434, 436, 438, 440, 442, 444, 439, 455, 461, 463, 464, 465, 468, 469, 470, 475, 480, 486, 495, 511, 517, 53. (N=52).

Table 11. 9: Taxa represented at land-based fossil occurrences from Sandstone 12 (1.2 - 1.4 Mya-old)

Taxa	**68**	**73**	**133**	**610**	**642**	**Scatters**	**General collections**
Artiodactyla (general)	68	113	55	133	274	66	34
Bovids (General)	43	23	78	209	731	45	85
Cervids	44	102	17	2	1	13	19
Suids	23	6	0	6	6	0	0
Equids	4	36	25	29	39	12	27
Rhinoceros	0	5	0	0	14	0	2
Proboscidean	0	1	0	0	0	84 (56)	63 (52)
Carnivora (general)	4	30	0	6	1	0	0
Hyaenideae	1	13	1	2	1	0	0
Felidae	0	8	0	0	0	1	0
Canidae	2	11	0		0	0	0
Ursidae	0	3		0	0	0	0
(All carnivores	**7**	**65**	**1**	**8**	**2**	**1**	**0**
Crocodylus	0	0	1	0	0	0	0
Turtle	0	0	0	0	0	0	0
Eggshell		0	15	1	0	20	39
Indeterminates	479	587	612	2841	3146	1088	777
Diagnostics (non-indet)	193	354	208	388	1073	242	279
less turtle, eggshell **	**193**	**354**	**193**	**387**	**1073**	**222**	**240**
Average	n/a	n/a	n/a	n/a	n/a	60.5	26.4
Total	**672**	**941**	**820**	**3229**	**4219**	**1330**	**1056**

Scatters: 21, 37, 45, 56, 60, 69, 109, 110, 115, 123, 136, 270, 604, 605, 626, 637, 641, 644, 645, 651, 668, 670, 675. (N=22)
General collections: 15, 24, 25, 26, 27, 52, 65, 67, 75, 76, 77, 78, 80, 112, 118, 137, 142, 146, 600, 606, 618, 619, 620, 622, 623, 627, 631, 635, 640, 643, 646, 647, 648, 674, 649, 652, 653, 662, 674, 678, 682. (N=40)

Table 11. 11: Taxa represented at fluvial fossil occurrences >1.6 Mya old

Taxa	Turtle-rich localities	Scatters with turtle	General collections with turtle
Artiodactyla (general)	251	61	52
Bovids (General)	279	105	70
Cervids	51	31	29
Giraffids	6	4	0
Suids	2	0	0
Hippopotamids	6	7	0
Equids	7	12	0
Rhinoceros	9	15	4
Proboscidean	43	93	55
Carnivora (general)	7	2	0
Hyaenideae	3	1	0
Felidae	2	0	0
Canidae	2	2	0
Ursidae	0	0	0
(All carnivores	14	5	0
Crocodylus	17	14	3
Turtle	949	358	134
Eggshell	0	21	4
Indeterminates	4662	1544	1027
Diagnostics	1640	729	351
less turtle, eggshell **	**691**	**350**	**213**
Average	787.8	145.9	114.8
Total	**6302**	**2273**	**1378**

Localities: 156, 427, 435, 437, 481, 487, 489, 491. (N = 8).
Scatters: 244, 248, 254, 260, 265, 271, 275, 284, 285, 413, 415, 433, 445, 446, 451, 466, 452, 454, 485, 515. (N= 19)

General collections: 203, 216, 236, 259, 276, 406, 479, 483, 497, 499, 503, 507. (N=12)

Table 11. 12. Taxa represented at fluvial fossil occurrences from Sandstone 12 (1.2 - 1.4 Mya-old)

Taxa	Turtle-rich localities	Scatters with turtle	General collections with turtle
Artiodactyla (general)	94	55	13
Bovids (General)	68	59	22
Cervid/Tragulids	129	77	13
Giraffids	0	0	0
Suids	8	3	0
Equids	15	14	2
Rhinoceros	0	3	2
Proboscidean	854 (833)	65	11
Carnivora (general)	7	0	0
Hyaenideae	0	0	0
Felidae	0	0	0
Canidae	1	1	1
Ursidae	0	0	0
(All carnivores)	**8**	**1**	**1**
Crocodylus	20	14	5
Turtle	585	571	36
Eggshell	39	25	1
Indeterminates	1608	1191	488
Diagnostics	1833	903	107
less turtle, eggshell **	**1239**	**307**	**70**
Average	688.2	95.2	74.4
Total	**3441**	**2094**	**595**

Elephas: figure in brackets denote tooth fragments.
Localities: 14, 83, 124, 128, 666. (N=5)
Scatters: 17, 19, 23, 30, 33, 34, 35, 36, 38, 42, 47, 48, 114, 117, 122, 125, 127, 140, 628, 650, 655, 656, 664. (N=22)
General collections: 49, 51, 74, 119, 126, 132, 144, 672. (N=8)

Table 11: 15: Total number of specimens of each taxa represented in localities, scatters, and general collections in the Pabbi Hills

Taxon	Localities	Scatters	General collections	Total
Artiodactyla	995	451	247	1693
Bovid - general	1487	558	398	2443
Bovid (Large)	106	34	44	184
Bovid (Medium, general)	513	86	46	645
Bovid (Small)	39	6	4	49
Cervid - general	350	271	139	760
Giraffid	13	9	8	30
Sivatherium	**2**	**0**	**0**	**2**
Suid, general	51	14	6	71
Anthracocere	0	2	0	2
Hippotamus*/*Hexaprotodon	**8**	**10**	**8**	**26**
Equus sivalensis	157	76	63	296
Rhinoceros sivalensis	41	25	21	87
Proboscidean	856	157	70	1083
Elephas hysudricus	60	567	286	913
Stegadon	0	2	0	2
Carnivores, general	57	5	3	65
Hyaenidae	18	2	0	20
Felidae	9	3	0	12
Canidae	14	3	3	20
Ursidae	3	0	0	3
Viverridae	1	0	0	1
Primate (1999)	**1**	**0**	**0**	**1**
Crocodylus	38	39	11	88
Chelonia	1536	1338	251	3125
Total	**6358**	**3661**	**1610**	**11629**

Table 11. 17: Immature specimens in the Pabbi Hills

Occurrence	Specimen	Type of occurrence	Taxon	Element	Part
30	3, 40	Scatter	Suid	Femur	Proximal
31	9	Scatter	Cervid	Ulna	Proximal
34	9	Scatter	Cervid	Vertebra	Thoracic
34	1	Scatter	Equid	Radius	Proximal
83	7	Locality	Indeterminate	Humerus	
83	8	Locality	Ungulata	Femur	Proximal
116	1	Scatter	Indeterminate	Humerus	
116	4	Scatter	Indeterminate	Vertebra	Centrum
128	102	Locality	Cervid	Metacarpal	Distal
133	107	Locality	Artio/Cervid	Metacarpal	Distal
208	5	General collection	Artiodactyla	Vertebra	Split
341	1	General collection	Bovid	2nd Phalange	Complete
343	6	Scatter	Artiodactyla	Metapodial	Barrel
362	6	Locality	Bovid (Medium)	Humerus	
370	5	General collection	*Elephas*	Ramus	
435	1	Locality	Indeterminate	Femur	Epiphysis
435	34	Locality	Indeterminate	Femur	
437	17	Locality	Bovid	Metapodial	Split
437	58, 99	Locality	Bovid	Metapodial	Split
437	18	Locality	Indeterminate	Vertebra	
445	3	Scatter	Artiodactyla	Femur	
445	2	Scatter	Artiodactyla	Tibia	
451	21	Scatter	Artiodactyla	Metapodial	Split
471	3, 8	Scatter	Indeterminate	Vertebra	
481	24	Locality	Artio (Small)	Vertebra	Complete
481	31	Locality	Artiodactyla	Humerus	
481	10	Locality	Artiodactyla	Metapodial	Split
481	8,25,26,39	Locality	Indeterminate	Vertebra	
483	26	General collection	Artiodactyla	Metapodial	Split
487	19	Locality	Bovid	Metapodial	Epiphysis
487	3	Locality	Indeterminate	Femur	
489	10	Locality	Artiodactyla	Radius	
489	15	Locality	Artiodactyla	Tibia	
489	15	Locality	Artiodactyla	Humerus	
489	4	Locality	Artiodactyla	Metapodial	Epiphysis
489	1	Locality	Artiodactyla	Tibia	
489	6,7	Locality	Artiodactyla	Vertebra	
489	27	Locality	Artiodactyla	Femur	
489	19, 20	Locality	Indeterminate	Vertebra	
491	6	Locality	Bovid	Femur	
491	1	Locality	Bovid	Metacarpal	
491	9	Locality	Indet/Artio	Humerus	
497	4	General collection	Artiodactyla	Calacaneum	
497	2	General collection	Bovid	Humerus	
664	23	Scatter	Bovid/Cervid	2nd Phalange	Complete
664	21	Scatter	Indet (Medium)	Humerus/Femur	Epiphysis

Table: 11. 19: Summary of surface weathering of specimens from the Pabbi Hills

Weathering stage	**1**	**2**	**2a**	**2b**	**3**	**4**	**5**	**Total**
Riverine localities >1.6 Mya-old			26	117	**228**	90	48	509
Locality 156			1	1	7	1	**8**	18
Locality 427				**4**	3	2	0	9
Locality 435			2	5	**33**	23	10	73
Locality 437			1	12	**40**	9	9	71
Locality 481			6	29	**31**	12	6	84
Locality 487			0	3	9	**20**	6	38
Locality 489			11	34	**74**	20	6	145
Locality 491			5	29	**31**	3	3	71
Carnivore localities > 1.6 Mya-old								
Locality 214			1	7	**9**	4	1	22
Locality 362* surface	80	16			3			
Locality 362* excavation	66	9	8	2	**2**			
Scatters >1.6 Mya-old			7	18	**91**	82	40	238
General collections >1.6 Mya			3	25	**71**	23	48	170
Scatters 1.2 - 1.4 Mya-old	13	25	8	80	155	131	**213**	625
General collections 1.2 - 1.4 Mya-old		5	1	5	17	20	**79**	125
Sandstone 12: fluvial (1.2- 1.4 Mya-old):								
Locality 14	1	**32**		2	8	3	1	47
Locality 83*		1	3	12	**70**	47	55	188
Locality 124	1		5	3	**10**	4	0	22
Locality 128	1	15	5	**11**	5	2	0	39
Locality 666				4	21	47	**181**	253
Sandstone 12: carnivore localities								
Locality 68	1	7	2	10	**142**	1		163
Locality 73 surface	27		**46**	29	19	6	4	131
Locality 73, excavation	12		**140**	59	43	8	2	264
Locality 133				4	18	**73**	44	139
Locality 610			5	22	**332**	301	93	753
Locality 642			3	22	272	704	**1497**	2498

Notes: weathering data for locality 362 provided by Coard. For locality 83, the totals given exclude 800 Proboscidean severely weathered (weathering stage 5) tooth/tusk fragments.

Post-script and personal summary

After 12 years of preparation, this volume is finally about to go the (patient) publishers. Now that the relentless grind of writing, editing, cross-checking, formatting and producing camera-ready copy is over, I will allow myself a brief statement as to what I think has, and has, not been achieved by the research in the Pabbi Hills.

First, the disappointments: most obviously, we failed to find a hominin fossil. The disappointment was often akin to being one step from the summit of a high mountain, but unable to reach it. Whoever looks for the next 40,000 fossils may have better luck, and find it in the brief interlude between exposure and destruction; it will be there somewhere. Secondly, we failed to find major archaeological sites – there is no equivalent of Olduvai Bed I in the Pabbi Hills, and, I suspect, in similar palaeo-landscapes. In a discipline that values its "flag-ship sites", the archaeological results were sadly very meagre.

Second, the successes – those that sustained my commitment to this volume: I have always believed that one of the most important building blocks of a discipline such as palaeoanthropology are its monographs and field reports. It is chiefly through these that researchers can evaluate what others have done, and can offer detailed criticism, and different interpretations what was found. Too often one hears the criticism that too little was published on a major piece of fieldwork beyond a couple of preliminary reports and perhaps a general synthesis that does not allow thorough examination. We all have a "book of shame", of major sites or assemblages that remain unknowable through inadequate publications. (Thankfully, there is also a "hall of fame" for those lengthy reports that will endure). I intended from the outset that this fieldwork would be published as fully as possible, just as Rendell and I did in our previous monograph on the Soan Valley in 1989. The fieldwork in the Pabbi Hills was one of the largest and longest-lasting field projects of its kind in recent years, and investigated a million-year slice of Early Pleistocene time in a region where very little was previously known in detail. As it may be the last of its kind in that area for some time, colleagues are entitled to as much information as possible. Hopefully, if and when others return to the Pabbi Hills, they can start where we left off.

There are some obvious successes to this work. First, Pakistan now has a large, well-dated Early Pleistocene (Upper Siwalik) fossil vertebrate reference collection that others can study at the Geological Survey head quarters in Islamabad. This collection includes many excellent fossil specimens, and detailed assemblages from the field surveys and the three excavations. With this volume, there is also a full catalogue of what was found, where, and with what.

I think the subdivision of the Upper Siwaliks into an earlier and later phase is robust, and makes a useful contribution to understanding the critical changes at the start of the Pleistocene.

This volume also places some important interpretative issues in the public domain. One arises from Coard's taphonomic investigations into the excavated assemblages of localities 73, 362 and 642. These are among the first accounts of Early Pleistocene fossil localities from southern Asia, and can now be discussed along side better-known localities in very different environmental settings in Europe and Africa. A second concerns the investigation of the fossil record of fluvial palaeo-landscapes. The Pabbi Hills produced a huge amount of new information that allows comparison of different types of environmental settings and periods. As commented early in this report, there are sadly few detailed accounts of these, despite their importance to early hominins, and this volume may help address the imbalance.

Although not found in large concentrations or indisputably primary contexts, I think the lithic material should be considered carefully. The study of low-level artefact discard (or "off-site" archaeology) is a major challenge that has not been extensively addressed in early hominin studies. An enormous effort went into investigating modern erosion rates, and the transport of bone (and fossil) in the survey areas. Hurcombe's chapter breaks new ground, and I think opens up many other possibilities for investigating not only the lithic record of fluvial palaeo-landscapes, but also the integrity of their fossil record. I think also her arguments considerably strengthen the case that at least some, if not most, of the stone artefacts probably came from the same horizons as the fossil remains. I still remain totally convinced that the artefact at 269 could only have come from the overlying sand deposit, and was thus >1.9 Mya old.

Curiously, I think I learnt more about *Homo erectus* by not finding its remains, or major archaeological sites than perhaps I would if I had been successful. Like the dog that didn't bark, its absence - or at most, only the most shadowy of traces – made me wonder as to why that might have been. I think now the reason is that it was very rare, and much less able to cope than I anticipated with extensive flood-plains that lacked abundant sources of stone, such as the palaeo-Pabbi Hills. That perhaps has more value to me than learning a little more about the unique or derived details of its molar cusp patterns.

Finally, a friend once offered the comment that science is an endeavour in its own right, but also a means of understanding oneself and others, and of furthering personal and social relationships. The fieldwork in the Pabbi Hills was not just hard work, but also probably one of the happiest and most stimulating times of my life, and I am forever indebted to the way it enabled me to appreciate Pakistan, and the many I have been privileged to know as a result of this work.

Bibliography

Agrawal, D.P. 1982. *The Archaeology of India.* London: Curzon.

Allchin, B. 1995. Early human cultures and environments in the Northern Punjab, Pakistan: an overview of the Potwar Project of the British Archaeological Mission to Pakistan (1981-1991). In Wadia, S., Korisettar, S.R. and Kale, V.S. *Quaternary Environments and Geoarchaeology of India – Essays in Honour of Professor S.N. Rajaguru,* 150-157. Bangalore: Geological Society of India.

Allchin, B. and Allchin, R. 1982. *The Rise of Civilization in India and Pakistan.* Cambridge: Cambridge University Press.

Amano, K. and Taira, A. 1992. Two-phase uplift of Higher Himalayas since 17 Ma. *Geology* 20: 391-394.

Andrews, P. and Cronin, J.E. 1982. The relationships of *Sivapithecus* and *Ramapithecus*, and the evolution of the orang-utan. *Nature* 297: 541-546.

Antón, S.C. 2002. Evolutionary significance of cranial variation in Asian *Homo erectus. American Journal of Physical Anthropology 118: 301-323.*

Arif, M. 1985. Stratigraphy and chronology of vertebrate fossil localities in Mirpur area, Azad Kashmir. Government of Pakistan: Geological Survey of Pakistan Information Release 264: 1-13.

Arif, M. and Vos, J. de 1989. Description of *Sivatherium giganteum* (Mammalia, Giraffidae) material from H.-GSP 8560 in the Upper Siwaliks of Pakistan. *Proceedings of the Koninkijke Nederlandse Akademie van Wetenschappen* 92B (3): 169-177.

Armand, J. 1983. *Archaeological Excavations in Durkadi Nala, An Early Paleolithic Pebble-Tool Workshop in Central India* New Delhi: Munshiram.

Ascenzi, A., Biddittu, I., Cassoli, P.F., Segre, A.G. and Segre-Naldini, E. 1996. A calvarium of late *Homo erectus* from Ceprano, Italy. *Journal of Human Evolution* 31: 409-423.

Asfaw, B., White, T., Lovejoy, O., Latimer, B., Simpson, S. and Suwa, G. 1999. *Australopithecus garhi*: a new species of early hominid from Ethiopia. *Science* 284: 629-635.

Ashfaque S.M. and Saleem-ul-Haq. 1988. Physiography of Pabbi Hills and prospects of Palaeolithic studies. *Pakistan Archaeology 1987-1988* 23: 1-42.

Ashfaque S.M. and Saleem-ul-Haq. 1991. Palaeolithic survey in Pabbi Hills field season: spring 1990. *Journal of Central Asia* 14 (1): 43-79.

Ashton, N., McNabb, J. and Parfitt, S. 1992. Choppers and the Clactonian: a reinvestigation. *Proceedings of the Prehistoric Society* 58: 22-29.

Azzaroli, A. and Napoleone, G. 1982. Magnetostratigraphic investigation of the Upper Siwaliks near Pinjor, India. *Rivista Italiana Paleontologica* 87 (4): 739-762.

Bacon, A.-M. and Vu The Long 2001. The first discovery of a complete skeleton of a fossil orang-utan in a cave of the Hoa Binh Province, Vietnam. *Journal of Human Evolution* 41 (3): 227-241.

Badam, G.L. 1979. *Pleistocene Fauna of India.* Pune: Deccan College Post-Graduate and Research Institute.

Badgley, C. 1986a. Taphonomy of mammalian fossil remains from Siwalik rocks of Pakistan. *Paleobiology* 12 (2): 119-142.

Badgley, C. 1986b. Counting individuals in mammalian fossil assemblages from fluvial environments. *Palaios* 5 (1): 328-338.

Badgley, C. and Behrensmeyer, A.K. 1980. Palaeoecology of Middle Siwalik sediments and faunas, northern Pakistan. *Palaeogeography, Palaeoclimatology, Palaeoecology* 30: 133-155.

Badgley, C. and Tauxe, L. 1990. Paleomagnetic stratigraphy and time in sediments: studies in alluvial Siwalik rocks of Pakistan. *Journal of Geology* 98: 457-477.

Barnt, J., Johnson, N.M., Johnson, G.D., Opdyke, N.D., Lindsay, E.H., Pilbeam, D. and Tahirkheli, R.A.K. 1978. The magnetic polarity stratigraphy and age of the Siwalik group near Dhok Pathan village, Potwar Plateau, Pakistan. *Earth and Planetary Science Letters* 41: 355-364.

Barry, J.C. 1987. The history and chronology of Siwalik Cercopithecids. *Human Evolution* 2 (1): 47-58.

Barry, J.C. 1995. Faunal turnover and diversity in the terrestial Neogene of Pakistan. In *Paleoclimate and Evolution*, ed. E.S. Vrba, G.H. Denton, T.C. Partridge and L.H. Burckle, 115-134. New Haven and London: Yale University Press.

Barry, J.C., Behrensmeyer, A.K. and Monoghan, M. 1980. A geologic and biostratigraphic framework for Miocene sediments near Khaur village, northern Pakistan. *Postilla* 183 (1): 1-19.

Barry, J.C., Lindsay, E.H. and Jacobs, L.L. 1982. A biostratigraphic zonation of the Middle and Upper Siwaliks of the Potwar Plateau of northern Pakistan. *Palaeogeography, Palaeoclimatology, Palaeoecology* 37: 95-130.

Barry, J.C., Morgan, M.E., Flynn, L.J., Pilbeam, D., Behrensmeyer, A.K., Raza, S.M., Khan, I.A., Badgley, C., Hicks, J. and Kelley, J. 2002. Faunal and environmental change in the late Miocene Siwaliks of northern Pakistan. *Paleobiology Memoirs* 3: 1-55 (Supplement to Paleobiology 28 (2)).

Bar-Yosef, O. 1995. The role of climate in the interpretation of human movements and cultural transformations in western Asia. In *Paleoclimate and Evolution*, ed. E.S. Vrba, G.H. Denton, T.C. Partridge and L.H. Burckle, 507-524. New Haven and London: Yale University Press.

Basu, P.K. 2004. Siwalik mammals of the Jammu Sub-Himalaya, India: an appraisal of their diversity and habitats. *Quaternary International* 117: 105-118.
Behrensmeyer, A.K. 1975a. The taphonomy and paleoecology of Plio-Pleistocene vertebrate assemblages east of Lake Rudolf, Kenya. *Bulletin of the Museum of Comparative Zoology, Harvard* 146 (10): 474-576.
Behrensmeyer, A.K. 1975b. The bones of Amboseli Park as a key to East African paleoecology. *National Geographical Society* 1975: 91-109.
Behrensmeyer, A.K. 1978. Taphonomic and ecologic information from bone weathering. *Paleobiology* 4 (2): 150-162.
Behrensmeyer, A.K. 1982. Time resolution in fluvial vertebrate assemblages. *Paleobiology* 8 (3): 211-227.
Behrensmeyer, A.K. 1988. Vertebrate preservation in fluvial channels. *Palaeogeography, Palaeoclimatology, Palaeoecology* 63: 183-199.
Behrensmeyer, A.K. and Tauxe, L. 1982. Isochronous fluvial systems in Miocene deposits of northern Pakistan. *Sedimentology* 29: 331-352.
Behrensmeyer, A.K., Gordon, K.D. & Yanagi, G.T. 1986. Trampling as a cause of bone surface damage and pseudo-cutmarks. *Nature* 319: 768-771.
Behrensmeyer, A.K. , Todd, N.E., Potts, R. and McBrinn, G.E. 1997. Late Pliocene faunal turnover in the Turkana Basin, Kenya and Ethiopia. *Science* 278: 1589-1594.
Berger L.R. and Tobias P.V. 1996. A chimpanzee-like tibia from Sterkfontein, South Africa and its implications for the interpretation of bipedalism in *Australopithecus africanus*. *Journal of Human Evolution* 30: 343-348.
Berggren, W.A., Kent, D.V., Swisher, C.C., III, and Aubry, M.-P., 1995. A revised Cenozoic geochronology and chronostratigraphy. *In* Berggren, W.A., Kent, D.V., Aubry, M.-P., and Hardenbol, J. (Eds.), *Geochronology, Time Scales and Global Stratigraphic Correlation.* Spec. Publ. Soc. Econ. Paleontol. Mineral. (Soc. Sediment. Geol.), 54: 129-212.
Biagi, P. and Cremaschi, M. 1988. The early Palaeolithic sites of the Rohri Hills (Sind, Pakistan) and their environmental significance. *World Archaeology* 19 (3): 421-433.
Binford, L.R. 1981. *Bones: Ancient Men and Modern Myths.* London, New York: Academic Press.
Binford, L R. 1988. Fact and fiction about the *Zinjanthropus* floor: data, arguments, and interpretations (with reply by Bunn and Kroll). *Current Anthropology* 29 (1): 123-149.
Blumenschine, R. J. and Selvaggio, M.M. 1988. Percussion marks on bone surfaces as a new diagnostic of hominid behaviour. *Nature* 333: 763-765.
Boaz, N.T. and Behrensmeyer, A.K. 1976. Hominid taphonomy: transport of human skeletal parts in an artificial fluviatile environment. *American Journal of Physical Anthropology* 45: 53-60.
Brown, F., Harris, J., Leakey, R. and Walker, A. 1985. Early *Homo erectus* skeleton from west Lake Turkana, Kenya. *Nature* 316: 788-792.
Brunet, M., Beauvillain, A., Coppens, Y., Heintz, E., Moutaye, A.H.E. and Pilbeam, D. 1995. The first australopithecine 2500 kilometres west of the Rift Valley (Chad) *Nature* 378: 273-275.
Brunet, M., Guy, F., Pilbeam, D., Mackaye, H.T., Likius, A., Ahounta, D., Beauvillain, A., Blondel, C., Bocherens, H., Boisserie, J.-R., de Bonis, L., Coppens, Y., Dejax, J., Denys, C., Duringer, P., Eisenmann, V., Fanone, G., Fronty, P., Geraads, D., Lehann, T., Lihoreau, F.,, Louchart, A., Mahamt, A., Merceron, G., Otero, O., Campomanes, P.P., Ponce de Leon, M., Rage, J.-C., Sapanet, M., Schuster, M., Sudre, J., Tassy, P., Valentin, X., Vignaud, P., Vitriot, L., Zazzo, A. and Zollikofer, C. 2002. A new hominoid from the Upper Miocene of Chad, Central Africa. *Nature* 418: 145-151.
Burbank, D.W. and Johnson, G.D. 1982. Intermontane-basin development in the past 4 Myr in the north-west Himalaya. *Nature* 298: 432-436.
Burbank, D.W. and Johnson, G.D. 1983. The late Cenozoic chronologic and stratigraphic devrelopment of the Kashmir intermontane basin, northwestern Himalaya. *Palaeogeography, Palaeoclimatology, Palaeoecology* 43: 205-235.
Burbank, D.W. and Raynolds, R.G.H. 1984. Sequential late Cenozoic structural disruption of the northern Himalayan foredeep. *Nature* 311: 114-118.
Butler, R.W. and Prior, D.J. 1988. Tectonic controls on the uplift of the Nanga Parbat Massif, Pakistan Himalayas. *Nature* 333: 247-250.
Cande, S.C. and Kent, D.V. 1995. Revised calibration of the geomagnetic polarity timescale for the Late Cretaceous and Cenozoic. *Journal of Geophysical Research* 100: 6093-6095.
Carbonell, E., Bermudez de Castro, J.M., Arsuaga, J.L., Díez, J.C., Rosas, A., Cuenca-Bescós, G., Sala, R., Mosquera, M and Rodríguez, X.P. 1995a. Lower Pleistocene hominids and artifacts from Atapuerca-TD6 (Spain) *Science* 269: 826-829.
Carbonell, E., Mosquera, M., Rodríguez, X.P. and Sala, R. 1995b. The first human settlement of Europe *Journal of Anthropological Research* 52 (1): 107-114.
Carbonell, E., Mosquera, M., Rodriguez X.P. and Sala, R. 1999. Out of Africa: the dispersal of the earliest technical systems reconsidered. *Journal of Anthropological Archaeology* 18: 119-136.
Cavallo, J.A. and Blumenschine, R.J. 1989. Tree-stored leopard kills: expanding the hominid scavenging niche. *Journal of Human Evolution* 18: 393-399.

Cerling, T.E. 1992. Development of grasslands and savannahs in East Africa during the Neogene. *Palaeogeography, Palaeoclimatology and Palaeoecology* 97: 241-247.
Cerling, T.E., Harris, J.M., MacFadden, B.J., Leakey, M.G., Quade, J., Eisenmann, V., Ehleringer, J.R.. 1997. Global vegetation change through the Miocene/Pliocene boundary. *Nature* 389: 153-158.
Chauhan, P. 2003. An overview of the Siwalik Acheulian and reconsidering its chronological relationship with the Soanian – a theoretical perspective. *Assemblage* 7.
Churcher, C.S. 1978. Giraffidae In *Evolution of African Mammals*, ed. V.J. Maglio and H.B.S. Cooke, 540-572. Cambridge (Mass): Harvard University Press.
Clark, J.D. 1958. The natural fracture of pebbles from the Bakota Gorge, Northern Rhodesia, and its bearing on the Kafuan industries of Africa. *Proceedings of the Prehistoric Society* 24: 64-77.
Coard, R. 1993. *A taphonomic study of a Plio-Pleistocene fossil assemblage in the Pabbi Hills, northern Pakistan: an experimental study as an aid to site interpretation*. Unpublished Ph.D. dissertation, Dept. of Archaeology, University of Sheffield.
Colbert, E.H. 1935. Siwalik mammals in the American Museum of Natural History. *Transactions of the American Philosophical Society* 26: 1-401.
Corvinus, G. and Nanda, A.C. 1994. Stratigraphy and palaeontology of the Siwalik Group of Surai Khola and Rato Khola in Nepal. *N. Jb. Geol. Palaont. Abh.* 191 (1): 250-268.
Corvinus, G. 1998. Lower Palaeolithic occupations in Nepal in relation to South Asia. In *Early Human Behavior in Global Context: The Rise and Diversity of the Lower Palaeolithic Record*, 391-417, edited by M. Petraglia and R. Korisettar. London: Routledge.
Corvinus, G. and Rimal, L.N. 2001. Biostratigraphy and geology of the Neogene Siwalik Group of the Surai Khola and Rato Khola areas in Nepal. *Palaeogeography, Palaeoclimatology, Palaeoecology* 165: 251-279.
Cruz-Uribe, K. 1991. Distinguishing hyaena from hominid bone accumulations. *Journal of Field Archaeology* 18: 467-486.
Darwin, C. 1871. *The Descent of Man and Selection in Relation to Sex*. Reprinted N.Y.: AMS Press, 1972.
Davidson, I. and Solomon, S. 1990. Was OH7 the victim of a crocodile attack? In *Problem Solving in Taphonomy*, ed. S. Solomon, I. Davidson and D. Watson. *Tempus* 2: 229-239.
Day, M.H. 1986. *Guide to Fossil Man* (4th edition). London: Cassell.
deMenocal, P.B. 1995. Plio-Pleistocene African climate. *Science* 270: 53-59.
Dennell, R.W. 1983. *European Economic Prehistory: A New Approach*. London and New York: Academic Press.
Dennell, R.W. 1989. Report of the British Archaeological Mission to Pakistan 1980-89. *Man and Environment* 14 (1): 129-131.
Dennell, R.W. 1991a. Report of the British Archaeological Mission's Potwar Project for the year 1989-90. *South Asian Studies* 7: 161-165.
Dennell, R.W. 1991b. *Pakistan's Prehistory: A Glimpse at the First Two Million Years.* British Archaeological Mission to Pakistan Series 3: Ancient India and Iran Trust, Cambridge. 44 pp.
Dennell, R.W. 1993. Evidence on human origins: a rediscovered source in the Upper Siwaliks of northern Pakistan. *Interdisciplinary Science Reviews* 18 (4): 379-389.
Dennell, R.W. 1995a. The early stone age of Pakistan: a methodological review. *Man and Environment:* 20 (1): 21-28.
Dennell, R.W. 1995b. Do human origins lie only in Africa? New evidence from northern Pakistan. *Cranium* 12 (1): 21-24.
Dennell, R. W. 1998. Grasslands, tool-making and the earliest colonization of south Asia: a reconsideration. In (M. Petraglia and R. Korisettar, Eds.) *Early Human Behavior in Global Context: The Rise and Diversity of the Lower Palaeolithic Record*, pp. 280-303. London: Routledge.
Dennell, R.W. 1999. The Palaeolithic and Pleistocene potential of the Indus drainage system: a review of recent work. In *The Indus River: Biodiversity, Resources, Humankind*, ed. A. and P. Meadows, 306-319. Linnaean Society, London/Karachi: Oxford University Press.
Dennell, R.W. 2003. Dispersal and colonisation, long and short chronologies: how continuous is the Early Pleistocene record for hominids outside East Africa? *Journal of Human Evolution* 45: 421-440.
Dennell, R.W and Allchin , B. 1988. 1987 report of the British Archaeological Mission to Pakistan palaeolithic project. *South Asian Studies* 10: 145-147.
Dennell, R.W. and Hurcombe, L.M. 1992. Paterson, the British Clactonian and the Soan Flake Industry: a re-evaluation of the early palaeolithic of northern Pakistan. In *South Asian Archaeology 1989*. (Proceedings of the International Conference of South Asian Archaeologists in Western Europe, Paris, July 1989), ed. C. Jarrige, pp. 69-72. Madison: Prehistory Press.
Dennell, R.W. and Roebroeks, W. 1996. The earliest colonisation of Europe: the short chronology revisited. *Antiquity* 70: 535-542.

Dennell, R.W., Rendell, H.R. and Hailwood, E. 1988. Early tool-making in Asia: two-million-year-old artefacts in Pakistan. *Antiquity* 62: 98-106.
Dennell, R.W., Hurcombe, L.M., Jenkinson, R. and Rendell, H.M. 1990. Preliminary results of the Palaeolithic programme of the British Archaeological Mission to Pakistan, 1983-1987. In *South Asian Archaeology 1987.* (Proceedings of the International Conference of South Asian Archaeologists in Western Europe, Venice, July 1987), ed. M. Taddei and P. Callieri, pp. 17-29.
Dennell, R.W., H. Rendell, M. Halim and E. Moth 1991. Site 55, Riwat: a 42,000 yr.-bp. open-air palaeolithic site from northern Pakistan. *Journal of Field Archaeology* 19: 17-33.
Dennell, R.W., Hurcombe, L.M., Coard, R., Beech, M., Anwar, M. and ul Haq, S. 1992. The 1990 field season of the British Archaeological Mission to Pakistan in the Baroth area of the Pabbi Hills, northern Pakistan. *South Asian Archaeology 1991* (Proceedings of the Conference of South Asian Archaeologists in Europe, Berlin, July 1991), 1-14.
Dennell, R.W. and Hurcombe, L.H. 1995. Comment on Pedra Furada. *Antiquity* 69: 604-605.
Dmitrieva, E. L. 1977. Tajikstan's and India's fossil Aleclaphinae. *Journal of the Palaeontological Society of India* 20: 97-101.
Dowsett, H., Thompson, R., Barron, J., Cronin, T., Fleming, F., Ishman, S., Poore, R., Willard, D. and Holtz, T. jnr. 1994. Joint investigations of the Middle Pliocene climate I: PRISM paleoenvironmental reconstructions. *Global and Planetary Change* 9: 169-95.
Dowsett, H.J., Barron, J.A., Poore, R.Z., Thompson, R.S., Cronin, T.M., Ishman, S.E. and Willard, D.A. 1999. Middle Pliocene paleoenvironmental reconstruction: PRISM2. US Geol. Survey Open File Report 99-535: 1-23.
Dubois, E. 1898. On *Pithecanthropus erectus*: a transitional form between man and the apes. *Scientific Transactions of the Royal Dublin Society* 6: 1-18.
Etler, D. 1996. The fossil evidence for human evolution in Asia. *Annual Review of Anthropology* 25: 275-301.
Etler, D.A. 1998. Implications of new fossil material attributed to Plio-Pleistocene Asian Hominidae. http://www.cruzio.com/~cscp/art1.
Falconer, H. and Cautley, P.T. 1845. *Fauna Antiqua Sivalensis, being the fossil zoology of the Sewalik Hills in the North of India.* London: Smith, Elder and Co.
Fernández-Jalvo, Y., Denys, C., Andrews, P., Williams, T., Dauphin, Y. and Humphreys, L. 1998. Taphonomy and palaeoecology of Olduvai Bed-I (Pleistocene, Tanzania). *Journal of Human Evolution* 34: 137-172.
Flynn, L.J., Pilbeam, D., Jacobs, J., Barry, J.C., Behrensmeyer, A.K. and Kappelman, J.W. 1990. Time and faunas in a Miocene terrestial setting. *Journal of Geology* 98: 589-604.
Foley, R. 1987. *Another Unique Species*. Harlow: Longman Scientific and Technical.
Gabunia, L., Vekua, A., Lordkipanidze, D., Swisher, C.C., Ferring, R., Justus, A., Nioradze, M., Tvalchrelidze, M., Antón, S.C., Bosinski, G., Jöris, O., de Lumley, M.A., Majsuradze, G. and Mouskhelishvili, A. 2000a. Earliest Pleistocene hominid cranial remains from Dmanisi, Republic of Georgia: taxonomy, geological setting, and age. *Science* 288: 1019-1025.
Gabunia, L., Vekua, A. and Lordkipanidze, D. 2000b. The environmental contexts of early human occupation of Georgia (Transcaucasia). *Journal of Human Evolution* 38: 785-802.
Gabunia, L. and Vekua, A. 1990. L'évolution de paléoenvironment au cours de l'anthropogène en Géorgie (Transcaucasie). Dmanisi. *L'Anthropologie* 90 (4): 643-650.
Gabunia, L. and Vekua, A. 1995. A Plio-Pleistocene hominid from Dmanisi, East Georgia, Caucasus. *Nature* 375: 509-512.
Gaillard, C. 1996. Processing sequences in the Indian Lower Palaeolithic: Examples from the Acheulian and Soanian *Indo-Pacific Prehistory Association Bulletin* 1: 57-67.
Gaillard, C. Raju, D.R Mishra, V.N. Rajaguru, S.N.1983. Acheulian occupation at Singi-Talav in the Thar dessert: a preliminary report on the 1982 excavation. *Man and Environment* 7: 112-130.
Gaillard, C. Mishra, V.N. Rajaguru, S.N. Raju, D.R. and Raghavan, H. 1985. Acheulian occupation at Singi-Talav in the Thar dessert: a preliminary report on the 1981 excavation. *Bulletin of the Deccan College Research Institute* 44: 141-152.
Gamble, C. 1995. The environmental background. In *The Earliest Occupation of Europe*, ed. W. Roebroeks and T.van Kolfschoten, pp. 279-295. Leiden: University of Leiden.
Ganjoo, R.K. 1984. Some vertebrate fossil discoveries from the Upper Siwaliks of Jammu (J and K). *Proceedings of the Symposium on Quaternary Episodes*: Dept. of Geology, M.S.U. Baroda. 1984: 135-138.
Ganjoo, R.K. 1985. Some new fossil Proboscidea from the Siwaliks of Jammu. *Publications of the Centre for Advanced Studies in Geology, Panjab University, Chandigarh.* 1: 177-184.
Gaur, R. 1987. *Environment and Ecology of Early Man in Northwest India*. New Delhi: B.R. Publishing Corporation.
Gaur, R. and Chopra, S.R.K. 1984. Taphonomy, fauna, environment and ecology of Upper Sivaliks (Plio-Pleistocene) near Chandigargh, India. *Nature* 308: 353-355.
Gentry, A.W. 1978. Bovidae In *Evolution of African Mammals*, ed. V.J. Maglio and H.B.S. Cooke, 540-572. Cambridge (Mass): Harvard University Press.

Gibert, J., Gibert, Ll. and Iglesias, A. 1998. Two 'Oldowan' assemblages in the Plio-Pleistocene deposits of the Orce region, southeast Spain. *Antiquity* 72: 17-25.
Gibert, J., Gibert, L., Ferràndez-Canyadell, C., Iglesias, A. and Gonzales, F. 2001. Venta Micena, Barranco Leon-5 and Fuentenueva 3: three archaeological sites in the Early Pleistocene deposits of Orce, south-east Spain. *In A Very Remote Period Indeed: Papers on the Palaeolithic presented to Derek Roe*, ed. S. Milliken and J. Cook, 144-152. Oxford: Oxbow Books Ltd.
Gifford, D.P.and Behrensmeyer, A.K. 1977. Observed formation and burial of a recent human occupation site in Kenya. *Quaternary Research* 8: 245-266.
Gifford-Gonzalez, D.P., Damrosch, D.B.,Damrosch, D.R., Pryor, J. 1985. The third dimension in site structure: an experiment in trampling and vertical dispersal. *American Antiquity* 50 (4): 803-818.
Goldberg, P., Nash, D, T. and Petraglia, M. D (ed.s) 1993. Formation Processes in Archaeological Context. *Monographs in World Archaeology*17. Madison: Prehistory Press,
Graziosi, P. 1964. *Prehistoric Research in Northwestern Punjab, Anthropological Research in Chitral, Italian Expeditions to the Karakoram (K2) and the Hindu Kush.* Leiden: Brill.
Green, S. 1988. Pontnewydd Cave: the selection of raw materials for artefact manufacture and the question of natural damage, in R.J.MacRae and N. Moloney *Non-Flint Stone Tools and the Palaeolithic Occupation of Britain* Oxford: British Archaeological Reports (British Series) 189: 223-242.
Harris, J.W.K. and Capaldo, S.D. 1993. The earliest stone tools: their implications for an understanding of the activities and behaviour of late Pliocene hominids. In A. Berthelet and J. Chavaillon (eds.) *The Use of Tools by Human and Non-human Primates* pp.196-220. Oxford: Clarendon.
Hay, R.L. 1976. *The Geology of Olduvai Gorge*. Berkeley: University of California Press.
Heinzelin, J.de, Clark, J.D., White, T., Hart, W., Renne, P., Wolde-Gabriel, Beyene, Y. and Vrba, E. 1999. Environment and behaviour of 2.5-million-year-old Bouri hominids. *Science* 284: 625-629.
Hill, A. and Ward, S. 1988. Origin of the Hominidae: the record of African large hominoid evolution between 14 My and 4 My. *Yearbook of Physical Anthropology* 31: 49-83.
Huffman, F. 2001. Geologic context and age of the Perning/Mojokerto Homo erectus, East Java. *Journal of Human Evolution* 40: 353-362.
Hurcombe, L.M. and Dennell, R.W. 1993. A Pre-Acheulean in the Pabbi Hills, northern Pakistan? In *South Asian Archaeology 1989* (Proceedings of the International Conference of South Asian Archaeologists in Western Europe, Paris, July 1989), ed. C. Jarrige, pp. 133-136.
Hussain, S.T., Bergh, G.D. van den, Steensma, K.J., Visser, J.A. de, Vos, J. de, Arif, M., Dam, J. van, Sondaar, P.Y. and Malik, S.B. 1992. Biostratigraphy of the Plio-Pleistocene continental sediments (Upper Siwaliks) of the Mangla-Samwal Anticline, Azad Kashmir, Pakistan. *Proceedings of the Koninkijke Nederlandse Akademie van Wetenschappen* 95 (1): 65-80.
Hyodo, M., Nakaya, H., Urabe, A., Saegusa, H., Shunrong, X., Jiyun, Y. and Xuepin, J. 2002. Paleomagnetic dates of hominid remains from Yuanmou, China, and other Asian sites. *Journal of Human Evolution* 43: 27-41.
Isaac, G. L. 1989. The archaeology of human origins: studies of the Lower Pleistocene in East Africa 1971-1981. In *The Archaeology of Human Origins: Papers of Glynn Isaac*, ed. B. Isaac, pp. 120-187. Cambridge: Cambridge University Press.
Islamov, Y.I. 1990. Sel'oungour, un nouveau site du paleolithique inferieur en Asie centrale. *L'Anthropologie* 94 (4): 675-688.
Jenkinson, R., Dennell, R.W., Rendell, H. M., Jah, A. and S. Sutherland, S. 1989. Upper Siwalik palaeoenvironments and palaeoecology in the Pabbi Hills, Northern Pakistan. *Zeitschrift für Geomorphologie* 33 (4): 417-428.
Johnson, G.D. and Vondra, C.F. 1972. Siwalik sediments in a portion of the Punjab re-entrant: the sequence at Haritalyangar, District Bilaspur, H.P. *Himalayan Geology* 2: 118-144.
Johnson, G.D., Zeitler, P., Naeser, C.W., Johnson, N.M., Summers, D.M., Frost, C.D., Opdyke, N.D. and Tahirkheli, R.A.K.. 1982. The occurrence and fission-track ages of late Neogene and Quaternary volcanic sediments, Siwalik group, northern Pakistan. *Palaeogeography, Palaeoclimatology, Palaeoecology* 37: 63-93.
Johnson, G.D., Opdyke, N.D., Tandon, S.K. and Nanda, A.C.l. 1983. The magnetic polarity stratigraphy of the Siwalik group at Haritalyangar (India) and a new last appearance for *Ramapithecus* and *Sivapithecus* in Asia. *Palaeogeography, Palaeoclimatology, Palaeoecology* 44: 223-249.
Johnson, G.D., Raynolds, R.G.H. and Burbank, D.W. 1986. Late Cenozoic tectonics and sedimentation in the north-western Himalayan foredeep: I. Thrust ramping and associated deformation in the Potwar region. In: Allen, P.A. and Homewood, P. (eds) Foreland Basins. *Special Publication International Association of Sedimentologists*, 8: 273-291.
Johnson, N.M., Opdyke, N.D., Johnson, G.D., Lindsay, E.H. and Tahirkheli, R.A.K. 1982. Magnetic polarity stratigraphy and ages of Siwalik group rocks of the Potwar Plateau, Pakistan. *Palaeogeography, Palaeoclimatology, Palaeoecology* 37: 17-42.

Johnson, N.M., Stix, J., Tauxe, L., Cerveny, P.F. and Tahirkheli, R.A.K. 1985. Paleomagnetic chronology, fluvial processes, and tectonic implications of the Siwalik deposits near Chinji village, Pakistan. *Journal of Geology* 93: 27-40.
Jones, P.R. 1994. Results of experimental work in relation to the stone industries of Olduvai Gorge, in M. Leakey with D. Roe. *Olduvai Gorge*, Vol. 5, 254-298. Cambridge: Cambridge University Press.
Kappelman, J. et al. 1997. Bovids as indicators of Plio-Pleistocene paleoenvironments in East Africa. *Journal of Human Evolution* 32: 229-256.
Keates, S.G. 2000. Early and Middle Pleistocene Hominid Behaviour in Northern China. *British Archaeological Reports (International Series)* 863: 1-387.
Keller, H.M. 1975. *The magnetic polarity stratigraphy of an Upper Siwalik sequence in the Pabbi Hills of Pakistan.* Unpublished MA thesis, Dartmouth College, N.H., USA.
Keller, H.M., Tahirkheli, R.A.K., Mirza, M.A., Johnson, G.D., Johnson, N.M. and Opdyke, N.D. 1977. Magnetic polarity stratigraphy of the Upper Siwalik deposits, Pabbi Hills, Pakistan. *Earth and Planetary Science Letters* 36: 187-201.
Kerbis-Peterhans, J.C. and Horwitz, L.K. 1992. A bone assemblage from a striped hyaena (Hyaena hyaena) den in the Negev Desert, Israel. *Israel Journal of Zoology* 37: 225-245.
Khan, I.A. 1994. Alluvial architecture and sandstone petrology of the stratotype Chinji Formation, Siwalik Group, northern Pakistan. *Himalayan Geology* 15: 123-141.
Khan, I.A., Bridge, J.S., Kappelman, J. and Wilson, R. 1997. Evolution of Miocene fluvial environments, eastern Potwar plateau, northern Pakistan. *Sedimentology* 44: 221-251.
Kimbel, W.H. 1995. Hominid speciation and Pliocene climatic change. In *Paleoclimate and Evolution*, ed. E.S. Vrba, G.H. Denton, T.C. Partridge and L.H. Burckle, 425-437. New Haven and London: Yale University Press.
Klein R.G. 1989. Why does skeletal part representation differ between smaller and larger bovids at Klaises River Mouth and other archaeological sites? *Journal of Archaeological Science* 16(4): 363-382.
Klein, R. G. and Cruz-Uribe, K. 1984. *The Analysis of Animal Bones from Archaeological Sites*. Chicago: Chicago University Press.
Knudson, R. 1979. Inference and imposition in lithic analysis. In B.Hayden (ed) *Lithic Use-wear Analysis,* pp. 269-281. New York: Academic Press.
Kumar, M. and Rishi, K.K. 1986. Acheulian elements from Hoshiharpur region (Punjab). *Man and Environment* 10: 141-142.
Langbroek, M. 2003. *Out of Africa: A Study into the Earliest Occupation of the Old World.* Leiden University: Faculty of Archaeology, unpublished Ph.D. dissertation.
Langbroek, M. and Roebroeks, W. 2000. Extraterrestrial evidence on the age of the hominids from Java. *Journal of Human Evolution* 38: 595-600.
Larkin, N.R., Alexander, J. and Lewis, M.D. 2000. Using experimental studies of recent faecal material to examine hyaena coprolites from the West Runton freshwater Bed, Norfolk, U.K. *Journal of Archaeological Science* 27: 19-31.
Larick, R. and Ciochon, R.L. 1996 .The African emergence and early Asian dispersal of the genus *Homo*. *American Scientist* Nov-Dec. 1996. http://www.amsci.org/amsci/articles/96articles.
Larick R., Ciochon, R.L., Zaim, Y., Suminto, S., Izal, Y., Aziz, F., Reagan, M. and Heizler, M. 2001. Early Pleistocene Ar-40/Ar-39 ages for Bapang Formation hominins, Central Java, Indonesia. *Proceedings of the National Academy of SciencesUSA* 98 (9): 4866-4871.
Leakey, L.S.B. 1931. *The Stone Age Cultures of Kenya Colony.* Cambridge: Cambridge University Press.
Leakey, M. 1971. *Olduvai Gorge* (vol. 3). Cambridge: Cambridge University Press.
Leakey, M. with D. Roe 1994. *Olduvai Gorge* (vol. 5). Cambridge: Cambridge University Press.
Leakey, M.G., Feibel, C.S., McDougall, I. and Walker, A. 1995. New four-million-year-old hominid species from Kanapoi and Allia Bay, Kenya. *Nature* 376: 565-571.
Leakey, M.G., Spoor, F.H., Brown, F.H., Gathogo, P.N., Kiarie, C., Leakey, L.N. and McDougal, I. 2001. New hominin genus from eastern Africa shows diverse middle Pliocene lineages. *Nature* 410: 433-440.
Leinders, J.J.M., Aziz, F., Sondaar, P.Y. and De Vos, J. 1985. The age of the hominid-bearing deposits of Java: state of the art. *Geologie en Mijnbouw* 64: 167-173.
Lewis, G.E. 1934. Preliminary notice of new man-like apes from India. *American Journal of Science* 161-179.
Lindsay, E.H., Opdyke, N.D. and Johnson, N.M. 1980. Pliocene dispersal of the horse *Equus* and late Cenozoic mammalian dispersal events. *Nature* 287: 135-138.
Lyman, R. L. 1994. *Vertebrate Taphonomy*. Cambridge: Cambridge University Press.
Lyman, R.L. and Fox, G.L. 1989. A critical evaluation of bone weathering as an indication of bone assemblage formation. *Journal of Archaeological Science* 16 (3): 293-318.
Marean C.W. 1991. Measuring the post-depositional destruction of bone in archaeological assemblages. *Journal of Archaeological Science* 18 (6): 677-694.
Matthew, W. D. 1915. Climate and evolution. *Annals of the New York Academy of Science* 24: 171-318.

Matthew, W. D. 1929. Critical observations upon Siwalik mammals (exclusive of Proboscidea). *Bulletin of the American Museum of Natural History* 56: 437-560.
McBrearty, S. et al. 1998. Tools underfoot: human trampling as an agent of lithic artifact edge modification. *American Antiquity* 63 (1): 108-129.
McCrae, L.E. 1990. Paleomagnetic isochrons, unsteadiness, and non-uniformity of sedimentation in Miocene fluvial strata of the Siwalik Group, Northern Pakistan. *Journal of Geology* 98: 433-456.
MacFadden, B. J. 1992. *Fossil Horses: Systematics, Paleobiology, and Evolution of the Family Equidae*. Cambridge University Press.
Medlicott, H.B. 1864. On the geological structure and relation of the southern portion of Himalayan ranges between the river Ganges and Ravi. *Memoirs of the Geological Survey of India* 3 (2): 1-212.
Marzke, M.W. 1992. Evolution of the power ("squeeze") grip and its morphological correlates in humans. *American Journal of Physical Anthropology* 89: 283-298.
Milliken, S. 1999. The earliest occupation of Italy. *Accordia* 7: 7-36.
Mishra, S. 1992. The age of the Acheulian in India: new evidence. *Current Anthropology* 33: 325-328.
Misra, V.N., Rajaguru, S.N. Raju, D.R. Raghavan, H. and Gaillard, C 1982. Acheulian occupation and evolving landscape around Didwana in the Thar Desert, India. *Man and Environment* 6: 72-86.
Mishra, S., Venkatesan, T.R., Rajaguru, S.N. and Somayajulu, B.L.K. 1995. Earliest Acheulean industry from peninsula India. *Current Anthropology* 36 (5): 847-851.
Misra, V.N., Rajaguru, S.N., Raghavan, H. 1988. Late Middle Pleistocene environment and acheulian culture around Didwana, Rajasthan. In D.V.S. Jain, D.P.Agrawal, P. Sharma and S.K. Gupta (ed.s) *Paleoclimatic and Plaeoenvironmental changes in Asia*, pp. 93-106. New Delhi: Indian National Science Academy.
Mohapatra, G.C. 1981. Acheulian discoveries in the Siwlaik frontal range. *Current Anthropology* 22: 433-435.
Mohapatra, G.C. and Singh. M. 1979. Stratified occurence of lithic artefacts in the Siwalik frontal range of western sub-Himalaya. *Punjab University Research Bulletin* 10: 65-77.
Moloney, N,. S.L.Olsen and V. Volosin 2001. Lower and Middle palaeolithic occupation in Central Kazakhstan: the Batpak Valey and environs, in S. Milliken and J. Cook (eds) *A Very Remote Period Indeed: Papers on the Palaeolithic Presented to Derek Roe*, pp 138-143. Oxford: Oxbow.
Morgan, M.E., Kingston, J.D. and Marino, B.D. 1994. Carbon isotope evidence for the emergence of C4 plants in the Neogene from Pakistan and Kenya. *Nature* 367: 162-165.
Moya-Sola, S. and Kohler, M. 1997. The Orce skull: anatomy of a mistake. *Journal of Human Evolution* 33 (1): 91-98.
Nanda, A.C. 1976. Some Proboscidean fossils from the Upper Siwalik subgroup of Ambala. *Himalyan Geology* 6: 1-26.
Nanda, A.C. 1977. A new Pentalophodon (Mammalia, Gomphotheridae) from the Tatrot Formation of Ambala, India. *Recent Researches in Geology* 3: 44-51.
Nanda, A.C. 1978a. Occurrence of *Camelus sivalensis* Falconer and Cautley from the Tatrot Formation of Ambala, India. *Journal of the Geological Society of India* 19 (4): 160-164.
Nanda, A.C. 1978b. Fossil equids from the Upper Siwalik Subgroup of Ambala, Haryana. *Himalayan Geology* 8: 149-177.
Nanda, A.C. 1978c. *Hexaprotodon sivalensis* Falconer and Cautley from the Upper Siwalik Subgroup, Ambala Haryana. *Recent Researches in Geology* 5: 163-174.
Nanda, A.C. 1979. Skull characters of *Hemibos acuticornis* (Falconer) (Mammalia Bovidae) from the Pinjor Formation of Ambala, Haryana. *Indian Journal of Earth Sciences* 6 (2): 175-185.
Nanda, A.C. 1981. Occurrence of pre-Pinjor beds in the vicinity of Chandigargh. *Procedings of the Neogene/Quaternary Boundary Field Conference, India 1979*, 113-116.
Nanda, A.C. 1988. A new species of *Stegodon* from the Upper Siwalik subgroup of Ambala, India. *Geological Survey of India Special Publication* 11: 365-370.
Nanda, A.C. 1992. Upper Siwalik mammalian faunas from Chandigarh and Jammu regions with comments on certain faunal discrepancies. *Proc. 1st South Asian Geological Congres,* pp. 39-45.
Nanda, A.C. 1997. Some biostratigraphic observations based on the Upper Siwalik mammalian faunas of the Siwalik Group of India and Nepal. In *Geology in South Asia: II*, ed. Wijayananda, N.P., Cooray, P.G. and Mosley, P., pp. 171-189. Sri Lanka: Geological Survey and Mines Bureau, Professional Paper 7.
Nanda, A.C. and Corvinus, G. 1992. A report on the occurrence of Tatrot and Pinjor faunas from the Siwalik Group in Nepal. *Journal of Himalayan Geology* 3 (2): 209-211.
Nanda, A.C. and Corvinus, G. 2000. Skull characteristics of two Proboscideans from the Upper Siwalik subgroup of Nepal. *N. Jb. Geol. Palaont Abh.* 217 (1): 89-110.
Nanda, A.C. and Sehgal, R.K. 1993. Siwalik mammalian faunas from Ramnagar (J. & K.) and Nurpur (H.P.) and lower limit of *Hipparion*. *Journal of the Geological Society of India* 42: 115-134.
Nash, D.T. 1993. Distinguishing stone artefacts from naturefacts created by rockfall processes. In P. Goldberg et al. (eds.) *Formation Processes in Archaeological Context*. Madison: Prehistory Press, Monographs in World Archaeology 17: 125-138.

Ninkovich, D. and Burckle, L.H. 1978. Absolute age of the base of the hominid-bearing beds in eastern Java. *Nature* 275: 306-308.
Ohel, M.Y. 1977. On the Clactonian: reexamined, redefined, and reinterpreted. *Current Anthropology* 18 (2): 329-331.
Olsen, J.W. and Miller-Antonio, S. 1992. The palaeolithic in southern China. *Asian Perspectives* 31: 129-160.
Oms, O., Parés, J.M., Martínez-Navarro, B., Agusti, J., Toto, I., Martínez-Fernández, G. and Turq, A. 2000. Early human occupation of western Europe: paleomagnetic dates for two paleolithic sites in Spain. *Proceeedings of the National Academy of Sciences USA* 97 (19): 10666-10670.
Opdyke, N.D. 1972. Paleomagnetism of deep-sea cores. *Review of Geophysics and Space Physics*, 10, 213.
Opdyke, N.M., Lindsay, E., Johnson, G.D., Johnson, N., Tahirkheli, R.A.K. and Mirza, M.A.I. 1979. Magnetic polarity stratigraphy and vertebrate palaeontology of the Upper Siwalik subgroup of northern Pakistan. *Palaeogeography, Palaeoclimatology, Palaeoecology* 27: 1-34.
Opdyke, N.M., Johnson, N.M., Johnson, G.D., Lindsay, E.H. and Tahikheli, R.A.K. 1982. Paleomagnetism of the Middle Siwalik formations of northern Pakistan and rotation of the Salt Range decollement. *Palaeogeography, Palaeoclimatology, Palaeoecology* 37: 1-15.
Paddayya, K. 1982. *The Acheulian Culture of the Hunsgi Valley (Peninsular India), A Settlement System Perspective* Poona: Deccan College.
Paddaya, K. and Petraglia, M.D. 1993. Formation processes of Acheulean localities in the Hunsgi and Baichbal Valleys, Peninsular India. In Goldberg et al. (eds.) *Formation Processes in Archaeological Context.* Monographs in World Archaeology 17: 61-82. Madison: Prehistory Press.
Paddayya, K., Jhaldiyal, R. and Petraglia, M.D. 2000. Excavation of an Acheulean workshop at Isampur, Karnataka (India). *Antiquity* 74: 751-752.
Pappu, S. 1996. Reinvestigation of the prehistoric archaeological record in the Kortallayar Basin, Tamil Nadu. *Man and Environment* 21: 1-23.
Pappu, S. 2001. A Re-examination of the Palaeolithic Archaeological Record of Northern Tamil Nadu, South India. *British Archaeological Reports (International Series)* 1003.
Pappu, S. Gunnell, Y., Taieb, M., Brugal, J.-P. and Touchard, Y. 2003. Excavations at the palaeolithic site of Attirampakkam, South India: preliminary findings. *Current Anthropology* 44 (4): 591-598.
Partridge, T.C., Granger, D.E., Caffee, M.W. and Clarke, R.J. 2003. Lower Pliocene hominid remains from Sterkfontein. *Science* 300: 607-612.
Patnaik, R. and Schleich, H.H. 1993. Fossil crocodile remains from the Upper Siwaliks of India. *Mitt. Bayer. Staatslg. Palaont. hist. Geol.* 33: 91-117.
Patriat, P. and Achache, J. 1984. India-Eurasia collision chronology has implications for crustal shortening and driving mechanism of plates. *Nature* 311: 615-621.
Patterson, L.W. 1983. Criteria for determining the attribute of man-made lithics. *Journal of Field Archaeology* 10: 297-307.
Peters, C.R. and Blumenschine, R.J. 1995. Landscape perspectives on possible land use patterns for Early Pleistocene hominids in the Olduvai Basin, Tanzania. *Journal of Human Evolution* 29: 321-362.
Petraglia, M.D., and Nash, D.T. 1987. The impact of fluvial processes on experimental sites, in D.T. Nash and M. D. Petraglia (eds.) *Natural Formation Processes and the Archaeological Record.* Oxford: British Archaeological Reports (International Series) 352: 108-130.
Petraglia, M.D. and Potts, R. 1994. Water flow and the formation of early Pleistocene artifact sites in Olduvai Gorge, Tanzania. *Journal of Anthropological Archaeology* 13: 228-254.
Petraglia, M.D., Schuldenrein, J. and Korisettar, R. 2003. Landscapes, activity and the Acheulean to Middle Paleolithic transition in the Kaladgi basin, India. *Eurasian Prehistory* 1: 3-24.
Pilbeam, D. 1997. Siwaliks. In *History of Physical Anthropology*, ed. Fr. Spencer, 934-937. London and New York: Garland Publishing Co.
Pilgrim, G.E. 1910. Preliminary note on a revised classification of the Tertiary freshwater deposits of India. *Geological Survey Records of India* 40: 185-205.
Pilgrim, G.E. 1911. The fossil Giraffidae of India. *Palaeontologia Indica* 4 (1): 1-29.
Pilgrim, G.E. 1913. The correlation of the Siwaliks with mammal horizons of Europe. *Geological Survey Records of India* 43: 264-326.
Pilgrim, G.E. 1915a. The dentition of the tragulid genus *Dorcabune. Records of the Geological Survey of India* 45: 226-238.
Pilgrim, G.E. 1915b. New Siwalik primates and their bearing on the question of the evolution of man and the Anthropoidea. *Records of the Geological Survey of India* 45: 1-74.
Pilgrim, G.E. 1927. A *Sivapithecus* palate and other primate fossils from India. *Palaeontologica Indica* 14: 1-24.
Pilgrim, G.E. 1932. The fossil carnivora of India. *Memoirs of the Geological Survey of India (Palaeontolgica Indica)* 18: 1-232.
Pilgrim, G.E. 1937. Siwalik antelopes and oxen in the American Museum of Natural History. *Bulletin of the American Museum of Natural History* 72: 729-874.

Pilgrim, G.E. 1939. The fossil Bovidae of India. *Memoirs of the Geological Survey of India (Palaeontologica Indica)* 26: 1-356.
Plummer, T.W. and Stanford, C.B. 2000. Analysis of a bone assemblage made by chimpanzees at Gombe National Park, Tanzania. *Journal of Human Evolution* 39: 345-365.
Pope, G.C. 1983. Evidence on the age of the Asian hominidae. *Proceedings of the New York Academy of Science, USA* 80: 4988-4992.
Pope, C.G. and Cronin, J.E. 1984. The Asian hominidae. *Journal of Human Evolution* 13: 377-396.
Pope, G., Barr, S., MacDonald, A. and Hahabanlang, S. 1986. Earliest radiometrically dated artefacts from S.E.Asia. *Current Anthropology* 27 (3): 275-279.
Potts, R. 1986. Temporal span of bone accumulations at Olduvai Gorge and implications for early hominid foraging behavior. *Paleobiology* 12 (1): 25-31.
Potts, R. and Shipman, P. 1981. Cutmarks made by stone tools on bones from Olduvai Gorge, Tanzania. *Nature* 291: 577-580.
Ranov, V. 1995. The 'Loessic Palaeolithic' in South Tadjikistan, Central Asia: its industries, chronology and correlation. *Quaternary Science Reviews* 14: 731-745.
Qian Wang and Tobias, P.V. 2001. An old species and a new frontier: some thoughts on the taxonomy of *Homo erectus*. *Przegląd Anthropogiczny - Anthropological Review* 64: 9-20.
Quade, J. and Cerling, T.E. 1995. Expansion of C4 grasses in the Late Miocene of northern Pakistan: evidence from stable isotopes in paleosols. *Palaeogeography, Palaeoclimatology and Palaeoecology* 115 (1-4): 91-116.
Quade, J., Cerling, T.E. and Bowman, J.R. 1989. Development of Asian monsoon revealed by marked ecological shift during the latest Miocene in northern Pakistan. *Nature* 342: 163-166.
Quade, J., Cerling, T.E., Bowman, J.R. and Jah, A. 1993. Paleoecologic reconstruction of floodplain environments using palaeosols from Upper Siwalik Group sediments, northern Pakistan. In *Himalaya to the Sea: Geology, Geomorphology and the Quaternary*, ed. J.F. Schroder, pp. 213-226. London and New York: Routledge.
Ranga Rao, A., Agarawal, R.P., Sharma, U.N., Bhalla, M.S. and Nanda, A.C. 1988. Magnetic polarity stratigraphy and vertebrate palaeontology of the Upper Siwalik subgroup of Jammu Hills, India. *Journal of the Geological Society of India* 31 (4): 361-385.
Ranga Rao, A., Nanda, A.C., Sharma, U.N. and Bhalla, M.S. 1995. Magnetic polarity stratigraphy of the Pinjor Formation (Upper Siwalik) near Pinjore, Haryana. *Current Science* 68: 1231-1234.
Ranov, V. 2001. Cleavers: their distribution, chronology and typology, in S. Milliken and J. Cook (eds) *A Very Remote Period Indeed: Papers on the Palaeolithic Presented to Derek Roe*, pp 105-113. Oxford: Oxbow.
Ranov, V., Carbonell, E. and Rodriguez, X.P. 1995. Kuldara: earliest human occupation in Central Asia and its Afro-Asian context. *Current Anthropology* 36 (2): 337-346.
Raynolds, R.G.H. and Johnson, G.D. 1985. Rates of Neogene depositional and deformational processes, north-west Himalayan foredeep margin, Pakistan. In *The Chronology of the Geological Record*, ed. N.J. Snelling. Oxford: Blackwells Scientific Publications Memoir 10, pp. 297-311.
Reed, K.E. 1997. Early hominid evolution and ecological change through the African Plio-Pleistocene. *Journal of Human Evolution* 32: 289-322.
Rendell, H. and Dennell, R.W. 1985. Dated lower Palaeolithic artefacts from northern Pakistan. *Current Anthropology* 26 (3): 393.
Rendell, H. M. and Dennell, R.W. 1987. The dating of an upper pleistocene archaeological site at Riwat, northern Pakistan. *Geoarchaeology* 1: 6-12.
Rendell, H., Hailwood, E. and Dennell, R.W. 1987. Magnetic polarity stratigraphy of Upper Siwalik Sub-Group, Soan Valley, Pakistan: implications for early human occupance of Asia. *Earth and Planetary Science Letters* 85: 488-496.
Rendell, H.M., Dennell, R.W. and Halim, M. 1989. Pleistocene and Palaeolithic Investigations in the Soan Valley, Northern Pakistan. *British Archaeological Reports International Series* 544: 1-346.
Reynolds, T.E.G. 1993. Problems in the stone age of South-east Asia. *Proceedings of the Prehistoric Society* 59: 1-16.
Rightmire, G.P. 1991. The dispersal of *Homo erectus* from Africa and the emergence of more modern humans. *Journal of Anthropological Research* 47 (2): 177-191
Rightmire, G.P. 1998. Evidence from facial morphology for similarity of Asian and African representatives of *Homo erectus*. *American Journal of Physical Anthropology* 106 (1): 61-85.
Roberts, M. and Parfitt, S.A. 1999. *Boxgrove: a Middle Pleistocene hominid site at Eartham Quarry, Boxgrove, West Sussex.* London: English Heritage.
Roche, H., Delagnes, A., Brugal, J.-P., Feibel, C., Kibunjias, M., Mourrel, V. and Texier, P.-J. 1999. Early hominid stone tool production and technical skill 2.34 Myr ago in West Turkana, Kenya. *Nature* 399: 57-60.
Roe, D. 1994. Summary and Overview, in M. Leakey with D. Roe Olduvai Gorge, vol.5, 299-310. Cambridge: Cambridge University Press.
Roebroeks, W. 1996. The English palaeolithic record: absence of evidence, evidence of absence and the first occupation of Europe. *The English Palaeolithic reviewed*, ed. C. Gamble and A. Lawson ,57-62. Trust for Wessex Archaeology.

Roebroeks, W. and Kolfschoten, T. van 1994. The earliest occupation of Europe: a short chronology. *Antiquity* 68: 489-503.
Roebroeks, W. and Kolfschoten, T. van (eds.) 1995. *The Earliest Occupation of Europe.* Leiden: Leiden University Press.
Ron, H. and Levi, S. 2001. When did hominids first leave Africa? New high-resolution paleomagnetic evidence from the Erk-El-Ahmar formation, Israel. *Geology* 29: 887-890.
Ronen, A. 1991. The Yiron-gravel lithic assemblage: artefacts older than 2.4 My in Israel. *Archaologisches Korrespondenzblatt* 21: 159-164.
Sahni, M.R. and Khan, E. 1988. *Pleistocene Vertebrate Fossils and Prehistory of India.* New Dehli: Books and Books.
Sankalia, H. D. 1974. (2nd edition) *The Prehistory and Protohistory of India and Pakistan* Poona, Deccan College.
Sankhyan, A.R. 1997. Fossil clavicle of a Middle Pleistocene hominid from the Central Narmada Valley, India. *Journal of Human Evolution* 32: 3-16.
Schaller, G.B. and Lowther, G.R. 1969. The relevance of carnivore behavior to the study of early hominids. *Southwestern Journal of Anthropology* 25 (4): 307-341.
Schick, K. 1986. Stone Age Sites in the Making. *British Archaeological Reports (International Series)* 189: 1-311.
Schick K.D. and Toth N. 1993. *Making Silent Stones Speak: Human Evolution & the Dawn of Technology* London: Phoenix.
Schick, K.D. and Zhuan 1993. Early paleolithic of China and eastern Asia. *Evolutionary Anthropology* 2 (1): 22-35.
Schrenk, F., Bromage, T.G., Bernier, C.G., Ring, U. and Juwayeyi, Y.M. 1993. Oldest *Homo* and Pliocene biogeography of the Malawi Rift. *Nature* 365: 833-36.
Schwartz, J.H. and Tattersall, I. 1996. Whose teeth? *Nature* 381: 201-202.
Selvaggio, M. 1994. Carnivore tooth marks and stone tool butchery marks on scavenged bones: archaeological implications. *Journal of Human Evolution* 27: 215-228.
Semaw, S., Renne, P., Harris, J.W.K., Feibel, C.S., Bernor, R.L., Fesseha, N. and Mowbray, K. 1997. 2.5-million-year-old stone tools from Gona, Ethiopia. *Nature* 385: 333-336.
Sonkalia, A. 1985. Skull cap of an early man from the Narmada Valley alluvium (Pleistocene) of Central India. *American Anthropologist* 87: 612-616.
Spencer, L.M. 1997. Dietary adaptations of Plio-Pleistocene Bovidae: implications for hominid habitat use. *Journal of Human Evolution* 32: 201-228.
Steensma, K.J. and Hussain, S.T. 1992. *Merycopotamus dissimilis* [Artiodactyla, Mammalia] from the Upper Siwalik Subgroup and its affinities with Asian and African forms. *Proceedings of the Koninkijke Nederlandse Akademie van Wetenschappen* 95 (1): 97-108.
Stern, N. 1993. The structure of the Lower Palaeolithic archaeological record. *Current Anthropology* 34 (3): 201-225.
Stiles, D. 1978. Palaeolithic artefacts in Siwalik and post-Siwalik deposits of northern Pakistan. Kroeber Anthropological Society Papers 53-54: 129-148.
Susman, R.L. 1988. Hand of *Paranthropus robustus* from Member 1, Swartkrans: fossil evidence for tool behavior. *Science* 240: 780-781.
Susman, R.L. 1991. Who made the Oldowan tools? Fossil evidence for tool behaviour in plio-pleistocene hominids. *Journal of Anthropological Research* 47 (2): 129-151.
Susman, R.L. and Stern, J.T. 1982. Functional morphology of *Homo habilis*. *Science* 217: 931-934.
Swisher III C.C. Curtis G.H. Jacob T. Getty A.G. Suprijo A. and Widiasmoro 1994. Age of the earliest known hominids in Java Indonesia. *Science* 263: 1118-1121.
Swisher, G.C., Rink, W.J., Antón, S.C., Schwarcz, H.P., Curtis, G.H., Suprijo, A. and Widiasmoro 1996. Latest *Homo erectus* of Java: potential contemporaneity with *Homo sapiens* in Southeast Asia. *Science* 274: 1870-1874.
Tandon, S.K., Kumar, R., Koyama, M. and Niitsuma, N. 1984. Magnetic polarity stratigraphy of the Upper Siwalik subgroup, east of Chandigarh, Punjab Sub-Himalaya, India. *Journal of the Geological Society of India* 25 (1): 45-55.
Tattersall, I. 1997. Out of Africa again...and again? *Scientific American* 276 (4): 46-53.
Tauxe, L. and Opdyke, N.D. 1982. A time framework based on magnetostratigraphy for the Siwalik sediments of the Khaur area, northern Pakistan. *Palaeogeography, Palaeoclimatology, Palaeoecology* 37: 43-61.
Tchernov, E. 1989. The age of the 'Ubeidiya Formation. *Israel Journal of Earth Sciences* 36: 3-30.
Terra, H. de and Chardin. P.T. de. 1936. Observations on the Upper Siwalik Formation and later Pleistocene deposits in India. *Proceedings of the American Philosophical Society* 76: 791-822.
Terra, H. de and Paterson, T.T. 1939. *Studies on the Ice Age in India and Associated Human Cultures*. Washington: Carnegie Institute Publications 493.
Tixier, J. 1956. Le hachereau dans l'Acheuléen nord africain, notes typologiques. *Congrès Préhistorique de France, 15ème session*, Poitiers, Angoulème, pp 1-10.
Todd, L.C. and Rapson, D.J. 1988. Long bone fragmentation and interpretation of faunal assemblages: approaches to comparative analysis. *Journal of Archaeological Science* 15 (3): 307-326.
Toth, N. 1987. The first technology. *Scientific American* 256 (4): 104-113.
Vasishat, R.N. 1985. *Antecedents of Early Man in Northwestern India*. New Delhi: Inter-India Publications.

Vekua, A. 1995. Die Wirbeltierfauna des Villafranchium vom Dmanisi und ihre biostratigraphische Bedeutung. *Jahrbuch des Römisch-Germanischen Zentralmuseums, Mainz.*
Visser, C.F. and Johnson, G.D. 1978. Tectonic control of Late Pliocene molasse sedimentation in a portion of the Jhelum Re-entrant, Pakistan. *Geologisches Rundschau* 67 (1): 15-37.
Vos, J. de 1984. Reconsideration of Pleistocene cave faunas from South China and their relation to the faunas from Java. *Courier Forschungs-Institut Senckenberg* 69: 259-266.
Vos, J. de and Sondaar, P. 1994. Dating hominid sites in Indonesia. *Science* 266: 1726.
Vrba , E.S. 1995. The fossil record of African antelopes (Mammalia, Bovidae) in relation to human evolution and paleoclimate. In *Paleoclimate and Evolution*, ed. E.S. Vrba, G.H. Denton, T.C. Partridge and L.H. Burckle, 385-424. New Haven and London: Yale University Press.
Vrba, E.S. 1996. Climate, heterochrony, and human evolution. *Journal of Anthropological Research* 52: 1-28.
Walker, A. and Shipman, P. 1996. *The Wisdom of Bones: In Search of Human Origins*. London: Weidenfeld and Nicholson.
Wallis, J. 1988. Natural damage in quarzite pebbles, in R.J.MacRae and N. Moloney *Non-Flint Stone Tools and the Palaeolithic Occupation of Britain* Oxford: British Archaeological Reports (British Series 189) pp. 155-157.
Wang, H., Ambrose, S.H., Liu, C.-L. J. and Follmer, L.R. 1997. Paleosol stable isotope evidence for early hominid occupation of East Asian temperate environments. *Quaternary Research* 48 (2): 228-238.
Wanpo H., Ciochon, R., Yumin, G., Larick, R., Qiren, F., Schwarcz, H., Yonge, C., Vos J. de and Rink, W. 1995. Early *Homo* and associated artefacts from Asia. *Nature* 378: 275-278.
Wheeler, Sir M. 1950. *Five Thousand Years of Pakistan.* London: Royal India and Pakistan Society.
Wheeler, Sir M. 1959. *Early India and Pakistan.* London: Thames and Hudson.
Wheeler, P.E. 1991. The thermoregulatory advantages of hominid bipedalism in open equatorial environments: the contribution of increased convective heat loss and cutaneous evaporative cooling. *Journal of Human Evolution* 21:107-115.
White, C. 1995. La Grotte du Vallonet. *Lithics* 16: 70-77.
White, T.D., Suwa, G. and Asfaw, B. 1994. *Australopithecus ramidus*, a new speices of early hominid from Aramis, Ethiopia. *Nature* 371: 306-312.
White, T.D. 1995. African omnivores: global climatic change and Plio-Pleistocene hominids and suids. In *Paleoclimate and Evolution*, ed. E.S. Vrba, G.H. Denton, T.C. Partridge and L.H. Burckle, 369-385. New Haven and London: Yale University Press.
White, T.D., Suwa, G. and Asfaw, B. 1995. Corrigendum: *Australopithecus ramidus*, a new species of early hominid from Aramis, Ethiopia. *Nature* 375: 88.
Willis, B.J. 1993. Evolution of Miocene fluvial systems in the Himalayan foredeep through a two kilometre-thick succession in northern Pakistan. *Sedimentary Geology* 88: 77-121.
Willis, B.J. and Behrensmeyer, A.K. 1995. Fluvial systems in the Siwalik Miocene and Wyoming Paleogene. *Palaeogeography, Palaeoclimatology, Palaeoecology* 115: 13-35.
Willoughby, P.R. 1987. Spheroids and Battered Stones in the African Early and Middle Stone Age. *British Archaeological Reports (International Series)* 321: 1-253.
Wood, B. 1984. The origin of *Homo erectus. Courier Forschungsinstitut Senckenberg* 69: 99-111.
Wood, B. 1997. The oldest whodunnit in the world. *Nature* 385: 292-293.
Wood, B. 1999.'*Homo rudolfensis*' Alexeev, 1986-fact or phantom? *Journal of Human Evolution* 36: 115-118.
Wood, B. and Collard, M. 1999. The human genus. *Science* 284: 65-71.
Wood, B. and Turner, A. 1995. Out of Africa and into Asia. *Nature* 378: 239-240.
Wood, J.M., Thomas, R.G. and Visser, J. 1989. Fluvial processes and vertebrate taphonomy: the upper Cretaceous Judith River Formation, South-Central Dinosaur Provincial Park, Alberta, Canada. *Palaeogeography, Palaeoclimatology, Palaeoecology* 66: 127-143.
Wu, X. 2000. Longgupo hominoid mandible belongs to ape. *Acta Anthropologica Sinica* 19: 1-10.
Yaowalak Chaimanee, Varavudh Suteethorn, Pratueng Jintasakul, Chavalit Vidthayanon, Bernard Marandat and J.-J. Jaeger 2004. A new orang-utan relative from the Late Miocene of Thailand. *Nature* 427: 439-441.
Zeitler, P.K., Johnson, N.M., Naeser, C.W. and Tahirkheli, R.A.K. 1982. Fission-track evidence for Quaternary uplift of the Nanga Parbat region, Pakistan. *Nature* 298: 255-257.
Zhu, R.X., Hoffman, K.A., Potts, R., Deng, C.L., Pan, Y.X., Guo, B., Shi, C.D., Guo, Z.T., Hou, Y.M. and Huang, W.W. 2001. Earliest presence of humans in northeast Asia. *Nature* 413: 413-417.

www.ingramcontent.com/pod-product-compliance
Ingram Content Group UK Ltd.
Pitfield, Milton Keynes, MK11 3LW, UK
UKHW061843041026
15591UKWH00015B/888

9 781841 713717